World Soils Book Series

Series editor

Prof. Alfred E. Hartemink
Department of Soil Science, FD Hole Soils Laboratory
University of Wisconsin–Madison
Madison
USA

Aims and scope

The *World Soils Book Series* brings together soil information and soil knowledge of a particular country in a concise and reader-friendly way. The books include sections on soil research history, geomorphology, major soil types, soil maps, soil properties, soil classification, soil fertility, land use and vegetation, soil management, and soils and humans.

International Union of Soil Sciences

More information about this series athttp://www.springer.com/series/8915

L.T. West · M.J. Singer
A.E. Hartemink
Editors

The Soils of the USA

Springer

Editors
L.T. West
USDA-Natural Resources Conservation Service
Lincoln, NE
USA

A.E. Hartemink
University of Wisconsin–Madison
Madison, WI
USA

M.J. Singer
University of California–Davis
Davis, CA
USA

ISSN 2211-1255 ISSN 2211-1263 (electronic)
World Soils Book Series
ISBN 978-3-319-82452-9 ISBN 978-3-319-41870-4 (eBook)
DOI 10.1007/978-3-319-41870-4

Printed on acid-free paper

This Springer imprint is published by Springer Nature
The registered company is Springer International Publishing AG
The registered company address is: Gewerbestrasse 11, 6330 Cham, Switzerland

Foreword

The land resource regions shown in this book are a convenient way to illustrate the diversity of natural resources of the USA. A century ago, we knew little about the properties and diversity of natural resources of the USA, including soil resources; how the land was used; and did it matter how the natural resources were managed. Over the past century, much progress has been made toward understanding the properties, genesis, distribution, and management of the soil resource, and this book highlights our current knowledge of the USA's soil resources.

Prof. Hans Jenny in his 1941 book noted that "clearly it is the union of the geographic and the functional methods that provides the most effective means of pedological research." Soil distributions result from formation and evolution of soils and their landscapes. The experiments of nature have already been done and the existing heterogeneity suggests that scales are important in understanding what and where soils occur. Soil functions deal with active biogeochemical processes and relate to use and management of soil. Experiments can be designed and implemented to evaluate various treatments and how soils respond to manipulations. The main purpose of combining scientific works about soil distributions and their functioning is to transfer technology to appropriate locations.

The history of the National Cooperative Soil Survey involves the ever-increasing cooperation of federal, state, county, and private soil scientists. Interdisciplinary research into soil processes and functions as they relate to the landscapes of the nation has continually evolved since the founding of agricultural experiment stations and land grant universities. Today, the expansion on this legacy in cooperative efforts such as the Critical Zone Observatories and similar expansive projects is truly phenomenal.

In the twenty-first century, there is an increasing concern about the degradation and loss of soil resources because these affect food and fiber supplies for civilization to survive and prosper. The authors of this book have graciously dedicated their knowledge and efforts to help us understand and act accordingly to the concept that *Sustainability is the acceptance of resource stewardship—locally and globally.*

2016

Richard Arnold
Director Emeritus, USDA-NRCS Soil Survey Division

Preface

This book reviews the properties, behavior, distribution, genesis, and management of soils across the USA. It highlights the diversity of soils and the importance of soils to ecosystem services and productivity, agriculture, forestry, and urban infrastructure. Soils are a critical component of natural and managed ecosystems and perform functions that support the needs and well-being of the global human population; these include nutrient and water storage and supply for plant growth, hydrologic buffering, disposal and renovation of anthropogenic wastes, habitat for soil organisms, and support for roads, buildings, and other infrastructure. Soils are a major reservoir of global carbon and can, with proper management, serve as a sink for atmospheric carbon to reduce greenhouse gasses. Although most soils are relatively resilient, they are subjected to degradation if managed improperly or otherwise disturbed. Thus, conservation of the soil resource and its continued use to perform ecosystem functions to support the ever-increasing global population depends on understanding the properties of and processes occurring in the soil at any point in the landscape.

Discussions of the soil resource in this book are stratified geographically, based on land resource regions (LRRs) and, within the LRRs, major land resource areas (MLRAs). The major part of the book consists of chapters that discuss soils that occur in each LRR in the USA and its territories. Several chapters address multiple LRRs that have similar landscape and environmental characteristics. The book includes chapters that provide background information that may be needed to better understand concepts presented including processes important to soil formation, concepts and products of soil survey, and the structure and nomenclature of *Soil Taxonomy*. Also included are chapters discussing changes in soil properties related to human activities, and challenges facing soil science and soil survey in the future.

The wide diversity of soil conditions across the USA precludes any one individual that can synthesize the large amount of knowledge about the soils across the nation. Thus, multiple authors, each with extensive understanding of soils within their region, have written the chapters. The content of each chapter varies somewhat and reflects the diversity in the regions as well as the authors' interest and experience. We express our appreciation to all of the contributing authors for their dedication and effort in preparation of their chapter. We also appreciate the cooperation we received from the USDA-Natural Resources Conservation Service in allowing many NRCS staff to devote time to authoring chapters, providing data on properties and extent of soils in each LRR, and assistance with preparation of map figures.

Systematic mapping of soils on the landscape and interpretation of their expected behavior has been the objective of the National Cooperative Soil Survey (NCSS) since the late 1800s. Thousands of soil scientists from universities, state and federal agencies, and the private sector have strived to understand the soil resource and to develop the inventory of soils that is now widely available. Without the efforts of all these scientists and soil practitioners, this book would not have been possible. We dedicate this book to this unnamed cadre of dedicated soil scientists. Thank you!

The comprehensive inventory and availability of soil information across the USA is an example for many countries. It will continue to enable scientists and producers to design and implement soil management systems that allow sustainable production of food, fiber, and fuel crops necessary to maintain the quality of life. The inventory of the soil resource will also allow research and direction on new challenges including sustaining and improving soil quality and health, soil sequestration of atmospheric carbon, renovation of wastewater, and storage and delivery of water and nutrients for plant production. It is our humble hope that the knowledge brought together in this book will be used to educate a generation that will continue to deliver solutions to the environmental challenges that we face, now and in the future.

Fayetteville, USA L.T. West
Davis, USA M.J. Singer
Madison, USA A.E. Hartemink

Contents

1 **Introduction** 1
L.T. West, M.J. Singer, and A.E. Hartemink

2 **Concepts of Soil Formation and Soil Survey** 9
Robert C. Graham and Samuel J. Indorante

3 **Soil Properties and Classification (Soil Taxonomy)** 29
Craig A. Ditzler

4 **Soils of the USA: The Broad Perspective** 43
James G. Bockheim

5 **Soils of the Pacific Coast Region: LRRs A and C** 77
Susan Burlew Southard, Randal J. Southard, and Scott Burns

6 **Soils of the Rocky Mountain and Inland Pacific Northwest Region: LRRs B and E** 101
Paul A. McDaniel

7 **Soils of the Western Range and Irrigated Land Resource Region: LRR D** 115
Craig Rasmussen, Rebecca A. Lybrand, Caitlin Orem, Jennifer Kielhofer, and Molly Holleran

8 **Soils of the Great Plains** 131
Daniel Richard Hirmas and Rolfe David Mandel

9 **Southwest Plateaus and Plains Range and Cotton Region and Southwestern Prairies Cotton and Forage Region: LRRs I and J** 165
David Emmanuel Ruppert

10 **Soils of the Northern Lake States Forest and Forage Region: LRR K** 191
Randall J. Schaetzl

11 **Soils of the Central Feed Grains and Livestock Region and Lake State Fruit, Truck Crop, and Dairy Region USA: LRRs M and L** 209
Zamir Libohova, Philip Schoeneberger, Doug Wysocki, and Phillip R. Owens

12 **East and Central Farming and Forest Region and Atlantic Basin Diversified Farming Region: LRRs N and S** 227
Brad D. Lee and John M. Kabrick

13 Soils of the Southeastern USA: LRRs O, P, and T 243
L.T. West, J.N. Shaw, and E.P. Mersiovsky

14 Northeastern Forage and Forest Region: LRR R 261
M.A. Wilson and R.K. Shaw

15 Soils of Florida and the Caribbean: LRRs U and Z 281
Mary E. Collins

16 Soils of Hawaii and the US Territories in the Pacific: LRRs Q and V 305
Michael P. Robotham, Robert T. Gavenda, and Christopher W. Smith

17 Soils of Alaska: LRRs W1, W2, X1, X2, and Y 329
C.L. Ping, M.H. Clark, D. D'Amore, G.J. Michaelson, and D.K. Swanson

18 Human Land-Use and Soil Change 351
Skye A. Wills, Candiss O. Williams, Michael C. Duniway, Jessica Veenstra, Cathy Seybold, and DeAnn Presley

19 Future Challenges for Soil Science Research, Education, and Soil Survey in the USA 373
Patrick J. Drohan

Index 385

Contributors

James G. Bockheim University of Wisconsin–Madison, Madison, WI, USA

Scott Burns Portland State University, Portland, OR, USA

M.H. Clark USDA-NRCS, Palmer, AK, USA

Mary E. Collins University of Florida, Gainesville, FL, USA

Craig A. Ditzler USDA-Natural Resources Conservation Service, Lincoln, NE, USA

Patrick J. Drohan Pennsylvania State University, State College, PA, USA

Michael C. Duniway US Geological Survey, Moab, UT, USA

D. D'Amore USDA Forest Service, Juneau, AK, USA

Robert T. Gavenda USDA-NRCS, Merizo, Guam

Robert C. Graham University of California, Riverside, CA, USA

A.E. Hartemink University of Wisconsin–Madison, Madison, WI, USA

Daniel Richard Hirmas University of Kansas, Lawrence, KS, USA

Molly Holleran The University of Arizona, Tucson, AZ, USA

Samuel J. Indorante USDA-Natural Resources Conservation Service, Carbondale, IL, USA

John M. Kabrick USDA Forest Service, Northern Research Station, Columbia, MO, USA

Jennifer Kielhofer The University of Arizona, Tucson, AZ, USA

Brad D. Lee University of Kentucky, Lexington, KY, USA

Zamir Libohova USDA-NRCS, Lincoln, NE, USA

Rebecca A. Lybrand The University of Arizona, Tucson, AZ, USA

Rolfe David Mandel Kansas Geological Survey, Lawrence, KS, USA

Paul A. McDaniel University of Idaho, Moscow, ID, USA

E.P. Mersiovsky USDA-NRCS, Little Rock, AR, USA

G.J. Michaelson University of Alaska Fairbanks, Palmer, AK, USA

Caitlin Orem The University of Arizona, Tucson, AZ, USA

Phillip R. Owens USDA-Agricultural Research Service, Booneville, AR, USA

C.L. Ping University of Alaska Fairbanks, Palmer, AK, USA

DeAnn Presley Kansas State University, Manhattan, KS, USA

Craig Rasmussen The University of Arizona, Tucson, AZ, USA

Michael P. Robotham USDA-NRCS, Lincoln, NE, USA

David Emmanuel Ruppert Texas A&M University-Kingsville, Kingsville, TX, USA

Randall J. Schaetzl Michigan State University, East Lansing, MI, USA

Philip Schoeneberger USDA-NRCS, Lincoln, NE, USA

Cathy Seybold USDA-NRCS, Lincoln, NE, USA

J.N. Shaw Auburn University, Auburn, AL, USA

R.K. Shaw USDA-NRCS, Somerset, NJ, USA

M.J. Singer University of California, Davis, Davis, CA, USA

Christopher W. Smith USDA-NRCS, Washington, DC, USA

Randal J. Southard University of California, Davis, Davis, CA, USA

Susan Burlew Southard USDA-NRCS, Davis, CA, USA

D.K. Swanson USDI-National Park Service, Fairbanks, AK, USA

Jessica Veenstra Flagler College, St. Augustine, FL, USA

L.T. West USDA-NRCS, Lincoln, NE, USA

Candiss O. Williams USDA-NRCS, Lincoln, NE, USA

Skye A. Wills USDA-NRCS, Lincoln, NE, USA

M.A. Wilson USDA-NRCS, Lincoln, NE, USA

Doug Wysocki USDA-NRCS, Lincoln, NE, USA

Abbreviations

Al	Aluminum
AWC	Available water capacity
B.P.	Before present
C	Carbon
Ca	Calcium
$CaCO_3$	Calcium carbonate
CCE	Calcium carbonate equivalent
CEC	Cation exchange capacity
CH_4	Methane
cm	Centimeter(s)
cmol	Centimole(s)
CNMI	Commonwealth of the Northern Mariana Islands
CO_2	Carbon dioxide
d	Day
dS	Decisemen(s)
EAA	Everglades Agricultural Area
EC	Electrical conductivity
ECT	Eastern Cross Timbers
ESP	Exchangeable sodium percentage
Fe	Iron
g	Gram(s)
GP	Grand Prairie
H	Hydrogen
h, hr	Hour
ha	Hectare
HCl	Hydrochloric acid
HCO_3	Bicarbonate
K	Potassium
ka	Thousand years
KCl	Potassium chloride
km	Kilometer
KOH	Potassium hydroxide
kPa	Kilopascal(s)
K_s	Saturated hydraulic conductivity
l	Liter
LRR	Land resource region
m	Meter(s)
MAAT	Mean annual air temperature
Ma	Million years
MAP	Mean annual precipitation
MAST	Mean annual soil temperature

MAT	Moist acid tundra
Mg	Magnesium
Mg	Megagram(s)
MJT	Mean July temperature
MLRA	Major land resource area
mm	Millimeter(s)
Mn	Manganese
MNT	Moist nonacid tundra
Na	Sodium
NaOH	Sodium hydroxide
NBP	Northern Blackland Prairie
NCP	Texas Clay Pan Area, Northern Part
NCSS	National Cooperative Soil Survey
NCT	Northern Cross Timbers
NLSFFR	Northern Lake States Forest and Forage Region
NRCS	Natural Resources Conservation Service
P	Phosphorus
PET	Potential evapotranspiration
SAR	Sodium adsorption ratio
SBP	Southern Blackland Prairie
SCP	Texas Clay Pan Area, Southern Part
SCS	Soil Conservaiton Service, now Natural Resources Conservation Service
Si	Silicon
SiO_2	Silicon dioxide
SMR	Soil moisture regime
SO_4	Sulfate
SOC	Soil organic carbon
SOM	Soil organic matter
SSD	Soil Science Division
SSSA	Soil Science Society or America
STR	Soil temperature regime
SWI	Summer warmth index
T	Ton, megagram
TCB	Texas Central Basin
USDA	United States Department of Agriculture
w/w	Weight per weight
WAT	Wet acid tundra
WCT	Western Cross Timbers
WNT	Wet nonacid tundra
y, yr	Year
^oC	Degrees Celsius

Introduction

1

L.T. West, M.J. Singer, and A.E. Hartemink

1.1 Introduction

This is a book about soils of the United States. Soils are a critical and often unappreciated resource because they are belowfoot and mostly out of sight. This book brings to you a comprehensive overview of the diversity, beauty, and vital importance of soils to ecosystems, agriculture, forestry, and urban infrastructure. It is intended to be a reference and learning tool that will enhance your knowledge, understanding, and appreciation of the soil resources in the USA. Soil supports all terrestrial life forms, and performs functions critical to the well-being of the global population including nutrient and water storage and supply for plant growth, partitioning of precipitation into ground and surface waters, disposal and renovation of anthropogenic wastes, habitat for soil organisms, and support for roads, buildings, and other infrastructure. Soils are a major reservoir of global carbon and can, with proper management, serve as a sink for atmospheric carbon to reduce greenhouse gasses. Soils are relatively resilient, but are subject to degradation if managed improperly. Only by understanding the properties of and processes occurring in the soil, can the soil resource be conserved and sustained for continued support of the Earth's population.

L.T. West retired from USDA-NRCS

L.T. West (✉)
USDA-NRCS, Lincoln, NE, USA
e-mail: larrywestar@gmail.com

M.J. Singer
University of California, Davis, Davis, CA, USA
e-mail: mjsinger@dcn.org

A.E. Hartemink
University of Wisconsin, Madison, Madison, WI, USA
e-mail: hartemink@wisc.edu

L.T. West et al. (eds.), *The Soils of the USA*, World Soils Book Series,
DOI 10.1007/978-3-319-41870-4_1

1.2 Diversity of Soils in the USA

Soil Taxonomy is the taxonomic system used to classify soils in the USA as well as in many other countries across the globe (Fig. 1.1). It defines soil as "a natural body comprised of solids (minerals and organic matter), liquid, and gases that occurs on the land surface, occupies space, and is characterized by one or both of the following: horizons, or layers, that are distinguishable from the initial material as a result of additions,losses, transfers, and transformations of energy and matter or the ability to support rooted plants in a natural environment" (Soil Survey Staff 2014). This is the definition that was used in this book. From the initial parent material (rock, sediment, etc.), soils become vertically differentiated by natural processes over time into layers called "horizons". The properties and arrangement of soil horizons reflect the parent material, vegetation, climate, and the interacting chemical, physical, and biological processes that have been important in development of the soil. The conglomeration of these processes is known as soil formation. The properties and arrangement of horizons defines the soil and make soils on different landscape segments different from those on other segments.

Properties of the soil at any point in the landscape are due to the action of and interaction among five environmental factors; climate, parent material, relief ortopography, biology, and time (Jenny 1941). Across the USA (land area of about 9,857,000 km^2), a nearly infinite range in climatic, geologic, geomorphologic, and vegetative conditions has resulted in a enormous range in soil morphological, chemical, physical, mineralogical, and biological characteristics. This requires compartmentalization of soils into groups as an aid in communication of soil properties so that soil information can be useful. This grouping of knowledge is one of the major roles of Soil Taxonomy.

Climate, specifically temperature and precipitation, influences the processes that convert parent material into soil. Air temperatures vary from those of the tundra in northern Alaska to those of Death Valley in the California

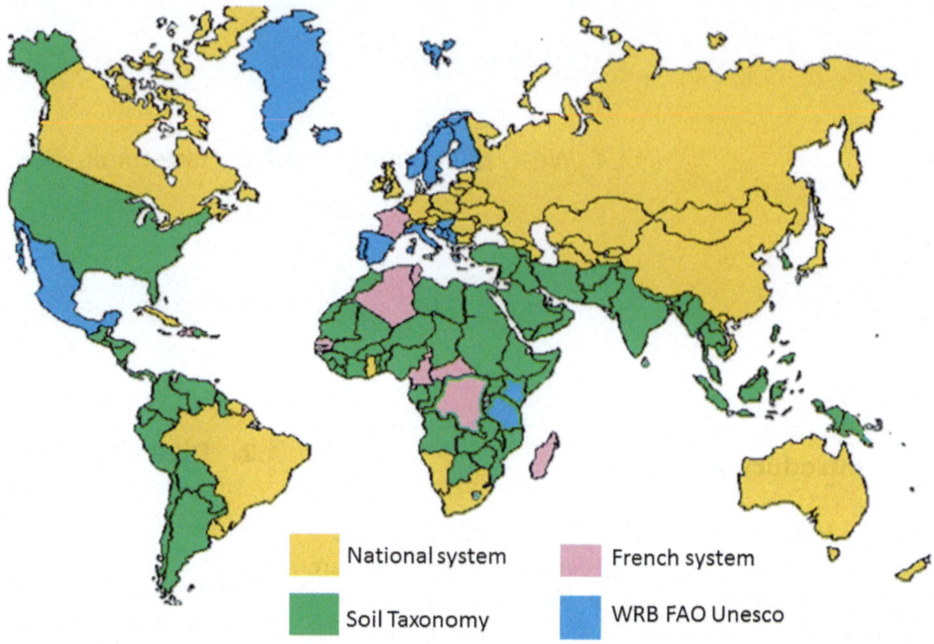

Fig. 1.1 Use of Soil Taxonomy and other soil classification systems across the world

Mojave Desert to those in the rainforest in tropical Puerto Rico. Annual precipitation varies from a few millimeters in the southwestern deserts and the lee side of the Hawaiian Islands to more than 10 m on the windward side of some of the Pacific Islands. Because of the wide ranges in climate found across the USA, native vegetation is extremely diverse and includes dense evergreen forests, hardwood forests, savannahs, mountain meadows, prairies of all types, mangrove swamps, coastal marshes, tropical rainforests, and many other communities.

A vast range of types and composition of rocks occurs across the country including granites, schists, gneisses, limestones, sandstones, shales, and many others. Continental and mountain glaciers have left large deposits of glacial sediments. Streams have built alluvial terraces and deltas. Volcanoes have deposited ash and lava over large areas, and wind has formed extensive dunes and blanketed vast areas with loess.Upland landforms vary from broad level plains to steep mountain slopes to closed depressions ponded with water for periods of the year.

1.3 Objective

The primary objective of this book is to present an overview of the soils of the USA including properties, genesis, and landscape relationships of important soils across the country as these relate to the five factors of soil formation. A secondary objective is to describe relationships between the soils and land use patterns and to highlight soil vulnerabilities related to human use of the soils.

This book is intended to serve as a reference for earth scientists, ecologists, biologists, foresters and all others interested in an overview of the properties of soils in the USA. Although use as a textbook in university settings was not a primary objective, all or parts of this text may be useful to educate students in geography, earth sciences, ecology, and other disciplines about the properties, genesis, and distribution of soils in the USA. The book will also aid development of students' understanding of the dynamics of the soil system, the influence of soils on ecosystem characteristics and dynamics, and the importance of soils and proper soil management for sustaining agricultural and silvicultural production while maintaining or improving environmental conditions.

1.4 Organization

Although the first soil maps in the USA were made in the 1820s (Brevik and Hartemink 2013), systematic mapping of different soils on the landscape and interpretation of their expected behavior has been ongoing since 1899. As part of the National Cooperative Soil Survey (NCSS), various state and federal agencies and universities across the country have contributed to this inventory of soils. These efforts have resulted in the Soil Survey of the USA, which contains soil information for more than 90 % of the country. The Soil Survey includes maps of soil distribution, descriptions of how the soils in an area relate to specific landforms, descriptions of the properties of the soils, interpretations of soil behavior, and suitability and limitations of soils for specific uses. This information serves as the basis for this overview of the soils across the USA and without it, assembling the information included in this text would not be possible.

Over 21,000 different soil series are now recognized in the USA, and it is virtually impossible to provide a meaningful overview of their great diversity and include specific information on the most extensive and/or important of these soils without a meaningful stratification of the soils. Since the publication of Soil Taxonomy (Soil Survey Staff 1975), several authors of texts discussing soil properties and genesis have used the 12 soil orders (the highest level of this system of soil classification) to stratify global soil resources into meaningful divisions, e.g. Wilding et al. (1983), West and Wilding (2011), Buol et al. (2011), Bockheim (2014) and Lal (2016). This approach has been well received, extremely useful, and has helped earth scientists across the world to develop an understanding of the basis for soil classification using Soil Taxonomy and the properties, genesis, and global distribution of soils in each order. This means of grouping soils has also proven to be useful for university Pedology courses as properties and genesis of soils can be presented in a systematic manner that emphasizes similarities and differences among soils worldwide.

In contrast to stratifying soils based on their classification, the discussions of the soil resource in this text are organized geographically, i.e. soil properties, genesis, and use of soils are described for different regions of the USA. Several publications are available that discuss distribution and properties of soils for specific states and regions of the USA, e.g. Buol (1973), Hole (1976), Montague (1982), Springer and Elder (1980), Daniels et al. (1999) and Brye et al. (2013), and between 1883 and the present, 129 state soil maps have been produced (Bockheim and Gennadiyev 2015). A text that attempts to describe and discuss the soil resource for all of USA and its territories, however, has not been published in recent years. The last comprehensive treatment of soil properties for the entire country was published in 1936 (Marbut 1936). More recently, Bockheim (2014) provided a country-wide overview of the distribution of diagnostic horizons.

1.4.1 The Base for This Book

The geographic basis chosen to stratify soils in the USA for this book is the land resource region (LRR) and within the LRRs, the major land resource area (MLRA) (USDA-NRCS 2006). These are regionallandscape units derived by aggregation of soil and landscape units from detailed soil maps up to the regional scale. Each LRR and MLRA has similar topography, climate, water resources, potential natural vegetation, land use, elevation, and soils (USDA-NRCS 2016). Relationships between LRRs and MLRAs, USEPA Level III Ecoregions (USEPA 2003; Omernik 1987), and U.S. Forest Service ecological sections (Cleland et al. 2005; McNab et al. 2005) as well as detailed discussions of the geology, physiography, water resources, and land use for each LRR and MLRA can be found in USDA-NRCS (2006).

The major part of this book is aseries of chapters that discuss the soils in the various LRRs recognized in the USA and its territories. Each LRR is designated by a capital letter and named for their location and primary land uses. LRR A is in the Pacific Northwest region of the USA, and the letters designating the LRRs generally increase in an easterly and southerly direction (Fig. 1.2). The LRRs range in area from

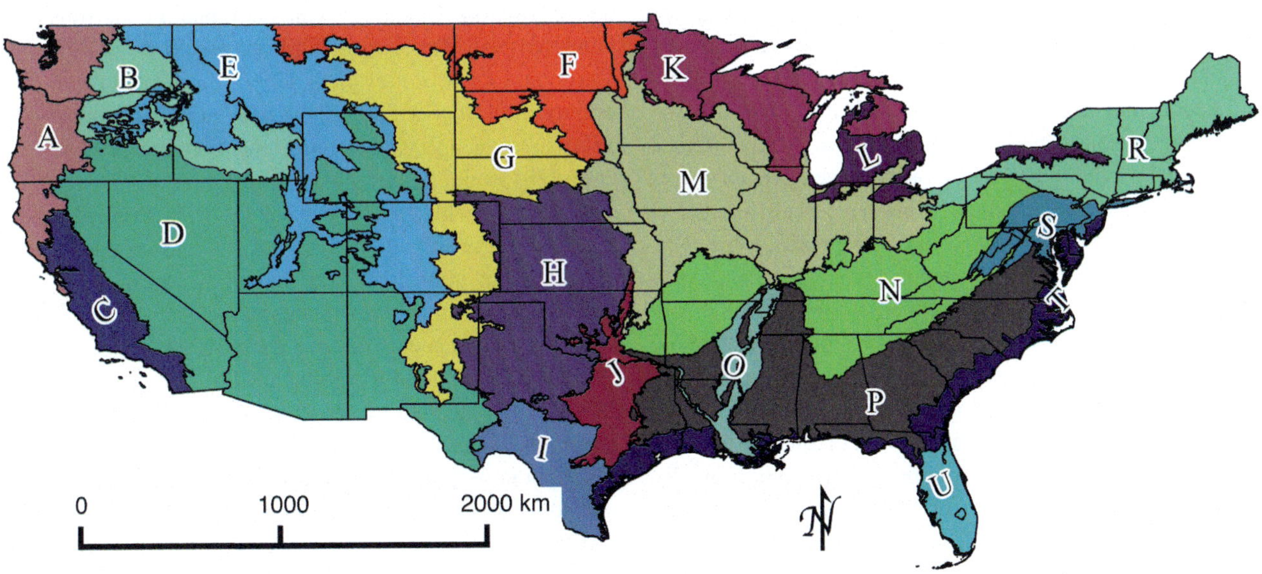

Fig. 1.2 LRRs of the conterminous USA

260 km^2 for the Hawaii Region to 1,400,000 km^2 for LRR D of the arid southwestern USA.

Several of the chapters include discussions of the soils in two or more LRRs. In many of the chapters that address multiple LRRs, a LRR with relatively small land area is combined with an adjoining larger LRR with similar soils, climate, topography, and soil parent materials. In other cases, LRRs with similar and related soils, topography, and parent materials are aggregated into a single chapter. For example, LRRs F, G, and H comprise the central and northern parts of the Great Plains of the USA and are combined into a single chapter. Similarly, parent materials, climate, topography, and soils for LRRs P, T, and O are similar and related, and these are combined into a single chapter to reduce redundancy.

1.4.2 About This Book

No one individual has the knowledge and experience to adequately describe and discuss the diversity of soils across the USA. Thus, a multi-author approach was used in order to have knowledgeable scientists who are familiar with the soils in the various regions of the country assemble the information in each chapter. As such, the emphasis of and information included in each chapter reflects the authors' interest and experience, and will thus vary somewhat. All chapters, however, discuss relationships of soil properties and distribution to the climatic, geomorphic, geologic, and vegetative conditions in each region. Similarly, the impact of soil use and management on the soil resource is included to a greater or lesser extent for the LRRs included in each chapter.

The information included and level of detail varies among the chapters. This variation is often due to the author's concepts of the types of information that should be included in the chapter. The availability of data and research specific to soils in the LRR(s) included in each chapter varies, however, and details may not be included because they are unknown. Research on soil genesis and behavior in other LRRs is extensive. For these, details of properties, genesis, and use of many extensive and important soils have been omitted or generalized to keep the size of the book manageable. References are included for each chapter to enable readers to find additional details concerning properties of particular soils and/or genesis and distribution of soils in particular regions.

While the chapters generally emphasize the most extensive soils in each LRR, most also have included discussions of important soils of limited extent, important soil features, and unique soil conditions. Additional information on distribution and properties of soils for specific local areas within the USA is available from theWeb Soil Survey (Soil Survey Staff 2016a; http://websoilsurvey.nrcs.usda.gov), from the SoilWeb app (UC-Davis California Soil Resource Lab 2015; http://casoilresource.lawr.ucdavis.edu/soilweb-apps), and from Purdue University's Integrated Spatial Education Experience (ISEE) site (Schulze et al. 2015; http://isee.purdue.edu). Additional data and information on soils across the USA is available from the USDA-NRCS soils website (http://www.nrcs.usda.gov/wps/portal/nrcs/site/soils/home/).

1.4.3 The Focus of This Book

The focus of this book is the chapters on characteristics, genesis, and use of the soils in each LRR. Because not all readers may be familiar with concepts and terminology related to soil formation and soil survey, Chap. 2 (Concepts of Soil Formation and Soil Survey) provides an overview of important processes that are involved in formation of soils. This chapter also includes a section describing important concepts related to the soil survey of the USA including types of soil maps, map scale and resolution, and products of the USA soil survey.

Soil classification is not a primary focus of this book. Because it is the language of soil science, however, names of taxa from Soil Taxonomy (Soil Survey Staff 2014) are used throughout the chapters to refer to soils that occur in the various LRRs and MLRAs. Thus, a general understanding of the structure and nomenclature of Soil Taxonomy is needed to understand the information presented in each chapter. Eachtaxon is defined by certain properties, and often these properties can be inferred from the name of the taxon without knowing the specifics of criteria used for placement. Chapter 3 (Soil Properties and Classification) provides an overview of Soil Taxonomy including its structure and derivation of names of taxa that can be used to infer properties. Additional details of Soil Taxonomy can be found in the Illustrated Guide to Soil Taxonomy (Soil Survey Staff 2016b), Soil Taxonomy (Soil Survey Staff 1999), the Keys to Soil Taxonomy (Soil Survey Staff 2014), and various textbooks, e.g. Wilding et al. (1983), Fanning and Fanning (1989), Buol et al. (2011) and Schaetzl and Thompson (2015).

Where possible, soil series names were avoided in discussions of the soils in each region since series names carry no additional information and have little meaning to those unfamiliar with the series. Series names are used in instances, however, because they are the basis for organizing data and other information within the NCSS. The "Official Series Descriptions" include description of the properties of each series used in the NCSS along with its classification, description of typical landforms and parent material, and related information (Soil Survey Staff 2016c).

Chapter 4 (Soils of the USA: The Broad Perspective) provides a broad overview of the soil resources for the USA. It includes general information on distribution of soil orders and suborders across the USA and its territories as well as an overview of important soil forming factors and process for each LRR. The chapter also includes brief discussions of the climate, geology, geomorphology, and soil resources in each LRR as well as an overview of the diversity of soils and relative abundance of various taxa for each LRR and MLRA.

Chapters 5 through 17 comprise the major focus of this work, which is a discussion of properties, genesis, and use of soils in one or more of the 28 LRRs that comprise the USA and its territories (Figs. 1.2, 1.3, 1.4).

Pedologists have long recognized that soils change over time. The changes in properties related to soil formation have been related to processes operating over centuries to millennia and changes in the landscape that have occurred over geologic time. Recently, there has been wider recognition that changes in soil properties are often related to human use of soil. To understand soil change and apply that knowledge to sustainable land management, we must understand how human activities impact the soil itself and soil's interactions with the wider environment. Chapter 18 (Human Land-Use and Soil Change) provides an overview of important concepts for evaluating and understanding soil change and presents examples of soils that have undergone major changes due to agricultural management, mining, and other types of drastic disturbance.

The final chapter of this work, Chap. 19 (Future Challenges for Soil Science Research, Education, and Soil Survey in the USA) attempts to discuss the current state of soil science and pedologic research and education in the USA. The chapter also offers solutions to a few of the issues that are limiting soil science education and research. The intent of the chapter is to highlight issues facing soil science and offer one scientist's concepts of potential actions that might overcome these issues. The concepts put forward in this chapter may encourage a continuing dialog among soil scientists, policy makers, agency administrators, and the public to develop a strategy to promote advancement in the study and teaching of the soil—one of the nation's most critical non-renewable natural resources.

1.5 Summary

It is the editors' and authors' desire that readers of this book will find a useful reference and learning tool and that it will enhance their knowledge of the USA's soil resources. Soil is

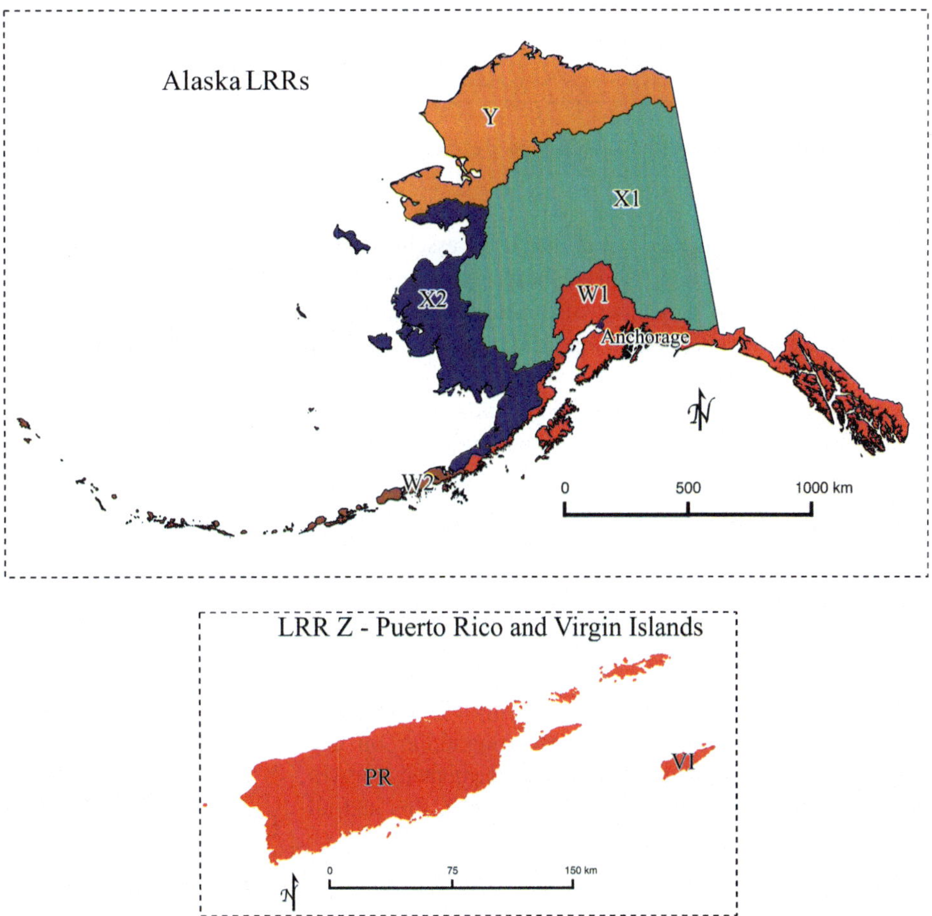

Fig. 1.3 LRRs of Alaska and the Caribbean Area

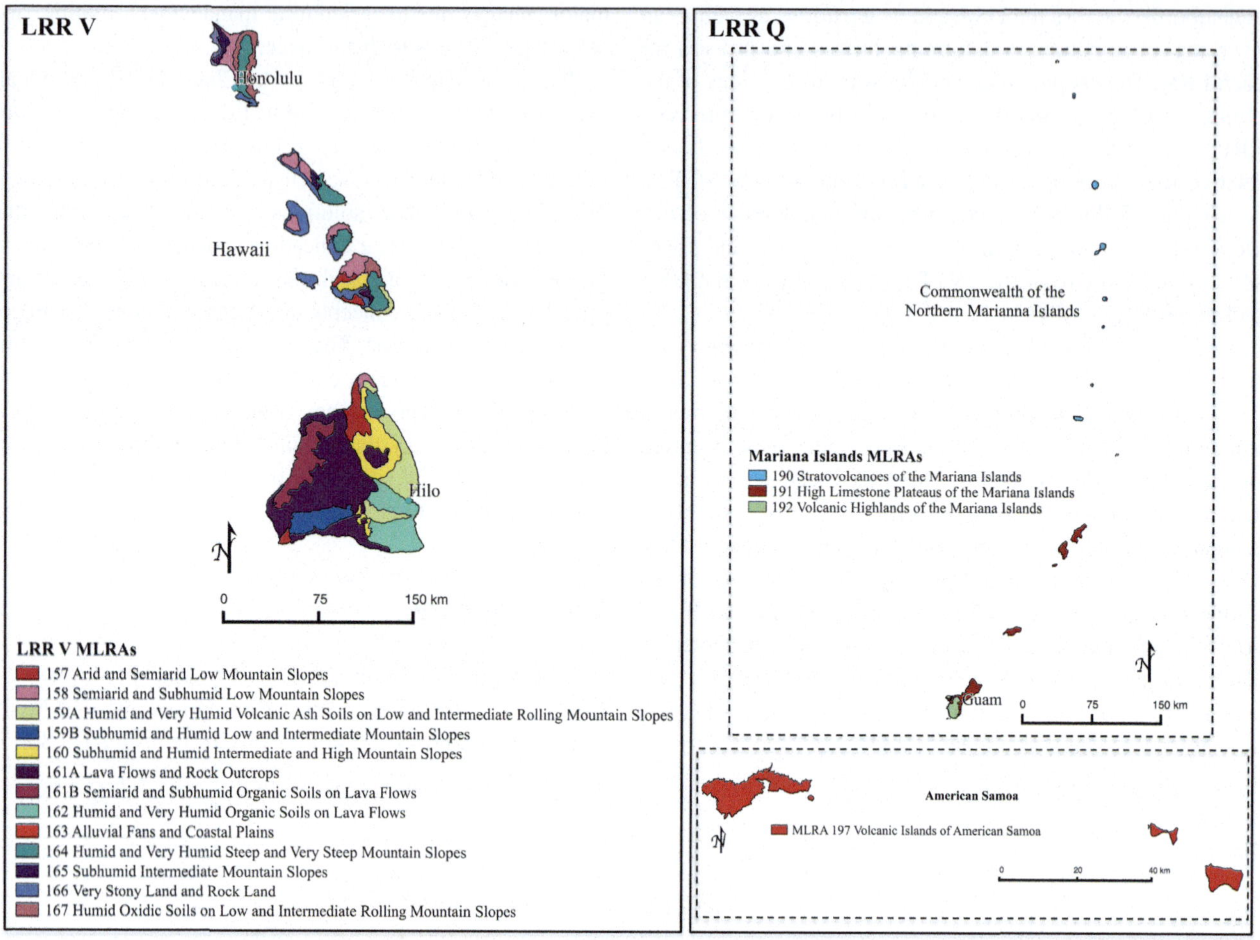

Fig. 1.4 LRRs and MLRAs of Hawaii and the Pacific Islands

the thin skin covering our planet that is an essential component of all ecosystems. Soil supports all terrestrial life forms, and performs a number of functions critical to the well-being of the global population including nutrient supply and storage for terrestrial and aquatic ecosystems, partitioning of rainfall and snow into ground and surface reservoirs, disposal and renovation of anthropogenic wastes, habit for soil organisms, and support for roads, buildings, and other infrastructure. Soils are also an important part of the global carbon cycle and potentially can sequester atmospheric carbon to reduce greenhouse gas concentrations. Soils are relativelyresilient and will perform these and other functions almost indefinitely with proper management. In contrast, improper use and management will quickly destroy this valuable non-renewable resource. Only by understanding the properties of and processes occurring in the soil, can management systems be designed and implemented that will ensure the long-term sustainability of the soil, this and all other nation's most valuable natural resource.

References

Bockheim JG (2014) Soil geography of the USA. Springer, Dordrecht

Bockheim J, Gennadiyev AN (2015) General state soil maps in the USA. Geoderma 253–254:78–89

Brevik E, Hartemink AE (2013) Soil maps of the United States of America. Soil Sci Soc Am J 77:1117–1132

Brye KR, Mersiovsky E, Hernandez L, Ward L (2013) Soils of Arkansas. Arkansas Agric Exp Stn, University of Arkansas Div of Agric, Fayetteville

Buol SW (ed) (1973) Soils of the southern states and Puerto Rico. Southern Coop. Series bull. no. 174

Buol SW, Southard RJ, Graham RC, McDaniel PA (2011) Soil genesis and classification, 6th edn. Wiley-Blackwell, Ames

Cleland DT, Freeouf JA, Keys JE, Nowacki GJ, Carpenter CA, McNab WH (2005) Ecological subregions: sections and subsections of the conterminous United States. U.S. Department of Agriculture, Forest Service, Washington

Daniels RB, Boul SW, Kleiss HJ, Ditzler CA (1999) Soil systems in North Carolina. Tech. Bull. 314, NC State University, Raleigh

Fanning DS, Fanning MCB (1989) Soil: morphology, genesis, and classification. Wiley, Hoboken

Hole FD (1976) Soils of Wisconsin. University of Wisconsin Press, Madison

Jenny H (1941) Factors of soil formation. McGraw-Hill, New York

Lal R (ed) (2016) Encyclopedia of soil science. Marcel-Dekker, New York

McNab WH, Cleland DT, Freeouf JA, Keys JE, Nowacki GJ, Carpenter CA (2005) Description of ecological subregions: sections of the conterminous United States. U.S. Department of Agriculture, Forest Service, Washington

Marbut CF (1936) Part III, soils of the United States. Atlas of American Agriculture, USDA, Washington

Montague C (1982) Soils of Montana. Bull 744, Montana Agric Exp Stn, Bozeman

Omernik JM (1987) Ecoregions of the conterminous United States. Ann Assoc Am Geogr 77:118–125

Schaetzl RJ, Thompson ML (2015) Soils: genesis and geomorphology, 2nd edn. Cambridge University Press, Cambridge

Schulze DG, Miller CC, Glotzbach RJ, Kocur LA (2015) Integrated spatial educational experience (ISEE). http://isee.purdue.edu

Springer ME, Elder JA (1980) Soils of Tennessee. Bull 596, Tenn Agric Exp Stn, Knoxville

Soil Survey Staff (1975) Soil Taxonomy: a basic system of soil classification for making and interpreting soil surveys. Handbook 436, USDA-Natural Resources Conservation Service. US Government Print Office, Washington

Soil Survey Staff (1999) Soil Taxonomy: a basic system of soil classification for making and interpreting soil surveys, 2nd ed. Handbook 436, USDA-Natural Resources Conservation Service. US Government Print Office, Washington

Soil Survey Staff (2014) Keys to Soil Taxonomy, 12th edn. USDA-Natural Resources Conservation Service, Washington

Soil Survey Staff (2016a) Web soil survey. USDA-Natural Resources Conservation Service. http://websoilsurvey.nrcs.usda.gov

Soil Survey Staff (2016b) Illustrated guide to Soil Taxonomy. USDA, Natural Resources Conservation Service. http://www.nrcs.usda.gov/wps/portal/nrcs/main/soils/survey/class/

Soil Survey Staff (2016c) Official series descriptions. USDA-Natural Resourrces Conservation Service. http://www.nrcs.usda.gov/wps/portal/nrcs/detailfull/soils/survey/class/data/?cid=nrcs142p2_053587

UC-Davis California Soil Resource Lab (2015) SoilWeb Apps. http://casoilresource.lawr.ucdavis.edu/soilweb-apps

USDA-NRCS (2006) Land resource regions and major land resource areas of the United States, the Caribbean, and the Pacific Basin. Handbook 296, USDA-Natural Resources Conservation Service. http://www.nrcs.usda.gov/wps/portal/nrcs/detail/soils/survey/geo/?cid=nrcs142p2_053624

USDA-NRCS (2016) Land resource regions and major land resource areas. In Part 649, National soil survey handbook, title 430-VI. http://www.nrcs.usda.gov/wps/portal/nrcs/detailfull/soils/ref/?cid=nrcs142p2_054242

United States Environmental Protection Agency (2003) Level III ecoregions of the conterminous United States. Revised from Omernik, 1987. USEPA National Health and Environmental Effects Research Laboratory (NHEERL). http://www.epa.gov/wed/pages/ecoregions/level_iii.htm

West LT, Wilding LP (2011) Classification of soils. In: Huang PM, Li Y, Sumner ME (eds) Handbook of soil science, properties and processes, 2nd edn. CRC Press, Boca Raton, pp 33-1–33-190

Wilding LP, Smeck NE, Hall GF (eds) (1983) Pedogenesis and Soil Taxonomy II. The soil orders. Elsevier, Amsterdam

2 Concepts of Soil Formation and Soil Survey

Robert C. Graham and Samuel J. Indorante

2.1 Soil Formation

2.1.1 Factors of Soil Formation

One of the earliest concepts in soil science is that soil formation is influenced by five factors: parent material, climate, topography, organisms, and time. These factors were first identified by V.V. Dokuchaev in the late 1800s as he inventoried soils of the Russian steppes. They were popularized in the USA by the book, *The Factors of Soil Formation*, in which Jenny (1941) sought mathematical expressions of soil formation based on the variables he referred to as cl, o, r, p, and t (climate, organisms, relief, parent material, and time). These variables provide an effective context for considering soil formation.

Humans have had considerable influence on soils for thousands of years, but this influence has vastly increased since the beginning of the twentieth century. Now extensive areas of soils worldwide have been irrevocably altered. Because of their huge impact on soils, humans have been considered as a specific agent of soil formation (Amundson and Jenny 1991) and soil change (Chap. 18).

2.1.1.1 Parent Material

The initial material from which soils form is considered the parent material. In the case of Histosols (organic soils), the parent material is plant debris, but for most soils, it is mineral matter. The parent material may be solid rock that weathers in place to form soil (Fig. 2.1), or it may be transported and deposited before having soils form in it (Fig. 2.2). Transported parent materials include those deposited by running water (alluvium), gravity (colluvium), glaciers (till), wind (loess, aeolian sand), and volcanic eruptions (tephra). In certain parts of the country, it is common to find soils with loess, tephra, or other deposits on top of soils formed from other parent materials. In these cases, the soils form in two parent materials: one below and one above (Fig. 2.3).

In any case, the parent material provides the geochemical foundation of the soil. The minerals that compose the parent material are the sources of elements that serve as nutrients for plants or precipitate as clay minerals. A parent material that lacks certain elements may limit the nutrient status of the soils that form from it as well as the kinds of minerals that can form by weathering. For example, ultramafic rock contains high levels of Fe, Mg, and Si, but very low levels of Ca, K, and Al. As a result, the soils support distinctive vegetation communities adapted to low Ca:Mg ratios and minimal K. The silicate clays that form in the soils are enriched in Mg or Fe rather than Al.

Likewise, the parent material limits the textural range of the soils derived from it. Soil textures are typically finer than the grain sizes of the original parent material because weathering reduces the size of the original grains and precipitates clay-size material, thus reducing the grain size overall. Coarse soil textures result from parent materials such as sandstones and granites (Fig. 2.4) that have sand-size quartz grains, which are resistant to weathering. Fine textures result from sediments that are already clay- and silt-rich, such as shale (Fig. 2.5), or rocks that are composed of easily weathered primary minerals, such as basalt.

2.1.1.2 Climate

The influence of climate on soil formation is largely through the combined effects of water and temperature, although wind and solar radiation also play important roles.

Water, delivered to the soil as rain, snowmelt, or fog drip, is required for weathering, biological activity, and the transport of materials through soils. The disposition of water in soils can be considered using the water balance relationship:

R.C. Graham (✉)
University of California, Riverside, CA, USA
e-mail: robert.graham@ucr.edu

S.J. Indorante
USDA-Natural Resources Conservation Service, Carbondale, IL, USA
e-mail: sam.indorante@il.usda.gov

L.T. West et al. (eds.), *The Soils of the USA*, World Soils Book Series,
DOI 10.1007/978-3-319-41870-4_2

Fig. 2.1 An Ultisol (Bethlehem series) formed from the in situ weathering of schist bedrock on the North Carolina Piedmont. The 10–60 cm depth zone is a kandic horizon and is red due the pigmentation of hematite, an iron oxide. The zone below 100 cm is saprolite (bedrock so thoroughly altered by weathering that it is soft and clay enriched). Photograph credit: John A. Kelley, Soil Scientist, USDA-NRCS

Fig. 2.2 An Entisol formed in a debris flow deposit in the San Bernardino Mountains, California. The pine needles, twigs, and other plant parts that compose the O horizon at the surface become more decomposed with depth. Photograph credit: Judith Turk

$$\text{Change in storage in the soil} = \text{Precipitation} - \text{Runoff} - \text{Evapotranspiration} - \text{Deep percolation}$$

The water stored in the soil is available for weathering reactions and biological activity, and the influx of water to fill the storage moves materials through the soil in solution and suspension.

Both runoff and deep percolation are dependent on properties of the soil itself. Runoff occurs when water is unable to infiltrate into the soil, either because all of the pores are already filled with water, or because the pores are too few, small, or unconnected to allow infiltration at a rate that can accommodate the rainfall or snowmelt. Deep percolation occurs when more water infiltrates into the whole soil volume than it can retain against the force of gravity. A soil's water-holding capacity varies with texture, such that sandy soils hold less water and have more deep percolation than clayey soils.

Fig. 2.3 An Aridisol near Quartzsite, Arizona. Desert pavement overlies a V horizon with columnar structure that formed in eolian dust (0–6 cm depth), which overlies a B horizon formed in alluvium. Photograph credit: Judith Turk

Fig. 2.4 An Entisol (Wake series) formed from weathering of granite on the North Carolina Piedmont. The soil textures above 50 cm are sandy, and below this depth is saprock (bedrock softened by weathering, but not thoroughly altered). Photograph credit: John A. Kelley, Soil Scientist, USDA-NRCS

Fig. 2.5 A Mollisol formed from the weathering of shale near Laramie, Wyoming. The A horizon is darkened by humified organic matter from the decomposition of grass roots and other plant parts. The dark, organic-rich soil material is redistributed to the 90 cm depth by burrowing animals (note the backfilled burrows). Photograph credit: R. C. Graham

Precipitation and evapotranspiration are the climatic variables in the water balance equation. The balance between the input of water (precipitation) and its loss to the atmosphere (evapotranspiration) defines the climatic effect on soil moisture. Four climate-dependent soil moisture regimes recognized by the USDA soil taxonomic system are shown in Fig. 2.6. A udic soilmoisture regime occurs mainly in the midwestern and eastern states where precipitation input is greater than evapotranspiration losses, such that deep leaching is expected. The ustic soil moisture regime is mostly in the Plains states where a moderate soil water deficit exists. The aridic soil moisture regime reflects a strong water deficit with no deep leaching and is common in the western states. The xeric soil moisture regime is unique in that it defines soils that are moist in the winter, even to the point of being strongly leached, but are dry during the summer. These soils are mostly in California and Oregon. The aquic soil moisture regime, also shown in Fig. 2.6, is more related to topographic conditions than strictly climate. Aquic soils are found in low-lying areas where the water table is high so that at some time during the year reducing conditions prevail. As a result, these soils generally have reducing conditions and concentrations of iron oxides and an accumulation of organic matter (Fig. 2.7).

Climate dictates soil temperature as well as soil moisture, and temperature affects the rates of chemical and biological reactions in soils. In general, chemical reactions increase exponentially with increasing temperature, so weathering of minerals to produce clays and iron oxides is vastly increased in warm climates (e.g., Hawaii) compared to cold climates (e.g., Alaska), assuming moisture is available. A notable exception is calcite ($CaCO_3$), a common soil mineral. Calcite is more soluble at low temperatures; thus, all else being equal, it is more likely to precipitate and be present in soils with high temperatures.

Soil biological activity is generally minimal below about 5 °C and increases rapidly to the 30–37 °C range, above

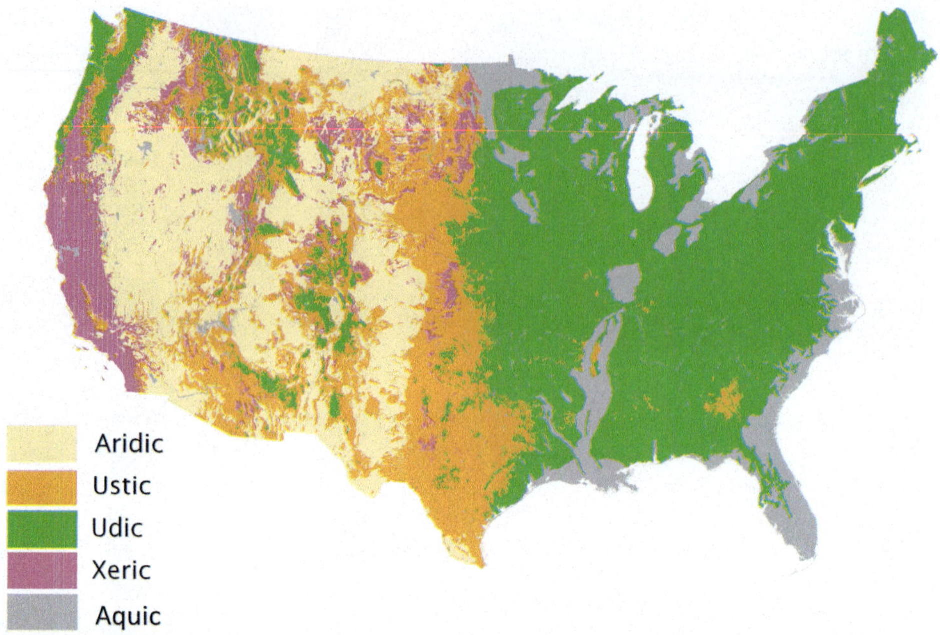

Fig. 2.6 Map of the distribution of soil moisture regimes in the contiguous USA (based on Winzeler et al. 2013)

Fig. 2.7 An Ultisol with an aquic soil moisture regime (Paxville series) formed in a poorly drained depression on a broad interstream divide on the middle Coastal Plain of North Carolina. The 43–120 cm depth is an argillic horizon with gray redox depletions and yellow–red redox concentrations. The A horizon (above the 27 cm depth) is darkened by humified organic matter. Photograph credit: R.C. Graham

which it drops precipitously. Such high temperatures are not sustained in soils, except those influenced by geothermal conditions, but very low soil temperatures are common at high latitudes and altitudes. The inhibitory effect of low temperatures on microbial activity decreases organic matter decomposition rates and results in organic matter accumulation in cold soils. Thus, soils of higher latitudes (and altitudes) generally contain more organic matter than those at low latitudes (and altitudes).

Temperatures within the upper 50 cm, or so, of soil are subject to diurnal fluctuation and are affected by variations in day-to-day weather conditions. Yet, it is the cumulative effect of soil temperature that determines the overall rates of organic matter decomposition and mineral weathering, thereby making a visible imprint on the soil morphology. Thus, soils of higher latitudes tend to be darker in color due to organic matter accumulation and soils of low latitudes tend to be redder due to enhanced weathering and iron oxide (hematite) formation (Fig. 2.1).

Wind is an aspect of climate that impacts soil formation through its role in erosion and deposition. Wind erosion occurs on dry, bare soils, such as those found on glacial outwash, playas, and post-wildfire landscapes. Of course, wind erosion can also occur on soils that are laid bare by human activities such as plowing and over grazing. Severe windstorms can uproot trees over large areas, especially when the soil has lost cohesive strength because it is saturated with water. When the trees fall over, a volume of soil is lifted up by the roots, mixing the soil horizons. Sometimes, large areas of forest are affected so that the soils are essentially plowed by this "tree throw."

Solar radiation is an important climatic factor in soil formation in that it drives photosynthesis that fixes carbon in organic matter, some of which is ultimately incorporated into the soil. Solar radiation is the overall source of heat flux to the soil.

2.1.1.3 Topography

The shape of the land's surface, or topography, influences how water flows onto and off of the soil, as well as how it moves into and through the soil. Consequently, it exerts a strong control on the balance between soil organic matter additions and decomposition, erosion and deposition, leaching and accumulation, and even oxidation and reduction. Topography is formed by depositional features, such as lava flows, moraines, alluvial fans, and dunes, and by geologic erosion, such as stream incision, glaciation, and mass wasting.

Slope shape determines the disposition of water on the landscape. On nearly level surfaces, water flow is largely vertical, into the soil, so runoff and run-on are minimal and so are erosion and deposition. Thus, nearly level land surfaces, such as stream terraces and broad hill summits, are stable locations for prolonged soil development. On the other hand, sloping land removes water from one part of the landscape and delivers it to another. A linear slope simply delivers water from upper slope positions to lower slopes, making these lower slope positions moister than one would infer from the climate. Slopes that are convex across and downslope, such as the nose of a ridge, disperse water flow and are the driest parts of the landscape (Fig. 2.8). Soils are less leached than on other landscape positions, and vegetation density is relatively low; hence, organic matter additions to the soil are also low. Slopes that are concave across and downslope, such as the head slope of a watershed, concentrate water flow so that the lower slopes of these places are the wettest parts of the landscape (Fig. 2.8). As a result, the soils tend to have the highest organic matter contents and may experience reducing conditions, at least periodically, qualifying them as having an aquic soil moisture regime or placing them in aquic subgroup classifications.

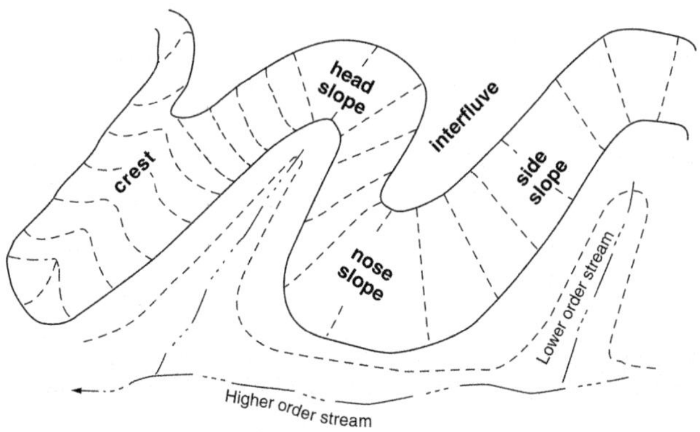

Fig. 2.8 Simplified diagram of landscape components (modified from Schoeneberger et al. 2012)

Topography also introduces localized variations in soil climate, most commonly in the form of slope aspect. In the Northern Hemisphere, south-facing slopes most directly intercept the sun's rays and, consequently, are warmer and drier than north-facing slopes. Because they are cooler and moister, soils on north-facing slopes tend to contain more organic matter than those on other slope orientations.

2.1.1.4 Organisms

Soils provide habitats for a multitude of organisms, and these plants, animals, and microbes strongly impact the soils as well. Plants are primary producers, acquiring carbon from the atmosphere by photosynthesis. They also scavenge the soil for mineral-derived elements (calcium, magnesium, potassium, phosphorus, etc.), and some have symbiotic relationships with bacteria that obtain nitrogen from the atmosphere. All of the elements that plants incorporate into their biomass are added to the soil via biocycling, so soil A horizons become enriched in carbon, nitrogen, and other nutrient ions. Plant roots also have direct impacts on soil physical properties as they create pores and promote soil aggregation.

Burrowing animals mix the soil, incorporating organic matter, homogenizing soil horizons, and converting parent material into soil (Fig. 2.5). Burrowing animals, including earthworms, ants, gophers, and ground squirrels, create porosity for preferential water flow and root growth. Certain animals create distinctive soil structures, such as earthworm and cicada casts.

The integrated effects of organisms can be observed by comparing surface soil horizons developed under coniferous forest with those developed under grass. Much of the organic matter in coniferous forest soils is derived from foliage that falls from the trees. This material accumulates on the soil surface as an O horizon (Fig. 2.2) and, because it is largely unpalatable to earthworms and burrowing rodents, it is only slowly mixed into the underlying mineral soil. As a result, A horizons are relatively thin. In contrast, grassland soils have virtually no O horizon, but a profusion of fine roots in the near-surface zone contributes abundant organic matter. Roots and above ground plant material serve as food for gophers and other burrowing rodents, which, together with earthworms, mix the soil and create thick, dark A horizons (Fig. 2.5).

Aside from its essential role in the decomposition of organic matter, microbial activity alters soil morphology by reducing iron oxides in water-saturated soils where free oxygen is absent. The yellowish and reddish colors of iron oxides are replaced by the background gray colors of the silicate minerals as the iron is reduced and solubilized (Fig. 2.7).

2.1.1.5 Time

The environmental factors of climate, organisms, topography, and parent material must interact over a period of time to produce soil. The longer these factors are able to act together, the more developed and vertically differentiated and distinctive the soil will become. Thus, the stability of the soil landscape governs the duration of soil formation, and the character of the soil reflects how long the other environmental factors have exerted their influence.

Most landscapes in the USA have been stable (e.g., neither extensively eroded or deposited upon) for much less than a million years. The youngest soils are those formed on recently deposited sediments, such as floodplain alluvium, or very steep slopes with ongoing erosion. The oldest soils are those on long stable geomorphic features, such as alluvial terraces and lava flows. Soils that are tens of thousands of years old and older have generally experienced pronounced changes in climate and organisms over the course of their formation.

Glaciation has affected much of the northern USA and farther south in alpine regions such as the Rocky Mountains, the Cascade Range, and the Sierra Nevada. Glaciers wiped away previously formed soils and deposited fresh parent material (known as drift). In a sense, the clock for soil formation on glaciated landscapes was reset to zero corresponding to the most recent glacial deposits. The last major period of glaciation ended 11,700 years ago (the end of the Pleistocene Epoch), so many soils of high latitudes and altitudes are younger than this.

2.1.2 General Soil Processes

The environmental conditions defined by the five soil-forming factors generate processes that determine how soils function and develop. These processes can be generally categorized as those that cause additions, losses, translocations, and transformations within soils (Simonson 1959). The balance among these processes determines the soil behavior and observable characteristics.

2.1.2.1 Additions

The most notable addition to most soils is organic matter derived from plant parts (Fig. 2.2). Plant litter that falls on the surface of soils is broken down by insects, earthworms, fungi, and other decomposers and mixed into the mineral soil by burrowing animals, including ants, earthworms, and gophers. Organic matter is also added directly to the subsurface by plant roots, which can make up a large proportion of the total plant biomass. Roots are an especially large contributor of organic matter to grassland soils. If the rate of organic matter loss by decomposition is less than the rate of addition, then organic matter will accumulate in the soil, and a thick, dark A horizon will result (Figs. 2.5, 2.7 and 3.2), or in very cold or wet soils, the organic matter will accumulate to form thick O horizons.

Dust is a ubiquitous addition to soils in arid regions. It blows off of dry lakebeds and intermittent stream channels, incorporating very fine sand- and silt-size particles as well as various salts into atmospheric suspension. These materials are blown onto upland parts of the landscape where they are trapped by rough surfaces and incorporated into the soil surface to form a V horizon (Fig. 2.3). Dust originates from any bare surface with fine earth materials. While these conditions are common in deserts, they are also common on the margins of active glaciation. Dust is blown between continents, so, to some degree, all soils on earth receive at least a background input of dust.

Volcanic ash is another material added from the atmosphere. Thin deposits are incorporated into the underlying soil by mixing processes, but thick deposits bury existing soils and soil formation starts anew in the fresh volcanic deposit. In the USA, the influence of volcanic ash is strongest in the Pacific Northwest, Alaska, Hawaii, and the Pacific Islands, but volcanic ash from Cascade Range volcanoes has been deposited at least halfway across the North American continent.

Alluvial deposits added to floodplains build the soils upward and, again, if thin enough, are mixed into the existing soils. These alluvial additions are common on floodplains of major rivers across the country.

2.1.2.2 Losses

The most obvious process leading to loss from a soil is erosion, a process that can occur in any part of the country. Erosion can be by either water or wind, but both processes are facilitated by bare, poorly aggregated soil. Erosion removes the soil surface materials preferentially, and these surface horizons are usually the most enriched in organic matter and nutrients. Erosion not only reduces the overall soil thickness, but specifically removes the soil nutrient supply.

As water percolates through the soil, it carries materials in suspension (colloids) and in solution (ions, dissolved organic matter). If the water balance is such to allow deep percolation, these waterborne materials are carried down into groundwater or otherwise out of the soil zone. This form of loss from soils is most common in the humid regions of the USA, where precipitation greatly exceeds evapotranspiration. Ultimately, this process is depleting the soil of weathering products and is likely to result in the formation of Ultisols.

A less visible loss occurs during soil organic matter decomposition. As microbes decompose organic matter, they respire CO_2 gas. Thus, the loss of carbon from the soil during soil organic matter decomposition occurs through the gas phase and can be substantial on an annual basis—on the order of 200–1000 g m^{-2}, depending on the ecosystem (Raich and Potter 1995). This loss through decomposition is so great in warm, humid regions that organic matter accumulation in soils is minimal. Conversely, in cool regions, the carbon loss via CO_2 respiration is less than the rate of organic matter additions, so the soils become enriched in stored organic matter. When soils are saturated with water, organic matter decomposes slowly by anaerobic processes, and methane (CH_4) is released, not CO_2. The evolution of carbon as gas from decomposing organic matter represents not only a loss from the soil, but a very significant addition of greenhouse gases to the atmosphere. More carbon is stored in soils as organic matter than is contained in the atmosphere or vegetation, so this particular loss from soils is of special interest to climate change modelers.

2.1.2.3 Translocations

Translocations of materials from one part of the soil to another can result in distinctive soil horizons. Materials are commonly moved by water as it percolates through the soil. The zone from which materials are removed is termed the eluvial zone and the zone into which materials are translocated is known as the illuvial zone. Some of the commonly translocated materials and the horizons they form are addressed below.

Clay

Soils are composed of variously sized mineral particles, but the smallest, the clay-sized (<2 µm diameter) particles are most easily moved through the soil matrix. These colloidal particles can be physically dislodged by flowing water, but they are most effectively mobilized for transport if they are chemically dispersed. Chemical dispersion means that the clays electrostatically repel each other so they remain thoroughly suspended in the soil solution. This dispersion is enhanced with increasing pH, Na on the cation exchange sites, dilute soil solutions, and dissolved organic compounds.

Once clay is in suspension, it moves with the water as it percolates through the soil. The clay is deposited when the water stops moving in a lower part of the soil. Often, the water moves preferentially through macropores, such as root channels or cracks between soil structural units. As the suspension flows down through these pores, capillary forces pull the water into the matrix of the pore walls. When this happens, the clay is filtered out and remains as a "clay film" or "clay lining" on the pore wall surface. Clay can move through the micropores of the soil matrix as well, but just not as far or as fast. Over time, enough translocated clay accumulates in the subsoil to form a distinct horizon labeled as a Bt horizon or, according to Soil Taxonomy, an argillic, natric, or kandic horizon (Figs. 2.1, 2.7). These horizons are common in landscapes that have been stable long enough for significant amounts of clay to accumulate, which is often on the order of thousands of years. Considering that in many soils the clay must be produced by mineral weathering before it can be moved and that mineral weathering is much slower at low temperatures, it is uncommon to find argillic horizons in soils less than ten thousand years old or in very cold climates.

Fe- and Al-Humus Complexes

In certain soils, organic compounds are leached from vegetation and organic material on the soil surface (O horizons). As these dissolved organic molecules move through the soil, they act as chelates to bind with Fe and Al cations, strip them from minerals in the upper part of the soil (E horizon), and transport them in solution to a lower zone. As the organic molecules move through the soil, they pick up more Fe and Al until all their chelate sites are full, at which point they precipitate. The zone in which they precipitate and accumulate takes on a dark reddish-brown color and is known as a Bhs horizon, or a spodic horizon (Fig. 2.9. The organic matter in these horizons undergoes decomposition, losing carbon through microbial respiration, and releasing the Fe and Al to reprecipitate as oxides. Subsequent percolating dissolved chelates become adsorbed on these oxides, furthering the accumulation of organic-metal complexes.

This process operates best in sandy soils, through which water leaches readily. While most prevalent in the northern latitudes or high elevations, it also operates in soils with a fluctuating water table even in warm climates. Soils with spodic horizons are common in the northeastern states, the northern midwestern states, the Pacific Northwest, and along the Atlantic and Gulf coasts in the southeastern USA.

Calcite

As water percolates through the soil, CO_2 from microbial respiration and Ca from mineral weathering, rainfall, dust input, and other sources are dissolved in it. Under certain conditions, the dissolved carbonate and Ca will precipitate as calcite ($CaCO_3$). The most common cause of precipitation is concentration of the solution by evapotranspiration. In arid regions, rainfall is insufficient to cause deep leaching. Instead, percolating water carrying dissolved carbonate and Ca ions only reaches the subsoil. As the soil solution dries, calcite precipitates. As this process is repeated over thousands of years, a distinct subsoil zone of calcite accumulation forms, known as a Bk horizon, or calcic horizon (Figs. 2.10, 3.2). In its initial stages of formation, the calcite occurs as isolated white masses or filaments or as concentrations on the undersides of gravel. With the passage of time for more accumulation, the whole horizon becomes white with powdery calcite (Bkk horizon), and with even more

Fig. 2.9 A Spodosol (Leon series) formed in sandy marine sediments on the South Carolina coast at Baruch North Island Reserve. Chelates leached from the O horizon (upper 10 cm) have stripped iron and aluminum from the E horizon (10–40 cm depth). The metal-organic complexes are precipitated in the spodic horizon (50–70 cm). Photograph credit: John A. Kelley, Soil Scientist, USDA-NRCS

Fig. 2.10 An Aridisol (Rotura series) near Las Cruces, New Mexico. An argillic horizon (30–70 cm depth) and a calcic horizon (70–110 cm) overlie a petrocalcic horizon (top at 110 cm). Photograph credit: Curtis Monger

time (hundreds of thousands of years), the horizon becomes cemented by calcite so that it is rock hard. At this stage, it is indicated as a Bkkm, or petrocalcic, horizon (Fig. 2.10).

In arid regions, calcite is often added to the soil surface as dust, blowing in from other parts of the landscape. This calcite can be dissolved by infiltrating water and reprecipitated in the subsoil as described above.

Calcium carbonate can be precipitated in another way in moist soils. If CO_2 is lost as a gas from the soil solution, calcite will precipitate, much as it does to form stalactites in caves. This can happen when CO_2 is degassed from a high water table in the soil. A calcic horizon will form in the zone of the water table. These are the kinds of calcic horizons that can occur in the more humid parts of the country.

Soluble Salts

In arid regions, the amount of water supplied by rainfall is much less than can be lost by evapotranspiration. As a result, leaching is insufficient to remove soluble salts from the soils. Furthermore, in some basins, capillary rise of groundwater reaches the surface and evaporation concentrates the dissolved salts so they precipitate as a fine powder or a crust on the soil surface (Fig. 2.11).

The ultimate source of the salts is the release of ions during rock weathering, often in more humid regions adjacent to the deserts. In the western US, mountain ranges receive much more precipitation than adjacent deserts. The dissolved ions are transported in solution either into groundwater or in surface streams, both of which flow to the topographic low position, e.g., a basin. Many basins in the western USA are dry lakebeds (playas) that were filled with water during the much wetter, cooler climate of the Pleistocene. The dissolved ions that reach the basins are precipitated as salts as the water dries, often from capillary rise to the surface. Salts on the surface in and around playas are blown with dust onto upland areas, such as alluvial fans and mountains, where infiltrating rainwater carries them into, but not out of, the soils. Over thousands of years, the salts accumulate in the subsoils of these upland soils. Sodium chloride is a common salt, but

Fig. 2.11 A salt crust on the surface of Black Rock Playa near Gerlach, Nevada. Note that the salt crust is only a few millimeters thick and can be scraped away to reveal moist soil material underneath

sulfates, nitrates, and carbonates are also common. When salts are sufficiently concentrated in the soil, the zone in which they are concentrated is called a salic horizon.

Amorphous Silica

Silica is released during the weathering of silicate minerals and volcanic ash. Under conditions of limited leaching, the silica is transported to the subsoil, but not leached entirely from the soil. Over thousands, or more often tens of thousands, of years, the silica precipitates and accumulates as opaline silica, cementing the subsoil to form a rock-hard duripan.

Mixing

While substances translocated by water can accumulate to form soil horizons, there are physical mixing processes in soils that can destroy or prevent those horizons from forming, while forming other types of horizons. Soils can be mixed by animals and plants (bioturbation), freezing and thawing (cryoturbation), and shrink–swell activity (pedoturbation).

Burrowing animals are very effective at mixing the soil. Earthworms, ants, termites, and crayfish are some invertebrates that burrow and move material around within the soil. Gophers and ground squirrels are mammals that move a large amount of soil during their activities. These animals often transport organic matter or A horizon material to some depth, thereby enriching the subsoil with organic matter (Fig. 2.5). Trees can mix the soil when they fall and their roots pull up a volume of soil material.

The expansion and contraction forces associated with the freezing and thawing of soil water mix the surface and subsurface materials of Gelisols into convoluted patterns. Likewise, clayey soils with a large component of swelling-type clays (smectite) can undergo on the order of 10–15 % volume change as they swell when wet and shrink when dry. The expansion–contraction forces involved, together with material falling into cracks that are open to the surface, are effective at mixing these Vertisols.

2.1.2.4 Transformation

During the course of soil formation, some materials are altered but not moved much within the soil. This is the case with mineral weathering. Weatherable minerals, such as feldspars, are often altered to clay minerals while still maintaining their original shape. Subsoil bedrock can be weathered such that its porosity increases and its hardness decreases, while its appearance as rock is maintained. The soft bedrock that results is called saprock (Fig. 2.4) or, if more thoroughly weathered, saprolite (Fig. 2.1).

Soil fabric can be altered by shrink–swell processes to produce distinct horizons of soil structure (Fig. 2.3). Soil color is altered by the production of iron oxides from the weathering of iron-bearing silicate minerals and by the concentrating effects of redox processes. When a soil is saturated with water, iron oxides are chemically reduced and dissolved, but they will reprecipitate in localized oxidized zones. The result is a matrix with gray zones (iron oxides reduced and removed) and reddish or orangish zones where iron oxides precipitated, often around root channels or sandy lenses (Fig. 2.12). Decomposing organic matter, as in soil O horizons, is altered in situ by microorganisms, converting plant parts to humus (Fig. 2.2).

2.2 Soil Survey

The imprint of the soil-forming factors and processes is reflected in the soils that occur on landscapes. The goal of soil survey is to model and record this imprint. The emphasis is on displaying the geographic distribution of soils and making predictions about the soils' behavior (Soil Survey Staff 1951; Soil Survey Division Staff 1993). Soil distributions are commonly displayed in hard copies or digital copies of soil maps, with associated soil descriptions, map

Fig. 2.12 Intact subsoil material from a soil formed in a wetland in Owens Valley, California. The gray matrix was depleted of iron oxides by reducing conditions, while iron oxides precipitated around the oxygenated root channel macropores

unit descriptions, and tabular data that can be used to make predictions about soil behavior. Other forms of presentation include hand-drawn and digital block diagrams that depict soil-landscape relationships and numerous forms of 3-D soil maps that can be generated as a combination of soil maps and digital elevation models.

2.2.1 Soils, Landscapes, and Soil-Landscapes—The Fundamentals of Soil Mapping and Soil Survey

Soils are landscapes as well as profiles (Soil Survey Staff 1951, pp. 5–8; Soil Survey Division Staff 1993, pp. 9–11), and an understanding of both are needed in mapping and depicting the geographic distribution of soils. For the purpose of this chapter, soil is defined as—the unconsolidated mineral or organic matter on the surface of the earth that has been subjected to and shows effects of genetic and environmental factors of climate, macro- and microorganisms, conditioned by relief, acting on parent material over a period of time (Soil Science Society of America 1997). The effects of the genetic and environmental factors are reflected in a soil profile, which is a vertical cut through the soil. A soil profile description (i.e., pedon description) is a record of the soil horizons and the soil properties associated with each horizon. The physical and chemical properties of the soil horizons and the occurrences of the soil horizons are used to classify and name the soils in *Soil Taxonomy* (Soil Survey Staff 1999). Applying *Soil Taxonomy* (Soil Survey Staff 1999) to label soil polygons in soil survey will be discussed in the next section of this chapter.

A landscape is a portion of the land surface that the eye can comprehend in a single view and is a collection of landforms (Ruhe 1969; Peterson 1981). The link between soils and landscapes are soil-landscape units. A soil-landscape unit can be thought of as a landscape unit (landscape, landform, or landform component) modified by one or more of the soil-forming factors (Soil Survey Staff 1999). Within a soil-landscape unit, the five factors of soil formation interact in a distinct manner, and as a result, areas of a soil-landscape unit have a relatively homogenous soil pattern (Soil Survey Staff 1999).

A soil surveyor observes and maps a geographic pattern of soils by grouping soils with similar genesis and separating soils where there is a change in one or more of the soil-forming factors. The soil groupings are shown as delineations (i.e., polygons) on a base map which is typically an aerial photograph or topographic map. There are numerous approaches to accomplishing the mapping and labeling of soil geographic patterns as polygons at various scales. The most common soil-landscape approaches used in US soil survey are presented in detail by Milne (1936), Jenny (1941), Hudson (1990, 1992), and Chap. 5 of *Soil Taxonomy* (Soil Survey Staff 1999).

2.2.2 Soil Survey: Scale, Labels, and Maps

Labeling soil map units (e.g., soil-landscape units) at the various scales is accomplished primarily by using the taxa within *Soil Taxonomy* (Soil Survey Staff 1999). The taxa for Soil Taxonomy are presented in Table 2.1. Chapter 3 of this book presents a more detailed coverage of soil properties, soil classification, and *Soil Taxonomy*.

Table 2.1 Soil classification taxa levels and the complete taxa for a phase of the Fayette Soil Series

Taxa	Example	Concept of group
Order	Alfisol	Soil-forming processes as indicated by the presence or absence of major horizons
Suborder	Udalf	May indicate soil features such as moisture condition, property of parent material or vegetation
Great Group	Hapludalf	Formulated by adding another syllable in front of the suborder to provide more information about the soil properties
Subgroup	Typic Hapludalf	Modifies the great group. Depicts a normal soil condition or special feature
Family	fine-loamy, mixed, mesic Typic Hapludalf	Soil properties such as texture, mineralogy, and temperature
Series	Fayette	Soils with very similar profiles
Type	Fayette silt loam	A subdivision of the soil series based on the texture of the surface soil
Phase	Fayette silt loam, nearly level	Groupings by soil features (e.g., rock fragments, slope) created to serve specific purposes of soil surveys

The Fayette soil series is a fine-loamy, mixed, mesic Typic Halpudalf
Refer to the Soil Survey Manual (Soil Survey Staff 1993) and Soil Taxonomy (Soil Survey Staff 1999) for detailed discussion

A map unit is a collection of areas defined and named the same in terms of their soil components (e.g., taxa) or miscellaneous areas or both. Each individual demarcation of a map unit on a map is a delineation. The three most common kinds of map units in soil surveys are as follows: *consociations, complexes*, and *associations*. In consociations, map units are dominated by a single taxon, and typically, the map unit is named for the dominant taxon, often a soil series. Complexes consist of two or more taxa that occur in a regularly repeating pattern, but cannot be separated at the scale mapped. Associations consist of two or more taxa that could be mapped separately, but it is deemed not necessary to do so because their occurrences are distinct and obvious within each delineation. Complexes are used at scales larger than about 1:24,000 (more detailed coverage), and associations are used at scales smaller than about 1:24,000 (broader coverage).

In all soil surveys, virtually every delineation includes areas of soil components (taxa) or miscellaneous areas that are not identified in the name of the map units. This emphasizes the importance of the map unit description that is included in almost all large- and small-scale soil maps. The map unit description gives more detailed information on the occurrence and distribution of the named soils (major components) in the map unit label, and the occurrence and distribution of the soils that are not named (minor components) in the map unit label. Map unit design, labeling, and description are all important steps in soil survey.

Mapping and labeling soil variability and soil survey are scale dependent (Table 2.2). The US General Soil Map is the smallest scale soil map listed in Table 2.2, and it is labeled with the soil order. As the map scale becomes larger, there is increasing specificity in the taxon or taxa used to name map units. The Digital General Soil Map of the USA or STATSGO2 is a broad-based inventory of soils and non-soil areas that occur in a repeatable pattern on the landscape and that can be cartographically shown at the scales listed in Table 2.2. Map units are named for the dominant soil order, which is the taxon also commonly used for naming map units in regional soil maps such as land resource region (LRR) and major land resource region (MLRA) maps (USDA-NRCS 2006).

State and county general soil maps are commonly produced and displayed at scales ranging from 1:100,000 to 1:500,000. These levels of mapping are still considered small scale and are designed for broad planning, and management uses covering state, regional, and multistate areas. Map units are commonly labeled with the dominant soil order, suborder, or great group or by the most common soil series that occurs in the unit. In the USA, detailed soil surveys are commonly produced at large and intermediate scales ranging from 1:12,000 in areas of intensive land use to 1:24,000 in areas of less intensive land use, and the map units are most commonly labeled by the dominant soil series, soil type, or soil phase consociations or associations.

The detail of a soil survey is primarily a function of the size of the survey area, the complexity of the soil-landscape, the detail required for the intended use of the soil survey, and the ability of the soil scientist to consistently identify the map units through the application of the available knowledge and tools within the constraints of cost and time (Soil Survey Staff 1999). It should also be emphasized that the assignment of taxonomic names, such as the name of a soil order or the name of a soil series, to label a map unit means that if we examine various delineations of the map unit, we expect most locations within the delineations to meet the criteria of the taxon or taxa (Holmgren 1988).

2.2.3 Soil Survey Products

In the USA, the most common soil map unit label is the soil series. The soil series is a foundational concept in soil survey. There are two primary definitions for soil series. Historically, it is defined as a group of soils having the same genetic horizons, that is, horizons that result from the same soil-forming factors and processes (Kellogg 1956). The

Table 2.2 Soil maps, soil map scales, and soil map labels

Map	Approximate scale range	Common taxonomic labels
US General Soil Map and Regional Soil Maps	1:7,500,000–1:3,500,000	Order
Regional Soil Maps and State General Soil Maps	1:1,500,000–1:500,000	Order, suborder, great group, association of soil series
Detailed Soil Survey General Soil Map (e.g., county)	1:250,000–1:63,360	Suborder, great group, associations of soil series
Detailed Soil Survey (Less Intensive Land Use)	1:63,360–1:31,860	Soil series, soil type, soil phase consociations and complexes
Detailed Soil Survey (Intensive Land Use)	1:31,860–1:12,000	Soil series, soil type, soil phase consociations, and complexes

more modern definition describes the series as a group of soils that have horizons similar in arrangement and in differentiating characteristics (Soil Survey Division Staff 1993, p. 20). The first definition has been widely used in the labeling of soil maps with soil series or other taxonomic labels (Table 2.2, e.g., order and suborder). Following are examples of common soil survey products using the various taxonomic labels.

2.2.3.1 General Soil Region/Association Maps

The US General Soil Region Map (Fig. 2.13) and the Illinois General Soil Region Map (Fig. 2.14) are considered schematic soil maps. Schematic soil maps are made by using many sources of information such as climate, vegetation, geology, landforms, detailed soil maps, and other factors related to soil (Soil Survey Division Staff 1993; Soil Survey Staff 1999). The labels for these General Soil Region/Association Maps use the dominant soil order.

Figure 2.15 shows a General Soil Association Map of Perry County, Illinois (Grantham and Indorante 1988). Associations of soil series are used to label this map, and each association has a distinctive pattern of soils, relief, and drainage. Each association is a unique natural landscape. Typically, an association consists of one or more major soil and some minor soils. It is named for the major soils. The soils making up one association can occur in another, but in a different pattern. The general soil map can be used to compare the suitability of large areas for general land uses. Areas of suitable soils can be identified on the map. Likewise, areas where the soils are not suitable can be identified. Because of its small scale, the map is not suitable for planning the management of a specific area. The soils in any one association differ from place to place in slope, soil depth, drainage, parent material, climate and microclimate, and other characteristics. These all affect management.

2.2.3.2 Detailed Soil Maps

The detailed soil map in Fig. 2.16 includes the Homen–Hickory–Bunkum Association, the Marine–Stoy–Pierron Association, and the Flood Plain Soils Association (Fig. 2.15). The legend for this soil map is presented in Table 2.3. Each map unit delineation on the detailed map

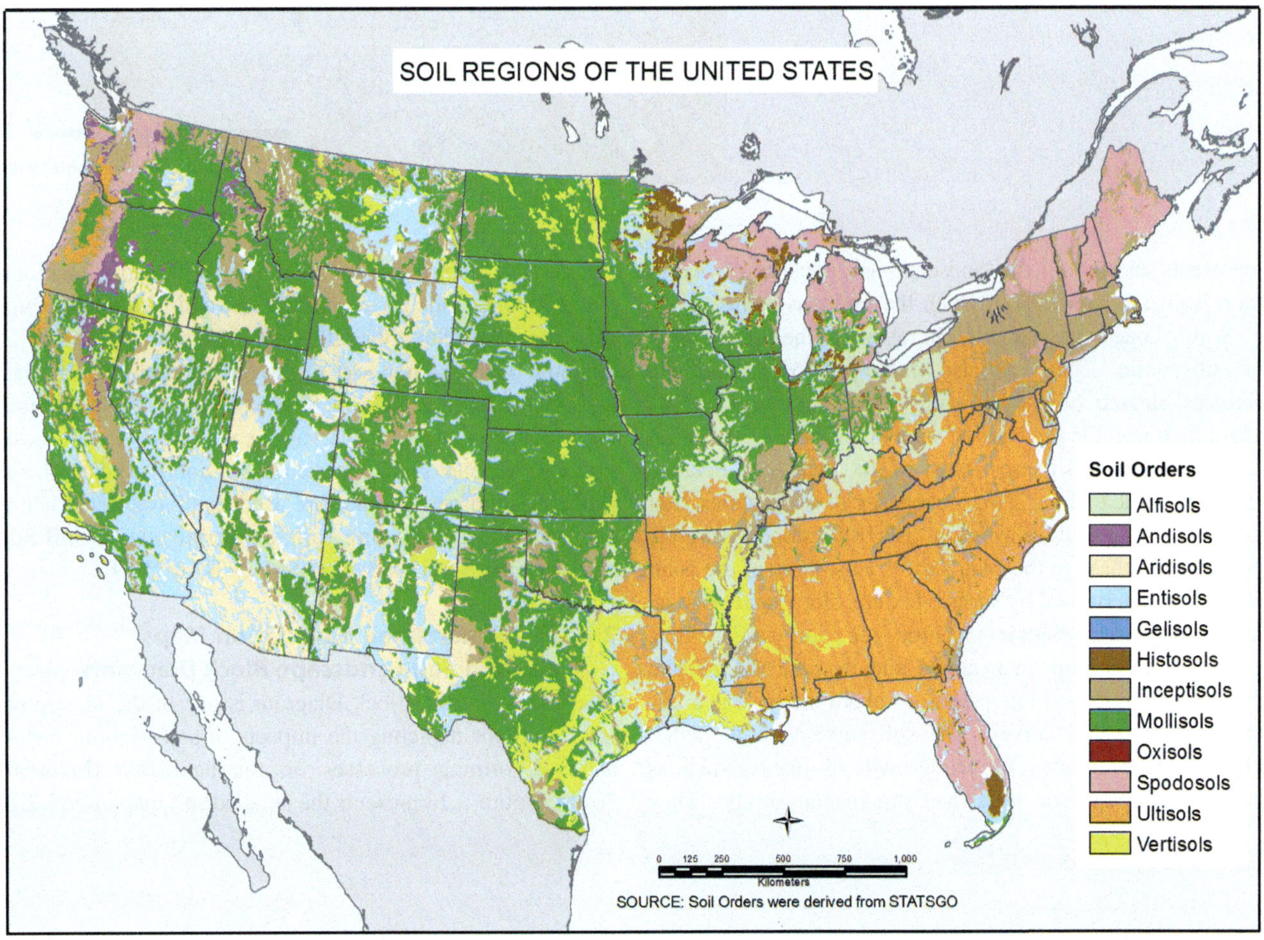

Fig. 2.13 A General Soil Map of Soil Regions of the contiguous USA. Soil orders are the taxonomic labels (Table 2.2) used in this map

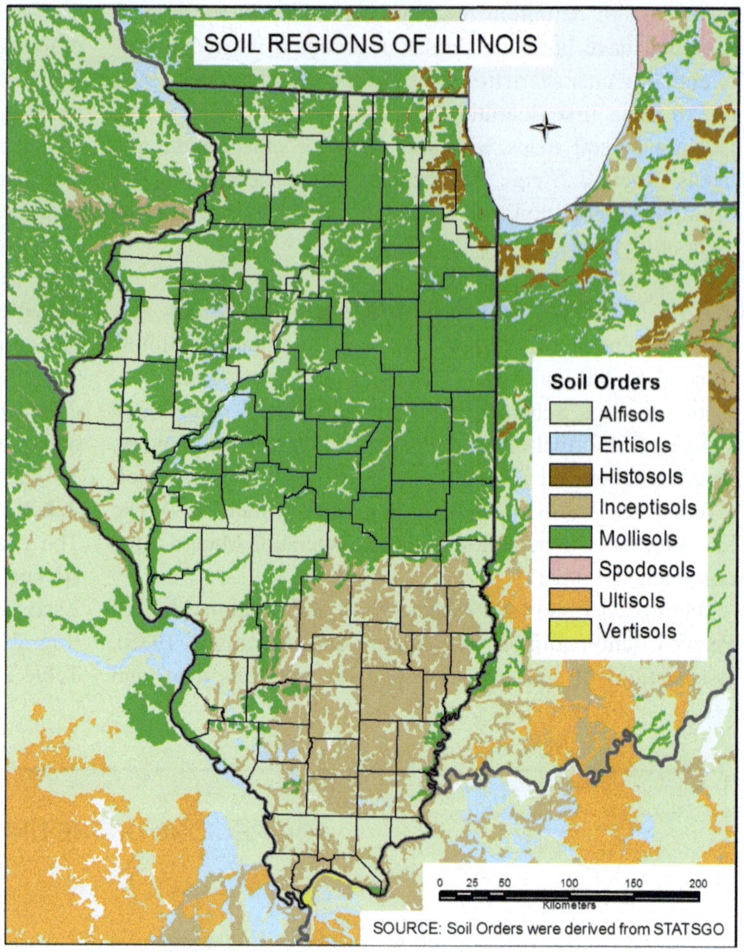

Fig. 2.14 A General Soil Map of Soil Regions of Illinois. The soil order is the taxonomic level of labels (Table 2.2) used in this map

represents an area on the landscape and consists of one or more soils (soil series) for which the unit was named.

Soils of one series can differ in texture of the surface layer or substratum. They can also differ in slope, stoniness, wetness, degree of erosion, and other characteristics that affect their use. On the basis of such differences, a soil series is divided into soil phases. Most of the areas on a detailed soil map, and in Fig. 2.16, are phases of soils series. For example, "Blair silty clay loam, 10–18 % slopes, severely eroded" is a phase of the Blair series. This phase of the Blair series was determined by surface texture, slope, and erosion.

A soil survey is both an inventory (i.e., soil maps) and an evaluation of the soils in an area. Using detailed soil maps to make predictions and interpretations about the soils is a key component of soil survey. The soil surveys in the USA include interpretations for the growth of plants, such as crops, forage species, trees, and ornamental shrubs. They also include interpretations for urban, rural, and recreational development and for conservation and wildlife habitat planning (Soil Survey Staff 1999). Figure 2.17 shows an example of a common soil survey interpretation, corn yield (*Zea mays* L.), for a farm field in Perry County, Illinois. Historical yield data on the various soil types were used to produce the map. Field experience and collected data on soil properties and performance are used as a basis in predicting soil behavior and creating various interpretations (Soil Survey Division Staff 1993; Soil Survey Staff 1999).

2.2.3.3 General Soil Association Maps and Soil-Landscape Block Diagrams

The soil-landscape Block Diagram is one of the most powerful tools for depicting the impacts of soil-forming factors and soil-forming processes on the landscape (Indorante 2011). Figure 2.18 presents the detailed soil map in Fig. 2.16

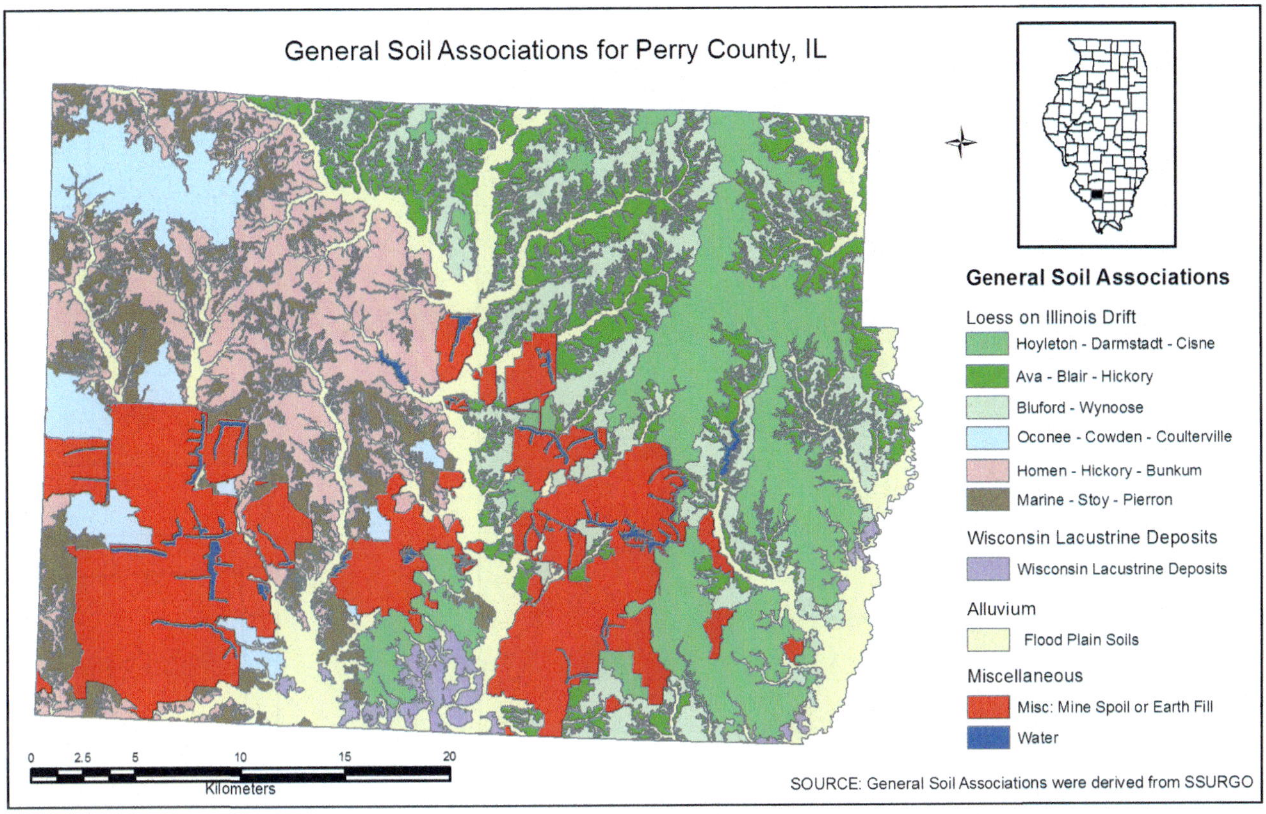

Fig. 2.15 General Soil Associations for Perry County, Illinois. Associations of soil series are the taxonomic labels (Table 2.2) used in this map. The associations are also grouped by parent material

as a General Soil Association Map and as a soil-landscape Block Diagram. This block diagram emphasizes the impact of parent material and topography on soil distribution. Even though parent material and topography are the two soil-forming factors that are emphasized in this diagram, climate and microclimate can be inferred by topographic position, and time can be inferred from parent material (e.g., alluvium as a relatively younger parent material). Soil-landscape block diagrams are available in many published (hard copy) and digital soil surveys or are available at http://www.nrcs.usda.gov/wps/portal/nrcs/detail/soils/survey/?cid=nrcs142p2_05431.

2.3 Chapter Summary

Soil formation reflects the impact of environmental conditions on the landscape. The soil-forming factors of topography, climate, and biology produce a variety of processes that imprint soil parent materials with distinct morphological and chemical properties. The expression of this imprinting is enhanced as the processes act over longer periods of time. Ultimately, the types of soil horizons that form and their degree of expression record the influences of the environments in which they formed and the consequent processes that have operated within them. Soil surveys are designed to

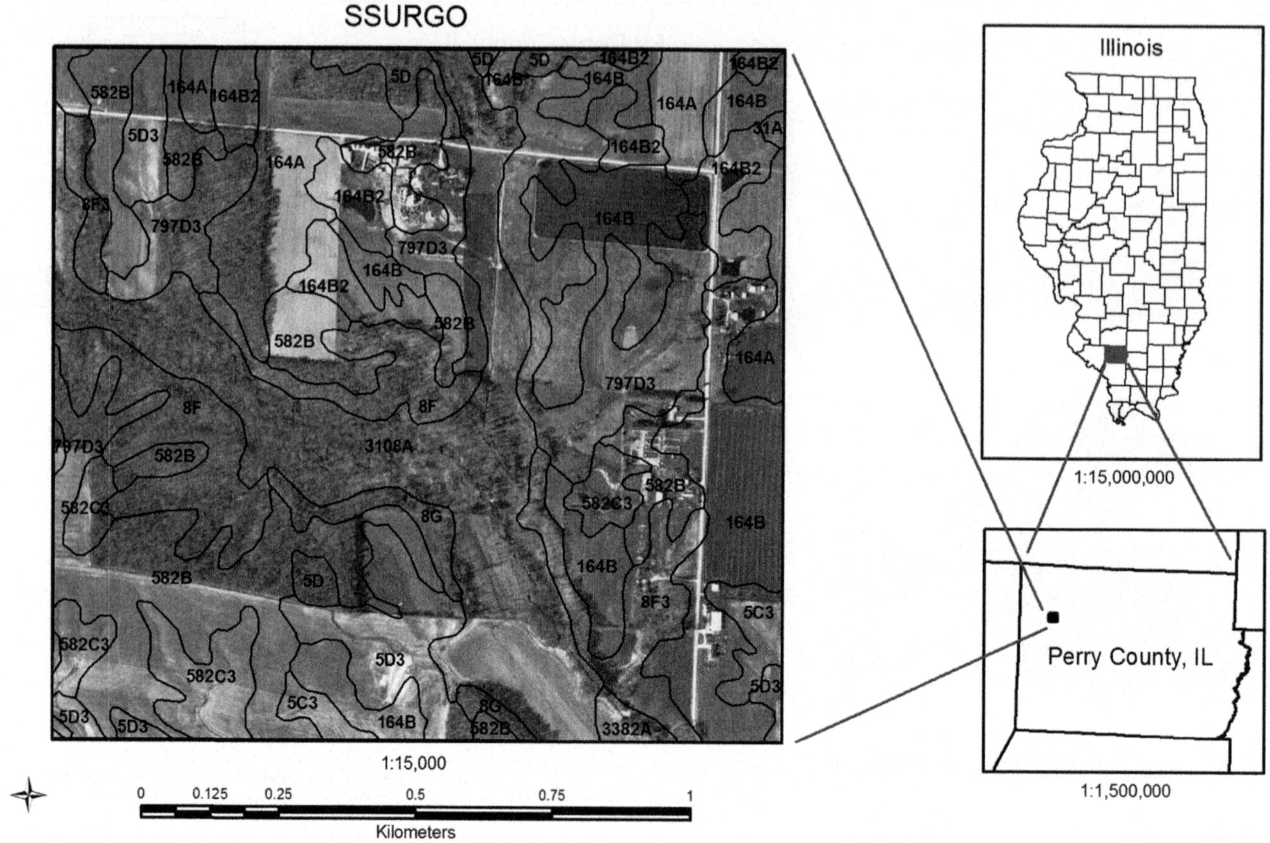

Fig. 2.16 An example of a detailed soil map from Perry County, Illinois. Soil series, soil type, soil phase consociations, and complexes are the taxonomic labels used in this map

Table 2.3 Legend* for detailed soil map in Fig. 2.16

Map unit symbol	Series	Map unit label (phase of soil series)
5C3	Blair	Blair silty clay loam, 5–10 % slopes, severely eroded
5D	Blair	Blair silty clay loam, 10–18 % slopes
5D3	Blair	Blair silty clay loam, 10–18 % slopes, severely eroded
8F	Hickory	Hickory silt loam, 18–35 % slopes
8F3	Hickory	Hickory clay loam, 18–35 % slopes, severely eroded
8G	Hickory	Hickory silt loam, 35–70 % slopes
31A	Pierron	Pierron silt loam, 0–2 % slopes
164A	Stoy	Stoy silt loam, 0–2 % slopes
164B	Stoy	Stoy silt loam, 2–5 % slopes
164B2	Stoy	Stoy silt loam, 2–5 % slopes, eroded
582B	Homen	Homen silt loam, 2–5 % slopes
582C3	Homen	Homen silt loam, 5–10 % slopes, severely eroded
797D3	Hickoy, Homen	Hickory–Homen silty clay loams, 10–18 % slopes, severely eroded
3108A	Bonnie	Bonnie silt loam, 0–2 % slopes, frequently flooded

*For more information refer to: Soil Survey Staff. Web Soil Survey. USDA Natural Resources Conservation Service, Available online at http://websoilsurvey.nrcs.usda.gov/. Accessed [07/09/2014]

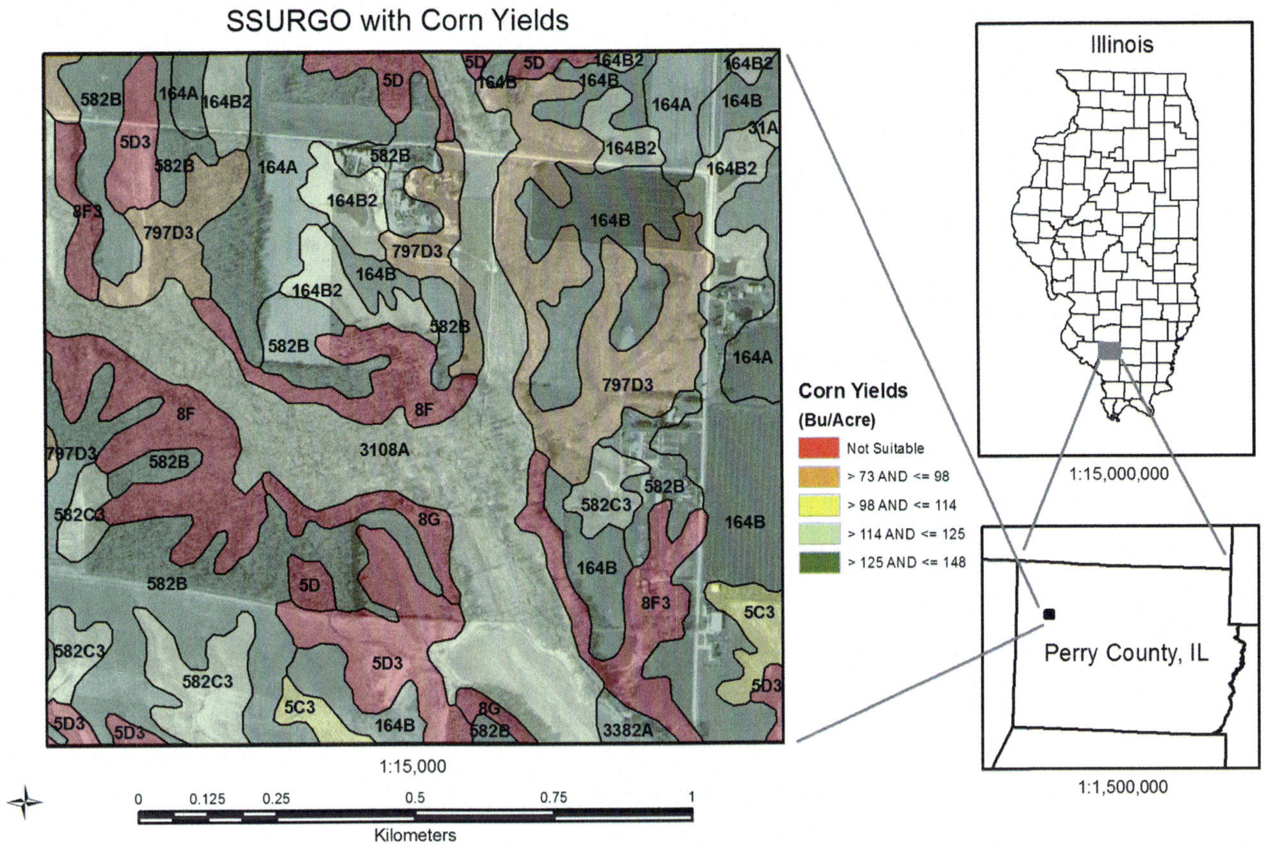

Fig. 2.17 Expected corn yields (*Zea mays* L.) by soil type for a field in Perry County, Illinois

spatially identify the imprint of soil formation across landscapes and to provide information on the properties and use potentials of the soils. Different scales of soil surveys, from highly detailed to very broad coverage, are used to serve different purposes. Parts of the landscape with similar soil taxa are grouped into map units, which serve as the basis for soil descriptions and interpretations.

Acknowledgments We thank Jon Bathgate (USDA-NRCS) for the preparation of Figs. 2.13, 2.14, 2.15, 2.16, 2.17, and 2.18, Edwin Winzeler (Department of Agronomy, Purdue University) for preparing Fig. 2.6, Doug Wysocki (USDA-NRCS) for Fig. 2.8, Amy Salvador for preparing photographic files, and the contributors of photographs as noted in the figure captions.

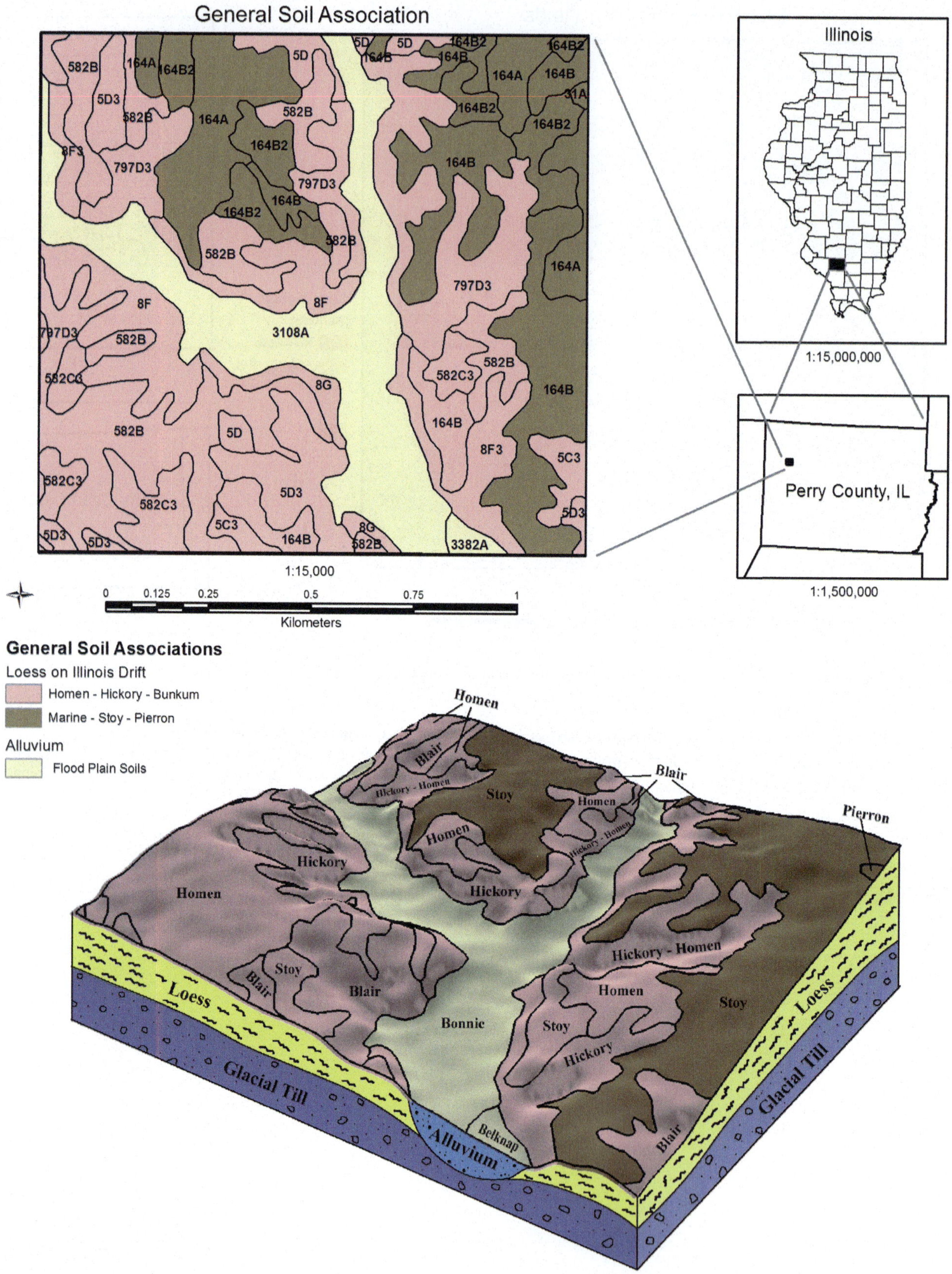

Fig. 2.18 The detailed soil map from Fig. 2.16 presented as a General Soil Association Map and as a soil-landscape block diagram

References

Amundson R, Jenny H (1991) The place of humans in the state factor theory of ecosystems and their soils. Soil Sci 151(99):109

Grantham DR, Indorante SJ (1988) Soil survey of Perry County Illinois. United States Department of Agriculture, Soil Conservation Service. US Gov Print Off, Washington, DC

Holmgren GGS (1988) The point representation of soil. Soil Sci Soc Am J 52:712–716

Hudson BD (1990) Concepts of soil mapping and interpretation. Soil Surv Horiz 31:63–72

Hudson BD (1992) The soil survey as paradigm-based science. Soil Sci Soc Am J 56:836–841

Indorante SJ (2011) The art and science of soil-landscape block diagrams: examples of one picture being worth more than 1000 words. Soil Surv Horiz 52:89–93

Jenny H (1941) Factors of soil formation. McGraw-Hill, New York

Kellogg CE (1956) The soils that support us. MacMillan, New York

Milne G (1936) A provisional map of East Africa. East African Agric Res Stn Amani Memoirs

Peterson FF (1981) Landforms of the Basin and Range Province defined for soil survey. Nevada Agric Exp Stn Tech Bull 28

Raich JW, Potter CS (1995) Global patterns of carbon dioxide emissions from soils. Global Biogeochem Cycles 9:23–36

Ruhe RV (1969) Quaternary landscapes in Iowa. Iowa State University Press, Iowa

Schoeneberger PJ, Wysocki DA, Benham EC, Soil Survey Staff (2012) Field book for describing and sampling soils. Version 3.0. Natural Resources Conservation Service, National Soil Survey Center, Lincoln

Simonson RW (1959) Outline of a general theory of soil genesis. Soil Sci Soc Am Proc 23:152–156

Soil Science Society of America (1997) Glossary of soil science terms

Soil Survey Division Staff (1993) Soil survey manual. U.S. Department of Agriculture Handbook 18. USDA Soil Conservation Service, Washington, DC

Soil Survey Staff (1951) Soil survey manual. U.S. Department of Agriculture Handbook 18. USDA Soil Conservation Service, Washington, DC

Soil Survey Staff (1999) Soil taxonomy: a basic system of soil classification for making and interpreting soil surveys. U.S. Department of Agriculture Handbook 436. USDA Natural Resources Conservation Service, Washington, DC

USDA-NRCS (2006) Land resource regions and major land resource areas of the United States, the Caribbean, and the Pacific Basin, U. S. Department of Agriculture Handbook 296. USDA Natural Resources Conservation Service, Washington, DC

Winzeler HE, Owens PR, Waltman SW, Waltman WJ, Libohova Z, Beaudette D (2013) A methodology for examining changes in soil climate geography through time: U.S. soil moisture regimes for the period 1971–2000. Soil Sci Soc Am J 77:213–225

3 Soil Properties and Classification (Soil Taxonomy)

Craig A. Ditzler

3.1 Documenting Soil Properties in the Field and Laboratory

The consistent use of standard procedures to describe soils in the field, analyze soil samples in the laboratory, and classify soils in a way that effectively organizes knowledge of the nation's soils is essential for carrying out an effective soil survey program.

3.1.1 Properties Used to Describe Soils in the Field

In the USA, hundreds of soil scientists working over a span of more than a century have collectively studied and recorded information from hundreds of thousands of soil profiles as part of the National Cooperative Soil Survey (NCSS). Such a large undertaking requires a standard approach so that soil descriptions can be compared, grouped, interpreted, and classified consistently. A soil profile is a 2-dimensional face observed in a pit, road cut, or other exposure large enough to reveal the boundaries of soil horizons and the largest structural units. The standards for describing soil profiles have been recorded in successive versions of the Soil Survey Manual (Soil Survey Division Staff 1993) and more recently, in the Field Book for Describing and Sampling Soils (Schoeneberger et al. 2012).

The process of describing the soil generally begins by exposing a soil profile over a lateral distance of about 1 m and to a depth of 1.5–2 m (Fig. 3.1). From the exposure depicted in Fig. 3.1, a sequence of layers or *horizons* can be observed ranging from the darkened *surface layer* ("A" horizon), to a lighter colored *subsurface layer* ("E" horizon), through the *subsoil* ("B" horizon), and, to the underlying, little-altered *parent material* ("C" horizon).

Soil scientists use a shorthand nomenclature to record the kind and thickness of horizons present in the soil profile. Master horizons are denoted with a capital letter (Table 3.1). Lower case letter suffixes can be used to differentiate within master horizons to describe some important aspects of the soil material or infer important soil-forming processes (Table 3.2).

Each horizon identified in the profile is labeled with the appropriate nomenclature. Once the kind and thickness of soil horizons is recorded, additional properties are noted for each horizon. Properties commonly recorded for each horizon include those shown in Table 3.3. A typical example of a standard narrative soil horizon description is as follows:

> Bt - 20 to 35 cm; yellowish brown (10YR 5/4), silt loam; 5 % gravel fragments; moderate medium subangular blocky structure; friable; few fine distinct light gray (10YR 7/2) redox depletions; common fine black (10YR 2/1) iron-manganese concretions; many dark yellowish brown (10YR 5/4) clay films on ped faces and in pores; common fine and medium roots; common fine tubular pores; slightly acid; clear smooth boundary.

3.1.2 Soil Properties Measured in the Laboratory

Soils have important properties that cannot be observed or measured by visual or tactile methods alone in the field. Therefore, laboratory analysis is needed to fully describe and classify soil profiles. Analyses include physical, chemical, and mineralogical properties that significantly impact the use and management of the soil. Because of their importance to understanding and managing soils, many of these properties are included as quantitative criteria in the definitions for various taxonomic classes in Soil Taxonomy (Soil Survey Staff 2014b). For example, the mollic epipedon (described below in Sect. 3.2.1) requires base saturation of 50 % or more throughout its thickness. Base saturation is an

C.A. Ditzler (✉)
USDA-Natural Resources Conservation Service, Lincoln, NE, USA
e-mail: craig.ditzler@earthlink.net

L.T. West et al. (eds.), *The Soils of the USA*, World Soils Book Series,
DOI 10.1007/978-3-319-41870-4_3

Fig. 3.1 An exposed soil profile reveals a sequence of soil horizons. Photograph courtesy USDA-NRCS Soil Survey Staff

Table 3.1 Common master soil horizons and the nomenclature (capital letters) used for each

Horizon symbol	General description
O	Composed dominantly of organic soil materials[a], relatively low bulk density
A	Surface mineral horizon, generally darkened by accumulating humus
E	Subsurface layer characterized by a loss of iron, aluminum, clay, or organic matter. Generally relatively light in color. (letter E denotes "eluvial")
B	Subsoil layer, generally characterized by either removals, transformations, or accumulations of constituents (such as iron, aluminum, silica, clay, humus, calcium carbonate, calcium sulfate, or sesquioxides) or by alterations resulting in clay and/or oxide formation and structure development
C	Unconsolidated, little-altered, underlying parent material
R	Consolidated bedrock

[a]Organic soil materials have a sufficiently large content of organic matter (as compared to mineral matter) so that the overall properties of the soil are dominantly influenced by the organic matter. See Soil Survey Staff (2014b) for the technical definition of "organic soil material"
See Schoeneberger et al. (2012) for additional, less commonly observed, master horizon kinds

important property impacting the fertility of the soil, but since it is a chemical property, it can only be measured in the laboratory.

Soil Taxonomy employs "operational definitions" for the criteria used. This means that along with the quantitative class limits for the property (e.g., "base saturation of 50 % or more"), the method of analysis is also specified because results measured for many properties will differ with the method used for the measurement. The criterion for base saturation of the mollic epipedon specifies measurement by the ammonium acetate method at pH 7.0.

Laboratory methods used for characterizing and classifying soils in the NCSS are described in the Kellogg Soil Survey Laboratory Methods Manual (Soil Survey Staff 2014a). Additional information about the use and importance of individual soil properties is in the Soil Survey Laboratory Information Manual (Soil Survey Staff 2011). Table 3.4 lists some of the properties requiring laboratory analysis that are used to characterize and classify soils.

3.2 Classifying Soils with Soil Taxonomy

The system of classification known as *Soil Taxonomy* is used in the USA, as well as several other countries. It was officially adopted by the United States Department of Agriculture (USDA) in 1975 and has remained the standard for soil classification ever since. Dr. Guy D. Smith, Director Soil Survey Investigations Division, led the development of the system during the 1950s to early 1970s. Specific objectives for the system included (Smith 1963):

- organizing the many kinds of soils known to exist around the world into useful classes whose members would share key properties considered important for understanding the soil's response to use and management and that reflect the major genetic processes thought to have formed the soils,
- developing definitions for diagnostic horizons and characteristics that use quantitative criteria that are properties of the soil and that can be observed in the field or measured in the laboratory. These criteria reflect, but are not directly defined by, soil genetic processes, and
- developing a classification system that can be used effectively by a diverse cadre of soil scientists in soil mapping projects of the NCSS.

Table 3.2 Common subordinate lower case suffixes describing physical characteristics of the soil (upper part of the table) and chemical or mineralogical characteristics (lower part of the table)

Describing	
Suffix	General description
Physical characteristics	
a	Highly decomposed organic matter (only used with O horizons)
b	Buried horizon that exhibits past soil-forming development
d	Dense, root-restrictive layer. Natural or artificially induced compaction
e	Moderately decomposed organic matter (only used with O horizons)
f	Permanently frozen layer
i	Slightly decomposed organic matter (only used with O horizons)
jj	Cryoturbated layer
m	Cemented layer
p	Artificial disturbance, commonly by plowing
r	Weathered bedrock, soft
ss	High shrink–swell potential (evidenced by slickenside presence)
t	Illuvial accumulation of clay
w	Weak color or structure development
x	Brittleness
Chemical or mineralogical properties	
g	Strong gleying (iron reduction)
h	Illuvial accumulation of organic matter
k	Accumulation of calcium carbonate, visible
n	Accumulation of sodium
o	Residual accumulation of sesquioxides
q	Accumulation of silica
s	Illuvial accumulation of iron and organic matter
v	Presence of plinthite
y	Accumulation of gypsum
z	Accumulation of salts (more soluble than gypsum)

See Schoeneberger et al. (2012) for additional, less commonly observed, subordinate symbols

The system was published in 1975 as Soil Taxonomy: A basic system of soil classification for making and interpreting soil surveys (Soil Survey Staff 1975). A second edition was published in 1999 (Soil Survey Staff 1999). In addition, as knowledge of soils around the world improved, 12 revised and updated versions of the taxonomic keys have been published (Soil Survey Staff 2014b).

Table 3.3 Features commonly described for soil horizons in the field

Property	General description
Color	Hue, value, chroma (e.g., 10YR 5/4) and adjective term (e.g., yellowish brown), as described in the Munsell® color system
Texture class	Based on % sand, silt, clay of material <2 mm diameter (e.g., sandy loam)
Rock fragments	Volume of fragments >2 mm diameter
Structure	Cohesive structural units formed by pedogenic processes (e.g., angular blocky)
Consistence	Degree of cohesion and resistance to deformation or rupture. (e.g., friable)
Redoximorphic (Redox) features	Features caused by reduction/oxidation processes in wet soils
Concentrations	Accumulations due to pedogenic processes (e.g., Fe nodules)
Ped surface features	Coatings on ped or pore surfaces due to translocation (e.g., clay films)
Roots	Size and number of roots in a specified area
Pores	Size, number, and shape of pores in a specified area
pH	Acidity/alkalinity of a sample, normally measured in a 1:1 soil–water mixture
Boundary	Topography and gradation of the transition between horizons (e.g., clear wavy)

See Schoeneberger et al. (2012) for additional properties described for soil horizons

3.2.1 Diagnostic Horizons and Characteristics

Soil Taxonomy is a morphogenetic system. It is based directly on the soil's morphology and indirectly on the inferred genetic history of the soil. The morphology of the soil is expressed in the kinds of horizons and other characteristic features that have formed through the interaction of the soil-forming factors as discussed in Chapter 2. Based on the observations of many soil profiles whose properties reflect the interaction of major soil-forming processes in different environments, it has been possible to define a set of key *diagnostic horizons and characteristics* to use as the basis for classifying soils. They are defined by morphological characteristics that are observed and measured in the field (color, texture, structure, thickness, etc.) and also by other properties that are measured in a laboratory (base saturation, cation-exchange capacity, organic carbon content, etc.). Because the criteria used to define diagnostic horizons and characteristics in Soil Taxonomy are based on the properties of soil profiles formed by major soil-forming processes, they effectively integrate pedogenic theory into soil classification. With this approach, the individual who is classifying a soil does not need to know the genetic history of the soil, but simply needs to identify the kinds of diagnostic horizons and characteristics that are present based on the existing properties of the soil.

The following two examples illustrate this concept. In humid to semiarid prairie regions dominated by grasslands, many soils have a thick, dark, humus-rich surface horizon with high native fertility. Over time, the grass roots and decaying surface residue have contributed large amounts of biomass deep into the soil. This results in a fairly high organic matter content and good overall soil tilth. The relatively high native fertility of the soil is maintained through nutrient cycling by the grasses and soil organisms such as earthworms and ants accompanied by only modest amounts of nutrients leaching out of the profile. This diagnostic surface horizon, illustrated in Fig. 3.2, is a *mollic epipedon* (L. *mollis*, soft; *Gr. epi,* over; and *pedon,* soil).

Some soils in desert or semidesert areas have a subsoil horizon into which significant amounts of calcium carbonate have accumulated. Water from precipitation dissolves calcium carbonate in the surface horizon and moves it into the subsoil. The water is subsequently taken up by plant roots, or it evaporates, leaving any dissolved minerals behind, such as calcium carbonate. This diagnostic subsoil horizon, also illustrated in Fig. 3.2, is a *calcic* horizon (L. *calcis*, lime).

The diagnostic horizons and characteristics defined in Soil Taxonomy (Soil Survey Staff 2014b) consist of eight surface horizons (epipedons), 22 subsoil horizons, and 36 characteristic features in the soil that do not constitute a horizon, but rather are a special feature within a horizon. Tables 3.5, 3.6, and 3.7 describe some of the more common diagnostic epipedons, subsoil horizons, and other characteristics, respectively, that occur in US soils.

3.2.2 How Soil Taxonomy Is Structured

The taxonomic categories of Soil Taxonomy are arranged in a hierarchy consisting of six levels. The six categories from highest to lowest level are as follows: *order*, *suborder*, *great group*, *subgroup*, *family*, and *series*. The highest level consists of 12 soil orders. They are the most broadly defined classes. Moving progressively lower in the hierarchy from suborder to series, the number of classes within each category increases and each class is progressively more narrowly defined.

In the USA, soils belonging to all 12 orders defined by Soil Taxonomy have been identified as well as all 68 suborders. Of 337 great groups defined in Soil Taxonomy, 271 are known to occur in the US. Of 2264 subgroups, 1565 are known to occur in the US. Complete records for families and series do not exist worldwide, but in the US, soils belonging to 10,895 families and 22,253 series had been documented in the Soil Series Classification Database (NRCS 2014a) at the time of this writing. As more is learned about soils in the US and around the world, new classes are periodically added to Soil Taxonomy.

The classes within each of the six categories reflect the major soil-forming processes responsible for the development of the soils and/or important characteristics of the soils. Soil orders have a limited number of important properties reflecting major soil-forming processes that are mostly operating over large regions. Suborders generally reflect the presence of a key soil property that exerts a controlling influence over currently active soil-forming processes. Commonly this consists of dynamic climatic controls such as the soil moisture or temperature regime. Great groups are often defined by one or more subordinate properties that exert additional controls over soil development. Commonly these are static properties such as the presence of a particular kind of diagnostic horizon. Subgroups are mostly defined either by having properties shared by similar soils in other categories, or by having unique properties not recognized in any other class. Soil families are defined by properties that are considered important for the practical use of the soil, such as those related to agronomic or engineering purposes. These include capacity factors such as soil texture, mineralogy, or depth, as well as intensity factors such as soil temperature, reaction, or cation-exchange activity. At the lowest level, soil series are separated by specifying a limited range of some key properties that impact soil use and management considerations and to facilitate field identification for making soil maps.

3.2.3 Soil Taxonomic Names

The names of the taxonomic classes are designed to be connotative, reflecting both their position in the hierarchy and key properties of the soils making up the class (Heller 1963). As a result, the names effectively communicate important information about the soils. This is accomplished through the use of short formative elements at each level (above the soil family) in particular ways to form the names. The formative elements are derived mostly from Greek or Latin roots.

Soil order names all begin with a syllable from the root word used to represent the order, and end with the syllable *sol* (L. *solum*, soil). The two syllables are connected with the

Table 3.4 Common laboratory analyses performed for characterizing and classifying soils grouped by physical properties (top), chemical properties (middle), and mineralogical properties (lower)

Property	General description	Importance
Physical properties		
Bulk density	Mass per unit volume at specified water content. (g cm^{-3} at 10 kPa, 33 kPa, and oven dry). Includes solids plus pore space	Required to convert from weight to a volumetric basis (e.g., kg carbon m^{-3}), to estimate saturated hydraulic conductivity, and in some studies of soil genesis. Used in definitions of mineral and organic soils, andic soil properties, and several taxonomic classes
Coefficient of linear extensibility (COLE)	Describes the proportional change in soil clod length in response to wetting and drying (moist and oven dry)	Higher values relate to increasing shrink/sell potential and soil movement hazard. Used to calculate linear extensibility for classifying vertic subgroups. Can be used to infer smectitic clay mineralogy
Particle-size analysis	Reports proportions of sand, silt, and clay for the <2 mm fraction. Sand fractions (5) determined by sieving; silt and clay fractions by differential settling in a water column	Used to determine textural class placement. Also used for taxonomic family particle-size class, some diagnostic horizon criteria, and taxonomic class placement
Water content	Water content of <2 mm fraction at specified soil–water tensions 33 kPa, (10 kPa for coarse textures), and 1500 kPa	Used to calculate water retention difference, available water capacity, pore-size distribution and porosity. Used in definitions of "vitric" and "hydric" taxonomic classes and in some "substitute" family particle-size classes
Chemical properties		
Cation-exchange capacity (CEC)	Quantity of readily exchangeable cations associated with negatively charged sites on soil particles. Reported as cmol charge kg^{-1} soil. Laboratory procedures vary by general soil types (acid, calcareous, saline) and interpretive purpose	Infers nutrient-holding capacity of the soil. Used in criteria for oxic and kandic diagnostic horizons, and in some taxonomic classes. When converted to a clay-only basis, can be used to infer clay mineralogy
Base saturation	Percent of cation-exchange sites occupied by exchangeable basic cations—Ca, Mg, Na, and K. Determined from results of CEC measured at pH 7.0 or pH 8.2	Used for separating Alfisols and Ultisols, mollic and umbric epipedons, and for some great groups and subgroups. Useful in soil genesis studies and for liming recommendations. Differences in results from the two methods are due to pH-dependent CEC
pH	Acidity or alkalinity of the soil (H-ion activity). Commonly measured in a 1:1 soil–water mixture, but several additional methods are used for various purposes	The pH (by specified methods) is used to identify sulfidic materials and their potential to form a sulfuric horizon. Also used in definitions of several great groups and subgroups, family reaction classes, and isotic mineralogy class. Used in criteria for andic soil properties. Useful for estimating nutrient availability, crop suitability, potential toxicity of elements, liming needs, and fertilizer response. Useful in soil genesis studies related to chemical processes and rate of chemical reactions. Useful in identifying soils where strong P-fixation makes phosphorous unavailable to crops
Phosphate retention	Percent phosphate fixed in the soil due to presence of short-range-order amorphous minerals (New Zealand P-retention test)	Used in criteria for andic soil properties. Useful in identifying soils of volcanic origin where strong P-fixation makes phosphorous unavailable to crops. Suggests dominance of active aluminum in amorphous clay minerals
Calcium carbonate equivalent (CCE)	A measure (weight percent) of the amount of carbonates in the soil relative to total dry sample weight. Manometric test of evolved CO_2 after acid treatment	Used in criteria for mollic and calcic diagnostic horizons, and in a few taxonomic classes from suborder to family. Carbonates affect fertility, susceptibility to erosion, and water-holding capacity. Useful in studies of soil genesis and terrestrial carbon cycling processes
Gypsum content	A measure (weight percent) of the amount of gypsum relative to total dry sample weight. Extraction with acetone and precipitation	Used in criteria for gypsic and petrogypsic diagnostic horizons, substitute particle-size classes, and mineralogy classes. Loss of gypsum due to solubility in irrigation or urban runoff water can result in subsidence and structural damage. Available water and CEC generally decrease as gypsum content increases

(continued)

Table 3.4 (continued)

Electrical conductivity (EC)	Conductivity of electricity (dS m^{-1}) through a saturated paste extract (or 1:5 soil–water (v/v) for subaqueous soil analysis). Infers level of salinity	Used to infer soluble salt levels in salt-affected soils. Used in criteria for the salic diagnostic horizon and for halic subgroups. Also used to differentiate fresh and brackish subaqueous soil classes. Salt content affects plant growth and plant available water
Exchangeable sodium percentage (ESP)	Exchangeable sodium as a percentage of CEC (at pH 7)	Used in criteria for the natric diagnostic horizon, and several sodium-related great groups and subgroups. Excessive levels of sodium can inhibit plant growth and adversely affect soil structure and hydraulic conductivity
Melanic index	Absorption spectrum test of humus acid extract to infer botanical origin	Used in criteria for the melanic diagnostic horizon in soils of volcanic origin. Used to infer grassland origin of humus vs. that of forest origin
Organic carbon	Dry combustion to determine total carbon, minus inorganic carbon	Widely used for taxonomic placement including definition of organic soil material, mollic, umbric, and histic diagnostic horizons, and for other taxa implying high organic matter content such as "Humic" suborders. Organic matter has significant effects on soil chemical properties such as CEC, and fertility; physical properties such as soil structure, aggregate stability, water-holding capacity, bulk density, and permeability. Important soil quality indicator. Important for terrestrial carbon cycle studies
Mineralogical properties		
Clay mineral species	X-ray diffraction or thermal analysis	Used in soil family mineralogy class for clayey soils. Useful in studies of soil genesis and weathering processes. Clay mineralogy strongly affects CEC and impacts nutrient-holding capacity. High content of smectitic clay can result in high shrink/swell, impacting roads, utilities, and building structures
Sand/silt fraction mineral species	Petrographic analysis for mineral identification and grain counts of sand fraction	Used in soil family mineralogy class for sandy and loamy soils. Useful in studies of soil genesis and weathering processes. Kinds of sand minerals, their size distribution, and total amounts can affect stability, erosivity, and suitability for engineering uses

See Keys to Soil Taxonomy appendix (Soil Survey Staff 2014b) for additional laboratory analysis required for classifying soils

vowel *i or o* (Table 3.8). For example, the order name *Gelisol* is based on the Latin word *gelare* (to freeze). This soil order contains the world's very cold, permafrost-affected soils. In the USA, these are predominantly found in Alaska. The name *Histosol* is based on the Greek word *histos* (tissue) and consists of soils formed in thick deposits of decaying organic material from plants. Histosols occur throughout much of the USA (except arid and semiarid areas), mostly in low-lying swamps or marshlands. The other orders are named following the same rules.

Below the order level, the names end with the short formative element syllable denoting the order as indicated in the second column of Table 3.8. For example, any taxonomic name ending in "alf" is an *Alfisol*, or ending in "and" is an *Andisol*.

Suborder names consist of two syllables, for example *Calcids*. The first syllable is connotative of either a major soil property or the moisture or temperature regime of the soil. The second syllable is the soil order's formative element. So a soil in the *Calcids* suborder is an Aridisol (denoted by the second syllable *id*) that has a calcic diagnostic horizon in the subsoil (denoted by the first syllable *calc*). Table 3.9 lists some of the common formative elements used in suborder names in the USA.

Great group names consist of one (or sometimes two) formative elements added to the beginning of the name of the suborder, for example *Petrocalcids*. In many cases, the formative element used at the great group level describes the presence of an additional soil property that exerts further control on the soil-forming process. So continuing with this example, a Petrocalcid is an Aridisol with a calcic horizon (Calcid suborder) and with a petrocalcic horizon (Petrocalcid great group). Table 3.10 lists some of the formative elements commonly used in great group names of soils in the USA.

Fig. 3.2 This soil profile from a semiarid grassland area of Texas has a mollic epipedon extending from the surface to about 38 cm and a calcic horizon in the subsoil below. The *dark color* of the mollic epipedon is due to the relatively high content of organic matter. The *white concentrations* scattered throughout the calcic horizon are accumulations of calcium carbonate (scale, decimeters *left*, feet *right*). Photograph courtesy USDA-NRCS Soil Survey Staff

Some formative elements used at the suborder level are also used at the great group level, generally with an "i" or "o" added, e.g., *calci*, *cryo*, *humi*, *torri*, and *vitri*. Great group names typically consist of three or sometimes four syllables.

Subgroup names consist of two words. An adjective term (ending with *ic*) precedes the great group name. The adjective term generally describes some important soil feature that is either shared by a higher class or that is unique and not shared by any other higher class. For example, the adjective *vertic* is used for subgroups that share the feature of relatively high shrink–swell potential with the Vertisols order. These kinds of subgroups are *intergrades*. In other cases, the adjective terms are unique to the subgroup and not shared with other classes of soils. For example, the adjective *lithic* is used for soils that are shallow to bedrock, a feature not used in any other higher category. These kinds of subgroups are *extragrades*.

The adjective descriptor is a separate word, not joined to the great group name, e.g., *Vertic Hapludalfs*, or *Lithic Hapludalfs*. Also, one subgroup within each great group is named with the adjective "*Typic*", for example "*Typic Hapludalfs*". These soils may be the most common of that group or have what was considered to be the most typical properties of the group, although this is frequently not the case. In all cases, however, they are the last subgroup listed in the key to subgroups within that particular great group.

Formative elements for subgroups that are intergrades are also used at the order, suborder, or great group level. As a subgroup term, they generally have "ic" added to the end, for example, *aquic, aridic, calcic, gypsic, kandic, natric, ustic, vertic*, and xeric. Table 3.11 lists some of the formative elements used in subgroup names that are not also used in the higher categories.

Soil family names do not utilize formative elements as described above for the higher categories. Rather they use three to five terms describing properties that are considered important to the use and management of soil. These include the following:

- calcareous and reaction class
- cation-exchange activity class
- class of coatings on sand grains
- human-altered and human-transported materials class
- mineralogy class
- particle-size class
- permanent cracks
- rooting depth class
- rupture resistance class
- soil temperature class.

Soil family names consist of the subgroup name (e.g., *Vertic Hapludalfs*) proceeded by the set of family terms specifically prescribed for this class (e.g., *very-fine, smectitic, thermic Vertic Hapludalfs*). In this example, family class terms for particle-size (*very-fine*), mineralogy (*smectitic*), and temperature (*thermic*) are used.

The rules for assigning family class terms are complex and in some cases vary between mineral and organic soils. The terms used in each family name vary with the kind of soil so that only some of the family class terms are applied.

Table 3.5 Epipedons commonly occurring in US soils

Horizon name	Major pedogenic processes	General concept
Histic	Accumulation of decaying organic matter	Poorly drained organic soil material. Occurs in a wide range of water-saturated settings
Melanic	Humus accumulation in association with poorly crystalline mineral species	Thick, black, organic-rich horizon with *andic soil properties*. Mostly associated with parent materials of volcanic origin modified by grassland vegetation
Mollic	Humus accumulation in association with high base saturation of the cation-exchange complex	Thick, dark-colored, organic-rich horizon, with high native fertility. Commonly associated with grassland soils
Ochric	Any illuvial or eluvial processes, but minimal in intensity	A surface horizon exhibiting pedogenic development, but that does not meet the criteria for any other epipedon. Typically thin or light colored
Umbric	Humus accumulation in association with low base saturation of the cation-exchange complex	Thick, dark-colored, organic-rich horizon, with low-to-moderate native fertility. Commonly associated with forestland soils

For a complete list of all eight epipedons and their criteria, see Keys to Soil Taxonomy (Soil Survey Staff 2014b)

Table 3.6 Diagnostic subsurface horizons commonly occurring in US soils

Horizon name	Major pedogenic processes	General concept
Albic	Leaching of clay and/or free iron oxides	Intensely leached layer with light color due to uncoated silt and sand grains
Argillic	Illuvial accumulation of phyllosilicate clay	Clay-enriched horizon, with evidence of illuviation such as clay films. Mostly humid, forested (or previously forested) environments
Calcic	Illuvial accumulation of calcium carbonate	Calcareous horizon with evidence of pedogenesis such as secondary forms of calcium carbonate. Typically in arid or semiarid environments

(continued)

Table 3.6 (continued)

Horizon name	Major pedogenic processes	General concept
Cambic	Any pedogenic alteration, but minimal in intensity	Pedogenically altered horizon, but not meeting criteria for any other diagnostic horizon
Duripan	Accumulation of silica (cemented)	Root-restrictive cemented layer, often associated with volcanic tephras. Slakes in KOH or NaOH (but not HCl)
Fragipan	Pedogenic processes are not well known, but result in a firm, brittle consistence	Firm, brittle, root-restrictive layer. Slakes in water (is not cemented). Often associated with transported materials such as loess, till, and colluvium
Glossic	Leaching of clay and iron oxides from a clay-enriched horizon	Degrading argillic, kandic, or natric horizon. Has separate distinct illuvial (B) and eluvial (E) parts. Typically in humid environments
Gypsic	Accumulation (or in situ transformation) of gypsum	A gypsum-enriched layer. Secondary gypsum is illuvial and/or has been dissolved and reprecipitated in place. Typically in arid or semiarid environments
Kandic	Accumulation (illuvial or in situ transformation) of low-activity clay	Horizon with low nutrient-holding capacity and significantly more clay than the overlying surface layer. High degree of weathering and predominance of low-activity clay minerals. Typically in warm, humid environments
Natric	Illuvial accumulation of clay along with sodium	Clay-enriched horizon, with evidence of illuviation and high sodium content. Often with columnar structure
Oxic	Intense leaching resulting in few remaining weatherable minerals and dominance of low-activity clay minerals	Intensely weathered horizon dominated by resistant minerals and low-activity clays. Very low nutrient-holding capacity. Dominantly on old stable landscapes

(continued)

Table 3.6 (continued)

Horizon name	Major pedogenic processes	General concept
		common in tropical environments
Petrocalcic	Illuvial accumulation of calcium carbonate (cemented)	Root-restrictive, calcareous horizon cemented with secondary calcium carbonate. Slakes with HCl. Advanced stage of calcic horizon. Typically in arid or semiarid environments
Petrogypsic	Accumulation (or in situ transformation) of gypsum (cemented)	Root-restrictive, horizon cemented with secondary gypsum. Advanced stage of gypsic horizon. Commonly in arid or semiarid environments
Salic	Accumulation of salts more soluble than gypsum	Highly saline horizon. Commonly in arid or semiarid environments
Spodic	Illuvial accumulation of organic matter in complex with aluminum and (commonly) iron	Reddish or dark-colored subsoil horizon enriched with amorphous organo-metal complexes. Commonly associated with an overlying albic horizon. Typically in warm to cool humid environments, mostly forested

For a complete list of all 22 subsoil horizons and their criteria, see Keys to Soil Taxonomy (Soil Survey Staff 2014b)

Table 3.7 Other diagnostic characteristics commonly occurring in US soils

Diagnostic characteristic	Major pedogenic processes	General concept
Abrupt texture change	Clay eluviation and illuviation	A considerable increase in clay content over a short vertical distance
Andic soil properties	Weathering and mineral transformation of primary alumino-silicate minerals (often glassy) to form poorly crystalline minerals and Al-humus complexes. Little or no illuviation	Weakly weathered soil with high volcanic glass content, or moderately weathered soil rich in short-range-order minerals such as allophane, imogolite, and ferrihydrite. Low bulk density, high p-adsorption.

(continued)

Table 3.7 (continued)

Diagnostic characteristic	Major pedogenic processes	General concept
		Commonly associated with cool, humid environments, often rich in tephra
Aquic conditions	Saturation and anaerobic conditions leading to biogeochemical reduction of Mn and Fe	Continuous or periodic saturation of sufficient length to result in oxygen depletion and formation of redox features. Wet soil environments
Densic materials	Non-pedogenic, geologic, or human-induced compaction	Root-restrictive, non-cemented, dense material. Commonly basal till, mudflows, or mechanically compacted layers
Durinodes	Silification, resulting in the formation of bodies cemented by SiO_2 such as opal or microcrystalline silica	Nodules or concretions 1 cm or more in size that slake in hot KCl, but not water or HCl. Often associated with soils containing volcanic glass as the silica source
Fibric soil materials	Organic matter accumulation and decay	Slightly decomposed organic soil materials, high in fiber content
Hemic soil materials	Organic matter accumulation and decay	Moderately decomposed organic soil material, moderate fiber content
Identifiable secondary carbonates	Translocation and precipitation of $CaCO_3$	Visible forms of secondary $CaCO_3$ such as soft masses, nodules, coatings, and filaments
Lamellae	Illuvial clay accumulation	Two or more thin layers (generally a few mm thick) enriched with illuvial clay in otherwise coarse-textured soil material
Lithic contact	Occurs below the depth of pedogenesis	2-dimensional horizontal contact with hard bedrock
Paralithic contact	Occurs below the depth of pedogenesis	2-dimensional horizontal contact with soft bedrock
Permafrost	Freezing	Layer with temperature <0°C for two or more consecutive years

(continued)

Table 3.7 (continued)

Diagnostic characteristic	Major pedogenic processes	General concept
Petroferric contact	Iron accumulation and cementation with little or no organic matter	2-dimensional horizontal contact with an iron-cemented, hard continuous layer of ironstone. Ironstone may be thick or in thin sheets
Plinthite	Accumulation of iron-oxide with clay and quartz	Reddish colored, firm, iron-oxide-rich, humus-poor mass. Irreversibly hardens to ironstone upon repeated wet/dry cycles
Redoximorphic (redox) features	Biogeochemical oxidation/reduction reactions	Morphological features with distinctive red/gray color patterns caused by wetness
Sapric soil materials	Organic matter accumulation and decay	Highly decomposed organic soil material, low fiber content
Slickensides	Shrink/swell of smectitic clays	Polished and grooved ped surface features indicating shear movement in response to wetting and drying cycles

For a complete list of all 36 diagnostic features and their criteria, see Keys to Soil Taxonomy (Soil Survey Staff 2014b)

Table 3.8 Soil order names, their formative elements, derivation, and general description of each order

Soil order	Formative element	Derivation	General description
Alfisols	-alf	Meaningless syllable (from early soil term "pedalfer")	Soils with high native fertility, high base saturation, and clay-enriched subsoil
Andisols	-and	Modified "ando". J. *An*, dark, *do*, soil	Soils characterized by short-range-order minerals, mostly of volcanic origin
Aridisols	-id	L. *aridus*, dry	Soils of desert regions (hot or cold)
Entisols	-ent	Meaningless, from "recent"	Young soils with minimal soil profile development
Gelisols	-el	L. *gelare*, to freeze	Soils with permafrost
Histosols	-ist	Gr. *histos*, tissue	Soils formed in thick organic deposits of decaying plants
Inceptisols	-ept	L. *inceptum*, beginning	Youthful soils with weak, but noticeable, profile development
Mollisols	-oll	L. *mollis*, soft	Dark-colored soils with high native fertility, mostly of grasslands
Oxisols	-ox	Fr. *oxide*, oxide	Highly weathered with low native fertility. Mostly in tropical regions
Spodosols	-od	Gr. *spodos*, wood ash	Acidic soils with low native fertility. Subsoil contains illuvial complexes of organic matter and iron and/or aluminum
Ultisols	-ult	L. *ultimus*, last	Soils with low native fertility, low base saturations, and clay-enriched subsoil
Vertisols	-ert	L. *verto*, turn	Very clayey soils that shrink and swell markedly with moisture change

See *Keys to Soil Taxonomy* (Soil Survey Staff 2014b) for a complete description of soil family classes and their rules of application.

The names for soil series are not appended to the higher category terms as described above for the other categories. Rather they are assigned names derived mostly from local places such as cities, towns, or geographic features near where the soil occurs. Examples of soil series names used in the USA include *Almirante*, *Cecil*, *Hilo*, *Holdrege*, *Narragansett*, *Ruston*, and *Yolo*. Although soil series names are easy to pronounce and remember locally, they do not contain the information included within the formative elements or family class terms used in the higher categories, and are therefore only effective for communicating among people who are familiar with the specific soil series.

Table 3.9 Some commonly used formative elements in suborder names of US soils, their derivation, and soil property connotation

Formative element	Derivation	Soil property connotation
Aqu	L. *aqua*, water	Water-saturated soils
Calc	L. *calcis*, lime	Has a horizon with calcium carbonate accumulation
Cry	Gr. *kryos*, cold	Cold soil temperatures
Fibr	L. *fibra*, fiber	Slightly decomposed organic matter
Fluv	L. *fluvius*, river	Formed in alluvial sediments
Gyps	L. *gypsum*, gypsum	Has a horizon with gypsum accumulation
Hem	Gr. *hemi*, half	Moderately decomposed organic matter
Hum	L. *humus*, earth	High in organic matter content
Orth	Gr. *orthos*, true	Common profile characteristics
Psamm	Gr. *psammos*, sand	Sandy texture
Sal	L. *sal*, salt	Has a horizon with salt accumulation
Sapr	Gr. *sapros*, rotten	Highly decomposed organic matter
Torr	L. *torridus*, hot and dry	Hot and dry (torric/aridic) moisture regime
Turb	L. *turbidis*, disturbed	Intense mixing by frost action (cryoturbation)
Ud	L. *udus*, humid	Udic moisture regime, (ample, well-distributed rainfall)
Ust	L. *ustus*, burnt	Ustic moisture regime, (somewhat limited moisture and a dry summer season)
Vitr	L. *vitrum*, glass	Contains volcanic glass
Xer	Gr. *xeros*, dry	Mediterranean type of climate: cool, moist winter, and a dry summer

Additional formative elements for suborder names are described in Soil Taxonomy (Soil Survey Staff 1999)

Table 3.10 Some commonly used formative elements in great group names of US soils, their derivation, and soil property connotation

Formative element	Derivation	Soil property connotation
Acr	Gr. *arkos*, at the end	Extremely weathered
Argi	L. *argilla*, white clay	Has an argillic horizon
Dystr	Gr. *dys*, ill, infertile	Low base saturation
Duri	L. *durus*, hard	Has a duripan
Endo	Gr. *endo*, within	Groundwater saturates the whole soil profile
Epi	Gr. *epi*, on, above	Groundwater perched on an impervious layer in the profile
Eutr	Gr. eu, good, fertile	High base saturation
Fragi	L. *fragilis*, brittle	Has a fragipan
Gloss	Gr. *glossa*, tongue	Has a glossic horizon
Hapl	Gr. *haplos*, simple	Minimal soil profile development
Hydr	Gr. *hydro*, water	Presence of water
Kand	Modified from *kandite*	Dominance of low-activity clay minerals
Moll	L. *mollis*, soft	Has a mollic epipedon
Natr	L. *natrium*, sodium	Accumulation of sodium in the subsoil
Pale	Gr. *paleos*, old	Excessive development
Petro	Gr. *petra*, rock	Has a cemented horizon
Plinth	Gr. *plinthos*, brick	Presence of plinthite
Quartzi	Ger. *Quarz*, quartz	High quartz content

Additional formative elements for great group names are described in Soil Taxonomy (Soil Survey Staff 1999)

The properties used to define soil series include characteristics such as kind and arrangement of soil horizons, and ranges of physical properties such as color, texture, and consistence, and chemical properties such as pH and salinity. Descriptions of all current US soil series are recorded in the Official Soil Series Database (NRCS 2014b). Seven examples of soil series and their taxonomic name at the various category levels are shown in Table 3.12.

Table 3.11 Some commonly used formative elements in subgroup names of US soils, their derivation, and soil property connotation

Formative Element	Derivation	Soil property connotation
Abruptic	L. *abruptum*, torn off	Has an abrupt change in texture
Aeric	Gr. *aerios*, air	Aeration is implied, the soil is drier than expected
Alic	From *alluminum*	High aluminum ion concentration
Arenic	L. *arena*, sand	Moderately thick, sandy, surface horizon
Cumulic	L. *cumulus*, heap	Over-thickened surface horizon
Grossarenic	L. *grossic*, thick; and arenic, *sand*	Thick, sandy, surface horizon
Halic	Gr. *hals*, salt	High salinity
Leptic	Gr. *leptos*, thin	A thin soil
Lithic	Gr. *lithos*, stone	Has a shallow contact with bedrock
Oxyaquic	Modified from *oxygen*, and L. *aqua*, water	Wetness, but with oxygenated water
Petronodic	Gr. *petra*, rock; and L. *nodulus*, little knot	Contains concretions or nodules (petronodes)
Rhodic	Gr. *rhodon*, rose	Dark red color
Sodic	From *sodium*	Contains sodium salts
Sulf	L. *sulfur*, sulfur	Has a sulfuric horizon or high content of oxidizable sulfur
Terric	L. *terra*, earth	Organic soils with underlying mineral soil material
Typic	Modified from *typical*	Either the central concept or simply the last subgroup listed in the classification key

Additional formative elements for subgroup names are described in Soil Taxonomy (Soil Survey Staff 1999)

3.3 Conclusions

Over more than 100 years, the NCSS has developed many standards and procedures for carrying out soil surveys throughout the USA. Standards for describing soil profiles in the field include nomenclature for recording the kinds and arrangements of master horizons and their subdivisions, as well as terms, definitions, and classes for key properties routinely included in descriptions for all horizons. For important soil properties that can only be measured in a laboratory, standard procedures for their measurement have been developed. The US system of soil classification, *Soil Taxonomy*, was developed to organize our knowledge about soils and facilitate communication about them by grouping soils with similar properties and genesis. It is a six-category, hierarchical system that uses a unique naming system to convey a wealth of information about each taxa. Utilizing these standards has proven effective for recording, comparing, classifying, and interpreting soil information consistently.

References

Heller JL (1963) The nomenclature of soils, or what's in a name. Soil Sci Soc Am Proc 27:212–215

Natural Resources Conservation Service, United States Department of Agriculture (2014a) Soil series classification database. http://www.nrcs.usda.gov/wps/portal/nrcs/detail/soils/survey/class/?cid=nrcs142p2_053583. Accessed 1 June 2015

Natural Resources Conservation Service, United States Department of Agriculture (2014b) Official soil series descriptions. http://www.nrcs.usda.gov/wps/portal/nrcs/detail/soils/scientists/?cid=nrcs142p2_053587. Accessed 1 June 2015

Schoeneberger PJ, Wysocki DA, Benham EC (eds) and Soil Survey Staff (2012) Field book for describing and sampling soils, Version 3.0. Natural Resources Conservation Service, National Soil Survey Center, Lincoln

Table 3.12 Examples of taxonomic names for seven US soil series

Series	Order	Suborder	Great group	Subgroup	Family
Almirante	Oxisol	Udox	Hapludox	Plinthic Hapludox	Very-fine, kaolinitic, isohyperthermic Plinthic Hapludox
Cecil	Ultisol	Udults	Kanhapludults	Typic Kanhapludults	Fine, kaolinitic, thermic Typic Kanhapludults
Hilo	Andisol	Udands	Hydrudands	Acrudoxic Hydrudands	Medial over hydrous, ferrihydritic, isohyperthermic Acrudoxic Hydrudands
Holdrege	Mollisol	Ustolls	Argiustolls	Typic Argiustolls	Fine-silty, mixed, superactive, mesic Typic Argiustolls
Narragansett	Inceptisol	Udepts	Dystrudepts	Typic Dystrudepts	Coarse-loamy over sandy or sandy-skeletal, mixed, active, mesic Typic Dystrudepts
Ruston	Ultisol	Udults	Paleudults	Typic Paleudults	Fine-loamy, siliceous, semiactive, thermic Typic Paleudults
Yolo	Entisol	Fluvents	Xerofluvents	Mollic Xerofluvents	Fine-silty, mixed, superactive, non-acid, thermic Mollic Xerofluvents

Smith GD (1963) Objectives and basic assumptions of the new classification system. Soil Sci 96:6–16

Soil Survey Division Staff (1993) Soil survey manual. Soil Conservation Service. U.S. Department of Agriculture Handbook 18. Washington, DC

Soil Survey Staff (1975) Soil taxonomy: a basic system of soil classification for making and interpreting soil surveys. Soil Conservation Service. U.S. Department of Agriculture Handbook 436. Washington, DC

Soil Survey Staff (1999) Soil taxonomy: a basic system of soil classification for making and interpreting soil surveys. 2nd edition. Natural Resources Conservation Service. U.S. Department of Agriculture Handbook 436. Washington, DC

Soil Survey Staff (2011) Soil Survey Laboratory Information Manual. Soil Survey Investigations Report No. 45, Version 2.0. R. Burt (ed.). U.S. Department of Agriculture, Natural Resources Conservation Service

Soil Survey Staff (2014a) Kellogg soil survey laboratory methods manual. Soil Survey Investigations Report No. 42, Version 5.0. R. In: Burt and Soil Survey Staff (ed) U.S. Department of Agriculture, Natural Resources Conservation Service

Soil Survey Staff (2014b) Keys to soil taxonomy, 12th edn. USDA-Natural Resources Conservation Service, Washington, DC

Soils of the USA: The Broad Perspective

4

James G. Bockheim

4.1 Soil Suborders by Land Resource Region

Soils form in response to climate, organisms, relief, parent materials, and time. In view of the heterogeneity of these factors in a country as large as the USA, the soils are extremely diverse. The purpose of this chapter is to provide a broad overview of soils in each Land Resource Region (LRR) as related to soil-forming factors. There are 28 LRRs in the USA and its territories that are named for their location and primary land uses (NRCS 2006; Figs. 1.1 to 1.3). Each LRR is divided into from one to 23 Major Land Resource Areas (MLRAs; total 278). The MLRAs are basedprimarily on physiography and geology but also on climate, soils, and vegetation type.

The USA (excluding its territories) has a total land area of 9.0 million km^2. The LRRs range in area from 260 km^2 for the Hawaii Region to 1.4 million km^2 for the Western Range and Irrigated Region (Table 4.1). The number of soil series in a LRR is governed not only by its area but also by the diversity of the soil-forming factors. The number of established soil series ranges from 84 in the relatively small Mississippi Delta Region (LRR O; area = 100,710 km^2) to 5467 in the massive Western Range and Irrigated Region (LRR D; area = 1424,480 km^2). The ratio of land area to number of soil series per LRR ranges from 116 to 185 for LRRs A, B, C, L, and T, which represent high pedodiversity indices, and 593–754 for LRRs F, H, and P, which represent low pedodiversity indices (the ratio of land area to number of soil series per LRR).

The number of soil orders per LRR ranges from 5 for various regions in Alaska to 10 in the Hawaii Region (LRR V) and averages 7 (Table 4.1). The number of suborders ranges from 9 in western Alaska to 44 in the Northwestern Forest, Range, and Specialty Crop Region (LRR A) and averages 21. Much of Alaska has been mapped only at a reconnaissance scale so that the number of soil series is not consistent with its large area. The number of soil great groups ranges from 24 in the Mississippi Delta to 118 in the Western Range and Irrigated Region (LRR D) and averages 60.

The dominant soil taxa by LRR or MLRA can be determined through three techniques: (i) by overlaying the LRR or MLRA map on the soil order and suborder maps, (ii) identifying all of the series in an MLRA using the NRCS soil classification database and the accompanying areal extent maps, and (iii) from the number of soil series per taxa. There is a highly significant correlation between the number of soil series and the areas of soil orders (Fig. 4.1).

From Table 4.1, it can be seen that from one to four suborders (those in bold face) account for half of the soil series in each LRR. Mollisols are the dominant order in the USA, accounting for 23 % of the land area and 31 % of the soil series (Table 4.2). Mollisols are most common in the Great Plains from Montana, North Dakota, and Minnesota down through South Dakota, Nebraska, Kansas, northeastern Colorado, and Oklahoma to Texas. However, they are also abundant in the five Far West states: Washington, Oregon, Idaho, Nevada, and California. Udolls are dominant in LRRs F, K, and M; and Ustolls are prevalent in E, F, G, H, I, J, V, and Z; Xerolls in A, B, C, D, and E; and Cryolls in E (Table 4.1).

Alfisols are the second most abundant order in the USA, accounting for 15 % of the land area and 17 % of the soil series (Table 4.2). Whereas Udalfs prevail in K, L, M, N, P, and S, Ustalfs are common in I and J, Xeralfs in C, and Aqualfs dominate LRRs T and U. Entisols comprise 12 % of the land area and of the soil series; they tend to be dispersed among the LRRs. Orthents comprise 11 of D and 13 % of G, and Psamments compose 10 % of U. Inceptisols account for 10 % of the land area and 13 % of the soil series. Udepts are common in LRRs A and Z, Ustepts in Q (Pacific Basin), Xerepts in A, and Aquepts in L and R.

Ultisols compose 10 % of the land area and 5.6 % of the soil series (Table 4.2). Udults are a predominant suborder in LRRs N, P, S, T, and Z, with Aquults abundant in LRR T.

J.G. Bockheim (✉)
University of Wisconsin, Madison, WI 53706-1299, USA
e-mail: bockheim@wisc.edu

L.T. West et al. (eds.), *The Soils of the USA*, World Soils Book Series,
DOI 10.1007/978-3-319-41870-4_4

Table 4.1 Dominant suborders by Land Resource Region

	Land Resource Region	Area (km^2)	Number				Dominant suborders[b]
			Soil series[a]	Orders	Suborders	Great groups	
A	Northwestern Forest, Forage, and Specialty Crop Region	233,635	2019	8	44	105	**Xerepts (260), Xerolls (252), Udepts (202), Xeralfs (173), Udands (162)**
B	Northwestern Wheat and Range Region	210,555	1729	8	37	82	**Xerolls (920)**, Cryolls (99), Calcids (92), Argids (83)
C	California Subtropical Fruit, Truck, and Specialty Crop Region	161,570	943	8	34	82	**Xerolls (297), Xeralfs (248)**, Xerepts (62), Orthents (58)
D	Western Range and Irrigated Region	1424,480	5467	8	37	118	**Xerolls (1023), Argids (762), Orthents (627), Calcids (455)**
E	Rocky Mountain Range and Forest Region	612,875	3152	8	37	108	**Ustolls (611), Cryolls (482), Xerolls (390), Cryalfs (210)**
F	Northern Great Plains Spring Wheat Region	368,535	621	6	23	52	**Ustolls (197), Aquolls (105), Udolls (104)**
G	Western Great Plains Range and Irrigated Region	554,395	1085	6	24	44	**Ustolls (407), Orthents (142)**, Argids (82)
H	Central Great Plains Winter Wheat and Range Region	569,420	755	6	19	45	**Ustolls (355), Ustalfs (108)**, Ustepts (57)
I	Southwest Plateaus and Plains Range and Cotton Region	187,460	334	6	16	34	**Ustolls (136), Ustalfs (68)**
J	Southwestern Prairies Cotton and Forage Region	154,695	336	6	17	31	**Ustalfs (119), Ustolls (88),** Ustepts (31), Usterts (28)
K	Northern Lake States Forest and Forage Region	307,795	1026	7	21	45	**Udalfs (285), Udolls (159), Orthods (124)**, Aquolls (106)
L	Lake States Fruit, Truck Crop, and Dairy Region	118,460	639	6	14	36	**Udalfs (167), Aquepts (84), Udepts (72),** Orthods (43)
M	Central Feed Grains and Livestock Region	731,905	2000	7	23	61	**Udolls (605), Udalfs (459)**, Aquolls (306), Aqualfs (173)
N	East and Central Farming and Forest Region	612,645	1059	8	20	45	**Udalfs (324), Udults (283)**, Udepts (183), Udolls (63)
O	Mississippi Delta Cotton and Feed Grains Region	100,710	84	5	11	24	**Aquerts (6)**, Aqualfs (12), Aquepts (12), Udolls (11)
P	South Atlantic and Gulf Slope Cash Crops, Forest, and Livestock Region	684,340	1041	8	20	55	**Udults (380), Udalfs (235)**, Aqualfs (77), Aquults (71)
Q	Pacific Basin Region	2585	71	8	18	33	**Perox (11), Ustepts (8)**, Udands (6)
R	Northeastern Forage and Forest Region	312,625	806	6	21	55	**Udepts (235), Orthods (74), Aquepts (141)**, Udalfs (81)
S	Northern Atlantic Slope Diversified Farming Region	105,905	593	8	18	45	**Udults (186), Udalfs (124)**, Udepts (94), Aquults (34)
T	Atlantic and Gulf Coast Lowland Forest and Crop Region	240,055	736	8	24	71	**Udults (152), Aquults (79), Aqualfs (76), Aquents (74)**
U	Florida Subtropical Fruit, Truck Crop, and Range Region	92,275	216	7	18	30	**Aqualfs (34), Aquods (26), Aquents (23), Psamments (22), Saprists (21)**
V	Hawaii Region	16,260	242	10	27	53	**Udands (56), Ustands (36), Ustolls (25), Humults (17)**
W1	Southern Alaska	246,710	58	5	13	14	**Cryods, Cryands,** Aquands, Cryepts, Gelepts, Saprists

(continued)

Table 4.1 (continued)

	Land Resource Region	Area (km²)	Number				Dominant suborders[b]
			Soil series[a]	Orders	Suborders	Great groups	
W2	Aleutian Alaska	27,645	nd	nd	nd	nd	**Cryands,** Fibrists
X1	Interior Alaska	671,835	94	6	15	21	**Orthels,** Turbels, Cryepts, Gelepts
X2	Western Alaska	236,585	18	5	9	11	**Turbels**, Orthels, Gelepts, Cryepts
Y	Northern Alaska	325,345	nd	nd	nd	nd	**Turbels**, Orthels, Histels
Z	Caribbean Region	9310	228	10	30	60	**Ustolls (28), Udepts (26), Udults (18), Udox (14)**

[a]A soil series generally occurs in more than one LRR
[b]Suborders in bold face are most extensive; the number of soil series in the suborder is in parentheses

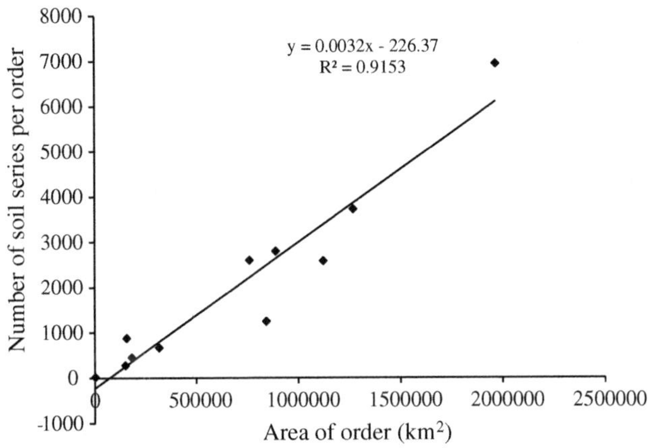

Fig. 4.1 Relation between number of soil series in a suborder and suborder area (km^2)

Gelisols occur only in LRRs X1, X2, and Y in interior, western, and northern Alaska. Although Gelisols account for 9.4 % of the land area, only 55 Gelisol soil series are included in the NRCS database and they account for only 0.2 % of the land area. Aridisols compose 9 % of the land area and nearly 12 % of the soil series in the USA. However, they are dominant only in LRR D. Spodosols occupy 3.8 % of the land area and account for 3.0 % of the soil series. Whereas Orthods are common in the northern Lake States and northeastern USA (LRRs K and R, respectively), Aquods are abundant in southern Florida (U), and Cryods are common in southern Alaska (W1).

Vertisols account for 2.0 % of the land area and 2.1 % of the soil series identified (Table 4.2). Aquerts are the dominant suborder in the Mississippi Delta (LRR O; Table 4.1). Andisols cover 1.8 % of the US land area but account for 3.9 % of the soil series. Udands are abundant in northwestern USA and Hawaii (LRRs A and V, respectively), Ustands occur in Hawaii, and Cryands are abundant in southeastern and southwestern Alaska (W1 and W2). Histosols occur on 1.8 % of the land surface in the USA and account for 1.2 % of the soil series identified. Oxisols occur to the least extent in the USA, occupying 0.02 % of the land area and 0.03 % of the soil series. Perox are common in the Pacific Basin Region (LRR Q) and Udox occur in the Caribbean (Puerto Rico, LRR Z).

4.2 Soil-Forming Factors

Soils of the USA occur in 12 orders, 65 suborders, 344 great groups, and numerous subgroups and families, yielding about 23,000 soil series. The distribution of soils in the USA is controlled primarily by climate (especially the Aridisols, Gelisols, and Oxisols), vegetation (especially the Histosols and Mollisols), parent materials (especially the Andisols, Entisols, and Vertisols), time (Inceptisols), or a combination of factors (Alfisols, Spodosols, and Ultisols). Figures 4.2, 4.3, and 4.4 are soil order maps of the USA and its territories (Commonwealth of the Northern Mariana Islands is not shown because of scale limitations) with the boundaries of the LRRs delineated. These figures show that Alfisols dominate in LRRs M, K, L, and J; Andisols in A and E; Aridisols in D; Entisols in D, B, C, and G; Gelisols in W, X, and Y; Histosols in K and L; Inceptisols in R, S, E, and A; Mollisols in B, C, E, F, G, H, I, and M; Oxisols in V and Z; Spodosols in R, K, and U; Ultisols, in N, P, and T; and Vertisols in O and P.

The ranking of suborders by abundance in the USA is Ustolls > Udults > Udalfs > Turbels > Orthents > Udolls > Udepts > Xerolls; these suborders comprise 53 % of the soils of the USA. There is a highly significant correlation (R^2 = 0.58) between the number of soil series per state and the area of the state (Fig. 4.5). Ninety-five percentage of the soils in the continental USA contain either an ochric or a mollic epipedon (Figs. 4.6, 4.7). Soils with an argillic horizon comprise 49 % of the soils of the continental USA; 24 % of the soils (Entisols and Histosols) lack a diagnostic subsurface horizon; and 18 % contain a cambic horizon (Figs. 4.8, 4.9).

Table 4.2 Distribution of soil suborders by area and number of soil series

Order	Suborder	Area (km^2)	No. of soil series		
			%		%
Alfisols	Aqualfs	215,497	2.6	575	2.6
	Cryalfs	69,129	0.8	235	1.1
	Udalfs	622,300	7.4	1609	7.2
	Ustalfs	281,586	3.3	682	3.1
	Xeralfs	80,771	1.0	625	2.8
			15.0		16.7
Andisols	Aquands	2299	0.0	27	0.1
	Cryands	87,531	1.0	215	1.0
	Gelands	0	0.0	0	0.0
	Torrands	134	0.0	12	0.1
	Udands	17,456	0.2	228	1.0
	Ustands	1149	0.0	39	0.2
	Vitrands	17,319	0.2	116	0.5
	Xerands	29,924	0.4	241	1.1
			1.8		3.9
Aridisols	Argids	322,921	3.8	991	4.5
	Calcids	221,716	2.6	634	2.8
	Cambids	103,009	1.2	388	1.7
	Cryids	562	0.0	13	0.1
	Durids	74,371	0.9	381	1.7
	Gypsids	14,061	0.2	126	0.6
	Salids	23,482	0.3	38	0.2
			9.0		11.6
Entisols	Aquents	124,331	1.5	345	1.5
	Arents			34	0.2
	Fluvents	178,579	2.1	564	2.5
	Orthents	570,811	6.8	1146	5.1
	Psamments	248,662	2.9	484	2.2
	Wassents	0	0.0	12	0.1
			13.3		11.6
Gelisols	Histels	87,440	1.0	7	0.0
	Orthels	124,975	1.5	18	0.1
	Turbels	580,530	6.9	27	0.1
			9.4		0.2
Histosols	Fibrists	22,058	0.3	20	0.1
	Folists	8795	0.1	32	0.1
	Hemists	29,159	0.3	70	0.3
	Saprists	89,543	1.1	156	0.7
	Wassists	0	0.0	0	0.0
			1.8		1.2

(continued)

Table 4.2 (continued)

Order	Suborder	Area (km^2)	No. of soil series		
			%		%
Inceptisols	Aquepts	123,753	1.5	629	2.8
	Cryepts	190,517	2.3	359	1.6
	Gelepts	72	0.0	14	0.1
	Udepts	348,528	4.1	890	4.0
	Ustepts	149,122	1.8	400	1.8
	Xerepts	77,521	0.9	510	2.3
			10.5		12.6
Mollisols	Albolls	15,314	0.2	71	0.3
	Aquolls	233,749	2.8	874	3.9
	Cryolls	136,132	1.6	696	3.1
	Gelolls	0	0.0	1	0.0
	Rendolls	3223	0.0	22	0.1
	Udolls	356,064	4.2	852	3.8
	Ustolls	884,688	10.5	1929	8.7
	Xerolls	337,287	4.0	2499	11.2
			23.3		31.2
Oxisols	Aquox	0	0.0	3	0.0
	Perox	15	0.0	16	0.1
	Torrox	0	0.0	3	0.0
	Udox	1048	0.0	22	0.1
	Ustox	603	0.0	15	0.1
			0.0		0.3
Spodosols	Aquods	55,269	0.7	154	0.7
	Cryods	85,873	1.0	166	0.7
	Gelods	0	0.0	5	0.0
	Humods	4612	0.1	8	0.0
	Orthods	171,911	2.0	334	1.5
			3.8		3.0
Ultisols	Aquults	75,472	0.9	153	0.7
	Humults	16,489	0.2	160	0.7
	Udults	742,749	8.8	891	4.0
	Ustults	823	0.0	11	0.0
	Xerults	6726	0.1	31	0.1
			10.0		5.6
Vertisols	Aquerts	55,469	0.7	103	0.5
	Cryerts	0	0.0	3	0.0
	Torrerts	10,025	0.1	37	0.2
	Uderts	30,934	0.4	70	0.3
	Usterts	74,120	0.9	146	0.7
	Xererts	10,934	0.1	92	0.4
			2.1		2.0
	Total	8,449,142	100.0	22,259	100

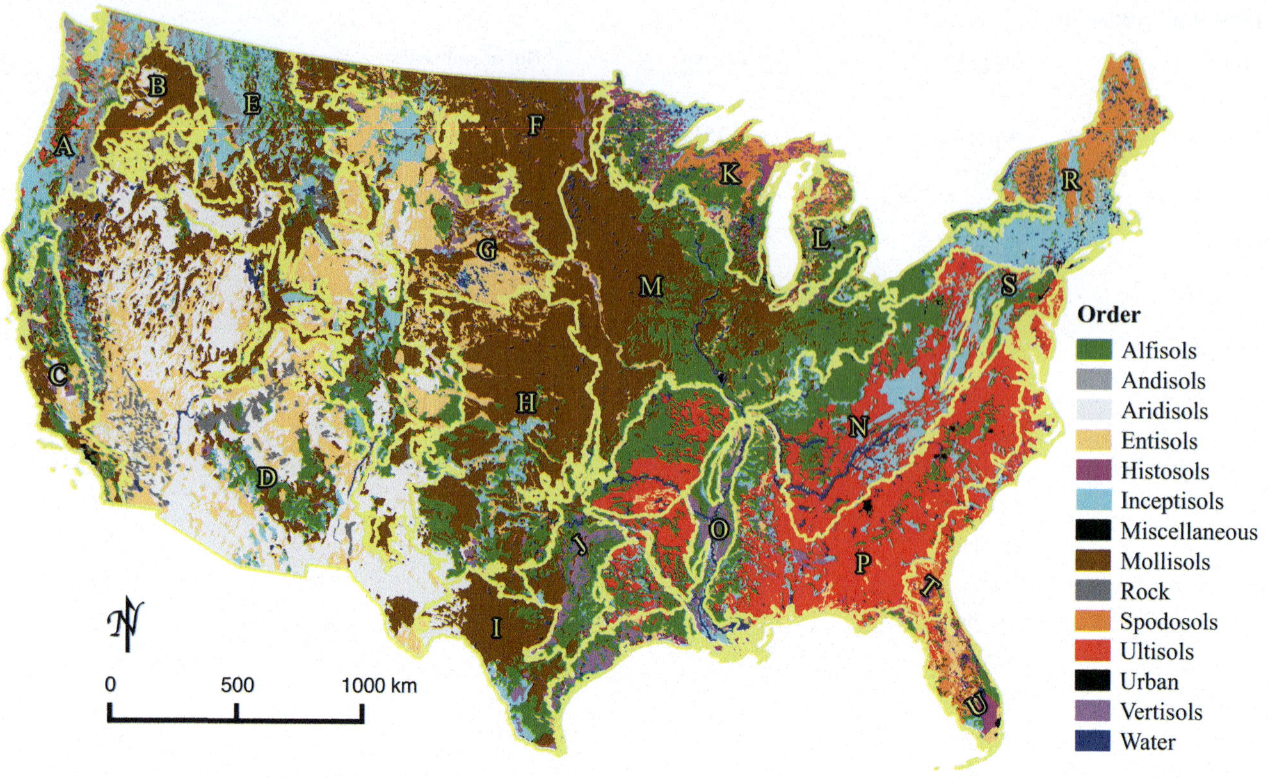

Fig. 4.2 Soil orders of the conterminous USA with Land Resource Regions delineated

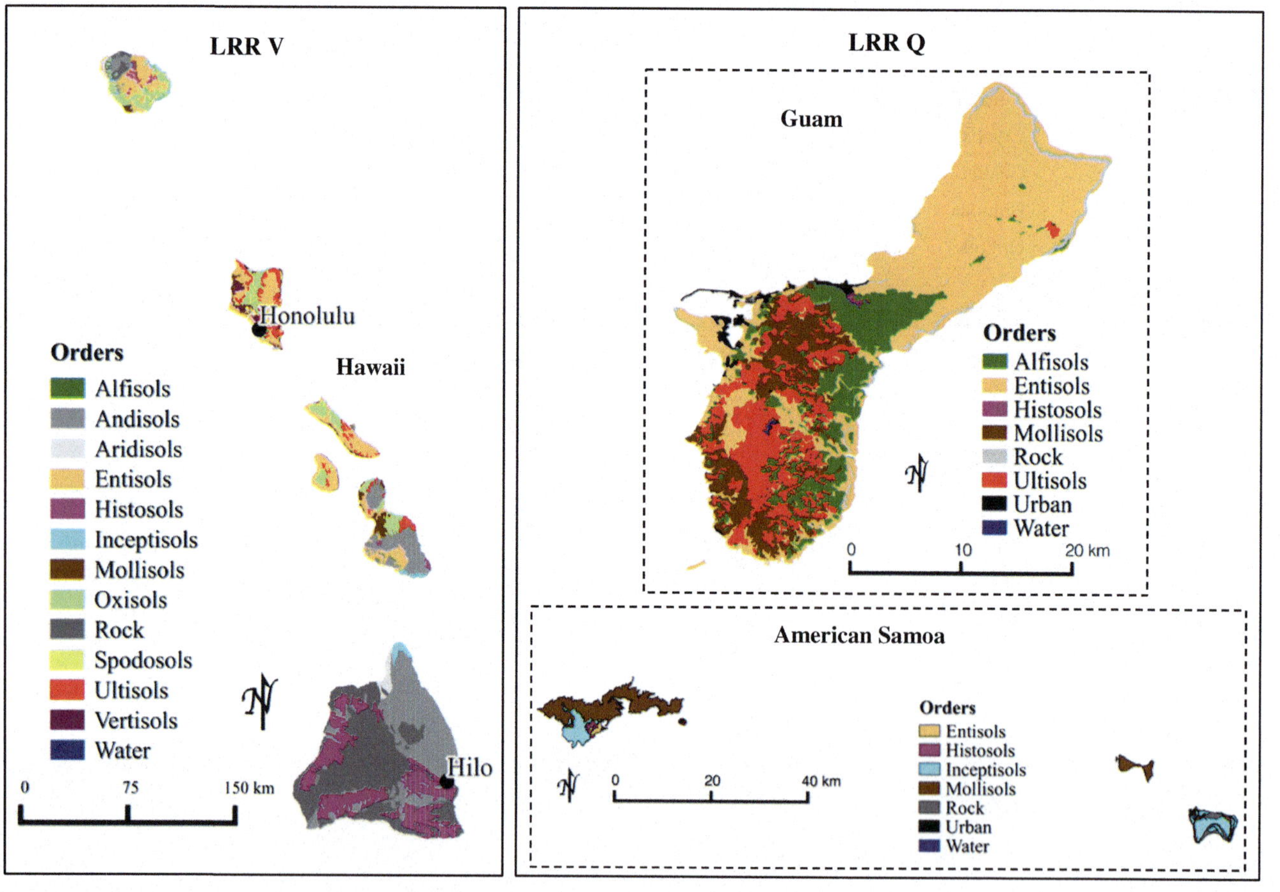

Fig. 4.3 Soil orders of Hawaii and the Pacific Island Territories (LRRs Q and V)

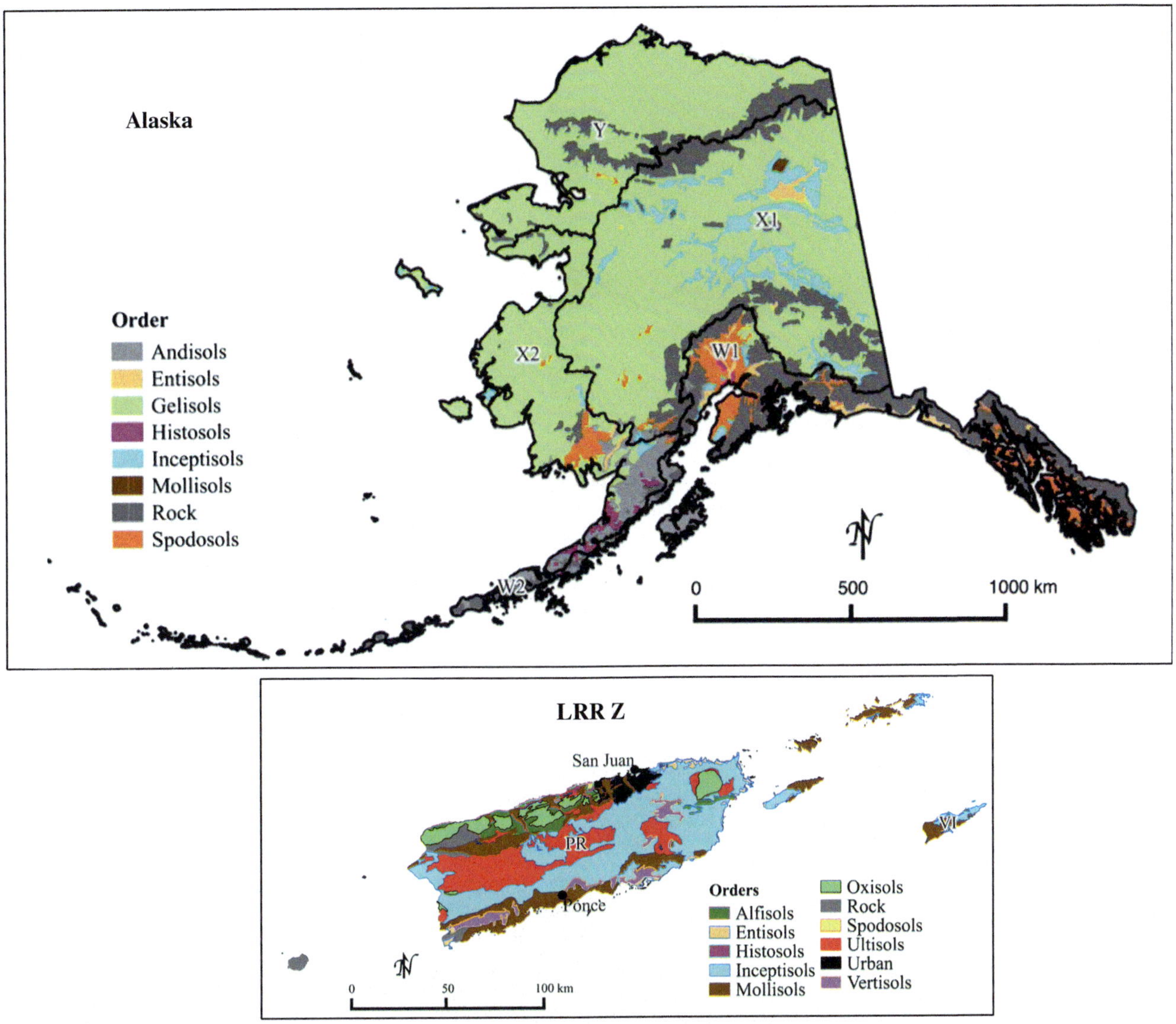

Fig. 4.4 Soil orders of Alaska and the Caribbean Area (LRR Z)

4.3 Soil-Forming Processes

The dominant soil-forming processes in soils of the continental USA are argilluviation, melanization, base-cation cycling, and cambisolization (Figs. 4.10, 4.11). Argilluviation, or clay translocation, is prominent in LRRs D, F, G, H, I, J, K, L, M, N, O, P, and T. Argillic, natric, and kandic horizons are present in 56 % of the soil area of the USA (Bockheim and Hartemink 2013). Melanization (also called humification) refers to the accumulation of dark-colored organic compounds in the upper mineral soil horizons. This process is particularly evident in LRRs B, C, E, F, G, H, I, and M. Soils with mollic, umbric, and melanic horizons cover 42 % of the USA land are (Bockheim 2014). Approximately 37 % of the soils reflect biological enrichment of base cations by grassland or deciduous forest vegetation that renders the soils productive for agriculture or pasture. These include Mollisols, Alfisols, and Vertisols that are particularly abundant in LRRs J, K, L, and M. More than a quarter (28 %) of the soils of the USA show weak forms of the various soil-forming processes, i.e., cambisolization. These soils are particularly dominant in LRRs A, D, E, R, and S.

Calcification, the accumulation of secondary carbonates, occurs in 10 % of the soils in the USA, especially in LRRs B, D, and E (Fig. 4.10). Gleization, processes that include gleying and development of redoximorphic features, occurs in hydric soils, i.e., aqu-suborders of Alfisols, Andisols, Entisols, Inceptisols, Mollisols, Oxisols, Spodosols, Ultisols, and Vertisols. This process occurs in 9.7 % of the soils, especially in LRRs M, O, T, and U. Ferralitization, the residual accumulation of Fe and Al and loss of Si

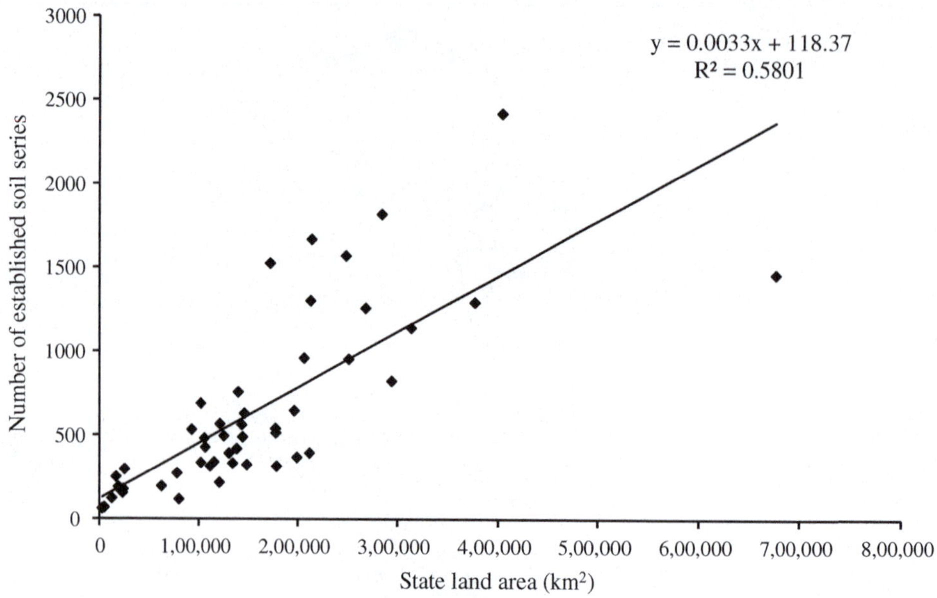

Fig. 4.5 Relation of the number of established soil series and the land area by state in the conterminous USA

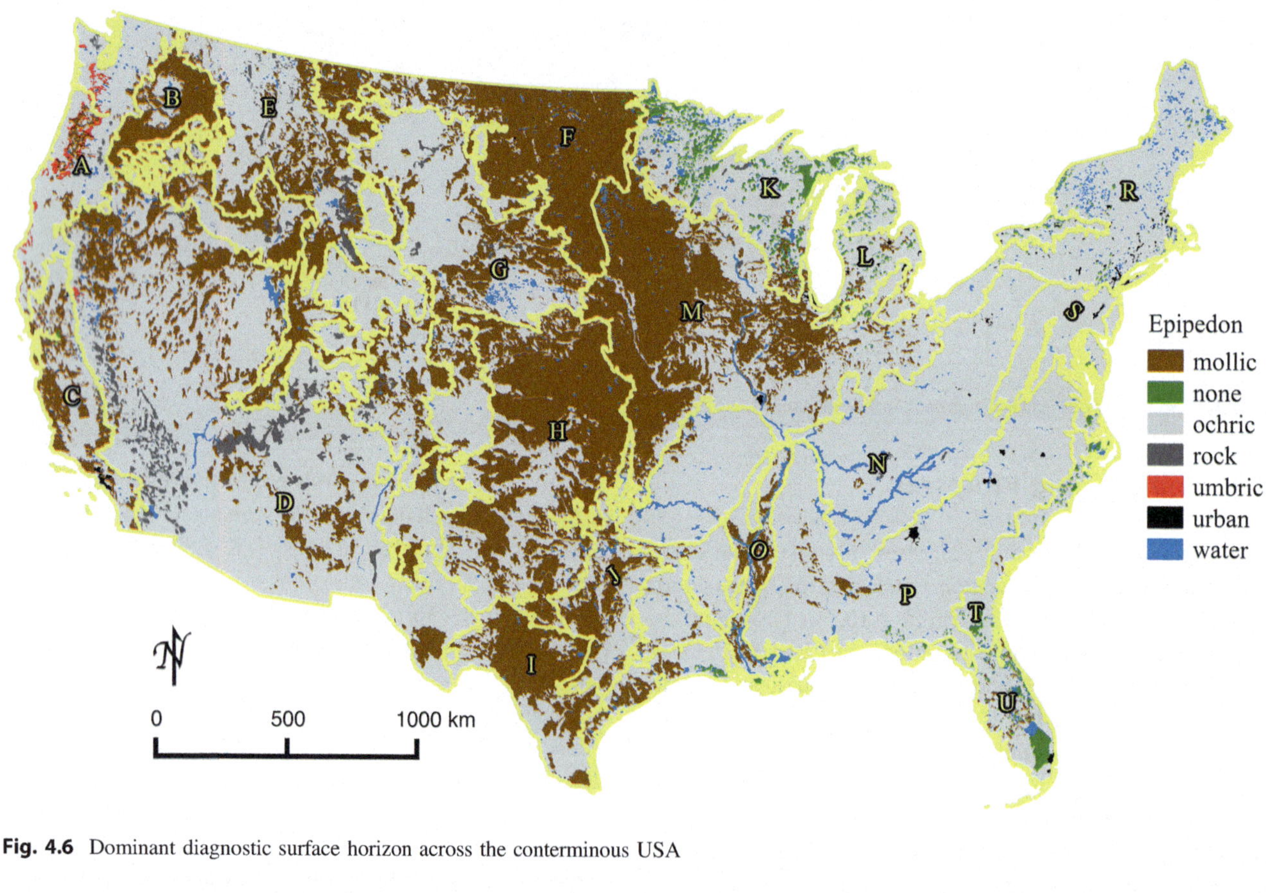

Fig. 4.6 Dominant diagnostic surface horizon across the conterminous USA

from intense tropical weathering, is a dominant process in Ultisols of LRRs N, P, S, and T and in Oxisols of LRRs V and Z, Hawaii and the Caribbean region. These soils cover about 9.2 % of the USA and its territories. Silicification, the accumulation of poorly crystalline silica in the profile, is common in 3.7 % of the soils, particularly in LRR D.

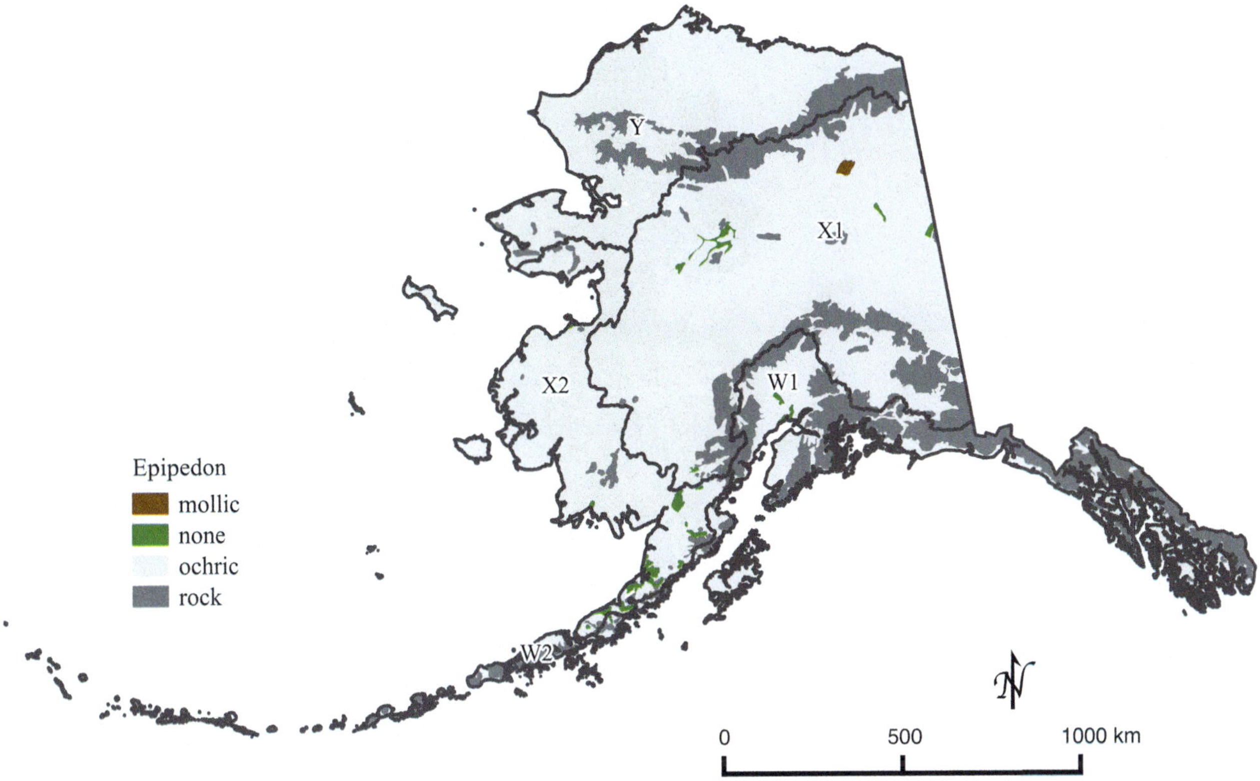

Fig. 4.7 Dominant diagnostic surface horizon across Alaska

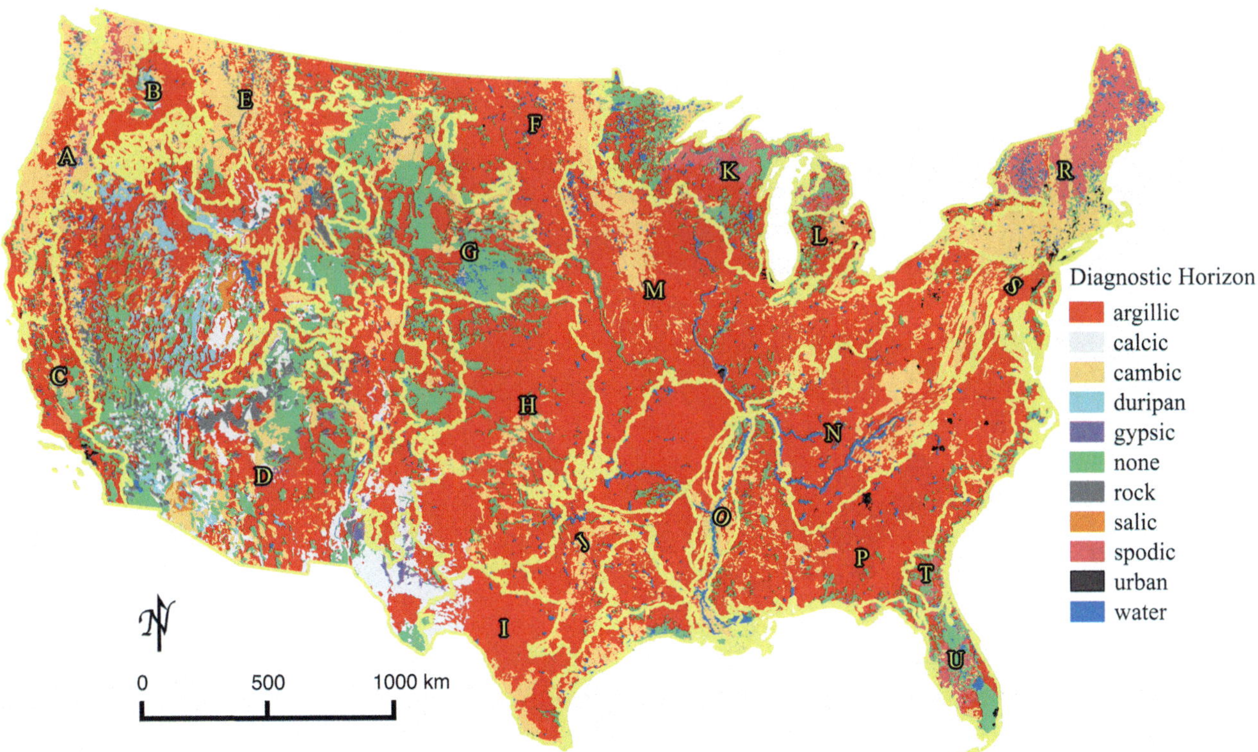

Fig. 4.8 Dominant diagnostic subsurface horizon across the conterminous USA

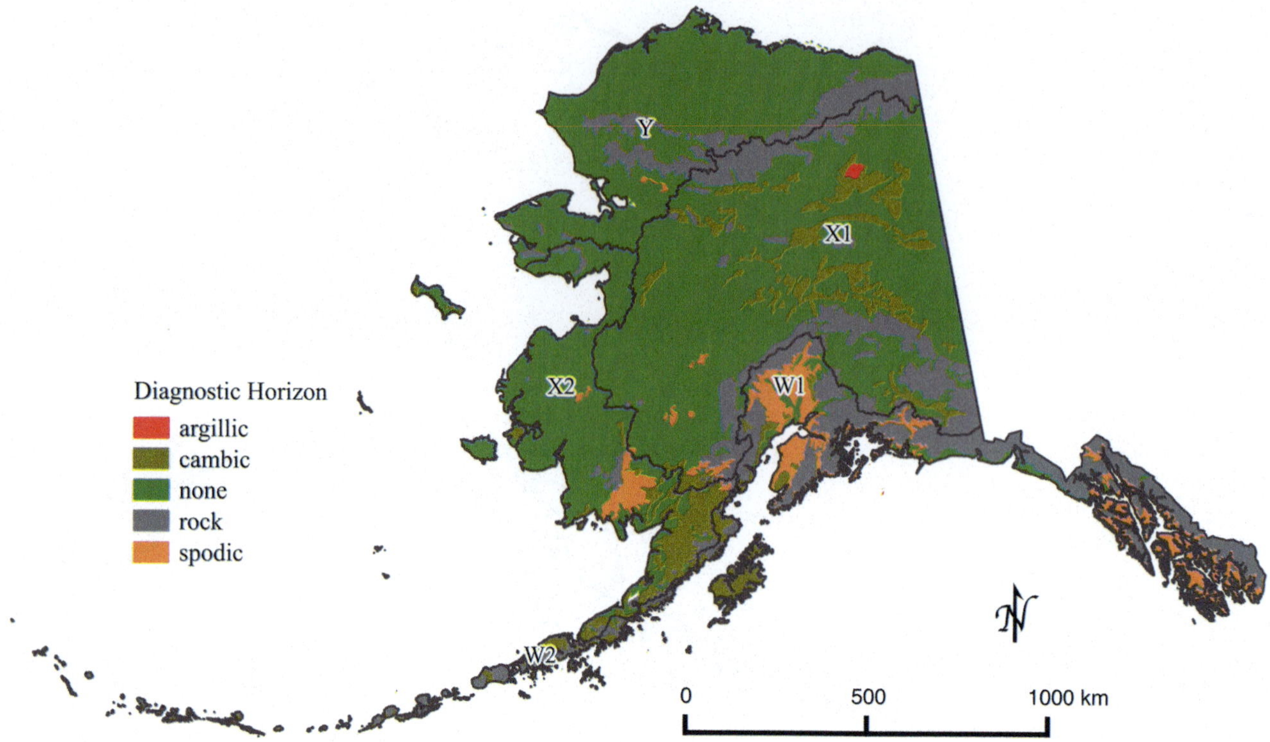

Fig. 4.9 Dominant diagnostic subsurface horizon across Alaska

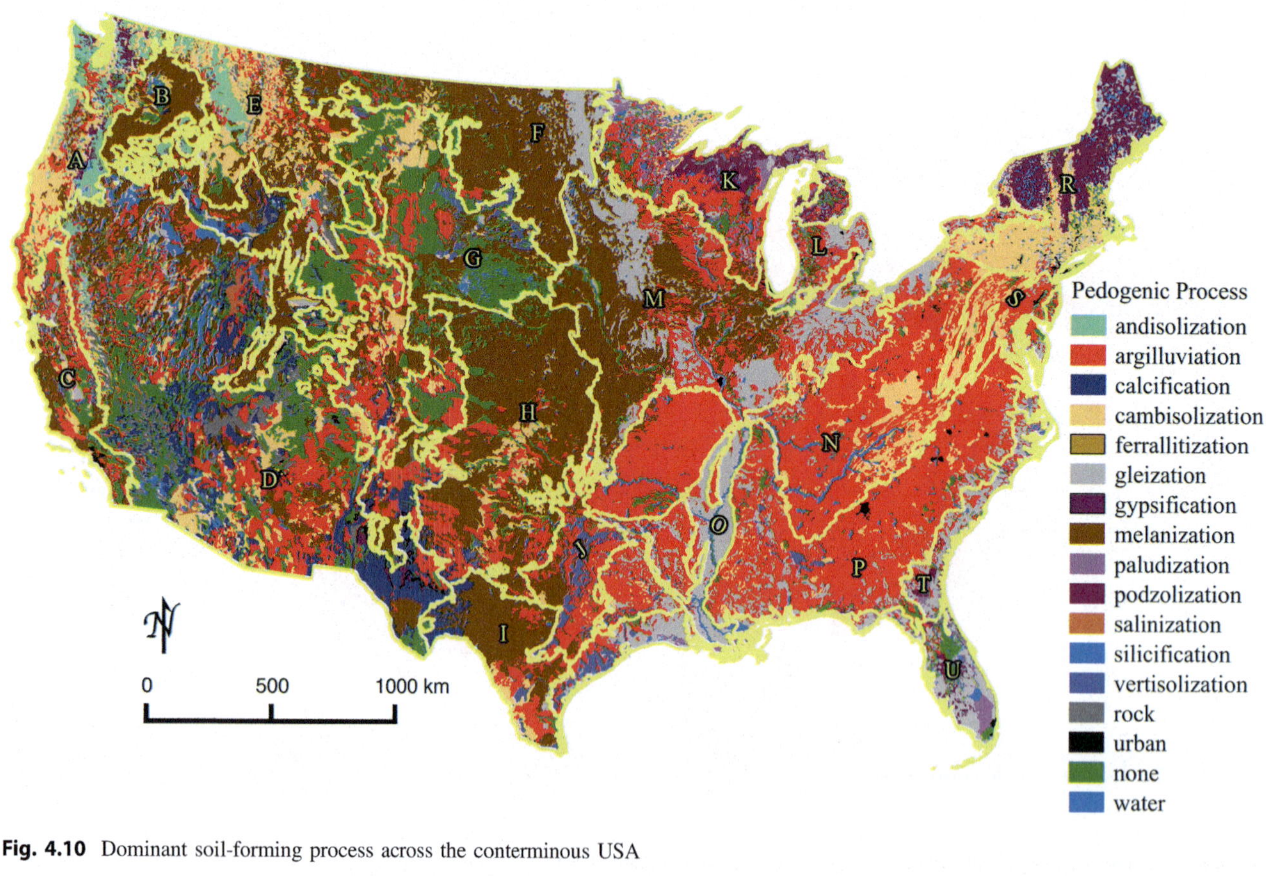

Fig. 4.10 Dominant soil-forming process across the conterminous USA

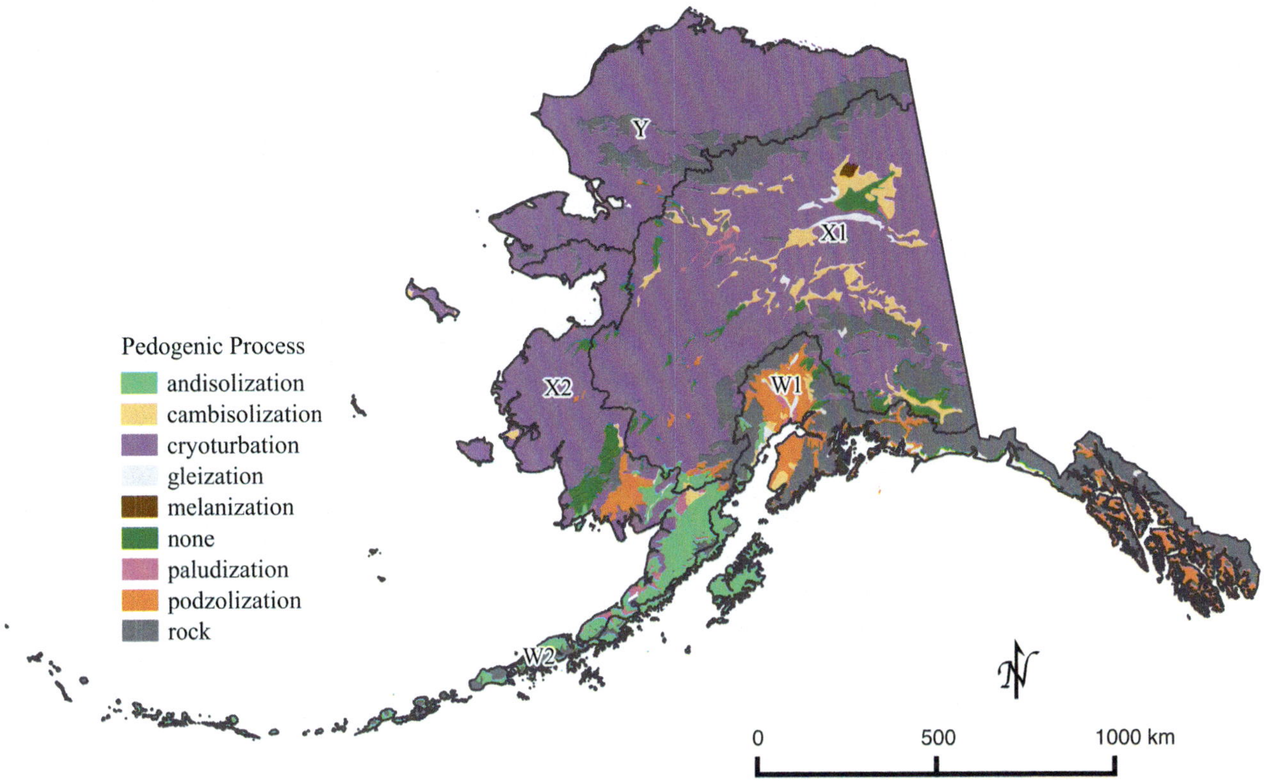

Fig. 4.11 Dominant soil-forming process across Alaska

Podzolization is a dominant process in LRRs K, L, R, and U in about 3.5 % of the USA soils. Vertization, or shrinking and swelling that leads to wedge-shaped aggregates and slickensides, is evident in 2 % of the soils, especially in LRRs J and O. Andisolization, the accumulation of amorphous compounds and development of andic properties, is dominant in LRRs A and E and occurs in 1.7 % of the USA soils. Paludification, the accumulation of organic soil materials, is evident in 1.6 % of the soils in the USA, especially in LRRs K and U.

4.4 Major Land Resource Areas

There are 278 Major Land Resource Areas (MLRAs) within the 28 LRRs. Each MLRA contains one or two dominant suborders that compose about 50 % of the area. These soils are shown in Table 4.3, which also provides information about the soil-forming factors: parent materials, vegetation, elevation, and mean annual precipitation and temperature. In the following section, soil distribution and soil-forming factors are considered for each LRR and its attendant MLRAs.

4.4.1 Northwest Forest, Range, and Specialty Crop Region (A)

This region contains 7 MLRAs that include the Willamette and Puget Sound Valleys (MLRA 2), the Olympic and Cascade Mountains (3 and 6), the Northern Pacific Coast and Siskiyou-Trinity Ranges (1 and 5), and the Pacific coastal temperate rain forest belts of Sitka Spruce (4A) and Coastal Redwoods (4B) (Figs. 1.1 and 5.1).

The valleys contain soils with a xeric soil moisture regime, mainly Xerepts and Xerolls, because moisture is intercepted by the mountains to the west. Soils in the Olympic and Cascade Mountains have a cryic soil temperature regime and are mainly Cryands and Cryods (Fig. 4.2, Table 4.3). The Sitka Spruce belt along the Washington coast contains Udands derived from volcanic ash from the Cascade Stratovolcanoes; and the Coastal Redwoods belt features Humults from thousands of years of litter accumulation and turnover in old-growth redwood stands. Whereas the North Pacific Coast Range has an udic soil moisture regime (Udepts and Udands), the more southerly and drier Siskiyou-Trinity Area and the eastern slopes of the Range have a xeric soil moisture regime (Xerands and Xerolls).

Table 4.3 Dominant suborders and soil-forming factors by Land Resource Region and Major Land Resource Area

Symbol	Land Resource Region and Major Land Resource Area	Dominant Suborders		Dominant Parent materials		Dominant Vegetation		Min. MAP (mm)	Max. MAP (mm)	Min. MAAT (°C)	Max. MAAT (°C)	Min. Elev (m)	Max. Elev (m)
A	Northwestern Forest, Forage, and Specialty Crop Region												
1	Northern Pacific Coast Range, Foothills, and Valleys	Udepts	Udands	Volcanic residuum	Till	Western conifers-alder		1525	5080	4	13	30	760
2	Willamette and Puget Sound Valleys	Xerepts	Xerolls	Alluvium	Till	Western conifers-alder		760	1525	6	12	0	500
3	Olympic and Cascade Mountains	Cryands	Cryods	Till	Outwash	Western conifers-alder		1525	3555	−3	12	200	1710
4A	Sitka Spruce Belt	Udands	Udepts	till	Alluvium	Sitka spruce		1320	1525	7	13	0	550
4B	Coastal Redwood Belt	Udalfs	Udepts	Marine terrace	Colluvium	Coastal redwoods		585	2490	10	15	0	795
5	Siskiyou-Trinity Area	Xerepts	Xeralfs	Sedi residuum	Igne-meta residuum	Pine savanna	Grasslands	355	510	5	17	100	1830
6	Cascade Mountains, Eastern Slope	Cryands	Xerepts	Till	Volcanic ash	Pine savanna	Grasslands	305	2210	0	12	275	2440
B	Northwestern Wheat and Range Region							911	2414	4	13	86	1226
7	Columbia Basin	Cambids	Calcids	Lacustrine	Alluvium	Desert shrub	Grasslands	150	255	9	12	90	610
8	Columbia Plateau	Xerolls	Xerolls	Loess	Volcanic ash	Desert shrub	Grasslands	255	405	8	12	395	1100
9	Palouse and Nez Perce Prairies	Xerolls	Xerolls	Loess	Volcanic ash	Desert shrub	Grasslands	330	710	8	12	610	1220
10	Central Rocky and Blue Mountain Foothills	Xerolls	Xerolls	Volcanic residuum	Alluvium	Desert shrub	Grasslands	205	405	2	12	395	2010
11	Snake River Plains	Calcids	Argids	Loess	Alluvium	Desert shrub	Grasslands	180	305	5	13	640	1525
12	Lost River Valleys and Mountains	Calcids	Xerolls	Alluvium	Colluvium	Desert shrub	Mixed conifers	180	635	2	7	1220	3660
13	Eastern Idaho Plateaus	Xerolls	Cryolls	Loess	Alluvium	Desert shrub	Grasslands	305	635	2	9	1370	2010
C	California Subtropical Fruit, Truck, and Specialty Crop Region							229	479	5	11	674	1734
14	Central California Coastal Valleys	Xerolls	Xeralfs	Alluvium	Alluvium	Oak savanna	Grasslands	280	1675	13	16	0	600
15	Central California Coast Range	Xerolls	Xeralfs	Colluvium	Colluvium	Oak savanna	Grasslands	150	510	10	19	0	810
16	California Delta	Aquents	Aquepts	Alluvium	Alluvium	Marsh	Marsh	305	535	15	16	0	200
17	Sacramento and San Joaquin Valleys	Xeralfs	Cambids	Alluvium	Alluvium	Grasslands	Oak savanna	125	305	15	20	0	200
18	Sierra Nevada Foothills	Xeralfs	Xerolls	Volcanic residuum	Igne-meta residuum	Grasslands	Oak savanna	455	1145	8	20	200	505
19	Southern California Coastal Plain	Xeralfs	Xerolls	Alluvium	Alluvium	Grasslands	Oak savanna	255	735	13	19	0	600
20	Southern California Mountains	Xeralfs	Orthents	Volcanic residuum	Igne-meta residuum	Mixed conifers	Grasslands	205	1295	15	23	305	2400
D	Western Range and Irrigated Region							254	886	13	19	72	759
21	Klamath and Shasta Valleys and Basins	Xerolls	Xerolls	Volcanic residuum	Colluvium	Grasslands	Pine savanna	305	760	4	11	795	1400

(continued)

Table 4.3 (continued)

Symbol	Land Resource Region and Major Land Resource Area	Dominant Suborders		Dominant Parent materials		Dominant Vegetation		Min. MAP (mm)	Max. MAP (mm)	Min. MAAT (°C)	Max. MAAT (°C)	Min. Elev (m)	Max. Elev (m)
22A	Sierra Nevada Mountains	Xerepts	Xeralfs	Igne-meta residuum	Colluvium	Mixed conifers	Grasslands	1015	2030	−4	17	455	2745
22B	Southern Cascade Mountains	Xerands	Xeralfs	-meta residuum	Colluvium	Pine savanna	Mixed conifers	380	2030	1	17	455	2500
23	Malheur High Plateau	Xerolls	Durids	Volcanic residuum	Colluvium	Desert shrub	Grasslands	150	305	4	11	1190	2105
24	Humboldt Area	Argids	Cambids	Alluvium	Loess	Desert shrub	Grasslands	150	305	3	12	1205	1800
25	Owyhee High Plateau	Durids	Xerolls	Alluvium	Loess	Desert shrub	Grasslands	180	405	2	12	915	2300
26	Carson Basin and Mountains	Xerolls	Argids	Igne-meta residuum	Colluvium	Desert shrub	Grasslands	125	915	3	12	1190	1995
27	Fallon-Lovelock Area	Argids	Orthents	Alluvium	Colluvium	Desert shrub	Desert shrub	125	255	6	12	1005	1800
28A	Great Salt Lake Area	Xerolls	Salids	Alluvium	Colluvium	Desert shrub	Grasslands	125	305	4	12	1205	3400
28B	Central Nevada Basin and Range	Durids	Orthents	Alluvium	Colluvium	Desert shrub	Desert shrub	100	305	1	11	1495	3630
29	Southern Nevada Basin and Range	Calcids	Orthents	Alluvium	Colluvium	Desert shrub	Desert shrub	75	305	−2	22	595	2865
30	Mojave Desert	Orthents	Calcids	Alluvium	Colluvium	Desert shrub	Desert shrub	50	205	6	25	−85	1205
31	Lower Colorado Desert	Orthents	Argids	Alluvium	Alluvium	Desert shrub	Desert shrub	75	560	12	24	−84	1205
32	Northern Intermountain Desertic Basins	Orthents	Argids	Alluvium	Colluvium	Desert shrub	Grasslands	150	305	4	9	1190	1800
34A	Cool Central Desertic Basins and Plateaus	Argids	Orthents	Alluvium	Colluvium	Desert shrub	Grasslands	180	305	0	8	1585	2285
34B	Warm Central Desertic Basins and Plateaus	Calcids	Orthents	Alluvium	Colluvium	Desert shrub	Grasslands	150	255	5	12	1250	2285
35	Colorado Plateau	Ustalfs	Argids	Alluvium	Colluvium	Desert shrub	Grasslands	150	455	2	19	1295	1510
36	Southwestern Plateaus, Mesas, and Foothills	Ustalfs	Ustolls	Alluvium	Colluvium	Grasslands	Desert shrub	205	785	3	14	1400	2835
38	Mogollon Transition	Ustolls	Ustalfs	Alluvium	Alluvium	Pine savanna	Grasslands	255	940	8	21	915	1675
39	Arizona and New Mexico Mountains	Ustolls	Ustalfs	Volcanic residuum	Colluvium	Grasslands	Desert shrub	380	760	2	13	1220	2135
40	Sonoran Basin and Range	Argids	Calcids	Alluvium	Colluvium	Desert shrub	Desert shrub	75	255	15	23	300	1100
41	Southeastern Arizona Basin and Range	Argids	Fluvents	Alluvium	Colluvium	Desert shrub	Pine savanna	230	510	8	20	1500	2715
42	Southern Desertic Basin, Plains, and Mountains	Calcids	Argids	Alluvium	Colluvium	Desert shrub	Grasslands	205	355	10	22	795	1510
E	Rocky Mountain Range and Forest Region							210	592	4	16	947	2122
43A	Northern Rocky Mountains	Vitrands	Udepts	Volcanic ash	Till	Mixed conifers	Mixed conifers	635	1525	0	11	550	2135
43B	Central Rocky Mountains	Cryolls	Cryepts	Till	Colluvium	Mixed conifers	Pine savanna	635	1525	−4	9	1830	2440

(continued)

Table 4.3 (continued)

Symbol	Land Resource Region and Major Land Resource Area	Dominant		Dominant		Dominant		Min.	Max.	Min.	Max.	Min.	Max.
		Suborders		Parent materials		Vegetation		MAP (mm)	MAP (mm)	MAAT (°C)	MAAT (°C)	Elev (m)	Elev (m)
43C	Blue and Seven Devils Mountains	Xerolls	Cryands	Volcanic ash	Colluvium	Pine savanna	Mixed conifers	305	1090	1	7	395	2410
44	Northern Rocky Mountain Valleys	Ustepts	Ustolls	Alluvium	Colluvium	Grasslands	Mixed conifers	305	405	1	9	535	2105
46	Northern Rocky Mountain Foothills	Ustolls	Orthents	Colluvium	Alluvium	Grasslands	Pine savanna	305	510	1	9	1000	2400
47	Wasatch and Uinta Mountains	Xerolls	Cryolls	Colluvium	Alluvium	Grasslands	Mixed conifers	380	760	−1	15	1495	4115
48A	Southern Rocky Mountains	Cryolls	Cryepts	Colluvium	Colluvium	Grasslands	Mixed conifers	180	1600	−3	12	1980	4390
48B	Southern Rocky Mountain Parks	Cryalfs	Cryepts	Alluvium	Outwash	Desert shrub	Grasslands	255	405	1	6	2395	3310
49	Southern Rocky Mountain Foothills	Cryolls	Cryalfs	Igne-meta residuum	Colluvium	Grasslands	Pine savanna	305	635	2	12	1525	2440
51	High Intermountain Valleys	Argids	Calcids	Alluvium	Alluvium	Desert shrub	Grasslands	180	255	4	8	2105	2700
F	Northern Great Plains Spring Wheat Region							348.5	871	0.2	9.8	1381	2844.5
52	Brown Glaciated Region	Ustolls	Ustalfs	Till	Alluvium	Grasslands	Grasslands	255	430	4	7	600	1400
53A	Northern Dark Brown Glaciated Plains	Ustolls	Ustepts	Till	Alluvium	Grasslands	Grasslands	305	380	3	7	595	900
53B	Central Dark Brown Glaciated Plains	Ustolls	Aquolls	Till	Alluvium	Grasslands	Grasslands	355	510	3	7	500	600
53C	Southern Dark Brown Glaciated Plains	Ustolls	Aquolls	Till	Alluvium	Grasslands	Grasslands	380	635	6	9	395	700
54	Rolling Soft Shale Plain	Ustolls	Orthents	Alluvium	Shale residuum	Grasslands	Grasslands	355	455	3	8	505	1100
55A	Northern Black Glaciated Plains	Udolls	Aquolls	Till	Alluvium	Grasslands	Grasslands	355	485	2	5	365	775
55B	Central Black Glaciated Plains	Udolls	Aquolls	Till	Alluvium	Grasslands	Grasslands	405	535	3	7	305	625
55C	Southern Black Glaciated Plains	Ustolls	Aquolls	Till	Alluvium	Grasslands	Grasslands	425	635	6	10	400	600
56	Red River Valley of the North	Aquerts	Aquolls	Glaciolacustrine	Outwash	Grasslands	Grasslands	455	585	2	7	200	305
G	Western Great Plains Range and Irrigated Region							366	517	4	7	429	778
58A	Northern Rolling High Plains, Northern Part	Ustepts	Orthents	Sedi residuum	Alluvium	Grasslands	Grasslands	205	560	5	10	900	1000
58B	Northern Rolling High Plains, Southern Part	Argids	Orthents	Alluvium	Colluvium	Grasslands	Grasslands	230	685	5	9	1800	2100
58C	Northern Rolling High Plains, Northeastern Part	Orthents	Ustolls	Alluvium	Sedi residuum	Grasslands	Grasslands	355	430	5	7	600	1000
58D	Northern Rolling High Plains, Eastern Part	Orthents	Ustolls	Sedi residuum	Alluvium	Grasslands	Grasslands	355	430	6	7	700	1200
60A	Pierre Shale Plains	Orthents	Ustalfs	Shale residuum	Alluvium	Grasslands	Grasslands	330	560	6	9	800	1100
60B	Pierre Shale Plains, Northern Part	Ustalfs	Orthents	Alluvium	Shale residuum	Grasslands	Grasslands	280	380	6	8	900	1005

(continued)

Table 4.3 (continued)

Symbol	Land Resource Region and Major Land Resource Area	Dominant		Dominant		Dominant		Min.	Max.	Min.	Max.	Min.	Max.
		Suborders		Parent materials		Vegetation		MAP (mm)	MAP (mm)	MAAT (°C)	MAAT (°C)	Elev (m)	Elev (m)
61	Black Hills Foot Slopes	Ustalfs	Orthents	Shale residuum	Alluvium	Grasslands	Oak savanna	355	610	5	9	900	1200
62	Black Hills	Ustalfs	Ustolls	Igne-meta residuum	Alluvium	Pine savanna	Grasslands	405	940	2	9	1100	2000
63A	Northern Rolling Pierre Shale Plains	Orthents	Ustolls	Shale residuum	Alluvium	Grasslands	Grasslands	355	485	6	10	500	900
63B	Southern Rolling Pierre Shale Plains	Orthents	Ustolls	Shale residuum	Alluvium	Grasslands	Grasslands	430	635	7	10	400	500
64	Mixed Sandy and Silty Tableland and Badlands	Orthents	Ustolls	Eolian	Sedi residuum	Grasslands	Grasslands	330	485	6	10	900	1200
65	Nebraska Sand Hills	Psamments	Psamments	Eolian	Eolian	Grasslands	Grasslands	380	660	8	10	600	1200
66	Dakota-Nebraska Eroded Tableland	Ustolls	Psamments	Eolian	Alluvium	Grasslands	Grasslands	455	635	8	9	600	900
67A	Central High Plains, Northern Part	Ustolls	Orthents	Sedi residuum	Eolian	Grasslands	Grasslands	305	485	6	10	975	1675
67B	Central High Plains, Southern Part	Ustolls	Ustalfs	Sedi residuum	Alluvium	Grasslands	Grasslands	305	455	7	13	915	2380
69	Upper Arkansas Valley Rolling Plains	Orthents	Ustalfs	Eolian	Alluvium	Grasslands	Grasslands	255	485	8	12	1100	1900
70A	Canadian River Plains and Valleys	Ustolls	Argids	Sedi residuum	Alluvium	Grasslands	Grasslands	255	535	8	14	1525	2135
70B	Upper Pecos River Valley	Argids	Calcids	Alluvium	Sedi residuum	Grasslands	Grasslands	330	380	12	16	1130	1615
70C	Central New Mexico Highlands	Calcids	Argids	Sedi residuum	Alluvium	Juniper-pinyon	Grasslands	280	380	8	16	1525	2255
70D	Southern Desert Foothills	Ustolls	Calcids	Sedi residuum	Alluvium	Grasslands	Grasslands	305	560	11	17	1220	1830
H	Central Great Plains Winter Wheat and Range Region							325	539	7	11	954.5	1454.75
71	Central Nebraska Loess Hills	Orthents	Ustolls	Loess	Alluvium	Grasslands	Grasslands	535	735	8	11	500	800
72	Central High Tableland	Ustolls	Orthents	Loess	Alluvium	Grasslands	Grasslands	355	635	8	14	795	1190
73	Rolling Plains and Breaks	Ustolls	Ustolls	Loess	Alluvium	Grasslands	Grasslands	485	760	9	14	505	915
74	Central Kansas Sandstone Hills	Ustolls	Ustolls	Sedi residuum	Alluvium	Grasslands	Grasslands	660	840	11	13	400	500
75	Central Loess Plains	Ustolls	Ustolls	Loess	Loess	Grasslands	Grasslands	585	915	10	15	505	600
76	Bluestem Hills	Ustolls	Udolls	Ls residuum	Colluvium	Grasslands	Grasslands	785	965	11	15	300	505
77A	Southern High Plains, Northern Part	Ustolls	Ustalfs	Loess	Eolian	Grasslands	Grasslands	380	560	12	14	885	1370
77B	Southern High Plains, Northwestern Part	Ustalfs	Ustalfs	Eolian	Eolian	Grasslands	Grasslands	355	455	10	14	1160	1585
77C	Southern High Plains, Southern Part	Ustalfs	Ustolls	Eolian	Lacustrine	Grasslands	Grasslands	405	560	13	17	795	1400
77D	Southern High Plains, Southwestern Part	Calcids	Ustalfs	Alluvium	Eolian	Grasslands	Grasslands	330	485	13	18	855	1585
77E	Southern High Plains, Breaks	Ustalfs	Ustolls	Sedi residuum	Alluvium	Grasslands	Grasslands	405	635	13	16	670	1465
78A	Rolling Limestone Prairie	Ustolls	Orthents	Ls residuum	Alluvium	Grasslands	Grasslands	585	735	17	19	370	695

(continued)

Table 4.3 (continued)

Symbol	Land Resource Region and Major Land Resource Area	Dominant		Dominant		Dominant		Min.	Max.	Min.	Max.	Min.	Max.
		Suborders		Parent materials		Vegetation		MAP (mm)	MAP (mm)	MAAT (°C)	MAAT (°C)	Elev (m)	Elev (m)
78B	Central Rolling Red Plains, Western Part	Ustolls	Ustepts	Sedi residuum	Alluvium	Grasslands	Grasslands	485	660	14	18	440	895
78C	Central Rolling Red Plains, Eastern Part	Ustalfs	Ustepts	Sedi residuum	Alluvium	Grasslands	Grasslands	560	760	13	18	305	610
79	Great Bend Sand Plains	Ustolls	Ustalfs	Alluvium	Eolian	Grasslands	Grasslands	610	890	13	15	505	795
80A	Central Rolling Red Prairies	Ustolls	Ustolls	Alluvium	Sedi residuum	Grasslands	Grasslands	635	965	14	18	260	455
80B	Texas North-Central Prairies	Ustalfs	Ustolls	Sedi residuum	Ls residuum	Oak savanna	Grasslands	660	840	17	19	200	705
I	Southwest Plateaus and Plains Range and Cotton Region							519	729	12	16	556	945
81A	Edwards Plateau, Western Part	Ustolls	Calcids	Ls residuum	Alluvium	Desert shrub	Grasslands	380	660	16	21	305	825
81B	Edwards Plateau, Central Part	Ustolls	Ustolls	Ls residuum	Alluvium	Grasslands	Desert Shrub	485	815	17	20	275	760
81C	Edwards Plateau, Eastern Part	Ustolls	Ustepts	Ls residuum	Alluvium	Grasslands	Grasslands	610	760	17	20	275	610
81D	Southern Edwards Plateau	Calcids	Orthents	Ls residuum	Alluvium	Desert shrub	Grasslands	255	380	17	21	365	1065
82A	Texas Central Basin	Ustalfs	Ustolls	Igne-meta residuum	Colluvium	Oak savanna	Grasslands	610	785	18	19	365	670
83A	Northern Rio Grande Plain	Ustalfs	Ustolls	Sedi residuum	Alluvium	Oak savanna	Grasslands	535	940	20	22	60	305
83B	Western Rio Grande Plain	Argids	Calcids	Sedi residuum	Alluvium	Grasslands	Grasslands	455	635	20	23	50	365
83C	Central Rio Grande Plain	Ustalfs	Ustolls	Ls residuum	Eolian	Grasslands	Grasslands	535	735	21	23	45	260
83D	Lower Rio Grande Plain	Ustalfs	Ustolls	Sedi residuum	Alluvium	Grasslands	Grasslands	560	685	22	23	85	185
83E	Sandsheet Prairie	Ustalfs	Aqualfs	Alluvium	Eolian	Grasslands	Grasslands	560	710	22	23	0	280
J	Southwestern Prairies Cotton and Forage Region							499	711	19	22	183	533
82B	Wichita Mountains	Ustolls	Ustolls	Igne-meta residuum	Colluvium	Grasslands	Grasslands	660	785	15	17	335	615
84A	North Cross Timbers	Ustalfs	Udalfs	Sedi residuum	Alluvium	Oak savanna	Grasslands	760	1015	13	17	300	395
84B	West Cross Timbers	Ustalfs	Udalfs	Sedi residuum	Alluvium	Oak savanna	Grasslands	660	1065	17	19	305	395
84C	East Cross Timbers	Ustalfs	Ustalfs	Sedi residuum	Colluvium	Oak savanna	Grasslands	865	1040	17	19	150	305
85	Grand Prairie	Ustolls	Usterts	Ls residuum	Alluvium	Grasslands	Oak savanna	685	1040	16	19	150	400
86A	Texas Blackland Prairie, Northern Part	Usterts	Ustolls	Alluvium	Alluvium	Grasslands	Grasslands	760	1170	17	21	90	185
86B	Texas Blackland Prairie, Southern Part	Usterts	Ustolls	Alluvium	Alluvium	Grasslands	Grasslands	890	1120	19	21	60	185
87A	Texas Claypan Area, Southern Part	Ustalfs	Ustalfs	Sedi residuum	Alluvium	Oak savanna	Grasslands	685	1145	18	21	60	230
87B	Texas Claypan Area, Northern Part	Udalfs	Aqualfs	Sedi residuum	Alluvium	Oak savanna	Oak savanna	990	1145	17	19	75	230

(continued)

Table 4.3 (continued)

Symbol	Land Resource Region and Major Land Resource Area	Dominant Suborders		Dominant Parent materials		Dominant Vegetation		Min. MAP (mm)	Max. MAP (mm)	Min. MAAT (°C)	Max. MAAT (°C)	Min. Elev (m)	Max. Elev (m)
K	Northern Lake States Forest and Forage Region							773	1058	17	19	169	327
57	Northern Minnesota Gray Drift	Udalfs	Aqualfs	Till	Till	Northern hardwoods	Mixed pines	585	735	3	6	300	500
88	Northern Minnesota Glacial Lake Basins	Udalfs	Saprists	Glaciolacustrine	Glaciolacustrine	Northern hardwoods	Mixed pines	510	735	2	5	275	410
89	Wisconsin Central Sands	Psamments	Psamments	Glaciolacustrine	Glaciolacustrine	Pine-oak	Pine-oak	760	840	6	7	270	425
90A	Wisconsin and Minnesota Thin Loess and Till, Northern Part	Udalfs	Psamments	Loess/till	Loess/till	Northern hardwoods	Northern hardwoods	660	865	3	7	335	595
90B	Wisconsin and Minnesota Thin Loess and Till, Souhern Part	Udalfs	Udalfs	Loess/till	Loess/till	Northern hardwoods	Northern hardwoods	685	840	4	8	205	470
91A	Central Minnesota Sandy Outwash	Udolls	Saprists	Outwash	Outwash	Pine-oak	Swamp conifers	585	735	4	6	250	450
91B	Wisconsin and Minnesota Sandy Outwash	Orthods	Psamments	Outwash	Outwash	Pine-oak	Swamp conifers	635	865	3	8	245	455
92	Superior Lake Plain	Udalfs	Orthods	Glaciolacustrine	glaciolacustrine	Mixed decid-evergreen	Mixed decid-evergreen	685	940	4	6	185	425
93A	Superior Stony and Rocky Loamy Plains and Hills, Western Part	Udepts	Hemists	Till	Till	Mixed decid-evergreen	Swamp conifers	635	760	2	4	185	700
93B	Superior Stony and Rocky Loamy Plains and Hills, Eastern Part	Orthods	Orthods	Till	Till	Mixed decid-evergreen	Swamp conifers	760	965	3	6	185	600
94A	Northern Michigan and Wisconsin Sandy Drift	Orthods	Psamments	TILL	Outwash	Northern hardwoods	Pine barrens	685	760	5	9	260	525
94B	Michigan Eastern Upper Peninsula Sandy Drift	Orthods	Udalfs	Outwash	Till	Northern hardwoods	Mixed decid-evergreen	760	915	4	6	175	425
94C	Michigan Northern Lower Peninsula Sandy Drift	Orthods	Udalfs	Outwash	Till	Northern hardwoods	Mixed decid-evergreen	710	865	5	7	175	295
94D	Northern Highland Sandy Drift	Orthods	Saprists	Outwash	Till	Northern hardwoods	Swamp conifers	760	890	4	5	455	565
95A	Northeastern Wisconsin Drift Plain	Udalfs	Udalfs	Till	Glaciolacustrine	Northern hardwoods	Mixed decid-evergreen	760	915	5	8	200	400
95B	Southern Wisconsin and Northern Illinois Drift Plain	Udalfs	Udolls	Till	Till	Northern hardwoods	Grasslands	760	965	6	9	200	300
L	Lake States Fruit, Truck Crop, and Dairy Region							683	849	4	7	244	471

(continued)

Table 4.3 (continued)

Symbol	Land Resource Region and Major Land Resource Area	Dominant Suborders		Dominant Parent materials		Dominant Vegetation		Min. MAP (mm)	Max. MAP (mm)	Min. MAAT (°C)	Max. MAAT (°C)	Min. Elev (m)	Max. Elev (m)
96	Western Michigan Fruit Belt	Orthods	Psamments	Till	Outwash	Northern hardwoods	Mixed pines	760	915	5	9	175	295
97	Southwestern Michigan Fruit and Truck Crop Belt	Orthods	Udalfs	Till	Outwash	Oak-hickory	Mixed hardwoods	890	1015	8	11	200	305
98	Southern Michigan and Northern Indiana Drift Plain	Udalfs	Aquolls	Till	Till	Oak-hickory	Mixed hardwoods	735	1015	7	10	175	335
99	Erie-Huron Lake Plain	Udalfs	Aqualfs	Glaciolacustrine	Glaciolacustrine	Oak-hickory	Mixed hardwoods	760	915	7	10	200	230
101	Ontario-Erie Plain and Finger Lakes Region	Udalfs	Udepts	Glaciolacustrine	Outwash	Northern hardwoods	Mixed decid-evergreen	735	1145	5	10	100	400
M	Central Feed Grains and Livestock Region							776	1001	6	10	170	313
102A	Rolling Till Prairie	Udolls	Aquolls	Till	Till	Grasslands	Grasslands	485	735	4	7	305	410
102B	Till Plains	Ustolls	Aquolls	Till	Till	Grasslands	Grasslands	585	660	6	9	350	575
102C	Loess Uplands	Ustolls	Aquolls	Loess	Loess	Grasslands	Grasslands	585	760	6	11	335	610
103	Central Iowa and Minnesota Till Prairies	Udolls	Aquolls	Till	Till	Grasslands	Grasslands	585	890	6	10	300	400
104	Eastern Iowa and Minnesota Till Prairies	Udolls	Udalfs	Till	Till	Grasslands	Grasslands	735	940	7	10	300	400
105	Northern Mississippi Valley Loess Hills	Udalfs	Udolls	Loess	Sedi residuum	Oak-hickory	Grasslands	760	965	6	10	200	400
106	Nebraska and Kansas Loess Drift Hills	Udolls	Udolls	Loess	Till	Grasslands	Grasslands	710	1015	10	13	300	505
107A	Iowa and Minnesota Loess Hills	Udolls	Aquolls	Loess	Till	Grasslands	Grasslands	660	790	7	9	340	520
107B	Iowa and Missouri Deep Loess Hills	Udolls	Udolls	Loess	Till	Grasslands	Grasslands	660	1040	8	13	185	475
108A	Illinois and Iowa Deep Loess and Drift, Eastern Part	Udolls	Udalfs	Loess	Till	Grasslands	Oak-hickory	890	1090	8	12	200	300
108B	Illinois and Iowa Deep Loess and Drift, East-Central Part	Udolls	Udalfs	Loess	Till	Grasslands	Oak-hickory	840	990	8	12	200	300
108C	Illinois and Iowa Deep Loess and Drift, West-Central Part	Udolls	Aquolls	Loess	Till	Grasslands	Grasslands	840	965	8	11	155	340
108D	Illinois and Iowa Deep Loess and Drift, Western Part	Udolls	Udalfs	Loess	Till	Grasslands	Oak-hickory	840	940	9	11	210	460
109	Iowa and Missouri Heavy Till Plain	Udolls	Aquolls	Till	Till	Grasslands	Grasslands	865	1040	9	12	200	300
110	Northern Illinois and Indiana Heavy Till Plain	Udolls	Aquolls	Till	Till	Grasslands	Grasslands	785	1015	7	11	200	250

(continued)

Table 4.3 (continued)

Symbol	Land Resource Region and Major Land Resource Area	Dominant Suborders		Dominant Parent materials		Dominant Vegetation		Min. MAP (mm)	Max. MAP (mm)	Min. MAAT (°C)	Max. MAAT (°C)	Min. Elev (m)	Max. Elev (m)
111A	Indiana and Ohio Till Plain, Central Part	Udalfs	Aquolls	Till	Till	Oak-hickory	Swamp hardwoods	915	1090	9	12	205	380
111B	Indiana and Ohio Till Plain, Northeastern Part	Udalfs	Aqualfs	Till	Till	Oak-hickory	Swamp hardwoods	760	990	8	11	190	470
111C	Indiana and Ohio Till Plain, Northwestern Part	Udalfs	Aquolls	Till	Till	Oak-hickory	Swamp hardwoods	890	990	9	11	190	285
111D	Indiana and Ohio Till Plain, Western Part	Udalfs	Aquolls	Till	Till	Oak-hickory	Grasslands	915	1090	10	12	160	320
111E	Indiana and Ohio Till Plain, Eastern Part	Aqualfs	Aquolls	Till	Till	Oak-hickory	Swamp hardwoods	890	1040	9	11	175	425
112	Cherokee Prairies	Ustolls	Udalfs	Ls residuum	Alluvium	Grasslands	Oak-hickory	865	1145	11	17	100	400
113	Central Claypan Areas	Aqualfs	Udalfs	Till	Alluvium	Grasslands	Swamp hardwoods	915	1170	11	14	200	300
114A	Southern Illinois and Indiana Thin Loess and Till Plain, Eastern Part	Udalfs	Aqualfs	Loess	Till	Oak-hickory	Swamp hardwoods	940	1170	9	14	100	380
114B	Southern Illinois and Indiana Thin Loess and Till Plain, Western Part	Udalfs	Aqualfs	Till	Loess	Oak-hickory	Swamp hardwoods	940	1170	11	14	105	365
115A	Central Mississippi Valley Wooded Slopes, Eastern Part	Udalfs	Aqualfs	Loess	Loess	Oak-hickory	Swamp hardwoods	1015	1195	11	14	100	310
115B	Central Mississippi Valley Wooded Slopes, Western Part	Udalfs	Aqualfs	Loess	Loess	Oak-hickory	Swamp hardwoods	965	1220	12	14	100	310
115C	Central Mississippi Valley Wooded Slopes, Northern Part	Udalfs	Aqualfs	Loess	Loess	Oak-hickory	Swamp hardwoods	865	1015	9	13	130	270
N	East and Central Farming and Forest Region							804	1004	8	12	205	387
116A	Ozark Highland	Udults	Udalfs	Ls residuum	Ls residuum	Oak-hickory	Mixed pines	965	1145	12	16	60	245
116B	Springfield Plain	Udults	Udalfs	Ls residuum	Ls residuum	Oak-hickory	Grasslands	1040	1145	13	15	305	520
116C	St. Francois Knobs and Basins	Udults	Udalfs	Ls residuum	Ls residuum	Oak-hickory	Mixed pines	1015	1170	12	14	135	540
117	Boston Mountains	Udults	Udepts	Sedi residuum	Alluvium	Oak-hickory	Mixed pines	1065	1395	13	16	200	800
118A	Arkansas Valley and Ridges, Eastern Part	Udults	Udults	Sedi residuum	Alluvium	Oak-hickory	Oak savanna	1040	1145	14	17	90	840
118B	Arkansas Valley and Ridges, Western Part	Udalfs	Udepts	Sedi residuum	Alluvium	Oak-hickory	Oak savanna	990	1170	15	17	170	455
119	Ouachita Mountains	Udults	Udalfs	Sedi residuum	Alluvium	Oak-hickory	Mixed decid-evergreen	1270	1675	14	17	100	800
120A	Kentucky and Indiana Sandstone and Shale Hills and Valleys, Southern Part	Udalfs	Udalfs	Loess	Sedi residuum	Oak-hickory	Mixed hardwoods	1145	1370	13	14	105	290

(continued)

Table 4.3 (continued)

Symbol	Land Resource Region and Major Land Resource Area	Dominant		Dominant		Dominant		Min.	Max.	Min.	Max.	Min.	Max.
		Suborders		Parent materials		Vegetation		MAP (mm)	MAP (mm)	MAAT (°C)	MAAT (°C)	Elev (m)	Elev (m)
120B	Kentucky and Indiana Sandstone and Shale Hills and Valleys, Northwestern Part	Udalfs	Udepts	Loess	Sedi residuum	Oak-hickory	Mixed hardwoods	1090	1220	11	13	105	290
120C	Kentucky and Indiana Sandstone and Shale Hills and Valleys, Northeastern Part	Udalfs	Udults	Loess	Sedi residuum	Oak-hickory	Mixed hardwoods	1040	1195	11	14	115	325
121	Kentucky Bluegrass	Udalfs	Udolls	Sedi residuum	Loess	Mixed hardwoods	Mixed hardwoods	1040	1145	10	14	200	300
122	Highland Rim and Pennyroyal	Udalfs	Udults	Ls residuum	Loess	Oak-hickory	Mixed hardwoods	1090	1600	11	16	200	300
123	Nashville Basin	Udalfs	Udolls	Ls residuum	Loess	Oak-hickory	Mixed hardwoods	1220	1450	14	16	135	405
124	Western Allegheny Plateau	Udults	Udepts	Sedi residuum	Colluvium	Oak-hickory	Mixed pines	940	1145	8	13	200	400
125	Cumberland Plateau and Mountains	Udults	Udepts	Sedi residuum	Colluvium	Mixed hardwoods	Oak-hickory	940	1145	10	15	200	1205
126	Central Allegheny Plateau	Udalfs	Udults	Sedi residuum	Colluvium	Oak-hickory	Mixed hardwoods	865	1145	9	13	200	400
127	Eastern Allegheny Plateau and Mountains	Udults	Udepts	Sedi residuum	Colluvium	Mixed hardwoods	Mixed pines	840	1725	6	12	300	1400
128	Southern Appalachian Ridges and Valleys	Udepts	Udults	Sedi residuum	Colluvium	Oak-hickory	Pine-oak	1040	1395	11	17	200	1500
129	Sand Mountain	Udults	Udepts	Sedi residuum	Colluvium	Oak-hickory	Pine-oak	1345	1525	13	17	100	700
130A	Northern Blue Ridge	Udepts	Udults	Sedi residuum	Colluvium	Oak-hickory	Mixed hardwoods	915	1145	9	14	250	1280
130B	Southern Blue Ridge	Udults	Udepts	Colluvium	Colluvium	Oak-hickory	Mixed hardwoods	915	1525	8	16	275	2010
O	Mississippi Delta Cotton and Feed Grains Region							1039	1313	11	15	174	715
131A	Southern Mississippi Alluvium	Aquerts	Aquepts	Alluvium	Alluvium	Swamp hardwoods	Swamp hardwoods	1170	1525	14	21	5	100
131B	Arkansas River Alluvium	Aquerts	Fluvents	Alluvium	Alluvium	Swamp hardwoods	Swamp hardwoods	1245	1475	16	18	15	75
131C	Red River Alluvium	Uderts	Aquepts	Alluvium	Alluvium	Swamp hardwoods	Swamp hardwoods	1195	1575	17	19	12	80
131D	Southern Mississippi River Terraces	Udalfs	Aqualfs	Alluvium	Alluvium	Mixed hardwoods	Grasslands	1245	1420	16	18	15	75

(continued)

Table 4.3 (continued)

Symbol	Land Resource Region and Major Land Resource Area	Dominant Suborders		Dominant Parent materials		Dominant Vegetation		Min. MAP (mm)	Max. MAP (mm)	Min. MAAT (°C)	Max. MAAT (°C)	Min. Elev (m)	Max. Elev (m)
P	South Atlantic and Gulf Slope Cash Crops, Forest, and Livestock Region							1245	1420	16	18	15	75
133A	Southern Coastal Plain	Udults	Udults	Coastal plain	Alluvium	Pine-oak	Mixed pine	1040	1525	13	20	25	200
133B	Western Coastal Plain	Udults	Udalfs	Coastal plain	Alluvium	Pine-oak	Mixed pine	990	1600	16	20	25	200
134	Southern Mississippi Loess	Udalfs	Udalfs	Loess	Loess	Pine-oak	Mixed pine	1195	1525	14	20	25	185
135A	Alabama and Mississippi Blackland Prairie	Uderts	Aquepts	Alluvium	Alluvium	Pine-oak	Pine-oak	1345	1550	16	18	30	180
135B	Cretaceous Western Coastal Plain	Udults	Aqualfs	Ls residuum	Alluvium	Mixed hardwoods	Mixed hardwoods	1040	1400	16	18	25	105
136	Southern Piedmont	Udults	Udepts	Igne-meta residuum	Igne-meta residuum	Mixed decid-evergreen	Pine-oak	940	1145	12	18	100	400
137	Carolina and Georgia Sand Hills	Udults	Psamments	Coastal plain	Coastal plain	Pine-oak	Pine-oak	1040	1345	15	18	50	200
138	North-Central Florida Ridge	Udults	Psamments	Igne-meta residuum	Coastal plain	Pine-oak	Pine-oak	1345	1525	19	21	25	50
Q	Pacific Basin Region							1117	1452	15	19	38	190
190	Stratovolcanoes of the Mariana Islands	Ustands	Ustepts	Lava flows	Volcanic ash	Tropical forest	Grasslands	2030	2030	26	26	0	970
191	High Limestone Plateaus of the Mariana Islands	Ustalfs	Ustolls	Ls residuum	Volcanic residuum	Tropical forest	Grasslands	2030	2540	26	26	0	500
192	Volcanic Highlands of the Mariana Islands	Ustox	Ustalfs	Volcanic residuum	Volcanic residuum	Tropical forest	Grasslands	2160	2540	26	26	0	400
193	Volcanic Islands of Western Micronesia	Perox	Perox	Volcanic residuum	Volcanic residuum	Tropical forest	Grasslands	3100	3685	27	27	0	240
194	Low Limestone islands of Western Micronesia	Rendolls	Rendolls	Ls residuum	Volcanic residuum	Tropical forest	Grasslands	3760	3760	27	27	0	80
195	Volcanic islands of Central and Eastern Micronesia	Perox	Udepts	Volcanic residuum	Lava flows	Tropical forest	Tropical forest	3685	5715	27	27	0	800
196	Coral Atolls of Micronesia	Orthents	Psamments	Coral	Volcanic residuum	Strand forests	Strand forests	1015	3685	27	27	0	1000
197	Volcanic Islands of American Samoa	Udolls	Udands	Volcanic residuum	Volcanic residuum	Tropical palms	Tropical forest	3175	6350	27	27	0	900
R	Northeastern Forage and Forest Region							2619	3788	27	27	0	611
139	Lake Erie Glaciated Plateau	Udalfs	Aqualfs	Till	Outwash	Northern hardwoods	Northern hardwoods	865	1270	7	10	200	305
140	Glaciated Allegheny Plateau and Catskill Mountains	Udepts	Aquepts	Outwash	Glaciolacustrine	Northern hardwoods	Mixed oak	760	1145	4	10	200	1100
141	Tughill Plateau	Orthods	Aquods	Till	Till	Northern hardwoods	Spruce-fir	1145	1600	4	8	300	600

(continued)

Table 4.3 (continued)

Symbol	Land Resource Region and Major Land Resource Area	Dominant		Dominant		Dominant		Min.	Max.	Min.	Max.	Min.	Max.
		Suborders		Parent materials		Vegetation		MAP (mm)	MAP (mm)	MAAT (°C)	MAAT (°C)	Elev (m)	Elev (m)
142	St. Lawrence-Champlain Plain	Aqualfs	Udepts	Till	Outwash	Northern hardwoods	Northern hardwoods	760	1145	5	9	25	305
143	Northeastern Mountains	Orthods	Aquepts	Till	Till	Northern hardwoods	Spruce-fir	815	1145	1	8	305	1500
144A	New England and Eastern New York Upland, Northern Part	Udepts	Orthents	Till	Outwash	Northern hardwoods	Oak-hickory	890	1145	6	12	0	305
144B	New England and Eastern New York Upland, Southern Part	Orthods	Udepts	Till	Till	Northern hardwoods	Spruce-fir	840	1145	4	9	0	900
145	Connecticut Valley	Udepts	Orthents	Alluvium	Alluvium	Northern hardwoods	Oak-hickory	840	1320	6	11	0	305
146	Aroostook Area	Orthods	Udepts	Till	Outwash	Northern hardwoods	Spruce-fir	890	1065	3	5	0	305
S	Northern Atlantic Slope Diversified Farming Region							867	1220	4	9	114	625
147	Northern Appalachian Ridges and Valleys	Udepts	Udults	SEDI residuum	Colluvium	Mixed oaks	Oak-hickory	785	1145	7	14	100	1350
148	Northern Piedmont	Udalfs	Udepts	igne-meta residuum	Colluvium	Mixed oaks	Mixed hardwoods	940	1320	9	14	100	500
149A	Northern Coastal Plain	Udults	Psamments	Coastal plain	Coastal plain	Mixed hardwoods	Mixed pines	1015	1195	11	14	0	100
149B	Long Island-Cape Cod Coastal Lowland	Udepts	Udalfs	Outwash	Till	Mixed decid-evergreen	Mixed pines	1040	1220	10	12	0	100
T	Atlantic and Gulf Coast Lowland Forest and Crop Region							945	1220	9	14	50	512.5
150A	Gulf Coast Prairies	Uderts	Aqualfs	Coastal plain	Alluvium	Grasslands	Oak savanna	1145	1600	19	22	0	50
150B	Gulf Coast Saline Prairies	Aqualfs	Psamments	Eolian	Alluvium	Grasslands	Grasslands	1145	1450	20	23	0	8
151	Gulf Coast Marsh	Saprists	Aquents	Alluvium	Alluvium	Saltwater marsh	Grasslands	1525	1650	19	21	0	3
152A	Eastern Gulf Coast Flatwoods	Aquults	Aquods	Coastal plain	Alluvium	Mixed pines	Mixed pines	1525	1725	18	21	0	25
152B	Western Gulf Coast Flatwoods	Aqualfs	Aquults	Coastal plain	Alluvium	Mixed decid-evergreen	Grasslands	1170	1525	19	20	25	100
153A	Atlantic Coast Flatwoods	Aquults	Aquods	Coastal plain	Alluvium	Mixed pines	Mixed pines	1120	1450	15	21	8	50
153B	Tidewater Area	Aquults	Aqualfs	Coastal plain	Alluvium	Mixed pines	Mixed pines	1015	1475	14	21	0	8
153C	Mid-Atlantic Coastal Plain	Aquults	Udults	Coastal plain	Alluvium	Mixed decid-evergreen	Mixed pines	1015	1120	12	14	0	25
153D	Northern Tidewater Area	Aquults	Udults	Coastal plain	Alluvium	Mixed decid-evergreen	Mixed pines	965	1145	11	15	0	60

(continued)

Table 4.3 (continued)

Symbol	Land Resource Region and Major Land Resource Area	Dominant Suborders		Dominant Parent materials		Dominant Vegetation		Min. MAP (mm)	Max. MAP (mm)	Min. MAAT (°C)	Max. MAAT (°C)	Min. Elev (m)	Max. Elev (m)
U	Florida Subtropical Fruit, Truck Crop, and Range Region							1181	1460	16	20	4	37
154	South-Central Florida Ridge	Psamments	Udults	Coastal plain	Coastal plain	Pine-oak	Pine-oak	1170	1420	20	23	0	100
155	Southern Florida Flatwoods	Aquods	Aqualfs	Coastal plain	Coastal plain	Mixed pines	Mixed pines	1120	1525	20	24	0	25
156A	Florida Everglades and Associated Areas	Aquents	Saprists	Coastal plain	Coastal plain	Swamp hardwoods	Mangroves	1015	1575	23	25	0	25
156B	Southern Florida Lowlands	Aqualfs	Saprists	Coastal plain	Coastal plain	Swamp hardwoods	Swamp conifers	1170	1575	22	24	2	25
V	Hawaii Region							1119	1524	21	24	1	44
157	Arid and Semiarid Low Mountain Slopes	Torrands	Ustands	Lava flows	Volcanic ash	Grass-shrub	Grass-shrub	255	890	13	24	0	1830
158	Semiarid and Subhumid Low Mountain Slopes	Ustox	Ustolls	Volcanic ash	Alluvium	Shrubs	Grasslands	255	760	21	24	0	490
159A	Humid and Very Humid Volcanic Ash Soils on Low and Intermediate Rolling Mountain Slopes	Udands	Udands	Volcanic ash	Volcanic ash	Tropical forest	Grasslands	3050	5080	12	23	0	1830
159B	Subhumid and Humid Low and Intermediate Mountain Slopes	Folists	Udands	Volcanic ash	Cinders	Tropical forest	Grasslands	1270	2540	12	23	0	1830
160	Subhumid and Humid Intermediate and High Mountain Slopes	Udands	Ustands	Volcanic ash	Lava flows	Grasslands	Shrubs	510	1905	10	22	305	2745
161A	Lava Flows and Rock Outcrops	Ustands	Folists	Lava flows	Lava flows	Grasslands	Shrubs	255	2540	3	25	0	4206
161B	Semiarid and Subhumid organic Soils on Lava Flows	Folists	Folists	Organics	Organics	Grasslands	Shrubs	760	2030	13	24	0	1830
162	Humid and Very Humid organic Soils on Lava Flows	Folists	Folists	Organics	Organics	Shrubs	Shrubs	1525	5970	12	23	0	1220
163	Alluvial Fans and Coastal Plains	Ustolls	Psamments	Coastal plain	Alluvium	Shrubs	Shrubs	510	760	22	24	0	60
164	Humid and Very Humid Steep and Very Steep Mountain Slopes	Saprists	Udepts	Volcanic residuum	Organics	Temperate rain forest	Wetlands	1905	6350	12	24	0	2130
165	Subhumid intermediate Mountain Slopes	Humults	Ustands	Igne-meta residuum	Volcanic ash	Grasslands	Shrubs	635	1525	16	23	120	1130
166	Very Stony Land and Rock Land	Orthents	Ustolls	Igne-meta residuum	volcanic ash	Grasslands	Shrubs	255	1525	10	24	0	2440
167	Humid Oxidic Soils on Low and Intermediate Rolling Mountain Slopes	Ustands	Humults	Igne-meta residuum	Volcanic ash	Grasslands	Montane forest	890	3050	19	24	0	610
W1	Southern Alaska							929	2687	13	24	33	1719
220	Alexander Archipelago-Gulf of Alaska	Cryods	Saprists	Colluvium	Till	Western conifers	Subalpine conifers	1525	3050	3	8	0	1420

(continued)

Table 4.3 (continued)

Symbol	Land Resource Region and Major Land Resource Area	Dominant		Dominant		Dominant		Min.	Max.	Min.	Max.	Min.	Max.
		Suborders		Parent materials		Vegetation		MAP (mm)	MAP (mm)	MAAT (°C)	MAAT (°C)	Elev (m)	Elev (m)
221	Kodiak Archipelago	Cryands	Aquepts	Colluvium	Till	Sitka spruce	Dwarf shrub	585	1525	4	7	0	1340
222	Southern Alaska Coastal Mountains	Cryods	Saprists	Colluvium	Till	Alpine	Dwarf shrub	3050	5080			0	5490
223	Cook Inlet Mountains	Cryods	Cryepts	Colluvium	Till	Alpine	Alpine	380	760	−3		760	6195
224	Cook Inlet Lowlands	Cryods	Cryepts	Till	Outwash	Mixed decid-evergreen	Subalpine conifers	380	1525	−3	2	0	1340
225	Southern Alaska Peninsula Mountains	Cryands	Cryepts	Volcanic ash	Colluvium	Tall scrub	Dwarf shrub	1525	2285	−3		0	2160
W2	Aleutian Alaska							1241	2371	0	6	127	2991
226	Aleutian Islands-Western Alaska Peninsula	Cryands	Fibrists	Volcanic ash	Cinders	Dwarf shrub	Alpine	535	1980	2	4	0	2800
X1	Interior Alaska												
227	Copper River Basin	Turbels	Cryepts	glaciolacustrine	Till	Mixed decid-evergreen	Mixed spruce	255	510	−5	−2	185	1160
228	Interior Alaska Mountains	Turbels	Cryepts	Till	Colluvium	Subalpine conifers	Alpine	380	510	−3		455	6195
229	Interior Alaska Lowlands	Turbels	Histels	Till	Glaciofluvial	Mixed spruce	Mixed decid-evergreen	255	380	−6	−4	30	580
230	Yukon-Kuskokwim Highlands	Turbels	Histels	Colluvium	Organics	Mixed spruce	Mixed decid-evergreen	255	1015	−7	−4	9	1370
231	Interior Alaska Highlands	Turbels	Histels	Colluvium	Organics	Mixed decid-evergreen	Dwarf shrub	255	1015	−7	−4	120	2000
232	Yukon Flats Lowlands	Turbels	Histels	Glaciofluvial	Alluvium	Mixed decid-evergreen	Mixed decid-evergreen	150	380	−7	−4	90	305
233	Upper Kobuk and Koyukuk Hills and Valleys	Turbels	Histels	Alluvium	Organics	Mixed spruce	Mixed spruce	255	1015	−7	−6	45	1450
234	Interior Brooks Range Mountains	Turbels	Orthents	Colluvium	Alluvium	Dwarf shrub	Dwarf shrub	255	760	−13	−9	490	2440
X2	Western Alaska							258	698	−7	−5	178	1938
235	Northern Alaska Peninsula Mountains	Cryands	Fibrists	Volcanic ash	Till	Tall scrub	Dwarf shrub	760	2540			4	2135
236	Bristol Bay-Northern Alaska Peninsula Lowlands	Cryands	Cryods	Volcanic ash	Alluvium	Dwarf shrub	Dwarf shrub	330	1270	−1	2	0	760
237	Ahklun Mountains	Turbels	Gelepts	Colluvium	Alluvium	Tall scrub	Tussock tundra	510	760	1		0	1420
238	Yukon-Kuskokwim Coastal Plain	Turbels	Histels	Alluvium	Alluvium	Sedge meadows	Dwarf shrub	380	760	−2	1	0	700
239	Northern Bering Sea Islands	Turbels	Histels	Alluvium	Marine	Sedge meadows	Dwarf shrub	255	635	−4		0	670
240	Nulato Hills-Southern Seward Peninsula Highlands	Turbels	Cryepts	Colluvium	Alluvium	Tall scrub	Tussock tundra	380	510	−3		0	1190

(continued)

Table 4.3 (continued)

Symbol	Land Resource Region and Major Land Resource Area	Dominant Suborders		Dominant Parent materials		Dominant Vegetation		Min. MAP (mm)	Max. MAP (mm)	Min. MAAT (°C)	Max. MAAT (°C)	Min. Elev (m)	Max. Elev (m)
Y	Northern Alaska							436	1079	−2	2	1	1146
241	Seward Peninsula Highlands	Turbels	Orthels	Colluvium	alluvium	dwarf shrub	dwarf shrub	255	380	−6		0	1430
242	Northern Seward Peninsula-Selawik Lowlands	Turbels	Histels	colluvium	Alluvium	Sedge meadows	Tussock tundra	255	760	−6		0	90
243	Western Brooks Range Mountains, Foothills, and Valleys	Turbels	Histels	Colluvium	Alluvium	Dwarf shrub	Dwarf shrub	255	380	−13	−9	6	2600
244	Northern Brooks Range Mountains	Turbels	Turbels	Colluvium	Alluvium	Dwarf shrub	Dwarf shrub	380	1015	−13	−9	600	2600
245	Arctic Foothills	Turbels	Histels	Colluvium	Alluvium	Dwarf shrub	Tussock tundra	255	510	−12	−8	200	610
246	Arctic Coastal Plain	Turbels	Histels	Alluvium	Eolian	Wet sedge	Tussock tundra	100	150	−15	−10	0	200
Z	Caribbean Region							250	533	−11	−9	134	1255
270	Humid Mountains and Valleys	Udepts	Humults	Volcanic residuum	Colluvium	Grasslands	Grasslands	2030	2160	21	23	50	1340
271	Semiarid Mountains and Valleys	Ustolls	Ustepts	Volcanic residuum	Colluvium	Grasslands	Grasslands	890	1145	26	26	50	395
272	Humid Coastal Plains	Udults	Udox	Marine terrace	Alluvium	Grasslands	Grasslands	1525	1650	25	25	0	700
273	Semiarid Coastal Plains	Aquolls	Usterts	Marine terrace	Volcanic residuum	Grasslands	Grasslands	760	1145	26	26	0	75

The parent materials in this region are varied and include primarily till, volcanic ash, colluvium, and alluvium. The vegetation is composed of coniferous forests, with pine savanna and grasslands in the more xeric areas. The mean annual precipitation (MAP) varies from as low as 300 mm along the eastern slopes of the Cascades to over 5000 mm on the western slopes of the northern Pacific Coast Range. The mean annual air temperature (MAAT) ranges from slightly below 0 °C in the high Cascades to over 15 °C in northern California. Elevations range from sea level to over 2400 m.

4.4.2 Northwest Wheat and Range Region (B)

This region is located to the east of LRR A (Fig. 1.1) and, therefore, is much drier. The region is exceptionally diverse and contains basins (Columbia Basin, MLRA 7), plateaus (Columbia, 8, and eastern Idaho, 13), prairies and plains (Palouse and Nez Perce, 9, and Snake River, 11), and valleys and mountains (Central Rocky and Blue Mountain Foothills, 10, and Lost River Valley and Mountains, 12).

The basins, plains, and valleys contain primarily Aridisols, especially Durids and Calcids; the plateaus, prairies, and foothills feature mainly Xerolls; and Cryolls occur at the higher elevations (Fig. 4.2, Table 4.3). Xerolls comprise 53 % of the soil suborders of the region.

Parent materials are generally alluvium or colluvium, but loess is common on the plateaus, prairies, and plains. The vegetation is primarily desert shrub and short-grass prairie. The MAP commonly ranges from 230 to 480 mm, the MAAT from 5 to 11 °C, and the elevation from 90 to over 3600 m.

4.4.3 California Subtropical Fruit, Truck, and Specialty Crop Region (C)

This region contains 7 MLRAs in central and southern California that include valleys (Central California Coastal Valley, 14, and Sacramento and San Joaquin Valleys, 17), the California Delta (16), the Southern California Coastal Plain (19), mountains (Central California Coast Range, 15, and Southern California Mountains, 20), and foothills (Sierra Nevada Foothills, 18) (Fig. 1.1).

Most of the soils have a xeric soil moisture regime and include Xerolls and Xeralfs, which comprise 31 and 25 % of the soils, respectively (Fig. 4.2, Table 4.3). The California Delta contains Aquents and Aquepts.

The parent materials are dominantly alluvium and colluvium. The vegetation is either short-grass prairie, chaparral, or oak savanna. The MAP commonly ranges from 250 to 900 mm, the MAAT from 13 to 19 °C, and elevations range from sea level to 2400 m.

4.4.4 Western Range and Irrigated Region (D)

This exceptionally large region contains much of southwestern USA from Oregon and Wyoming down to California, Arizona, New Mexico, and western Texas (Fig. 1.1). It has 23 MLRAs that include valleys and basins (Klamath and Shasta, 21); mountains (Sierra, 22A; south Cascades, 22B; and Arizona and New Mexico, 39); high plateaus (Colorado, 35; Malheur; 23, and Owyhee, 25); basins and mountains (Humboldt, 24; and Carson, 26); basin and range (central Nevada, 28B; southern Nevada, 29; Sonoran, 40, and southeastern Arizona, 41); basin, plains, and mountains (Fallon-Lovelock, 27; and southern Desertic, 42); plateaus, mesas, and foothills (southwestern, 36); desert basins and plateaus (cool central, 34A and warm central, 34B); deserts (Mojave, 30 and lower Colorado, 31; desert basins (Great Salt Lake, 28; and northern Intermountain, 32); and plateaus, trenches, and mountains (Mogollon Transition, 38).

Physiographic province plays an important role in the distribution of soil suborders in LRR D, with Xerolls in the valleys; Argids and Calcids in basins (also called "sinks") and deserts, Orthents in associated ranges, Argids and Orthents on plateaus, Xerolls and Durids on high plateaus, Ustalfs on footslopes, and Xerepts, Xeralfs, and Xerands in the mountains (Fig. 4.2, Table 4.3). The Great Salt Lake Region contains Salids. Xerolls comprise 19 % of the suborders in region D.

Colluvium and alluvium are the most common parent materials. Desert shrubs and short-grass prairie are prevalent in most regions, except for conifer forests in the mountains. The MAP ranges from less than 100 mm in the deserts to over 2000 in the Sierra Nevada and southern Cascade Mountains. The MAAT commonly ranges from 4 to 16 °C but may be below 0 °C in the mountains and up to 24 °C in the lower Colorado Desert. Elevations range from 85 m below sea level in the Mojave and lower Colorado deserts to over 3600 m in the central Nevada Range. The Malheur and Owyhee high plateaus range up to 2300 m in elevation.

4.4.5 Rocky Mountain Range and Forest Region (E)

There are 10 MLRAs in the LRR E, including valleys (Northern Rocky Mountains, 44, and High Intermountain, 51), foothills (Northern Rocky Mountains, 46, and Southern Rocky Mountains, 49), parks (Southern Rocky Mountains, 48B), and mountains (Northern Rocky, 43A; Southern

Rocky, 43B; Blue and Seven Devils, 43C; Wasatch and Uinta, 47, and Southern Rocky, 48A) (Fig. 1.1).

The valleys feature Ustepts, Ustolls, Argids, and Calcids; the foothills contain Ustolls, Orthents, Cryolls, and Cryalfs; the parks are dominated by Cryalfs and Cryepts; the mountains contain Vitrands, Cryands, Udepts, Cryepts, Ustolls, Xerolls, Cryolls, and Orthents. Ustolls and Cryolls comprise 19 and 15 % of the soils in region E (Fig. 4.2, Table 4.3).

Colluvium and alluvium are the most common parent materials, with till and volcanic ash occurring in the mountains. Although grasslands are present in valleys, foothills, and parks, mixed conifers are common in the mountains. The MAP commonly ranges from 350 to 870 mm, but may be less than 200 mm in the High Intermountain Valleys (51) and up to 1600 mm in the Southern Rocky Mountains (48A). The MAAT is generally between 0.2 and 10 °C, but may be below 0 °C at elevations above 4000 m. Elevations range from around 400 to 4390 m.

4.4.6 Northern Great Plains Spring Wheat Region (F)

This region contains 9 MLRAs in north-central USA that have been glaciated and are divided by color of the surface, including Brown (MLRA 52), Dark Brown (53A, 53B, and 53C), and Black (55A, 55B, and 55C); the region also contains the Rolling Soft Shale Plain (54) and the Red River Valley of the North (56; to distinguish it from the Red River in the Mississippi Delta Region) (Fig. 1.1).

The soils are mostly Mollisols, including Ustolls in the west and Udolls in the east, with Aquolls in depressions (Figs. 4.2, Table 4.3). Aquerts are common in the Red River Valley that runs along the North Dakota and Minnesota border. Other common soil suborders in the region include Ustalfs, Ustepts, and Orthents. Ustolls, Udolls, and Aquolls comprise 32, 17, and 17 % of the soils in region F, respectively.

Till is ubiquitous in the region and is crossed by alluvium. Shale residuum is exposed in the Rolling Soft Shale Plain. Grasslands cover nearly all of the region, with grasses of intermediate height in the west and tallgrass prairie in the east. Because the elevation ranges narrowly between 200 and 1400 m, there are fewer climatic variations than in the previously described regions. The MAP ranges commonly between 350 and 500 mm; and the MAAT is between 4 and 7 °C.

4.4.7 Western Great Plains Range and Irrigated Region (G)

This region is divided into high plains (Northern Rolling High Plains, 58A through 58D; Pierre Shale Plains, 60A and 60B; Northern Rolling Pierre Shale Plains, 63A and 63B; Central High Plains, 67A and 67B; Upper Arkansas Valley Rolling Plains, 69; Canadian River Plains and Valleys, 70A and 70B); valleys (Upper Pecos River Valley, 70B); foothills (Black Hills Foot Slopes, 61, and Southern Desert Foothills, 70D); hills (Black Hills, 62, and Nebraska Sand Hills, 65); tablelands (Mixed Sandy and Silty Tableland and Badlands, 64, and Dakota-Nebraska Eroded Tableland); and highlands (Central New Mexico Highlands, 70C) (Fig. 1.1).

The high plains contain mainly Ustolls and Orthents, with some Ustepts, Ustalfs, and Usterts (Fig. 4.2, Table 4.3). The foothills feature Ustalfs and Orthents; the hills have Ustalfs, Ustolls, and Psamments; and the tablelands have Ustolls, Orthents, and Psamments. The southern portion of the region is arid so that the Canadian and Pecos River valleys and Central New Mexico Highlands have mainly Calcids and Argids. Ustolls occupy 38 % of region G.

The high plains have a variety of parent materials, particularly residuum from shale and other sedimentary rocks, alluvium, and eoliansands. The vegetation is predominantly short-grass prairie with pines in the Black Hills and juniper-pinyon pine on the Central New Mexico Highlands. Mean annual precipitation varies narrowly between 325 and 540 mm; the MAAT is generally between 7 and 11 °C; and the elevation commonly ranges from 1000 to 1500 m.

4.4.8 Central Great Plains Winter Wheat and Range Region (H)

Southeast of region G, region H is centered over western Kansas and extends from the Nebraska Sand Hills to northern Texas (Fig. 1.1). This region contains 17 MLRAs that include prairies (Rolling Limestone, 78A; Central Rolling Red, 80A; Texas North-Central, 80B); plains (Rolling Plains and Breaks, 73; Central Loess, 75; Southern High, 77A through 77E; Central Rolling Red, 78B and 78C; and Great Bend Sand, 79); hills (Central Nebraska Loess, 71; Central Kansas Sandstone, 74; and Bluestem, 76); and tablelands (Central High, 72).

The prairies have primarily Ustolls, along with Ustalfs and Orthents; the plains have primarily Ustolls, but also Ustalfs, Ustepts, and Calcids in the southwest; the hills have

Ustolls, Udolls, and Orthents; and the tablelands have Ustolls and Orthents (Fig. 4.2, Table 4.3). Ustolls comprise about 47 % of the soils in the region.

The parent materials are alluvium, loess, eolian sands, and residuum from limestone and other sedimentary rocks. Grasslands of intermediate height occupy nearly all of the region. The MAP commonly ranges between 520 and 730 mm; the MAAT varies mostly between 12 and 16 °C; and the elevation from 440 to 950 m.

4.4.9 Southwest Plateaus and Plains Range and Cotton Region (I)

This region is limited to southern Texas and is divided into 10 MLRAs, including basins (Texas Central, 82A), a Sandsheet Prairie (83E), plains (Rio Grande, 83A through 83E), and plateaus (Edwards Plateau, 81A through 81D) (Fig. 1.1).

The basins contain Ustalfs and Ustolls; the sand prairie is composed of Ustalfs and Aqualfs; the plains feature especially Ustalfs and Ustolls but also Argids and Calcids; and the plateaus contain especially Ustolls and Calcids but also Ustepts and Orthents (Fig. 4.2, Table 4.3). Ustolls comprise 41 % of the area.

The dominant parent materials in region I are limestone and other sedimentary rock residuum and alluvium, with smaller areas of eolian sands. The vegetation is short-grass prairie, with oak savanna and desert shrubs in some areas. The MAP commonly ranges between 500 and 700 mm; the MAAT from 19 to 22 °C; and the elevation is from sea level to over 1000 m, but it normally ranges between 185 and 535 m.

4.4.10 Southwestern Prairies Cotton and Forage Region (J)

This region includes east-central Texas, a strip through central Oklahoma, and a small portion of southeastern Kansas (Fig. 1.1). Composed of 9 MLRAs, region J includes the Grand Prairie (85), blackland prairies (Texas Blackland, 86A and 86B), forested regions (North, West, and Cross Timbers, 84A through 84C), the Wichita Mountains (82B), and the Texas Claypan Area (87A and 87B).

The Grand and Texas Blackland Prairies contain Usterts and Ustolls; the Texas Claypan Area has Ustalfs, Udalfs, and Aqualfs; the Wichita Mountains have Ustolls; and the Cross Timbers area feature Ustalfs and Udalfs (Fig. 4.2, Table 4.3). Ustalfs and Ustolls comprise 35 and 25 % of region J, respectively.

The parent materials are primarily residuum from sedimentary materials and alluvium. Grasslands of intermediate height cover much of the area, with oak savanna and grasslands in the Cross Timbers and Claypan areas. The MAP normally ranges from 770 to 1060 mm, the MAAT from 17 to 19 °C, and the elevation from 170 to 330 m.

4.4.11 Northern Lake States Forest and Forage Region (K)

This region contains the upper Great Lakes states, including Minnesota, Wisconsin, and Michigan, with a small area in northern Illinois (Fig. 1.1). Composed of 16 MLRAs, region K is separated into drift plains (Northern Minnesota Gray Drift, 57; Superior Stony and Rocky Loamy Plains and Hills, 93A and 93B; Northern Michigan and Wisconsin Sandy Drift, 94A; Michigan Eastern Upper Peninsula Sandy Drift, 94B; Michigan Northern Lower Peninsula Sandy Drift, 94C; Northern Highland Sandy Drift, 94D; Northeastern Wisconsin Drift Plain, 95A; and Southern Wisconsin and Northern Illinois Drift Plain, 95B), glacial lake basins (Northern Minnesota Glacial Lake Basins, 88; Wisconsin Central Sands, 89; Lake Superior Lake Plain, 92), areas of thin loess over till (Wisconsin and Minnesota Thin Loess and Till, 90A and 90B), and areas of sandy outwash (Central Minnesota Sandy Outwash, 91A; and Wisconsin and Minnesota Sandy Outwash, 91B).

The MLRAs are strongly differentiated on soil texture, including sands (89, 91, 94), silt loams (90 and 95B), loams (57, 88, 93, and 95A), and clays (92). Drift plains feature primarily Udalfs and Orthods, but also have Aqualfs, Udolls, Udepts, and Hemists (Fig. 4.2, Table 4.3). Glacial lake basins contain Udalfs, Orthods, Psamments, and Saprists. Areas of sandy outwash contain mainly Orthods and Saprists, but also Udolls, Psamments, and Udalfs. The dominant suborders are Udalfs, Udolls, and Orthods, which comprise 28, 15, and 12 % of the area, respectively.

Till and glaciolacustrine sediments are the major parent materials, with lesser areas of outwash or thin loess over till. Biologic materials have accumulated in depressions. The vegetation is closely related to soil texture with pine-oak on extremely sandy materials, mixed pines on loamy sand materials, and northern hardwoods on sandy loam and other finer-textured materials. Grasslands occur to a limited extent, especially in the Southern Wisconsin and Northern Illinois Drift Plain. The MAP varies narrowly between 680 and 850 mm, the MAAT from 4 to 7 °C, and the elevation mostly from 245 to 470 m.

4.4.12 Lake States Fruit, Truck Crop, and Dairy Region (L)

This small region includes southern Michigan and small portions of northern Ohio, Indiana, and Illinois (Fig. 1.1). There are 5 MLRAs in region L that include drift plains (Southwestern Michigan Fruit and Truck Crop Belt, 97; and Southern Michigan and Northern Indiana Drift Plain, 98), lake plains (Erie-Huron, 99, and Ontario-Erie, 101), and outwash plains (Western Michigan Fruit Belt, 96).

Whereas the lake plains have Udalfs, Aqualfs, and Udepts, the drift plains have Orthods, Udalfs, and Aquolls, and the outwash plains have Orthods and Psamments (Fig. 4.2, Table 4.3). Udalfs are the dominant suborder, accounting for 26 % of the soil cover.

The three main parent materials are till, glaciolacustrine, and outwash. The vegetation is mainly broad-leaved forest types, including northern hardwoods, mixed hardwoods, and oak-hickory; mixed pines are also present. The MAP ranges from 780 to 1000 mm; the MAAT ranges from 6 to 10 °C; and the elevation ranges from 170 to 315 m.

4.4.13 Central Feed Grains and Livestock Region (M)

This region lies to the south of regions K and L and is the second largest region (Fig. 1.1). Land Resource Region M includes 27 MLRAs identified as loess prairies (Cherokee Prairies, 112), till prairies (Rolling, 102A; Central Iowa and Minnesota, 103; and eastern Iowa and Minnesota, 104), loess uplands and hills (Loess Uplands, 102C; Northern Mississippi Valley Loess Hills, 105; Nebraska and Kansas Loess Drift Hills, 106; Iowa and Minnesota Loess Hills, 107A; and Iowa and Missouri Deep Loess Hills, 107B), drift plains (Till Plains, 102B; Iowa and Missouri Heavy Till Plain, 109; Northern Illinois and Indiana Heavy Till Plain, 110; and Indiana and Ohio Till Plain, 111), loess and drift plains (Illinois and Iowa Deep Loess Drift, 108; and Southern Illinois and Indiana Thin Loess and Till Plain, 114), Central Clay Pan Areas (113), and the Central Mississippi Valley Wooded Slopes (115).

The prairies, loess hills and uplands, drift plains, and loess and drift plains contain primarily Udolls, with Aquolls in depressions and Udalfs on wooded slopes (Fig. 4.2, Table 4.3). The clay pan area is dominated by Aqualfs, and the Central Mississippi Valley Wooded Slopes feature Udalfs, with Aqualfs in depressions. Udolls, Udalfs, and Aquolls comprise 30, 23, and 15 % of region M.

The parent materials are predominantly till, with a variable thickness of loess capping many soils. Residuum from limestone is common in the Cherokee Prairies. The vegetation is mainly grasslands in the prairies and loess hills. Oak-hickory forests are present on the Indiana, Ohio, and southern Illinois till plains; and swamp hardwoods occur in depressions with Aqualfs throughout the region. The MAP normally ranges between 800 and 1000 mm; the MAAT from 8 to 12 °C; and the elevation from 200 to 400 m.

4.4.14 East and Central Farming and Forest Region (N)

This region occurs to the south and east of region M as two lobes bisected by the Mississippi River (Fig. 1.1). The region has not been glaciated and is composed of 21 MLRAs, including basins (St. Francois Knobs and Basins, 116C; and Nashville Basin, 123), plains (Springfield Plain, 116B), hills (Kentucky Bluegrass, 121), long ridges (Northern and Southern Blue Ridges, 130), valley and ridges (Arkansas Valley and Ridges, 118; Kentucky and Indiana Sandstone and Shale Hills and Valleys, 120; and Southern Appalachian Ridges and Valleys, 128), highlands and plateaus (Ozark Highland, 116A; Highland Rim and Pennyroyal, 122; Western Allegheny Plateau, 124; and Central Allegheny Plateau, 126), plateaus and mountains (Cumberland Plateau and Mountains, 125; and Eastern Allegheny Plateau and Mountains, 127), and mountains (Boston, 117; Ouachita, 119; and Sand Mountain, 129).

The basins are composed of Udalfs, with Udults and Udolls as accessories (Fig. 4.2, Table 4.3). The hills and hills and valleys contain primarily Udalfs. The Springfield Plain has Udults and Udalfs. The Blue Ridge and the valleys and ridges areas are composed mainly of Udults and Udepts. The plateaus, plateau and mountains, and mountains are commonly Udults, associated with Udepts.

The parent materials are residuum from sedimentary and metamorphic rocks, alluvium in valleys, and loess in protected areas. The vegetation is mainly oak-hickory, along with mixed hardwoods and mixed pines. Oaks savanna and grasslands occur in the plains and valleys. The MAP commonly varies between 1000 and 1300 mm; the MAAT from 11 to 15 °C; and the elevation from 175 to 715 m, but extending to over 2000 in the Southern Blue Ridge.

4.4.15 Mississippi Delta Cotton and Feed Grains Region (O)

The Mississippi Delta Region (O) is the second smallest region in the conterminous USA (Fig. 1.1). The region is divided into 4 MLRAs that are all composed of alluvium (Southern Mississippi, 131A; Arkansas River, 131B; and Red River, 131C) or river terraces (Southern Mississippi, 131D).

The soils are primarily Aquerts, Aquepts, and Uderts on alluvium and Udalfs and Aqualfs on the higher river terraces

(Fig. 4.2, Table 4.3). The alluvium supports primarily swamp hardwoods, and the river terraces contain mixed hardwoods and some grasslands. The MAP commonly ranges from 1200 to 1400 mm; the MAAT from 16 to 18 °C; and the elevation from 15 to 75 m.

4.4.16 South Atlantic and Gulf Slope Cash Crops, Forest, and Livestock Region (P)

This region is the third largest and includes the piedmont and upper coastal plain of the southeastern USA (Fig. 1.1). The region is divided into 8 MLRAs, including the coastal plain (Southern, 133A; Western, 133B; Cretaceous Western, 135B; and North-Central Florida Ridge, 138), Southern Mississippi Loess (134), Alabama and Mississippi Blackland Prairie (135A), the Carolina and George Sand Hills (137), and the Southern Piedmont (136).

The coastal plain contains mainly Udults, along with Udalfs and Psamments (Fig. 4.2, Table 4.3). The piedmont contains Udults, along withUdepts; the loess-covered area has Udalfs; the sand hills have Udults and Psamments; and the blackland prairie has Uderts and Aquepts.

The parent materials are coastal plain sediments and residuum from igneous and metamorphic rocks and alluvium. The vegetation is mainly pine-oak or mixed pine, with mixed hardwoods in some areas. The MAP is between 1100 and 1450 mm; the MAAT ranges from 15 to 19 °C; and the elevation is from 19 to 38 m.

4.4.17 Pacific Basin Region (Q)

The Pacific Basin includes territories managed by the USA, including the Mariana Islands (Guam and the Commonwealth of the Northern Mariana Islands) and American Samoa (Fig. 1.3). The 8 MLRAs include stratovolcanoes of the Mariana Islands (190), High Limestone Plateaus of the Mariana Islands (191), volcanic islands or highlands (192 and 197), and coral atolls (196).

The stratovolcanoes of the Mariana Islands have mainly Ustands and Ustepts (Table 4.3). The Limestone Plateaus of the Mariana Islands containUstalfs, Ustolls, and shallow Entisols. The volcanic islands of American Samoa features Udolls and Udands (Fig. 4.3). As mentioned earlier, because of the differences in age and composition of parent materials and climate associated with elevation change, the soils of the Pacific Basin are highly diverse.

The parent materials are primarily residuum from volcanic materials, with lava flows, volcanic ash, and coral also being represented. Limestone residuum occurs on high plateaus of the Mariana Islands. The vegetation is tropical forest, strand forests (also called linear swamp forests), tropical palms, with abundant grasslands. The MAP commonly ranges between 2600 and 3800 mm; the MAAT is around 27 °C; and the elevation ranges from 0 to 1000 m.

4.4.18 Northeastern Forage and Forest Region (R)

This region contains 9 MLRAs, including the Connecticut Valley (145), glaciated plains (St. Lawrence-Champlain, 142), glaciated plateaus (Lake Erie, 139; Allegheny Plateau and Catskill Mountains, 140; and Tughill Plateau, 141), uplands and highlands (New England and Eastern New York, 144; and Aroostook Area, 146), and the Northeastern Mountains (143) (Fig. 1.1).

The Connecticut Valley contains mainly Udepts and Orthents; the glaciated plains have Aqualfs and Udepts; the uplands and highlands contain primarily Udepts, with Orthods and Orthents; the glaciated plateaus have Udalfs, Aqualfs, Udepts, Aquepts, Orthods, and Aquods; and the Northeastern Mountains have Orthods and Aquepts (Fig. 4.2, Table 4.3). Udepts and Aquepts comprise 31 and 17 % of the soils in region R.

The parent materials are dominantly till and outwash, with alluvium in the Connecticut Valley. The vegetation is northern hardwoods and spruce-fir, with some oak-hickory. The MAP commonly ranges between 850 and 1200 mm, but may be as high as 1600 mm on the Tughill Plateau; the MAAT ranges narrowly between 4 and 9 °C; and the elevation ranges from 110 to 625 m, but may be above 1500 m in the Northeastern Mountains Area.

4.4.19 Northern Atlantic Slope Diversified Farming Region (S)

This small but exceedingly diverse region is divided into 4 MLRAs: the Northern Appalachian Ridges and Valleys (147), the Northern Piedmont (148), the Northern Coastal Plain (149A), and the Long Island-Cape Cod Coastal Lowland (149B) (Fig. 1.1).

Whereas Udults are common in the northern piedmont and coastal plain, Udepts and Udalfs are most abundant in the ridges, valleys, and lowlands (Fig. 4.2, Table 4.3). Udults, Udalfs, and Udepts comprise 31, 21, and 16 % of the area, respectively.

The parent materials include residuum from sedimentary, igneous, and metamorphic rocks, coastal plain sediments, till, outwash, and colluvium. The vegetation is comprised of mixed oaks, mixed hardwoods, mixed hardwoods and pines, and mixed pines. The MAP varies between 950 and 1200 mm; the MAAT from 9 to 14 °C; and the elevation from 50 to 500 m, but with elevations up to 1350 m in the Northern Appalachian Ridges and Valleys.

4.4.20 Atlantic and Gulf Coast Lowland Forest and Crop Region (T)

This region is divided into 9 MLRAs, including coastal prairies (Gulf Coast, 150A; and Gulf Coast Saline, 150B), coastal marshes (Gulf Coast, 151), flatwoods (Eastern Gulf Coast, 152A; Western Gulf Coast, 152B; and Atlantic Coast, 153A), tidewater (153B and 153D), and coastal plain (Mid-Atlantic, 153C) (Fig. 1.1).

The coastal prairies have Aqualfs, along with Uderts and Psamments; coastal marshes have Saprists and Aquents; the flatwoods have Aquults, Aqualfs, and Aquods; and the tidewater and coastal plain areas have Aquults, Udults, and Aqualfs (Fig. 4.2, Table 4.3). Soils with an aquic moisture regime and Histosols account for 51 % of the soils; Udults (21 %) are the most abundant soil suborder.

The parent materials are primarily alluvium and coastal plain deposits. The vegetation is mixed pines, with some grasslands, and mixed deciduous-coniferous forests. The MAP ranges between 1200 and 1450 mm; the MAAT is between 16 and 20 °C; and the elevation ranges narrowly between 0 and 100 m.

4.4.21 Florida Subtropical Fruit, Truck Crop, and Range Region (U)

This small region has 4 MLRAs, including the South-Central Florida Ridge (154), the Southern Florida Flatwoods (155), the Florida Everglades and Associated Areas (156A), and the South Florida Lowlands (156B) (Fig. 1.1).

Saprists and Aqualfs dominate the soils of this region, with lesser areas of Aquods in the flatwoods and Psamments and Udults along the dolomitic South-Central Florida Ridge (Fig. 4.2, Table 4.3). The parent materials are coastal plain sediments. The vegetation includes pine-oak, mixed pines, swamp hardwoods, swamp conifers, and mangroves. The MAP ranges between 1100 and 1500 mm; the MAAT is 21–24 °C, the warmest in continental USA; and the elevation ranges from 0 to 100 m on the South-Central Florida Ridge.

4.4.22 Hawaii Region (V)

This small region (16,260 km^2) is extremely diverse in terms of physiography, climate, topography, soils, and land use (Fig. 1.3). Region V is divided into 13 MLRAs, firstly by parent material and physiography, and secondarily by climate. These include mountain slopes (Arid and Semiarid Low, 157; Semiarid and Subhumid Low, 158; Subhumid and Intermediate, 165; Subhumid and Humid Low and Intermediate, 159B; Subhumid and Humid Intermediate and High, 160; Humid and Very Humid Steep and Very Steep, 164); volcanic ash soils (Humid and Very Humid, 159A); lava flows and rock outcrops (161A); organic soils on lava flows (Semiarid and Subhumid, 161B; Humid and Very Humid, 162); alluvial fans and coastal plain (163); very stony and rockland (166); and humid oxidic soils on rolling mountain slopes (167).

On mountain slopes, arid to semiarid soils are Torrands and Ustands, semiarid to subhumid soils are Ustox and Ustolls, subhumid soils are Humults and Ustands, subhumid to humid soils are Folists and Udands at low elevations and Udands and Ustands at higher elevations, and humid to very humid soils are Saprists and Udepts (Fig. 4.3, Table 4.3). Soils on alluvial fans and coast plains are Ustolls and Psamments; soils on volcanic ash are Udands; organic soils on lava flows are Folists; humid oxidic soils on rolling mountain slopes are Udox and Humults. Soils on lava flows with rock outcrops are Ustands and Folists; soils on very stony land and rockland are Orthents and Ustands. Udands, Ustands, and Ustolls account for 23, 15, and 10 % of the Hawaiian soils, respectively.

Parent materials include volcanic ash, lava flows, residual materials from igneous and metamorphic rocks, coastal plain sediments, and organic materials. The vegetation includes tropical forest, montane forest, tropical shrubs, and tropical grasslands. The MAP normally ranges between 930 and 2700 mm, but may be as low as 255 mm and as high as 6350 mm. The MAAT ranges between 13 and 25 °C, but may be considerable on the high volcanic peaks. Elevations range from 0 m to over 4200 m on Mauna Kea and Mauna Loa.

4.4.23 Southern Alaska (W1)

Southern Alaska includes 6 MLRAs, including archipelagos (Alexander, 220; and Kodiak, 221), mountains (Southern Alaska Coastal Mountains, 222; Cook Inlet Mountains, 223; and Southern Alaska Peninsula Mountains, 226), and the Cook Inlet Lowlands (224) (Fig. 1.2).

Unlike soils elsewhere in Alaska (except for the Aleutian Islands and Western Alaska Peninsula), the soils of southern Alaska lack permafrost. Soils in the archipelagos are Cryods, Saprists, Cryands, and Aquepts (Fig. 4.4, Table 4.3). In the mountains, Cryods, Saprists, Cryepts, and Cryands are dominant. Cook Inlet has Cryods and Cryepts.

The parent materials include till, volcanic ash, and extensive colluvium. The vegetation includes western conifers, Sitka spruce, subalpine conifers, and alpine tundra. The MAP ranges between 380 mm in the Cook Inlet region to over 5000 mm in the Coastal Mountains. Elevations range from sea level to nearly 6200 m.

4.4.24 Aleutian Alaska (W2)

The western Alaska Peninsula and Aleutian Islands extend 2700 km from the mainland into the northern Pacific Ocean (Fig. 1.2). MLRA 226 contains primarily Cryands and Cryods (Fig. 4.4, Table 4.3). The soils are derived from volcanic ash, cinders, and colluvium. The vegetation is dwarf shrubs and grassland tundra. The MAP ranges from 535 to 1980 mm; and the MAAT is between 2 and 4 °C. Elevations range from sea level to 2800 m.

4.4.25 Interior Alaska (X1)

This region is divided into 8 MLRAs, including a basin (Copper River, 227), lowlands (Interior Alaska, 229 and Yukon Flats, 232), hills and valleys (Upper Kobuk and Koyukuk, 233), highlands (Yukon-Kuskokwim, 230 and Interior Alaska, 231), and mountains (Interior Alaska, 228; and Interior Brooks Range, 234) (Fig. 1.2). Permafrost covers over 70 % of the area, and Turbels are the dominant suborder (Fig. 4.4, Table 4.3). Cryepts are abundant in areas either lacking permafrost or having permafrost below 2 m. Parent materials include till, colluvium, alluvium, and organic materials. The vegetation is mixed spruce, mixed deciduous and coniferous species, subalpine conifers, and dwarf shrubs. Mean annual precipitation varies from 150 mm in the Yukon Flats to over 1000 mm in the highlands. The MAAT ranges from −2 °C in the Copper River basin to −13 °C in the Brooks Range. Elevations range from 9 to 500 m, with areas in the Brooks Range above 2400 m.

4.4.26 Western Alaska (X2)

Western Alaska includes 6 MLRAs, including the Bering Sea Islands (239), the Yukon-Kuskokwim Coastal Plain (238), the Bristol Bay-Northern Alaska Peninsula Lowlands (236), the Nulato Hills-Southern Seward Peninsula Highlands (240), and the mountains (Northern Alaska Peninsula, 235; Ahklun, 237) (Fig. 1.2).

Whereas Cryands, along with Fibrists, are predominant on the Northern Alaska Peninsula and Mountains, Turbels and Histels occur in the other areas (Fig. 4.4, Table 4.3). Parent materials are primarily colluvium and alluvium, but include volcanic ash in areas 235 and 236. The vegetation ranges from tussock tundra and sedge meadow tundra to tall shrubs. The MAP ranges from 255 mm in the northern Bering Sea islands to over 2500 mm in the Alaska Peninsula Mountains. The MAAT ranges from 2 °C near Bristol Bay to −4 °C in the northern Bering Sea. Elevations range from 0 m to over 2100 m in the Alaska Peninsula Mountains.

4.4.27 Northern Alaska (Y)

The 6 MLRAs in northern Alaska include the Arctic Coastal Plain (246), the Northern Seward Peninsula-Selawik Lowlands (242), the Arctic Foothills (245), the Seward Peninsula Highlands (241), and the Northern Brooks Range Mountains (243 and 244) (Fig. 1.2). The soils are mainly Turbels, with some Histels. Orthels occur to a limited extent in the Seward Peninsula Highlands (Fig. 4.4, Table 4.3). The parent materials are mainly alluvium and colluvium. The vegetation includes dwarf shrub tundra, sedge meadow tundra, tussock tundra, and wet sedge tundra. The MAP ranges from 100 mm along the Arctic Coastal Plain to over 1000 mm in the northern Brooks Range. This is the coldest MLRA, with MAATs ranging from −8 to −15 °C. Elevations range from sea level to 2600 m in the Brooks Range.

4.4.28 Caribbean Region (Z)

The Caribbean Region includes the island of Puerto Rico and the Virgin Islands to the east (Fig. 1.2). This region is divided into coastal plains (Semiarid, 273 and Humid, 272) and mountains and valleys (Semiarid, 271 and Humid, 270). Despite its small area, the Caribbean region contains a high diversity of soils. The coastal plains contain primarily Udults, Udox, Usterts, and Aquolls; the mountains and valleys have Udepts, Humults, Ustolls, and Ustepts (Fig. 4.4, Table 4.3). The parent materials are volcanic residuum, marine terrace deposits, colluvium, and alluvium. Most of Puerto Rico has been cleared for agriculture. The native vegetation included grasslands, forest and shrubland, and tropical rain forest. The MAP ranges from 760 mm to over 2000 mm; the MAAT is from 21 to 26 °C; and the elevation ranges from sea level to 1340 m.

4.5 Conclusions

There are 28 Land Resource Regions (LRR) in the USA delineated on the basis of land use and 278 Major Land Resource Areas that are based primarily on physiography and parent materials but also have specific climates and soils. There are from 5 to 10 soil orders, 9 to 44 suborders, 24 to 118 great groups, and 84 to 5467 soil series per LRR. From one to four suborders account for half of the soil series in each LRR. The distribution of soils in the USA is controlled primarily by climate (Aridisols, Gelisols, and Oxisols), vegetation (Histosols and Mollisols), parent materials (Andisols, Entisols, andVertisols), time (Inceptisols), or a combination of factors (Alfisols, Spodosols, and Ultisols). The dominant soil-forming processes are argilluviation, melanization, base-cation enrichment, and cambisolization, which comprise 56, 42, 37, and 28 % of the soil area of the USA, respectively. The ranking of suborders by abundance is Ustolls > Udults > Udalfs > Turbels > Orthents > Udolls > Udepts > Xerolls; these suborders comprise 53 % of the soils of the USA. Each of these LRRs is discussed in more detail in the following chapters.

References

Bockheim JG (2014) Soil geography of the USA: a diagnostic-horizon approach. Springer, New York

Bockheim JG, Hartemink AE (2013) Distribution and classification of soils with clay-enriched horizons in the USA. Geoderma 209–210:153–160

Natural Resources Conservation Service (2004) Land Resource Regions and Major Land Resource Areas of Alaska. USDA-NRCS Alaska, Palmer, AK

Natural Resources Conservation Service (2006) Land resource regions and major land resource areas of the United States, the Caribbean, and the Pacific Basin. USDA Handbook 296

Soil Survey Staff (1999) Soil taxonomy: a basic system of soil classification for making and interpreting soil surveys, 2nd edn. United States Department of Agriculture Natural Resources Conservation Service. Agriculture Handbook No. 436, Superintendent of Documents, U.S. Government Printing Office, Washington, DC, p 20402

Soil Survey Staff (2014) Keys to soil taxonomy, 12th edn. USDA, National Resources Conservation Service, National Soil Survey Center, Lincoln

5 Soils of the Pacific Coast Region: LRRs A and C

Susan Burlew Southard, Randal J. Southard, and Scott Burns

Land Resource Region (LRR) A is located in the Pacific Northwest of the United States and is dominated by four major mountain ranges (Cascades, Klamath, Coast Range and the Olympics) and two main coastal areas (Puget Sound and Pacific Coast). The Region includes two large valley systems (Willamette Valley and the Puget Lowland) and the adjoining eastern and western mountain foot slopes. The region covers over 234,000 km^2. It includes a rich agricultural area due to a mild coastal climate, deep soils formed in alluvium and adequate precipitation for crops. The parent material of the Cascade Mountains is mainly volcanic tephra and residuum, the valleys are primarily mixed alluvial sediments, and the parent materials of the Coast Range, Klamath and Olympic Mountains are mixtures of alluvium, residuum and colluvium derived from sedimentary rocks that are uplifted sea floors and volcanic rocks. Elevations range from sea level to over 4,400 meters in the Cascade Mountains.

Land Resource Region (LRR) C covers about 161,500 km^2, lies entirely within California, and consists of the Central Valley, Sierra Foothills, Central Coast Range, Coastal Valleys, California Delta, Southern Mountains, and Southern Coastal Plain. Valley soils formed predominantly in alluvium and eolian materials; soils in the mountains and foothills formed predominantly in residuum and colluvium derived from mixed metamorphic and sedimentary rocks and granite. Volcanic rocks occur in the northern part, and volcanic activity in the Cascades of LRR A has draped many high-elevation soils in LRR C with tephra. California Delta soils formed from in situ accumulation of organic plant materials and episodic alluvial deposition. Elevations in LRR C range from below sea level in the Delta to 3500 m in the mountains and, as in LRR A, elevation has a strong influence on soil temperature and moisture regimes.

In both regions, regional plate tectonics and volcanism have contributed to a wide range of rock types, resulting in a wide range of chemical and physical soil properties inherited from the parent material. Soil hydrologic regimes have been altered by dams, canals, drainage structures, irrigation, mining, and in California, by significant regional soil subsidence. The two regions have soil climates that vary as a function of latitude, aspect, and elevation. LRR A has a general range of soil climate from udic and mesic in the north that trends to xeric and mesic in the south. In LRR C, soil climate is regionally xeric and thermic in the north grading to aridic and thermic in the south. Soil climate plays a major role in the greater sequestration of soil organic carbon in northern latitudes. In the drier southern latitudes of LRR C, soil carbon contents are lower than at higher latitudes due to lower biomass input rates and higher organic matter decomposition rates. Soil climates of coastal areas in both regions are moderated by a strong maritime effect. Most low-lying basins, depressions, and sinks have an aquic soil moisture regime, which may result in soil carbon sequestration due to in situ accumulation of organic matter, if not artificially drained.

S.B. Southard (✉)
USDA-NRCS, Davis, CA, USA
e-mail: susan.southard@lin.usda.gov

R.J. Southard
University of California, Davis, Davis, CA, USA
e-mail: rjsouthard@ucdavis.edu

S. Burns
Portland State University, Portland, OR, USA
e-mail: burnss@pdx.edu

L.T. West et al. (eds.), *The Soils of the USA*, World Soils Book Series,
DOI 10.1007/978-3-319-41870-4_5

5.1 Northwestern Forest, Forage, and Specialty Crop Region (LRR A)

5.1.1 Puget Sound Valley

The Puget Sound Valley is a northern extension of the Willamette Valley (Fig. 5.1). It is a synclinal valley filled with glacially derived sediment that lies between the Cascade Mountains and the Coast Range of Washington. It

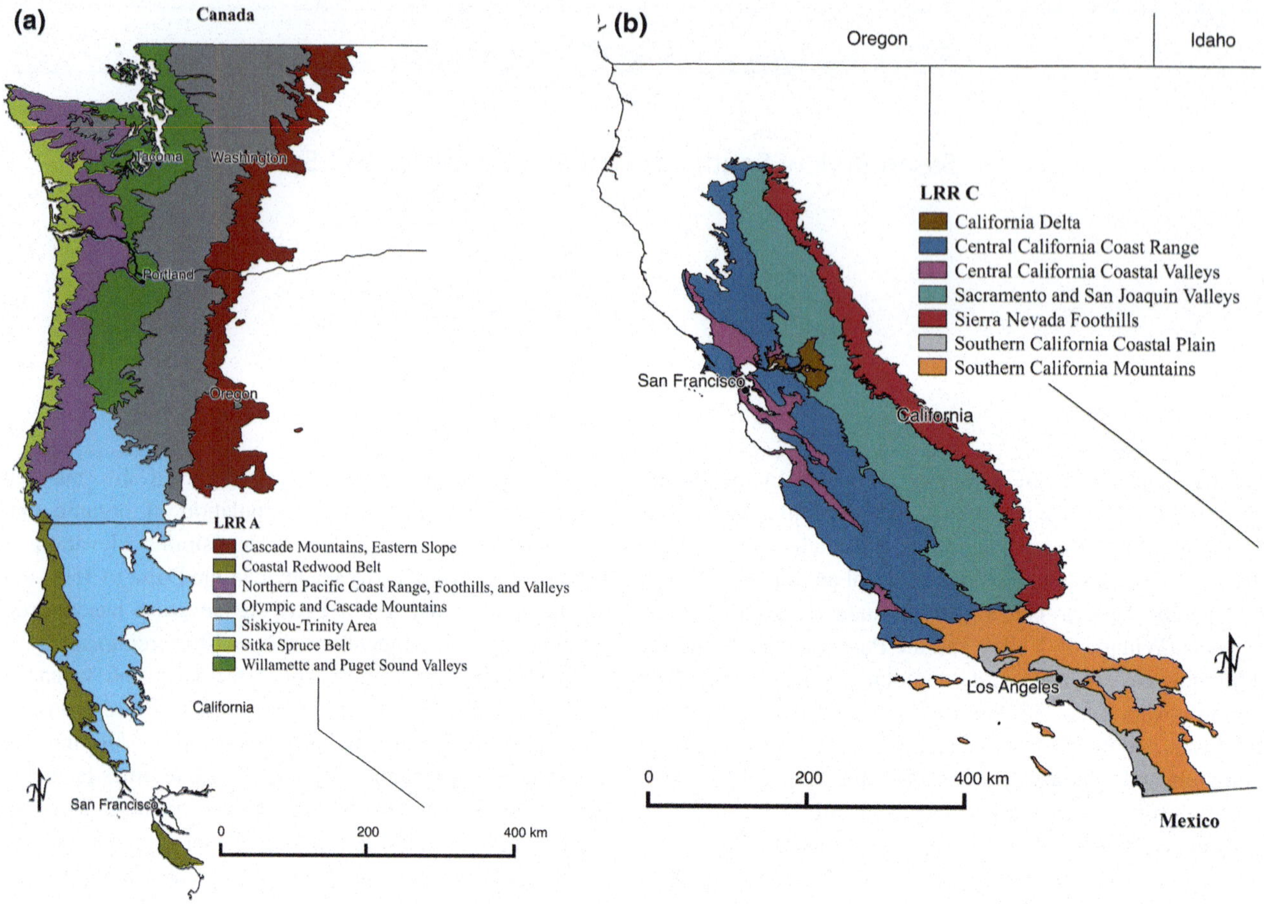

Fig. 5.1 Maps of Land Resource Regions A and C and physiographic areas discussed in this chapter. **a** Northwestern Forest, Forage, and Specialty Crop (LRR A); and **b** California Subtropical Fruit, Truck, and Specialty Crop Region (LRR C)

extends 250 km from Chehalis to Bellingham in Washington and covers an area of about 19,000 km^2. The Puget Sound Valley differs from the Willamette Valley in that the Puget Sound Valley contains mainly outwash, till, and glacial marine sediment, whereas the Willamette Valley contains mainly Missoula Flood deposits. The topography is undulating depending upon the glacial landforms. Elevation ranges from sea level to 500 m. In places, the till and outwash have been dissected by deep stream valleys. Annual precipitation ranges between 75 and 150 cm and falls mainly in the late fall, winter, and spring. The average temperature ranges from 6 to 12 °C. Soils support a dominant vegetation of Douglas-fir (*Pseudotsuga menziesii*) and western hemlock (*Tsuga heterophylla*).

Soils are mainly Dystroxerepts, Argixerolls, and Haploxerands formed in valley glaciofluvial outwash sediments on terraces, moraines, and till plains (Fig. 5.2). Some soils are formed from outwash over root- and water-restrictive dense glaciomarine deposits. During the wetter months of the year (October through May), a perched water table (episaturation) develops in these soils. In nearly level to gently sloping areas, the water table reaches the surface, and water might pond on the soils during periods of frequent precipitation.

Some soils of the valley have additions of eolian material, small amounts of volcanic ash, and/or relatively high amounts of organic matter in the surface layer. These types of additions can have a positive influence on the available water holding capacity for plants. Soils on terraces and toe slopes of adjacent foothills and mountains formed in loess and slope alluvium have significant additions of volcanic ash that may result in the development of andic soil properties.

5.1.2 Willamette Valley

The Willamette Valley extends 230 km from Eugene, Oregon north to Longview, Washington. It is a synclinal valley whose soils formed from alluvium. The valley lies between the Cascade and Coast Range Mountains and covers an area of about 13,000 km^2. Elevation ranges from near sea level to 500 m on valley side slopes. The majority of the soils developed in sandy and silty Missoula Flood sediments deposited between 15,000 and 18,000 years ago. These

Fig. 5.2 Many soils of the Puget Sound Valley, such as this Argialboll, formed in loamy glacial deposits over compact, clayey glaciomarine sediments. The change in particle size distribution with depth results in perched water over the dense layers (image by Toby Rodgers, NRCS; NRCS file photograph released to the public domain)

sediments were deposited when Missoula floodwaters in the Columbia River backed up into the Willamette Valley. The deposits are about 80 m thick in the northern part of the valley and range in texture from silt to gravel. Farther south, the deposits are silty, about 40 m thick, and are locally known as Willamette silts (Gannett and Caldwell 1998). Soils formed in the valley reflect the physical properties and geographic distribution of the Missoula flood deposits. Soils on the valley sides are mostly developed on Miocene Columbia River basalt or uplifted marine sandstones and siltstones. Today, agriculture dominates the use of the region's fertile soils.

The region has a mesic soil temperature regime and a xeric soil moisture regime; soils with an aquic soil moisture regime occur in low-lying areas. Annual rainfall ranges between 50 and 150 cm with dry summers. The valley bottoms are a mix of moderately well-drained Argixerolls, somewhat poorly drained Argialbolls, and poorly drained Albaqualfs, similar to some soils of the Puget Sound Valley. On the valley sides are mostly Haplohumults and Palehumults with relatively high contents of subsoil organic C, formed in basaltic colluvium and residuum and Haploxeralfs on sedimentary rocks (Figs. 5.3, 5.4).

Fig. 5.3 Jory, a Palehumult, is the state soil of Oregon. It is well-drained, very deep, and formed in clayey colluvium derived from mafic igneous rock. The soil temperature regime is mesic and the moisture regime is xeric. Depths are in cm (NRCS file photograph released to the public domain)

5.1.3 Northern Pacific Coast Range, Foothills, and Valleys

The Northern Pacific Coast Range, Foothills, and Valleys are inland from the Pacific Coast and extend from southern Oregon to the Olympic Mountains in northwest Washington. This long, narrow area covers about 27,000 km ^{2}. Elevations range from 30 to 1220 m. Bedrock is mostly uplifted marine sediments composed of Tertiary sandstones and siltstones mixed with basalt. The Washington Coast Range soils formed mainly in large deposits of outwash and till. The central and northern Oregon Coast Range is mainly sedimentary rocks, and the southern Oregon Coast Range is accreted terrane of uplifted marine sediments that are partially metamorphosed. Precipitation occurs mainly in fall, winter, and spring and ranges from 150 to 500 cm per year.

Fig. 5.4 The Jory landscape (foreground) consists of the foothills surrounding the Willamette and Umpqua Valleys. Jory generally supports Douglas-fir and Oregon white oak (*Quercus garryana*) although areas have been cleared for agriculture. The Jory soils and the climate of the Willamette Valley provide an ideal setting for the production of many crops, including Christmas trees, berries, hazelnuts, sweet corn, wheat, and grass seed. Jory is also suitable for the production of wine grapes (NRCS file photograph released to the public domain)

Soils support dense stands of Douglas-fir, western hemlock, and red alder (*Alnus rubra*) with some western red cedar (*Thuja plicata*) and grand fir (*Abies grandis*).

The soil temperature regime is mostly mesic with frigid zones near summits. The predominant soil moisture regime is udic. Hapludands typically have formed in colluvium over residuum, and Fulvudands are found in colluvium derived from basalt. Dystrudepts and Eutrudepts are the primary soils that develop on sedimentary rocks, but on more stable sites, Hapludalfs are common.

Uplifted, unconsolidated geologic materials that form soils of the Coast Ranges (in Washington, Oregon, and California) are prone to landslides and slippage (Fig. 5.5).

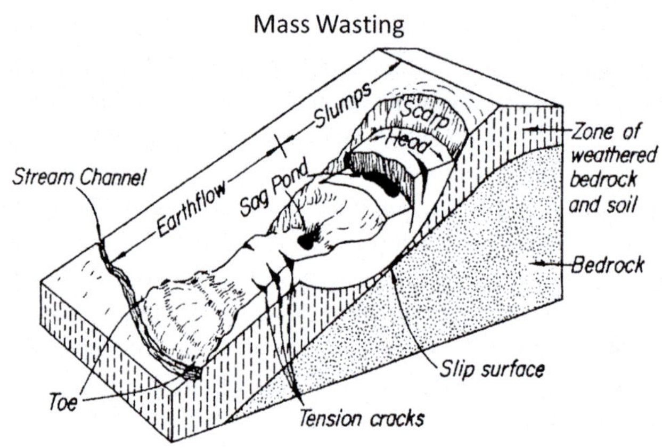

Fig. 5.5 Soils formed in colluvium in the Coast Range are often altered by mass wasting. Toe slopes are often cut by a channel resulting in destabilization of slopes when soils become saturated, which results in slope failure (NRCS file photograph released to the public domain)

5.1.4 Sitka Spruce Belt

This 13,800-km^2 area borders the Pacific Ocean in Oregon and Washington and consists of marine terraces, estuaries, sand sheets, and the western Coast Range foothills. Elevations typically range from sea level to 90 m, but in the foothills of the Coast Range elevations can reach 550 m. The soil climate is moderated by the ocean, and coastal fog drip

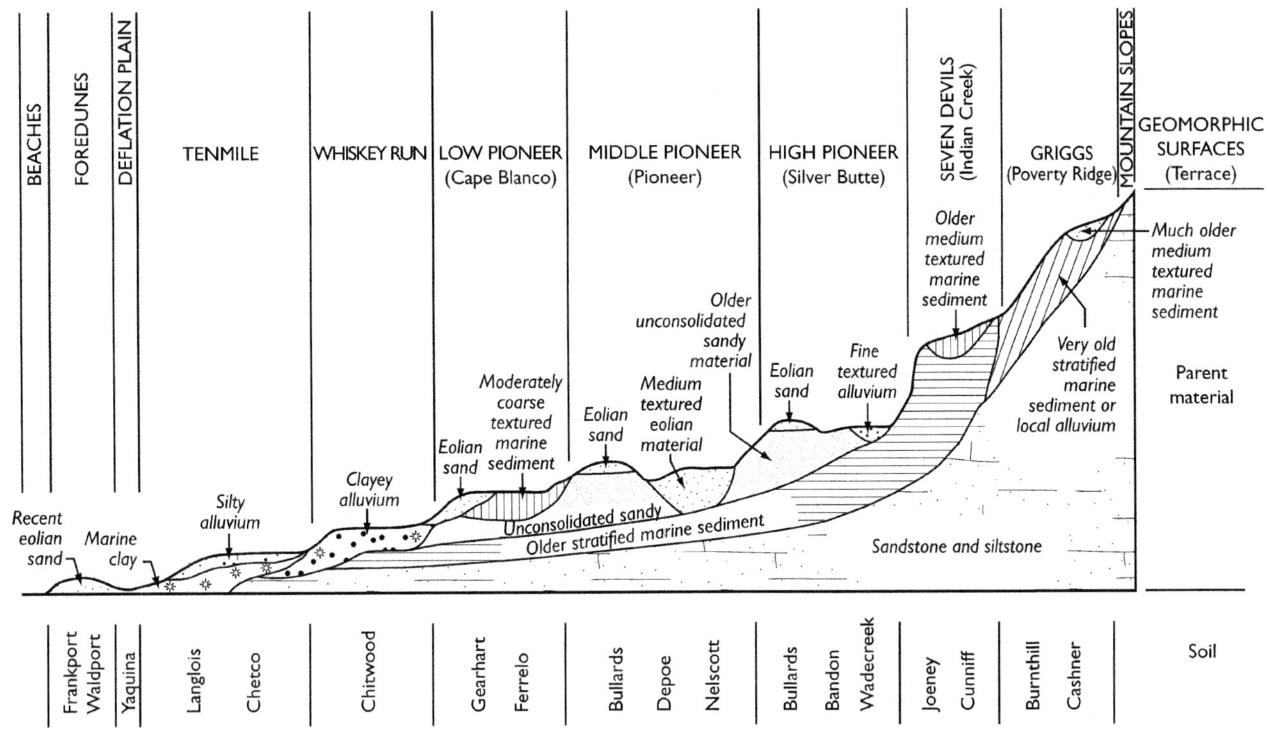

Fig. 5.6 Stylized soil geomorphology cross section of the narrow coastal Sitka Spruce Belt. Soils (named along *bottom* of diagram) are correlated with named terraces (*top* of the diagram) in Curry County, OR (Fillmore 2005). The Pacific Ocean lies to the *left* of the diagram. Parent material type is shown with *arrows*. Soil series names can be queried at the NRCS official soil series description site: https://soilseries.sc.egov.usda.gov/osdname.asp

supplies supplemental moisture to forested soils throughout the narrow area.

The area is dominated at lower elevations by Sitka spruce and at upper elevations by a combination of Sitka spruce (*Picea sitchensis*), Douglas-fir, western hemlock, western red cedar, and red alder. The area includes numerous coves and bays with estuaries, headlands, stacks, and small islands. In northern Washington, Tertiary sedimentary rocks and glacial deposits dominate the soil parent material, whereas soils formed in alluvium and eolian beach dune deposits are more prominent in the southern part of Washington. A sequence of dunes with Udipsamments and interdunal areas with Psammaquents parallels the shoreline. In the northern part of the Oregon coast, marine claystones and headlands of basalt are mixed with estuarine sediments of sands and silts. In the southern Oregon portion of this area lies the edge of the Klamath Mountain accreted terrane. Annual precipitation ranges from 150 cm near the beach to 500 cm in the highlands. Snowfall is rare.

The soil temperature regimes are mainly isomesic and isofrigid, and the regional moisture regime is udic. Soils are generally acidic throughout. The hills and steep uplands to the east side of the area are dominated by Fulvudands and Dystrudepts. Soils of the foothill areas are prone to erosion if deforested. Marine and glacial outwash terraces are mainly Fulvudands, Durudands, Placaquands, Haplorthods, and Duraquods. The flat floodplains and estuaries are mainly poorly drained Fluvaquents, Humaquepts, and Haplohemists in low-lying areas, with better drained Udifluvents and Dystrudepts on terraces and natural levees (Fig. 5.6). Low-lying soils are prone to flooding, ponding, and sedimentation.

5.1.5 Coastal Redwood Belt

This 12,100-km^2 area is mainly in California between the Klamath Mountains and the coastline. Like the Sitka Spruce Belt, it is made up mainly of marine terraces with some large estuaries and bays. Steep slopes dominate the eastern edge. Elevations range from sea level to 800 m in the foothills of the Klamath Mountains to the east. Gently sloping marine terraces border the coast, with a few broad valleys extending inland through the mountains. The coast is rugged and eroding. The marine terraces consist of gravels and sands that were originally deposited as beach systems and then were uplifted. Holocene eolian deposits cap the marine terrace beach materials in many locations. The coastal area is still being uplifted. Inland, the rocks are part of the Klamath uplifted seafloor terranes of contorted metamorphic rocks with some igneous intrusives. Serpentinite, which result in many landslides in the area. The variability of the rock types

Fig. 5.7 Palehumults, such as this very deep soil in Redwoods National Park have thick subsoil horizons of clay accumulation. These soils are common in redwood groves. These soils formed in colluvium and residuum on sandstone and schist (Seney and Frazier 2008). Depths are in cm (image by Joseph Seney, NRCS; NRCS file photograph released to the public domain)

has resulted in soils with a wide range of soil physical and chemical properties.

Annual precipitation ranges from 50 to 250 cm. Snowfall is rare along the coast. Heavy fog is common in the summer. Average temperature ranges from 10 to 15 °C. Vegetation is dominated by coast redwood (*Sequoia sempervirens*) with Douglas-fir, grand fir, western red cedar, and Bishop pine (*Pinus muricata*).

The soil temperature regime is mainly isomesic adjacent to the ocean and mesic inland. The soil moisture regime is mainly udic and trends from udic to ustic to xeric away from the coast. Summer coastal fog drip supplies supplemental moisture to soils under redwoods throughout the area, especially in the isomesic, udic zone along the ocean. Soils are Hapludalfs, Dystroxerepts, Dystrudepts, Palehumults (Fig. 5.7), Haploxerults, and Haplustalfs. Floodplain soils are mainly Udifluvents. On alluvial plains near the ocean, soils on backswamps, depressions, and meander scars are mainly Endoaquepts. There are also dune and interdune soils similar to those in the Sitka Spruce Belt.

5.1.6 Cascade and Olympic Mountains

The Cascade Mountains make up over 60,000 km^2, the largest area in the region. The majority of the Cascade Mountains consist of volcanic rocks, mostly andesite and basalt flows, with some tuffs and debris flows. Elevations range from 200 to 4400 m on Mount Rainier. Annual precipitation ranges from 150 cm on the east side to over 350 cm on the west side. The rain and snow fall mainly in the late fall, winter, and spring. Dense forests dominated by Douglas-fir and western hemlock with silver fir (*Abies alba*) and noble fir (*Abies procera*) in northern reaches cover the landscape up to tree line, above which alpine tundra occurs. The soil temperature regimes in the area are mostly mesic at lower elevations and frigid or cryic at higher elevations. Udic is the dominant soil moisture regime, but the moisture regime grades to xeric in California's southern Cascades.

The northern Cascades consist of sedimentary rocks on the western slopes and metamorphic rocks in the east that have been cut by intrusive igneous rocks (Fig. 5.8). Layered tephra, till, and outwash cover large areas of the Cascades and result in the development of soils with water and root-restrictive dense till and duripans (Fig. 5.9). Active composite volcanoes are found the entire length of the Cascade Range from Mount Rainier and Mount Baker in Washington and Oregon to the dacite plug dome volcano of Lassen Peak in northern California. Eruptions of Mount Mazama (Crater Lake in Oregon, part of the southern Cascades) 7700 years ago, Lassen Peak from 1914 to 1917, and Washington's Mount Saint Helens in 1980 have dramatically affected, and are still affecting, soil forming processes over a large area of the west. Dated ash deposits in soils help to understand their pedologic history and age (Fig. 5.10).

Vitricryands form mainly in volcanic ash, pumice, and cinders at high elevations in the Cascades. Dystrudepts, Humudepts, Humicryods, Haploxerands, Haplocryepts, Haplocryods, and Haplocryands, form in colluvium mixed with volcanic ash (Fig. 5.11). Fulvudands and Durudands form in residuum of volcanic rocks and soils overlain by volcanic ash and loess. Wet meadow areas on toe slopes are dominated by Histosols (Cryohemists and Cryosaprists) (Fig. 5.12).

The Olympic Mountains lie to the west of the Puget Valley in Washington and cover about 2000 km^2. They range in elevation from sea level to over 2400 m. The highest peaks

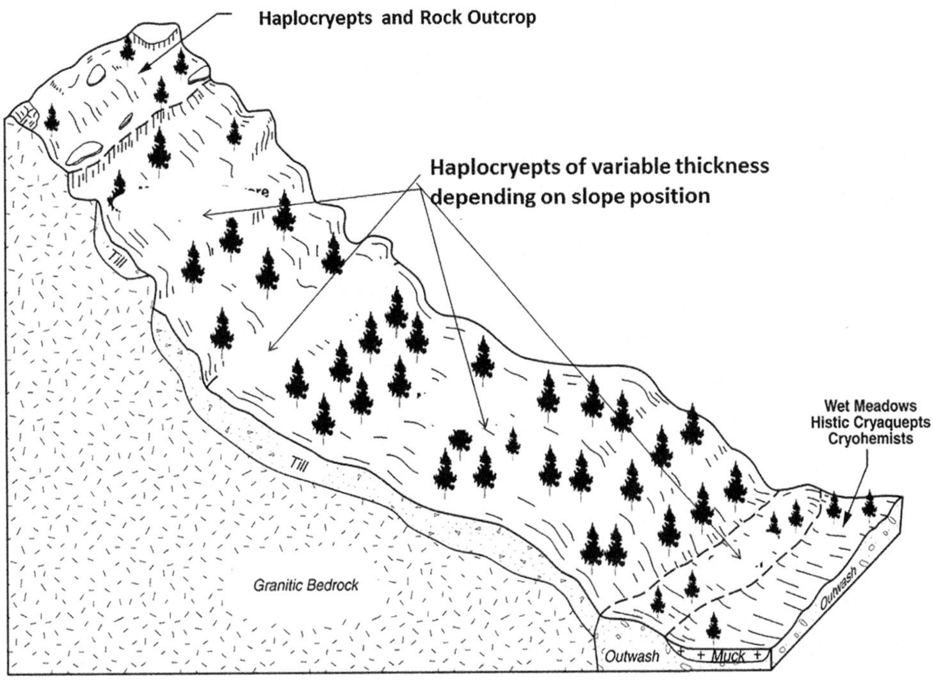

Fig. 5.8 Granitic bedrock underlies till and outwash on the eastern side of the North Cascades and controls slope shape and soil development. Minimally developed Haplocryepts form on eroding slopes; Histic Cryaquepts and Cryohemists form in organic muck deposits on wet lowland positions (NRCS file released to the public domain)

are glaciated. These mountains are mainly uplifted Tertiary ocean bottom composed of pillow basalts in the east and sandstones and siltstone turbidites to the west. Turbidites are geologic deposits of layered particles that grade upward from coarser to finer sizes and are thought to have originated from turbidity currents in oceans (Snavely et al. 1980). Annual precipitation ranges from 150 cm on the east side to over 700 cm on the west side, mostly falling in the late fall, winter, and spring. Dominant vegetation is Douglas-fir, silver fir, and western hemlock, and alpine tundra above tree line. Steep mountains, narrow valleys, narrow divides, and U-shaped glaciated valleys dominate the landscape.

Soils have dominantly frigid and cryic soil temperature regimes and an udic soil moisture regime. The majority of the soils are Dystrudepts and Humudepts on steep slopes. In lower elevations with high rainfall, there is intensive weathering of siltstones, sandstones, and basalts, and Haplohumults and Palehumults are common.

5.1.7 Cascade Mountains, Eastern Slope

This 34,100-km^2 area is the transition between the Cascade Mountains to the west and dry plateaus to the east. Strongly sloping mountains and U-shaped glaciated valleys are dominant in the north, and eroded basalt plateaus are more typical in the south. Many streams dissect this region. Landforms and soil forming factors are typical of those found in both the mountain and plateau areas. The geology in the northern part of the Cascades eastern slope consists of Mesozoic metamorphic rocks cut by younger igneous intrusive rocks. There is also a region of tilted marine sedimentary rocks in the far north.

Most of the eastern slopes to the center and south are Miocene Columbia River basalt with some younger continental, stream-deposited sediments on top of the basalt. Some areas have thick loess or ash deposits. South of Bend, Oregon, a thick blanket of ash and pumice from the eruption of Mount Mazama (which formed Crater Lake 7700 years ago) lies on top of basalt (Fig. 5.13).

Elevations range from about 275 to 2440 m from east to west. Annual precipitation increases from east to west from 30 to 220 cm and falls mainly in winter, spring, and fall. Average annual temperature ranges from 0 to 12 °C and decreases with increasing elevation. The climate, which is drier than the Cascades, dictates the vegetation, which transitions from forests of Ponderosa pine (*Pinus ponderosa*) and lodgepole pine (*Pinus contorta*) in the west to grasslands on the eastern edge.

The three dominant soil orders are Alfisols and Andisols in the west (similar to the Cascade Mountains) and Mollisols in the east. Vitricryands and Haplocryands formed from ash and other pyroclastics dominate the highest elevations. The area has cryic, frigid, or mesic soil temperature regimes (warmer trends from west to east) and a predominant xeric moisture regime that grades to aridic to the east.

At lower elevations, Vitrixerands and Haploxerands formed in loess and ash parent material occupy hills and benches, whereas Haploxerepts formed in residuum and colluvium occupy steep mountain slopes and ridges. Haploxeralfs also occur on older terraces and upland sites. Haploxerolls and Argixerolls formed in till and ground moraines occur in

Fig. 5.9 Tokul, a Vitrixerand, is the state soil of Washington, formed from volcanic ash and loess over dense glacial till cemented with iron and manganese. Gray till parent material is at 2.5 feet (about 75 cm). Tokul soils are limited for home sites because water perches above the gray dense till during wet periods, creating wet basements and making steep slopes unstable. Depths are in feet (NRCS file photograph released to the public domain)

Fig. 5.10 Past eruptions in the Cascade Mountains are present in the soils of Mount Rainier National Park. This Cryaquand preserves distinguishable tephra layers: Mount St. Helens tephra in the A horizon (2–10 cm), Mount Rainier C tephra (10–14 cm), Mount St. Helens P tephra (14 to 26 cm), Mount St. Helens Yn tephra (26–37 cm), Mount Rainier F tephra (44–51 cm), Mount Rainier D tephra (51–60 cm), and Mount Mazama O tephra (60–66 cm) (Rodgers and Roberts 2015) (image by Toby Rodgers and Phil Roberts, NRCS; NRCS file photograph released to the public domain)

the foothills. Grassland soils include Haploxerands, Vitrixerands, and Haploxeralfs formed in tephra, loess, residuum, and alluvium in the eastern part of the area.

5.1.8 Siskiyou–Trinity Area

This area is mainly in California (~62 %), but also extends into southern Oregon (~38 %), and covers about 52,000 km^2. It includes the Trinity and Siskiyou Ranges, a region of steep mountains and narrow valleys between ridges, and the Klamath Mountains in the western part. Other mountain ranges in the area are the Marble Mountains, the Trinity Alps, Scott Mountains, Salmon Mountains, and the northern Yolla Bolly Mountains. The area is mostly accreted terrane. Accreted terrane consists of ocean bottom and island arcs attached to oceanic crust that were scraped off of the edge of the North American plate as the oceanic crust was subducted. The region contains a very wide array of rock types including sandstones and shales with small regions of marble in the northern part (Fig. 5.14). Geologic formations have been folded and faulted and have variable degrees of metamorphism that result in a wide variety of soils.

Intrusions of granite, granodiorite, and gabbro are also common in southern Oregon. Some small areas of ultramafic rocks with seafloor origin, such as serpentinite, also occur in the region. The high Mg and low Ca contents of the ultramafic rocks form unusual soils that often support rare, endemic plants. These plants are adapted to soil conditions that include a low calcium-to-magnesium ratio and low contents of essential nutrients such as nitrogen, potassium, and phosphorus.

In the Klamath region, elevations range from 100 to 1900 m. Annual precipitation ranges from 50 cm at lower elevations to over 500 cm at the higher elevations. Average annual temperature ranges from 5 to 17 °C, decreasing with

Fig. 5.11 The northern Cascade Mountains are cold and wet resulting in the formation of Cryohemists or Cryosaprists (*left*, depth in inches) in organic deposits over alluvium in low-lying meadow areas. Haplocryods (*right*, depth in cm) formed in volcanic ash and glacial deposits under forest vegetation (Rodgers and Roberts 2012) (images from North Cascades National Park by Toby Rodgers and Phil Roberts, NRCS; NRCS file photograph released to the public domain)

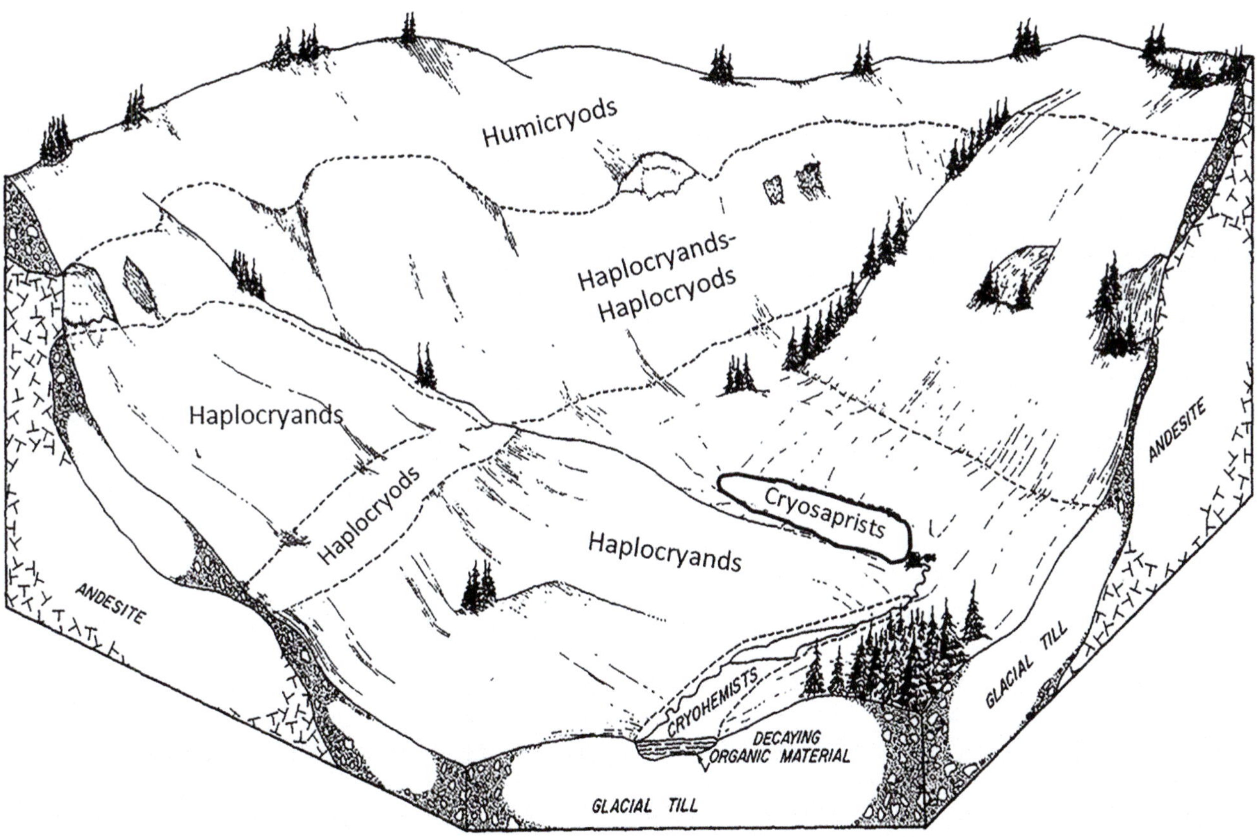

Fig. 5.12 Soils of the southern Cascade Mountains are predominantly in the cryic soil temperature regime and dominated by Andisols, Spodosols, and Histosols. Humicryods dominate areas of landscape stability and high leaching rates; a mix of Haplocryands and Haplocryods are on sideslopes. Cryosaprists are found in low-lying, wet landscape positions (NRCS file photograph released to the public domain)

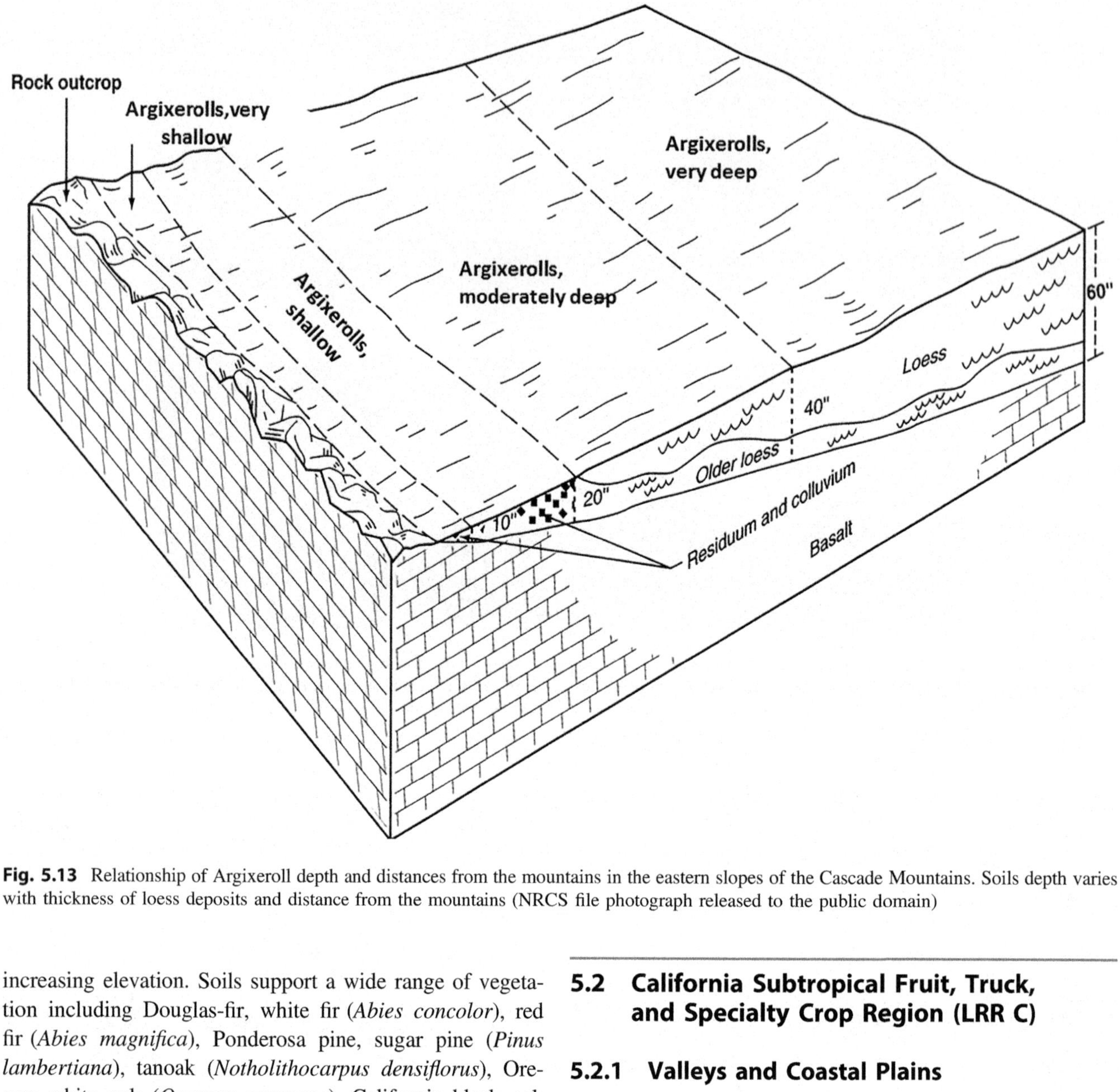

Fig. 5.13 Relationship of Argixeroll depth and distances from the mountains in the eastern slopes of the Cascade Mountains. Soils depth varies with thickness of loess deposits and distance from the mountains (NRCS file photograph released to the public domain)

increasing elevation. Soils support a wide range of vegetation including Douglas-fir, white fir (*Abies concolor*), red fir (*Abies magnifica*), Ponderosa pine, sugar pine (*Pinus lambertiana*), tanoak (*Notholithocarpus densiflorus*), Oregon white oak (*Quercus garryana*), California black oak (*Quercus kelloggii*), canyon live oak (*Quercus chrysolepis*), Pacific madrone (*Arbutus menziesii*), and diverse grasslands.

Soils in the Klamath–Siskiyou area are mainly in the mesic soil temperature regime and the xeric moisture regime. In the mountains, deep Dystroxerepts formed on colluvium over residuum are found on lower slopes, whereas shallow Dystroxerepts formed in residuum are generally upslope of the deeper soils. On more stable landscape positions, colluvium-over-residuum soils include Haploxeralfs and Haploxerults. The lower elevation grassland areas are occupied by Argixerolls and Haploxerolls also formed in colluvium over residuum (Fig. 5.15). Landslides are a dominant soil forming process and hazard in the Klamath Mountains, as in the Coast Ranges.

5.2 California Subtropical Fruit, Truck, and Specialty Crop Region (LRR C)

5.2.1 Valleys and Coastal Plains

5.2.1.1 Sacramento and San Joaquin Valleys

This mostly flat area is about 720 km long by 70 km wide (about 48,000 km^2). Elevations range from sea level to about 200 m. The San Joaquin River watershed drains most of the southern Central Valley (San Joaquin Valley). It flows westward to the valley from the Sierra Nevada, then northward to the California Delta. The southern part of the San Joaquin Valley is mostly internally drained. The Sacramento River drains the northern Central Valley (Sacramento Valley) and originates in the southern Cascades, Klamath Mountains (LRR A) and Modoc Plateau (LRR D).

Soils are derived predominantly from alluvium deposited by the Sacramento and San Joaquin Rivers and their tributaries. The alluvium is derived from granitic, volcanic,

Fig. 5.14 The steep Siskiyou Mountains are dominated by Dystroxerepts and Haploxerults formed on metamorphic rocks. An exposure of marble, in the foreground, is responsible for cave formation at Oregon Caves National Monument (image by Susan Burlew Southard, NRCS; NRCS file photograph released to the public domain)

metasedimentary, sedimentary, and metamorphic rock sources. Anthropogenic soils, formed from alluvium derived from hydraulic mining deposits, are extensive in some areas. The overall soil climate in the Central Valley varies from xeric and thermic in the north to predominantly aridic and thermic in the south. Soils in low-lying basins and sinks have an aquic soil moisture regime (Fig. 5.16).

Floodplains in the valley consist of a wide meander belt of very deep Xerofluvents and Fluvaquents with variable flooding frequency and depths to water table. Haploxerepts typically lie along the outer margin on the distal ends of alluvial terraces that border the floodplains (Burkett and Conlin 2006).

The flood basins are occupied by Vertisols with variable drainage. Low-lying basin Endoaquerts and Haploxererts require maintenance leveling for agriculture because the soils shrink and swell during periods of drying and wetting. Most farmers and ranchers in these areas do not invest in fencing because of high maintenance costs due to subsequent heaving of fence posts. On the east side of the Sacramento Valley, many Vertisols have a duripan, probably due to volcanic glass weathering from the Southern Cascade volcanics to the north, and silicate mineral weathering from the Sierra Nevada granites to the east (Burkett and Conlin 2006).

Alluvial fans of the eastern Sacramento Valley are composed of materials derived from southern Cascade volcanics (Tuscan Formation) and Sierra Nevada granites. Fans of the western Sacramento Valley are derived from Coast Range and Klamath sedimentary and metamorphic rocks. Farther south, the eastside fans are derived from Sierra Nevada granites and volcanic mudflows of the Mehrten Formation. The San Joaquin Valley fans are derived mostly from Sierran granitic rocks to the east, and from Coast Range sedimentary and metamorphic rocks to the west (Arroues 2006). The variability of the source material influences the properties of the soils, such as texture, cation exchange capacity, pH, calcium carbonate content, gravel content, and color. Native vegetation in the valley was predominantly valley oak (*Quercus lobata*), perennial and annual grasses, and forbs in the north, which contributed to relatively high soil organic matter content when compared to the southern Central Valley. In the drier San Joaquin Valley, native vegetation was dominated by desert shrubs, which has resulted in relatively low soil organic matter content.

Based on soil survey data, soils formed on fan terrace are more strongly weathered than soils on the alluvial fans. Lower, younger fan terrace soils are Argixerolls; higher, older terrace soils are Haploxeralfs, Durixeralfs, and Palexeralfs (Fig. 5.17).

Throughout the valley, soils on extensive river terraces often have root-restrictive soil horizons, such as duripans, petrocalcic horizons, and dense, clayey argillic horizons. Many of these soils have been deeply subsoiled in order to increase rooting depth and irrigation water penetration for agriculture. To increase irrigation water efficiency, thousands of hectares of soils have been precision-levelled throughout the valley for the production of a wide variety of crops. Higher terrace positions have a rolling topography and are usually irrigated by drip or micro-sprinkler methods.

Soils with a mound–swale topography (Mima mounds) are common on some of the highest elevations of the terraces (Fig. 5.18). Vernal pools also occur on the terraces in association with the mounds and are host to rare and endangered plants and invertebrates.

The Central Valley has been altered extensively by agriculture and water control features. The entire watershed has been extensively dammed, leveled, ditched, and leveed, which has changed the rates and types of soil forming processes. Soil hydrologic properties have been changed within the entire valley, reducing flooding and deposition of new soil-forming sediments. Hydraulic gold mining in the 1800s

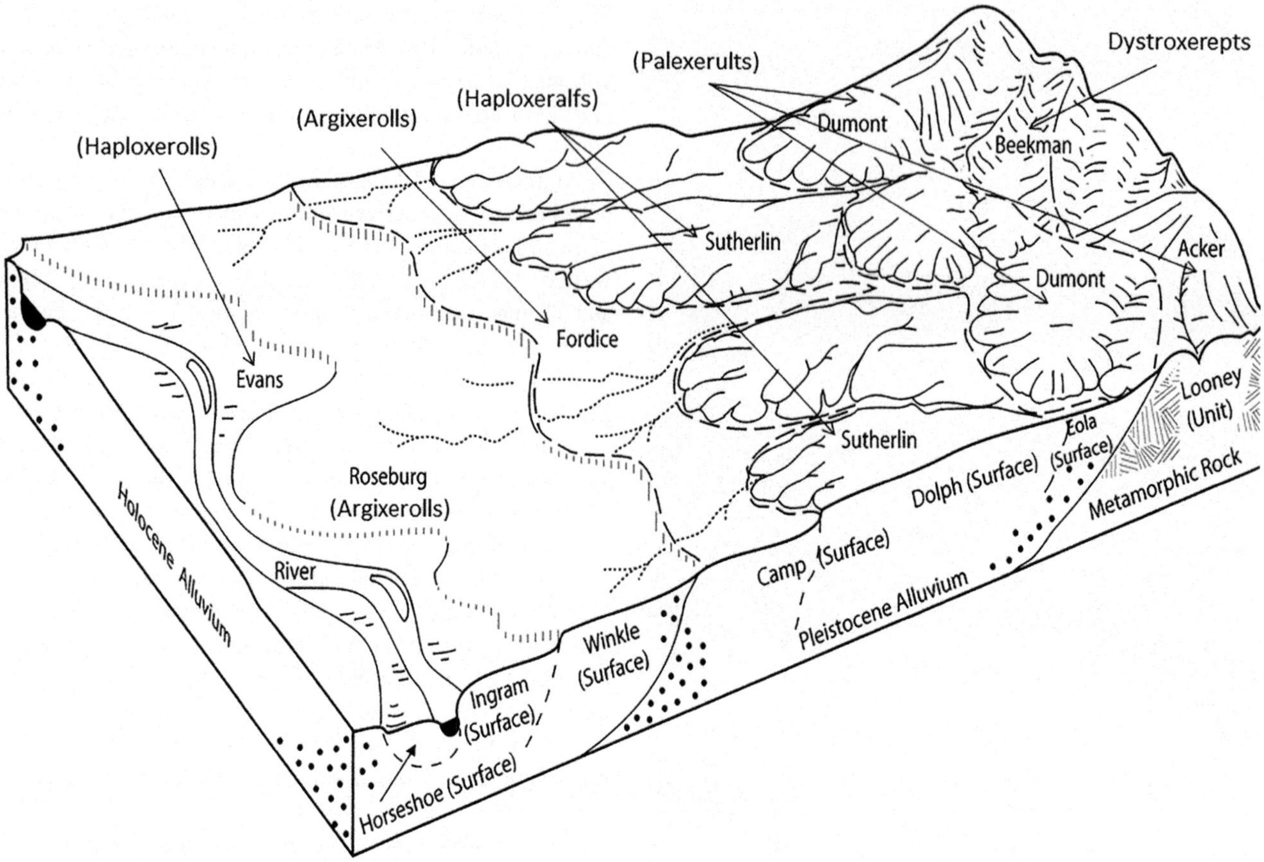

Fig. 5.15 Relationship of geomorphic surfaces, landform, parent material and named soil series (Roseberg, Fordice, Sutherlin, Dumont, Beekman, and Acker) in the Umpqua Valley of south-central Oregon. On lowest and youngest terrace (Ingram), Haploxerolls are dominant. Soils generally are more strongly weathered as reflected in argillic horizon formation and reduced base saturation from left to right, with Palexeralfs (the Eola surface) being the most developed. Dystroxerepts on the extreme right are on steep, eroding slopes and have little profile development. This general geomorphic sequence (with different soil series) is also common in the central valley of California as the valley grades to the foothills to the east and west (NRCS image adapted by Susan Burlew Southard)

deposited silty material (slickens) in the eastern Sacramento area, aggrading the present-day surface and altering hydrologic regimes (Burkett and Conlin 2006). Placer mining east of Sacramento completely changed the soil surface. Mining has created many new anthropogenic soils, which have been identified in soil mapping of the valley.

Salinization has increased due to the lack of seasonal flushing of salts by floodwaters, and by evaporative loss of irrigation water wicking salts to the soil surface. Fire was a common occurrence, but is now rare in the valley due to irrigated agriculture. Large land areas in the San Joaquin Valley have subsided due to ground water withdrawal, and rates of subsidence increase during periods of drought when more pumping occurs (Arroues 2006). Drainage and tillage have created sources for fugitive dust, a contributor to violations of United States EPA air quality standards (Madden et al. 2008).

5.2.1.2 California Delta

The California Delta is a relatively small area, about 3000 km^2, adjacent to the western edge of the Central Valley. The Delta formed during the last 10,000 years as sea level rose during the last deglaciation, causing San Joaquin and Sacramento River sediments to accumulate east of the narrow Carquinez Strait (USDA 2006). The strait is the only outlet from the Central Valley to San Francisco and San Pablo Bays. The California Delta is not a typical delta, but an inverted river delta, that formed inward rather than splayed outward. The Delta consists of around 60 peat and tule islands, surrounded by natural levees (now reinforced), and narrow sloughs. It was a freshwater marsh. Organic soils are mostly Haplosaprists and Haplohemists that formed in situ from plant residue when sea level stopped rising. Mineral soils include Endoaquepts and Endoaquolls formed in alluvium in Delta basins, sloughs, and salt marshes. Many

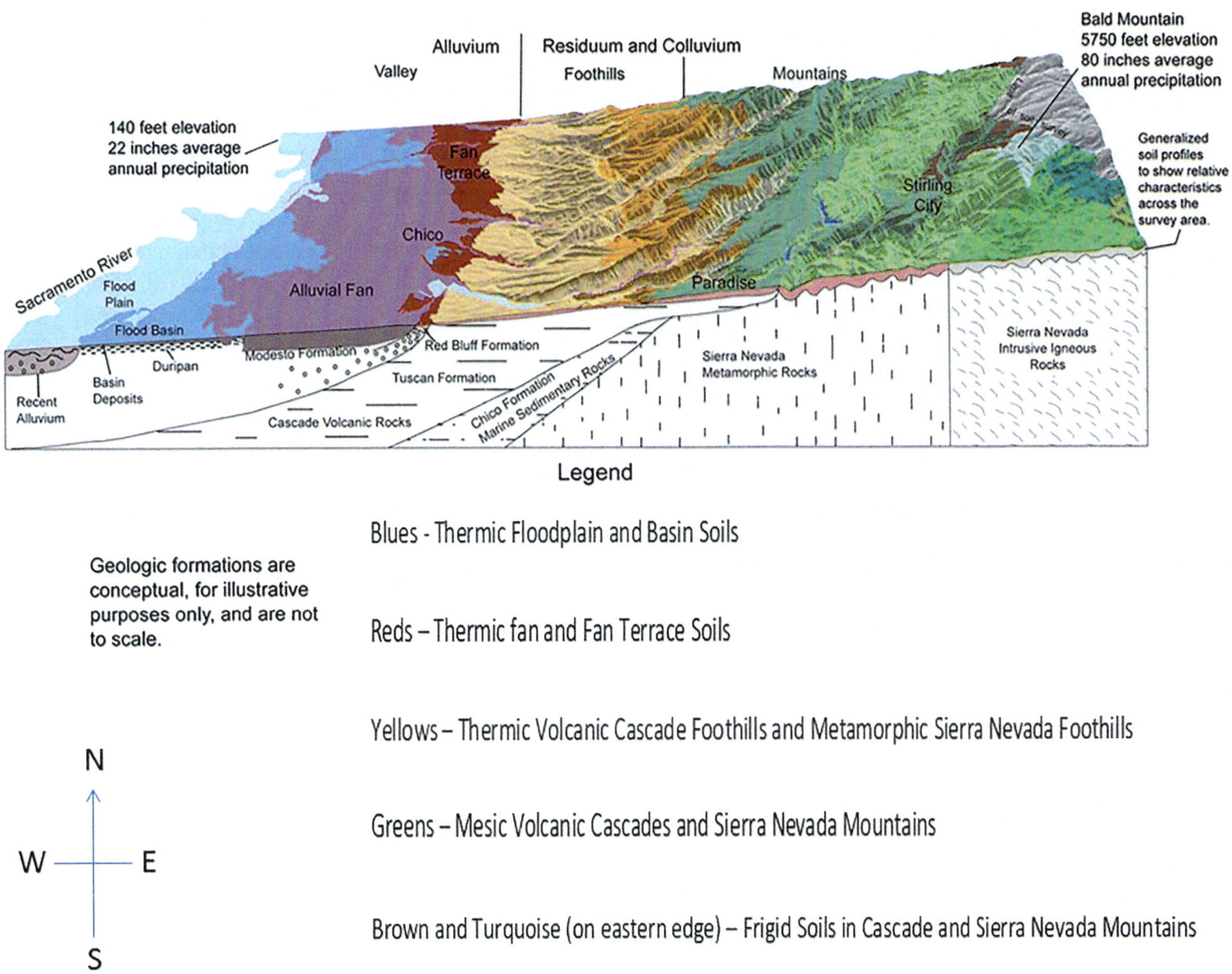

Fig. 5.16 Typical relationship of soils, elevation, landform, and geology in the eastern Sacramento Valley, Sierra Nevada Foothills, and northern Sierra Nevada Mountains (Burkett and Conlin 2006) (Diagram adapted from NRCS file photograph released to the public domain)

of these mineral soils were probably Histosols that have lost their surface organic horizons through oxidation of organic matter due to soil drainage, contributing to the overall subsidence of the Delta. Soil organic matter content remains highest in the Delta, as compared to any other area in LRR A and C (Fig. 5.19a, b).

Fluvaquents formed in alluvium occupy natural levees and floodplains. Many Delta soils have alternating mineral and organic layers illustrating changes in hydrologic regimes and episodic valley flooding. The soils have thermic soil temperature regimes, but are close to mesic due to the insulating effects of organic matter, the cool runoff waters contributing to the aquic moisture regime, seasonal maritime fog, cool air drainage, and evening summer breezes ("Delta breeze") due to thermal rise in the Central Valley.

The Delta was cleared and drained for agriculture during the California Gold Rush, when farmers planted orchards to provide fresh fruit for mining camps in the Sierra Nevada (USDA 2006). Flat terrain and year-round availability of freshwater made irrigation cheap and easy. Today, a network of artificial levees protects the islands from flooding (USDA 1997). Two, routinely dredged, deep-water ports (Sacramento and Stockton) are used for the import and export of agricultural products. The Delta is also a conduit of freshwater from northern California to southern California and to the Central Valley for both irrigation and domestic water via a system of pumping stations and aqueducts. Pumping has resulted in saltwater intrusion from the bays, and increased subsidence, peat fires, wind erosion, and soil organic carbon oxidation. Due to subsidence, most of the western and central Delta averages about 5 m below sea level.

5.2.1.3 Central California Coastal Valleys

This area covers 8500 km^2, and is predominately near the San Francisco Bay area, trending north and south along major fault lines (USDA 2006). Elevations range from sea level to about 600 m. Soils mostly have a thermic–xeric soil climate, although there is a marine influence near the ocean

Fig. 5.17 A Durixeralf on stable terraces, on the eastern side of the San Joaquin Valley, has well-expressed structure. Dark-brown loam abruptly overlies a reddish brown prismatic-structured, clayey argillic horizon (at 11 in. or 28 cm) over a duripan (at 18 in. or 46 cm). The soil pictured is similar to the state soil of California, the San Joaquin. Depths are in inches (image by Randal Southard, University of California, Davis)

that keeps some soils in isomesic or mesic temperature regimes. Vegetation is mostly annual grasses and forbs with scattered valley oak. Salt-tolerant plants grow in the tidal areas. Steep drainages have stands of coast redwoods. Many of the soils with an isomesic temperature regime have an ustic moisture regime. Other soils have udic soil moisture regimes and are associated with fog drip from redwoods and associated vegetation. Soils with aquic soil moisture regimes are in coastal marsh areas or low-lying valley positions.

Adjacent to the ocean, many well-developed soils are formed from eolian materials deposited over consolidated terrace beach deposits and alluvium. Soils are typically Xeropsamments or Haploxerolls on stabilized dunes and more strongly weathered Alfisols and Ultisols on stable terraces.

Soils formed from unconsolidated alluvial floodplain and alluvial fan deposits from Coast Range sources. Argixerolls, Haploxerolls, Haploxeralfs, and Haploxererts are common. Most soils in the coastal valleys are clayey, loamy, or silty and contain few rock fragments. As in the Central Valley, terrace soils are typically Durixeralfs, Palexeralfs, and Palexerolls.

Fluvaquents, Haplosaprists, and Sulfaquents, all of which formed in marshy conditions, are typically ditched, drained, or filled. These wet soils formed in drowned river valleys associated with post-glacial sea level rise. Areas of Sulfaquents that have been leveled and drained for salt production occur at the mouth of rivers that drain into San Francisco Bay. Some marsh soils are now Sulfaquepts, having become acid-sulfate soils, due to the oxidation of sulfides and the development of extremely low pH when artificially drained.

Many areas contain human-transported fill (Anthraltic Xerorthents, formerly Xerarents) that are prone to liquefaction and subsidence during earthquakes (Fig. 5.20). Mining has also altered the landscape in some areas, and soils may be contaminated with methyl mercury.

5.2.1.4 Southern California Coastal Plain

This mostly urbanized area is about 11,000 km^2 and includes the cities of Los Angeles, San Diego, and Ventura. Elevations range from sea level to 600 m. Most soils have a thermic soil temperature regime and xeric soil moisture regime. The area is mostly level, but some areas are strongly sloping. Vegetation includes annual grasses with coast live oak *(Quercus agrifolia)* and brush of ceanothus (e.g., *Ceanothus cuneatus*), chamise (*Adenostoma fasciculatum*), and coastal scrub oak *(Quercus dumosa)*.

The major soils are deep Haploxerolls and Haploxeralfs on alluvial fans and terraces, Durixeralfs on dissected terraces, Xerofluvents and Xeropsamments on fans and floodplains, and shallower Haploxerolls and Xerorthents in the rolling hills.

Soils are derived from sandy and loamy coastal plain alluvial and eolian sediments. Along the coast, terraces have formed via tectonic uplift and sea level changes associated with glacial periods. Some of these terraces are strongly dissected (Fig. 5.21).

The coastal plain begins where alluvial fans and colluvial slopes from the Southern California Mountains begin to coalesce with the plains (USDA 2006). The distance of transport of the alluvium from the hills to the plains affects soil texture in a slope gradient. Coarser-textured soils are nearer the mountains and finer-textured soils are closer to the ocean. Well-developed soils, such as Haploxeralfs and Palexerolls, occur on incised drainages (USDA 2006).

Because of urban development, the hydrologic and depositional regimes of the soils have been changed. Canals,

Fig. 5.18 Mound–swale and vernal pool landscapes exist in a narrow belt of level and sloping terraces in the eastern Sacramento Valley. The hydrologic function of these soils contributes to a wide variety of native plant associations (image by Susan Burlew Southard; NRCS file photograph released to the public domain)

ditches, dams, levees, and concrete-line flood-control channels divert floodwaters from industrialized and residential areas and retain water for urban uses. Alluvial deposition of sand from the Santa Clara River (that supplied Los Angeles beaches and stabilized dune soils) has been diminished due to gravel and sand mining operations.

Fire plays a major role in the ecosystems of Southern California. Many soils become hydrophobic from either fire or plant exudates. Soil organic carbon content is generally low in all the soils of this area, and rapid erosion of soils in the hills results in thin, weakly developed Xerorthents. In basin positions, salinity and encroachment of seawater are soil management concerns.

5.2.2 Mountains and Foothills

5.2.2.1 Sierra Nevada Foothills

The foothills lie along the eastern edge of the Central Valley. This soil region is approximately 720 km long and has an area of about 54,000 km^2 (USDA 2006). Elevations range from about 200 to 1200 m. Soils have mostly a thermic soil temperature regime. The soil moisture regime is predominantly xeric in the northern half and aridic in the south. Some north aspects have mesic soil temperature regimes, especially at higher latitudes and higher elevations in the northern half of the foothills. Precipitation varies from 50 to 100 cm in the very northern foothills to 20 to 50 cm in the extreme southern portion. The entire foothills area is tilted to the west by uplift of the Sierra Nevada, and drainages are bedrock-controlled (USDA 2006). Soil depth throughout the foothills ranges from very shallow to deep, depending on bedrock type and soil forming processes. Many of the soils formed in a combination of residuum and colluvium. Some soils formed in alluvium along narrow drainages. Vegetation is dominantly blue oak (*Quercus douglasii*) savanna on deeper soils; shallower soils are dominated by chamise.

The northern part of the foothills consists of low-elevation volcanic plateaus of the southern Cascades, composed of mudflows (lahars) and pyroclastic rocks that overlap the Sierran granites. Soils are commonly shallow Xerorthents over basalt or consolidated lahar and ash

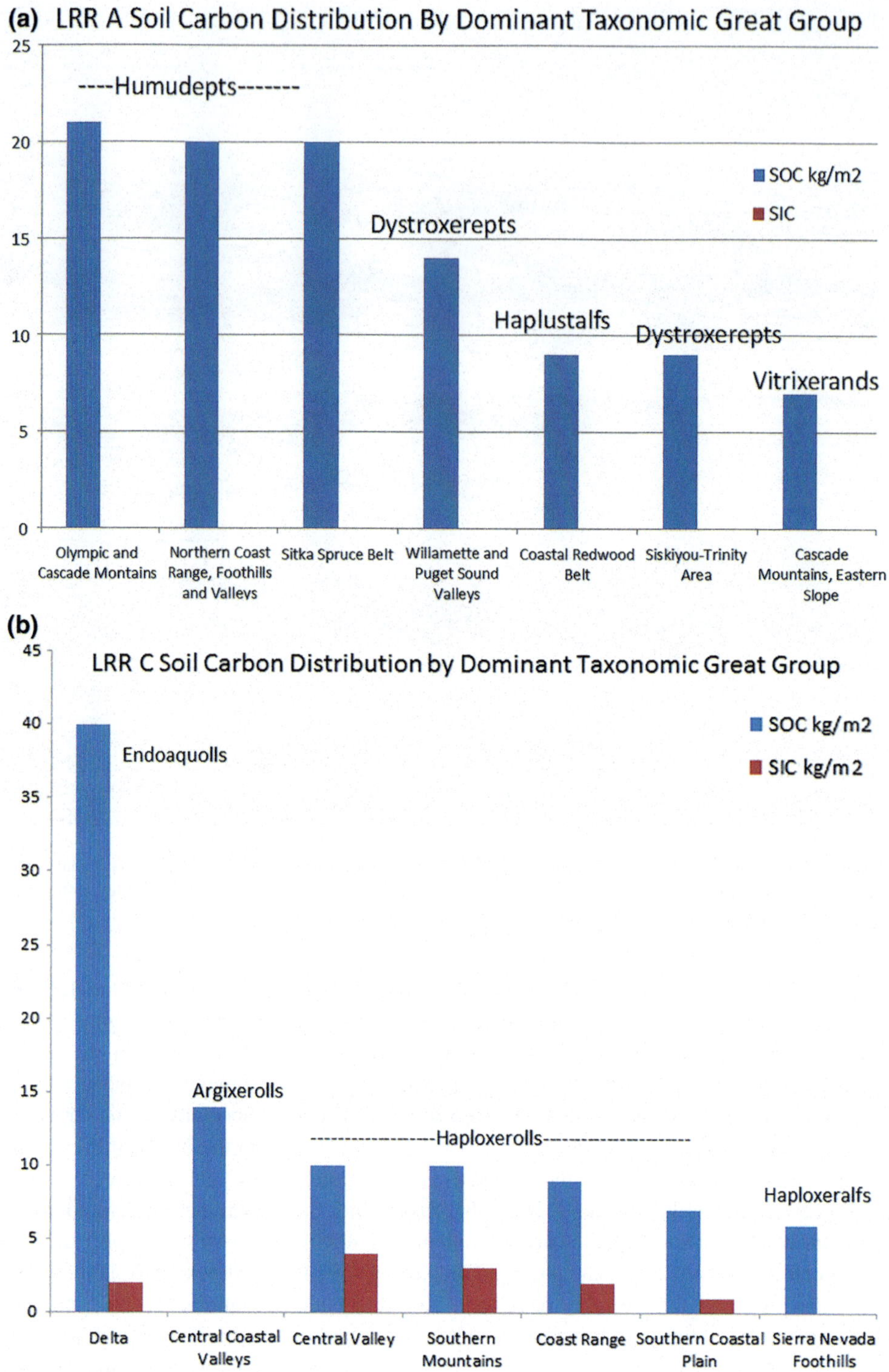

Fig. 5.19 Distribution of soil organic carbon (SOC) by the dominant soil great group mapped in LRR A (**a** *top*) and LRR C (**b** *bottom*). In LRR A, the trend of soil temperature regime shifts from cool mesic to thermic and soil moisture regime shifts from udic to ustic to xeric (left to right). More carbon is stored in cool, wet climates than in drier climates. None of the dominant great groups in LRR A has measurable soil inorganic carbon (SIC). The shift to drier, warmer soil climates in LRR C, as compared to LRR A, results in overall lower SOC. The most southern extent of the Central Valley, the Coast Range, and the Southern Mountains have calcium carbonate in soils with aridic and driest xeric climates, where leaching of carbonates from the profile is limited. Most soils of the present-day Delta have lost considerable amounts of soil organic carbon through human-induced subsidence. However, SOC levels remain very high when compared to the rest of the dominant soil great groups throughout both LRRs

deposits. Other soils include Haploxeralfs, at higher elevation, which are moderately deep or deep to either paralithic or lithic materials of mudflow breccia.

The central foothills, from about Oroville, CA, to south of the Merced River near Merced, CA, are underlain by Mesozoic-age, metamorphosed marine sedimentary and volcanic rocks associated with a subduction trench that existed just west of the present-day Sierra Nevada. Most soils in the central foothills are Haploxerepts and Haploxeralfs.

The southern foothills section is south of Merced and is underlain by granites of the Sierra Nevada, with occasional exposures of gabbro, and mafic metavolcanic rocks (USDA 1997). Most soils in the southern foothills are Haploxerolls

Fig. 5.20 Fort Baker, near the northern end of the Golden Gate Bridge, was built on leveled fill and mapped as Xerarents. The tectonically active coastline of California has no naturally deep ports and building areas are steep. Dredged channel materials are often used as fill, which can be unstable and have low soil strength (Southard 2013b) (images by Susan Burlew Southard, NRCS. NRCS file photograph released to the public domain)

or shallow Xerorthents, but more strongly weathered soils include Haploxerepts, Haploxeralfs, Palexeralfs, and Rhodoxeralfs on more stable landforms. In alluvial settings in the southern foothills, the soils are Torripsamments and coarse-textured Haplargids. The granite in this region is generally deeply weathered. The hardness of the weathered granite varies seasonally and is moisture dependent. During the winter when the soils are moist, the light-colored, well-drained soils and underlying weathered granite can be easily dug with a spade. During dry summer months, the soils and weathered rock zone are very hard, and excavation is extremely difficult. Under these conditions, the weathered rock material may be identified as a paralithic contact.

Soil color is often dependent on the type of parent material and weathering intensities. For example, throughout the foothills, Rhodoxeralfs typically occur on mafic plutonic rocks such as gabbro. A wet, warm climate zone at higher foothill elevations allows for maximum weathering, as illustrated by very high iron oxide accumulation in the red soils. The soils do not freeze, are below the snowline, and remain moist for much of the year, allowing for maximum microbial growth and chemical weathering of soil minerals. At higher latitudes, the foothill "red soil zone" occurs at lower elevations than in the southern foothill section.

Soils on ultramafic rocks are mostly dark-colored Haploxeralfs and Argixerolls. These soils have relatively high organic carbon contents. The soil organic matter is stabilized by magnesium and other metals weathered from the ultramafic rocks.

Some foothill soils, weathered from ultramafic rocks and metamorphosed gabbro, contain naturally occurring asbestos minerals. These minerals include fibrous serpentine minerals and fibrous amphiboles. Asbestos minerals have been mined for a variety of industrial uses (e.g., insulation materials, brake linings). Exposure to asbestos minerals has been linked to the occurrence of cancer and lung disease in humans. This region has also been extensively mined for gold and other metals, leaving behind a variety of

Fig. 5.21 Soils on tectonically uplifted marine terraces along the Pacific in LRR A and C preserve clues to past climates. In this image, taken at Cabrillo National Monument near San Diego, reticulate mottling has been preserved along an uplifted terrace in the park. Images *a* and *b* were taken in area designated by the arrow. Redoximorphic features (*red* and *gray* mottles) have become lithified and indicate a previous warm and wet climate. Image *c* illustrates present-day redoximorphic features forming in a soil in warm, humid North Carolina. Image *d* is the modern day soil profile in North Carolina (images by Susan Burlew Southard, and Dave Kelley, NRCS. NRCS file photograph released to the public domain)

anthropogenic land forms, including abandoned placer deposits, and hydraulic mine tailings, and mine shafts.

5.2.2.2 Central California Coast Range

This area extends 740 km from north to south and covers about 46,000 km^2 (USDA 2006). Elevations range mostly from sea level to 800 m, but some areas are as high as 1500 m on mountain summits. The area has annual grasses, grass–oak and shrub vegetation. Trees are dominantly blue oak, valley oak, and canyon live oak; brush is often coyotebrush (*Baccharis pilularis*), chamise, ceanothus, and common manzanita (*Arctostaphylos manzanita*).

Soils are mostly Haploxerepts, Argixerolls, Haploxerolls, Haploxererts, Haploxeralfs, and Palexeralfs. Southern areas of the Coast Range are drier and warmer than the northern areas. Soils close to the coast generally have an ustic soil moisture regime and an isomesic soil temperature regime (e.g., Haplustalfs, Haplustolls), which results in higher soil carbon content in the soil. Transpiration and evaporation rates along the coast are low due to summer fog, and there is little seasonal variation in soil temperature and moisture. Fog and cool temperatures near the coast contribute to a water balance that favors the development of soils with thick, darkened surface horizons with higher carbon content than soils inland and increased soil mineral weathering.

The landscape is strongly controlled by uplift and right-lateral strike-slip movement along the San Andreas Fault and other faults (USDA 2006). The northern part is dominated by the Franciscan Formation, consisting of Mesozoic-aged marine deposits. The Franciscan deposits were moved eastward during subduction, then sheared, deformed, and faulted along the strike-slip San Andreas Fault system during the last 30 million years (Elder 2001). Highly variable rock types with highly variable degrees of deformation, faulting and differential erosion result in extremely variable soils on the landscape (Fig. 5.22).

Between the Franciscan Formation and the uplifted Great Valley sequence of marine sediments to the east are

Fig. 5.22 Uplifted and tilted red radiolarian cherts that originated on the ocean floor (*left*) lend their color to red soils forming along the California coast in the Marin Headlands north of San Francisco (*right*). Unconsolidated uplifted marine sediments are susceptible to sliding as seen along the road (Southard 2013a) (images by Susan Burlew Southard, NRCS. NRCS file photograph released to the public domain)

serpentine-bearing ophiolites. These rock have undergone variable amounts of hydrothermal alteration resulting in a wide diversity of soils, some with magnesic mineralogy and some with naturally occurring asbestos, similar to those found in the foothills of the Sierra Nevada on the east side of the Central Valley. The overall landscape and soil forming processes include mass movement (Fig. 5.23), erosion and deposition from slopes, and in situ weathering of the highly variable rock types (USDA 1997).

In the Central California Coast Range, Haploxerepts occur on more resistant rock types, such as breccias and rhyolite. Haploxerolls occur on granites and on colluvial toe slopes. Uplifted mudstones and shale have contributed to the development of sloping Haploxererts that shrink and swell due to drying and wetting. Very weakly weathered Xerorthents commonly occur on sandstones (Figs. 5.24, 5.25). Palexerolls and Haploxerolls support blue oak woodland, which is effective in recycling nutrients as the leaves decompose on and in the soils.

Andesite weathers to finer-textured soils; rhyolite weathers to coarser-textured soils. Clayey Haploxerolls typically occur on northern aspects where the evaporation rate is lower than on southern aspects and soil moisture is conserved. Prolonged moist conditions on north aspects maximize the weathering of the parent material into soil, and the grass understory with numerous fine roots, is particularly effective in enriching the soils with organic matter.

Sedimentary rocks in the foothills include fanglomerates, sandstones, and shales. These rocks are acidic, are low in nutrients, and are often highly fractured. Soils formed from these rock types have low water storage capacity, and therefore are easily saturated and become susceptible to

Fig. 5.23 Mass movement or mass wasting as evidenced by hummocky soil surfaces and hill slope slumps are common in the entire Coast Range Mountains. Unstable, uplifted landforms have limited urban development in some areas. These areas are clayey Argixerolls and Haploxeralfs formed on unconsolidated Franciscan Formation (USDA 2013b) (image by Susan Burlew Southard, NRCS; NRCS file photograph released to the public domain)

erosion. Fertility is low because of (1) the low amount of nutrients supplied by the parent material, (2) the low amount of organic matter contributed by chamise vegetation, and (3) high nutrient losses due to erosion (Oster 2007).

5.2.2.3 Southern California Mountains

Comprising about 25,000 km^2, the northern part of the Southern Mountains includes the Transverse Range (of mixed marine and non-marine sediments and granite) that merges with the southern Coast Range (USDA 1997). Extending to the south are the Peninsular Mountains of mixed lithology. Both ranges lie near, but do not border the ocean. Because elevations reach 3550 m, soil temperature regimes range from thermic to frigid; aspect plays a large role in localized soil climates. The soils mostly have a xeric soil moisture regime. Vegetation is mostly brush and shrub-grass mixtures. Brushy chaparral areas include manzanita, ceanothus, sumacs (*Rhus integrifolia*) and mixed oaks at lower elevations. Mixed-conifer forests (mostly fir and pine) dominate at higher elevations.

Major soils are Haploxerolls, Haploxeralfs, and Xerorthents that tend to be sandy or loamy. Xerorthents are typically on steep slopes and are shallow over weathered granite or schist. More stable slopes develop Haploxeralfs.

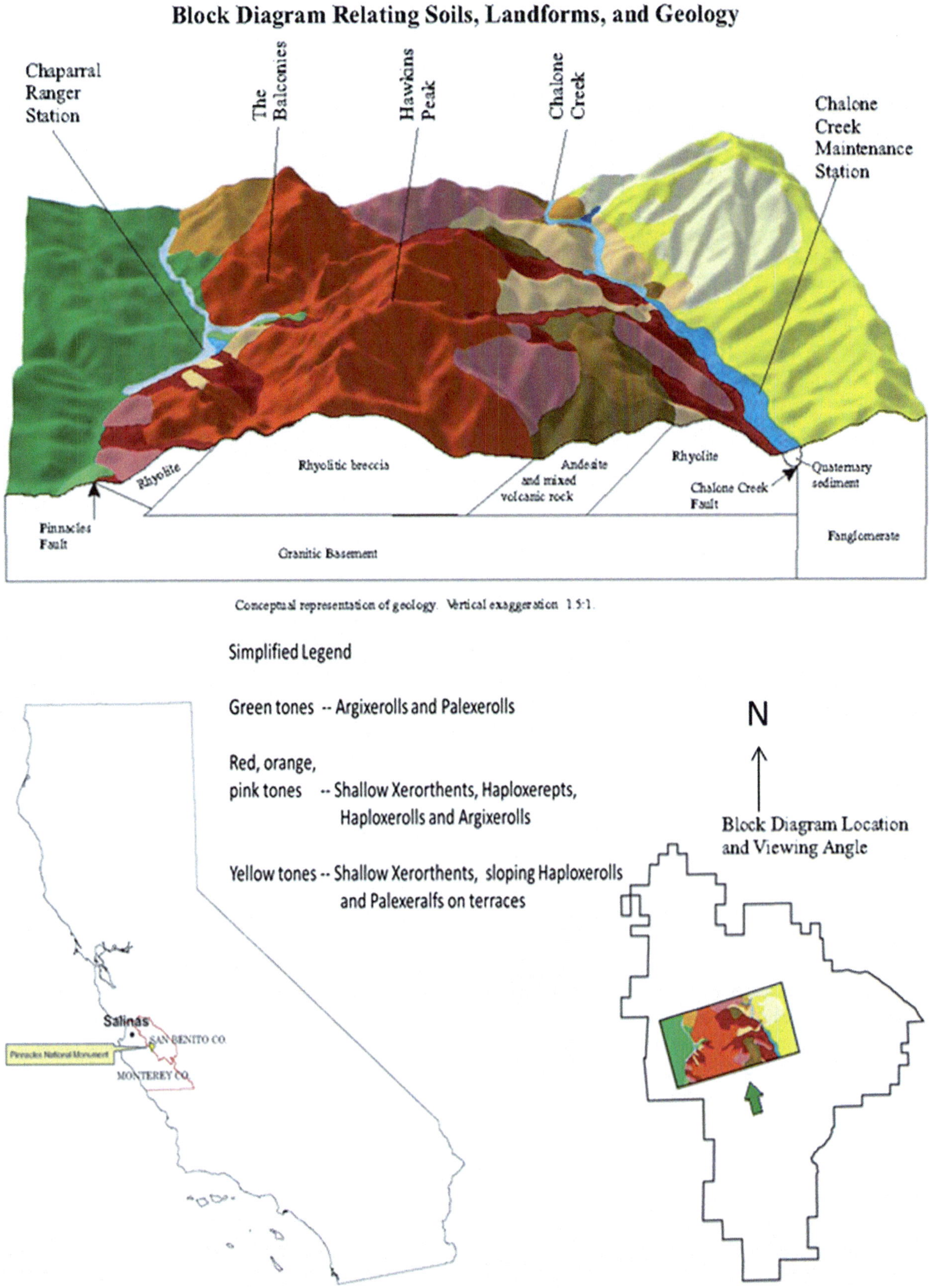

Fig. 5.24 Pinnacles National Monument illustrates tectonics, soil forming factors, and soil diversity in the Coast Range. The variable rock types associated with faulting and soil formation processes contribute to complex soil landscapes. Mesozoic granite has been carried north hundreds of kilometers on the San Andreas Fault to the Pinnacles National Monument and beyond to the southern San Francisco peninsula (Oster 2007)

Fig. 5.25 In the Central Coast Range, moderately deep, marginally developed soils, such as this Haploxerept (*left*) formed on breccia in Pinnacles National Monument. Erosion limits soil depth. Vegetation is typically mixed chaparral with chamise, buckbrush (*Ceanothus cuneatus*), and bigberry manzanita (*Arctostaphylos glauca*) (images by Susan Burlew Southard, NRCS; NRCS file photograph released to the public domain)

Haploxerolls are commonly on colluvial slope positions where soil has accumulated from eroding slopes above them (Fig. 5.26).

The overall climate and terrain make the area susceptible to wildfires, which can influence soil forming processes. After a fire, erosion may be accelerated by the loss of vegetation and surface ground cover. Heavy precipitation loosens rock and soil on slopes that lack the stabilizing effect of plant roots. Unconsolidated rock and soils that are suddenly saturated with water can detach and slide downslope, causing a slump or flow.

Fire often creates a plant mosaic of different ages and species. As a result, fire often increases the diversity of habitats (Southard 2013a). When burned, chaparral vegetation may result in areas of fire-induced soil hydrophobicity, which causes the soils to repel water. A thin layer of soil at or below the mineral soil surface can become hydrophobic after intense heating. The hydrophobic layer is the result of a waxy substance that is derived when volatile oils and other organic compounds from plants are vaporized during a hot fire and then re-condense in the cooler soil just beneath the surface. Water absorption is prevented due to a waxy subsurface layer that reduces soil porosity. Because the ground cover is often destroyed by the fire, hydrophobic soils have increased runoff and erosion. As a result, chaparral soils are commonly shallow. Clayey soils are the most resistant to developing hydrophobicity. The predominantly sandy or loamy soils in this area are much more susceptible to hydrophobicity when vegetated with chaparral (Southard 2013a).

The Southern California Mountains are susceptible to earthquakes and landslides. Earthquakes can cause landslides, but landslides are also caused by saturated soils that slip and slide. This is a natural geomorphic process that contributes to the distribution of soil types in the entire region.

Fig. 5.26 Typical landscape of Southern California Mountains northeast of San Diego with common exposures of rounded, eroded granite. Dominant soils are shallow Xerorthents associated with rock outcrop, with deeper Haploxerepts and Haploxerolls in colluvial positions. Development of soil hydrophobicity and high surface runoff after fires is common in this environment. (image by Susan Burlew Southard, NRCS; NRCS file photograph released to the public domain).

5.3 Summary

Land Resource Regions A and C have diverse soils due to wide ranges of soil forming factors. Climate varies with latitude and elevation giving rise to different soil weathering intensities. Soil carbon content exemplifies the diversity of soil forming factors and soil types. Due to the orographic effect of mountains and the range of latitude, drier climates on the eastern and southern part of the region result in less effective leaching and some accumulation of soil carbonates (Fig. 5.19). Parent materials include tephra deposits, alluvium derived from massive regional flood deposits, large rivers, and small creeks, a wide array of rock types, plant residues, constantly eroding colluvial slopes, eolian sources from beaches and riverbeds, and vast glacial deposits. Many soil profiles have two or more parent material types. Tectonics and differential erosion have played significant roles in landscape stability (or instability) and constrain the age of soil landscapes. Landscape stability (as proxy for time) has influenced the degree of soil profile development and has affected local soil hydrologic regimes in basins, sinks, canyons, and valleys. Vegetation ranges from grasses and shrubs to a wide diversity of coniferous and deciduous trees to high alpine meadows. The vegetation types contribute to soil chemical, physical, and morphological properties in the regions.

References

Arroues KD (2006) Soil survey of Fresno County, Western Part, California. USDA Natural Resources Conservation Service, Washington

Burkett DW, Conlin AE (2006) Soil Survey of Butte Area, California, Parts of Butte and Plumas Counties. USDA Natural Resources Conservation Service, Washington

Elder WP (2001) Geology of the Golden Gate headlands. In: Stoffer P, Gordon J (eds) Geology and natural history of the San Francisco Bay Area: a 2001 NAGT field-trip guidebook, p 61–86. http://pubs.usgs.gov/bul/b2188/b2188ch3.pdf. Accessed 15 Jan 2014

Fillmore MH (2005) Soil survey of Curry County, Oregon. USDA Natural Resources Conservation Service, Washington

Gannett MW, Caldwell RC (1998) Geologic framework of the Willamette lowland aquifer system, Oregon and Washington. US geological survey professional paper 424-A

Madden NM, Southard RJ, Mitchell JP (2008) Conservation tillage reduces PM10 emissions in dairy forage rotations. Atmos Environ 42:3795–3808

Oster K (2007) Soil survey of Pinnacles National Monument, California. USDA Natural Resources Conservation Service. http://soils.usda.gov/survey/printed_surveys/. Accessed 20 Jan 2014

Rodgers T, Roberts P (2012) Soil survey of North Cascades National Park Complex, Washington. USDA Natural Resources Conservation Service and USDI National Park Service. http://www.nrcs.usda.gov/wps/portal/nrcs/surveylist/soils/survey/state/?stateId=WA. Accessed 20 Jan 2014

Rodgers T, Roberts P (2015) Soil survey of Mount Rainier National Park. USDA Natural Resources Conservation Service, Washington

Snavely Jr PD, Niem AR, Macleod NS, Pearl JE, Rau WW (1980) Makah Formation-a deep marginal-basin sequence of late Eocene and Oligocene Age in the Northwestern Olympic Peninsula, Washington. US Geological Survey Professional Paper 1162-B

Seney JP, Frazier AC (2008) Soil survey of Redwood National and State Parks, California. USDA Natural Resources Conservation Service, Washington

Southard SB (2013a) Soil survey of Cabrillo National Monument, California. USDA Natural Resources Conservation Service and USDI National Park Service. http://soils.usda.gov/survey/printed_surveys. Accessed 20 Jan 2014

Southard SB (2013b) Soil survey of Golden Gate National Recreation Area, California. USDA Natural Resources Conservation Service and USDI National Park Service. http://soils.usda.gov/survey/printed_surveys. Accessed 20 Jan 2014

USDA United States Forest Service (1997) Ecological subregions of California. R5-EM-TP-005

United States Department of Agriculture (2006) Land Resource Regions and Major Land Resource Areas of the United States, the Caribbean and the Pacific Basin. US Department of Agriculture Handbook 296

Soils of the Rocky Mountain and Inland Pacific Northwest Region: LRRs B and E

6

Paul A. McDaniel

6.1 Northwestern Wheat and Range Region: Land Resource Region B (MLRAs 7, 8, 9, 10, 11, 12, 13)

The Northwestern Wheat and Range Region makes up just over 210,000 km^2 in Idaho, Washington, and Oregon and is often referred to as the Inland Northwest Region (Fig. 6.1a). Because this region lies on the leeward side of Cascade Range, the climate is relatively dry with mean annual precipitation (MAP) ranging from approximately 150 to 550 mm across much of the Inland Northwest. Mollisols and Aridisols are the dominant soils of the region, which includes some of the most productive wheat and small grain production systems in the USA. More than 40,000 km^2 of wheat and barley were harvested from Idaho, Oregon, and Washington in 2012; nationally, this constitutes 8 % of the small grain acreage and more than 13 % of the total yield (Eigenbrode et al. 2013). Livestock grazing is also a major land use.

6.1.1 Columbia Basin and Columbia Plateau (MLRAs 7, 8)

The Columbia Basin and Columbia Plateau MLRAs together encompass 65,085 km^2 almost entirely in Washington and north-central Oregon (Fig. 6.1a) (U.S. Dept. of Agriculture 2006). The location of these MLRAs in the rain shadow of the Cascade Range coupled with relatively low elevations results in some of the driest and warmest areas found in the region. Mean annual precipitation ranges from 150 to 255 mm across the Columbia Basin to over 400 mm in parts of the Columbia Plateau (U.S. Dept. of Agriculture 2006).

The Columbia Basin represents the relatively low-lying (~90–600 m elevation) part of this area. Entisols and Aridisols are the dominant soils and occur in a variety of parent materials—loess, alluvium, lacustrine sediments, outwash, wind-deposited sands, and glaciofluvial deposits. On younger landscapes, the texture and mineralogy of these parent materials strongly influence soil properties.

The relatively low rainfall also influences soil properties in two main ways. First, vegetative production is relatively low and soils consequently do not receive large additions of organic matter. Soil A horizons typically contain <1.0 % organic C and lack the dark A-horizon colors found in Mollisols of the higher-precipitation zones of the Northwest Range and Forest Region LRR (Fig. 6.2). Also, little leaching has occurred in these soils and horizons in which $CaCO_3$ has accumulated are common. On some older landscape positions, $CaCO_3$- and SiO_2-cemented duripans have formed in Haplodurids (Fig. 6.2). These duripans, by definition, only allow root penetration along vertical fractures that have of a horizontal spacing of at least 10 cm (Soil Survey Staff 1999). In addition to limited root growth, vertical movement of water is also restricted. These restrictions can have a major influence on plant community structure and productivity in the water-limited environment (aridic and xeric soil moisture regimes) of the Columbia Basin and Plateau MLRAs.

Where not irrigated, soils support shrubs and grasses, including sagebrush (*Artemisia tridentata*), bitterbrush (*Purshia tridentata*), bluebunch wheatgrass (*Pseudoroegneria spicata*), and needle-and-thread grass (*Hesperostipa comata*). Development of irrigation projects in the Columbia Basin has resulted in approximately 40 % of this MLRA being used as irrigated cropland (U.S. Dept. of Agriculture 2006).

The Columbia Plateau represents the higher-elevation counterpart to the Columbia Basin MLRA. As such, it generally has higher mean annual precipitation (255 to over 400 mm) and slightly lower mean annual temperatures (8–12 °C) (U.S. Dept. of Agriculture 2006). It consists of a basalt plateau that is mantled with varying amounts of loess and smaller amounts of volcanic ash. In some areas, the loess may be as much as 75 m thick, while in other areas

P.A. McDaniel (✉)
University of Idaho, Moscow, ID 83844-2339, USA
e-mail: paulm@uidaho.edu

L.T. West et al. (eds.), *The Soils of the USA*, World Soils Book Series,
DOI 10.1007/978-3-319-41870-4_6

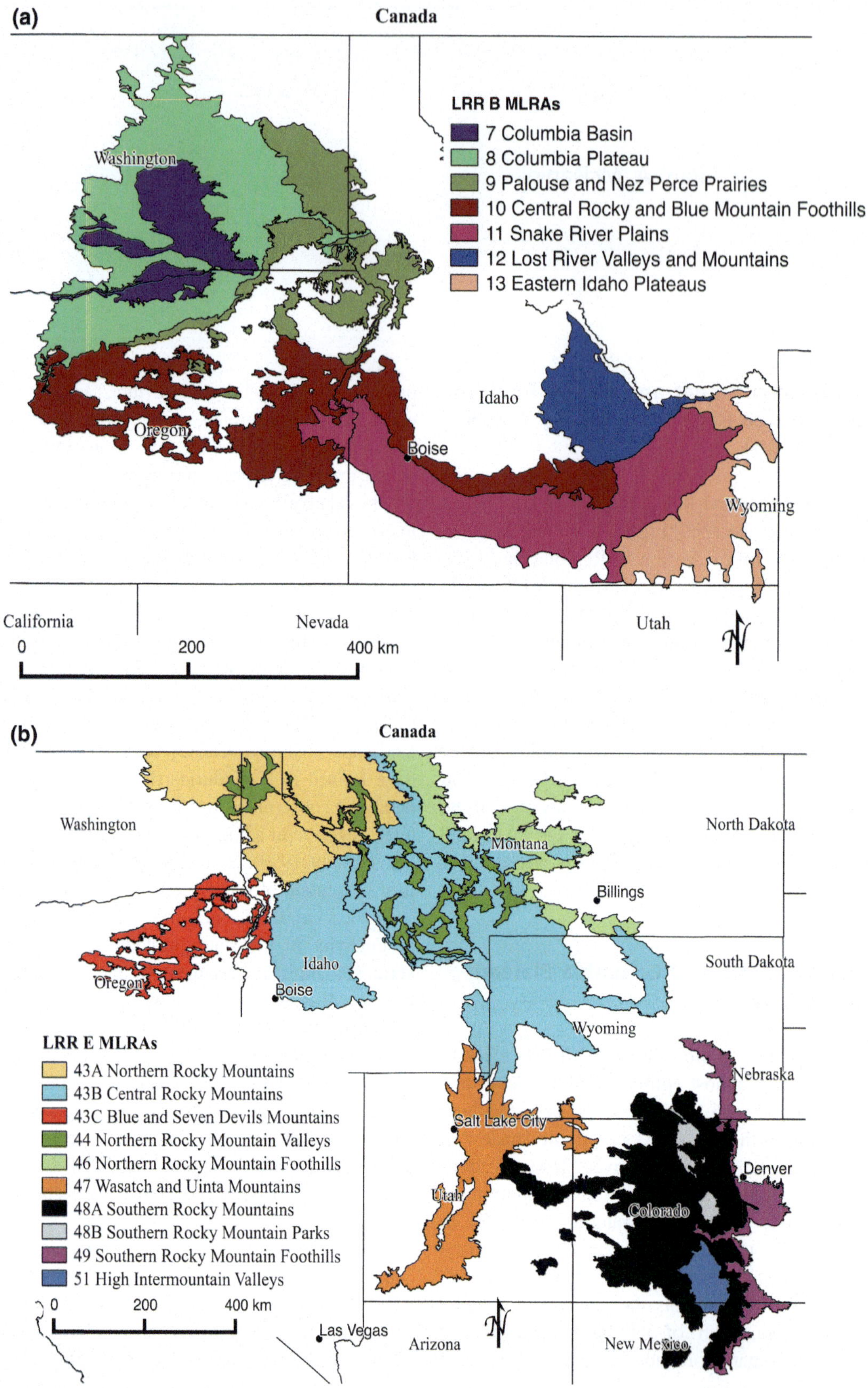

Fig. 6.1 Maps of Land Resource Regions and component MLRAs discussed in this chapter. **a** Northwestern Wheat and Range Region (LRR B), and **b** Rocky Mountain Range and Forest Region (LRR E)

Fig. 6.2 Relatively light-colored A horizon and thin SiO_2- and $CaCO_3$-cemented duripan (*arrow*) in a very gravelly Haplodurid. Cementation occurs between approximately 40 and 55 cm. *Scale* is in decimeters

such as the Channeled Scabland of eastern Washington, the loess mantle has been completely stripped off by catastrophic flooding from Glacial Lake Missoula. Mollisols, primarily Haploxerolls and Argixerolls, are the dominant soils of the Columbia Plateau. Organic C contents are considerably higher than those of soils found in the Columbia Basin (MLRA 7) as a result of the greater precipitation and associated vegetative production.

The majority of the Columbia Plateau MLRA is used for cropland (43 %) and rangeland (48 %) (U.S. Dept. of Agriculture 2006). The deeper, silty soils are preferred for agricultural production. In areas receiving more than ~355 mm of precipitation, winter wheat, peas, and lentils can be produced annually. Where precipitation is less, wheat is produced using a crop-fallow system. Grain is grown following a fallow year in which soil moisture is recharged. Only small areas of irrigated cropland exist along streams. Both wind and water erosion and the associated sedimentation of waterways have been problems since cultivation began in the early 1900s. Conservation practices have focused mainly on crop residue management and crop rotations.

6.1.2 Palouse and Nez Perce Prairies (MLRA 9)

The Palouse and Nez Perce Prairies represent an area of 22,825 km^2 characterized by deep loess hills overlying thick flows of the Columbia River Basalts (Fig. 6.1a). A pronounced east–west climate gradient exists across the MLRA. Mean annual precipitation ranges from less than 300 to approximately 1000 mm; mean annual temperatures range from 5 to 12 °C (U.S. Dept. of Agriculture 2006). These conditions support a wide range of native plant communities. On the drier, warmer western edge of the MLRA big sagebrush, Idaho fescue (*Festuca idahoensis*), and bluebunch wheatgrass dominate; on the moister, cooler eastern margin, ponderosa pine (*Pinus ponderosa*), Douglas-fir (*Pseudotsuga menziesii*), and grand fir (*Abies grandis*) forests are supported.

This bioclimatic gradient is well reflected in soils of the MLRA, which are dominantly Mollisols. Calcic diagnostic horizons are present in soils of the lower MAP areas, whereas $CaCO_3$ is typically absent from the upper meter of soils from the higher MAP areas (McDaniel and Hipple 2010). Vegetative productivity increases across the gradient with increasing MAP and is reflected by increasing organic matter content and increasing thickness of A horizons (Stevenson et al. 2005). This relationship is illustrated by comparison of two uncultivated soils of the MLRA with differing MAP (Table 6.1). The Palouse soil contains 18.4 kg C m^{-2}, nearly twice the amount in the Oliphant soil.

Table 6.1 Comparison of selected properties of uncultivated Haploxerolls (Oliphant[a] and Palouse[b] series) from the Palouse Region of MLRA 9

Soil series	Classification	Latitude/longitude	MAP (mm)	Depth to $CaCO_3$ (cm)	Mollic epipedon thickness (cm)	Organic C (kg m^{-2})[c]
Oliphant	Calcic Pachic Haploxeroll	46°46′52″ 117°47′35″	360	81	81	9.6
Palouse	Pachic Ultic Haploxeroll	46°53′29″ 117°4′20″	560	140	107	18.6

[a]Pedon 85WA075005 from the USDA-NRCS National Cooperative Soil Survey Soil Characterization Database
[b]Pedon 85WA075009 from the USDA-NRCS National Cooperative Soil Survey Soil Characterization Database
[c]Calculated to 1 m depth

Both mollic epipedon thickness and depth to $CaCO_3$ are also significantly greater in the Palouse soil.

In the higher MAP zones of the MLRA, many soils have dense water- and root-restrictive subsoil horizons that are classified as fragipans (Fragixeralfs) and/or argillic horizons (Argixerolls). Pedogenically, these horizons are part of a regional paleosol that is believed to have formed during the period between ~15 and 35 ka ago (McDaniel and Hipple 2010; Sweeney et al. 2004). These horizons have bulk density values of ~1.65 g cm^{-3} and have a poorly developed network of macropores. As a result, saturated hydraulic conductivities average only 0.1 and 0.06 cm day^{-1} for the argillic and fragipan horizons, respectively (McDaniel et al. 2001). Because of their limited ability to transmit water, these horizons perch water for as much as 7 months during the winter and spring and exert a profound influence on near-surface landscape hydrology. Almost 90 % of the incident precipitation and snowmelt during the spring exits hillslopes of the region as subsurface lateral flow at rates as high as 15 m day^{-1} (Brooks et al. 2012). This can result in rapid downslope transport of applied fertilizers and agrichemicals and degraded water quality of local streams and groundwater.

Where loess thickness is greatest (up to ~75 m), a sequence of at least 19 paleosols that formed over the past 1.5–2 million years can be observed; as such, the Palouse loess contains a unique pedostratigraphic record of the Quaternary in North America (Busacca 1989). Stable isotopic characterization of soil components obtained from these paleosols shows promise for reconstructing past shifts in climate and vegetation that have occurred across the region during the Holocene and Pleistocene (Stevenson et al. 2005).

Soils of this MLRA produce some of the highest non-irrigated winter wheat yields in the world. The deep silty Haploxerolls and Argixerolls provide excellent water-holding capacity and, coupled with well-timed spring rains, will produce winter wheat yields up to 8 metric tons ha^{-1} (119 bushels per acre) (Papendick 1996). The agricultural productivity of these soils has resulted in extensive conversion of the native prairie to agriculture since the late 1800s. It is estimated that <1 % of the original Palouse prairie vegetation remains today (Noss et al. 1995).

The combination of hilly topography, silty textures, and lack of significant plant cover in winter when the majority of precipitation is received results in some of the highest historical rates of water erosion in the USA. On steeper slopes, annual losses of as much as 450 metric tons ha^{-1} have been observed (U.S. Dept. of Agriculture 1978). Under poor management, historic annual erosion averaged 112 metric tons ha^{-1} and it is estimated that between 1939 and 1978, over 800 metric tons of soil were lost from every hectare of cropland in the Palouse Basin (U.S. Dept. of Agriculture 1978). Adoption of conservation practices such as contour tillage, minimal tillage, the use of cover crops, and spring cropping has substantially reduced modern-day erosion rates relative to those of the earlier 20th century.

6.1.3 Central Rocky and Blue Mountain Foothills (MLRA 10)

This MLRA (Fig. 6.1a) encompasses 43,385 km^2 of rolling hills, plateaus, and low mountains extending from the east side of Cascade Range in Oregon to southeastern Idaho. It resembles a gerrymandered congressional district, stretching an east–west distance of over 700 km (Fig. 6.1a). MAP is generally low across the area, typically ranging from 205 to 405 mm, and supports mainly shrub–grassland communities dominated by big sagebrush, Idaho fescue, and bluebunch wheatgrass (U.S. Dept. of Agriculture 2006). These communities are most commonly associated with Mollisols. Argillic horizons have formed in Palexerolls and Argixerolls on more stable landscape positions in a variety of parent materials. Loess is ubiquitous across the MLRA and is at least a minor component of most soils. It is generally thicker on older surfaces.

Most of the soils are used for grazing. Weed invasion has degraded rangeland quality in many areas where overgrazing has occurred. Wildfires are also a concern because of the very dry summer conditions—soil moisture regimes are either xeric or aridic. Many wildfires are generated by lightning during summer thunderstorms across the MLRA. These fires can greatly increase the potential for localized debris flows and flooding as vegetative cover is destroyed.

6.1.4 Snake River Plains (MLRA 11)

This MLRA occurs mostly in Idaho on the Snake River Plain (Fig. 6.1a). Mean annual precipitation ranges from approximately 180 to 305 mm over much of the area and supports mostly sagebrush–grass communities under native conditions. Soils of Snake River Plains are primarily Aridisols and have formed in loess overlying basalt bedrock (Fig. 6.3). Loess deposition has been extensive in this MLRA and several loess units have been identified (Bettis et al. 2003). These units are likely correlated with regional alpine glaciations and are thicker on older lava flows. On lava flows younger than ~13 ka, relatively little loess has accumulated and shallow Histosols (Folists) have developed in plant debris that accumulates in depressions and cracks in the basalt (Vaughan et al. 2011).

Because of the dry climate, most of the soils contain significant quantities of $CaCO_3$. On older geomorphic surfaces, Aridisols with argillic horizons (Argids) are present

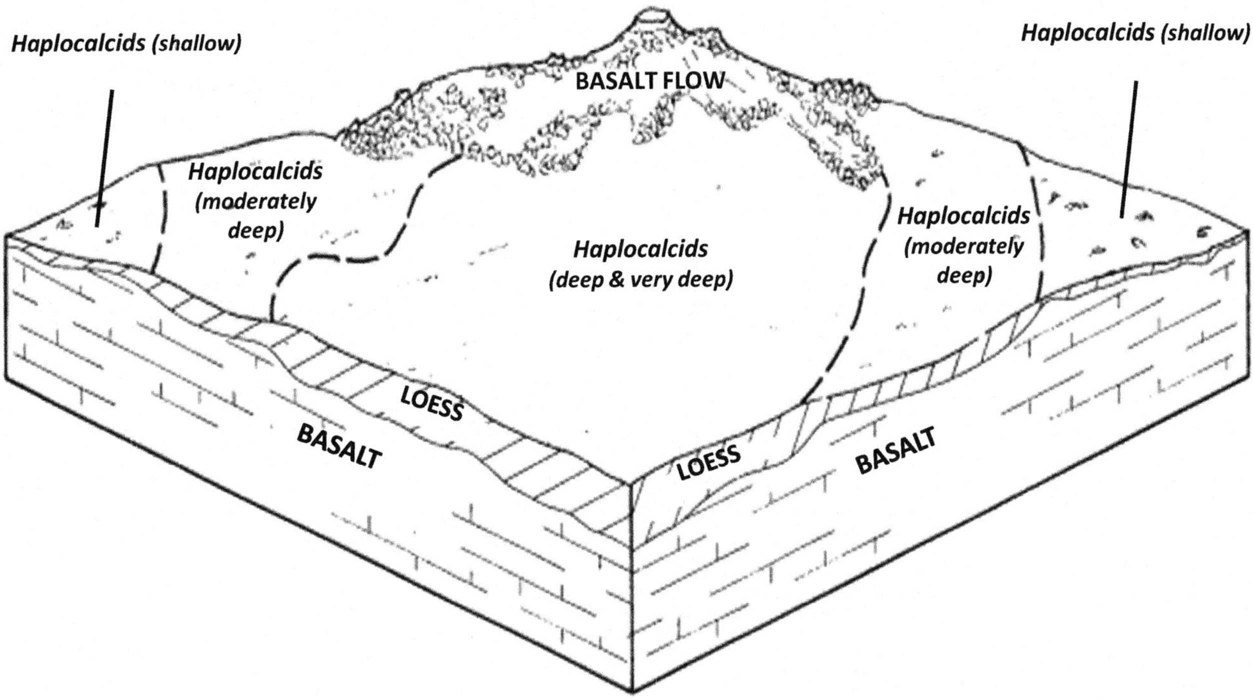

Fig. 6.3 Generalized pattern of soils formed in loess over basalt on the Snake River Plains. Haplocalcids are the dominant soils, but soil depth varies from shallow (25–50 cm) to very deep (>150 cm) depending on loess thickness. Adapted from Soil Survey of Bingham Area, Idaho (Salzman and Harwood 1973)

where $CaCO_3$ has been leached from upper horizons, thereby facilitating the movement and subsoil accumulation of clays. Duripans also occur in soils on older surfaces where sufficient SiO_2 has accumulated to cement at least half of the soil matrix. This likely occurs under conditions of high pH and may be facilitated by the presence of volcanic glass, which provides a good source of soluble SiO_2 (Soil Survey Staff 1999). Some of these Durids have duripans that may be up to 1 m thick (Fig. 6.4). Research suggests these thick duripans may be as old as 1.7 Ma (Othberg et al. 1997). Typical morphology is that of plates, each with a thickness ranging from a few millimeters to a few centimeters. These well-developed duripans appear to be the result of polygenesis, during which they were cyclically exposed at the soil surface and then reburied by eolian dust during the Pleistocene. Evidence for this includes: (1) a stromatolitic fabric within some duripan plates that suggests the presence of a microphytic crust; and (2) a distinct layer of air fall volcanic ash in the middle of a duripan plate (Blank et al. 1998).

The use and management of the soils are largely determined by loess thickness. Where loess is thicker and irrigation water is available, these soils represent some of the most important agricultural soils in the Snake River Plain. Approximately 25 % or 10,650 km^2 of this MLRA are irrigated (U.S. Dept. of Agriculture 2006). High yields of a variety of crops—corn, sugar beets, beans, wheat, barley, potatoes, alfalfa hay—are produced. Where irrigation is not available or practical, soils are used primarily for rangeland. Wind erosion and proliferation of invasive species are major concerns when vegetation is disturbed or removed through overgrazing, cultivation, or fire.

6.1.5 Lost River Valleys and Mountains and Eastern Idaho Plateaus (MLRAs 12, 13)

Mollisols and Aridisols dominate these MLRAs, which comprise 34,575 km^2 almost entirely in eastern Idaho (Fig. 6.1). There is a considerable variation in physiography and climate, with elevations ranging from approximately 1200 to over 3600 m and MAP ranging from 180 to over 1200 mm (U.S. Dept. of Agriculture 2006). At the drier end of the MAP range, Aridisols (Argids, Calcids, and Cambids) have formed. Limited leaching has resulted in significant quantities of $CaCO_3$ at or near the surface of these soils. Wyoming big sagebrush (*Artemisia tridentata* ssp. *wyomingensis*), shadscale (*Atriplex confertifolia*), Indian ricegrass (*Achnatherum hymenoides*), needle-and-thread grass, and bluebunch wheatgrass are among the dominant species. These soils are used primarily for rangeland unless irrigation water is available.

Fig. 6.4 Exposed duripan (*arrows*) in a quarry near Boise, Idaho. The duripan forms a visible ledge and is ~1 m thick (note shovel for scale)

As MAP increases, Mollisols are the most extensive soils and support grass–shrub communities. Xerolls occur in lower-elevation parts of the MLRA that receive most of the annual precipitation during the fall, winter, and spring; Cryolls occur in the colder, higher-elevation areas. There is substantial accumulation of soil organic matter in many of these Cryolls because of the cool temperatures and the proliferation of fine and very fine roots from grasses and forbs. Many Cryolls have developed very thick (>40 cm) mollic epipedons and are classified as Pachic subgroups of Mollisols (Fig. 6.5). Higher-elevation Cryolls and some Cryalfs support Douglas-fir forest with an understory of shrubs and various grasses. Because of a short frost-free season (<65 days), these soils are very limited for agricultural production. They are, however, used for timber production and grazing and provide excellent habitat for mule deer, elk, and other wildlife.

6.2 Rocky Mountain Range and Forest Region (MLRAs 43A, 43B, 43C, 44, 46, 47, 48A, 48B, 49, and 51)

The Rocky Mountain Range and Forest Region encompasses 612,875 km^2 in eight states and extends from northern New Mexico almost 1900 km in a northwesterly direction to the Canadian border (Fig. 6.1b). It includes the highest elevations of the region with many mountainous areas exceeding 3000 m. The highest elevation is Mount Elbert in Colorado at 4401 m. Because of the wide range of climatic regimes and associated plant communities, parent materials, and physiography, a tremendous diversity of soils exists in this region. Soil temperature regimes range from mesic at lower elevations to gelic at the highest elevations. Mean annual precipitation ranges widely from 230 to over 2000 mm (U.S. Dept. of Agriculture 2006), resulting in soils forming under aridic, xeric, ustic, and udic soil moisture regimes. Alfisols, Andisols, Aridisols, Entisols, Inceptisols, and Mollisols are all well represented. Histosols, Vertisols, Spodosols, and possibly Gelisols can also be found in more localized environments.

6.2.1 Northern Rocky Mountains, Valleys, and Foothills (MLRAs 43A, 44, and 46)

The Northern Rocky Mountains MLRA encompasses 81,460 km^2 of northeastern Washington, northern Idaho, and western Montana and extends to the Canadian border (Fig. 6.1b) (U.S. Dept. of Agriculture 2006). Much of this MLRA was glaciated during the last Ice Age, resulting in

Fig. 6.5 Pachic Haplocryoll from Bannock County, Idaho. The mollic epipedon extends from the surface to a depth of over a meter. The soil supports Douglas-fir and a pinegrass (*Calamagrostis rubescens*) understory. Elevation is ~2000 m, and MAP is 610 mm. *Scale* is in decimeters

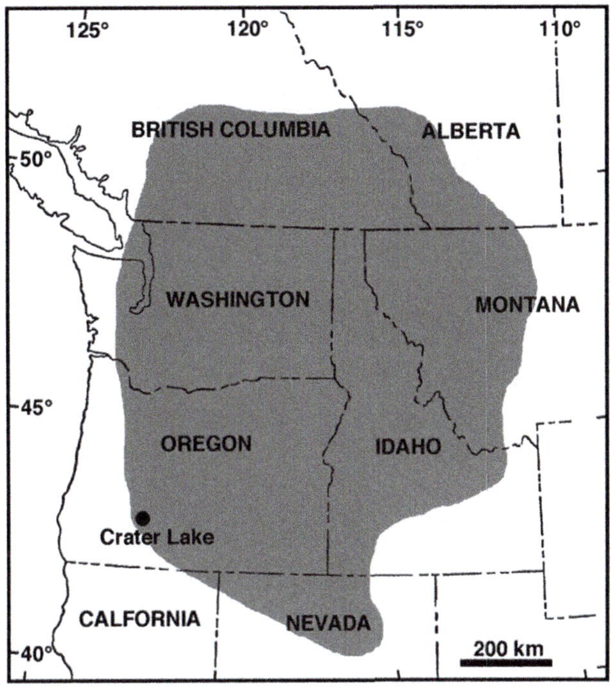

Fig. 6.6 Generalized distribution of volcanic ash deposition following the eruption of Mount Mazama (now Crater Lake, OR) approximately 7600 years ago. Adapted from Fryxell (1965)

rugged mountains and small valleys filled with glacial till, outwash, and glaciolacustrine sediments. As a result, soils are generally young and reflect many of the properties of their parent materials, which include alluvium, colluvium, glacial till and outwash, and eolian deposits.

Many soils of the Northern Rocky Mountains have been mantled by volcanic ash from Holocene eruptions in the Cascade Range. The largest of these eruptions was that of Mount Mazama (now Crater Lake, OR) approximately 7600 years ago, in which an estimated 116 km^3 of tephra fell across the entire Pacific Northwest Region (Zdanowicz et al. 1999; Bacon 1983) (Fig. 6.6). Although much of this original blanket of ash has been eroded away, Andisols are commonly found in mid- to high-elevation forests across these MLRAs where litter layers and canopy cover have helped protect the ash from erosion.

Characteristics of soil horizons formed in volcanic ash differ considerably from those of horizons formed in other parent materials. Several of these differences are illustrated in Table 6.2, which shows data for a representative Andisol of the Northern Rocky Mountains MLRA. The top three mineral horizons—A, Bw1, and Bw2—have formed in volcanic ash and exhibit andic soil properties. These include low bulk density, high glass content, high New Zealand P retention, and high concentrations of oxalate-extractable Al and Fe (Soil Survey Staff 2014). Note the change in these properties across the boundary between the volcanic ash and the underlying parent material at 49 cm. The New Zealand P retention values of the andic horizons indicate that at least 85 % of added phosphorus is strongly sorbed by soil colloids and retained in forms that are unavailable to plants. Oxalate-extractable Al and Fe serves as an indicator of the weathering products that form as volcanic glass weathers—these include allophane, imogolite, ferrihydrite, and metal humus complexes. It is these colloidal weathering products that give Andisols their unusual properties, such as high water retention, high P retention, and the ability to suddenly liquefy when pressure is applied (Buol et al. 2011).

Andisols across the MLRA such as the Bonner series typically have a 30- to 60-cm-thick mantle of weathered volcanic ash (Fig. 6.7). This loamy-textured mantle can greatly increase the water-holding capacity of a site (note coarse-textured horizons underlying ash mantle in the

Table 6.2 Selected properties of an Andisol[a] from the Northern Rocky Mountains MLRA

Oxyaquic Udivitrand (Threebear series)

Location: Latah County, Idaho; Latitude: 46.8628; Longitude: −116.5336

Elevation: 913 m; MAP: 991 mm

Horizon	Depth (cm)	Sand (%)	Silt (%)	Clay (%)	Bulk density (g cm^{-3})	Water retention (%)		pH H_2O (%)	Carbon (%)	Glass[b] (%)	Al + 0.5Fe[c] (%)	New Zealand P retention (%)
						33 kPa	1500 kPa					
Oi	0–3	–[d]	–	–	–	–	97.8	–	31.4	–	–	–
A	3–9	14.6	78.2	7.2	0.72	49.2	14.9	5.7	5.3	29	2.39	88
Bw1	9–30	18.8	76.4	4.8	0.77	43.4	11.2	6.2	1.8	32	3.12	91
Bw2	30–49	17.6	77.5	4.9	0.86	48.0	11.7	6.1	1.4	24	2.41	85
2E	49–77	11.8	75.1	13.1	1.63	19.6	5.3	5.5	0.2	1	0.33	17
2EB	77–100	12.3	70.3	17.4	1.63	20.1	6.8	5.6	0.2	1	0.40	20
2Btx	100–117	12.8	65.0	22.2	1.67	18.6	9.0	5.6	0.2	tr[e]	0.41	22

[a]Pedon 07ID057006 from the USDA–NRCS National Cooperative Soil Survey Characterization database
[b]In coarse silt fraction
[c]Extracted using acid ammonium oxalate
[d]Not determined
[e]Trace

Bonner profile). In the water-limited environments characterizing much of the region, this increased water-holding capacity translates into greater site productivity. A 25-cm-thick ash mantle can increase the height of a 50-year-old Douglas-fir by as much as 4 m compared with a Douglas-fir growing on a soil without an ash mantle (Kimsey et al. 2008). At lower elevations and in areas where disturbance (fire, logging, disease) has occurred, the ash is not present as a discrete mantle but rather has been mixed to varying degrees with other soil parent materials (Fig. 6.8). Depending on the degree of mixing, these soils are either classified as Andisols (less mixing) or vitrandic or andic subgroups of Andisols and Inceptisols (more mixing) (McDaniel and Hipple 2010). In extreme cases, loss of the ash mantle resulting from erosion following stand-replacing disturbance significantly decreases soil water-holding capacity. This has led to desertification, whereby affected sites are no longer able to support the forest community that previously existed. Many such sites can be found in previously forested areas burned by the catastrophic fires of 1910. These sites are now shrub fields with little or no conifer regeneration.

Alfisols and Inceptisols are also found throughout this MLRA. Inceptisols typically occur on the steeper mountain slopes, where greater runoff and erosion do not allow significant subsoil development. Alfisols occur on the gentler slopes where increased movement of water into the soil and greater slope stability promote the subsoil accumulation of layer silicate clays. Spodosols (Cryods) occur at higher elevations where increased snowmelt and organic compounds derived from coniferous litter promote the process of podzolization—the formation of a light-colored Fe- and Al-depleted E horizon overlying a reddish-brown Fe- and Al-enriched B horizon. The E horizons of these soils are extremely acidic with pH values of $\sim$3.5–4.0 and high concentrations of exchangeable Al^{3+} (McDaniel et al. 1993). It is possible that Gelisols occur in some higher-elevation areas of this MLRA where glaciers and permanent snowfields are found, but their existence needs to be confirmed.

The Northern Rocky Mountain Foothills MLRA is limited to Montana on the eastern edge of the Rocky Mountains. Most of the area is privately owned rangeland, and Mollisols and Entisols dominate the landscape. Mollisols are associated with the moister, more productive grasslands and forests that occur at higher elevations. Entisols and some Aridisols are found in the warmer, drier areas. On steeper slopes, many Entisols are shallow to parent material, commonly sandstones and shales.

The majority of the Northern Rocky Mountain Valleys MLRA occurs in western Montana, with smaller areas in northern Idaho and northeastern Washington (Fig. 6.1b). Inceptisols and Mollisols are the dominant soils, but Andisols are present where forest has helped retain the volcanic ash mantle. Many of the valleys have been modified by glacial activity and are filled with outwash and glaciolacustrine sediments. In many valleys, glacial activity has disrupted pre-existing drainage systems, resulting in localized areas of poorly drained Histosols.

The relationships between soil morphology, aboveground vegetative production, and the environmental gradients found within the foothill and valley rangelands of western Montana are illustrated in Fig. 6.9. Aboveground production

Fig. 6.7 Typic Vitrixerand (Bonner series) from Bonner County, Idaho. The volcanic ash mantle (*arrow*) is approximately 40 cm thick and overlies gravelly sand glacial outwash. The soil supports mixed conifer forest. Elevation is ~800 m, and MAP is 760 mm. *Scale* is in decimeters

increases in response to the moister, cooler conditions found at higher elevations, and this is reflected in two major changes in soil morphology (Munn et al. 1978). Thickness of the mollic epipedon—the dark, organic-rich mineral surface horizon—is highly correlated ($r = 0.89$) with site productivity. In these soils, the mollic epipedon serves as a long-term record of the balance between vegetative production, primarily in the form of roots, and decomposition. The combination of greater production and lower decomposition rates associated with moister, cooler sites results in a thicker and darker mollic epipedon. Total productivity is also highly correlated ($r = 0.71$) with depth to free carbonates (Munn et al. 1978). As production increases along a moisture gradient in response to greater precipitation, $CaCO_3$ is leached to greater depths. In this regard, the depth to $CaCO_3$ serves as a long-term site record of both productivity and precipitation regime.

As previously mentioned, Andisols support some of the most productive forests of the region, and timber production has historically been the primary economic driver in the Northern Rocky Mountains. In addition, the forested areas in these MLRAs are used for grazing, recreation, wildlife habitat, and watershed. In the forested areas of the Valleys and Foothills MLRAs, some timber production occurs. However, rangeland is the dominant land use. The rangeland supports beef cattle and sheep production and also serves as habitat for a variety of wildlife species including elk, deer, antelope, bobcat, badger, coyote, fox, and grouse (U.S. Dept. of Agriculture 2006). Lack of water is the major constraint for agricultural use, and irrigation in the Valleys and Foothills is generally restricted to the valleys. Some dryland wheat is produced along the northeastern side of the Foothills MLRA.

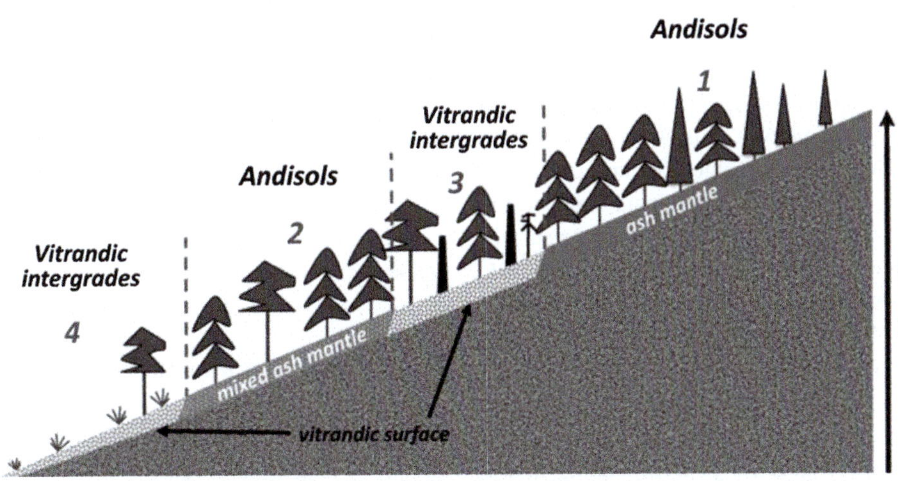

Fig. 6.8 Generalized landscape distribution of volcanic ash mantle and soil classification in the Northern Rocky Mountains MLRA. Andisols are present where the mantle is undisturbed (indicated by *1*) or only slightly mixed (*2*). Vitrandic intergrades to other soil orders are found where there has been significant loss of the mantle due to erosion following major disturbance (e.g. fire) (*3*) or by lack of canopy and litter cover (*4*)

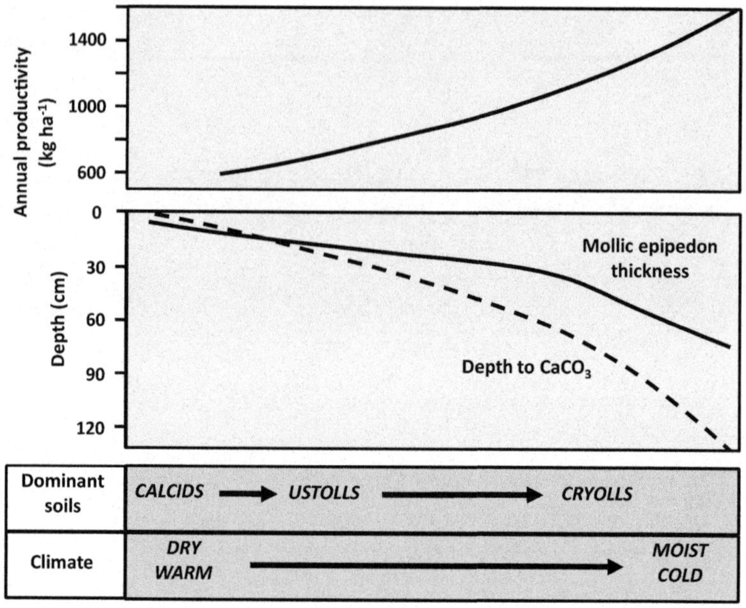

Fig. 6.9 Relationship between annual productivity, mollic epipedon thickness, depth to $CaCO_3$, soil classification, and climate in mountain and foothill range sites in western Montana. Adapted from Munn et al. (1978)

6.2.2 Central Rocky Mountains and Blue and Seven Devils Mountains (MLRAs 43B, and 43C)

These MLRAs together encompass just under 233,000 km^2 of rugged mountains, plateaus, canyons, and valleys in central Idaho, western Montana, and western Wyoming (U. S. Dept. of Agriculture 2006). Many of the mountains have been glaciated, and as a result, soils are relatively young and therefore exhibit limited development and weathering. Because of the general lack of weathering, parent materials strongly affect soil properties throughout these MLRAs. Alfisols, Mollisols, and Inceptisols are the major soils of the Central Rocky Mountains (U.S. Dept. of Agriculture 2006). In addition, Andisols and Andic and Vitrandic intergrades of other soil orders occur in forested areas where at least some volcanic ash has been retained.

Selected soil properties of a forested Alfisol from Wyoming are given in Table 6.3. This Alfisol represents a relatively well-developed soil in which organic matter and clay accumulation have been the dominant soil-forming processes. In addition to the forest litter layer at the surface, there is a 6-cm-thick dark A horizon containing 3.1 % C. Substantial clay movement has occurred with almost 40 % clay in upper part of the argillic horizon. This and similar soils that occur at high elevations have short growing seasons and cryic temperature regimes, both of which limit vegetative production. However, these soils have relatively high native fertility. pH values range from slightly acid to near neutral, and base saturation is relatively high. Coupled with adequate cation exchange capacity (CEC), these properties indicate good availability of Ca, Mg, K, and other macro- and micronutrients. Quantities of oxalate-extractable Fe and Al > 0.4 % in the upper horizons indicate the presence of poorly crystalline weathering products. This and the presence of some glass suggest some influence of volcanic ash. However, the ash influence is not sufficient for the soil to meet requirements for a Vitrandic intergrade (Soil Survey Staff 2014).

Land use in the Central Rocky Mountains is 31 % forest and 65 % grassland (U.S. Dept. of Agriculture 2006). In addition to timber production and livestock grazing, soils provide watershed and important habitat for a wide variety of wildlife, including big game species such grizzly bear, wolf, moose, cougar, and bighorn sheep.

The Blue and Seven Devils Mountains MLRA occurs primarily in northeastern Oregon (Fig. 6.1b). The geology of this area is quite complex, resulting in a wide suite of parent material lithology that ranges from Mesozoic limestones to Cenozoic volcanics (U.S. Dept. of Agriculture 2006). Eolian additions of loess and volcanic ash have also affected many soils of the area. Most of the soils are classified as Mollisols or Andisols. Andisols are dominant at higher elevations where greater canopy cover and litter production have helped retain the volcanic ash mantle.

6.2.3 Wasatch and Uinta Mountains (MLRA 47)

Most of this MLRA occurs in Utah and consists of steep mountain slopes and high plateaus. Elevations range from approximately 1500 to 4115 m, and MAP is highly variable, ranging from a little as 150 mm in lower-elevation valleys to well over 1800 mm at higher elevations; average annual

Table 6.3 Selected properties of a Eutric Haplocryalf[a] from the Central Rocky Mountains MLRA

Eutric Haplocryalf

Location: Fremont County, Wyoming; Latitude: 43.6746; Longitude: −109.3771

Elevation: 2579 m; MAP: ~825 mm

Horizon	Depth (cm)	Sand (%)	Silt (%)	Clay (%)	Rock fragments (% by wt)	Bulk density (g cm^{-3})	pH H_2O	Carbon (%)	CEC_7 $cmol_+$ kg^{-1}	Base saturation (%)	Glass[b] (%)	Al + 0.5Fe[c] (%)
Oi	0-4	–[d]	–	–	1	–	–	–	–	–	–	–
A	4-10	30.1	44.6	24.6	6	1.13	6.3	3.1	28.3	69	4	0.54
Bt1	10-39	31.3	28.8	39.9	16	1.20	6.1	0.6	40.4	77	4	0.55
Bt2	39-66	51.8	24.0	24.2	32	1.21	6.7	0.2	27.6	84	tr[e]	0.41
C1	66-89	51.5	27.7	20.8	48	1.17	6.9	0.3	25.4	86	5	0.30
C2	89-110	56.0	24.4	19.6	53	–	7.3	0.2	25.1	83	tr	0.27

[a]Pedon S10WY056001 from the USDA–NRCS National Cooperative Soil Survey Characterization database
[b]In coarse silt fraction
[c]Extracted using acid ammonium oxalate
[d]Not determined
[e]Trace

temperatures are between −1 and 15 °C (U.S. Dept. of Agriculture 2006). Several vegetation zones are associated with this elevational/climatic gradient: alpine meadows, coniferous forest, aspen forest, Gambel oak (*Quercus gambelii*) and other shrub–grass communities, and sagebrush–grass communities at the lowest elevations. Consequently, soils have a wide range of properties and are predominantly Aridisols, Mollisols, Entisols, and Inceptisols.

A generalized distribution of soils on steep, high-elevation mountainsides in northern Utah is shown in Fig. 6.10. Soils of this landscape tend to have high rock fragment content and are predominantly Palecryolls, which support quaking aspen (*Populus tremuloides*), scattered conifers, and a grass/forb understory. The cryic temperature regime has facilitated the accumulation of soil organic matter and formation of a mollic epipedon. In addition, on the more stable parts of the landscape sufficient translocation of clays has resulted in formation of an argillic horizon in the Palecryolls and Glossocryalfs. The Glossocryalfs support Douglas-fir and have an discontinuous litter layer at the surface rather than a mollic epipedon.

Another soil–landscape pattern is seen in the Gambel oak–grass communities of northern Utah. Mollisols are the dominant soils and occur in association with Vertisols (Graham and Southard 1983). It is believed that loss of the Gambel oaks and the subsequent establishment of mule ears (*Wyethia amplexicaulis*) has allowed the mollic epipedon to be eroded away, exposing a clayey argillic horizon at the surface. The shrink/swell activity of the clays results in surface cracks that help define the Vertisol order. This proposed pathway of soil evolution underscores the importance of vegetation in the formation and stability of soils.

6.2.4 Southern Rocky Mountains, Foothills, High Intermountain Valleys, and Parks (MLRAs 48A, 49, 51, and 48B)

These MLRAs comprise a total of 164,310 km^2 primarily in Colorado, but extending into Wyoming, Utah, and northern New Mexico (Fig. 6.1b). Physiography and climate are extremely varied, ranging from steep mountain slopes to nearly level intermountain valleys. Elevations range from 1980 to 4400 m and serve as proxy for MAP and MAAT; MAP ranges from 180 to over 1600 mm and MAAT ranges from −3 to 12 °C (U.S. Dept. of Agriculture 2006). Winter snowpack throughout the higher elevations of these MLRAs provides substantial water for local and regional needs. Mollisols, Alfisols, Inceptisols, Entisols, and Aridisols are the dominant soils, and it is possible some Gelisols occur at the highest elevations in close proximity to permanent snowfields and glaciers.

The Southern Rocky Mountains MLRA 48A occurs primarily in Colorado and includes what is often referred to as the Front Range on the eastern edge and the Precambrian igneous and metamorphic mountain core to the west. Many of the soils form entirely or partially in transported parent materials—alluvial, colluvial, and glacial deposits are common. Just over half of the MLRA is forested. Mollisols and Alfisols dominate the more stable landscape positions, while Inceptisols and Entisols occur on the steeper mountain slopes and summits.

Properties of a Cryalf (Leadville series) from this MLRA in Colorado are given in Table 6.4. This soil has formed in colluvium under mixed conifer and hardwood forest at an

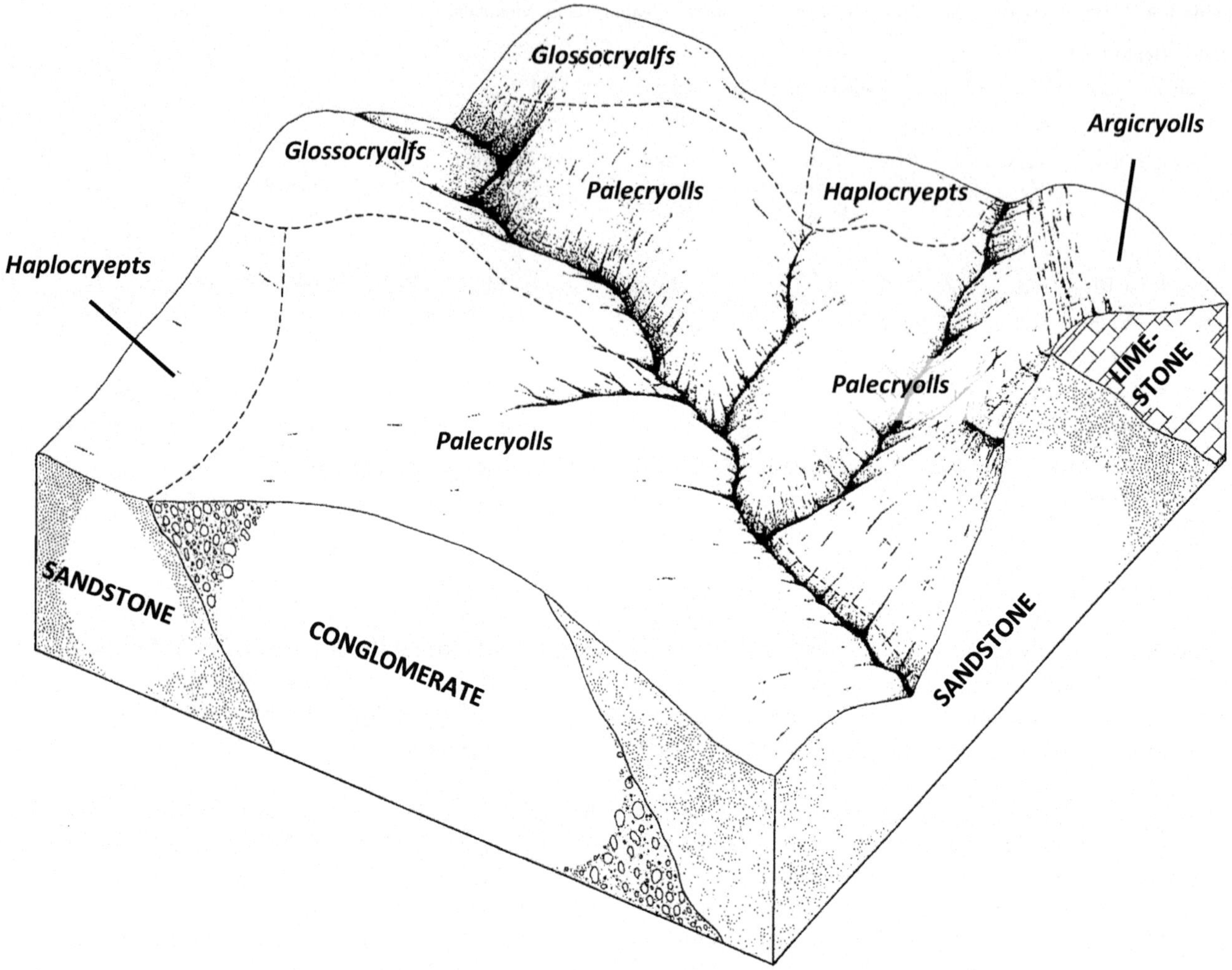

Fig. 6.10 Generalized soil pattern on steep mountainsides in the Wasatch and Uinta Mountains MLRA. Approximate elevation is 2400 m; MAP is ~825 mm. Palecryolls support aspen. Glossocryalfs support Douglas-fir and subalpine fir (*Abies lasiocarpa*). Adapted from Soil Survey of Rich County, Utah (Campbell and Lacey 1982)

elevation of 2775 m with approximately 430 mm MAP and a cryic soil temperature regime. As is typical of many of the mountainous landscapes, this soil has a high rock fragment content, which limits its water-holding capacity. Other than the presence of an O horizon, there is relatively little organic C accumulation when compared to forested soils of the Northern and Central Rocky Mountains (see Tables 6.2, 6.3). This is reflected in the horizon designations, as the surface mineral horizon is described as an E rather than an A horizon. The dominant soil-forming process has been the translocation of clay from the E horizons to the Bt horizons. This process is driven by snowmelt, because much of the MAP comes as snow. Although clay content of the argillic horizon is relatively low (~15–18 %), it nevertheless represents a significant accumulation relative to the E horizons (Table 6.4).

The Southern Rocky Mountain Parks MLRA consists of relatively high-elevation parks and valleys that are surrounded by higher mountains (Fig. 6.1b). The parks and valleys consist largely of deep deposits of materials derived from the mountains that have been deposited by rivers, streams, and glacial outwash. The dominant native vegetation of the MLRA is mountain big sagebrush and a variety of grasses including Idaho fescue, bluebunch wheatgrass, needle-and-thread grass (*Hesperostipa* spp.), and western wheatgrass (*Pascopyrum smithii*) (U.S. Dept. of Agriculture 2006). These grasses have favored the formation of Mollisols, which are the dominant soils of this MLRA. The annual input of fine and very fine roots and their subsequent dieback and decomposition results in substantial quantities of belowground soil organic matter. This process, referred to as melanization, gives rise to the thick, dark mollic epipedons that characterize many soils of the MLRA. Approximately 85 % of this MLRA is rangeland (U.S. Dept. of Agriculture 2006). Some irrigated pastures and hayfields are

Table 6.4 Selected properties of an Ustic Glossocryalf[a] from the Southern Rocky Mountains MLRA

Ustic Glossocryalf (Leadville series)

Location: Las Animas County, Colorado; Latitude: 37.1750; Longitude: −105.0728

Elevation: 2775 m; MAP: 432 mm

Horizon	Depth (cm)	Sand (%)	Silt (%)	Clay (%)	Rock fragments (% by wt)	Bulk density (g cm^{-3})	pH H_2O	Carbon (%)	CEC_7 ($cmol_+$ kg^{-1})	Base saturation (%)
Leadville										
Oe	0–8	–[b]	–	–	–	–	–	–	–	–
E	8–43	66.2	28.8	5.0	50	–	5.4	1.0	7.4	63
E/B	43–63	65.9	28.2	5.9	70	–	5.1	0.4	6.7	74
Bt1	63–119	51.3	33.5	15.2	86	1.84	6.2	0.3	10.9	84
Bt2	119–180	52.8	29.3	17.9	87	1.72	7.8	0.2	12.7	95

[a] Pedon 99CO071005 from the USDA–NRCS National Cooperative Soil Survey Characterization database
[b] Not determined

found along streams, but the short growing season precludes most other agricultural production.

The Southern Rocky Mountain Foothills MLRA lies along the eastern edge of the Southern Rocky Mountains and extends from Wyoming southward into Colorado and northern New Mexico (Fig. 6.1b). Most elevations range from 1500 up to 2440 m, and MAP increases from approximately 300 to 635 mm with increasing elevation (U. S. Dept. of Agriculture 2006). Plant communities are primarily grassland at lower elevations, shrub-grassland at middle elevations, and forest at higher elevations. Mollisols and Alfisols represent the more developed soils of the MLRA. Mollisol formation is favored under grassland vegetation, while Alfisols form at the higher elevations under forest. Entisols and Inceptisols are well represented on steeper slopes and areas that are shallow to bedrock. Because of aridity at lower elevations and cryic temperature regimes at higher elevations, agricultural land use is very limited. Common land uses are watershed, wildlife habitat, livestock grazing, and some limited timber production.

The High Intermountain Valleys MLRA consists of a high-elevation valley in southern Colorado and northern New Mexico known as the San Luis Valley (Fig. 6.1b). The valley consists of nearly level to gently sloping alluvial valley fill derived from the surrounding mountains. The climate is generally drier and warmer than found elsewhere throughout the Southern Rocky Mountains and supports desert shrub-grasslands. Aridisols and Entisols dominate the landscape. Haplargids form on the older terraces and alluvial fans. Less-developed Haplocalcids, Haplocambids, and Entisols occur on younger surfaces. Approximately 75 % of the valley is grassland and 16 % is irrigated cropland (U.S. Dept. of Agriculture 2006).

6.3 Summary

Soils of the Rocky Mountain and Inland Pacific Northwest Region together encompass almost 700,000 km^2 in eight western states and are among the most diverse found anywhere in the United States—every soil order except Oxisols is represented. This diversity of soils is due to the wide range of parent materials, vegetation, climate, and topography that exists across the region. Because these soil-forming factors often vary considerably over short distances, markedly different soil properties can be found within a relatively small geographic area.

In the Inland Pacific Northwest Region, Aridisols and Mollisols dominate the seasonally dry landscapes. Calcium carbonate-containing Aridisols support desert plant communities in the driest areas, but where irrigated, are used to produce a variety of crops. In the moister parts of the region, Mollisols support some of the most productive dryland wheat and small grain production systems in the USA. Grazing is also a common land use throughout the region.

Much of the Rocky Mountain Region is characterized by rugged landscapes that are dominated by Alfisols, Inceptisols, Mollisols, and Entisols. These soils support a variety of forest, grassland, and shrubland communities that extend from northern New Mexico to the Canadian border. Timber production, recreation, wildlife habitat, grazing, and watershed are the primary uses for these soils.

References

Bacon CR (1983) Eruptive history of Mount Mazama and Crater Lake caldera, Cascade Range, USA. J Volcanol Geotherm Res 18:57–115

Bettis EA III, Muhs DR, Roberts HM, Wintle AG (2003) Last glacial loess in the conterminous USA. Quat Sci Rev 22:1907–1946

Blank RR, Cochran B, Fosberg MA (1998) Duripans in southwestern Idaho: polygenesis during the Quaternary deduced through micromorphology. Soil Sci Soc Am J 62:701–709

Brooks ES, Boll J, McDaniel PA (2012) Hydropedology in seasonally dry landscapes: the Palouse region of the Pacific Northwest USA. In: Lin H (ed) Hydropedology: synergistic integration of soil science and hydrology. Academic Press, Elsevier BV, London, pp 329–350

Buol SW, Southard RJ, Graham RC, McDaniel PA (2011) Soil genesis and classification, 6th edn. Wiley-Blackwell, Chichester

Busacca AJ (1989) Long Quaternary record in eastern Washington, U. S.A., interpreted from multiple buried paleosols in loess. Geoderma 45:105–122

Campbell LB, Lacey CA (1982) Soil survey of Rich County, Utah. USDA-Soil Conservation Service, US Govt Print Office, Washington

Eigenbrode SD, Capalbo SM, Houston LL, Johnson-Maynard J, Kruger C, Olen B (2013) Agriculture. In: Dalton M, Mote P, Snover A (eds) Climate change in the Northwest: implications for our landscapes, waters, and communities. Island Press, Washington, pp 149–180

Fryxell RH (1965) Mazama and Glacier Peak volcanic ash layers: relative ages. Science 147:1288–1290

Graham RC, Southard AR (1983) Genesis of a Vertisol and associated Mollisol in northern Utah. Soil Sci Soc Am J 47:552–559

Kimsey MJ, Moore J, McDaniel PA (2008) A geographically weighted regression analysis of Douglas-fir site index in north central Idaho. For Sci 54:356–366

McDaniel PA, Fosberg MA, Falen AL (1993) Expression of andic and spodic properties in tephra-influenced soils of northern Idaho. Geoderma 58:79–94

McDaniel PA, Gabehart RW, Falen AL, Hammel JE, Reuter RJ (2001) Perched water tables on Argixeroll and Fragixeralf hillslopes. Soil Sci Soc Am J 65:805–810

McDaniel PA, Hipple KW (2010) Mineralogy of loess and volcanic ash eolian mantles in Pacific Northwest (USA) landscapes. Geoderma 154:438–446

Munn LC, Nielsen GA, Mueggler WF (1978) Relationships of soils to mountain and foothill range habitat types and production in western Montana. Soil Sci Soc Am J 42:135–139

Noss RF, LaRoe ET III, Scott MJ (1995) Endangered ecosystems of the United States: a preliminary assessment of loss and degradation. Biological report, vol 28. US National Biological Service, Washington

Othberg K, McDaniel PA, Fosberg MA (1997) Soil development on a Pleistocene terrace sequence, Boise Valley, Idaho. Northwest Sci 71:318–329

Papendick RI (1996) Farming systems and conservation needs in the Northwest Wheat Region. Am J Alternat Agric 11:52–57

Salzman RA, Harwood JO (1973) Soil survey of Bingham area, Idaho. USDA-Soil Conservation Service, US Govt Print Office, Washington

Soil Survey Staff (1999) Soil taxonomy—a basic system of soil classification for making and interpreting soil surveys, 2nd edn. USDA-Natural Resources Conservation Service, Washington

Soil Survey Staff (2014) Keys to soil taxonomy, 12th edn. USDA-Natural Resources Conservation Service, Washington

Stevenson BA, Kelly EF, McDonald EV, Busacca AJ (2005) The stable carbon isotope composition of soil organic carbon and pedogenic carbonates along a bioclimatic gradient in the Palouse region, Washington State, USA. Geoderma 124:37–47

Sweeney MR, Busacca AJ, Richardson CA, Blinnikov M, McDonald E (2004) Glacial anticyclone recorded in Palouse loess of northwestern United States. Geology 32:705–708

U.S. Dept. of Agriculture (1978) Palouse Cooperative River Basin Study. U.S. Department of Agriculture–Soil Conservation Service, Forest Service, Economics, Statistics, and Cooperatives Service, Washington

U.S. Dept. of Agriculture (2006) Land resource regions and major land resource areas of the United States, the Caribbean, and the Pacific Basin. U.S. Department of agriculture handbook, vol 296. USDA-Natural Resources Conservation Service, Washington

Vaughan KL, McDaniel PA, Phillips WM (2011) Episodic soil succession on basaltic lava fields in a cool, dry environment. Soil Sci Soc Am J 75:1462–1470

Zdanowicz CM, Zielinski GA, Germani MS (1999) Mount Mazama eruption: calendrical age verified and atmospheric impact assessed. Geology 27:621–624

7 Soils of the Western Range and Irrigated Land Resource Region: LRR D

Craig Rasmussen, Rebecca A. Lybrand, Caitlin Orem, Jennifer Kielhofer, and Molly Holleran

7.1 Western Range and Irrigated Land Resource Region

The arid and semiarid ecosystems of the Western Range and Irrigated Region (WRIR) represent critical agricultural and economic areas, occupying over 1,424,480 km^2 across the states of Oregon, California, Idaho, Nevada, Arizona, Utah, New Mexico, Wyoming, Colorado, and Texas (Fig. 7.1a). These areas encompass a desert and semi-desert region of plateaus, plains, basins, and isolated mountain ranges. The mean annual precipitation (MAP) spans an order of magnitude ranging from <150 mm year^{-1} on some of the plains and basins up to >1500 mm year^{-1} on some of the higher mountains (Fig. 7.1c). The native vegetation consists largely of shrubs, interspersed grasses and scattered trees in the low lying areas, with areas of forest in the cooler, wetter mountain ranges. Much of the low lying land in this region is used for grazing with areas of irrigated agricultural production where water is available and soils are suitable. One of the major resource management concerns on croplands includes soil salinity and sodium content as soluble salts accumulate rapidly in this generally warm, dry region with relatively salty irrigation water. Overgrazing is a concern on areas used for rangeland.

The soils in this region are dominantly Aridisols, Entisols, and Mollisols. The dominant suborders include Argids and Calcids on alluvial fans, plains, and basins; Orthents and Fluvents on alluvial fans, plains, plateaus, and valleys; and Xerolls and Ustolls on mountain slopes. The soils in the region dominantly have mesic and thermic soil temperature regimes (Fig. 7.1b), and aridic or ustic soil moisture regime, with mixed mineralogy. Soils of the WRIR display unique characteristics reflecting the water-limited climate and the complex geologic and geomorphologic conditions in which they form. The formation of soils may be related to factors such as climate, biologic activity, topography, geologic substrate, and soil age (Jenny 1941). In the water-limited conditions of the WRIR, climate and its variation over time is one of the dominant soil-forming factors. Climate moderates the availability of water that controls plant productivity, the movement of soil and sediment across the landscape, and the accumulation and loss of soluble materials from the soil profile. Modern climate in this region varies substantially with local variation in elevation and landform (Fig. 7.1b, c) and, in addition, has varied tremendously over the recent geologic past following global patterns in glaciation. This region is also characterized by a complex geologic history that has led to the formation of numerous mountain ranges, basins, and plateaus. These geologic features interact with modern and past climate to control the variation in soils that we observe in these landscapes today. The soils in this area provide critical ecosystem services, such as watershed and groundwater recharge, carbon sequestration, and maintenance of plant and animal diversity to a large area of the Western US.

7.2 Physiography

The WRIR includes a long and complex geologic history that has direct impacts on the soils and ecosystems observed. The main physiographic areas of this region include the Basin and Range Province, the Colorado Plateau Province, and the Cascade-Sierra Province, with smaller areas located in the Wyoming Basin and Columbia Plateau provinces. The

C. Rasmussen (✉) · R.A. Lybrand · C. Orem · J. Kielhofer · M. Holleran
The University of Arizona, Tucson, AZ, USA
e-mail: crasmuss@cals.arizona.edu

R.A. Lybrand
e-mail: rlybrand@email.arizona.edu

C. Orem
e-mail: orem@email.arizona.edu

J. Kielhofer
e-mail: jkeilhofer@email.arizona.edu

M. Holleran
e-mail: meholleran@email.arizona.edu

L.T. West et al. (eds.), *The Soils of the USA*, World Soils Book Series,
DOI 10.1007/978-3-319-41870-4_7

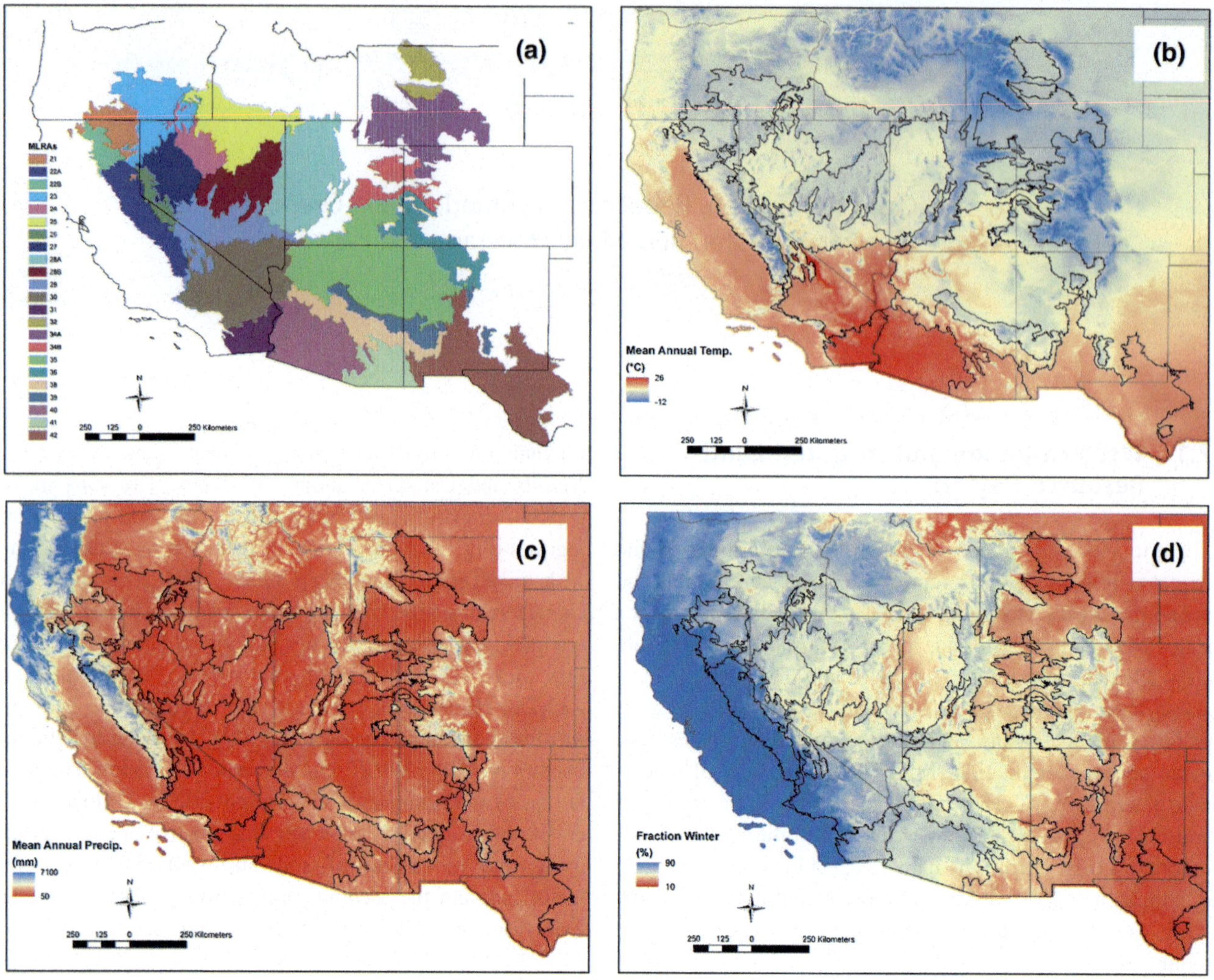

Fig. 7.1 The Major Land Resource Areas (MLRAs) for the Western Region Irrigated and Range Land Resource Region (**a**), and the mean annual temperature (**b**), mean annual precipitation (**c**), and fraction of mean annual precipitation that arrives during winter months (**d**). The climate data are derived from the 30 year normal for 1981–2010 from PRISM climate group available at http://www.prism.oregonstate.edu/

formation of these areas reflects a rich and complex geologic history and has a direct impact on present-day soil distribution and properties.

7.2.1 Basin and Range

The largest portion, over 47 % of the WRIR, is located within the Basin and Range Province. This province is characterized by extensive faulting resulting from crustal extension that leads to the development of numerous mountain ranges that rise abruptly from broad sediment-filled valleys and basins. The valleys and basins contain alluvial deposits of various ages, including alluvial fans, stream and river terraces, and lake deposits. Soil properties often vary as a direct function of deposit and landscape age, and landscape position (e.g., Parker 1995).

Typical rock units in the Basin and Range Province span a large range of geologic time, with rock units ranging from as old as the early proterozoic (~ 1–2 billion years) to late Holocene (<10,000 years) volcanic deposits. The oldest rocks include metamorphosed sedimentary and intrusive igneous rocks, typically found in uplifted mountain blocks (Gile et al. 1981; Palais and Peacock 1990; Wooden and Miller 1990). During the Paleozoic and Mesozoic Eras (spanning the time period of roughly 540–66 million years ago), many portions of the WRIR were affected by near shore and shallow water deposition resulting in numerous sedimentary units across large swaths of the Western USA (Stoffer 2004). These ancient, uplifted rocks subsequently erode and serve as the parent material for soils that now fill the intermountain valleys and basins.

Much of the topography characterizing the WRIR is the result of processes occurring in the Cenozoic Era that spans roughly 65 million years ago (Ma) to present. Beginning at ~30 Ma, Basin and Range style extension began across the region, first in the south portion of the region that includes Arizona, New Mexico, and the eastern deserts of southern California, following the complete subduction of the Farallon plate beneath the North American plate near Baja California. Over time, the extension propagated northward into Nevada and Utah as the Pacific plate began its subduction beneath the North American plate. This large-scale regional plate tectonic activity drove extension of the North American plate and resulted in pervasive normal faulting, mountain uplift and basin formation, and the formation of more complex structures such as metamorphic core complexes (Davis et al. 2004; Stoffer 2004). The normal faulting resulted in a horst and graben landscape where blocks of rock alternate between uplift (horst) and downward movement (graben) and generated the regular north–south-trending mountain ranges and basins seen in this region today. Additionally, volcanic activity was pervasive during the extensional period. In the greater southern Arizona and New Mexico regions, many calderas and other eruptive systems were active during the early Basin and Range extensional period (Dubray and Pallister 1991). In New Mexico and west Texas, the Basin and Range extension interacted with continental rifting of the Rio Grande Rift to create large mountain ranges and basins (Monger et al. 2009). Associated volcanism was mostly intermediate to felsic in composition and included many dike, sill, and plug features (Gile et al. 1981).

The basins in the Basin and Range Province can be internally or externally drained. Many of the basins are hydrologically closed and do not drain externally. This facilitated the formation of large lakes during cool, wet glacial periods in the recent geologic past. Following warming and drying of the climate in the modern period, these lakes evaporated and lead to the formation of playa deposits at the basin bottoms and beach terraces at the lake margins. The formation of playas and associated features is present across the region, with the largest lake features found in Nevada and Utah, e.g., the Great Salt Lake is the modern remnant of a much larger fluvial system. Other desert landforms, such as sand dunes, occur where sand material is available and wind directions are favorable. These features are found throughout the region, indicating the importance of eolian processes to soil–landscape development in this region.

7.2.2 Colorado Plateau

As summarized in Hendricks (1985), the Colorado Plateau region encompasses thick sequences of flat to gently undulating ancient sedimentary deposits. The Colorado Plateau extends over northern Arizona and New Mexico and into southern Utah and southwestern Colorado. The ancient Paleozoic aged sedimentary deposits that make up the plateau have been eroded into expansive plateaus and dissected by deep canyons. Whereas the Basin and Range is characterized by uplifted mountain ranges and sediment-filled basins, Colorado Plateau topography resulted from stream and river incision into the sedimentary deposits, generating deep, steep sided arroyos and canyons in an otherwise generally flat terrain. In this sense, the Colorado Plateau is structurally unique in the Western USA, because it is only moderately deformed compared to the surrounding regions.

The Colorado Plateau appears to have undergone substantial uplift over a period of 40 million years between the Late Cretaceous and middle Eocene time, leaving the region approximately 1.5 km above sea level (Hendricks 1985). The uplift resulted in a significant elevation gradient between the land surface and regional base level near sea level in the Gulf of California, leading to deep incision and canyon formation by regional river systems. The most well known of these canyons is the Grand Canyon in northwestern Arizona that cuts into the highest part of the plateau.

The Colorado Plateau region also experienced more recent volcanism over the last several million years. The San Francisco Peaks volcanic field in northern Arizona near the town of Flagstaff is one of the most extensive areas of volcanism on the plateau, covering more than 5000 km^2 (Hendricks 1985). Volcanism in this area spanned from the late Pliocene to the late Holocene, with the most recent eruption occurring roughly 900 years ago to form the cinder cone named Sunset Crater.

7.2.3 Sierra Nevada and Southern Cascades

The Sierra Nevada and Southern Cascade Ranges form a nearly continuous barrier along the eastern edge of California. The soils and vegetation across the ranges are generally similar; however, the two ranges vary substantially in their formation and geologic composition. The Sierra Nevada is primarily comprised of Mesozoic aged granitic rocks, emplaced during the Jurassic ~200 million years ago. During the Late Cretaceous, roughly 70 million years ago, the once deeply buried granitic rocks were exposed at the Earth's surface, but it was less than five million years ago that the range began to rise along its eastern margin. The modern Sierra Nevada is essentially an enormous tilted block with a long, gentle slope westward to California's Central Valley and steep eastern slope. The modern soils and landscapes in the Sierra Nevada have largely been shaped by erosion and glaciation that have occurred since uplift (http://geomaps.wr.usgs.gov/parks/province/pacifmt.html).

The southern Cascade Range and Sierra Nevada meet in northern California. The Cascade Province is comprised of a chain of explosive volcanic centers. Mountain building in this region was largely associated with volcanic activity and deposition of erupted materials.

7.3 Modern and Past Climates

The modern climatic regime of this region is highly variable and characterized by seasonal to inter-decadal variability in temperature and precipitation due to Pacific Ocean sea surface temperature patterns (e.g., El Niño-Southern Oscillation), giving rise to extended wet and dry periods (Trenberth et al. 2003). Further, this region exhibits a strong west to east gradient in the seasonality of rainfall, with, for example, the Mojave Desert in the west dominated by winter precipitation, the Sonoran Desert in the center that exhibits a bimodal precipitation regime with both winter and summer rainfall, and the Chihuahuan Desert in the east that is dominated by summer rainfall (Fig. 7.1d). Annual rainfall amounts in the desert systems range from <100 mm year^{-1} at the lowest and driest elevations to over 800 mm year^{-1} in the wettest high-elevation areas. Potential water loss to evaporation is typically in excess of 2500 mm year^{-1} for the low lying desert regions.

A distinct seasonality in precipitation source and characteristics exists when comparing winter and summer precipitation (Neilson 1987). Winter rainfall derives from large-scale Pacific frontal systems delivering low-intensity precipitation to broad areas. These storms are the dominant source of moisture in the western portion of the region. In contrast, the summer monsoon season is dominated by convective rainfall related to thunderstorm activity and pulses of subtropical moisture and contributes over half the annual precipitation to the Sonoran Desert and the majority of precipitation to the Chihuahuan Desert (Fig. 7.1d). Monsoon thunderstorms can produce large amounts of localized rainfall over very short periods of time that can have dramatic and long-lasting impacts on soil processes and plant production and diversity (Goodrich et al. 1995). The Sonoran Desert, which receives nearly equal proportions of winter and summer precipitation, contains some of the most diverse and abundant vegetation cover in the region, owing largely to the bimodal nature of the precipitation pulses. The soil moisture regimes in the WRIR are predominantly aridic owing to the relative lack of precipitation and high potential evapotranspiration rates. At higher elevations with greater precipitation and cooler temperatures, soil moisture regimes may be xeric in the western portion and ustic in the central and eastern portions of the WRIR, with a concomitant shift in vegetation from desert scrub and grassland to pine-oak woodlands and mixed conifer forests at the highest elevations.

Regional climate is heavily influenced by local topography in the WRIR. In general, annual precipitation increases and temperature decreases with increasing elevation. This variation in climate leads to a significant variation in water availability, plant productivity and composition, and soil-forming processes over relatively short distances. For example, the precipitation, temperature and vegetation range encompassed by the Santa Catalina Mountain range just outside of Tucson, AZ in the Sonoran Desert are equivalent to the environmental gradient spanned between the Sonoran Desert to forest in Canada (Whittaker and Niering 1975). These local elevation gradients have been termed "Sky Islands" in that they represent "islands" of forested ecosystems surrounded by "oceans" of desert. The Sky Islands serve to provide a wide range of soil-forming environments and soil variability in addition to providing critical ecosystems services related to groundwater recharge, carbon cycling, wildlife habitat and migration routes, and recreational opportunities.

7.3.1 Paleoclimate

The WRIR has experienced significant variation in climate over the recent geological past in concert with the glacial–interglacial cycles of the last ~2 Ma. These fluctuations have exerted substantial control over soil development, with many soil properties observed today relics of past climates. It is therefore important to understand these past fluctuations to better understand the distribution and range of soils in this region. Global atmospheric circulation patterns have considerable impacts on climate and subsequent landscape evolution processes, particularly in arid environments where small fluctuations in water availability can significantly impact soil process type and rate (Spaulding 1991b). At least four glacial–interglacial cycles occurred during the Pleistocene. During the last glacial maximum (~20 ka), the Laurentide continental ice sheet, which covered large regions of North America including Canada and the upper mid-west, created high pressure cells that forced the jet stream over the southwestern United States (Tchakerian and Lancaster 2002). This southward shift in the jet stream led to increased winter precipitation, cooler temperatures, and reduced evapotranspiration rates that significantly increased water availability. Increased water availability is evident through shifted vegetation boundaries as recorded in pack rack middens, increased pluvial lake formation as recorded in lake terraces and lacustrine deposits, and isotopic records from speleothems in caves and pedogenic carbonates. Following deglaciation, the jet stream shifted northwards leading to a general warming of temperatures and decreasing precipitation during the Pleistocene–Holocene transition that leads to paleolake desiccation and changed vegetation distributions toward those observed today. Glacial periods also led to the formation of extensive alpine glaciers in the high-elevation mountains of the region that dramatically carved and shaped the landscape into features we observe today, such as the dramatic U-shaped Yosemite Valley in the Sierra Nevada.

In the southern desert regions, packrat midden records dated to the late Pleistocene (25–12 ka) indicate extensive cover of pinyon–juniper woodlands across the region at elevations that today are dominated by desert scrub. For example, midden data from the Mojave, Sonoran, and Chihuahuan deserts all indicate extensive cover of coniferous evergreens during the last full glacial period of the late Pleistocene (McAuliffe and Van Devender 1998; Spaulding 1991a). A much wetter, cooler climate than the present is required to support these plant species. Modern-day desert scrub distribution is recorded in early Holocene middens and appears to have been well established by the early to mid-Holocene, ~4000 years ago. During the Pleistocene, there were over 100 pluvial lake systems across the WRIR with many showing maximum lake extents between 15 and 13.5 ka (Minnich 2007). In addition to confirming greater water availability, data from lacustrine sediments also document the shift toward greater summer rainfall and establishment of the monsoon at the Pleistocene–Holocene transition (Metcalfe et al. 2002). These findings are further supported in speleothems sampled from the Cave of the Bells in southern Arizona that record large deviations in the oxygen isotope content at approximately 15 ka, marking the major shift in climate from the wetter, cooler Pleistocene conditions to the warmer, drier conditions associated with Holocene conditions (Wagner et al. 2010).

7.4 General Summary of Regional Soils

7.4.1 Basin and Range Province (MLRAs 26, 27, 28A, 28B, 29, 30, 31, 32, 40, and 41)

The Basin and Range Province includes a number of MLRAs and dominates the WRIR with a combined land area of over 664,800 km^2. The dominant soils include Entisols, Aridisols, and Mollisols with the entire area generally dominated by Torriorthents and Torrifluvents, Haplocalcids, Haploargids and Haplocambids, and Argixerolls. The area includes large tracts of federally owned and managed lands, with the dominant land use of rangeland, areas of irrigated agriculture, and significant areas of urban and suburban development.

The modern topography of the Basin and Range Province is represented by widely spaced generally north–south-oriented mountain ranges, with large sediment-filled basins in-between. alluvial fan and pediment features are prevalent in the basins along with the imprint of cut and fill sequences that reflect past sediment aggradation followed by intervals of erosion. These alluvial fan cycles lead to the formation of progressively higher and older surfaces that express varying degrees of soil development (Fig. 7.2) (Gile 1975b; McAuliffe 1994; McFadden 1978). In general, older surfaces tend to contain well-developed soils that exhibit

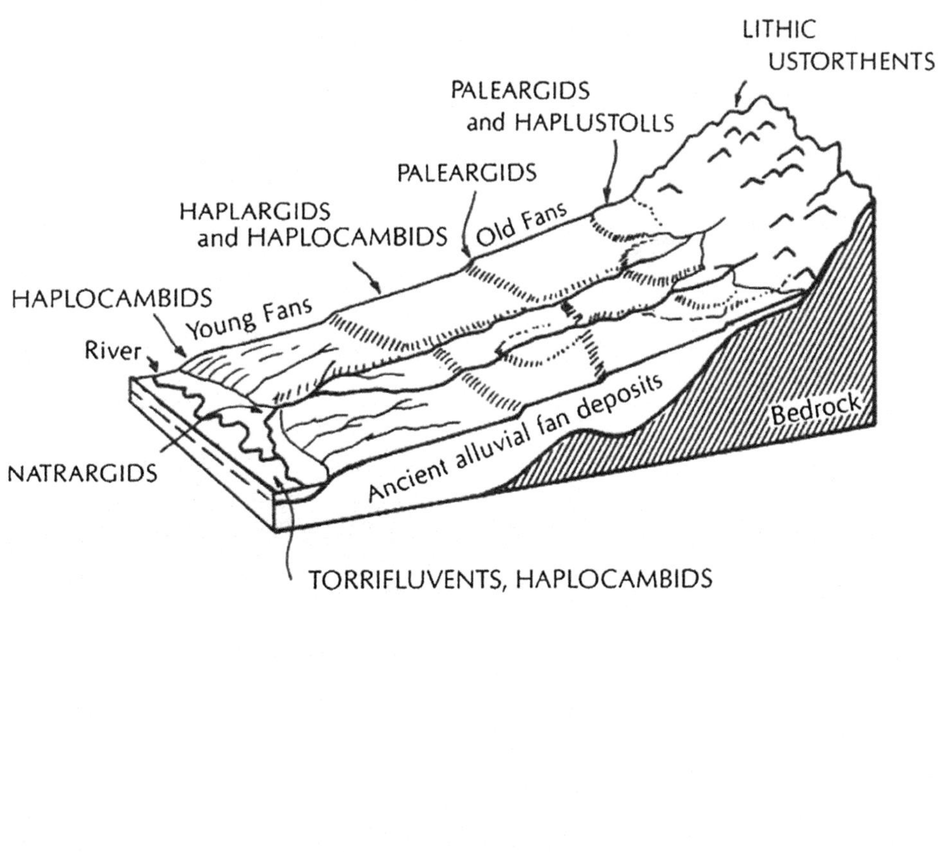

Fig. 7.2 Block diagram showing the typical soil–landscape relationships for the Basin and Range Province. Soils progress from shallow weakly developed Entisols and the mountainous uplands with varying degrees of soil development on alluvial fan deposits and basin floor sediments depending on landscape age. The relationships create a complex assemblage of soils in the sediment-filled basins. Image taken from Buol et al. (2011)

significant reddening due to Fe oxidation and chemical weathering, clay accumulation from both dust deposition and in situ chemical weathering, and the presence of diagnostic subsurface horizons such as argillic, calcic, or petrocalcic horizons (Fig. 7.3) (Mayer et al. 1988; McAuliffe 1994). In contrast, the relatively young, Holocene aged surfaces tend to contain poorly developed soils that lack any significant degree of soil formation. The Holocene aged soils tend to be relatively deep, can span a range of soil texture and rock fragment content depending on landscape position and depositional regime, and lack subsurface diagnostic horizons. Repeated cycles of aggradation and erosion related to past changes in climate have yielded a complex mosaic of soils and degree of soil development on these alluvial surfaces (Fig. 7.2).

7.4.1.1 Soil Development as a Function of Landscape Age

One major change observed over time in warm arid and semiarid environments is the formation and changing expression of two distinct soil horizons: (1) the calcic horizon and (2) the argillic horizon (Gile 1975a; Gile et al. 1966; Knuepfer and McFadden 1990; McAuliffe 1994; McFadden 1978; Monger et al. 2009). In fact, carbonate ($CaCO_3$) morphology in these systems advances through defined stages of increasing development (stages I through V) that correspond with greater carbonate accumulation as soils age (Gile et al. 1966) (Fig. 7.4). Soils on the youngest Holocene surfaces generally show little accumulation of secondary carbonate or clay. Mid-Holocene soils are generally characterized by weak pedogenesis in the form of stage I carbonate accumulation, cambic horizons, and/or structural development. Over time, these young soils undergo weathering and accumulate clay and carbonate from precipitation, weathering, and dust influx. Therefore, soils on stable Pleistocene surfaces generally are red in color and contain well-developed calcic and argillic horizons, with degree of horizon expression increasing with landscape age (Fig. 7.3). The characteristics of these older soils, such as carbonate content and clay content, differ dramatically from the original parent material.

Carbonate precipitation in soils requires a source of calcium and carbonate/bicarbonate that are then concentrated and precipitated in the soil matrix. The sources of the

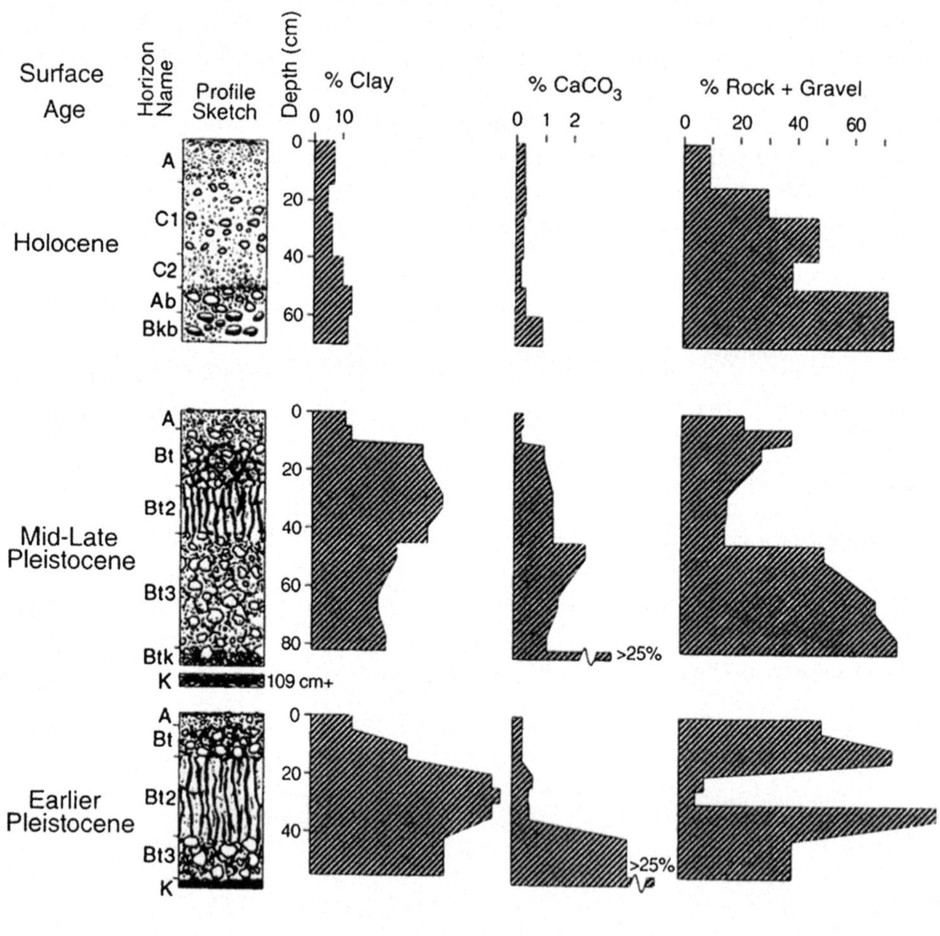

Fig. 7.3 Variation in soil profile morphology, clay and calcium carbonate content, and rock and gravel content for alluvial fans of varying age in southern Arizona. The data indicate increasing development in terms of morphology and clay and carbonate accumulation. The image is taken from McAuliffe (1994)

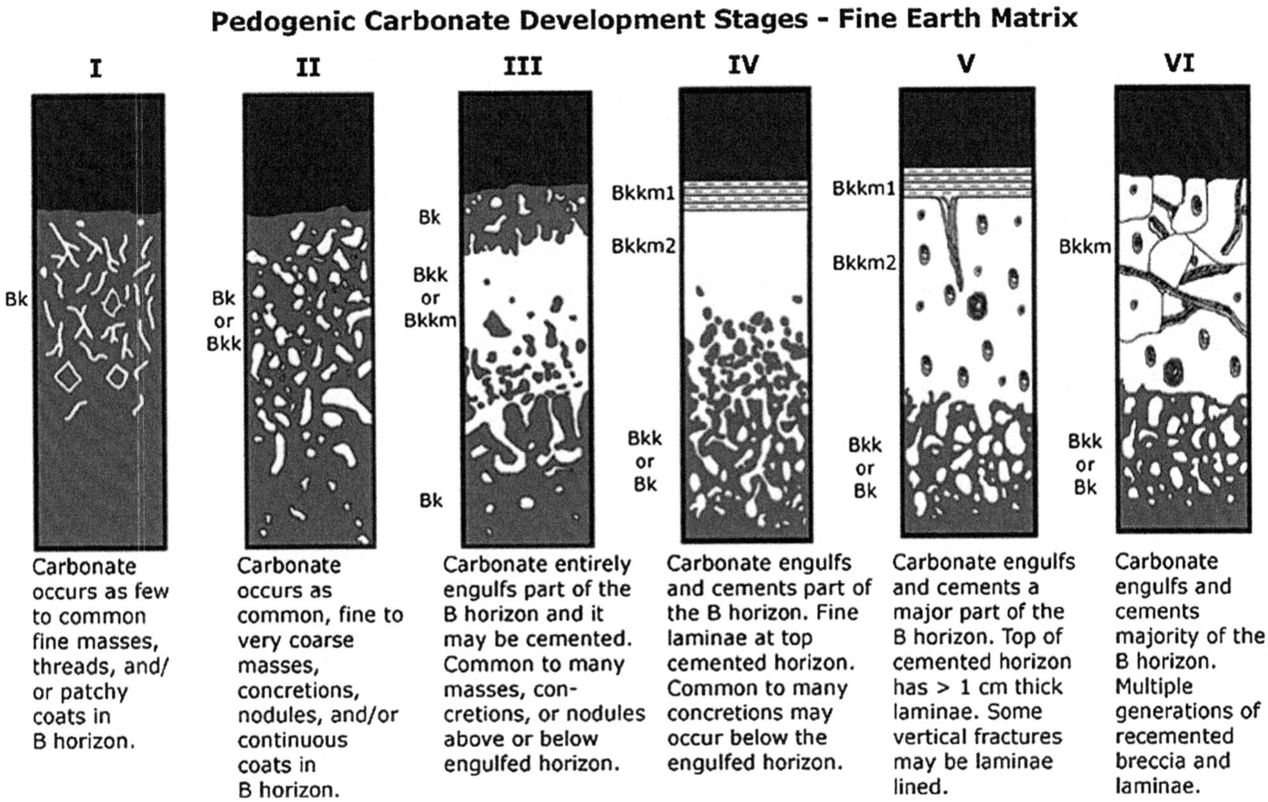

Fig. 7.4 Model of carbonate accumulation over time and associated carbonate stages (I, II, III, etc.) and horizon nomenclature and morphology. The image is taken from the Natural Resource Conservation Service Field Book for Describing and Sampling Soils v3.0 and based on the work of Gile et al. (1966)

carbonate reactants in Basin and Range soils include Ca inputs from precipitation, direct $CaCO_3$ influx associated with dust and eolian deposition, and both Ca and HCO_3 from the chemical weathering of primary minerals in the soil profile. The dominant source of reactants varies across the region, but is generally dominated by a combination of precipitation and dust inputs. For example, studies from the Chihuahuan Desert indicate that up to 90 % of the carbonate in a cemented carbonate horizon may be sourced from atmospheric inputs (Capo and Chadwick 1999). Dust data from the regions of southeastern California and Nevada indicate dust can contribute on the order of 0.1 to 12 g m^{-2} $year^{-1}$ of $CaCO_3$ (Reheis et al. 1995). In comparison, the dominant source of Ca in the Chihuahuan Desert appears to be precipitation, where carbonate precipitation resulting from Ca influx from rain is perhaps two to three times greater than carbonate formed via calcareous dust input (Gile et al. 1981). Carbonate accumulation likely is predominant during warm, dry interglacial periods that tend to provide the water-limited climatic conditions that favor accumulation of soluble salts in the soil.

Clay accumulation, translocation, and argillic horizon formation result from both in situ mineral weathering and input of fine-grained eolian materials. Much of the clay accumulation by mineral weathering and subsequent translocation likely occurred during the periods of continental glaciation, when these systems were cooler and dominated by winter rainfall, allowing for greater water availability to drive chemical weathering and colloid transport. However, there is also evidence that a large fraction of silicate minerals in subsurface soils may be derived from eolian materials, e.g., Capo and Chadwick (1999) estimated that up to 70 % of silicates derived from atmospheric deposition in the form of dust in a Chihuahuan Desert profile and Crouvi et al. (2013) estimated that up to 80 % of the fine earth fraction in a Mojave Desert landscape was derived from eolian materials. Eolian materials are generally dominated by clay and silt sized particles. These fine-grained materials both add clay sized mineral material to the soil that may be translocated into the subsurface and generate greater water holding capacity that facilitates longer periods of increased soil moisture that fosters in situ chemical weathering and clay formation (Harden et al. 1991).

The morphological differences associated with progressive soil development over time have a dramatic impact on the classification of younger versus older soils. In many study areas, the Holocene soils belong to one of two soil orders, the Aridisols and Entisols, and generally fall under the classification of Torriorthents or Torrifluvents in mid to late Holocene soils. Early Holocene soils, which often exhibit incipient development of calcic and argillic horizons, are classified as Haplocalcids and Haplocambids in the Aridisol order. Finally, Pleisotcene soils are generally Haplocalcids, Calciargids, and/or Paleargids in the Aridisol order due to the presence of well-defined argillic and calcic horizons. However, landscape evolution can also result in the obliteration of soil horizons and limit soil development. Generally, the oldest surfaces are the most dissected and eroded, as they have experienced the greatest number of cut and fill cycles. In many cases, only remnants of early Pleistocene alluvial fans remain, and their associated soils have been truncated. Such dissection events have the greatest impact on argillic horizons, which may be eroded away or reduced in thickness. Therefore, while argillic horizons might be expected to continually develop over time, this may not be the case on active alluvial surfaces. In addition to erosion and dissection, carbonate engulfment also reduces the thickness of the argillic horizon. With greater carbonate accumulation and cementation, the calcic horizon transitions to a petrocalcic horizon and begins to engulf the lower portions of the argillic horizon.

7.4.1.2 Eolian Deposition and Desert Pavement

The influx of eolian materials is a major factor in the genesis of soils in the arid portions of the Basin and Range with modern dust flux rates ranging from 2 to 20 g m^{-2} $year^{-1}$ (Reheis 2006). The warm, dry interglacial periods in this region resulted in paleolake desiccation, playa exposure, and reduced vegetation cover, conditions that favor higher rates of dust production and transport into the atmosphere, as well as increased eolian deposition rates. Eolian influx can alter rates of soil development and general soil morphology as demonstrated across chronosequences in the Mojave Desert (Harden et al. 1991). Eolian influx contributes to rapid accumulation of carbonate and clay in Holocene soils that leads to increased rates of soil development. The relatively fine-grained eolian material enhances soil water holding capacity and as a result, more water is available to drive mineral weathering, silicate clay formation, and iron oxidation. Eolian influx also contributes to the accumulation of carbonates and other soluble salts, with dust contributing a large fraction of the Ca for carbonate formation, particularly in the Mojave and Sonoran deserts.

Desert pavements are another important surface feature developed in arid Basin and Range soils as a result of eolian influx. Desert pavements consist of closely packed, interlocked gravels in a smooth surface layer that can be several centimeters thick (Fig. 7.5). This smooth interlocking layer of stone rests on top of, or is partially embedded in, a layer of fine-grained soil that is relatively free of stones (Laity 2011). The fine-grained soil material below the pavement is typically derived in large part from eolian materials (Fig. 7.5). Pavements may include surfaces with anywhere from ~15 % surface stone cover to greater than 90 % surface coverage; typically in the Basin and Range Province, a well-developed pavement is considered one where greater than 65 % of the surface is covered with stones (Wood et al. 2002).

A generalized model of pavement formation includes a number of processes as summarized by Laity (2011) and based on the pioneering work of Wells et al. (1985) in the Mojave Desert (Fig. 7.5). Briefly, pavement formation requires a stable geomorphic surface where aggradation of eolian materials outpaces erosive processes such as concentrated surface wash or other fluvial activity. Common stable surfaces that favor pavement formation include old alluvial fans and areas underlain by Quaternary basalt flows. The formation of the smooth interlocked pavement surface is driven by entrapment and infiltration of dust into the soil surface. Rocky surfaces, in particular, serve as effective dust traps. The entrapped dust accumulates in the spaces between the rocks, accelerating the mechanical weathering and break down of the surface rocks into smaller fragments. As discussed above, accumulation of fine-grained dust enhances the soil water holding capacity and increases the rate of clay formation. Dust and clay accumulate until the soil has sufficient strength and shrink-swell potential to raft stones upward. This process continues, concentrating stones at the soil surface while the accumulated mantle of dust and clay increases in thickness (Anderson et al. 2002). Essentially, desert pavements are "born at the land surface" and rafted upwards on an accretionary mantle of eolian materials (McFadden et al. 1987).

Soils beneath the pavement are typically fine-grained and stone free and exhibit some degree of pedogenesis. In particular, it is common for pavement soils to have a surface horizon with vesicular pores. Vesicular pores are spherical in shape and are disconnected from each other. Vesicular pores are present in soils that cover over 156,000 km^2 in the Western USA, largely in the Basin and Range Province (Turk and Graham 2011). They are found predominantly in soils that receive eolian inputs and typically are associated with a surface pavement or biological crust that can effectively capture fine-grained wind transported materials. Their formation likely results from air that cannot escape when dry soils are rapidly wetted during rainfall events. Some data suggest the formation of vesicles may be favored by relatively rapid heating of the soil surface after a rainfall event that causes the air to expand (McFadden et al. 1998). Vesicular horizons are typically shallow, on the order of <10 cm in depth, and located directly beneath the surface stone layer or biological crust. They are important in terms of surface hydrology in that they have been found to have slow infiltration rates (Young et al. 2004), and in terms of

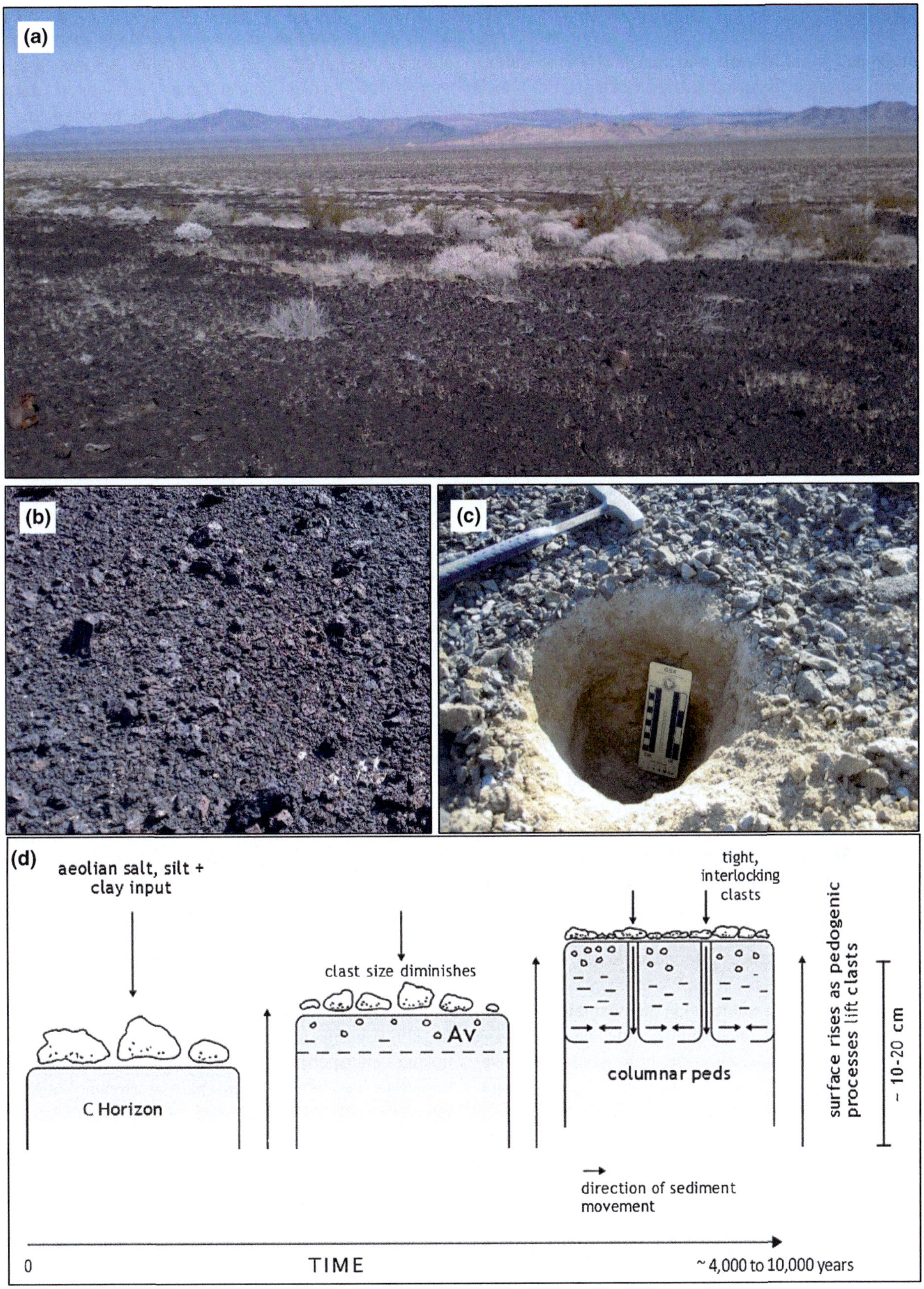

Fig. 7.5 Example of a desert pavement landscape on Quaternary basalt flows in the Mojave Desert of California (**a**), an example of a dense interlocked pavement surface (**b**), and associated eolian deposited layers beneath the pavement (**c**). Conceptual model of desert pavement formation taken from Laity (2011) and based on the work of Anderson et al. (2002) and Young et al. (2004)

dust production. Given that vesicular horizons predominantly form in eolian materials, they are highly susceptible to subsequent wind erosion and transport if the surface pavement or crust is disturbed, particularly by human activity (Goossens and Buck 2009).

7.4.1.3 Interrelation of Soils and Vegetation

The degree of soil development controls potential and actual soil water availability leading to a significant variation in vegetation density and assemblage with soil properties (e.g., Hamerlynck et al. 2002). A simplified comparison may be made between a young Holocene aged soil that lacks substantial subsurface development and clay accumulation with a mid- to early Pleistocene aged soil that has a well-developed argillic and/or calcic horizon in the subsurface. The hydraulic properties of the Holocene aged soil favor more rapid and deeper infiltration of precipitation, favoring deep rooted species such as creosote (*Larrea tridentata*) and mesquite (*Prosopsis velutina*). In contrast, the hydraulic properties of the Pleistocene aged soil limit deep water infiltration and limit water residence time in the soil, favoring shallow, laterally rooted plant species such as cacti, grasses, and certain shrubs. Indeed, research in the Mojave suggests that intermediate soil development and the associated hydraulic conditions may form an optimal setting for some plant species production and survival (Hamerlynck et al. 2002). The relationship between soil age, subsurface soil texture, pedogenic development, and vegetation patterns may be one of the dominant mechanisms responsible for observed variation in modern vegetation distributions and aboveground productivity in these water-limited landscapes (Shepard et al. 2015).

7.4.1.4 "Sky Island" Elevation Gradients

The mountainous topography of the Basin and Range generates elevation gradients that encompass large gradients in precipitation, temperature, and vegetation over relatively short distances in sequences referred to as Sky Islands. In general, elevation gradients in the Western United States that extend across >1500 m of elevation gain exhibit a substantial increase in water availability with elevation. The increase in water availability results from greater precipitation, decreased temperature, and decreased evapotranspiration demands at higher elevations. In many cases, such gradients span the transition from water-limited to energy-limited ecosystems where water and energy limitations are defined as the ratio of mean annual precipitation (MAP) to potential evapotranspiration (PET), or a MAP/PET ratio <1 in water-limited sites and MAP/PET >1 in energy-limited sites. This transition can manifest as significant variations in primary production and soil development, with strong climate-related gradients in degree of soil development (Whittaker et al. 1968).

One example of a typical Basin and Range elevation gradient includes the Santa Catalina Mountains (SCM) environmental gradient in southeastern Arizona (Fig. 7.6). Across this gradient, mean annual temperature decreases from 18 to 9 °C and mean annual precipitation increases from 450 mm year^{-1} at low elevations of 1092 m asl to 950 mm year^{-1} at high elevations of 2408 m asl. The SCM elevation gradient spans the water to energy limitation spectrum, with MAP/PET values of 0.53 in low-elevation locations that increased up to 1.5 in high-elevation locations. Vegetative communities change concurrently with climate variation across the SCM gradient, ranging from desert scrub at low elevation to mixed conifer at high elevation, and are well characterized in terms of aspect, climate, and net primary productivity (Whittaker and Niering 1975). Along with climate and vegetation, soil properties vary significantly following the increase in water availability and vegetation production with elevation across the SCM gradient (Lybrand and Rasmussen 2015). In particular, the amount of organic carbon stored in the soil and the degree of chemical weathering increases with elevation, whereas soil pH decreases with increasing elevation. These differences then manifest as changes in soil taxonomic classification with soils grading from Typic Torriorthents in the desert scrub locations, to Aridic Haplustolls in the mid-elevation grassland–oak woodland locations, and to Typic Humustepts in the conifer locations.

7.4.2 Southern Cascade and Sierra Nevada Region and Klamath and Shasta Valleys (MLRAs 21, 22A, and 22B)

These MLRAs comprise over 93,620 km^2 extending from southern Oregon to east-central California in landscapes dominated by mountainous uplands. The Klamath and Shasta Valley areas (MLRA 21) are in the transition zone between the Basin and Range Province to the southeast, the Cascade and Klamath Mountains to the west and northwest, and the Sierra Nevada Mountains to the south. The area is largely volcanic terrain characterized by a vast volcanic upland interspersed with isolated volcanic peaks and numerous reservoirs, lakes, narrow stream valleys, and internally drained basins with lakes or periodically dry lakebeds. Soils in this area are predominantly Mollisols with small areas of Inceptisols and Histosols. The dominant Mollisols include Argixerolls and Haploxerolls that cover 41 and 19 % of MLRA 21. Approximately 50 % of the area is federally owned and largely used for grazing.

The Sierra Nevada and Cascade MLRAs (MLRAs 22A and 22B) are characterized by large north–south-oriented mountain ranges that generate steep environmental gradients over short distances. The dominant soils include Alfisols,

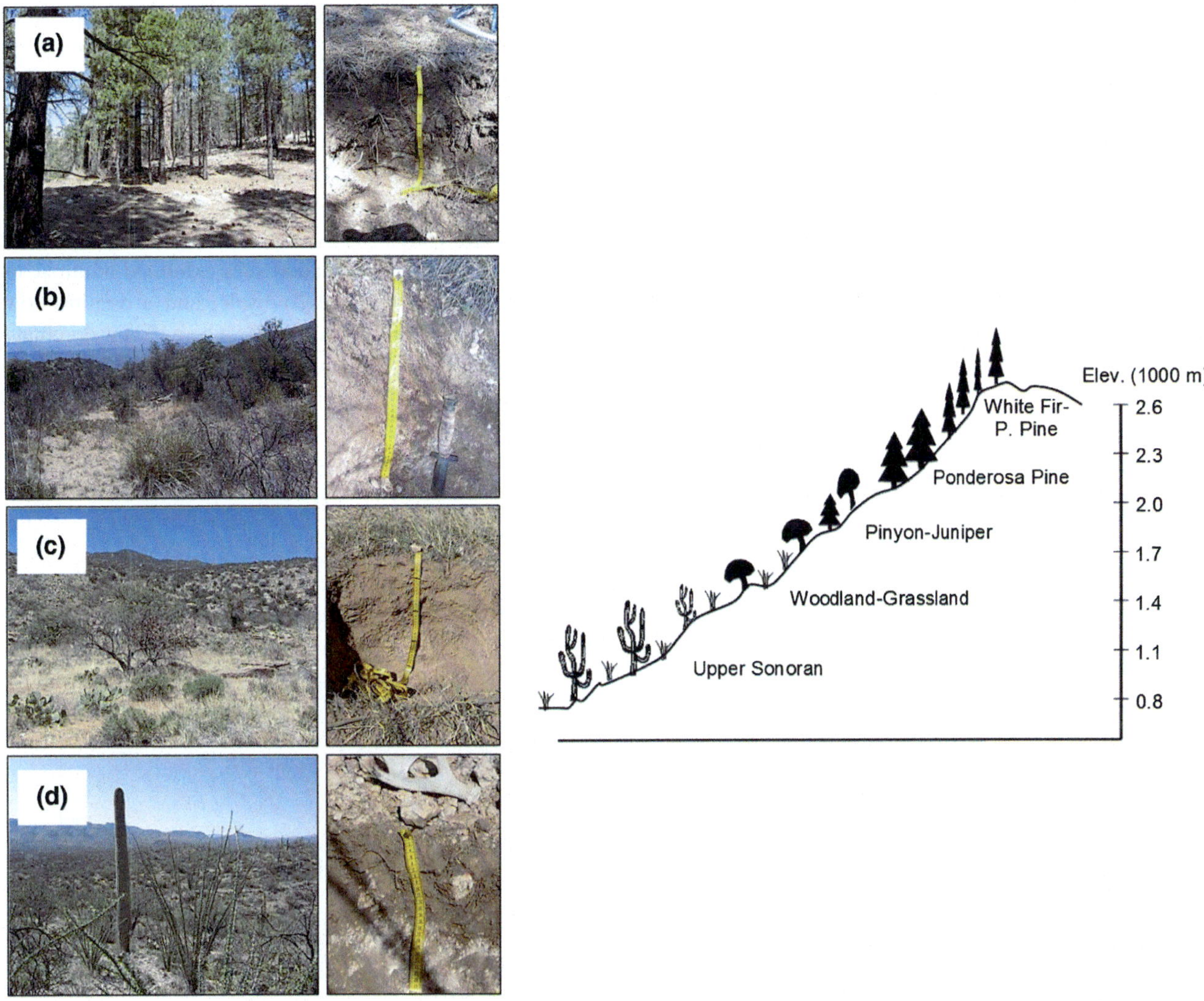

Fig. 7.6 Typical variation in soils along a Sky Island elevation gradient in the Basin and Range Province. These images are from the Santa Catalina and Rincon Mountains in southern Arizona. The steep environmental gradients span vegetation from pine and mixed conifer (**a**), pinyon–juniper woodlands (**b**), desert grasslands (**c**), and desert scrub (**d**). Soils tend to increase in depth, carbon content, and degree of weathering from low elevation to high elevation with changes in climate and vegetation (Lybrand and Rasmussen 2015)

Entisols, Inceptisols, Andisols, Mollisols, and Ultisols. The soils in the area exhibit clear trends in soil properties with local variation in elevation, climate, and vegetation. Soils dominantly have a mesic, frigid, or cryic soil temperature regime, a xeric soil moisture regime, and mixed mineralogy. The majority of soils formed in residuum and colluvium on hills and mountains; upland soils are generally very shallow to deep well drained with loamy or sandy particle-size classes, whereas soils in basins can be somewhat poorly to poorly drained with aquic soil moisture regimes. Roughly three-fourths of the Southern Cascade and Sierra Nevada region area is federally owned land, which is primarily in national forests and parks. The rest of the area is privately owned forestland, farms, and ranches. Approximately 80 percent of the area is forested and used for timber, recreation, wildlife habitat, and watershed. The area has very little agriculture, with agricultural land use largely consisting of range and pastureland, with grazing confined to mountain meadows and areas with open stands of timber, and small areas of cropland used mainly for deciduous fruits, grain, or hay.

The dominant soils of the Sierra Nevada include Haploxeralfs (23 %), Haplohumults (17 %), and Haploxerults (16 %), whereas the Cascade soils have a greater proportion of Andisols with the dominant soils including Haploxeralfs (33 %), Vitrixerands (19 %), and Haploxerands (14 %). A number of studies have quantified variation in soil properties across the Sierra Nevada and Cascade elevation gradients that indicate a number of consistent climate-controlled trends in soil development with increasing elevation (Dahlgren et al. 1997; Rasmussen et al. 2007, 2010). The transects all span broad environmental gradients on the western slope of the Sierra Nevada or southern Cascade ranges, with variation in mean annual soil temperature, ranging from 3 to 17 °C, and mean annual precipitation, ranging from 330 to 1500 cm year^{-1} (Fig. 7.7). With increasing elevation, temperature decreases and mean annual precipitation increases; the current permanent winter snowline occurs in the mid-elevations near 1300 to 1600 m asl depending on the latitude of the transect location. Soil moisture regimes are predominantly xeric (winter precipitation dominated), although the high-elevation zone may remain sufficiently moist due to late melting of the snowpack and summer thunderstorm activity to classify as udic. Soil temperature regime ranges from thermic to cryic with increasing elevation. Following elevation-driven changes in climate, vegetation communities progress from oak woodlands at low elevation through mixed conifer systems at mid-elevations, to subalpine mixed conifer and alpine grassland ecosystems at high elevations.

The degree of soil development and chemical weathering varies significantly with elevation, coincident with changes in mean annual precipitation and temperature (Fig. 7.7). In particular, all of the transect data indicate a threshold type change in soil development at the transition between zones where precipitation is dominated by rain or snow. Snow-dominated high-elevation sites exhibited minimal soil development (Entisols and Andisols) and a dominance of soil fractions by primary minerals, whereas rain dominated low-elevation sites were characterized by well-developed soils (Alfisols and Ultisols) and secondary kaolins and Fe oxyhydroxides. A threshold in pedogenesis and weathering may be observed in mid-elevation sites at the rain–snow transition in that the rain dominated soils tend to be highly

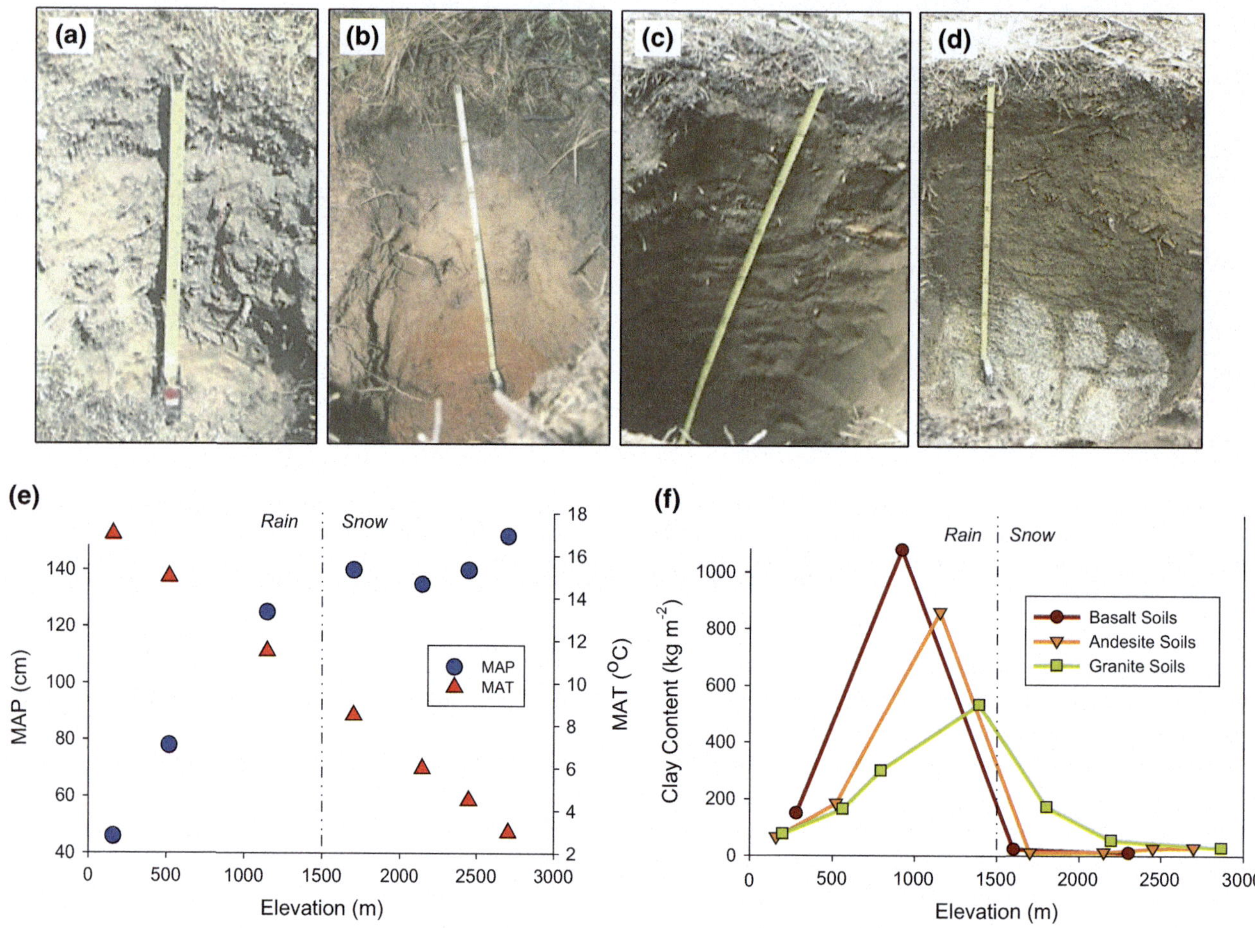

Fig. 7.7 Variation in soil development across the elevation and climate gradients that span the western slopes of the southern Cascade and Sierra Nevada Mountains. The soils grade from warm and dry low elevations (**a**), to warm and wet mid-elevations with precipitation dominantly in the form of rain (**b**), to cool and wet mid-elevations dominated by snowfall (**c**), and to cold, wet high elevations (**d**). The mean annual temperature (MAT) and mean annual precipitation (MAP) exhibit relative decrease and increase, respectively, with increasing elevation (**e**). Soil development and clay accumulation increase with elevation in the rain dominated soils and decreases with elevation in the snow-dominated soils suggesting a climate threshold in soil weathering (**f**). Data from Dahlgren et al. (1997), Rasmussen et al. (2007, 2010)

weathered Ultisols with a maximum of clay accumulation, whereas the snow-dominated soils are generally Andisols and Inceptisols with limited chemical weathering and clay accumulation. The weathering threshold is likely an interaction between soil moisture and temperature. Low-elevation soils remain relatively warm during the wet winter months, promoting in situ mineral weathering. In contrast, high-elevation soils experience cold temperatures when the soil profile is moist, followed by dry conditions during the warm summer period. The data suggest soil development limited by temperature in high-elevation sites and by water availability in low-elevation sites.

7.4.3 Colorado Plateau and Southwestern Plateaus, Mesas, and Foothills (MLRAs 35 and 36)

These MLRAs span the four corners region and occupy 133,630 km^2, largely northern Arizona and northwestern New Mexico, southeastern Utah, and a small portion of southwest Colorado. The Colorado Plateau has been structurally uplifted generating a surface that consists of gently sloping to strongly sloping plains. Rivers flowing across the area cut down into the bedrock as it was being uplifted, resulting in spectacular geologic scenery as best expressed in the Grand Canyon. Areas of shale, sandstone, limestone, dolomite, and volcanic rock outcrop are extensive. Rocks representing almost the entire geologic time span are exposed from the bottom of the Grand Canyon up to the present-day surface. Quaternary and Tertiary lava flows and cinder cones occur in the southwest part of this area. Older flows cap plateaus and mesas, and isolated volcanic cones and eroded volcanic necks occur throughout the area. On the order of three quarters of this area is rangeland used for sheep and cattle production. Greater than 10 % of the area is juniper and pinyon–juniper woodland that are used for firewood and pinyon nut production in addition to grazing. Severe gullying, overgrazing, and the lack of a dependable water supply represent serious land use issues. Small areas of irrigated cropland are located and used to grow alfalfa, hay, and corn. Soils in this area are predominantly Alfisols, Aridisols, Entisols, and Mollisols with dominant Great Groups including Torriorthents and Ustorthents, Haplocalcids and Calciargids, and Haplustalfs.

Grand Canyon soils are mostly shallow, with extensive areas of rock outcrop due to the rapid removal of material from canyon surfaces and uplands. There are areas of alluvial deposits and terraces along the Colorado River that largely classify as Entisols (Hendricks 1985). The uplands surrounding the Grand Canyon consist of broad plateaus with soils formed in alluvium in washes and material reworked by wind on the flat sedimentary rock surfaces. These soils also generally lack extensive soil development. There are also areas that exhibit substantial soil development, particularly on areas of the Kaibab Plateau that is underlain by limestone. This plateau has north–south-trending faults that generate ridges and gentle slopes that are geomorphically stable and have soils with extensive clay accumulation through in situ weathering and accumulation of fine-grained eolian sediments (Hendricks 1985).

Most of the area associated with the Southwestern Plateaus, Mesas, and Foothills MLRA is characterized by horizontal beds of Jurassic, Cretaceous, and Tertiary sedimentary rocks that have been uplifted and eroded into plateaus, mesas, hills, and canyons. The area includes thick deposits of Pleistocene age eolian material on top of the mesas in some areas. This MLRA also includes small areas of Tertiary and Quaternary volcanic rocks, including cinder cones and lava flows, as well as wide valleys associated with the Rio Grande rift basin that have accumulated deep alluvial sediments. Nearly all of this area is used as grazing or forestland. Substantial land area is also used for irrigating agriculture and the production of wheat, barley, beans, oats, alfalfa, and hay. Areas of pinyon–juniper woodlands are a source of fuel wood, and at higher elevations, ponderosa pine and Douglas fir are commercially harvested for timber. The dominant soils include Alfisols, Inceptisols, Mollisols, Entisols, and Aridisols.

7.4.4 Mogollon Transition and Arizona and New Mexico Mountains (MLRAs 38 and 39)

These MLRAs span over 88,450 km^2 in central Arizona and New Mexico. The Mogollon Transition zone occurs at the boundary between the Basin and Range in the south and the Colorado Plateau in the north. The Arizona and New Mexico Mountains occur throughout Arizona and New Mexico, with a large section of mountains associated with the Mogollon Transition. The Mogollon Transition consists largely of canyons and structural troughs and valleys filled by deep alluvium washed in from adjacent mountains. This MLRA is also an area of intensive volcanism with a combination of old intrusive igneous rocks as well as more recent andesite and basalt flows. Some outcrops of Paleozoic sediments are associated with the uplift in the vicinity of the older intrusive rock units. Some of these sediments have been metamorphosed. Most of the area is used for livestock grazing with many tracts of rangeland subdivided for community development. The dominant soil orders in this area include Aridisols, Alfisols, and Mollisols that predominantly have a thermic or mesic soil temperature regime, an aridic or ustic soil moisture regime. There is substantial diversity in soils in this region as evidenced by a lack of several dominant Great Groups, but rather an even distribution of land area among many that include Argiustolls, Haplargids, Haplusterts, Ustorthents and Torriorthents, and Haplustolls.

The Arizona and New Mexico Mountains are characterized by volcanic fields, uplifted crystalline, and metamorphic rocks, along with gently dipping sedimentary rocks that have been eroded into plateaus, valleys, and deep canyons. The vegetation ranges from grasslands to mixed conifer vegetation, with most areas covered by forest and woodlands. The area is largely used for timber and livestock grazing with growing areas of land subdivided for community development. The dominant soils in this region include Inceptisols, Mollisols, Alfisols, and Entisols and span mesic to cryic soil temperature regimes with ustic soil moisture regimes. The dominant Great Groups include Argiustolls and Haplustalfs that cover 43 and 11 % of the MLRA.

Soil studies in this region indicate significant variation in soil properties with parent material (Heckman and Rasmussen 2011). This particular study focused on a lithosequence of four parent materials that included rhyolite, granite, basalt, and dolostone under ponderosa pine (*Pinus pondersa*) (Fig. 7.8). Results indicated significant variation in profile characteristics and chemical weathering loss and transformation according to parent material. The soils ranged in taxonomic classification from Entisols to Alfisols, with generally greater reddening and clay accumulation in the basalt soils followed by the soils on dolostone, rhyolite, and granite. Geochemical and mineralogical data indicated that a large fraction of the basalt (>50 %) soils were derived from eolian materials and volcanic cinders, highlighting the role of eolian inputs even in these relatively wet systems. Rhyolite and granite soils exhibited large differences in degree of weathering and mass flux regardless of the nearly identical elemental and mineralogical compositions of the respective parent materials that could be attributed to differences in parent material grain size and bulk density. The rhyolitic materials were relatively lower in bulk density that facilitates ready water movement into the parent material that can then participate in weathering reactions, that in conjunction with smaller mineral grain size, provides greater reactive surface area for chemical reactions to occur.

7.4.5 High Plateau Region (MLRAs 23, 24, and 25)

The area occupies over 121,105 km^2 in southwestern Oregon, southern Idaho, and northern Nevada. The general physiography of this area includes nearly level to moderately steep volcanic plateaus, basins and valleys filled with gently sloping alluvial fans and lacustrine deposits, bordered by north–south-trending fault-block mountain ranges. Playas occur in the lowest areas in valleys with closed drainage systems. Much of the area and volcanic plateaus are underlain by young andesite and basalt layers with older volcanic rocks and marine and continental sediments exposed in the mountain ranges. The basins between the mountains and lava plateaus are filled with a mixture of Quaternary alluvium, continental sediments, and volcanic ash. The dominant soil orders in this area include Aridisols and Mollisols, with smaller areas of Entisols and Inceptisols. The dominant Great Groups include Argixerolls and Haplargids with areas of Argidurids and Haplodurids on older

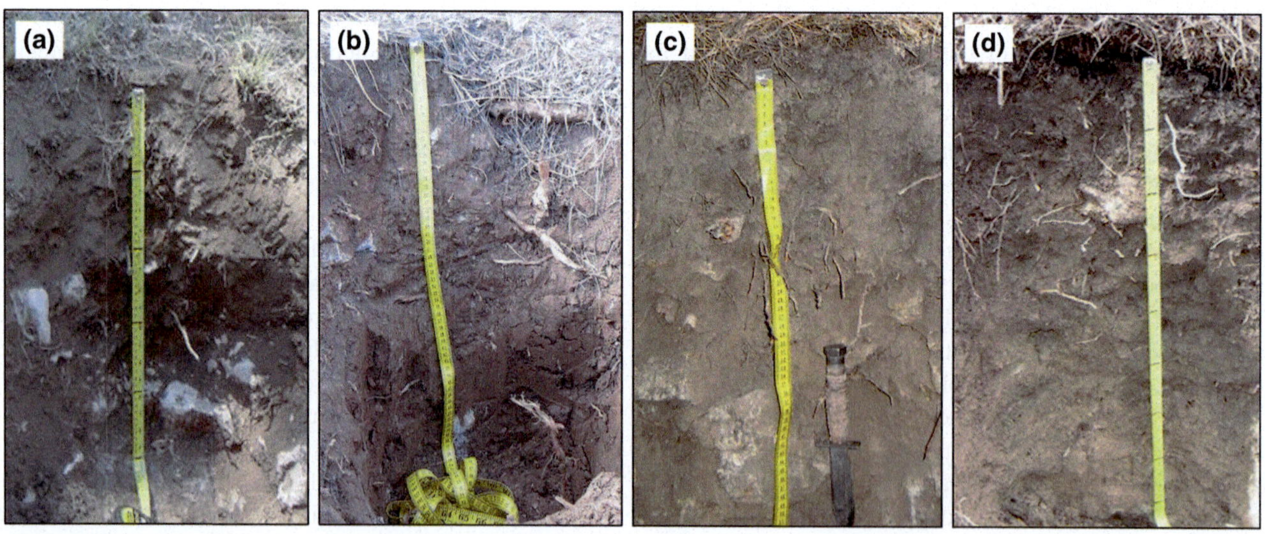

Fig. 7.8 Example of soil variation with parent materials in the Arizona and New Mexico Mountains region. Soils are developed from dolostone (**a**), basalt (**b**), granite (**c**), and rhyolite (**d**) and exhibit varying degrees of soil development and morphology with soils classified as Lithic Argiustolls on the dolostone, Typic Paleustolls on basalt, Typic Ustorthents on granite, and Typic Haplustept on rhyolite (Heckman and Rasmussen 2011)

landscapes, and Haplocambids and Torriorthents on younger landscapes. The soils dominantly have a mesic or frigid soil temperature regime, an aridic or xeric soil moisture regime, and mixed or smectitic mineralogy. The soils on uplands generally are well drained, loamy or clayey, and shallow or moderately deep, whereas the soils in basins generally are poorly drained to well drained, loamy or clayey, and very deep. Locally, large areas have an ashy particle-size class and glassy mineralogy.

7.4.6 Desertic Basins and Plateaus (MLRAs 32, 34A, and 34B)

The Desertic Basins and Plateaus region occupies over 141,895 km^2, in Wyoming, northwestern Colorado and northeastern Utah. The Northern Intermountain Desertic Basins (MLRA 32) and Cool Central Desertic Basins and Plateaus (MLRA 34A) are largely located in the Wyoming Basin encompassing uplifted mountains and large sediment-filled basins. The area is drained by three large rivers with the Green River to the south, the Big Horn River to the north, and the North Platte River to the east. The climate is generally arid to semiarid with vegetation dominated by salt desert species in the lower lying arid regions, sagebrush steppe in the semiarid basins, and a transition to juniper woodlands and conifer forest in higher elevations. The land is predominantly federally owned with private land largely parceled into ranches and farms, and land use is primarily livestock grazing. Soils in these two MLRAs are dominantly Entisols and Aridisols with mesic to frigid soil temperature regimes and aridic soil moisture regimes. MLRA 32 is dominated by Torriorthents and Torrifluvents that cover 56 and 12 % of the land area, respectively. The soils in MLRA include a greater proportion of Aridisols with Haplargids (28 %), Calciargids (18 %), and Torriorthents (15 %).

The Warm Central Desertic Basins and Plateaus (MLRA 34B) is mainly located in Utah and western Colorado and encompasses the Canyon Lands in the south and the Uinta Basin area in the north. The MLRA consists of broad intermountain sediment-filled basins bounded by plateaus and steep escarpments consisting mainly of sedimentary shale and sandstone. Similar to MLRAs 32 and 34A, this region spans salt desert, semiarid steppe, and upland shrub and conifer vegetation zones. The largest area is covered by salt desert that includes species such as saltbush (various species of *Atriplex* such as *A. canescens*, *A. corrugate*, and *A. gardneri*), greasewood (*Sarcobatus vermiculatus*), and shadscale (*Atriplex confertifolia*). The semiarid steppe is dominated by sagebrush, whereas the upland shrub and conifer zones transition to pinyon pine and juniper forests. Nearly three quarters of the land is federally owned and most of the area is used for recreation and livestock grazing. The dominant soils include Aridisols and Entisols, with small areas of Mollisols at higher elevations. The dominant Great Groups include Torriorthents, Haplocalcids, and Haplocambids that occupy 35, 24, and 12 % of the MLRA, respectively.

7.5 Conclusion

The Western Range and Irrigated Region encompasses a large area that has a long and complex geologic history, combined with complex topography and climate history, that directly control the soils and ecosystems that comprise this large region of the Western USA. The main physiographic areas of this region include the Basin and Range Province that dominates the land area of this region, the Colorado Plateau Province, and the Cascade-Sierra Province, with smaller areas located in the Wyoming Basin and Columbia Plateau provinces. The region is comprised of desert and semi-desert ecosystems on plateaus, plains, basins, and isolated mountain ranges providing areas of forested habitat. The soils in this region are dominantly Aridisols, Entisols, and Mollisols and provide key ecosystem services such as watershed and groundwater recharge, carbon sequestration, and maintenance of plant and animal diversity over a large area of the Western USA.

References

Anderson K, Wells S, Graham R (2002) Pedogenesis of vesicular horizons, Cima Volcanic Field, Mojave Desert, California. Soil Sci Soc Am J 66(3):878–887

Buol SW, Southard RJ, Graham RC, McDaniel PA (2011) Soil genesis and classification. John Wiley & Sons, Inc

Capo RC, Chadwick OA (1999) Sources of strontium and in desert soil and calcrete. Earth Planet Sci Lett 170(1–2):61–72

Crouvi O, Pelletier JD, Rasmussen C (2013) Predicting the thickness and aeolian fraction of soils in upland watersheds of the Mojave Desert. Geoderma 195:94–110

Dahlgren RA, Boettinger JL, Huntington GL, Amundson RG (1997) Soil development along an elevational transect in the western Sierra Nevada, California. Geoderma 78(3–4):207–236

Davis GH, Constenius KN, Dickinson WR, Rodriguez EP, Cox LJ (2004) Fault and fault-rock characteristics associated with Cenozoic extension and core-complex evolution in the Catalina-Rincon region, southeastern Arizona. Geol Soc Am Bull 116(1–2):128–141

Dubray EA, Pallister JS (1991) An ash flow caldera in cross-section: ongoing field and geochemical studies of the Midtertiary Turkey Creek Caldera, Chiricahua Mountains, Se Arizona. J Geophys Res Solid Earth Planets 96(B8):13435–13457

Gile LH (1975a) Causes of soil boundaries in an arid region. 1. Age and parent materials. Soil Sci Soc Am J 39(2):316–323

Gile LH (1975b) Holocene soils and soil-geomorphic relations in an arid region of southern New-Mexico. Quatern Res 5(3):321–360

Gile LH, Hawley JW, Grossman RB (1981) Soils and geomorphology in the Basin and Range area of southern New-Mexico. New Mexico Bureau of Mines and Mineral Resources, Socorro

Gile LH, Peterson FF, Grossman RB (1966) Morphological and genetic sequences of carbonate accumulation in desert soils. Soil Sci 101:347–360

Goodrich DC, Faures JM, Woolhiser DA, Lane LJ, Sorooshian S (1995) Measurement and analysis of small-scale convective storm rainfall variability. J Hydrol 173(1–4):283–308

Goossens D, Buck B (2009) Dust emission by off-road driving: experiments on 17 arid soil types, Nevada, USA. Geomorphology 107(3–4):118–138

Hamerlynck EP, McAuliffe JR, McDonald EV, Smith SD (2002) Ecological responses of two Mojave Desert shrubs to soil horizon development and soil water dynamics. Ecology 83(3):768–779

Harden JW, Taylor EM, Hill C, Mark RK, McFadden LD, Reheis MC, Sowers JM, Wells SG (1991) Rates of soil development from four soil chronosequences in the southern Great-Basin. Quatern Res 35 (3):383–399

Heckman K, Rasmussen C (2011) Lithologic controls on regolith weathering and mass flux in forested ecosystems of the southwestern USA. Geoderma 164(3–4):99–111

Hendricks DM (1985) Arizona soils. University of Arizona Press, Tucson

Jenny H (1941) Factors of soil formation; a system of quantitative pedology, vol xii. McGraw-Hill book company, Inc., New York

Knuepfer PLK, McFadden LD (1990) Soils and landscape evolution. In: Proceedings of the 21st Binghamton symposium in geomorphology, held 6–7 October, 1990, vol viii. Elsevier, Amsterdam; New York, pp 197–575

Laity JE (2011) Pavements and stone mantles. In: Thomas D (ed) Arid zone geomorphology—process, form and change in drylands. Wiley, Oxford, pp 181–208

Lybrand RA, Rasmussen C (2015) Quantifying climate and landscape position controls on soil development in semiarid ecosystems. Soil Sci Soc Am J 79(1):104–116

Mayer L, Mcfadden LD, Harden JW (1988) Distribution of calcium-carbonate in desert soils: a model. Geology 16(4):303–306

McAuliffe JR (1994) Landscape evolution, soil formation, and ecological patterns and processes in Sonoran Desert Bajadas. Ecol Monogr 64(2):111–148

McAuliffe JR, Van Devender TR (1998) A 22,000-year record of vegetation change in the north-central Sonoran Desert. Palaeogeogr Palaeoclimatol Palaeoecol 141(3–4):253–275

McFadden LD (1978) Soils of the Canada del Oro Valley, southern Arizona, vol ix. Thesis (MS, Geosciences) Thesis, University of Arizona

McFadden LD, McDonald EV, Wells SG, Anderson K, Quade J, Forman SL (1998) The vesicular layer and carbonate collars of desert soils and pavements: formation, age and relation to climate change. Geomorphology 24(2–3):101–145

McFadden LD, Wells SG, Jercinovich MJ (1987) Influences of eolian and pedogenic processes on the origin and evolution of desert pavements. Geology 15(6):504–508

Metcalfe S, Say A, Black S, McCulloch R, O'Hara S (2002) Wet conditions during the last glaciation in the Chihuahuan Desert, Alta Babicora basin, Mexico. Quatern Res 57(1):91–101

Minnich RA (2007) Climate, paleoclimate, and paleovegetation, In: Barbour, M. (ed.), Terrestrial Vegetation of California, 3rd Edition, University of California Press, pp 43–70

Monger HC, Cole DR, Buck BJ, Gallegos RA (2009) Scale and the isotopic record of C-4 plants in pedogenic carbonate: from the biome to the rhizosphere. Ecology 90(6):1498–1511

Neilson RP (1987) Biotic regionalization and climatic controls in western North America. Vegetatio 70(3):135–147

Palais DG, Peacock SM (1990) Metamorphic and stratigraphic constraints on the evolution of the Santa-Catalina Mountains metamorphic core complex, Arizona. J Geophys Res Solid Earth Planets 95(B1):501–507

Parker KC (1995) Effects of complex geomorphic history on soil and vegetation patterns on and alluvial fans. J Arid Environ 30(1):19–39

Rasmussen C, Dahlgren RA, Southard RJ (2010) Basalt weathering and pedogenesis across an environmental gradient in the southern Cascade Range, California, USA. Geoderma 154(3):473–485

Rasmussen C, Matsuyama N, Dahlgren RA, Southard RJ, Brauer N (2007) Soil genesis and mineral transformation across an environmental gradient on andesitic lahar. Soil Sci Soc Am J 71(1):225–237

Reheis MC (2006) A 16-year record of eolian dust in Southern Nevada and California, USA: controls on dust generation and accumulation. J Arid Environ 67(3):487–520

Reheis MC, Goodmacher JC, Harden JW, McFadden Leslie D, Rockwell TK, Shroba RR, Sowers Janet M, Taylor EM (1995) Quaternary soils and dust deposition in southern Nevada and California. Geol Soc Am Bull 107(9):1003–1022

Shepard C, Schaap MG, Crimmins MA, van Leeuwen WJ, Rasmussen C (2015) Subsurface soil textural control of aboveground productivity in the US Desert Southwest. Geoderma Reg 4:44–54

Spaulding WG (1991a) A middle Holocene vegetation record from the Mojave Desert of North-America and its paleoclimatic significance. Quatern Res 35(3):427–437

Spaulding WG (1991b) Pluvial climatic episodes in North-America and North-Africa—types and correlation with global climate. Palaeogeogr Palaeoclimatol Palaeoecol 84(1–4):217–227

Stoffer P (2004) Desert landforms and surface processes in the Mojave National Preserve and vicinity. USGS Open File Report 2004–1007

Tchakerian VP, Lancaster N (2002) Late Quaternary arid/humid cycles in the Mojave Desert and western Great Basin of North America. Quatern Sci Rev 21(7):799–810

Trenberth KE, Dai A, Rasmussen RM, Parsons DB (2003) The changing character of precipitation. Bull Am Meteorol Soc 84 (9):1205–1217

Turk JK, Graham RC (2011) Distribution and properties of vesicular horizons in the Western United States. Soil Sci Soc Am J 75 (4):1449–1461

Wagner JDM, Cole JE, Beck JW, Patchett PJ, Henderson GM, Barnett HR (2010) Moisture variability in the southwestern United States linked to abrupt glacial climate change. Nat Geosci 3(2):110–113

Wells SG, Dohrenwend JC, McFadden LD, Turrin BD, Mahrer KD (1985) Late Cenozoic landscape evolution on lava flow surfaces of the Cima volcanic field, Mojave Desert, California. Geol Soc Am Bull 96(12):1518–1529

Whittaker RH, Buol SW, Niering WA, Havens YH (1968) A soil and vegetation pattern in Santa Catalina Mountains Arizona. Soil Sci 105(6):440–451

Whittaker RH, Niering WA (1975) Vegetation of Santa Catalina Mountains, Arizona. 5. Biomass, production, and diversity along elevation gradient. Ecology 56(4):771–790

Wood YA, Graham RC, Wells SG (2002) Surface mosaic map unit development for a desert pavement surface. J Arid Environ 52 (3):305–317

Wooden JL, Miller DM (1990) Chronological and isotopic framework for Early Proterozoic crustal evolution in the eastern Mojave Desert Region, Se California. J Geophys Res Solid Earth Planets 95 (B12):20133–20146

Young MH, McDonald EV, Caldwell TG, Benner SG, Meadows DG (2004) Hydraulic properties of a desert soil chronosequence in the Mojave desert, USA. Vadose Zone J 3(3):956–963

Soils of the Great Plains

8

Daniel Richard Hirmas and Rolfe David Mandel

8.1 General Description of the Great Plains

The Great Plains region encompasses an area of approximately 2,900,000 km^2—roughly equivalent to one-third of the land area of the USA—making it one of the largest physiographic provinces in North America (Wishart 2011). The Great Plains, as defined by Fenneman (1931), lies between the Rocky Mountains to the west, the Central Lowlands to the east, the Gulf Coastal Plain to the south, and the Canadian boreal forest to the north. The soils in the portion of the Great Plains between the US–Canada border and the Edwards Plateau of central Texas are considered in this chapter. This area includes the Northern Great Plains Spring Wheat Region (Land Resource Region, LRR, F), the Western Great Plains Range and Irrigated Region (LRR, G), and the Central Great Plains Winter Wheat and Range Region (LRR, H) (Fig. 8.1).

8.1.1 Physiography and Regional Geology

Compared to the Rocky Mountains or Appalachians, the Great Plains seems flat and somewhat featureless. Some areas in this region, such as the High Plains of the Texas Panhandle, eastern New Mexico, eastern Colorado, and western Oklahoma, Kansas, and Nebraska, are, indeed, flat. If, however, comparisons with the mountainous regions of North America are set aside, it is clear that the Great Plains region contains diverse landscapes and many prominent landforms. The Black Hills of South Dakota and the Bear Paw, Big Snowy, and Judith mountains of Montana, for example, rise 450–1200 m above the surrounding plains. Low rolling hills, such as those of the Flint Hills and Smoky Hills in eastern Kansas and the Rolling Plains of north-central Texas, also are common, as are deep valleys and canyons, such as the Canadian Breaks and Palo Duro Canyon in the Texas Panhandle and Ladder Creek valley in western Kansas.

During the early Pliocene, some 5 million years ago, the Great Plains, with the exception of the volcanic and strongly uplifted areas, was a broad and gently sloping plain that extended from the foot of the Rocky Mountains eastward almost to the Central Lowlands (Trimble 1980). The High Plains is the largest physiographic subprovince of the Great Plains and forms most of the western one-third of the region. Although the edge of the High Plains is heavily dissected, it is essentially a plateau characterized by broad reaches of flat uplands with poorly developed surface drainage. Late Quaternary landscape evolution, including soil formation, however, has greatly modified the High Plains surface (Frye 1946).

As the Rocky Mountains were slowly uplifted during the Laramide orogeny (80–35 million years ago), large volumes of rock were eroded from its slopes and transported eastward by streams. By the end of the Pliocene, the upper surface of a large sheet of sand and gravel formed a gently eastward-sloping plain extending from the eastern front of the Rockies to areas hundreds of kilometers to the east. The High Plains represents the uneroded remnants of this extensive plain, and the deposits of pre-Quaternary sand and gravel that lie below the surface comprise the Ogallala Formation (Frye et al. 1956; Ludvigson et al. 2009).

The Ogallala Formation is a major part of the High Plains aquifer system; springs are common along the High Plains escarpment and where the Ogallala is exposed in stream valleys. Given that these springs were probably reliable sources of water for game and people during the past 12,000 years, it is not surprising that archaeological sites are often located at or near them (Mandel 2006). Today, the Ogallala Aquifer is one of the world's most important sources of groundwater for irrigation. In 2000, the Ogallala yielded about 30 % of the groundwater used for irrigation in

D.R. Hirmas (✉)
University of Kansas, Lawrence, KS 66045, USA
e-mail: hirmas@ku.edu

R.D. Mandel
Kansas Geological Survey, Lawrence, KS 66047, USA
e-mail: mandel@kgs.ku.edu

L.T. West et al. (eds.), *The Soils of the USA*, World Soils Book Series,
DOI 10.1007/978-3-319-41870-4_8

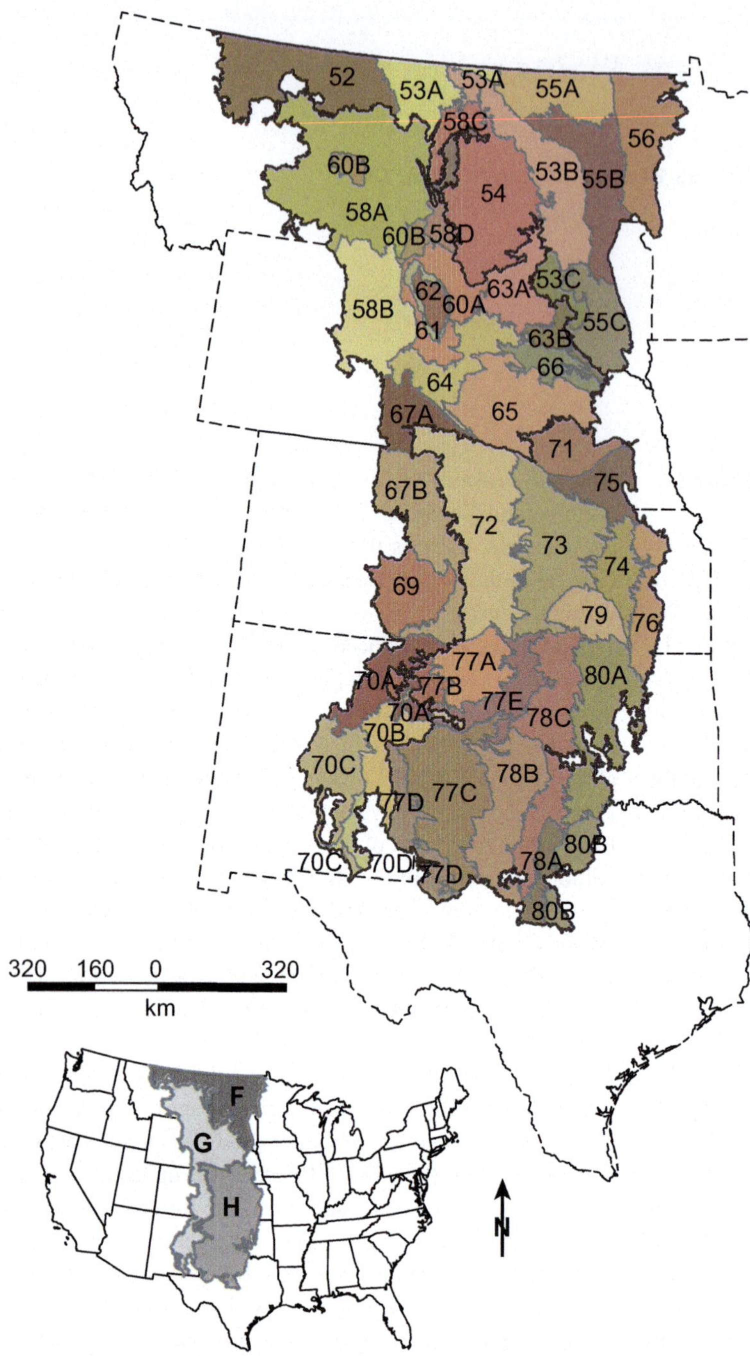

Fig. 8.1 Map of the Major Land Resource Areas (MLRAs) within Land Resource Regions (LRRs) *F*, *G*, and *H* that make up the Great Plains in the USA. Shapefiles were downloaded from the Natural Resource Conservation Service (http://www.nrcs.usda.gov/wps/portal/nrcs/detail/soils/survey/?cid=nrcs142p2_053624) and projected using the North American Albers Equal Area projection

the USA, which was approximately 23 % of total groundwater used that year in the USA (Maupin and Barber 2005).

A sheet of late-Quaternary wind-blown sediment, or *loess*, mantles much of the High Plains surface (Bettis et al. 2003). The loess directly overlies Pleistocene alluvium or the Ogallala Formation. Fertile soils with high base saturation have developed in this extensive mantle of loess making the High Plains an important part of the USA wheat belt. Sand dunes are also common on the High Plains with most of the dunes concentrated on the south side of major streams, such as the Platte, Arkansas, and Cimarron rivers.

Within the Great Plains, there are many large areas, such as the Nebraska Sand Hills, the Black Hills, and the Nebraska and South Dakota Badlands that differ greatly from the High Plains. The Nebraska Sand Hills—the largest dune field in the western hemisphere, covers 50,000–60,000 km^2 and mostly consists of stabilized, grass-covered sand dunes (Major Land Resource Area, MLRA, 65; Fig. 8.1). This dune field extends from the White River in South Dakota southward beyond the Platte River almost to the Republican River in western Nebraska and to the Loup River in central Nebraska (Bleed 1990).

The Black Hills of western South Dakota and extreme eastern Wyoming is one of the most conspicuous regions in the Great Plains. It was uplifted about 60–65 million years ago and formed an isolated dome that covers about 13,000 km^2 and stands high above the surrounding plains (Darton and Paige 1925; Carter et al. 2002). Land-surface altitudes range from 2207 m above sea level at Harney Peak to about 915 m at the foot of the dome. The geology of the Black Hills is complex, which contributes to a mosaic of different soils. Igneous rocks associated with late Cenozoic volcanism occur in the northern Black Hills (DeWitt et al. 1989). Precambrian granite, pegmatite, and metamorphic rocks comprise the core of the entire Black Hills uplift, and sedimentary rocks dating to the Paleozoic, Mesozoic, and Cenozoic form a rim around the core (MLRA 61 and 62; Figs. 8.1, 8.2; Table 8.1).

The Badlands of northwestern Nebraska and western South Dakota (MLRA 64) have a unique topography characterized by sharply eroded buttes, deep gullies, and prominent ridges. This landscape is a product of differential erosion of bedrock, including Pierre Shale, non-marine limestones, and lithified volcanic ash (Harris and Tuttle 1990). The combination of wind erosion and deep incision by streams formed the dissected terrain of the South Dakota and Nebraska Badlands.

8.1.2 Climate

Surface geology has influenced the physical and chemical properties of soils of the Great Plains, but climate has strongly controlled the pedogenic pathways and spatial

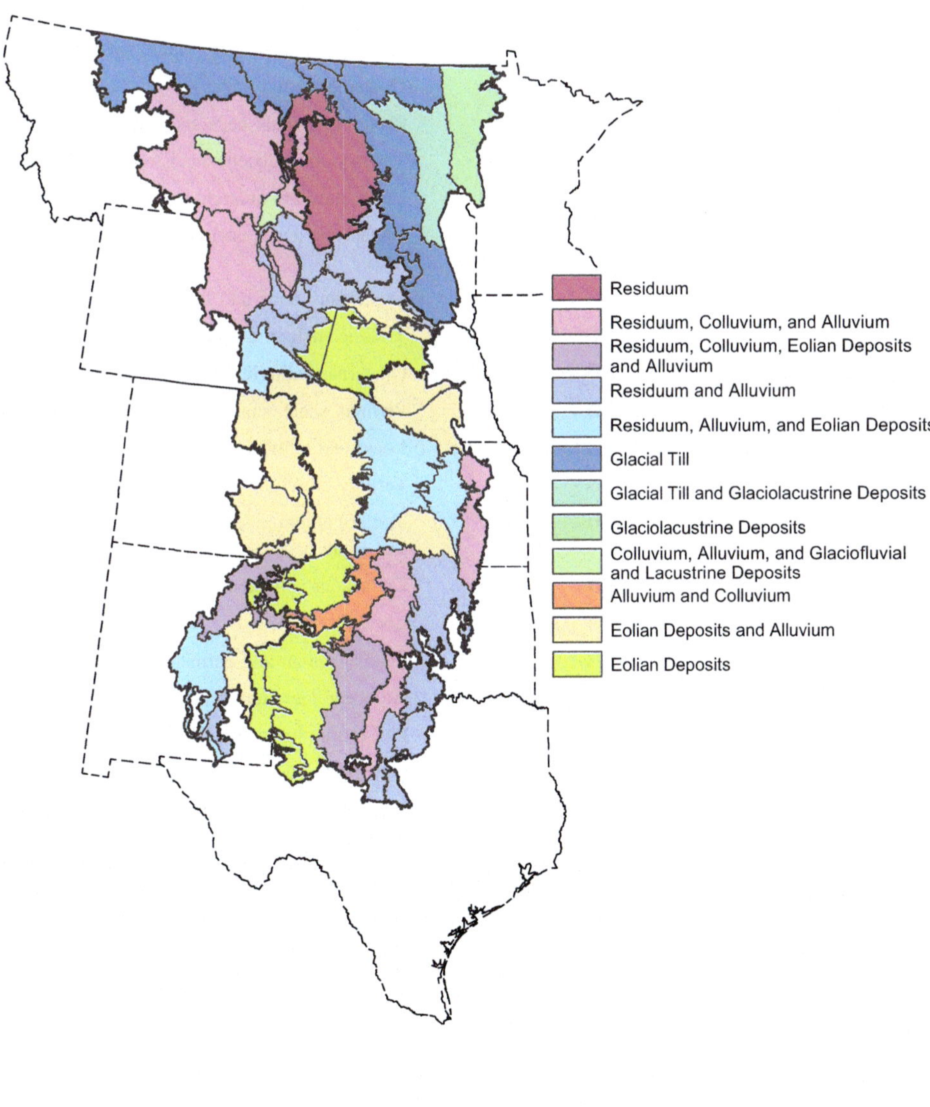

Fig. 8.2 Distribution of dominant soil parent materials across the Great Plains. Boundaries of the MLRAs are shown as *solid gray lines* and LRRs are shown as *solid black lines*. Shapefiles were downloaded from the Natural Resource Conservation Service http://www.nrcs.usda.gov/wps/portal/nrcs/detail/soils/survey/?cid=nrcs142p2_053624) and projected using the North American Albers Equal Area projection. Parent materials were determined from the dominant soil series in each MLRA as presented in Natural Resources Conservation Service (2006)

Table 8.1 Description of the dominant lithologies occurring at or near the surface within each MLRA

LRR[a]	MLRA[a]	Dominant lithology
F	52	Glacial till
F	53A	Glacial till
F	53B	Glacial till
F	53C	Glacial till
F	54	Tertiary shale, siltstone, and sandstone
F	55A	Glacial till
F	55B	Glacial till and glaciolacustrine deposits
F	55C	Glacial till
F	56	Glaciolacustrine deposits
G	58A	Tertiary and cretaceous shale, siltstone, and sandstone
G	58B	Tertiary and cretaceous shale, siltstone, and sandstone
G	58C	Tertiary sediments, shale, siltstone, and sandstone
G	58D	Cretaceous sediments, shale, siltstone, and sandstone
G	60A	Cretaceous shale
G	60B	Cretaceous sediments
G	61	Cretaceous sediments and sandstone, permian limestone and shale, and triassic shale
G	62	Granite, slate, schist, mississippian and permian limestone
G	63A	Cretaceous shale
G	63B	Cretaceous shale
G	64	Tertiary sediments, claystone, siltstone, and sandstone, loess, and alluvium
G	65	Quaternary sand dunes
G	66	Cretaceous chalk beds and limestone sediment
G	67A	Tertiary and miocene sandstone and conglomerate
G	67B	Cretaceous and quaternary sediments, sand dunes, and eolian deposits and alluvium
G	69	Cretaceous and quaternary sediments and cretaceous shale, sandstone, and chalk beds
G	70A	Cretaceous shale and limestone and tertiary basalt and other volcanic rocks
G	70B	Triassic shale and sandstone
G	70C	Quaternary alluvium and sand dunes and permian shale, sandstone, and dolomite
G	70D	Permian shale, sandstone, limestone, and dolomite
H	71	Loess and alluvium
H	72	Tertiary alluvium, claystone, and sandstone, sand dunes
H	73	Tertiary alluvium, claystone, and sandstone, sand dunes, loess, and cretaceous shale and chalk beds
H	74	Cretaceous sandstone and loess
H	75	Loess
H	76	Pennsylvanian and permian shale and limestone
H	77A	Holocene loess and sand dunes
H	77B	Pleistocene loess and Holocene Eolian deposits
H	77C	Pleistocene loess
H	77D	Pleistocene loess and quaternary eolian deposits
H	77E	Tertiary sediments
H	78A	Permian shale and limestone
H	78B	Quaternary alluvium, triassic sediments, sandstone, shale, clay, and conglomerate, permian shale, sandstone, gypsum, and dolomite
H	78C	Permian shale, siltstone, sandstone

(continued)

Table 8.1 (continued)

LRR[a]	MLRA[a]	Dominant lithology
H	79	Late Pleistocene and Holocene eolian deposits and alluvium and sand dunes
H	80A	Pleistocene alluvium, permian sandstone and shale
H	80B	Pennsylvanian shale and limestone and cretaceous sandstone

[a]*LRR* Land Resource Region, *MLRA* Major Land Resource Area
Data taken from Natural Resources Conservation Service (2006)

patterns of soil types. The climate of the Great Plains is continental, which is characterized by hot summers, cold winters, and considerable variation in precipitation and temperature (Fig. 8.3). A distinct east-to-west precipitation gradient exists, with mean annual precipitation ranging from about 840 mm at the eastern edge of the region to less than 250 mm along the western edge. Mean annual temperature decreases from south to north, ranging from between 17 and 19 °C in west-central Texas to between 2 and 5 °C along the US–Canada border (Fig. 8.3). However, the topography causes anomalies within these precipitation and temperature gradients. For example, the mountains of the Black Hills in western South Dakota have greater amounts of precipitation and lower temperatures than the surrounding plains.

The precipitation and temperature gradients described above account for broad patterns of soil types that occur across the Great Plains (Fig. 8.4). For example, Mollisols and Alfisols dominate the more mesic eastern Plains, and Aridisols, Inceptisols, and Entisols are common in the arid and semiarid western Plains. Both Mollisols and Aridisols occur in the driest parts of the Great Plains with frigid temperatures, which includes an area stretching from western South Dakota to north-central Montana and an isolated tract in Colorado—the Black Forest. In general, as the climate gets progressively drier from east to west across the Great Plains, the depth to secondary carbonates decreases because there is less effective moisture to leach carbonates deeper in the soil profile.

The Great Plains receives approximately 75 % of its precipitation from April through September, largely as a result of frontal activity. The collision of pacific and polar air masses often produces intense rainfalls of short duration along the zone of convergence. Periodic intensification of westerly (zonal) airflow, however, prevents moist Gulf air from penetrating the Plains. This condition and the development of strong anticyclonic (high-pressure) activity in the upper atmosphere over the midcontinent tend to promote drought in the region (Borchert 1950; Bryson and Hare 1974; Namias 1982, 1983; COHMAP 1988; Laird et al. 1996; Smith and Hollander 1999).

Severe droughts have afflicted the Great Plains roughly every 20 years causing dramatic changes in the composition of grassland communities and significant (>75 %) losses of vegetative cover (Albertson and Weaver 1942; Albertson and Tomanek 1965; Tomanek and Hulett 1970; Borchert 1971; Frison 1978). Such droughts contribute to severe wind-driven soil erosion; this was observed across most of the Great Plains during the 1930s 'Dust Bowl.'

8.1.3 Vegetation

The Great Plains are within the Interior Grasslands region of North America, an area with distinct east–west- and north–south-trending grassland communities that are closely related to climate (Küchler 1964). The increase in elevation and the decrease in mean annual precipitation from east to west influence the composition and overall appearance of these communities. Short-grass prairie and bunchgrass steppes extend eastward from the foot of the Rocky Mountains in Colorado and Wyoming into Kansas, Nebraska, and South Dakota. These grasslands are dominated by blue grama (*Bouteloua gracilis*) and buffalo grass (*Bouteloua dactyloides*), but include larger plants such as yucca (*Yucca* spp.) and prickly pear cactus (*Opuntia* spp.), as well as woody shrubs such as sagebrush (*Artemisia tridentate*).

Mixed-grass prairies dominated by big bluestem (*Andropogon gerardi*), little bluestem (*Schizachyrium scoparium*), blue grama (*Bouteloua gracilis*), and sideoats grama (*Bouteloua curtipendula*) gradually replace short grasses along the eastern edge of the High Plains. Where the soils are sandy, sandsage-bluestem prairies dominate. Sand prairies include bluestem (*Andropogon gerardi* and *Schizachyrium scoparium*), sandreed (*Calamovilfa longifolia*), and switchgrass (*Panicum virgatum*), and these cover the Sand Hills on the north side of the Platte River in Nebraska and the Great Bend prairie on the south side of the Arkansas River in Kansas.

Mixed-grass prairies is replaced by tall-grass prairies in the wetter eastern portion of the Great Plains. The open tall-grass prairie is dominated by bluestem (*Andropogon gerardi* and *Schizachyrium scoparium*), switchgrass (*Panicum virgatum*), and Indian grass (*Sorghastrum nutans*). Trees are more common along the eastern fringe of the Great Plains, with oak (*Quercus* spp.), hickory (*Carya* spp.), elm (*Ulmus* spp.), sugar maple (*Acer saccharum*), and black

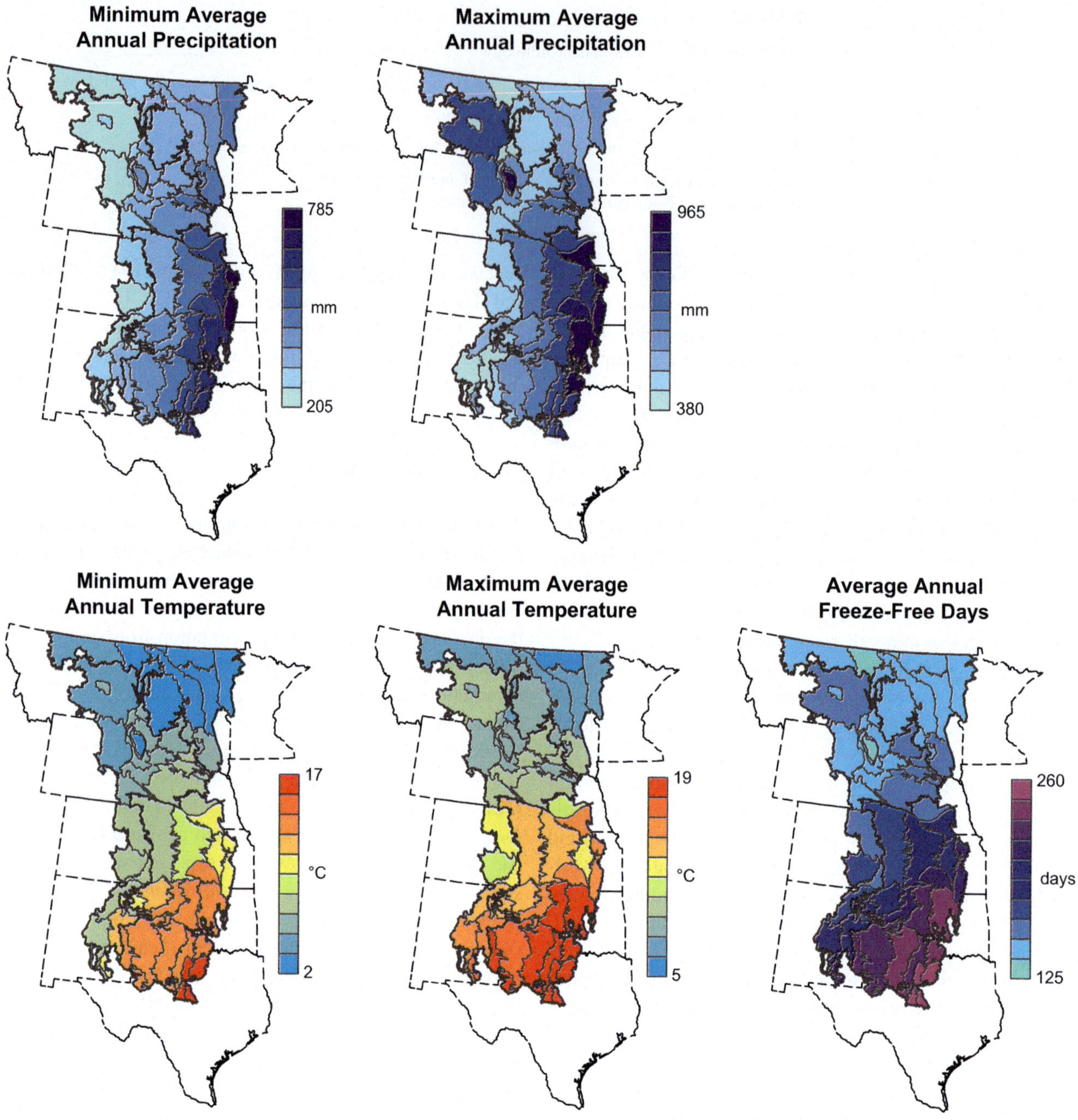

Fig. 8.3 Distribution of climate properties across the Great Plains reported in Natural Resources Conservation Service (2006). Boundaries of the MLRAs are shown as *solid gray lines* and LRRs are shown as *solid black lines*. Shapefiles were downloaded from the Natural Resource Conservation Service (http://www.nrcs.usda.gov/wps/portal/nrcs/detail/soils/survey/?cid=nrcs142p2_053624) and projected using the North American Albers Equal Area projection

walnut (*Juglans nigra*) dominating wooded areas on the steep slopes of stream valley. Also, riparian forests occur in narrow bands along major streams throughout the Great Plains. Hackberry (*Celtis occidentalis*), cottonwood (*Populus deltoides*), willow (*Salix* spp.), and American elm (*Ulmus Americana*) dominate these riparian woodlands.

In the Northern Great Plains, including areas of North Dakota, South Dakota, and Montana, cool-season grasses such as fescue (*Festuca* spp.), western wheatgrass (*Pascopyrum smithii*), and needlegrass (*Stipa* spp.), dominate the prairies. Also, wetlands are common in this region, and many have saline soils that support halophytic graminoids such as alkali bulrush (*Scirpus maritimus*), inland salt grass (*Distichlis spicata*), and Nuttall's alkali grass (*Puccinellia nuttalliana*), and shrubs such as black greasewood (*Sarcobatus vermiculatus*).

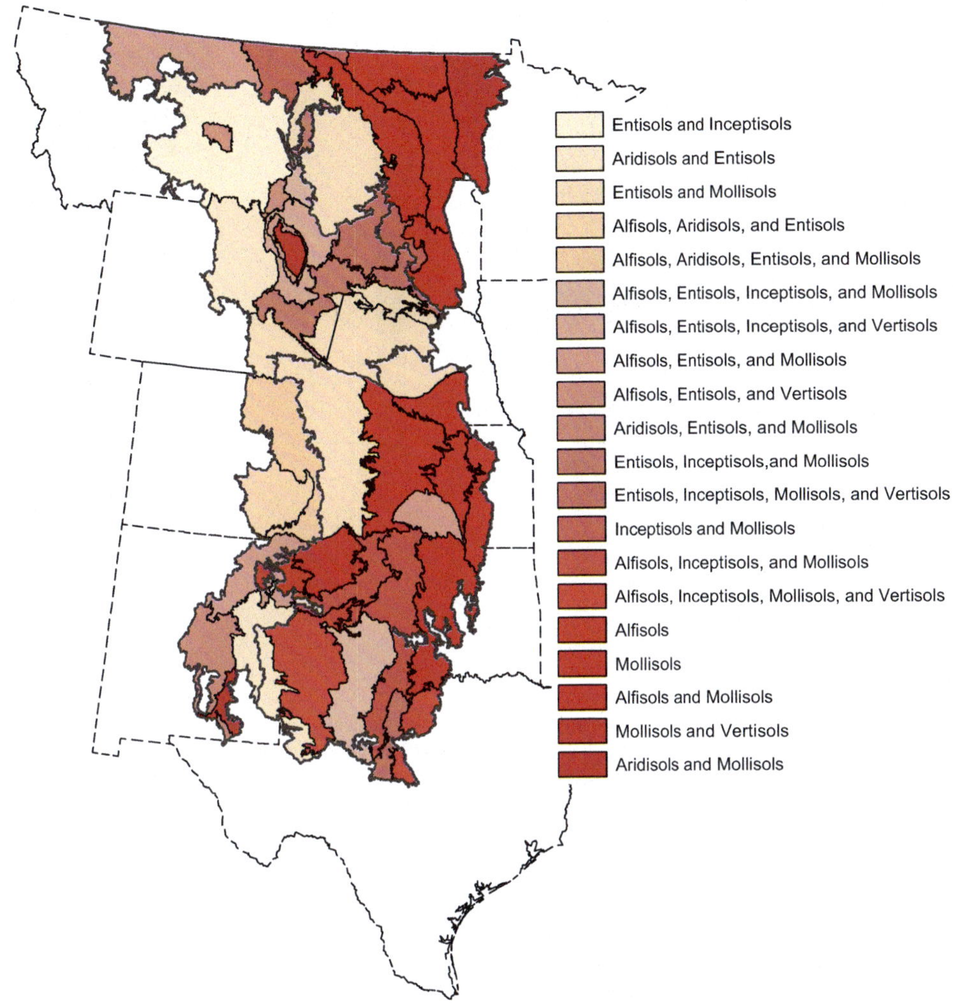

Fig. 8.4 Distribution of dominant soil orders across the Great Plains. Boundaries of the MLRAs are shown as *solid gray lines* and LRRs are shown as *solid black lines*. Shapefiles were downloaded from the Natural Resource Conservation Service (http://www.nrcs.usda.gov/wps/portal/nrcs/detail/soils/survey/?cid=nrcs142p2_053624) and projected using the North American Albers Equal Area projection. Dominant soil orders in each MLRA are presented in Natural Resources Conservation Service (2006)

Over the past 150 years, the grasslands of the Great Plains have been greatly modified by humans in general and agriculture in particular (Fig. 8.5). The short-grass prairie of the western Plains, the mixed-grass prairie of the central Plains, and tall-grass prairie of the eastern Plains largely correspond to the western rangelands, the wheat belt, and the corn and soybean region, respectively. Groundwater-supported irrigation has allowed crop production, including corn, to expand westward into the semiarid and dry-subhumid portions of the High Plains. Currently, more than 70 % of the land area in the Great Plains is used for agriculture (Karl et al. 2009). The organic-rich Mollisols associated with the prairies are especially suitable for a range of crops; hence, millions of hectares of natural grassland have been converted to cropland. In the eastern Plains, the once vast tall-grass prairie survives only in areas unsuited to plowing, such as the rocky landscape of the Kansas Flint Hills. Also, grasslands persist in areas of the central and western Plains with very sandy soils that are too droughty for crop cultivation such as the Nebraska Sand Hills and the Cimarron Grasslands of southwestern Kansas.

8.2 Soils Developed in Glacial Drift

8.2.1 General Characteristics of Glacial Drift in the Great Plains

At various times during the Pleistocene, the Laurentide ice sheet covered portions of the Northern and Central Great Plains, including the northern third of Montana, all but the southwest corner of North Dakota, most of the eastern two-thirds of South Dakota, most of the eastern quarter of Nebraska, and a portion of the northeastern quarter of Kansas. The areas of Nebraska and Kansas that were directly affected by the Laurentide ice sheet experienced glaciation

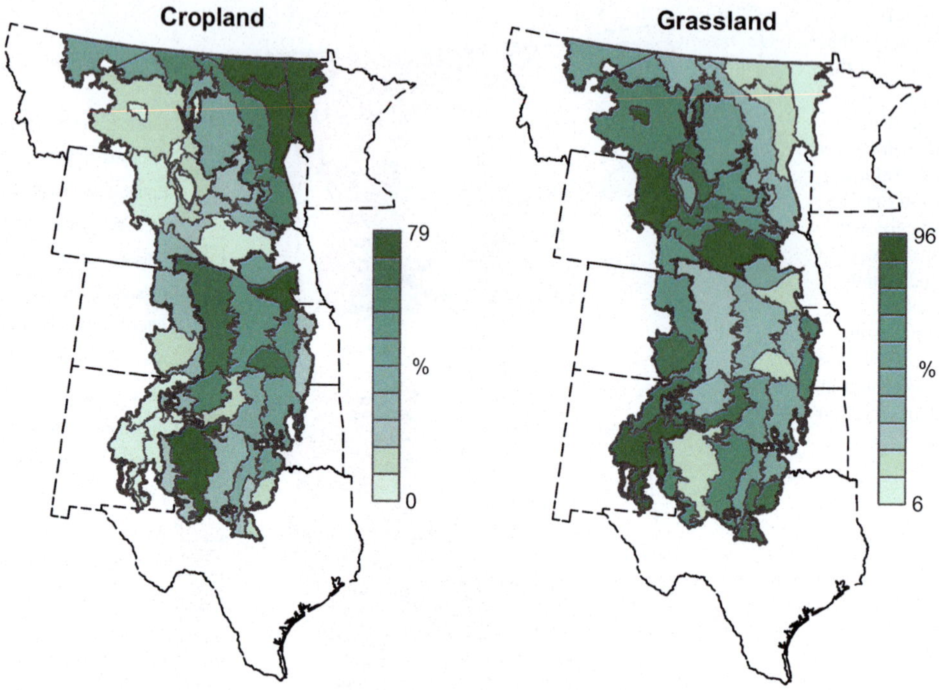

Fig. 8.5 Percentages each MLRA that are covered by cropland and grassland. Forest, urban, and other land uses make up only a minor fraction in most of the Great Plains except in the Black Hills (MLRA 62) and associated foothills (MLRA 61) where forests make up 47 and 24 % of the area, respectively. Boundaries of the MLRAs are shown as *solid gray lines* and LRRs are shown as *solid black lines*. Shapefiles were downloaded from the Natural Resource Conservation Service (http://www.nrcs.usda.gov/wps/portal/nrcs/detail/soils/survey/?cid=nrcs142p2_053624) and projected using the North American Albers Equal Area projection. Land use data are taken from Natural Resources Conservation Service (2006)

before the Illinoian Glacial Stage (~191,000–130,000 years ago), but not during subsequent glacial episodes of the Pleistocene. Other glaciated areas of the Great Plains were directly affected by continental ice during the last glacial episode—the Wisconsin Stage (~85,000–11,000 years ago). The disparity between the age of the Pre-Illinoian and Wisconsinan drift is reflected in the soils developed in these glacial deposits. The age of these deposits is important for understanding soils in glaciated landscapes of the Great Plains.

8.2.2 Age of Pleistocene Glaciated Landscapes

Pleistocene landscapes in northeastern Kansas and eastern Nebraska are products of a host of glacial, periglacial, and interglacial geomorphic processes. Paleosols within these deposits mark episodes of landscape stability, and erosion surfaces indicate former episodes of landscape instability. In some areas, the drift is exposed at the surface, but in many areas it is mantled by late Quaternary loess. The advance of continental glaciers into the region during the early Pleistocene left a complex stratigraphic record that indicates that the glaciers covered portions of the landscape on more than two occasions during the Pleistocene (Dort 1985). There were at least seven glacial advances across the nearby Till Plain of southern Iowa (Boellstorff 1978a, b; Hallberg and Boellstorff 1978; Hallberg 1986).

The Pleistocene stratigraphy of the dissected till plains beyond the Wisconsin and Illinoian glacial limits is based on a framework of Pre-Illinoian glacial tills and intercalated volcanic ashes, and younger loesses. These deposits are regional in extent and, thus, provide references to which more localized fluvial and colluvial units can be stratigraphically related. The available evidence, including paleomagnetic, radiometric, and biostratigraphic data, constrains the age of Pre-Illinoian drift (till, outwash, and glaciolacustrine deposits) in the Great Plains to the period about between 0.62 and 0.78 million years ago. Hence, surface soils developed in the drift are ancient relict paleosols that probably experienced multiple cycles of erosion and formation (i.e., they are *polygenetic*).

By contrast, the Wisconsinan drift and associated surface soils that occur on the Northern Plains are relatively young. The Laurentide Ice Sheet reached its maximum southern extent around 21,000 years ago, barely crossing the Missouri River near Vermillion, South Dakota. By ~13,000 years ago, the ice sheet had retreated across the US–Canada border, leaving behind a thick mantle of glacial drift on portions of the Northern Plains. The drift is time-transgressive, decreasing in age from south to north.

Two types of drift and associated soils are discussed in this section: glacial till and galcio lacustrine deposits. Glaciofluvial deposits and associated soils are described in Sect. 8.4.

8.2.3 Soils Developed in Glacial Till

Glacial till is unsorted and unstratified drift, generally unconsolidated and deposited directly by a glacier without subsequent reworking by meltwater; it consists of a heterogeneous mixture of clay, silt, sand, pebbles, cobbles, and boulders (Natural Resources Conservation Service 2014a). In northeastern Kansas and eastern Nebraska (northern and eastern parts of MLRA 75 and 76), soils developed on the dissected Pre-Illinoian till plain are either calcareous or noncalcareous. The calcareous soils are Argiustolls with thick A–Bt–Bk or A–Bt–Btk–Bk profiles, matrix hues of 10YR or 2.5Y, and clay loam or clay textures. Texture of the Bt horizon (argillic horizon) ranges from 20 to 50 % clay, and soils with clayey Bt horizons have shrink-swell potential. Bt horizons contain secondary carbonates although the upper part of the horizon may be noncalcareous. Redoximorphic features often occur in the argillic horizon, which may indicate a previously wetter soil hydrology. Two of the most common calcareous till-derived soil series in the region are the Burchard and Pawnee series (Fig. 8.6).

The noncalcareous till-derived soils are typically rubified (reddened), with matrix hues of 5YR or 7.5YR, and are formed in a *diamicton*—sediment that contains pebbles, cobbles, and boulders embedded in a fine-grained matrix. The coarse fraction of the till is dominated by granite,

Fig. 8.6 Examples of several soils developed in glacial till: Burchard series (Typic Argiudolls), Pawnee series (Oxyaquic Vertic Argiudolls), Morrill series (Typic Argiudolls), Williams series (Typic Argiustolls), Elloam series (Aridic Natrustalfs), Hamerly series (Aeric Calciaquolls), Houdek series (Typic Argiustolls), and Scobey series (Aridic Argiustolls). Photograph of the Morrill, Williams, and Houdek series courtesy of the Natural Resources Conservation Service. All other photographs were reproduced from Soils of the Great Plains: Land Use, Crops, and Grasses by Andrew R. Aandahl with the permission of the University of Nebraska Press. Copyright 1980 by the University of Nebraska Press

diorite, and Sioux quartzite. Fragments of weathered limestone also are common, but they generally comprise less than 50 % of the >2 mm sediment fraction.

Soils developed over noncalcareous till commonly have mollic epipedons and argillic horizons and are Argiudolls. The argillic horizon typically has redder hues than soils over calcareous tills and commonly has relict Fe concentrations (redox features) in the lower part of the horizon. Karlstrom (1994) noted that some of the rubified, noncalcareous till-derived soils in northeastern Kansas have morphological and genetic properties resembling those of Paleudalfs and Paleustalfs of subtropical to tropical climates. Although it is likely that the Morrill series and similar deeply weathered soils developed in Pre-Illinoian till are products of a wetter and warmer climate than exists today in the region, the long period of soil development played a role in shaping these soils. The Morrill series (Typic Argiudolls) is one of the most common noncalcareous till-derived soils in the region (Fig. 8.6).

The till-derived soils on the Northern Plains are much younger and, therefore, not as thick or well developed as soils formed on the dissected Pre-Illinoian till plain of eastern Nebraska and northeastern Kansas. Many of these Northern Plains soils are Calciustepts and Calciustolls that lack Bt horizons and have subsoil accumulations of carbonates (calcic horizons). Some of the till-derived soils in the Dakotas and northern Montana, however, do have well-developed profiles with Bt and Btk horizons. The Williams series is an example of these Argiustolls (Fig. 8.6) that has common carbonate masses in the argillic horizon. This soil formed in matrix-supported calcareous till on glacial till plains and moraines in north-central South Dakota, central and northwestern North Dakota, and northeastern Montana.

As with other soils in the Northern Plains, till-derived soils transition from Udolls to Ustolls westward across the region in response to decreasing mean annual precipitation. Also, secondary carbonate accumulation increases and the depth to carbonate decreases in the till-derived soils along the westward transect. The Hamerly series (Aeric Calciaquolls), for example, occurs on till plains in north-central and northeastern Montana and has secondary carbonate accumulation as shallow as 20 cm below the surface (Fig. 8.6). In addition, soils such as the Elloam series (Aridic Natrustalfs) in the most arid areas of the Northern Plains, especially northern Montana, tend to have accumulations of gypsum and other salts at shallow depths (Fig. 8.6). Elloam soils are mostly used as dryland farmland and as rangeland, and cultivated areas of the Hamerly soil are used for growing small grains, flax, and hay.

Some of the till-derived Mollisols in the Northern Plains are among the most productive agricultural soils in the world. On level to gently rolling areas, the Williams soil (Fig. 8.6) is used to grow wheat, barley, oats, flax, and sunflowers, whereas the rolling and hilly areas mantled by the Williams soil are used as rangeland. The Houdek series (Typic Argiustolls) formed in till has a thick mollic epipedon with 2–4 % organic matter (Fig. 8.6). This soil is fertile and is used to produce wheat, barley, oats, corn, alfalfa, and feed grains. The Scobey series (Aridic Argiustolls) consists of very deep, well-drained soils on till plains, hills, and moraines in the north-central part of Montana (Fig. 8.6). These soils occur on more than 280,000 ha and are among the most productive soils in Montana's 'Golden Triangle,' an area formed between the cities of Great Falls, Conrad, and Havre, Montana, known for its ideal climatic conditions for growing high-quality wheat.

8.2.4 Soils Developed in Glaciolacustrine Deposits

Glaciolacustrine deposits consist of sediments that were deposited by glacial meltwater in lakes. In the Northern Plains, these lakes include ice margin lakes formed by the damming action of a moraine or ice dam during the retreat of the melting Laurentide ice sheet, or by meltwater trapped against the ice sheet due to isostatic depression of the crust around the ice. Also, glacial erosion of bedrock and till created large depressions that collected meltwater and became lakes. Sediments in the bedload (mostly sands and gravels) and suspended load (mostly silts and clays) of meltwater streams were carried into the glacial lakes and deposited. Such lakes gradually evaporated during the warming period that followed the Ice Age, exposing thick glaciolacustrine deposits to subaerial weathering and soil formation.

In the Great Plains, glaciolacustrine deposits are exposed only in the Northern Plains; such deposits are deeply buried beneath late Pleistocene loess or Pre-Illinoian till in eastern Nebraska and northeastern Kansas. Some soils developed in Wisconsinan glaciolacustrine deposits have coarse textures and are formed in sands and gravels deposited along the margins of glacial lakes (e.g., the apex of deltas), while soils with fine textures are formed in suspended load that was deposited distal to the shorelines of lakes. For example, the Ulen series (Aeric Calciaquolls) formed in sandy glaciolacustrine deposits on the western shore of the glacial Lake Agassiz basin in eastern North Dakota. The Hamar series (Typic Endoaquolls) is associated with sandy shoreline facies of former glacial lakes in northeastern South Dakota and eastern North Dakota, and the Lohnes series (Entic Hapludolls) formed in coarse and medium sands on glacial lake plains in north-central and eastern North Dakota.

Typic (e.g., Tansem series) and Pachic Haplustolls (e.g., Roseglen and Makoti series) are associated with fine-grained glaciolacustrine deposits. These soils are formed in silty glaciolacustrine deposits on glacial lake plains in northwestern and west-central North Dakota and north-central South Dakota and have moderately developed A–Bw–Bk–C profiles with mostly silt loam or loam texture.

Most soils developed in the sandy glaciolacustrine deposits tend to be poorly drained because of perched water tables. Where drainage is problematic, the soils are used for pasture and range. Otherwise, they are mostly cropped to small grains, and in some places, potatoes, corn, and hay. By contrast, soils developed in fine-grained glaciolacustrine deposits tend to be well drained and are primarily used to produce small grains, flax, and hay.

8.3 Soils Developed in Eolian Sediments

Eolian sediments on the Great Plains take the form of extremely thick (>50 m) to very thin deposits of silt-size loess mantling large areas of southwestern North Dakota, central South Dakota, southern Nebraska, northern Kansas, and the Texas panhandle. They also include thick and extensive dune fields and sand sheets occurring from North Dakota to the Southern High Plains in Texas and New Mexico (Figs. 8.7, 8.8). The character, origin, and depositional timings of these eolian sediments vary. Loess in the northern portions of the Central Great Plains (LRR H) occurs on interfluves and Pleistocene terraces and can be distin-

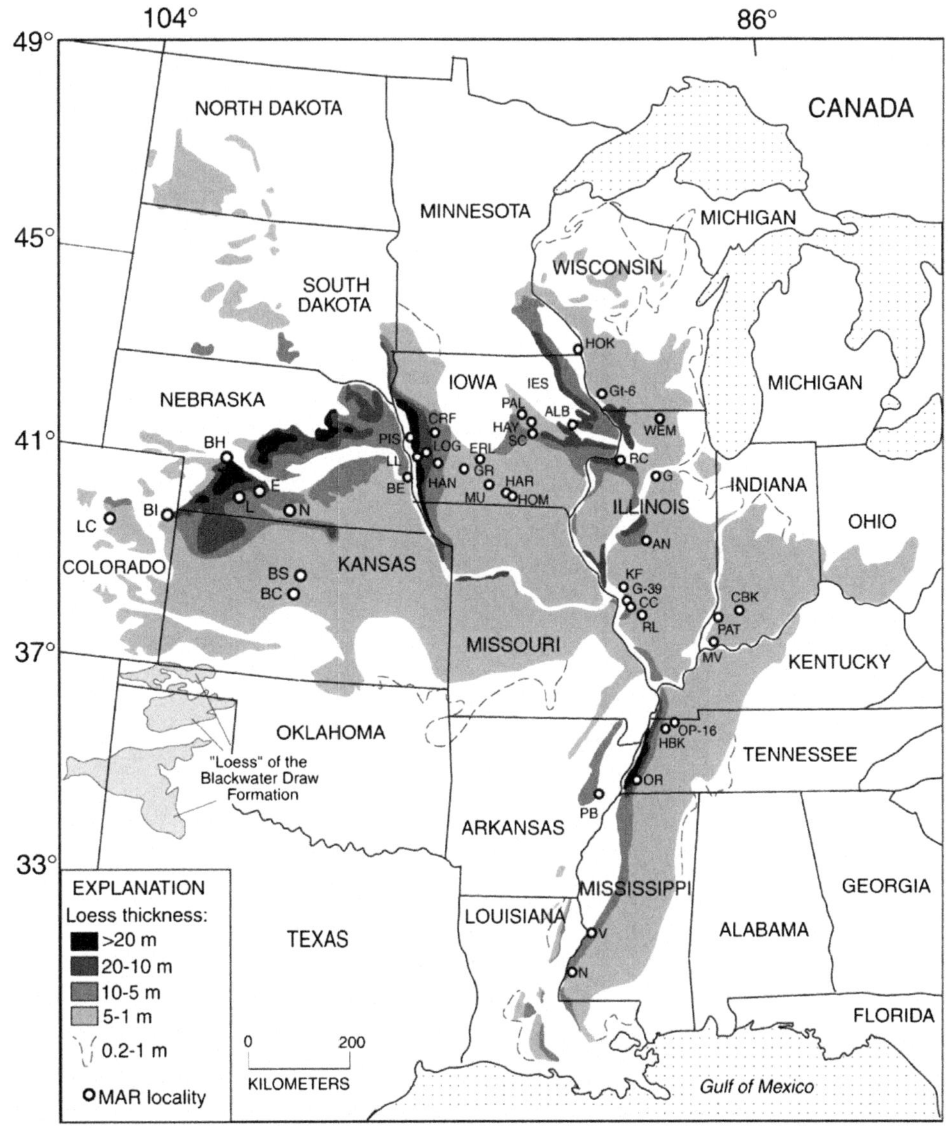

Fig. 8.7 Distribution and thickness of loess across the Great Plains and Central Lowlands. After Bettis et al. (2003) with permission

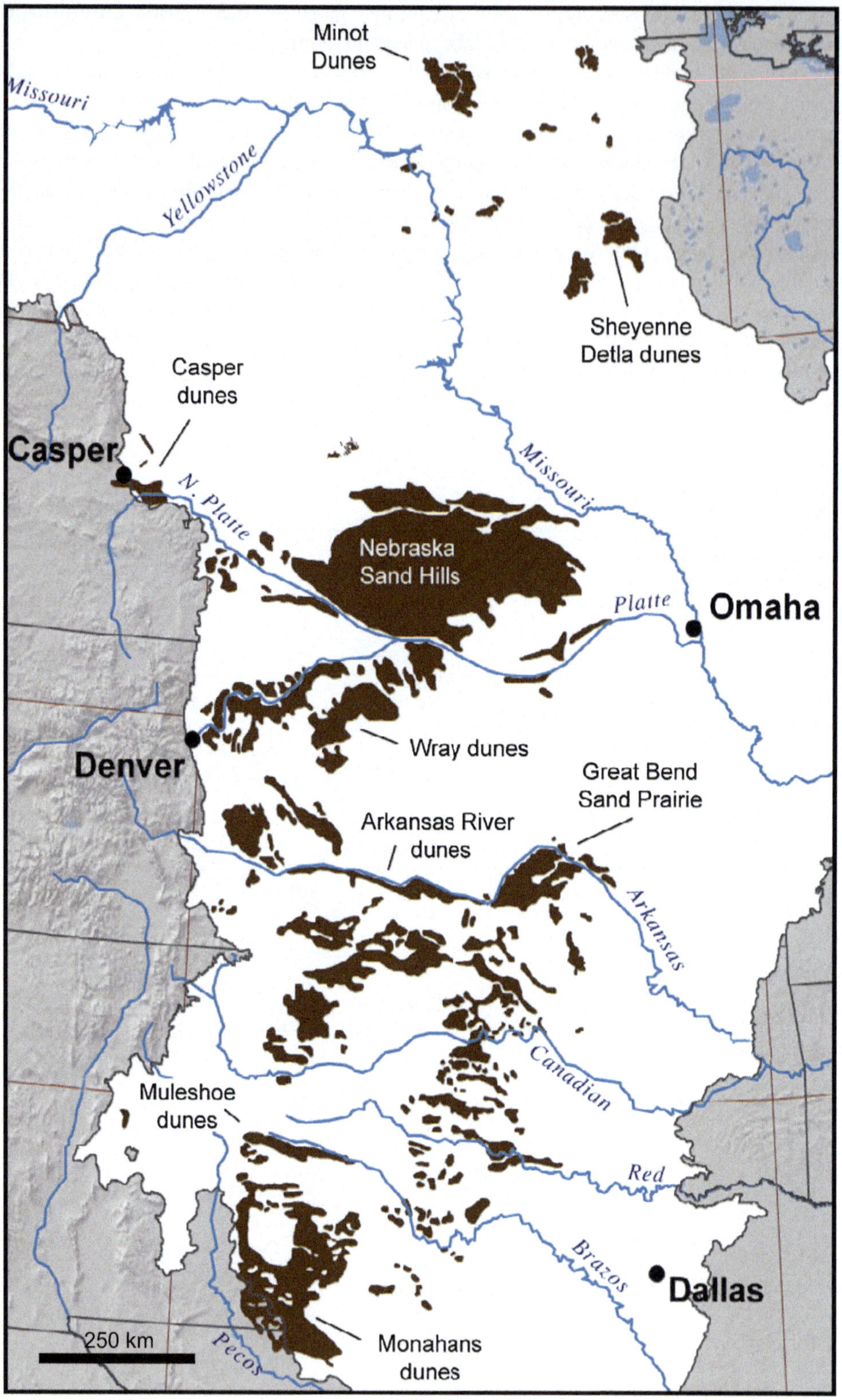

Fig. 8.8 Distribution of sand dunes across the Great Plains. Modified from Halfen and Johnson (2013)

guished as at least four stratigraphically superposed loesses —Loveland, Gilman Canyon, Peoria, and Bignell—ranging in age from Holocene to Illinoian (Johnson et al. 2007). Although the combined loess deposits can exceed 50 m (e.g., Fig. 8.9), loess thicknesses are generally less than 6 m, and in many areas, the Loveland and Gilman Canyon loesses have been eroded from the uplands leaving only the Peoria Loess in these landscape positions. However, in western

Fig. 8.9 Hay Canyon system in the Central High Tableland (MLRA 72) in northwestern Kansas showing the thick sequences of Quaternary loess in this area. Note the truck in the upper center of the photograph for scale. Photograph courtesy of A. N. Koop

South Dakota and North Dakota within the Northern Great Plains (LRR F), loess deposits mantling interfluves and Pleistocene terraces are mostly of Holocene age. These deposits comprise the lithostratigraphic unit known as the Oahe Formation (Clayton 1972). Eolian sand deposits tend to be concentrated in the Western Great Plains (LRR G) as well as the central, southern, and eastern Central Great Plains (LRR F) (Fig. 8.8). The most impressive dune field in the Great Plains and the largest in North America is the Nebraska Sand Hills (MLRA 65) (Figs. 8.1, 8.8). Dunes in the Sand Hills are typically 10–20 m high with some exceeding 100 m in height (Bleed 1990). The fragile sandy soils on the dunes are very susceptible to erosion, and dune activation may occur when drought, fire, overgrazing, or cultivation reduce the protective plant cover. This dune field was active at least 17,000–14,000 years ago if not earlier and possibly as recently as the Medieval Warm Period—a warm, dry climatic episode that occurred 1000–750 years ago (Mason et al. 2004, 2011; Miao et al. 2007; Halfen and Johnson 2013). Most of the dune fields in the Great Plains are reported to have been active during the latest Pleistocene and early Holocene at the earliest (Halfen and Johnson 2013).

8.3.1 Soils Developed in Loess

Soils formed in glacially derived loess deposits, especially in the northern Central Great Plains (e.g., MLRAs 72–75), occur along a steep precipitation gradient (Fig. 8.3) and have properties that strongly correlate with climate. For instance, Aridic Ustorthents (e.g., the Colby series), which occur in the Central High Tableland (MLRA 72) of eastern Colorado and western Kansas and receive approximately 450 mm of annual rainfall, are weakly developed (A–AC–C profiles) with relatively thin A horizons ($\sim$10 cm; Fig. 8.10). By contrast, Typic Argiustolls (e.g., the Holdrege series), occur further to the east (MLRA 73) and receive approximately 580 mm of annual rainfall, have thick (>30 cm) A horizons and well-developed argillic (i.e., clay-enriched) horizons (Fig. 8.10). Figure 8.11 illustrates the correlations that exist between organic carbon and average annual precipitation for a transect across the northern Central Plains. Soils derived in loess are often calcareous although the depth of the carbonate-free zone at the surface of the profile tends to be a function of rainfall due to leaching of the carbonates. These soils tend to be fertile owing to their high base saturation and silt loam to silty clay loam textures. They are mostly used for rangeland in the west and cultivated and irrigated for sorghum and corn toward the east.

In the southwestern portion of the Central Great Plains, the south part of the Southern High Plains (MLRA 77C)—also known as the *Llano Estacado*—is mantled with Pleistocene Blackwater Draw loess derived from alluvial sediments in the Trans-Pecos region of Texas and New Mexico. This mantle displays a fining gradient in the particle-size distribution of the sediment from southwest to northeast across the plateau following prevailing winds. Soils formed

Fig. 8.10 Examples of several soils developed in loess. Holdrege series (Typic Argiustolls) developed in calcareous loess and are extensive in south-central Nebraska and north-central Kansas (MLRA 71, 73, and 75). Colby series (Aridic Ustorthents) occur on plains and hillslopes of the Central High Tableland (MLRA 72). Amarillo series (Aridic Paleustalfs) occur on nearly level and gently sloping plains and playa slopes of the Llano Estacado (Southern High Plains, Southern Part—MLRA 77C) and have developed in Blackwater Draw loess derived from the Trans-Pecos region of New Mexico and Texas. Reproduced from Soils of the Great Plains: Land Use, Crops, and Grasses by Andrew R. Aandahl with the permission of the University of Nebraska Press. Copyright 1980 by the University of Nebraska Press

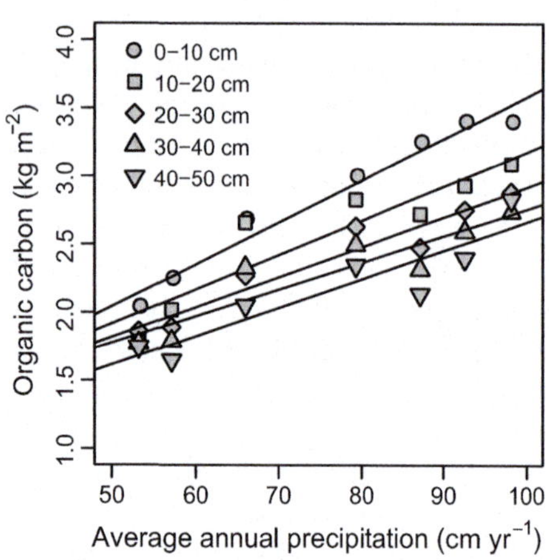

Fig. 8.11 Total OC as a function of average annual precipitation for 10-cm depth intervals of soils sampled along a transect across northern Kansas. Samples were taken from undisturbed pioneer cemeteries to minimize confounding effects from land use. After Klopfenstein et al. (2015)

in loess of the Blackwater Draw are commonly Mollisols and Alfisols and contain thick argillic and calcic horizons. Common crops grown are cotton, grain sorghum, and wheat. An excellent example of an extensive soil in this region is the Amarillo series (Aridic Paleustalfs) shown in Fig. 8.10.

8.3.2 Soils of Sand Dunes and Sheets

Soils developed in thick sand sheets and dune fields are commonly classified as Entisols (usually Ustipsamments) with A–AC–C profiles (e.g., Valentine series) or A–C profiles (e.g., Tivoli series). These soils have high, saturated hydraulic conductivities and excessive drainage classes due to the sandy textures throughout the profiles. Clay lamellae, which are relatively thin, horizontal but wavy accumulations of clay within a sandy substrate, can form in these soils from rapid translocation and concentration of clay at deeper depths through a process of *argilluviation* (Bockheim and Hartemink 2013); these lamellae can vertically coalesce to form well-developed argillic horizons over time (Birkeland 1999). The Brownfield series (Arenic Aridic Paleustalfs—MLRA 77C) is an example of a soil developed on eolian sand with a thick, red, iron oxide-rich and highly structured argillic horizon (Fig. 8.12). These soils are used for growing crops and livestock grazing.

8.3.3 Soils of Lunettes

Occurring on the lee-side of some upland depressions (especially playas and salinas described below) on the High Plains (i.e., MLRAs 67, 77, and 72) are isolated, often

Fig. 8.12 Examples of several soils developed in eolian sand. Valentine series (Typic Ustipsamments) occur on interdunes and dunes within the Nebraska Sand Hills (MLRA 65). Brownfield series (Arenic Aridic Paleustalfs) occur on nearly level to gently sloping plains of the Llano Estacado (Southern High Plains—MLRA 77C). Reproduced from Soils of the Great Plains: Land Use, Crops, and Grasses by Andrew R. Aandahl with the permission of the University of Nebraska Press. Copyright 1980 by the University of Nebraska Press

clay-rich eolian dunes, or *lunettes*, with sediment derived, in part, from the associated basin (Frederick 1998; Bowen and Johnson 2012; Rich 2013). Soils in these settings are Calciustepts (e.g., Drake series) in the Southern High Plains (Fig. 8.13) and Haplustolls (e.g., Ulysses series) in the Central and Northern High Plains. Soils and buried paleosols

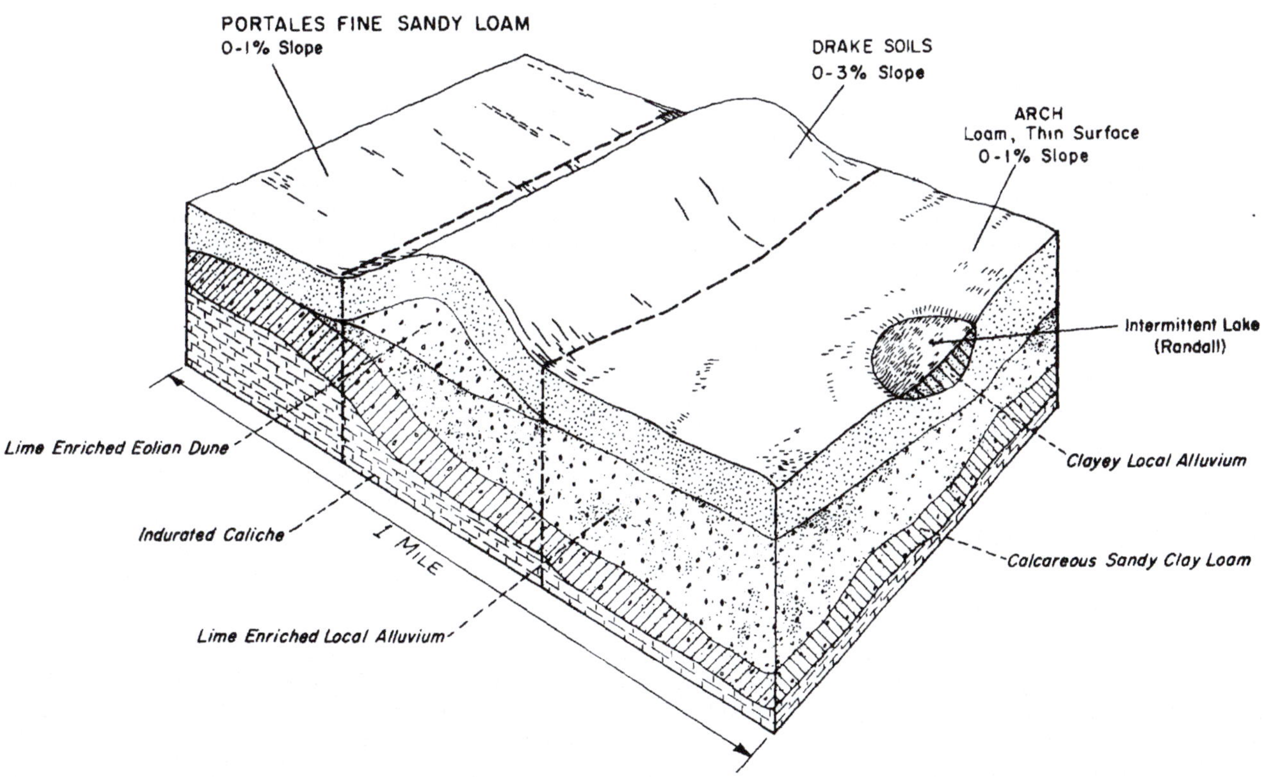

Fig. 8.13 Idealized block diagram illustrating the landscape-parent material-soil relationships of a playa-lunette system and associated Randall and Drake series. After Dittemore and Hyde (1964)

in these lunettes record the paleoclimatic and hydrologic changes of the region (e.g., Bowen and Johnson 2012; Rich 2013).

8.4 Alluvial Soils

8.4.1 General Characteristics of Alluvial Soils in the Great Plains

Alluvial soils are formed in materials transported and deposited by streams. These soils occupy a large area and are important because they represent some of the most productive agricultural soils in the Great Plains, often being designated by the USDA as prime farmland. These soils tend to have high concentrations of organic matter and plant nutrients and occur on relatively flat landscapes that are easily cultivated. Because the valley floors of streams receive runoff and sediments from various sources, however, alluvial soils also capture and store contaminants, including organic wastes and nitrates from animal feedlots, and pesticides and herbicides from croplands.

Alluvial soils exhibit characteristics imparted by the depositional environment and pedogenesis. Properties, such as color, texture, chemical composition, and mineralogy, are often inherited from the sediments comprising alluvial deposits. For example, alluvial soils on the High Plains of western Kansas and Nebraska tend to be brown (10YR 5/3) to yellowish brown (10YR 5/4), calcareous silt loams because they are developed in alluvium mostly derived from brown to yellowish brown, silty, calcareous Peoria Loess. By contrast, alluvial soils on the Rolling Plains of west-central Oklahoma tend to be red (2.5YR 5/6 and 5/8) to reddish brown (2.5YR 5/3 and 5/4) sandy loams because they are developed in alluvium mostly derived from red to reddish brown Permian sandstones.

Most alluvial deposits are stratified and exhibit upward-fining sequences. Consequently, bedding often occurs in alluvial soils, and it is not unusual for A horizons to have high clay contents because fine-grained alluvium is typically deposited at the top of the sedimentary sequence. Sudden changes in texture that occur in stratified alluvium can dramatically affect pedogenesis (Schaetzl and Anderson 2005). This is because the flow of water carrying dissolved and suspended materials tends to be retarded due to matric potential differences that form along lithologic discontinuities (i.e., the boundaries separating beds of fine- and coarse-grained sediment) resulting in precipitation and deposition of the materials. Hence, soils developed in calcareous alluvium often exhibit accumulations of calcium carbonate along contacts between fine-grained overbank (flood) deposits and underlying coarse-grained channel and near-channel deposits.

Many of the alluvial soils in the Great Plains are products of *cumulization*, a pedogenic and sedimentologic process involving the slow upward growth of the soil surface due to additions of sediment (Reicken and Poetsch 1960; Schaetzl and Anderson 2005). In stream valleys, cumulic soils receive influxes of alluvium while pedogenesis is occurring, but the rate of sedimentation is so slow that soil development keeps up with deposition (Nikiforoff 1949; Birkeland 1999; Mandel and Bettis 2001b; Mandel 2008). In such soils, the A horizon builds up through time and becomes *overthickened*, and it is not unusual for the A horizon to look stratified.

Although the physical and chemical properties of alluvial soils vary within and between subregions of the Great Plains, there are close relationships between alluvial soils and alluvial landforms. In the plains, alluvial soils are associated with four general types of landform sediment assemblages: floodplains, terraces, alluvial fans, and alluvial plains.

8.4.2 Soils of Floodplains

Areas of floodplains closest to alluvial channels tend to have thick, poorly developed soils with relatively coarse textures, whereas more distal floodplain areas usually have better developed soils with relatively fine textures due to the sedimentation rates and energy levels that decrease across a floodplain with distance from the channel. Abandoned channels, flood chutes, oxbow lakes, and low, backwater areas become traps for clay-rich alluvium and soils associated with such depressions on floodplains often are poorly drained and have redoximorphic and Vertic properties. Vertic Epiaquolls, for instance, occur in depressions and abandoned channels on floodplains in the Central Kansas Sandstone Hills (MLRA 74), Central Loess Plains (MLRA 75) and Bluestem Hills (MLRA 76) and are moderately developed with gleyed matrix colors (10YR 2/1 and 10YR 3/1), common to few strongly cemented iron and manganese oxide concretions, and common Pressure faces (e.g., Solomon series). Ponding is common with these soils, so they can be difficult to cultivate, especially during the rainy season from March through June.

Because modern floodplains are young landform sediment assemblages with surfaces that are often less than 100 years old, most floodplain soils in the Great Plains are weakly developed Entisols and Mollisols with A–C profiles. For example, Cumulic Hapludolls (e.g., Kennebec series), which occur on floodplains in stream valleys and on drainage ways that extend into uplands in the Bluestem Hills of the Eastern Plains (MLRA 76), have A–C profiles; the A horizon is thick as a result of cumulization (Fig. 8.14). This thick (>100 cm), organic-rich mollic epipedon of the Kennebec series makes it one of the most productive agricultural soils in the Great Plains.

Fig. 8.14 Examples of an overthickened mollic epipedon in the Kennebec series (Cummulic Hapludolls) and reddish hues of the Westola series (Typic Ustifluvents) derived from the alluvium of the Permian Redbeds. Photograph of the Westola series courtesy of the Natural Resources Conservation Service. Photograph of the Kennebec series reproduced from Soils of the Great Plains: Land Use, Crops, and Grasses by Andrew R. Aandahl with the permission of the University of Nebraska Press. Copyright 1980 by the University of Nebraska Press

On the High Plains of western and central Kansas, central Nebraska, and southeast South Dakota (MLRA 73), Cumulic Haplustolls, such as the Tobin and Roxbury series, typically occur on floodplains. These soils are developed in silty alluvium derived from loess, and both soils have cumulic A horizons that generally are 75–90 cm thick, over stratified C horizons.

Further west in the semiarid High Plains region of southwestern Wyoming, eastern Colorado and extreme western Kansas and Nebraska, Aridic Ustifluvents are the dominant soils on floodplains. These soils typically have thin, high-chroma A horizons (ochric horizon) above stratified, sandy, and/or gravelly C horizons. The A horizons tend to be slightly alkaline, whereas the C horizons are moderately to strongly alkaline. The coarse texture of these soils is related to the source of the alluvium—the Ogallala Formation, which largely consists of sand and gravel and has been deeply dissected by streams on the western High Plains. Aridic Ustifluvents have low organic matter contents, tend to be excessively drained, and are highly erodible; hence, they are mostly under native grass or used for grazing livestock. However, some of these soils, such as the Pathfinder series, produce corn and sugar beets when heavily irrigated.

Aridic Ustifluvents are dominant floodplain soils in the Northern Plains, including northeastern Wyoming, western North Dakota and South Dakota, and eastern Montana. These soils are mostly formed in stratified loamy or clayey alluvium derived from Cretaceous salt-rich shale, especially the Pierre Shale. These soils tend to be calcareous and saline, and some have secondary accumulations of gypsum and other salts (e.g., Lostriver series). These soils are used mainly as rangeland, but a few areas with lower salinity are used for irrigated and non-irrigated cropland.

In the Central Rolling Red Plains (MLRA 78B and 78C) of Texas, Oklahoma, and south-central Kansas, and throughout most of the Southern Plains, Typic Ustifluvents are the dominant soils on floodplains. These soils (e.g., Westola series) are formed in calcareous, silty, or sandy alluvium derived from siltstone or sandstone, respectively, comprising the Permian Redbeds. These weakly developed floodplain soils (A–C profiles) tend to have reddish hues of 2.5YR and 5YR (Fig. 8.14). The Typic Ustifluvents are mostly used for range, but a few areas are cultivated with wheat and sorghum.

8.4.3 Soils Formed in Terraces

8.4.3.1 Characteristics of Alluvial Terraces in the Great Plains

In the Great Plains, fill terraces (i.e., terraces created by episodes of stream aggradation followed by incision) as well as bedrock and fill strath terraces (i.e., erosional surfaces created by lateral migration of a stream during incision) often comprise stair-step-like flights of geomorphic surfaces, with the surfaces increasing in age with increasing elevation in the valley landscape. Hence, the magnitude of soil development increases on the surfaces with increasing elevation, forming a *chronosequence*—a series of soils where time (soil age) varies while all other soil-forming factors are held relatively constant by comparison (Stevens and Walker 1970; Yaalon 1975; Schaetzl and Anderson 2005).

Soil chronosequences also occur across terraces where channel fill is inset into older alluvial fill, and where laterally inset fills aggraded to approximately the same elevation. For example, a long cutbank that formed perpendicular to the axis of the South Fork Big Nemaha River valley of southeastern Nebraska revealed at least three laterally inset fills occurring beneath the T-1 terrace; radiocarbon dating of the fills indicated significant time-transgression going from the proximal to distal portions of the T-1 terrace (Fig. 8.15; Mandel and Bettis 2001a). Hence, soil variability on the terrace represents the effect of time, with the least developed soils occurring at the top of the youngest fill, proximal to the channel, and the best developed soils at the top of the oldest fill, distal to the channel.

8.4.3.2 Regional Differences in Surface Soils of Terraces

In the prairies of the Central Plains, one Holocene age alluvial terrace typically occurs above the elevation of the modern floodplain, and in the valleys of a few streams there are two Holocene terraces. Unlike the Pleistocene terraces that occur high in the valley landscape, the Holocene terraces lack a loess mantle; hence, soils on the Holocene terraces are formed entirely in alluvium. Going from east to west across the Great Plains, there is an increase in calcium carbonate accumulation in the subsoil and a decrease in organic matter content in A horizons of soils formed on similar-age Holocene terraces. The depth of secondary carbonates and the thickness of the A horizons also decrease from east to west. This east-to-west soil change is related to the steep climatic gradient, characterized by reduced mean annual precipitation, across the region.

Soils formed on moderately to well-drained Holocene terraces in the tall-grass prairies of the Eastern Plains (MLRA 76) typically have thick mollic epipedons, well-developed argillic horizons, and little or no secondary accumulation of calcium carbonate. Most of these soils are Pachic (e.g., Reading and Mason series) or Aquertic Argiudolls (e.g., Chase series). Soils in poorly drained backwater areas of Holocene terraces also have thick mollic epipedons, but lack argillic horizons. Instead, the high clay content of the alluvium and concomitant poor drainage in these low-energy depositional environments favors the development of poorly and very poorly drained soils with gleyed subsoil layers (Bg horizons) that have Fe and Mn masses and concretions. A horizons are often thick and the soils commonly occur on segments of Holocene terraces that are distal to the stream channel.

In the Central Plains, including central Kansas and southern and central Nebraska (MLRA 71, 73, 74, and 75), Cumulic Haplustolls with A–Bw or A–Bk profiles (e.g., Muir series) are the most common soils on Holocene ter-

Fig. 8.15 Photograph of the north bank of the South Fork Big Nemaha River in southeastern Nebraska showing a series of laterally inset fills beneath the T-1 terrace. The magnitude of soil development varies across the terrace and is directly related to the age of each fill. All of the radiocarbon ages were determined on charcoal

races. Further west in the dry subhumid to semiarid areas of the High Plains region, Fluventic, and Cumulic Haplustolls (e.g., Bridgeport and Roxbury series, respectively) are the dominant soils on Holocene terraces. In the same region, Cumulic Haplustolls with A–Bw–BC profiles (e.g., Hord series) and Pachic Argiudolls with A–Bt–BC or A-Bt-BCk profiles (e.g., Goshen series) occur on middle and early Holocene terraces that are 2–4 m higher than the late Holocene terraces. With some exceptions, the soils on the Holocene terraces in the Central and Western Plains are silt loams reflecting the silty Peoria Loess as a source of the alluvium.

Alluvial soils associated with stream terraces in the Northern Plains fall into two general categories: those formed in local Holocene alluvium and those formed in Pleistocene age glacial outwash. Most of these soils are Mollisols, with Hapludolls occurring in the eastern third of the Northern Plains and Haplustolls in the western two-thirds of the region. However, Entisols, primarily Ustifluvents, are common on low terraces in the western quarter of the region.

Buried alluvial soils are common in Holocene and late Pleistocene terrace fills throughout the Great Plains, and they are especially common in the Central Plains (see May 1989; Mandel 1992, 1994, 1995, 2006, 2008; Beeton and Mandel 2011; May and Holen 2014) and Southern Plains (see Holliday and Allen 1987; Ferring 1990; Blum et al. 1992; Murphy et al. 2014). Late Quaternary terrace fills in the valleys of large and small streams often contain multiple buried soils (Fig. 8.16).

8.4.4 Soils Developed in Alluvial Fans

Alluvial fans occur throughout the Great Plains and are common landforms on the margins of valley floors and basins and in the Western Great Plains in areas near the Front Range of the Rocky Mountains. In the valleys, alluvial fans form where small streams enter the valleys of larger streams. As the gradient of a small stream decreases, it deposits alluvium at the base of the valley wall of the larger stream. This reduces the capacity of the small stream channel and forces it to change direction and gradually build up a slightly mounded, fan-shaped landform. These same processes operate to form alluvial fans where small, intermittent streams flowing out of mountainous areas, such as the Rocky Mountains and Black Hills, deliver sediment to adjacent basin floors or the High Plains surface.

Sediment source strongly influences the lithology of alluvial fan deposits, which in turn has an effect on soil formation. For example, alluvial fans in the High Plains region of western Kansas and Nebraska tend to be fine-grained, mostly consisting of silt and/or fine sand, because the alluvium comprising the fans was derived from loess and/or eolian sand on the uplands. Hence, soils developed on these fans are predominantly loams, silt loams, and silty clay loams, and lack coarse clasts. The Atchison series (Aridic Haplustepts) is a good example of this and often occurs on alluvial fans in areas of the Central High Tablelands (MLRA 72) with both loess and eolian sand on the uplands. This soil has textures ranging from silt loam to fine sandy loam and lacks pebbles and cobbles. By contrast, alluvial fans in areas of the Great Plains bounded by mountains usually have coarse-grained sediment, including pebbles and cobbles derived from bedrock comprising the mountains. Soils formed on these fans are often gravelly. For example, the Judith series (Typic Calciustolls), associated with alluvial fans in central Montana (MLRAs 53A and 58A), typically has gravel throughout the solum with the highest concentration of gravel in the subsoil.

Alluvial fans in the Great Plains harbor well-preserved records of sedimentation and soil formation produced by landscape response to climatic change (Mandel and Bettis 2013). On the High Plains of western Kansas and Nebraska, fans began to form around 11,000 ^{14}C years BP Sedimentation initially was slow and punctuated by multiple episodes of soil development between ca. 13,500 and 9000 ^{14}C years BP (Mandel 2008). Organic-rich cumulic soils dating to the Younger Dryas Chronozone (12,900–11,700 years BP) are common and are mantled by thick early through middle Holocene fan deposits (Fig. 8.17); late Holocene fan deposits are generally rare, except in the Nebraska Sand Hills (Sweeney and Loope 2001) and Smoky Hills of north-central Kansas (Mandel 1992).

Buried soils are common in the early through middle Holocene fan deposits (Mandel 1995, 2006; Bettis and Mandel 2002; Faulkner 2002; Mandel and Bettis 2013), and since well-developed soils imply prolonged landscape stability, soils that are buried by alluvial fan deposits are evidence of a shift, at least locally, toward increased sediment yields (Faulkner 2002). It is likely that a change in regional climatic conditions, especially increased aridity, triggered these shifts.

8.4.5 Soils of Alluvial Plains

An alluvial plain is a relatively flat landform created by the deposition of sediment over a long period by one or more streams. In the Great Plains, there are three general types of alluvial plains: (1) a large assemblage of alluvial deposits that form low gradient, regional ramps along the flanks of the Rocky Mountains and extend great distances eastward from the sources of the alluvium (i.e., the High Plains), (2) alluvial deposits that aggraded on broad floodplains of ancient coalescing streams and were left high in the upland landscape when the streams deeply incised their valleys (i.e.,

Fig. 8.16 Photographs of late Quaternary terrace fills and associated buried soils in the valleys of high-order streams (>5th-order) on the Central Plains. All of the radiocarbon ages were determined on soil organic matter. **a** Lower Hackberry Creek in southwestern Kansas; **b** Little Blue River in south-central Nebraska; **c** Smoky Hill River in north-central Kansas; **d** Pawnee River in southwestern Kansas

Fig. 8.17 Photograph of The Willems Ranch alluvial fan formed at the mouth of an ephemeral, first-order stream that delivers sediment to valley floor of South Beaver Creek on the High Plains of northwestern Kansas, as shown above in **a**. The Younger Dryas paleosol is developed in the lower 1.5 m of the fan, as shown above in **b**

the Rolling Plains), and (3) glacial outwash plains consisting of glaciofluvial deposits that accumulated during the Pleistocene.

8.4.5.1 High Plains

The Ogallala Formation mostly consists of Miocene and Pliocene deposits of fluvial sand and gravel (Ludvigson et al. 2009). There are many petrocalcic (Bkm or Bkkm) horizons in the Ogallala Formation; the most distinct of these horizons is the resistant *caprock* caliche that forms the upper surface of the Ogallala (Fig. 8.18; Reeves 1976).

In the driest areas of the Southern High Plains, including northwestern Texas and eastern New Mexico, these soils are mostly Petrocalcids with A–Bw–Bkm, A–Bw–Bk–Bkm, or A–Bk–Bkm profiles. A series of strongly cemented to indurated petrocalcic horizons that are typically 15–90 cm thick and separated by gravelly and cobbly soil material often occur beneath the indurated Bkm or Bkkm horizons.

On convex, gently sloping to steep knobs and erosional hillslopes in the Canadian Breaks and margins of the Southern High Plains (MLRA 77E), Calciustolls, such as the Tascosa and Mansic soil series, are developed in sandy and

Fig. 8.18 Examples of several soils and horizons developed on alluvial plains. The Ogallala caprock is an ancient (Miocene to Pliocene in age) thick ($\sim$3 m in this photograph) petrocalcic (Bkkm) horizon formed in the upper Ogallala formation. Pastura series (Ustic Petrocalcids) formed on the Southern High Plains of eastern New Mexico and northwestern Texas. Rotan and Tillman series (Pachic Paleustolls and Vertic Paleustolls, respectively) are examples of relict paleosols. Renshaw series (Calcic Hapludolls) occur on glacial outwash plains in northeastern South Dakota and eastern North Dakota. Photograph of the Ogallala caprock is from Hirmas and Allen (2007). Photograph of the Raton soil courtesy of the Natural Resources Conservation Service. Photograph of the Renshaw soil reproduced from Soils of the Great Plains: Land Use, Crops, and Grasses by Andrew R. Aandahl with the permission of the University of Nebraska Press. Copyright 1980 by the University of Nebraska Press

gravelly alluvium comprising the Ogallala Formation. These soils have mollic epipedons above calcic horizons. Surface horizons typically have 20–40 % quartzite gravel, and gravel content increases with depth. Bk and Ck horizons have common to many secondary carbonate accumulations. These soils are primarily used for rangeland.

Further north, in the Southern High Plains Breaks (MLRA-77E) of Kansas and adjacent parts of Oklahoma,

Calciustepts with A–Bk–C profiles (e.g., case series) are developed in loamy, calcareous alluvium comprising the Ogallala Formation. The calcic horizons typically have a carbonate morphology characterized by common to many hard concretions and soft masses of calcium carbonate. These soils are mostly cultivated, with wheat and grain sorghums as the principal crops.

8.4.5.2 Rolling Plains

In the Rolling Plains of northwest Texas and southwest Oklahoma, including the Central Rolling Red Plains (MLRA-78B and 78C) and Rolling Limestone Prairies (MLRA-78D), alluvial soils are common on nearly level to gently sloping uplands. These soils formed in calcareous alluvium presumably deposited by ancient streams during the Pleistocene, though deposition during the Pliocene cannot be ruled out. Subsequent stream entrenchment left the former floodplain deposits high in the landscape, creating inverted topography; soils that formed on the alluvial plains are now paleosols since the morphology of these soils no longer reflects the current soil-forming environment (see Birkeland 1999:339). The Rotan and Tillman soil series, a Pachic Paleustoll and Vertic Paleustoll, respectively, are examples of these alluvial paleosols (Fig. 8.18). Both soils have thick, well-developed A–Bt–Btk–BCk profiles. Rotan soils have a thick mollic epipedon underlain by a yellowish red to red Btk horizon, which contains 35–45 % clay and about 25 % visible soft masses and concretions of calcium carbonate. The Tillman series is similar but has slightly higher clay contents in the subsoils. Parent material for both is Permian Redbed clays and claystones. Other common soils formed in ancient alluvium high in the landscape include Typic Paleustalfs (e.g., Miles and Wichita series), Pachic Argiustolls (e.g., Abilene series), Typic Haplusterts (e.g., Hollister series), and Vertic Calciustolls (e.g., Rowena series).

8.4.5.3 Glacial Outwash Plains

During the Wisconsin glacial stage, outwash plains formed in portions of the Northern Great Plains that were affected by the Laurentide ice sheet, including the northern third of Montana, all but the southwestern quarter of North Dakota, and most of the eastern half of South Dakota. The Laurentide ice sheet contained large amounts of sediment, and meltwater transported the sediment away from the ice sheet and deposited it on broad plains, or *sandars*, in front of the ice. Instead of being characterized by a single broad, flat geomorphic surface, outwash plains often have multiple levels, which are commonly terraces formed by incision of the glaciofluvial deposits due to changes in hydrology, sediment availability, or a drop in local base level (Benn 2009).

Typically, large amounts of sediment were transported as coarse bedload. The material in the outwash plains was often size-sorted by the runoff of the melting ice sheet, with fine-grained sediment deposited farthest from the ice, and coarse-grained sediment, including boulders, deposited close to the ice front. Consequently, as the ice front retreated, fine-grained outwash (mostly silt and fine sand) was typically deposited above coarse-grained outwash (mostly gravel and coarse sand). This pattern of sedimentation is reflected in many soils developed in glacial outwash in the Northern Plains, with abrupt textural changes occurring in the sola. For example, the Renshaw series (Calcic Hapludolls) commonly developed on outwash plains in northeastern South Dakota and eastern North Dakota has an A–Bw–2Bk–2C profile (Fig. 8.18). An abrupt boundary separates the Bw horizon, typically a loam, from the 2Bk horizon, typically consisting of very gravelly loamy sand with 15–60 % gravel. Further west, in northwest South Dakota, western North Dakota, and eastern Montana (MLRA 54), soils on outwash plains primarily are Haplustolls developed in upward-fining sequences, and many have abrupt textural changes. For example, the Bowdle series (Pachic Haplustolls) has an abrupt boundary that separates the lower Bk horizon with gravelly loam to sandy loam textures from the 2C horizon that contains gravelly to very gravelly loamy sand to sand textures. Other similar soils that frequently occur on outwash plains include the Wabek series (Entic Haplustolls) and Appam series (Typic Haplustolls).

8.5 Bedrock-Derived Soils

Soils derived from bedrock residuum are concentrated in areas of the Great Plains that have not been mantled with glacial till (e.g., the Rolling Soft Shale Plain—MLRA 54), lacustrine/glaciolacustrine deposits, eolian sediments, or extensively buried by thick alluvium (e.g., Central Rolling Red Plains—MLRA 78C) (Figs. 8.1, 8.2). These soils also occur in areas where chemical weathering and surface erosion are minimal because of resistant lithologies (e.g., the Black Hills and the Flint Hills—MLRAs 62 and 76) or because of a relatively dry climate (e.g., the Central New Mexico Highlands—MLRA 70C). The minimal weathering and erosion has maintained distinct stable upland topographies that are less prone to burial from alluvial, eolian, or glacial processes. The lack of significant surface deposition and the geomorphic stability of the level upland surface topography have maintained the presence of bedrock near the surface for extensive periods of time allowing for the development of strongly horizonated and, in some cases, deep soils to form in relatively resistant parent materials.

Bedrock-derived soils inherit properties of the bedrock from which they form. For example, the textures and hydraulic conductivities of the Rhoades series (Leptic Vertic Natrustolls) are markedly different from the textures and conductivities of the Flasher series (Typic Ustipsamments) despite both occurring under the same climate in the same MLRA (Fig. 8.19). Rhoades soils have developed in fine-grained shale, siltstone, and mudstone giving rise to silty clay textures, a slow permeability, strongly alkaline and sodic conditions, and the presence of carbonate, gypsum, and soluble salt masses, whereas Flasher soils have developed in sandstone and are characterized by fine-sand textures and a moderately rapid to rapid permeability. Despite being derived from bedrock, many of these soils have received influxes of eolian sediments (especially fine-sand and silt-sized loess), alluvium, and colluvium and/or have been reworked by wind, water, or gravity at the surface (e.g., Travessilla, Grant, and Wells series). In general, bedrock-derived soils are not used for cropland to a considerable degree owing largely to the large quantity of stones or abundance of shallow soils such as in the Flint Hills (Fig. 8.20).

Fig. 8.19 Several soil profiles developed in bedrock residuum. The very slowly permeable Rhoades series (Leptic Vertic Natrustolls) formed in soft shale, siltstone, and mudstone in upland swales. By contrast, the Flasher series (Typic Ustipsamments) is excessively drained with a moderately rapid to rapid permeability and developed from soft sandstone within the same MLRA (54). Similarly, the extremely gravelly and cobbly Florence (Udic Argiustolls) and thin Sogn (Lithic Haplustolls) series formed from cherty and hard limestones, respectively, within the same MLRA (76). Reproduced from Soils of the Great Plains: Land Use, Crops, and Grasses by Andrew R. Aandahl with the permission of the University of Nebraska Press. Copyright 1980 by the University of Nebraska Press

Fig. 8.20 Flint Hills (Bluestem HIlls—MLRA 76) in eastern Kansas after a fire showing the differential weathering of the cherty limestone and interbedded shale. After Beeton and Mandel (2011) with permission. Photograph by R. Swan

8.6 Soils Developed in Colluvium

Colluvium is unconsolidated sediment moved downslope primarily under the influence of gravity, though sheetwash and soil creep may be involved in the movement of the sediment. This term is also used to refer to sediments deposited at the base of hillslopes by unconcentrated surface runoff or sheetflow (Schaetzl and Anderson 2005; Natural Resources Conservation Service 2014a). Colluvium is often a poorly sorted mixture of sediments of various sizes that forms gently sloping aprons on footslopes and toeslopes. Colluvial deposits may bury and truncate pre-existing soils (buried paleosols), or bury older colluvium (Schaetzl and Anderson 2005); hence, crude bedding may occur in a colluvial apron, although individual beds rarely have internal stratification. Lenses or beds of alluvium may co-occur with colluvium, forming co-alluvial fans or aprons (Cremeens et al. 2003).

The texture of the colluvium, as well as its composition (calcareous vs. noncalcareous material), affects the physical and chemical properties of the soils developed in these materials. For example, soils such as the Elkader (Torriorthentic Haplustolls) and Hord series (Cumulic Haplustolls) that are associated with colluvial aprons in the High Plains region of Kansas and Nebraska tend to be calcareous silt loams because they are developed in sediment derived from calcareous loess on the uplands. In areas of the Central and Southern High Plains where the Ogallala Formation is at or near the upland surface, colluvial aprons and associated soils tend to be sandy and calcareous because of the contribution of sediment from the Ogallala; these soils are classified as Haplustepts (e.g., Mobeetie series).

Soils developed in a mix of coarse- and fine-grained colluvium (often with *skeletal* textures, or particle-size distributions with greater than 35 % rock fragments) tend to occur in regions of the Great Plains where the climate favors a mechanical weathering regime (i.e., cold and dry); the parent materials are resistant to chemical weathering, and/or in gravelly or coarser deposits that have not had sufficient time to weather to finer textures. Although they occur throughout the Great Plains, these soils are more common in the Black Hills and associated foot slopes (MLRAs 61 and 62), Canadian River Plains and Valleys (MLRA 70A) and the Flint Hills (MLRA 76) (Figs. 8.1, 8.2). These soils develop on footslope positions of mountains, hills, and ridges. Depending on the stability of the landform and its age, these soils range from poorly developed Entisols to relatively well-developed Alfisols and Mollisols. They can be difficult to cultivate due the high gravel content and are commonly used for rangeland or timber production.

In the Flint Hills of Kansas, footslope positions composed of a mix of fine- and coarse-grained colluvium derived from the upslope resistant cherty and hard limestone bedrock and interbedded shales have developed Pachic Argiustolls (e.g., Tully series) (Fig. 8.21). These soils have thick mollic epipedons (50–100 cm thick) that result from continuous, slow inputs of surface material from higher slope positions and thick argillic horizons that reflect the relative stability of the

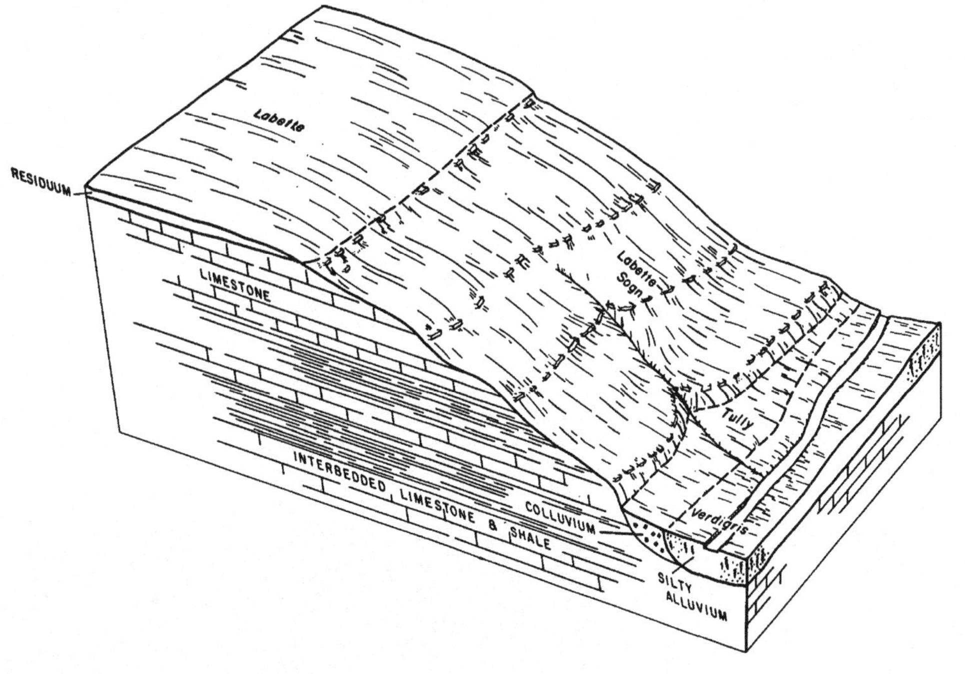

Fig. 8.21 Idealized block diagram of colluvium–soil–landscape relationships for the Tully series. After Penner et al. (1975)

surface and the more humid eastern Great Plains climate. Chert or limestone gravels can reach 35 % by volume in these soils (Natural Resources Conservation Service 2014b).

8.7 Soils Formed in Upland Depressions

8.7.1 Soils of Playas and Salinas

One of the striking features of the Great Plains region is the seemingly ubiquitous presence of small, semicircular, shallow basins known as *playas* (Bowen and Johnson 2012) (Figs. 8.13, 8.22). Prior to the extensive cultivation in this region, these ephemeral lakes were native prairie wetlands (Tsai et al. 2012) and used by Paleoindians and later groups making them important for the preservation of cultural materials (e.g., LaBelle et al. 2003; Mandel and Hofman 2003; Hurst et al. 2010). These natural basins contain a diversity of wetland plants, invertebrates, and amphibians and serve as stopover areas for shorebirds migrating through the region (Guthery and Bryant 1982; Davis and Smith 1998). Playas are found in significant numbers in Texas, New Mexico, Oklahoma, Kansas, Nebraska, and Colorado (Osterkamp and Wood 1987; Cariveau and Pavlacky 2008, 2009) and are concentrated on the Southern and Central High Plains with the number of playas estimated at approximately 20,000 on the Llano Estacado of Texas (Sabin and Holliday 1995; Wood 2002; Howard et al. 2003) and over 22,000 on the Central High Plains of western Kansas (Bowen et al. 2010). Area measurements of playas range from 0.03 to 483 ha corresponding to diameters of 20 m–2.5 km, although average areas are reported to be between 0.6 and 7.6 ha (Holliday et al. 2008; Cariveau and Pavlacky 2009; Bowen et al. 2010). Depths of these depressions range from almost 0 to over 14 m (Holliday et al. 2008).

Soils found in playas are commonly Epiaquerts and Haplusterts (e.g., Randall, Ness, and Lipan series) in the Southern and Central High Plains and Argiaquolls and Argialbolls with Vertic subgroups (e.g., Lodgepole, Filmore, Scott, and Massie series) in the Northern High Plains (Aandahl 1982; Gurdak and Roe 2009; LaGrange et al. 2011). Organic carbon typically ranges from 0.5 to over 2 % in the upper horizons (generally lower than what is expected from the dark and very dark gray colors) likely due to the high rates of decomposition from the frequent wetting and drying cycles (Allen et al. 1972; Smith 2003; Holliday et al. 2008). Slickensides and wedge-shaped soil structures are common in the subsurface horizons of these soils and reflect shrink-swell processes (Hovorka 1997; Natural Resources Conservation Service 2014b). Redox concentrations in the form of iron–manganese masses and concretions are also frequently found in the upper horizons of these soils (e.g., Randall series) owing to the cycles of saturation and drying in the upper horizons as the smectitic clays swell causing otherwise open macropores to close (Fig. 8.23).

Despite making up less than 1 % of the landscape (Bowen et al. 2010), playas contribute a significant amount to the recharge of the High Plains Aquifer (Wood et al. 1997; Gurdak and Roe 2009). This recharge happens as preferential flow through faunal burrows or dissolution pores in the underlying calcic and petrocalcic horizons of the Ogallala and/or Blackwater Draw formations especially in the Southern High Plains (Wood et al. 1997; Hirmas and

Fig. 8.22 Playa lakes in Meade County southwestern Kansas. Photograph courtesy of W.C. Johnson

Fig. 8.23 Randall series (Ustic Epiaquerts) found in playa basins of the Llano Estacado in the Texas panhandle (i.e., Southern High Plains, Southern Part—MLRA 77C). Reproduced from Soils of the Great Plains: Land Use, Crops, and Grasses by Andrew R. Aandahl with the permission of the University of Nebraska Press. Copyright 1980 by the University of Nebraska Press

Allen 2007). Perhaps more importantly, however, desiccation cracks contribute significantly to the recharge especially at early stages of infiltration (Wood et al. 1997; Gurdak and Roe 2009) where these cracks can open to approximately 1 m deep in playa soils (Hovorka 1997). Sedimentation of these playas as a result of various agricultural practices in the region has significantly impacted both the recharge of the High Plains Aquifer and the biodiversity of the avian, vertebrate, invertebrate, and floral communities of these wetlands (Haukos and Smith 1994; Luo et al. 1997; Gurdak and Roe 2009; LaGrange et al. 2011; Tsai et al. 2012).

In addition to the numerous semicircular playas, approximately 40 large and irregularly shaped ephemeral saline lakes, or *salinas*, occur on the Southern High Plains (Wood and Osterkamp 1987; Holliday 1990; Stout 2003). Salinas represent discharge areas for groundwater (Holliday et al. 1996); thus, soils mantling salina floors are saline due to the limited drainage in combination with the high evaporative demand in this semiarid area (Stout 2003). Soils on the basin floor are Halaquepts (e.g., Cedar Lake series), salt-encrusted at the surface, strongly saline (electrical conductivity, EC, between 16 and 32 dS m^{-1}), moderately sodic with sodium adsorption ratios (SARs) between 13 and 40 in the upper horizons, and contain calcic horizons in the lower subsoil with calcium carbonate equivalents (CCEs) between 40 and 60 % (Stout 2003; Natural Resources Conservation Service 2014b). Natrustolls can be found on the surrounding alluvial terraces 1–3 m above the basin floor (e.g., Yellow Lake series) which contain slightly lower salt concentrations in the upper horizons due to the enhanced drainage (EC between 4 and 16 dS m^{-1}), clay-enriched sodic subsoil horizons with SARs between 13 and 75, and gypsic horizons (Natural Resources Conservation Service 2014b). Because these soils are devoid of vegetation due to the high concentrations of salts and exchangeable sodium, salinas can be sources of dust emissions when the surface of the basin floor becomes dry (Stout 2003).

8.7.2 Soils of Prairie Potholes

The Northern Great Plains contains millions of natural, internally drained, depressional wetlands known as *prairie potholes* or *kettles* that were created by the deposition of thick but uneven glacial drift by retreating late Pleistocene glaciers (Shjeflo 1968; Mills et al. 2011). These potholes are found in an extensive area known as the Prairie Pothole Region (PRR) and comprise an area between 775,000 and 900,000 km^2 from the north-central USA to southern Canada including eastern South and North Dakota, and northern Montana (Sharratt et al. 1999; Euliss et al. 2006). The PRR has been extensively used for agriculture, which has led to the draining of more than half of the area in the USA (Tiner 1984). The region is also considered one of the most important breeding grounds in North America for waterfowl (Smith et al. 1964).

The genesis of soils that occur in many prairie potholes is controlled by subsurface hydrology (Richardson 1996) and consists mostly of Mollisols and Entisols. Figure 8.24 illustrates potholes in three different positions with respect to the surface topography and the flow of groundwater. Recharge potholes typically recharge groundwater in spring and early summer and contain soil profiles that are well-horizonated and fine-textured with significant argillic horizon development (Arndt and Richardson 1988). Argiaquic Argialbolls (e.g., Tonka series) are pothole soils in these topographically higher depressions with relatively thin

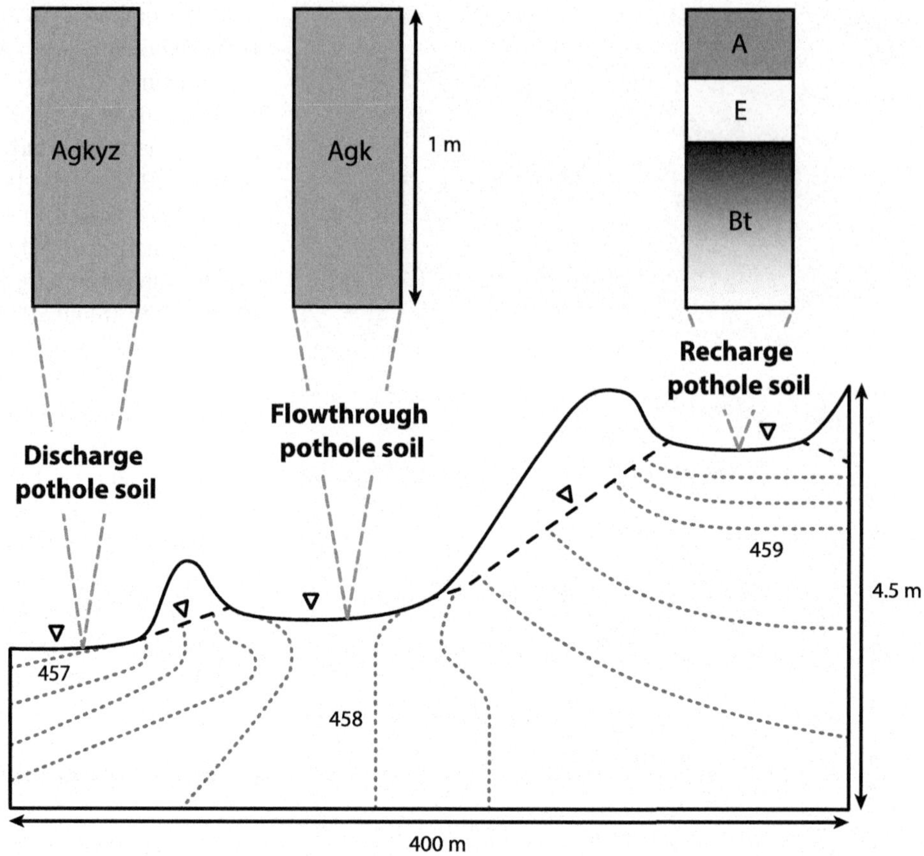

Fig. 8.24 Topographic cross section and flownet through three prairie potholes in different topographic positions illustrating the subsurface movement of water and the resulting soil profiles. (Equipotential values given as hydraulic head in meters above sea level.) The discharge and flow-through pothole soils contain thick-gleyed *A horizons* with carbonates where as the recharge pothole soil contains relatively thin *A horizon* and *E horizons*, and a thick zone of accumulation (Bt horizon). The discharge pothole soil also contains accumulations of gypsum and soluble salts. Modified from Richardson (1996)

A and E horizons and thick argillic horizons; soils in these positions can also form cumulic (overthickened) A horizons that result from high rates of organic matter production and constant sedimentation from surrounding areas (Arndt and Richardson 1988). Flow-through pothole soils can show effects of varying degrees of groundwater recharge and discharge depending on catchment characteristics and relative topographic position (Arndt and Richardson 1988). These soils tend to be more saline and contain higher saturated-paste extract concentrations of Na^+, Mg^{2+}, and SO_4^{2-} than recharge pothole soils; they are often gleyed throughout the profile (Arndt and Richardson 1988, 1989; Richardson 1996). Discharge potholes contain soils such as the poorly drained Playmoor series (Cumulic Endoaquolls) with concentrations of soluble salts and gypsum; gypsum has been reported to occur on the margins of the potholes (Steinwand and Richardson 1989). In these positions, salts have been concentrated by evapotranspiration and freezing which removes water from solution allowing salts such as the sodium sulfate mineral mirabilite to precipitate (Arndt and Richardson 1989; Steinwand and Richardson 1989). Figure 8.25 illustrates the landscape relationships between the Tonka (recharge), Parnell (flow-through), and Playmoor (discharge) soils for a prairie–pothole system in North Dakota.

8.7.3 Soils of Solution Basins

In addition to playas and salinas in the Southern and Central High Plains and potholes in the Northern High Plains, several sizable solution basins and a number of sinkholes occur on uplands throughout the region. An example of these solution basins is Big Basin in the Eastern Part of the Central Rolling Red Plains, south-central Kansas (Fig. 8.26). This and associated basins formed as a result of Quaternary dissolution of the underlying Permian age Hutchison Salt Member of the Wellington Formation (Anderson et al. 1998) and are mantled by Mollisols, Vertisols, and Entisols (Fig. 8.26).

The steep-walled sides of the basin are prone to erosion and dissection. These areas are comprised of weakly developed and shallow Ustorthents (e.g., Canlon series) on steeply sloping surfaces while Calciustolls (e.g., Campus series) occur on more stable portions of the wall where greater effective precipitation due to more gently sloping surfaces has formed shallow, well-drained soils with moderately developed calcic horizons in the underlying lime-cemented sandstone residuum. The gentlest sloping portions of the basin floor are composed of Haplustolls (e.g., Bippus series) with overthickened A horizons (reaching depths of 30–76 cm below the soil surface) likely the result

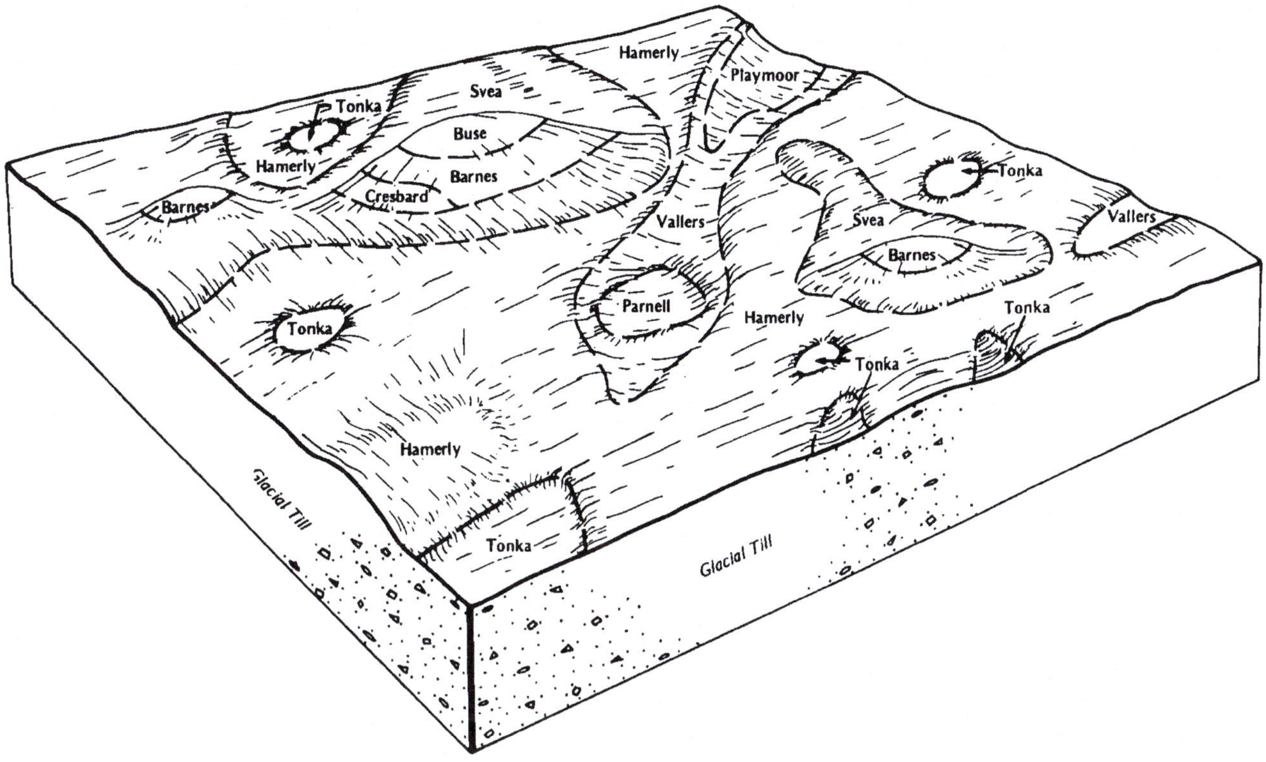

Fig. 8.25 Idealized block diagram of soils and parent materials in the Hamerly–Svea–Tonka association. After Heidt et al. (1989)

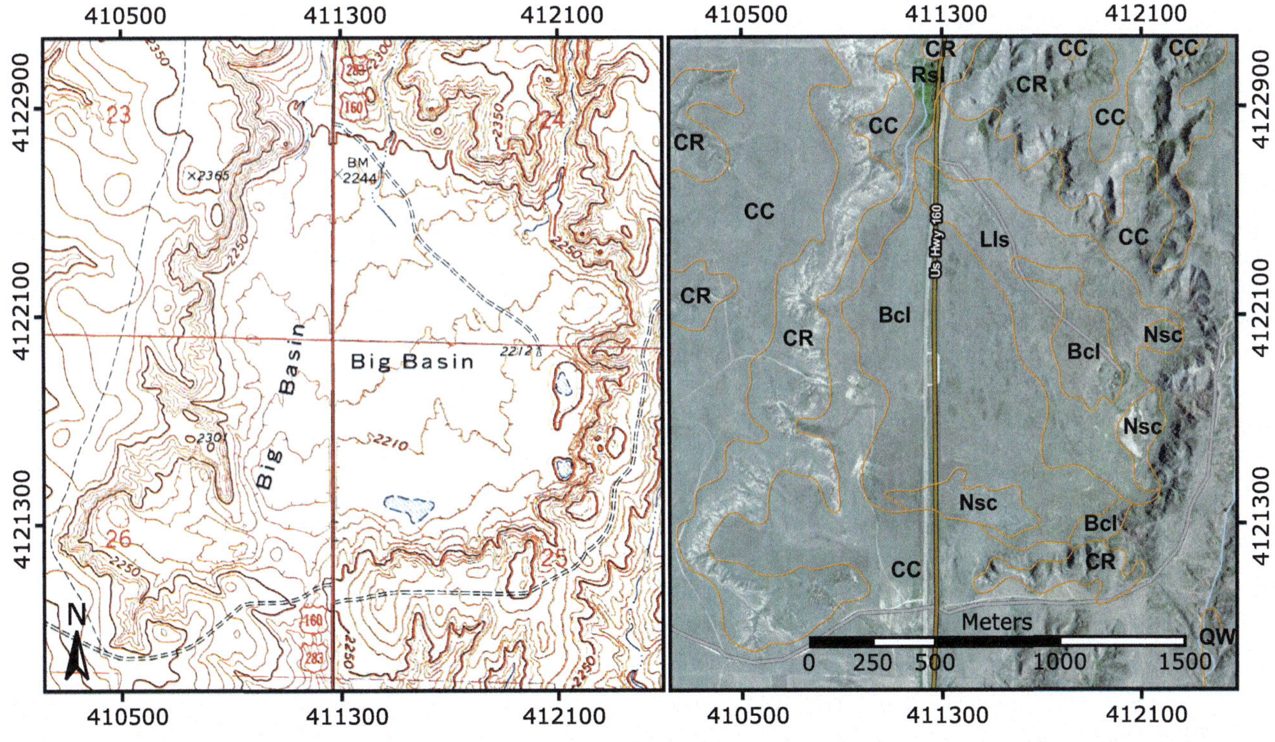

Fig. 8.26 Topographic (*left*) and soil (*right*) map of Big Basin, south-central Kansas. Contour intervals are 3.05 m. Soil series displayed are Bippus clay loam (*Bcl* Cumulic Haplustolls), Campus (Typic Calciustolls)–Canlon (Lithic Ustorthent) complex (CC), Canlon–Rock outcrop complex (CR), Likes loamy sand (*Lls* Aridic Ustipsamments), Ness silty clay (*Nsc* Ustic Epiaquerts), Quinlan (Typic Haplustepts)–Woodward (Typic Haplustepts) loams (QW), and Roxbury silt loam (*Rsl* Cumulic Haplustolls). Universal Transverse Mercator (UTM) coordinates correspond to Zone 14; UTM coordinates around the edge of the topographic map are approximate. Topographic quadrangles were downloaded from the USGS (http://store.usgs.gov); base NRCS soil map was downloaded from Web Soil Survey (http://websoilsurvey.sc.egov.usda.gov)

of periodic influxes from the surrounding walls of the basin. These calcareous soils have clay loam textures in the surface horizons and sandy clay loam textures in the subsurface. They are highly fertile and moderately permeable. Poorly drained Epiaquerts (e.g., Ness series) are also found on basin floors with smectitic clays with a high potential for shrink swell in the surface and subsurface. In addition to these fine-textured soils, deep (>2 m) and excessively drained Ustipsamments (e.g., Likes series) occur on basin floors near wall footslopes. These soils developed in sandy alluvial and colluvial parent material at the base of solution basin walls and contain visible secondary carbonates in subsurface horizons. The development of sinkholes in the Rolling and High Plains is ongoing and may be correlated with enhanced solution-subsidence processes from oil and gas exploration in the region (Steeples et al. 1986) although there is disagreement about the extent of this anthropogenic effect (Walters 1978; Anderson et al. 1998).

8.8 Conclusions

The Great Plains contains a variety of soils that have developed largely in response to the dominant soil-forming factors of parent material, climate, and time. Parent material varies across the region, with glacial drift (till, outwash and glaciolacustrine deposits) concentrated in the Northern Great Plains, loess mantling large areas of the Central Great Plains and Southern High Plains, eolian sand occurring in the Western Great Plains and several large areas in Kansas and Nebraska, alluvial deposits occurring beneath floodplains, terraces, fans, and plains throughout the region, and a scatter of colluvium and residuum occurring on footslopes and bedrock-dominated landscapes, respectively, across the plains. The soils that have formed in these deposits reflect the physical and chemical properties and age of the parent material as well as the strong regional east–west precipitation and latitudinal temperature gradients. These factors have given rise to the numerous agriculturally productive soils found in this region.

References

Aandahl AR (1982) Soils of the Great Plains: land use, crops, and grasses. University of Nebraska Press, Lincoln

Albertson FW, Tomanek GW (1965) Vegetation changes during a 30-year period in grassland communities near Hays, Kansas. Ecology 46:714–720

Albertson FW, Weaver JE (1942) History of the native vegetation during of western Kansas during seven years of continuous drought. Ecol Monogr 12:23–51

Allen BL, Harris BL, Davis KR, Miller GB (1972) The mineralogy and chemistry of High Plains playa lake soils and sediments. WRC-72-4, Water Resources Center, Texas Tech University, Lubbock

Anderson NL, Martinez A, Hopkins JF, Carr TR (1998) Salt dissolution and surface subsidence in central Kansas: a seismic investigation of the anthropogenic and natural origin models. Geophysics 63:366–378

Arndt JL, Richardson JL (1988) Hydrology, salinity and hydric soil development in a North Dakota prairie-pothole wetland system. Wetlands 8:93–108

Arndt JL, Richardson JL (1989) Geochemistry of hydric soil salinity in a recharge-throughflow-discharge prairie-pothole wetland system. Soil Sci Soc Am 53:848–855

Beeton JM, Mandel RD (2011) Soils and Late-Quaternary landscape evolution in the Cottonwood River Basin, east-central Kansas: implications for archaeological research. Geoarchaeology 26:693–723

Benn D (2009) Glaciofluvial sediments. In: Gornitz V (ed) Encyclopedia of paleoclimatology and ancient environments. Springer, The Netherlands

Bettis EA III, Mandel RD (2002) The effects of temporal and spatial patterns of Holocene erosion and alluviation on the archaeological record of the central and eastern Great Plains, USA. Geoarchaeology 17:141–154

Bettis EA III, Muhs DR, Roberts HM, Wintle AG (2003) Last glacial loess in the conterminous USA. Quatern Sci Rev 22:1907–1946

Birkeland PW (1999) Soils and geomorphology, 3rd edn. Oxford University Press, Oxford

Bleed AS (1990) Atlas of the Sand Hills. Conservation and Survey Division, University of Nebraska, Lincoln

Blum MD, Abbott JT, Valastro S Jr (1992) Evolution of landscapes on the Double Mountain Fork of the Brazos River, West Texas: implications for preservation and visibility of the archaeological record. Geoarchaeology 7:339–370

Bockheim JG, Hartemink AE (2013) Classification and distribution of soils with lamellae in the USA. Geoderma 206:92–100

Boellstorff J (1978a) A need for redefinition of North American Pleistocene stages. Trans Gulf Coast Assoc Geol Soc 28:65–74

Boellstorff J (1978b) Chronology of some Late Cenozoic deposits from the central United States and the ice ages. Trans Neb Acad Sci 6:35–49

Borchert JR (1950) The climate of the central North American grassland. Ann Assoc Am Geogr 40:1–30

Borchert JR (1971) The dust bowl in the 1970's. Ann Assoc Am Geogr 61:1–22

Bowen MW, Johnson WC (2012) Late Quaternary environmental reconstructions of playa-lunette system evolution on the central High Plains of Kansas, United States. Geol Soc Am Bull 124:146–161

Bowen MW, Johnson WC, Egbert SL, Klopfenstein ST (2010) A GIS-based approach to identify and map playa wetlands on the High Plains, Kansas, USA. Wetlands 30:675–684

Bryson RA, Hare FK (1974) Climates of North America. Elsevier, New York

Cariveau AB, Pavlacky D (2008) Assessment and conservation of playas in eastern Colorado. Rocky Mountain Bird Observatory, Brighton

Cariveau AB, Pavlacky D (2009) Biological inventory and evaluation of conservation strategies in southwest playa wetlands. Rocky Mountain Bird Observatory, Brighton

Carter JM, Driscoll DG, Williamson JE (2002) Atlas of water resources in the Black Hills area, South Dakota. U.S. Geological Survey, Hydrologic Investigations Atlas HA-747, Washington

Clayton L (1972) Geology of Mountrail County, North Dakota, Bulletin 55, Part IV. North Dakota Geological Survey, Grand Forks

Members COHMAP (1988) Climatic change of the last 18,000 years—observation and model simulations. Science 241:1043–1052

Cremeens DL, MacDonald DH, Lothrop JC (2003) Holocene colluvial soils and geoarchaeology in the unglaciated Appalachian Plateau: two examples from West Virginia, USA. Geoarchaeology 18:799–826

Darton NH, Paige S (1925) Central Black Hills, South Dakota. U.S. Geological Survey, Atlas of the United States, Folio 219, Washington

Davis CA, Smith LM (1998) Ecology and management of migrant shorebirds in the playa lakes region of Texas. Wildlife Monographs 140:3–45

DeWitt E, Redden JA, Buscher D, Wilson AB (1989) Geologic map of the Black Hills area, South Dakota and Wyoming. U.S. Geological Survey, Miscellaneous Investigations Series, Map I-1910, Washington DC

Dittemore WH Jr, Hyde HW (1964) Soil survey of Yoakum County, Texas. US Department of Agriculture, Soil Conservation Service, Washington

Dort W Jr (ed) (1985) Field evidence for more than two early Pleistocene glaciations of the central Plains. TER-QUA symposium series, vol 1. Nebraska Academy of Sciences, Lincoln

Euliss NH Jr, Gleason RA, Olness A, McDougal RL, Murkin HR, Robarts RD, Bourbonniere RA, Warner BG (2006) North American prairie wetlands are important nonforested land-based carbon storage sites. Sci Total Environ 361:179–188

Faulkner DJ (2002) Alluvial fans as records of Holocene sediment yield in the central Great Plains of southwest Nebraska. Phys Geogr 23:343–365

Fenneman NM (1931) Physiography of Western United States. McGraw-Hill, New York

Ferring CR (1990) Archaeological geology of the Southern Plains. In: Lasca NP, Donahue J (eds) Archaeological Geology of North America, vol 4. Geological Society of America Centennial Special, Boulder, pp 253–266

Frederick CD (1998) Late Quaternary clay dune sedimentation on the Llano Estacado, Texas. Plains Anthropologist 43:137–155

Frison GC (1978) Prehistoric hunters of the High Plains. Academic Press, New York

Frye JC (1946) The High Plains surface in Kansas. Transactions of the Kansas Academy of Science 49:71–86

Frye JC, Leonard AB, Swineford A (1956) Stratigraphy of the Ogallala Formation (Neogene) of northern Kansas. Bulletin 118, Kansas Geological Survey, Lawrence

Gurdak JJ, Roe CD (2009) Recharge rates and chemistry beneath playas of the High Plains Aquifer—a literature review and synthesis. Circular 1333, United States Geological Survey, Reston, Virginia

Guthery FS, Bryant FC (1982) Status of playas in the Southern Great Plains. Wildl Soc Bull 10:309–317

Halfen AF, Johnson WC (2013) A review of Great Plains dune field chronologies. Aeolian Research 10:135–160

Hallberg GR (1986) Pre-Wisconsinan glacial stratigraphy of the central plains region in Iowa, Kansas, and Missouri. Quatern Sci Rev 5:11–15

Hallberg GR, Boellstorff JD (1978) Stratigraphic "confusion" in the region of the type areas of Kansan and Nebraskan deposits. Geol Soc Am Abstr Programs 10:255–256

Harris A, Tuttle E (1990) Geology of the National Parks. Kendall/Hunt Publishing, Dubuque

Haukos DA, Smith LM (1994) The importance of playa wetlands to biodiversity of the Southern High Plains. Landsc Urban Plan 28:83–98

Heidt CJ, Moen DE, Morrison MW (1989) Soil survey of Nelson County Area, North Dakota. US Department of Agriculture, Soil Conservation Service, Washington

Holliday VT (1990) Soils and landscape evolution of eolian plains: the Southern High Plains of Texas and New Mexico. Geomorphology 3:489–515

Holliday VT, Allen BL (1987) Geology and soils. In: Johnson E (ed) Lubbock Lake: Late Quaternary Studies on the Southern High Plains. Texas A&M Press, College Station, pp 14–21

Holliday VT, Hovorka SD, Gustavson TC (1996) Lithostratigraphy and geochronology of fills in small playa basins on the Southern High Plains, United States. Geol Soc Am Bull 108:953–965

Holliday VT, Mayer JH, Fredlund GG (2008) Late Quaternary sedimentology and geochronology of small playas on the Southern High Plains, Texas and New Mexico, USA. Quatern Res 70:11–25

Hovorka SD (1997) Quaternary evolution of ephemeral playa lakes on the Southern High Plains of Texas, USA: cyclic variation in lake level recorded in sediments. J Paleolimnol 17:131–146

Howard T, Wells G, Prosperie L, Petrossian R, Li H, Thapa A (2003) Characterization of playa basins on the High Plains of Texas. Report 357, Texas Water Development Board, Austin

Hirmas DR, Allen BL (2007) Degradation of pedogenic calcretes in West Texas. Soil Sci Soc Am J 71:1878–1888

Hurst S, Johnson E, Holliday VT, Butler S (2010) Playa archeology on the Southern High Plains of Texas: a spatial analysis of hunter–gatherer occupations at Tahoka-Walker (41LY53). Plains Anthropol 55:195–214

Johnson WC, Willey KL, Mason JA, May DW (2007) Stratigraphy and environmental reconstruction at the middle Wisconsinan Gilman Canyon formation type locality, Buzzard's Roost, southwestern Nebraska, USA. Quatern Res 67:474–486

Karl T, Melillo J, Peterson T (eds) (2009) Global climate change impacts in the United States. Cambridge University Press, New York

Karlstrom E (1994) Paleoclimatic interpretation of a highly-weathered, relict paleosol in till at Topeka, Kansas. In: Dort W Jr (ed) TER-QUA symposium series, vol 2. Lincoln, Nebraska Academy of Sciences

Klopfenstein ST, Hirmas DR, Johnson WC (2015) Pedogenesis along a climosequence in loess-derived soils of the Central Great Plains, USA. Catena 133:25–54

Küchler AW (1964) Potential natural vegetation of the conterminous United States. Special publication no. 36, American Geographical Society, New York

LaBelle JM, Holliday VT, Meltzer DJ (2003) Early Holocene paleoindian deposits at Nall Playa, Oklahoma Panhandle, USA. Geoarchaeology 18:5–34

LaGrange T, Stutheit R, Gilbert M, Shurtliff D, Whited PM (2011) Sedimentation of Nebraska's playa wetlands: a review of current knowledge and issues. Lincoln, Nebraska Game and Parks Commission

Laird KR, Fritz SC, Maasch KA, Cumming BF (1996) Greater drought intensity and frequency before AD 1200 in the Northern Great Plains, USA. Nature 384:552–554

Ludvigson GA, Sawin, RS, Franseen EK, Watney WL, West RR, Smith JJ (2009) A review of the stratigraphy of the Ogallala Formation and revision of Neogene ("Tertiary") nomenclature in Kansas. Current Research in Earth Sciences, Bulletin 256, pt. 2, Kansas Geological Survey, Lawrence

Luo H-R, Smith LM, Allen BL, Haukos DA (1997) Effects of sedimentation of playa wetland volume. Ecol Appl 7:247–252

Mandel RD (1992) Soils and Holocene landscape evolution in Central and Southwestern Kansas: implications for archaeological research. In: Holliday VT (ed) Soils in archaeology: landscape evolution and human occupation. Smithsonian Institution Press, Washington, pp 41–100

Mandel RD (1994) Holocene landscape evolution in the Pawnee River Basin, Southwestern Kansas. Bulletin 236, Kansas Geological Survey and the Kansas State Historical Society, Lawrence
Mandel RD (1995) Geomorphic controls of the archaic record in the central plains of the United States. In: Bettis EA III (ed) Archaeological geology of the archaic period in North America. Special paper 297, Geological Society of America, Boulder
Mandel RD (2006) The Effects of late Quaternary landscape evolution on the archaeology of Kansas. In: Hoard RJ, Banks WE (eds) Kansas archaeology. University Press of Kansas, Lawrence, pp 46–75
Mandel RD (2008) Buried Paleoindian-age landscapes in stream valleys of the central plains, USA. Geomorphology 101:342–361
Mandel RD, Bettis EA III (2001a) Late Quaternary landscape evolution in the South Fork of the Big Nemaha River valley, southeastern Nebraska and northeastern Kansas. Guidebook no. 11, University of Nebraska Conservation and Survey Division, Lincoln
Mandel RD, Bettis EA III (2001) Use and analysis of soils by archaeologists and geoscientists: a North American Perspective. In: Goldberg P, Holliday VT, Ferring CR (eds) Earth science in archaeology. Kluwer Academic/Plenum, Norwell, pp 173–204
Mandel RD, Bettis EA III (2013) A 15,000 year record of landscape response to climate change preserved in alluvial fans in midcontinental North America. Geol Soc Am Abstr Programs 45:40
Mandel RD, Hofman JL (2003) Geoarchaeological investigations at the Winger Site: a Late Paleoindian bison bonebed in southwestern Kansas, USA. Geoarchaeology 18:129–144
Mason JA, Swinehart JB, Goble RJ, Loope DB (2004) Late Holocene dune activity linked to hydrologic drought, Nebraska Sand Hills, USA. The Holocene 14:209–217
Mason JA, Swinehart JB, Hanson PR, Loope DB, Goble RJ, Miao X, Schmeisser RL (2011) Late Pleistocene dune activity in the central Great Plains, USA. Quatern Sci Rev 30:3858–3870
Maupin MA, Barber NL (2005) Estimated withdrawals from principal aquifers in the United States, 2000. Circular 127, US Geological Survey, Washington DC
May DW (1989) Holocene alluvial fills in the South Loup Valley, Nebraska. Quatern Res 32:117–120
May DW, Holen SR (2014) Early Holocene alluvial stratigraphy, chronology, and Paleoindian/Early Archaic geoarchaeology in the Loup River Basin, Nebraska, USA. Quatern Int 342:73–90
Miao X, Mason JA, Swinehart JB, Loope DB, Hanson PR, Goble RJ, Liu X (2007) A 10,000-year record of dune activity, dust storms, and drought in the central Great Plains. Geology 35:119–122
Mills CT, Goldhaber MB, Stricker CA, Holloway JM, Morrison JM, Ellefsen KJ, Rosenberry DO, Thurston RS (2011) Using stable isotopes to understand hydrochemical processes in and around a Prairie Pothole wetland in the Northern Great Plains, USA. Appl Geochem 26:S97–S100
Murphy LR, Hurst SC, Holliday VT, Johnson E (2014) Late Quaternary landscape evolution, soil stratigraphy, and geoarchaeology of the Caprock Canyonlands, Northwest Texas, USA. Quatern Int 342:57–72
Namias J (1982) Anatomy of Great Plains protracted heatwaves (especially in the 1980 U.S. summer drought). Mon Weather Rev 110:824–838
Namias J (1983) Some causes of United States drought. J Clim Appl Meterol 22:30–39
Natural Resources Conservation Service (2006) Land resource regions and Major Land Resource Areas of the United States, the Caribbean, and the Pacific Basin. Handbook 296, United States Department of Agriculture, Washington
Natural Resources Conservation Service (2014a) National soil survey handbook. Title 430-VI. United States Department of Agriculture. http://www.nrcs.usda.gov/wps/portal/nrcs/detail/soils/survey/?cid=nrcs142p2_054242. Accessed 1 Dec 2014
Natural Resources Conservation Service (2014b) Official Series Description. National Cooperative Soil Survey, United States Department of Agriculture. https://soilseries.sc.egov.usda.gov/OSD_Docs. Accessed 22 Jul 2014
Nikiforoff CC (1949) Weathering and soil evolution. Soil Sci 67:219–223
Osterkamp WR, Wood WW (1987) Playa-lake basins on the Southern High Plains of Texas and New Mexico: Part I. Hydrologic, geomorphic, and geologic evidence for their development. Geol Soc Am Bull 99:215–223
Penner HL, Ekart SC, Ewing DA, Schmidt G, Smith J (1975) Soil survey of Butler County, Kansas. US Department of Agriculture, Soil Conservation Service, Washington
Reeves CC Jr (1976) Quaternary stratigraphy and geologic history of the Southern High Plains. In: Mahaney WC (ed) Quaternary stratigraphy of North America, Dowden, Hutchison, and Ross. Pennsylvania, Stroudsburg, pp 213–234
Reicken FF, Poetsch E (1960) Genesis and classification considerations of some prairie-formed soil profiles from local alluvium in Adair County, Iowa. Proc Iowa Acad Sci 67:268–276
Rich J (2013) A 250,000-year record of lunette dune accumulation on the Southern High Plains, USA and implications for past climates. Quatern Sci Rev 62:1–20
Richardson JL (1996) Mixing soil and water: the geology of wetlands. Geotimes 41:26–29
Sabin TJ, Holliday VT (1995) Playas and lunettes on the Southern High Plains: morphometric and spatial relationships. Ann Assoc Am Geogr 85:286–305
Schaetzl RJ, Anderson S (2005) Soils: genesis and morphology. Cambridge University Press, Cambridge
Sharratt B, Benoit G, Daniel J, Staricka J (1999) Snow cover, frost depth, and soil water across a prairie pothole landscape. Soil Sci 164:483–492
Shjeflo JB (1968) Evapotranspiration and the water budget of prairie potholes in North Dakota. Geological Survey Professional Paper 585-B, US Government Printing Office, Washington, DC
Smith LM (2003) Playas of the Great Plains. University of Texas Press, Austin
Smith MA, Hollander DJ (1999) Historical linkage between atmospheric circulation patterns and the oxygen isotopic record of sedimentary carbonates from Lake Mendota, Wisconsin, USA. Geology 27:589–592
Smith AG, Stoudt JH, Gollop JB (1964) Prairie potholes and marshes. In: Linduska JP (ed) Waterfowl tomorrow. US Fish and Wildlife Service, Washington, pp 39–50
Steeples DW, Knapp RW, McElwee CD (1986) Seismic reflection investigations of sinkholes beneath Interstate Highway 70 in Kansas. Geophysics 51:295–301
Steinwand AL, Richardson JL (1989) Gypsum occurrence in soils on the margin of semipermanent prairie pothole wetlands. Soil Sci Soc Am J 53:836–842
Stevens PR, Walker TW (1970) The chronosequence concept and soil formation. Q Rev Biol 45:333–350
Stout JE (2003) Seasonal variations in saltation activity on a High Plains saline playa: Yellow Lake, Texas. Phys Geogr 24:61–76
Sweeney M, Loope D (2001) Holocene dune-sourced alluvial fans in the Nebraska Sand Hills. Geomorphology 38:31–46
Tiner RW Jr (1984) Wetlands of the United States: current status and recent trends. National Wetlands Inventory, US Fish and Wildlife Service, US Government Printing Office, Washington
Tomanek GW, Hulett GK (1970) Effects of historic droughts on grassland vegetation in the Central Great Plains. In: Dort W Jr,

Jones JK Jr (eds) Pleistocene and recent environments of the Central Great Plains. The University Press of Kansas, Lawrence, pp 203–210
Trimble D (1980) The geologic story of the Great Plains. U.S. Geological Survey Bulletin 1493, Washington
Tsai J-S, Venne LS, McMurry ST, Smith LM (2012) Local and landscape influences on plant communities in playa wetlands. J Appl Ecol 49:174–181
Walters RF (1978) Land subsidence in central Kansas related to salt dissolution. Bulletin 214, Kansas Geological Survey, Lawrence
Wishart DJ (2011) Encyclopedia of the Great Plains. University of Nebraska, Center for Great Plains Studies, Lincoln
Wood WW (2002) Role of ground water in geomorphology, geology, and paleoclimate of the Southern High Plains, USA. Ground Water 40:438–447
Wood WW, Osterkamp WR (1987) Playa-lake basins on the Southern High Plains of Texas and New Mexico: Part II. A hydrologic model and mass-balance arguments for their development. Geol Soc Am Bull 99:224–230
Wood WW, Rainwater KA, Thompson DB (1997) Quantifying macropore recharge: examples from a semi-arid area. Ground Water 35:1097–1106
Yaalon DH (1975) Conceptual models of soil formation: Can soil-formation functions be solved? Geoderma 14:189–205

9 Southwest Plateaus and Plains Range and Cotton Region and Southwestern Prairies Cotton and Forage Region: LRRs I and J

David Emmanuel Ruppert

9.1 Introduction

Land resource regions (LRR) J (Fig. 9.1) and I (Fig. 9.2) are zones of transition. LRR J may be considered the beginning of the pre-European settlement transition zone between the more-or-less unbroken forests to the east and prairies to the west in the southern USA. The transition in LRR I is from grass and scrubland to peri-desert in often thinner or more droughty soils than in LRR J. The collective region corresponds roughly to the transition zone (95°–99°W) between eastern and western plant species identified by Macroberts and Macroberts (2003) and the line separating pedocals and pedalfers by Marbut (1928).

9.2 Land Resource Region J

Early sojourners noted stark changes in vegetation in transects across NE Texas (Bryan 1951). Leaving the forested landscapes of east Texas (MLRA 133B), paths would transect alternating zones of oak savannah and prairie if traveling west (Table 9.1; Fig. 9.3). In time, these different zones would define different economic opportunities and spell the difference between prosperity and poverty for homesteaders (Yelderman 1998), due to soil differences.

In such a transect (Fig. 9.3), underlying geologies transition from sandier to finer and more calcareous materials several times. The boundaries between these different deposits correspond to the boundaries between oak savannah and prairie and to different MLRAs. The central theme of LRR J is that prior to European colonization, non-calcareous, often sandy parent materials, having developed sandy surfaces over clayey subsoil, supported savannahs or woodlands, and often calcareous, finer-textured parent materials with loamy or clayey surfaces supported prairies (Table 9.1).

D.E. Ruppert (✉)
Texas A&M University-Kingsville, Kingsville, TX, USA
e-mail: david.ruppert@tamuk.edu

The digital and not smooth transition from forested east to prairie west[1] occurs because of geology. During the Cretaceous period, a shallow sea developed through the middle of the continent, depositing sandy, clayey, or calcareous materials depending on the energy associated with the depositing waters. Looking along the previously discussed transect (Fig. 9.3), the transition from the Paluxy (a sandy formation at the top of the Trinity Group) to the (fine and calcareous) Washita, and from the (sandy) Woodbine to the (calcareous) Austin Chalk corresponds to two incursions of the Cretaceous sea into the continent separated by a period of erosion.[2] Encroaching seas typically produce a sand-clay-carbonate signature as one ascends through the rock profile (Fig. 9.4). In the Tertiary period, deposition occurred mostly subaerially through fluvial, deltaic, or coastal processes as the sea retreated. As a result, many, Tertiary materials are sandier than marine Cretaceous deposits.

9.2.1 Oak-Alfisol-Dominated MLRAs

Considering the correlation between coarse textured surfaces and woodland in Texas, Tharp (1939) observes and speculates:

> they have the same woody dominants; namely, Post oak, blackjack and in some localities hickory.[3] [In addition to MLRAs 84BC and 87AB] Post oak is spattered out over [MLRAs 78ABC, 80B, 81ABC, 85] in localities where sandy or gravelly soil, with the right admixture of subsoil clay and sand

[1]To the west of LRR J, in MLRA 80B (LRR H), oak savannah vegetation continues; prairie becomes dominant further to the west in MLRA 78C (LRR H), though even here post oak and shinnery oak (*Quercus havardii*) may be found on coarser textured soils.

[2]A third incursion of the Cretaceous sea is recorded within the Trinity Group between the basal sands of the Trinity and the Glen Rose Limestone.

[3]Post oak (*Quercus stellata*); Blackjack oak (*Quercus marilandica*); Hickory (*Carya sp.*)

L.T. West et al. (eds.), *The Soils of the USA*, World Soils Book Series,
DOI 10.1007/978-3-319-41870-4_9

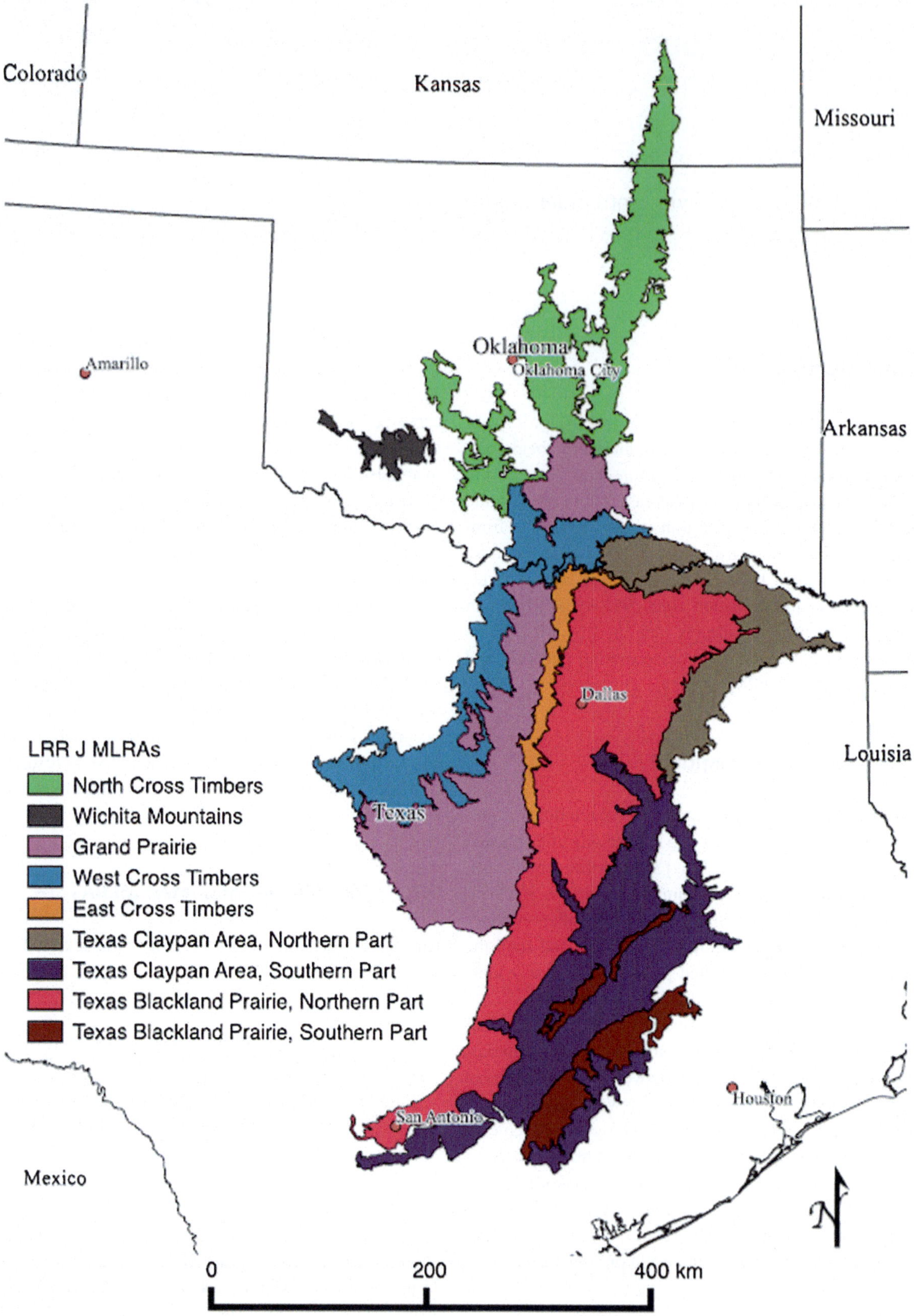

Fig. 9.1 MLRAs within LRR J

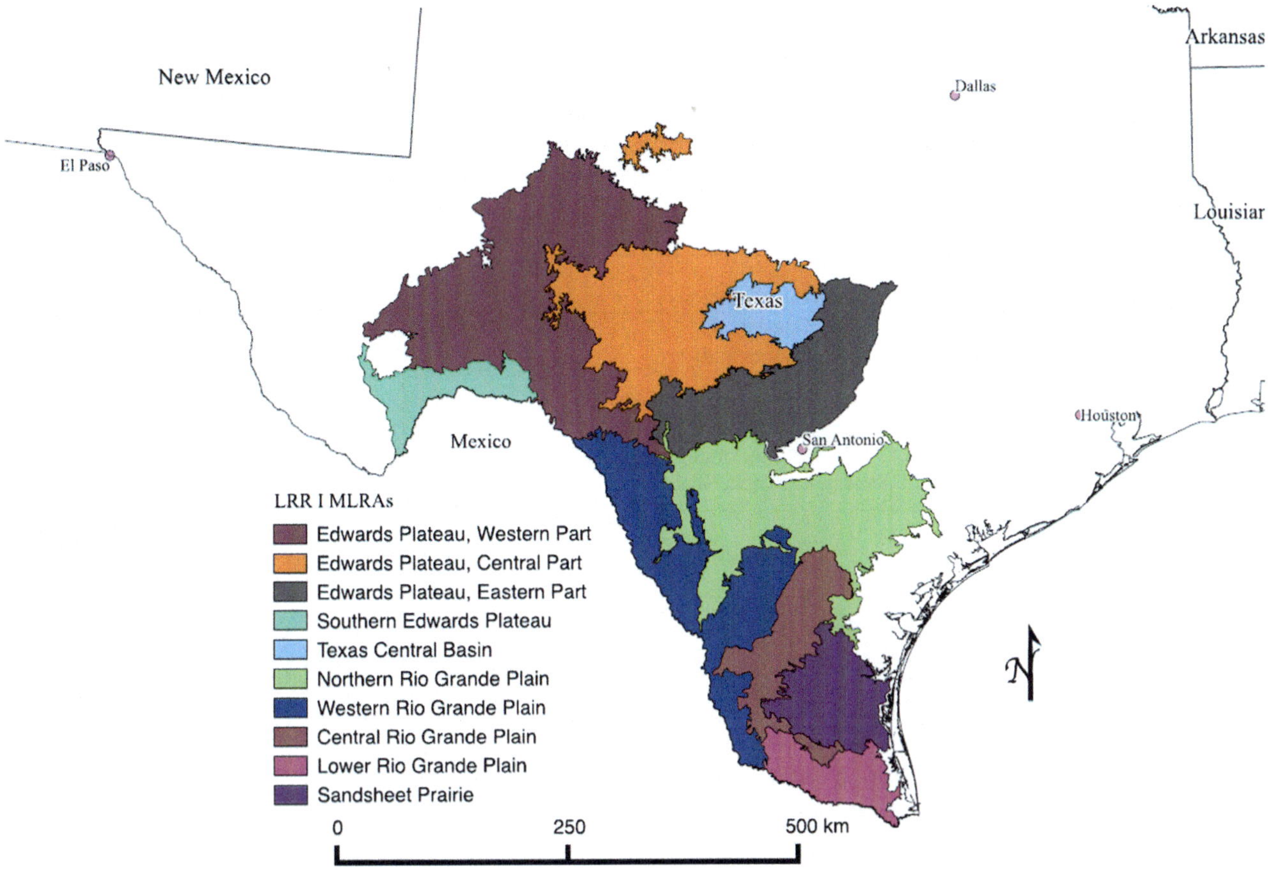

Fig. 9.2 MLRAs within LRR I

> to make a good water reservoir, occurs covered by sandy top soil adapted to a high degree of absorption in time of rain and of protection against surface evaporation in time of drouth. Such a soil is the best possible conservator of moderate or scant precipitation. Postoak and blackjack are well adapted to moderate supplies of water and, hence, are found on such soils farther west than any other trees of the eastern coastal plains forest.

The hypothesis of Tharp (1939) involving sand-maximized infiltration and sand-minimized evaporation has not been strictly tested for oak savannah vegetation in this region, but work by Sala et al. (1988), Alizai and Hulbert (1970), and Noy-Meir (1973) is consistent with it.

Tharp (1939) adds, regarding the presence of oak savannah in central Texas:

> Always they are found on sandy or gravelly soil with a reddish clay and sand sub-soil.

In effect, Tharp was observing a correlation between oak savannah and Alfisols (Table 9.1). Alfisols must contain a subsurface horizon higher in clay than overlying eluvial horizons, and this clay increase must result from eluviation-illuviation of clay (Soil Survey Staff 2014). Soils of oak savannahs are underlain overwhelmingly by Alfisols (Tables 9.1, 9.2). Therefore, the clay increase from surface to argillic horizon must be appreciable and hence the observation of Tharp (1939) that oak savannah 'always' occurs over a sandy surface and clayey subsurface. Unlike the oak savannah MLRAs, prairie-predominant MLRAs contain Mollisols (Grand Prairie, MLRA 85), Vertisols (North Blackland Prairie, MLRA 86A), or a balance of soil orders in which the majority of the soils themselves express vertic features (South Blackland Prairie, MLRA 86B) (Tables 9.1, 9.3). Relative to the Blackland Prairies the Alfisols of LRR J are not particularly fertile and are typically used for grazing (Table 9.1). The relative wealth and prospects associated with homesteader farming of the sandy Alfisols of the East Cross Timbers (MLRA 84C) versus the fertile Vertisols of the North Blackland Prairie (MLRA 86A) is provided by Yelderman (1998). Oak savannah Alfisols of North East Texas are well suited to orchards and can be used for vegetable crops as well as peanuts although grazingland is still the predominant agricultural land use.

9.2.1.1 Paleudalfs/Paleustalfs of LRR J

Often accompanying clay illuviation is the oxidation of reduced iron present in primary minerals. Argillic horizons often contain more iron than horizons above or below, hence

Table 9.1 Characteristics of MLRAs in LRR J

MLRA		Native Veg.[a]	Parent Material[b] Age	Parent Material[b] Type	Precip. (cm)	B-Horizon Munsell Hue	Surface Texture	Typical Soil Depth	Typical Ag. Use	Most common soil(s) and abundance (%)
Cross Timbers	North (84A)	Post/Blackjack Oak Savannah	Permian, Pennsylvanian	Sandstone, Shale	76-102	5YR – 10YR	Fine Sandy Loam	Mod. Deep	Grazing (often wooded)	Haplust/udalfs 60
	West (84B)		Cretaceous	Sandstone	66-107			Deep		Pale/Haplustalfs 88
	East (84C)			Fluvial/Deltaic/ Coastal Sandstones and Shales	87-104			Mod. Deep		Paleustalfs 76
Claypan	South (87A)		Tertiary		69-115			Very Deep		Paleustalfs 79
	North (87B)				99-115		Mix		Grazing	Pale/Hapludalfs 84
Wichita Mountains (82B)			Precambrian, Cambrian	Granite, Limestone, Sandstone, Chert, Shale	66-79		Loam	Mix		Argi/Haplustoll 89
Grand Prairie (85)		Mid/Tallgrass Prairie	Cretaceous	Limestone	69-104	7.5YR – 2.5Y	Clay Loam & Clay	Shallow		Calci/haplustolls 56 Haplusterts 17
Blackland Prairie	North (86A)			Marl, Chalk, Shale	76-117		Clay	Mix	Crops	Haplusterts 51 Paleustalfs 16
	South (86B)		Tertiary	Calcareous Clays, Sandstones and Marls	89-112		Mix	Very Deep	Crops and Grazing	Paleustalfs 19 Hapl/Calciusterts 38 Mollisols 28

[a]Predominating pre-European settlement plant communities occupying residual soils on uplands
[b]Predominating parent materials of residual soils

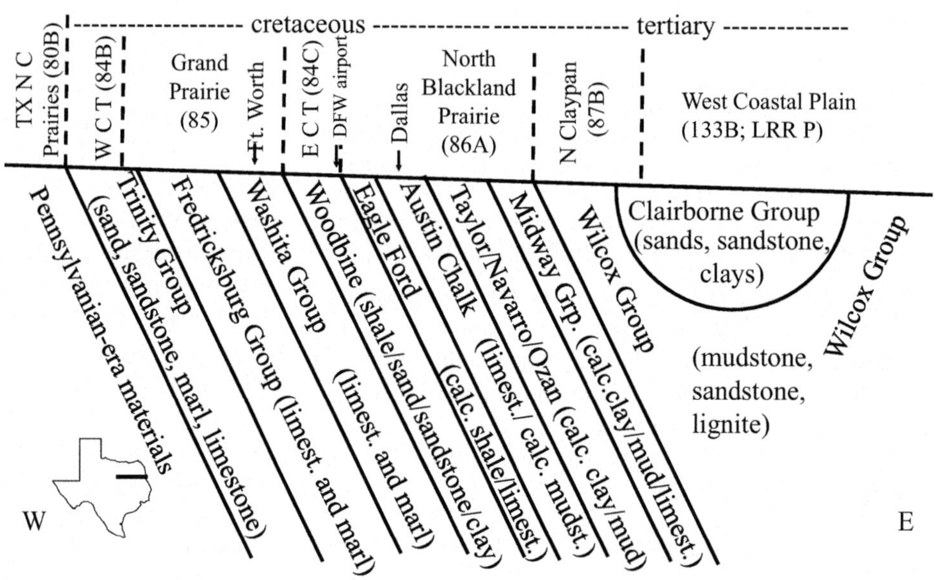

Fig. 9.3 Geologic groups and formations correlated to boundaries between Major Land Resource Areas of Central North and East Texas, land resource region J

the observation of 'reddish' subsoil underlying oak savannah by Tharp (1939). Subsurface 'B' horizons of residual soils of MLRAs typically associated with oak savannah are redder and brighter than B horizons of soils endemic to prairie biomes in LRR J (Table 9.1). Paleustalfs with red hues and bright chromas in the argillic horizon are abundant in the oak savannah MLRAs (Tables 9.1, 9.2), and soils in this great group typically have an abrupt and substantial increase in clay accompanying the transition from eluvial (A and E) to illuvial (Bt) horizons (Soil Survey Staff 2014). Such 'claypan' Paleudalfs and Paleustalfs are behind the names of the Texas Claypan MLRAs (87AB) of LRR J, but Alfisols with this morphology are found in the other MLRAs in the region (Table 9.4).

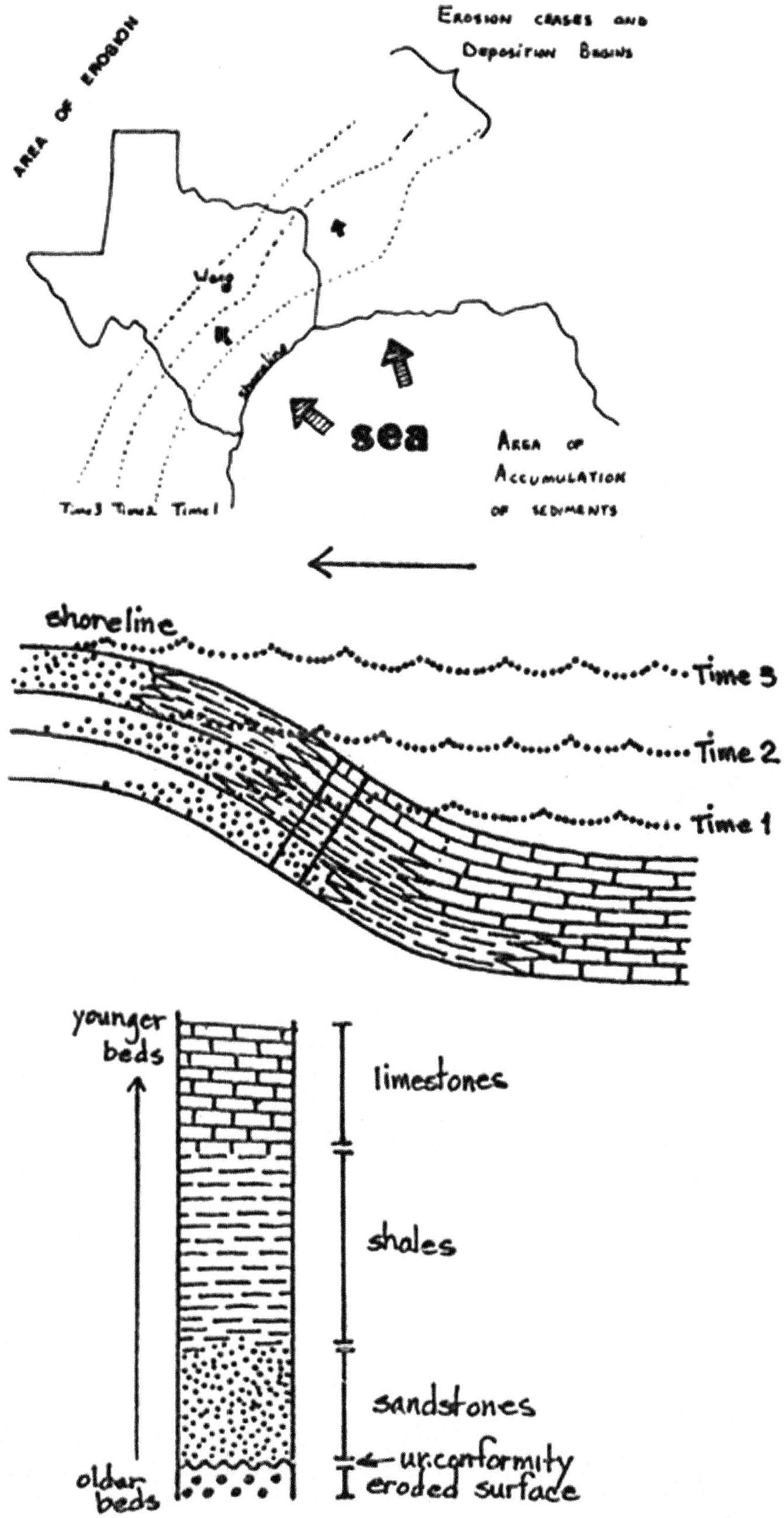

Fig. 9.4 Transgressing marine waters leaving an ascending sand-clay-carbonate profile in deposited materials. Modified from the original figure found in Baylor Geological Society (1979). By permission of Baylor Geological Society, Waco, Texas

Table 9.2 Distribution of major (> 5 %) soil great groups among the MLRAs of LRR J

Soil great groups	MLRA								
	84A (%)	84B (%)	84C (%)	87A (%)	87B[a] (%)	86A (%)	86B (%)	85 (%)	82B (%)
Haplust/udults			6						
Haplust/udalfs	60	17	5	5	40	11			
Paleust/udalfs	8	71	76	79	44	16	19		
Glossudalfs					5				
Glossaqualfs					6				
Haplust/udolls						10	6	15	12
Argiust/udolls							7	7	77
Calciustolls							10	41	8
Haplust/uderts						51	21	17	
Calciusterts							17		
Haplustepts	10							10	
Eutrudepts	7								
Ust/udifluvents	5	7	7						
Quartzipsamments	5								
Whole MLRA area (10^6 hectares)	1.61	1.15	0.30	2.13	0.53	3.86	0.81	2.47	0.07

[a]In 87B, the majority of soils are of the 'udic' moisture regime; in all others, the moisture regime is predominantly 'ustic'

Table 9.3 Percentage of soils in East and North East Texas MLRAs with vertic character

MLRA	Soils with 'vertic'[a] character excluding Vertisols (%)	Vertisols (%)	Paleustalfs with vertic character (%)
86A	27	51	97
86B	36	43	100
87A	17	2	20
87B	51	1	93

[a]Taxon includes the letters 'ertic'

Table 9.4 Differences among Paleustalfs in the Paleustalf-dominated MLRAs of LRR J

MLRA	Percentage of Paleustalfs with			
	likely claypan morphology[a] (%)	'vertic' character[b] (%)	'aquic' character[c] (%)	sandy character[d] (%)
84B	61	0	28	4
84C	48	24	2	10
87A	65	20	36	20

[a]'Fine', 'very-fine, or 'clayey' particle size control sections
[b]Subgroups containing the string 'ertic'
[c]Subgroups containing the string 'aqu'
[d]Subgroups containing the string 'arenic' or 'lamellic' or 'psammentic' subgroups

9.2.1.2 The Development of Alfisols and not Ultisols in LRR J

The line delineating LRRs J and P separates Alfisol- and Ultisol-dominated regions (Fig. 9.3). This would appear to be the case for two reasons: (1) Many of the soils in the coastal plain of LRR P (abundant Ultisols) are developed from sediments derived from upstream soils which were not base rich. In contrast, parent materials for soils (dominantly Alfisols) in LRR J are of marine or fluvial origin in which the fluvial sediments were derived from previously unweathered materials from the Rocky Mountains (Jurena 2005) or upstream carbonate-rich marine deposits (Greenwade 1996; Hyde et al. 1992). (2) Increased rainfall from west to east across the region results in moister soils with a higher likelihood of leaching in LRR P. Most upland soils in LRR J have an ustic moisture regime as opposed to soils in

Table 9.5 Percentage of soils, in East and North East Texas MLRAs with aquic or udic moisture regimes, ustic moisture regimes, Alfisols, or Ultisols

LRR	MLRA	Soils with 'aquic or 'udic' moisture regimes (%)	Soils with 'ustic' moisture regimes (%)	Alfisols (%)	Ultisols (%)
J	87B	98	2	95	1
	87A	7	94	85	2
	86A	12	88	28	tr[a]
P	133B	100	tr	36	49

[a]'tr' indicates a percentage which does not round up to 0.1 or more

LRR P which have a (moister) udic regime (Table 9.5). This indicates that the majority of soils in LRR J experience soil moisture levels in which leaching is limited. As a result, subsoils in LRR J have not been leached of exchangeable base elements (Na, K, Ca, Mg) as in Ultisols (Soil Survey Staff 2014). Limited leaching is cited by Gray and Roozitalab (1976) as an important mechanism for the maintenance of Alfisols in sub-humid environments.

9.2.1.3 Northern Cross Timbers (MLRA 84A)

The Northern Cross Timbers (NCT) extends from south central Oklahoma to Kansas and is comprised of three arms embedded within Permian (the smaller and westward arms) or Pennsylvanian (the eastern and uppermost arm) materials. Slopes can reach as high as 45 %, higher than in other MLRAs of the LRR. Many soil series in the region can be found in both colluvial and residual settings. Sandstone alone or sandstone interbedded with shale is the most common residual parent material (Table 9.1) and tends to yield Alfisols. Soils derived from sandstone and shale tend to be moderately deep and deep, respectively. Vertic properties can result where a sufficient fraction of the parent material is shale. In addition to Alfisols, sandstone can produce shallow loamy or sandy Inceptisols or Entisols (Table 9.2), especially when slopes are high (Fig. 9.5). Where interbedding of sandstone and shale occurs on a landscape scale, ridges are formed in sandstone with valleys in shale. The Pennsylvanian arm of the MLRA receives higher rainfall than the western Permian arms, which results in soils over Pennsylvanian age parent materials typically having an udic moisture regime. The NCT generally has Alfisols considered to be less weathered (Hapludalfs or Haplustalfs) than those of the other oak savannah-Alfisol MLRAs of LRR J (Paleustalfs) (Table 9.2).

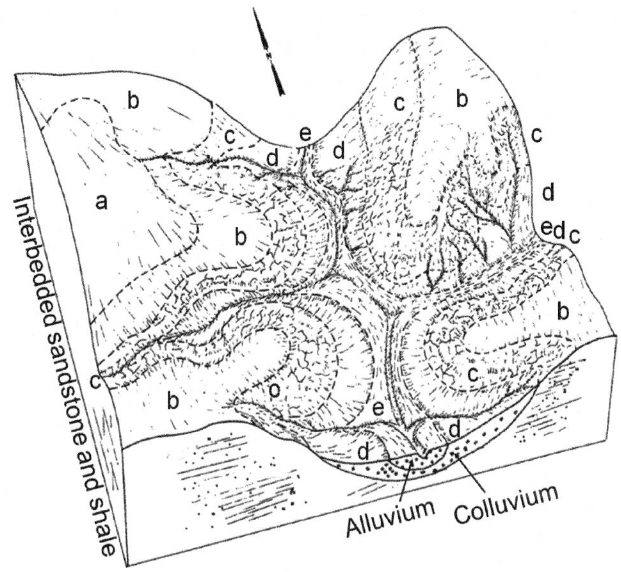

Fig. 9.5 Soil-landscape relationships in the North Cross Timbers. Haplustalfs (*a*), Haplustalfs-Haplustepts complex (*b*), Haplustalfs-Haplustepts or Haplustalfs-Quartzipsamments complex (*c*), Haplustepts or Paleustalfs (*d*), Usti- and Udifluvents (*e*). Modified from Bourlier et al. (1987)

9.2.1.4 Western Cross Timbers (MLRA 84B)

Soils in the Western Cross Timbers (WCT) are deeper than in the NCT (Table 9.1), which is likely a result of not only lower slopes than in the NCT but also higher temperatures and a longer growing season which would produce more extensive weathering (Schaetzl and Anderson 2005). In addition, in many places the sandy parent materials of the WCT occur as uncemented 'pack' sands, offering less resistance to erosion than (cemented) sandstones. More extensive weathering in the WCT is reflected in the dominance of Paleustalfs versus Haplustalfs (Table 9.2), with Paleustalfs in the WCT occupying a diversity of landscape positions (Fig. 9.6). More extensive weathering is also reflected in the lack of shallow Inceptisols and Entisols in uplands (Table 9.2). While many of the same soil series (Haplustalfs) that occur in the NCT also occur in the WCT, these soils are largely restricted to backslope positions (Fig. 9.6) where weathering would be relatively limited.

In the WCT the majority of terrace and floodplain soils formed in loamy sediments. Terraces typically contain Paleustalfs. Floodplains are dominantly Ustifluvents if the parent materials are non-calcareous (Fig. 9.6) with Haplustolls over calcareous sediments. The WCT is comprised of sandy formations of the Trinity Group (Sellards et al. 1932; Fig. 9.3). A significant number of soils in the MLRA are

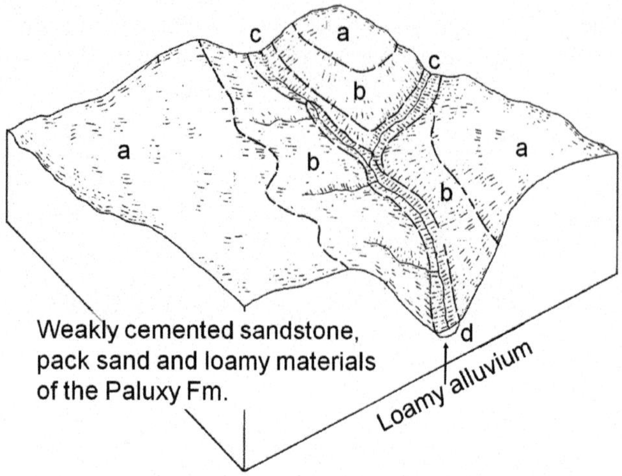

Fig. 9.6 Soil-landscape relationships in the West Cross Timbers, formed from the Paluxy and Trinity Sands of the Trinity Group. Paleustalfs (*a*), Haplustalfs (*b*), Paleustalfs (*c*), Ustifluvents (*d*). Modified from Ressel (1981)

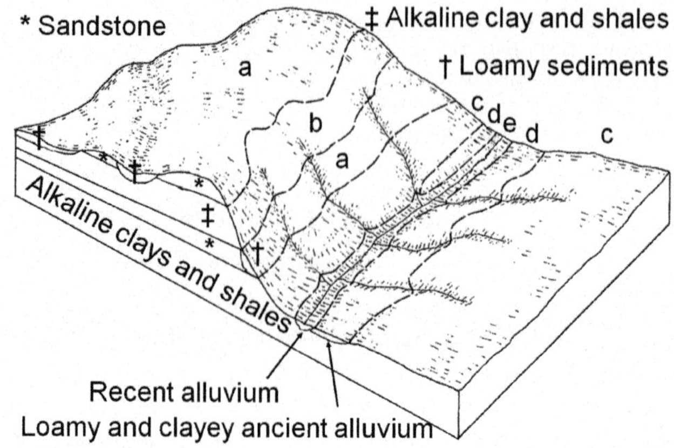

Fig. 9.7 Soil-landscape relationships in the East Cross Timbers, formed from the Woodbine Sands. Mostly Ultic and (some) Arenic Paleustalfs on loamy sediments and sandstone (*a*), Haplustults on acid shale or Haplustalfs on clayey high slope areas (*b*), Udertic Paleustalfs on alkaline clays and shale(*c*), Aquic Paleustalfs on loamy and clayey ancient alluvium; Haplustalfs when textures are sandier (*d*), and Ustifluvents (*e*). Modified from Ressel (1981)

formed from residual or terrace materials that show signs of having been reworked by wind. In the Texas portion of the WCT, agricultural production is largely supportive of cattle either directly in grazing lands and improved pasture, or indirectly in hay and small grain production for dairies (T. Riley Dayberry, NRCS, pers. comm.).

9.2.1.5 The Eastern Cross Timbers (MLRA 84C)

Seventy-six percent (Table 9.2) of the East Cross Timbers (ECT) is underlain by Paleustalfs (Fig. 9.7) developed over the Upper Cretaceous Woodbine 'Sand'. The Paleustalfs that make up the ECT are largely the same as in the WCT with an important exception. The most common Paleustalfs in the ECT are Udertic Paleustalfs, i.e., relatively non-droughty Paleustalf with shrink–swell characteristics. These soils are not found in the WCT. The Woodbine includes a number of shaley members (Hawkins et al. 1974; Main and Noto 2012) of sufficient clay to have shrink–swell properties. These geologic members and their resulting clayey Paleustalfs are transitional to the clayey soils of the Northern Blackland Prairie, which lie just to the east. The Woodbine is a fluvial/deltaic/near-shore marine deposit (Hawkins et al. 1974; Main and Noto 2012), which explains the presence of a former acid sulfate soil (e.g., 'Aubrey' series; Carson and Dixon 1983). Sulfuricization (Fanning and Fanning 1993) and elevated rainfall (Table 9.1) may also explain the greater abundance of Ultic Paleustalfs (relatively base poor) in the ECT (31 % of Paleustalfs) versus the WCT (8 %).

Vegetation species are reported by Marcy (1982) to be largely the same between the ECT and WCT, with post oak (*Quercus stellata*) and blackjack oak (*Quercus marilandica*) the dominant trees, and little bluestem (*Schizachyrium scoparium*) the dominant grass. In 1982, intact characteristic vegetation was found to occupy only 33 % of the original land area in the ECT and 43 % in the WCT (Marcy 1982). While the species present are largely the same, the occurrence of vegetation is not. Marcy (1982) reports that characteristic landscapes of the WCT are more parklike, with wider separation between trees (likely due to drier conditions). In contrast, denser stands (savannah or forest) occur in the ECT.

9.2.1.6 The Texas Claypan Area, Southern Part (MLRA 87A)

Even among its Paleustalf-dominated neighbors (84BC), the Texas Claypan Area, Southern Part (SCP), has the greatest abundance of Paleustalfs (mapped on nearly 79 % of the surface of the MLRA). While deep and sustained clay translocation coupled with extreme mineral weathering may place a soil in the category of Paleustalfs (Soil Survey Staff 2014), the term 'claypan' implies that many of the soils in this MLRA contain an abrupt texture change to a clayey (Bt) horizon, which is another property common to Paleustalfs.

The sudden clay increase at the top of the Bt horizon retards water infiltration; the resulting perching of water (episaturation) produces redoximorphic features in the upper parts of many SCP soils (Rehage 1985). Episaturation when wet and extreme drying in summer limits the agronomic utility of many SCP soils for row cropping (Rehage 1985). Instead, the primary usage of SCP soils is pasture or forage (Neitsch et al. 1989). The episaturation of Paleustalfs is not

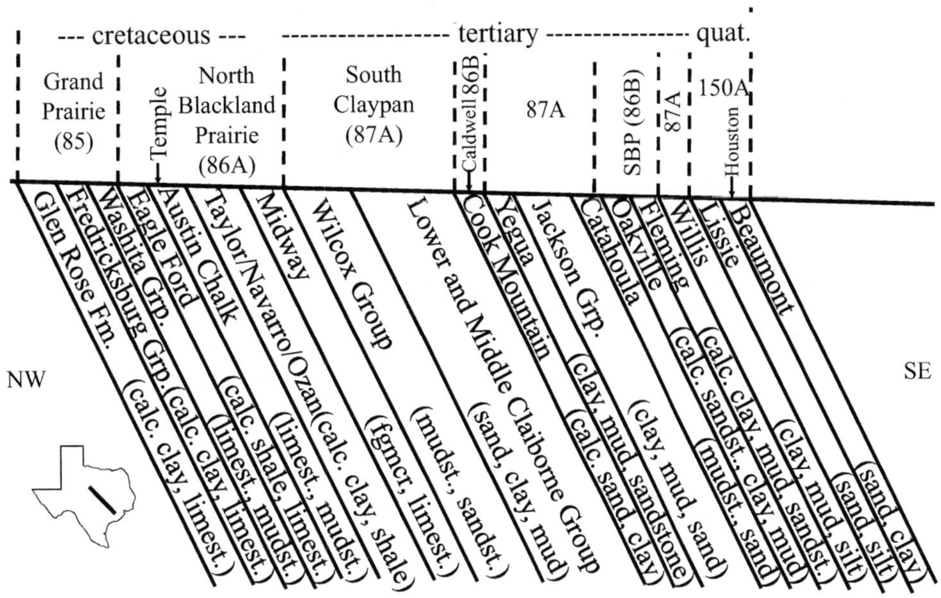

Fig. 9.8 Geologic groups and formations correlated to boundaries between Major Land Resource Areas (MLRAs) of Central and Southeast Central Texas, Land Resource Region (LRR) J. MLRA 150A occurs in LRRT. 'calc.' calcareous; 'limest.' limestone; 'mudst.' mudstone; 'Fgmcr' Fine-grained mixed clastic rock; 'sandst.' sandstone

unique to this MLRA but commonly occurs in the WCT and on minor soils in the ECT. Many SCP soils feature relict redox features left over from periods when groundwater levels were higher (Greenberg 1994). Greenberg (1994) measured the water table in three soil profiles displaying redoximorphic features and found that in none of them did the water table reach within 5 m of the surface.

The geologic formations that give rise to the residual soils of the SCP are tertiary in origin and parallel the present coastline (Table 9.1; Fig. 9.1). The SCP may be thought of as different regions subdivided by the presence of the South Blackland Prairies (Fig. 9.8) with many different parent materials represented. Soils overlying the Wilcox and Lower and Middle Claiborne tend to be 'deep' or very deep, but those over the Jackson group are generally only 'moderately deep' (Jurena 2007). Sandy parent materials can produce Paleustalfs with thick, sandy eluvial horizons (arenic and grossarenic subgroups). Siltstone and shale produce Paleustalfs with shrink-swell behavior (vertic, Udertic subgroups). Loamy or silty materials interbedded with sandstone produce (Ultic) Paleustalfs relatively low in base saturation (Fig. 9.9). While these subgroups of Paleustalfs can be found in other oak savannah-Alfisol MLRAs, the diversity of these subgroups is unique (Table 9.4) to the SCP.

Southwest of the San Antonio Prairie (see below) is an intriguing subset of the SCP called the 'lost pines'. Here, a forest of loblolly pine (*Pinus taeda*) incongruously occupies the landscape, displaced 100 miles west from the stronghold of the pine-dominated landscapes in East Texas (MLRA 133B). These stands are frequently observed on soils with thick sandy A and E horizons (grossarenic subgroups) (Jason Morris, NRCS, pers. comm.) and can be found in additional areas of the MLRA.

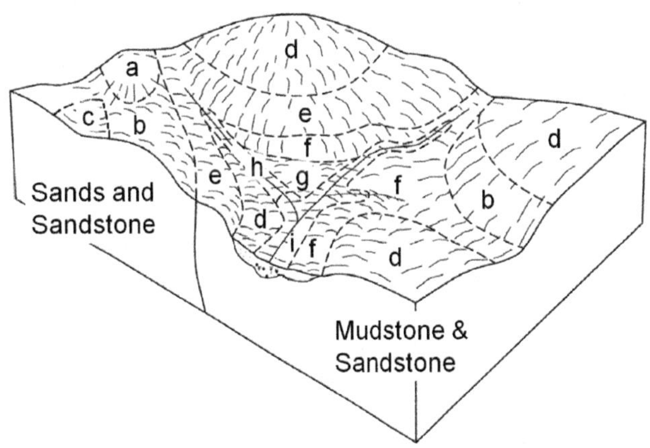

Fig. 9.9 Soil-landscape relationships in the Texas Claypan Area, Southern Part (MLRA 87A). Arenic Paleustalfs (*a*), Grossarenic Paleustalfs (*b*), Aquic Arenic Paleustalfs (*c*), Udic Paleustalfs (*d*), Ultic Paleustalfs (*e*), Vertic Paleustalfs (*f*), Aquic Paleustalfs (*g*), Aquic Haplustepts (*h*), Udifluventic Haplustepts (*i*). Modified from Jurena (2007)

9.2.1.7 The Texas Claypan Area, Northern Part (MLRA 87B)

The Texas Claypan Area, Northern Part (NCP), similar to the SCP, contains mostly Alfisols (Table 9.1). However, unlike any other MLRA in LRR J, the NCP contains a large amount (98 %) of soils with ('udic' or 'aquic' moisture regimes) (Table 9.5). In this sense, it is similar to MLRA 133B, which it abuts to the east; however, in MLRA 133B, Ultisols are the dominant soil order (Table 9.5). Also, unlike many other MLRAs in this LRR, the NCP contains no geological formations unique or largely unique to itself; instead, parent materials for NCP soils are shared with adjacent MLRAs. As a result, the NCP shares soils with its adjacent MLRAs and contains only one endemic series, a Glossudalf ('Hicota'), which forms in ancient terraces of the Red River.

9.2.2 Prairie-Dominated MLRAs of LRR J

9.2.2.1 Vertisols

Vertisols are the second most abundant soil order found in LRR J, occupying approximately 22 % of the surface area of the LRR. Vertisols need to be high in clay throughout and to exhibit physical signs of soil shrinking and swelling that are induced by moisture changes (Soil Survey Staff 2014). East Central Texas provides an ideal climate for substantial moisture changes in its soils. Rainfall occurs throughout the year, albeit with rain most likely in late spring and early fall (Fig. 9.10). The occurrence of periods with relatively little rainfall, especially during the height of summer heat, typically opens large cracks in Vertisols. Subsequent rewetting swells the soil, closing cracks. Soil swelling means that some elements of the profile are thrust upwards to alleviate longitudinal forces, inducing 'gilgai' microtopography, which helps to promote water storage (Fig. 9.11). Vertisol-dominated landscapes are extensively cropped (Table 9.1).

The North Blackland Prairie (MLRA 86A)

The North Blackland Prairie (NBP) can be split into four smaller regions running west to east: the Eagle Ford Prairie, the White Rock Cuesta, the Taylor Black Prairie, and the Eastern Marginal Prairie (Montgomery 1993; Hill 1901). The parent material of these different divisions corresponds, running west to east and of older to younger age, to the Eagle Ford Formation, the Austin Chalk Formation, the Taylor group and the Navarro and Midway groups, respectively (Figs. 9.3, 9.8). These different geologies give rise to clearly different soils: Vertisols, Mollisols, Vertisols, and Alfisols, respectively (Hallmark 1993).

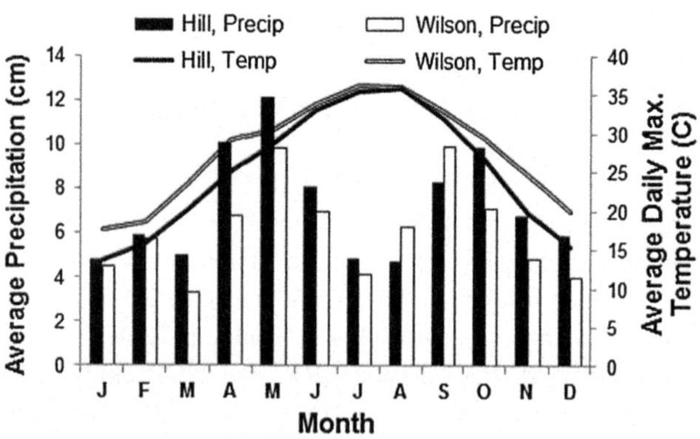

Fig. 9.10 Precipitation and high temperature averages in Hill (Brooks 1978) and Wilson (Taylor 1977) Counties, Texas

This region of Texas is home to the cities of Dallas, Waco, San Antonio, Austin, and Temple. The area is almost entirely used for agriculture (row cropping and grazing) or urban sprawl. The I-35 corridor connecting these cities (and preceding routes, e.g., the Chisolm trail) largely follows the relatively thin, stony, non-vertic soils overlying the Austin Chalk (Baylor Geological Society 1994). This MLRA, as a whole, is historically known for cotton production (Brooks et al. 1964), but current cropping also includes corn, oats, sorghum, soybeans, and wheat (Texas County Estimates 2015). Less than one hundredth of one percent of the original NBP soil-vegetation landscape remains (unplowed) (Brice Glidewell, The Nature Conservancy, pers. comm.). Many of these remnant parcels require intensive management to maintain the native plant community against invasives (James Eidson, pers. comm.).

Haplusterts are the predominant residual soils of the Eagle Ford and Taylor/Navarro groups, which are characterized by shale, clay, or marl (Fig. 9.12). Shallow and moderately deep, fine-textured Haplustolls and shallow loamy-textured Ustorthents and Haplustepts characterize residual soils of the Austin Chalk (Fig. 9.12). In many places, the effective delineation between the Blackland Prairie and the NCP occurs where Paleustalfs with vertic properties are mapped versus Paleustalfs without vertic features NBP, NCP (Table 9.3).

Clayey alluvium on terraces and floodplains forms Vertisols or soils with vertic properties unless slopes are relatively high. Calciustolls are associated with loamy calcareous alluvium, and if the loamy alluvium is non-calcareous, Haplustolls are common.

The South Blackland Prairie (MLRA 86B)

Surrounded on almost all margins by the SCP, the South Blackland Prairie (SBP) is displaced from the NBP (Fig. 9.1). Unlike the majority of the NBP, SBP lithologies are Tertiary and not Cretaceous in origin and somewhat sandier (Table 9.1). As a result of this higher sand content relative to parent materials of the NBP, SBP parent materials do not support the same high proportion of Vertisols; instead, a more balanced suite of great groups is found (Table 9.1). However, many of these soils have vertic features (Klich et al. 1990; Table 9.3).

The SBP occurs as two islands embedded within the SCP. In a transect through these two MLRAs (Fig. 9.8), the transition from SCP into the SBP occurs when clayier and often times more calcareous materials are encountered. The northern of the two sections is sometimes referred to as the 'San Antonio' prairie and occurs over portions of the (clayey) Cook Mountain and (loamy) Yegua formations; the southern of the two sections is known as the 'Fayette' prairie and occurs over Fleming and Oakville and in places over the Catahoula, Largarto, and Goliad formations.

Fig. 9.11 Gilgai microrelief of a Texas Vertisol. Note the substantial water retention. Anonymous source courtesy of the College of Agriculture and Life Sciences, University of Idaho

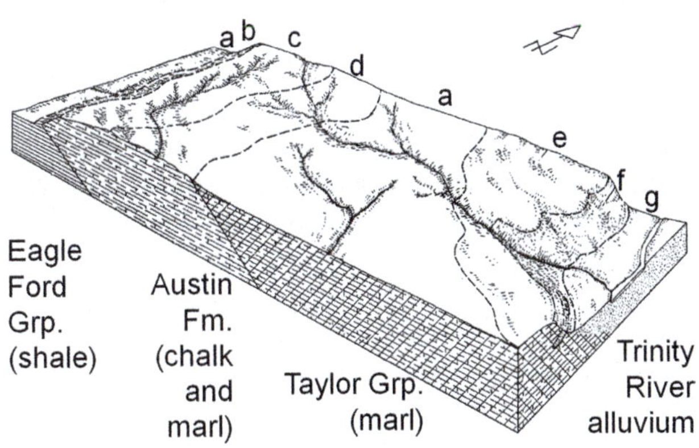

Fig. 9.12 Soil-landscape-geological relationships in the North Blackland Prairie (MLRA 86A) spanning the Eagle Ford, Austin, and Taylor Prairies. Haplusterts-Hapluderts complex (*a*), Haplusterts-Ustorthents complex (*b*), Ustorthents-Haplustolls complex (*c*), Haplustolls (*d*), Paleustalfs-Haplustalfs complex (*e*), Haplusterts-Eutrudepts complex (*f*), Hapluderts-Haplustolls complex (*g*). Modified from Brooks et al. (1964)

Vertic Paleustalfs dominate in loamy formations of the San Antonio Prairie, whereas in clayier formations Argiustolls and Haplusterts are also found (Fig. 9.13). Haplustalfs typically occupy terrace sediments. In the Fayette Prairie, the presence of Paleustalfs is diminished relative to that of Mollisols and Vertisols, with Calciustolls and Calciusterts occurring somewhat uniquely in this subset of the SBP (Fig. 9.14). Similar to the North Blackland Prairie, undisturbed landscapes in the SBP are rare (Smeins and Diamond 1983).

9.2.2.2 The Grand Prairie (MLRA 85)

The Grand Prairie (GP) spans Texas and Oklahoma and may be thought of as three or more distinct regions. The Oklahoma portion of the MLRA corresponds to the Arbuckle Mountains and surroundings; in Texas, it consists of the Washita Prairie to the North and East and the Lampasas Cut Plain/Glen Rose Prairie to the south and west. All regions share largely limestone parent materials (Table 9.1).

A complex geological history has produced many parent materials in the Arbuckles (Watterson et al. 1984). Limestones dominate, forming deposits more than a mile deep (Easterling 1972). Shales, sandstones, and conglomerates formed from sediment eroded from ancient, uplifted mountains (Watterson et al. 1984). Shaley materials weathering deeply result in Haplustert-Haplustoll-Haplustert (summit-backslope-footslope) catenas. Conglomerate tends to

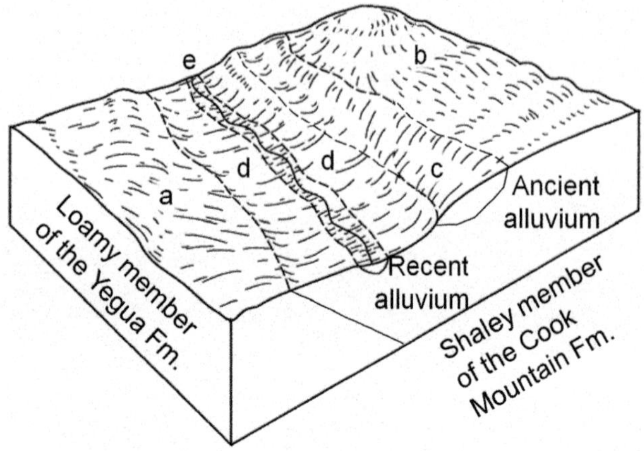

Fig. 9.13 Soil-landscape relationships in the San Antonio Prairie of the South Blackland Prairie (MLRA 86B). Paleustalfs (*a*), Haplusterts and Paleustalfs (*b*), Haplustalfs (*c*), Argiustolls (*d*), Haplustepts or Haplustolls (*e*). Modified from Jurena (2005)

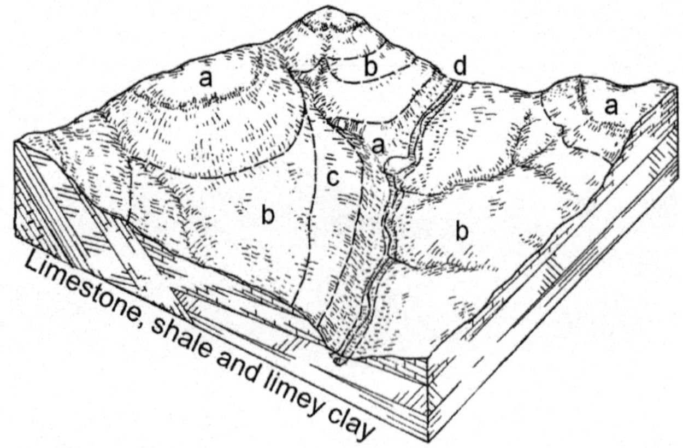

Fig. 9.15 Soil-landscape relationships in the Arbuckle Mountains of Oklahoma, in the Grand Prairie (MLRA 85). Calciustolls-rock outcrop complex (*a*), Calciustolls-Argiustolls complex (*b*), Argiudolls (*c*), Haplustolls (*d*). Modified from Bogard (1973)

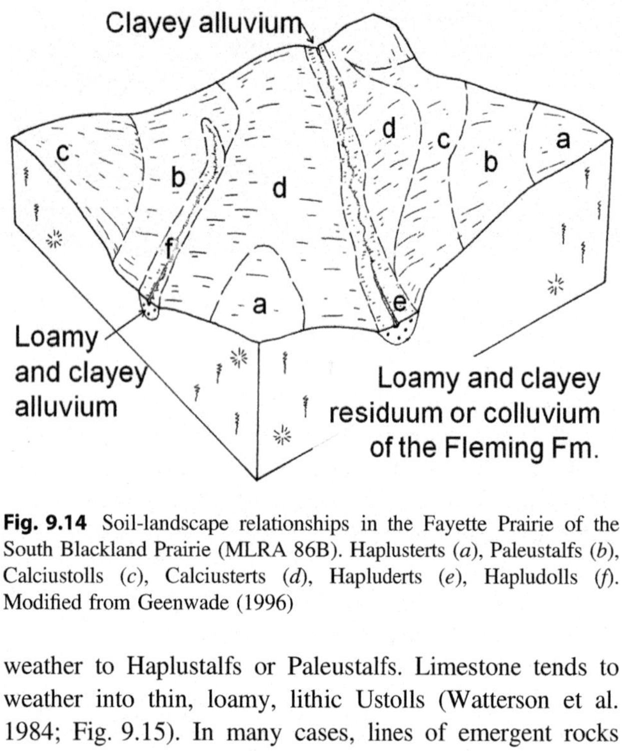

Fig. 9.14 Soil-landscape relationships in the Fayette Prairie of the South Blackland Prairie (MLRA 86B). Haplusterts (*a*), Paleustalfs (*b*), Calciustolls (*c*), Calciusterts (*d*), Hapluderts (*e*), Hapludolls (*f*). Modified from Geenwade (1996)

weather to Haplustalfs or Paleustalfs. Limestone tends to weather into thin, loamy, lithic Ustolls (Watterson et al. 1984; Fig. 9.15). In many cases, lines of emergent rocks occur along the fall line of the hill (Fig. 9.16). Paleustalfs are correlated with granite outcrops (Burgess 1977). Haplusterts are common that have linear gilgai microrelief occurring in rows oriented downslope (Fig. 9.17). This is a typical feature of Vertisols in moderately sloped environments.

In the Washita Prairie, the landscape is carved from limestones and marls of the Washita Group (Fig. 9.18). Limestone parent materials often weather to shallow and very shallow (lithic subgroups) Calciustolls and Ustorthents even if slopes are low. Shallow Haplustepts are commonly included when interbedding with marl occurs, forming a stair-stepped landscape when slopes are high. Deep to very deep Haplusterts can occur in concave positions in either uplands or toe slopes. Floodplain soils are Hapluderts or Haplustolls (Fig. 9.18). Row cropping in much of the Texas portion of the Grand Prairie is restricted to these rich bottom lands.

In the Lampasas Cut Plain (LCP) portion of MLRA 85, erosion has removed the Washita Group leaving formations of the Fredericksburg Group to control the topography (Hayward and Allen 1987). This occurs in the southwest portion of the Grand Prairie where the Edwards Limestone, an extremely resistant formation which thickens southward, forms mesa-tops in which lithic Haplustolls or Argiustolls have formed. Below the Edwards lie more impermeable but weatherable formations with higher shale and clay contents that form the sides and bases of the LCP table lands. The soils developing in these materials on sideslopes are shallow Calciustolls and Haplustepts. Footslopes below the mesas are occupied by Haplustepts or Calciustolls depending on slope and type (residuum versus colluvium) of parent material (Fig. 9.19). In this region of the Grand Prairie, dissection has occurred to such a high degree that some of the dissection itself has been erased (Hayward and Allen 1987). This results in sweeping vistas of broad valleys between widely separated mesas.

In this same general region of the Grand Prairie, erosion through the Fredericksburg and into the Trinity Group has exposed the Glen Rose Formation, in which limestone is interbedded with marl. The resulting ‘Glen Rose Prairie’ landscape is stair-stepped, depending on the resistance of the near-surface material; here Haplustepts, Haplustolls, and Calciustolls occur in complexes with limestone scarps (Fig. 9.20).

Fig. 9.16 Haplustolls-rock outcrop complex in the Arbuckle Mountain of Oklahoma; within the Grand Prairie (MLRA 85). Modified from Watterson et al. (1984)

Fig. 9.17 Linear downslope gilgai of some Vertisols in the Arbuckle Mountains of Oklahoma; within the Grand Prairie (MLRA 85). Modified from Moebius and Maxwell (1979)

The Cretaceous limestones of the Texas portion of the Grand Prairie are the same as those of the Edwards Plateau. However, in Cretaceous times, the seas in which these limy materials were deposited thinned northward resulting in thinner limestone deposits in the Texas portion of the Grand Prairie than in the more southerly Edwards Plateau. With less material to erode through in the LCP and GRP erosion has occurred to a greater degree through these materials than in the Edwards. The GRP/LCP represents Lower Cretaceous materials in an advanced erosional state relative to that of the Edwards Plateau (Hayward and Allen 1987). Erosion through the Glen Rose Limestone exposes basal Cretaceous sands, which are the parent material of much of the West Cross Timbers (Hayward and Allen 1987; Fig. 9.21).

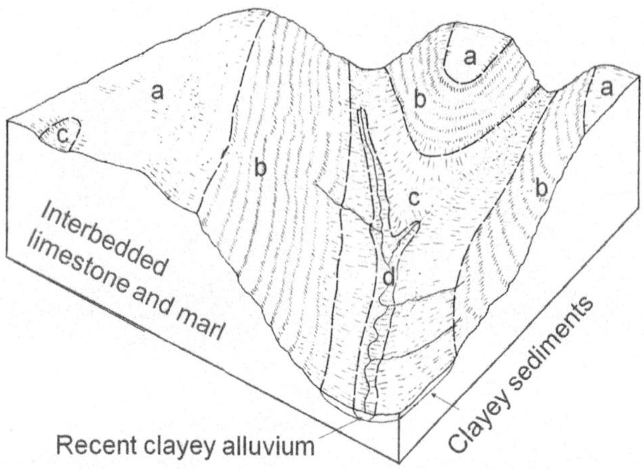

Fig. 9.18 Soil-landscape relationships in the Washita Prairie of the Grand Prairie (MLRA 85), Texas, formed from parent materials derived from the Washita Group. Calciustolls (*a*), Calciustolls, Ustorthents and Haplustepts (*b*), Haplusterts (*c*), Haplustolls (*d*). Modified from Ressel (1981)

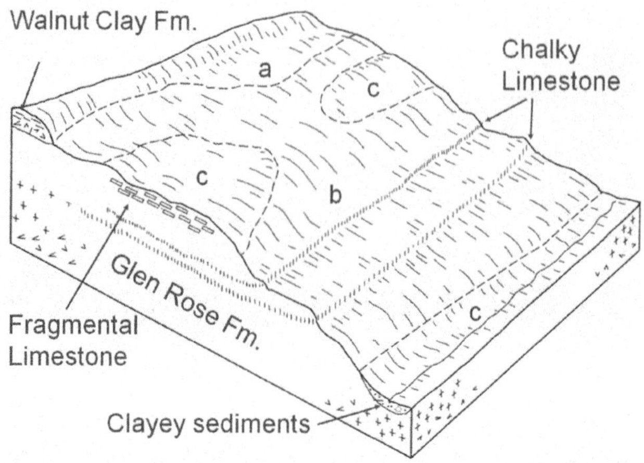

Fig. 9.20 Soil-landscape relationships in the Glenrose Prairie of the Grand Prairie (MLRA 85), and the Edwards Plateau (MLRA 81C), Eastern Part, Texas, formed from parent materials derived from the Glen Rose Formation of the Trinity Group. Calciustolls (*a*), Haplustolls (*b*), Haplustepts (*c*). Modified from Allison (1991)

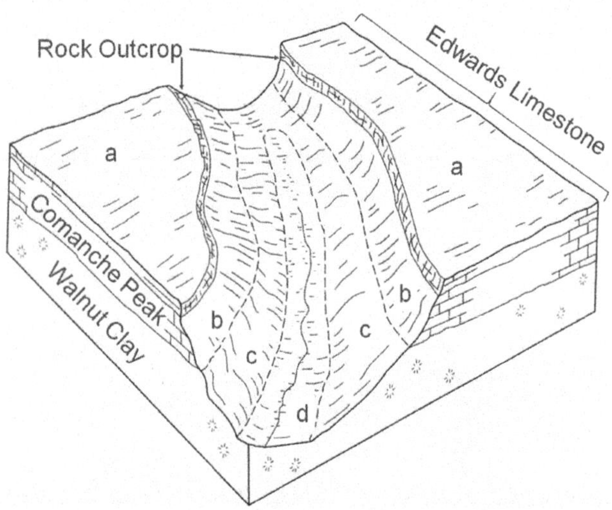

Fig. 9.19 Soil-landscape relationships in the Lampasas Cut Plain of the Grand Prairie (MLRA 85), and the Edwards Plateau, Eastern Part (MLRA 81C), Texas, formed from parent materials derived from the Fredericksburg Group. Lithic Haplustolls and Argiustolls (*a*), shallow Calciustolls (*b*), Calciustolls and shallow Haplustepts (*c*), Haplustolls (*d*). The distances separating mesa-tops in this figure are minimal relative to those found in the field. Modified from Allison (1991)

9.2.2.3 The Wichita Mountains (MLRA 82B)

The Wichita Mountains present the apparent contradiction of granite (typically low in bases) spires surrounded by Mollisols (higher in bases) (Fig. 9.22). The present landscape consists of the roots of mountains uplifted at the same time as the Arbuckle Mountains in MLRA 85. This Pennsylvanian orogeny was followed by Permian erosion and burial under Permian sands and shales (Johnson 2008). Present-day soils are derived from either Permian materials, granite outwash, or both. The soils associated with the granite outcrops are commonly shallow (mapped as 'stony rock land') (Mobley and Brinlee 1967). Soils surrounding the outcrops likely have significant influence of the Permian material which may help to explain the abundance of Mollisols in a landscape generally associated with outcroppings of rocks not typically considered to be high in base content.

9.3 Land Resource Region I

9.3.1 Texas Central Basin (MLRA 82A)

Unlike any other MLRA in this LRR, The Texas Central Basin (TCB) features the exposure of Precambrian igneous and metamorphic rocks (Table 9.6). Granite batholiths intermingled with schist and gneiss form the heart of the TCB (Fig. 9.23). Cambrian sandstones lie at the margin. More recent Paleozoic limestones lie at even more distant radii and are usually considered to be part of the Edwards Plateau, which is largely made up of Cretaceous limestones (see below). In terms of parent materials, the MLRA has much greater similarity to 82B (Wichita Mountains) of LRR J than to any other MLRAs in LRR I. An excellent review of the geology of the region, sometimes called the Llano uplift, is provided by Spearing (1991). Residual soils in the area are almost universally well drained with sandy loam the dominant surface texture. Subsoils are dominantly loamy or fine-loamy, and permeability is dominantly moderate.

In the droughty central Texas environment, limestone is highly resistant to weathering and tends to resist erosion more than granite (Baylor Geological Society 1983). As a result, the steepest slopes and shallowest soils are developed

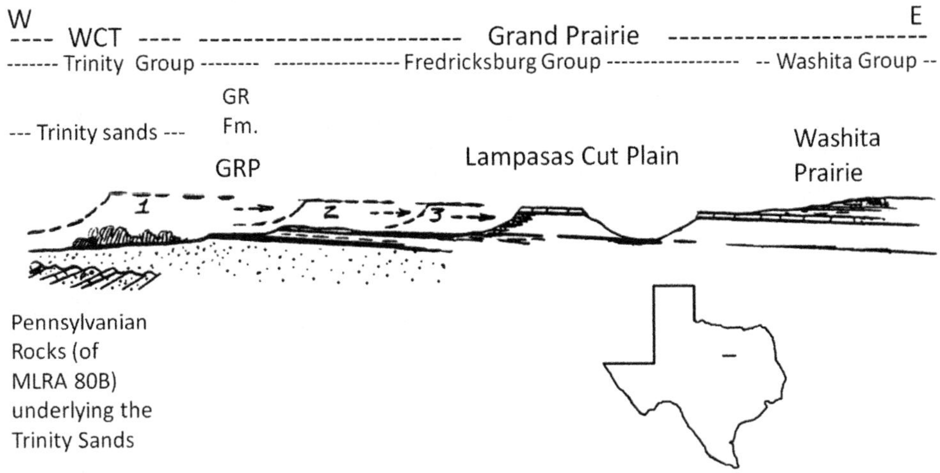

Fig. 9.21 Geography and geology of the southern Grand Prairie and Western Cross Timbers (WCT). Erosion of the Washita Group exposes the Fredericksburg Group and the capping rock of the Edwards limestone. Erosion through the Edwards creates separated mesas of the Lampasas Cut Plain. Erosion into the Trinity Group exposes the marls and limestones of the Glen Rose Formation (GR Fm.) creating the Glen Rose Prairie (GRP). Sandy deposits in the Trinity Group form the parent material for the WCT. The LCP has migrated eastward as Fredericksburg rocks have been eroded. Further erosion will expose Pennsylvanian materials characteristic of MLRA 80B (Texas North Central Prairies). Modified from Hayward and Allen (1987). By permission of Baylor Geological Society, Waco, Texas

on limestone and sandstone. In the TCB Limestone-developed landscapes tend to produce lithic subgroup soils on all but the gentlest depositional slopes and softest parent materials, and soil-landscape relationships bear many similarities and soil series with the Lampasas Cut Plain (MLRA 85) and the Eastern Edwards Plateau (MLRA 81C). Over sandstone parent materials, Ustorthents of escarpments are often bracketed by Haplustalfs on gentler surfaces. Paleustalfs occur on footslopes or flatter undulating lower landscapes either because of deep deposition or deep weathering into less resistant sandstone members (Fig. 9.24).

Among Precambrian parent materials, gneiss-developed landscapes feature Haplustepts on steep to gentle ridges and Paleustalfs on lower, flatter portions of the landscape.

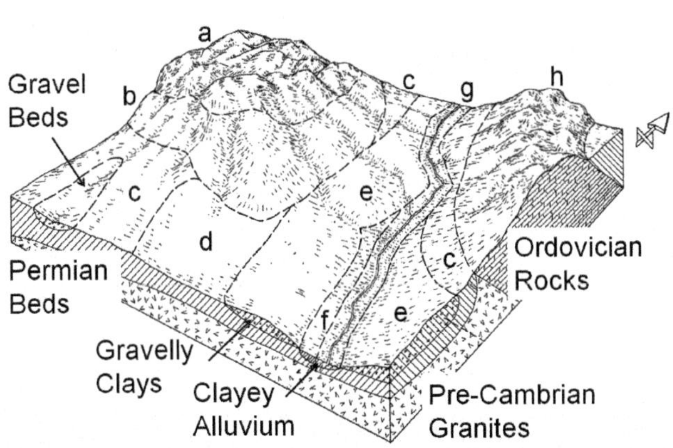

Fig. 9.22 Soil-landscape relationships in the Wichita Mountains, Oklahoma (MLRA 82B). Granite outcrop (*a*), 'stony rock land' (*b*), Paleustolls (*c*), Natrustolls (*d*), Argiustolls (*e*), Haplusterts (*f*), Haplustolls (*g*), Calciustolls (*h*). Modified from Mobley and Brinlee (1967)

Granitic landscapes are similar, but feature Haplustalfs as well on moderate and gentle ridges and sideslopes (Fig. 9.24). Schist is especially weatherable and typically does not develop ridges with slopes as steep as other parent materials (Mutis-Duplat and Gray 2011; Stenzel 1935). In addition to the Haplustalfs and Haplustepts found on summits and sideslopes, Rodustalfs or Argiustolls also occur depending on the nature of the schist. Natrustalfs commonly develop in drainage ways of landscape of Precambrian rocks (Fig. 9.24). Floodplains of small streams of Precambrian parent material landscapes typically have developed Haplustolls. At least 66 % of soils in the TCB are Alfisols (Table 9.7).

The most unique soils in the TCB occur at the top of Enchanted Rock, a granite batholith in which moderately deep to extremely shallow soils have developed in weathering pits that have been a sink for eolian materials and granite slope alluvium (Petersen and Duttmann 2009). The weathering pits are in several different stages of development (Fig. 9.25), and illuvial gypsum has been observed at the bottom of some pits (Duttmann and Petersen 2009). Flooded pits support a 'fairy shrimp' branchiopod despite elevations of tens of meters above the surrounding valley.

9.3.2 The Edwards Plateau

MLRAs of the Edwards Plateau and the Texas portion of the Grand Prairie largely correspond to the surficial occurrence of Lower Cretaceous limestones and marls. Major soils in these regions are shallow (<50 cm) to bedrock or petrocalcic horizons, well drained, and have moderate to moderately

Table 9.6 Characteristics of MLRAs in LRR I

MLRA		Current Vegetation	Parent Material† Age	Parent Material† Type	Precip. (cm)	Soil Temp. Class	Typical Soil Depth	Typical Ag. Use	Most common soils and abundance (%)	
Texas Central Basin (82A)		Mixed oak savannah with mid/tallgrasses	Precambrian Cambrian Cretaceous	Many	61-79	Thermic	Mod. Deep	Grazing	Ustalfs Haplustepts	66 16
Edwards Plateau	Eastern (81C)	Trees, shrubs, short/mid grasses	Cretaceous	Limestone	61-76	Thermic	Shallow	Grazing	Mollisolls Inceptisols	70 23
Edwards Plateau	Central (81B)	Trees, shrubs, short/mid grasses	Cretaceous	Limestone	49-82	Thermic	Shallow	Grazing	Calciustolls Haplustolls	74 14
Edwards Plateau	Western (81A)	Shrubs, short/mid grasses	Cretaceous	Limestone	38-66	Thermic	Shallow	Grazing	Calciustolls Haplocalcids	52 25
Edwards Plateau	Southern (81D)	Low shrubs, short/mid grasses	Cretaceous	Limestone	25-38		Shallow	Grazing	Calcids Torriorthents	73 23
Rio Grande Plain	Northern (83A)	Grassland, shrubs		Fluvial/Deltaic/Coastal Sandstones and Shales	54-94	Hyperthermic	Very Deep	Grazing and Crops	Ustolls Paleustalfs	51 25
Rio Grande Plain	Western (83B)	Grassland, shrubs	Tertiary	Fluvial/Deltaic/Coastal Sandstones and Shales	46-64	Hyperthermic	Shallow to Very Deep	Grazing	Ustepts Haplusterts Calciustolls	26 26 18
Rio Grande Plain	Central (83C)	Grassland, shrubs	Tertiary	Fluvial/Deltaic/Coastal Sandstones and Shales	54-74	Hyperthermic	Shallow to Very Deep	Grazing	Ustolls Paleustalfs Calciustepts	44 28 20
Rio Grande Plain	Lower (83D)	Grassland, shrubs, trees	Tertiary	Fluvial/Deltaic/Coastal Sandstones and Shales	56-69	Hyperthermic	Very Deep	Crops	Ustolls Calciustepts Haplusterts	49 20 10
Sandsheet Prairie (83E)		Grassland, shrubs, trees	Holocene	Aeolian	56-71	Hyperthermic	Very Deep	Grazing	Ustalfs Ustipsamments	81 13

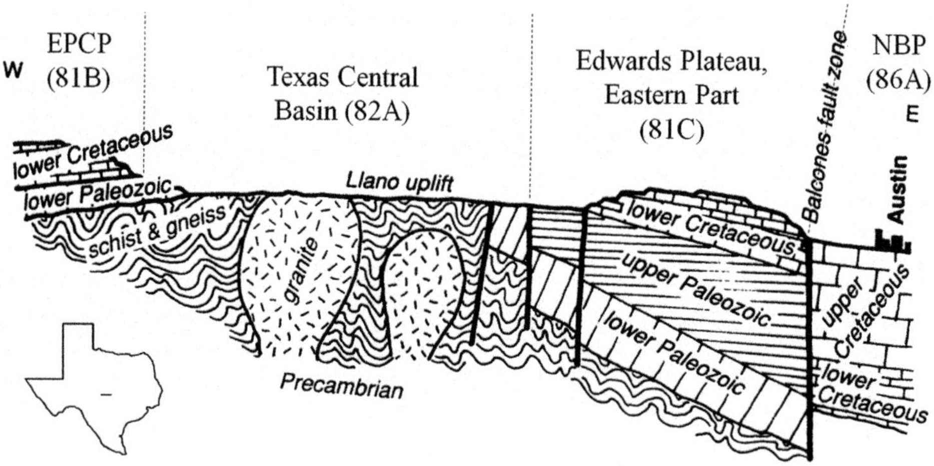

Fig. 9.23 Cross section through the geologies of the Central Mineral Region (MLRA 82A) and adjoining MLRAs. Precambrian schist, gneiss, and granite in the center, surrounded by marginal Paleozoic and Cretaceous deposits. From Spearing (1991). By permission of Mountain Press Publishing Company, Missoula, Montana

slow permeability. Many are skeletal or stony at the surface. Rangeland predominates as the major land use (Table 9.6). Climate and geology account for many of the differences that can be observed in these soil landscapes.

The Southern, Western, and Eastern Edwards Plateau MLRAs are highly dissected. A major source of this dissection is the uplift associated with the Balcones Fault Zone, a system of faults that has resulted in the relative uplift of

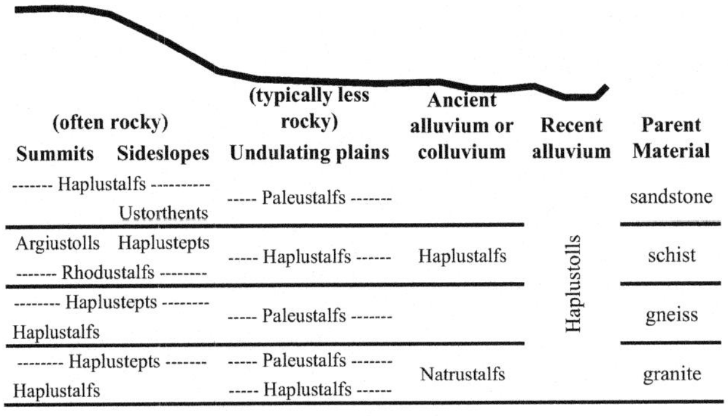

Fig. 9.24 Soil-landscape relationships in the Central Mineral Region (MLRA 82A) as a function of parent material. Based on information obtained from Soil Survey Staff (2015)

Lower Cretaceous versus Upper Cretaceous parent materials.[4] This uplift is relatively recent in geologic time ($\sim 10^6$ year; Spearing 1991), and created the gradient for much of the dissection occurring on the south and eastern margins of the Eastern Edwards Plateau (The Balcones Fault Zone) and likely some of the dissection occurring on the southern margin of the Western Edwards Plateau. An additional source of dissection in the Western Edwards Plateau is provided by the Rio Grande and Pecos River drainage incisions, which occur on the southern portion of the Western Edwards Plateau. Having carried high volumes of water from the Rocky Mountains, the Rio Grande and Pecos River basins incised deeply into the Lower Cretaceous limestone. On the northern side of the Western Edwards Plateau, such dissection has occurred due to Concho River drainage.

In contrast, the Central Edwards Plateau has remained relatively less affected by dissection and has a smoother surface. It has a central location away from the margins of the Edwards Plateau, the faults of the Balcones, and is less dissected by tributaries of major river systems.

While air temperatures are relatively constant across the Edwards Plateau, rainfall is not, dropping from east to west (Table 9.6). This correlates with increasing occurrence of calcic horizons, increasing strength of expression of calcic horizons, disappearance of argillic horizons, a switch from clayey to loamy control sections, and a switch from Mollisols to Aridisols (Tables 9.7, 9.8; Fig. 9.26). Rabenhorst and Wilding (1986b) suggest that a rough boundary between calcareous and non-calcareous soils in the surface layer occurs along a curve defined by 38-inch Thornthwaite Potential Evapotranspiration (Fig. 9.26).

Multiple models exist to explain the formation of calcic and petrocalcic horizons (Schaetzl and Anderson 2005). A model for petrocalcic horizon formation in Aridisols in southern New Mexico featuring allochthonous rain and dust inputs is discussed by Monger et al. (2009). However, current dust inputs (Rabenhorst et al. 1984) or dust retention (Rabenhorst and Wilding (1986a) in the Edwards Plateau may not be sufficient to explain the presence of calcic or petrocalcic horizons in the Edwards. An alternative model based on observations within the Edwards Plateau is provided by Rabenhorst and Wilding (1986c) which describes in situ secondary carbonate precipitation directly from the primary carbonates of the parent rock (with no illuviation). West et al. (1988) have made similar observations in the Grand Prairie.

Where petrocalcic horizons formed in skeletal parent materials containing limestone clasts or fragments, a third model may be more appropriate. Such skeletal materials are abundant in the highly dissected Western and Southern Edwards Plateau and in the Rio Grande Plains, which lie downstream. In this model, illuviation takes place with carbonate fragments forming both the parent material and fabric for petrocalcic horizon formation. In many cases, the limestone clasts which gave rise to the petrocalcic horizons have been completely dissolved above the petrocalcic leading to non-calcareous soils above the petrocalcic (Wayne Gabriel, NRCS, pers. comm.).

Current soil morphologies in the Edwards may not result from current illuvial or in situ calcite neoformation but might rather display the preserved or weathered remains of calcic or petrocalcic horizons which were developed in earlier times, with the differential rainfall across the Edwards plateau playing a role in preserving petrocalcic features in the west and destroying such features in the east (Loomis and Gabriel 1991). Petrocalcic-bearing soils in the Lampasas Cut Plain of the Grand Prairie are discussed as relict by Hayward and Allen (1987).

Ground-based identification of MLRAs can be performed using the relative drought tolerance of plants present in different portions of the landscape. In the Southern Edwards Plateau, for example, creosote bush (*Larrea tridentata*) is observed in all portions of the landscape, but juniper (*Juniperus*) species are only found in (relatively moist) drainage ways; in the next (and wetter) MLRA to the east (the Edwards Plateau, Western Part; MLRA81A), juniper is observed on all portions of the landscape, but Live Oak (*Quercus virginiana*) is only observed in the drainage ways, etc. (Table 9.9) (Wayne Gabriel, NRCS, pers. comm.).

9.3.2.1 Southern Edwards Plateau (MLRA 81D)

The Southern Edwards Plateau has the highest concentrations of Aridisols and shallow soils in the Edwards Plateau (Table 9.8). Subsoil colors are lighter (higher value) than in other MLRAs of the Edwards Plateau. Haplocalcids are found on relatively massive Edwards limestone, and Torriorthents have developed on flaggy Boquillas limestone

[4]This statement applies to the southern margin of the Edwards. To the north the Balcones represents displacement between different Upper Cretaceous formations, see Baylor Geological Society (1994).

Table 9.7 Distribution of major (>5 %) soil great groups among the MLRAs of LRR I

Soil great groups	MLRA									
	81D (%)	81A (%)	81B (%)	81C (%)	82A (%)	83A (%)	83B (%)	83C (%)	83D (%)	83E (%)
Haplustalfs					26	9				19
Paleustalfs					29	25		28		47
Rhodustalfs					11					
Natrustalfs							7			14
Haplocalcids	59	25								
Haplocambids		7								
Petrocalcids	14	9								
Haplustolls		5	14	33		6			10	
Paleustolls						9				
Calciustolls		52	74	23		15	18	19	27	
Argiustolls			6	13		21		22	11	
Haplusterts						7	26		12	
Haplustepts				18	16		12			
Calciustepts				5			14	18	20	
Ustorthents					8					
Ustifluvents									5	
Torriorthents	23									
Ustipsamments										13
Whole MLRA area (10^6 hectares)	0.26	3.16	3.13	1.49	0.48	2.57	2.16	0.82	0.61	0.99

(Fig. 9.27). The Edwards landscape is more common land would feature shallow Haplocalcids on the Edwards residuum, shallow Petrocalcids on gravelly (often Pleistocene aged) colluvium or alluvium, and deeper Haplocambids to Haplocalcids on more recently deposited materials (Fig. 9.28).

9.3.2.2 Edwards Plateau, Western Part (MLRA 81A)

This MLRA receives higher rainfall (Table 9.6) than its 'southern' counterpart (81D), and Aridisols and Mollisols in this MLRA are of the most comparable abundance in all of the Edwards Plateau (Table 9.8). Approximately 42 % of this MLRA is occupied by shallow to bedrock (lithic subgroup) Calciustolls that lack petrocalcic horizons. Such soils have developed on limestone residuum under diverse slopes. Frequently, these soils are mapped as complexes with rock outcrops. Catenas in these landscapes can be very similar to those in MLRA 81D (Fig. 9.28). Another 25 % of this MLRA is occupied by Haplocalcids also without petrocalcic horizons. Most of these are loamy in texture. On vast flats in this MLRA, these Haplocalcids are usually developed on eolian parent materials (Soil Survey Staff 2015), but in filled valleys (Fig. 9.28) are often of old alluvium.

9.3.2.3 Edwards Plateau, Central Part (MLRA 81B)

This MLRA may be distinguished from the Edwards Plateau, Western Part (MLRA 81A) in a number of ways: (1) Drainage to the deeply downcutting Pecos River is largely avoided (USDA NRCS 2006), which results in less dissected landscapes (Coburn 2007), and (2) rainfall is increased. Flatter landscapes and increased rainfall may contribute to (3) moister subsoils (lack of Aridisols; Table 9.8), (4) increased dominance of Calciustolls (Table 9.7), (5) the emergence of Haplustolls in residual (and not only depositional environments as in MLRA 81A), and (6) a switch from loamy to clayey textures (Table 9.8). Shallow Calciustolls occur over bedrock or over petrocalcic horizons developed in colluvium and slope alluvium. Deep Calciustolls are typical on older alluvial terraces (Fig. 9.29).

9.3.2.4 Edwards Plateau, Eastern Part (MLRA 81C)

The Eastern Part of the Edwards Plateau is classic 'Texas Hill Country'. Here the Balcones Fault System has resulted in Lower Cretaceous materials rising a few tens of meters above those of the Upper Cretaceous. The dramatic relief, coupled with the porous nature of the Edwards limestone, which is exposed in the interior of this MLRA, results in a series of springs that drain the MLRA near its outer margins.

Fig. 9.25 Summit of Enchanted Rock, Fredericksburg, Texas. Vernal pools in early stages of development (foreground) and later stages of development (background). Author's daughter in foreground provides scale (0.7 m)

Soils in this region reflect greater precipitation (Table 9.6), featuring very few calcic and petrocalcic horizons, and more argillic horizons, than other regions in the Edwards (Fig. 9.26). Clay contents are comparable to those of the Central Part (Table 9.8).

Soils in this portion of the Edwards are very similar to soils of the Lampasas Cut Plain and Glenrose Prairie of the Grand Prairie, due to similar geology (as discussed above with respect to the Grand Prairie). However, instead of the broad vistas of the Lampasas Cut Plain, the Eastern Part of the Edwards Plateau features a more intimately dissected landscape in which narrow ravines give way to valleys that are modest in size. Rivers are often lined with stately bald cypress.

9.3.3 The Rio Grande Plains

In the south of LRR I, all but one of the MLRAs are named 'Rio Grande Plain[s]'. This should not imply that the parent

Table 9.8 Differences in prevalence of selected soil taxa and soil properties in the Edwards Plateau

Selected soil taxa and properties	Portion of the Edwards Plateau			
	Southern (%)	Western (%)	Central (%)	Eastern (%)
Mollisols	tr[a]	56	93	70
Aridisols	77	42	1	0
Have argillic horizons	tr	tr	6	21
Have 'calcic' in subgroup	73	87	75	28
Have 'petrocalcic' in subgroup	14	15	17	2
<50 cm depth[b]	87	61	79	75
'Loamy'[c] in PSCS	98	67	22	39
Clay $\geq$ 35 % in PSCS	tr	5	70	60

PSCS particle size control section (typically the upper part of the B horizon or above a cemented horizon or rock contact)
[a]Trace to nonexistent
[b]To petrocalcic or bedrock
[c]Clay <35 % and textures not especially sandy or silty

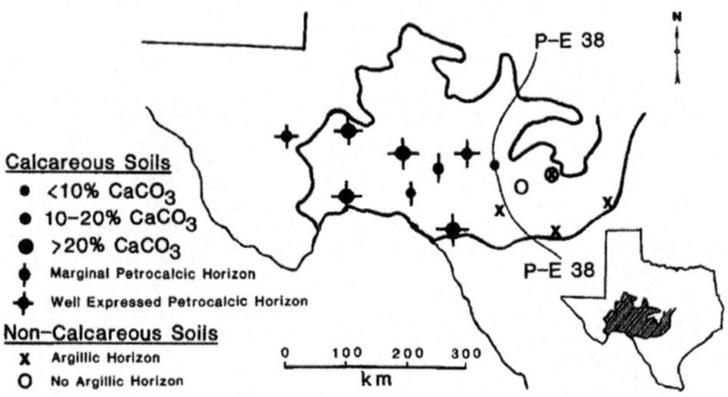

Fig. 9.26 Trends in occurrence of argillic and petrocalcic horizons and the occurrence of calcium carbonate in the Edwards Plateau. From Rabenhorst and Wilding (1986). Reprinted with the Permission of ASA, CSSA, SSSA

material for the soils in these regions was originally fluvial sediments of the Rio Grande. While the name provides a sense of place, many of the geologic formations making up the parent materials of these soils are continuous with Tertiary materials to the north and east. While most of these materials are indeed fluvial deposited, with the exception of those nearest the Rio Grande, they should not be considered materials deposited by the Rio Grande itself, but rather other rivers that may or may not be identifiable today. This region is perhaps better known as 'South Texas'.

Soils of South Texas classify as 'hyperthermic' (mean soil temperature at 50 cm >22 °C; Table 9.6), making them among the warmest soils in Texas and the country. Predominating uplands soils are well drained and saturated hydraulic conductivities are moderate (Soil Survey Staff 2015). From east to west, the region experiences a steep reduction in rainfall (Table 9.6). The western terminus of dryland row cropping occurs here. MLRAs in this region can be distinguished through the predominating land use (row cropping versus rangeland), the presence of petrocalcic horizons, the presence of salt-affected soils, the appearance of soils with moisture regimes typical of deserts ('aridic', 'torric'), and the presence of highly sandy soils (Table 9.10). In contrast to the relief found in the Edwards Plateau, much of South Texas is very flat. Southbound travelers transitioning from the Edwards to the South Texas Plains descend through dramatic roadcuts in the limestone bedrock before spilling onto the Plain's expanse. While major rivers and spring-fed streams originating in the Edwards flow continuously through the Rio Grande Plains, streams originating in the RGP are nothing more than ephemeral drainages.

Table 9.9 Vegetation differences which can be used for identification of MLRAs in the Edwards Plateau

Landscape position where plants[a] are observed furthest west	Southern	Western	Central	Eastern
All other than drains	Creosote[b]	Red Berry and Ashe Juniper	Live Oak	Texas Oak Chinkapin Oak
Drains	Red Berry Juniper	Live Oak	Texas Oak	Bald Cypress

[a]Creosote bush (*Larrea tridentata*); Red Berry Juniper (*Juniperus pinchotii*); Ashe Juniper (*Juniperus ashei*); Live Oak (*Quercus virginiana*); Texas Oak (*Quercus buckleyi*); Chinkapin Oak (*Quercus muehlenbergii*); Bald Cypress (*Taxodium distichum*)
[b]Creosote is not found in other MLRAs

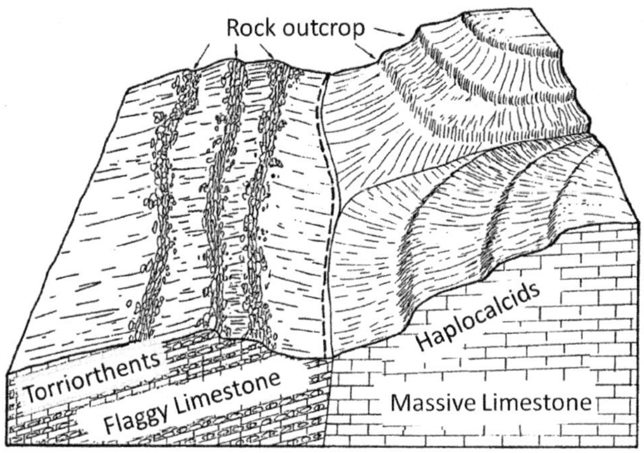

Fig. 9.27 Parent material-soil-landscape relationships in the Southern Edwards Plateau (MLRA 81D). Modified from Cochran and Rives (1985)

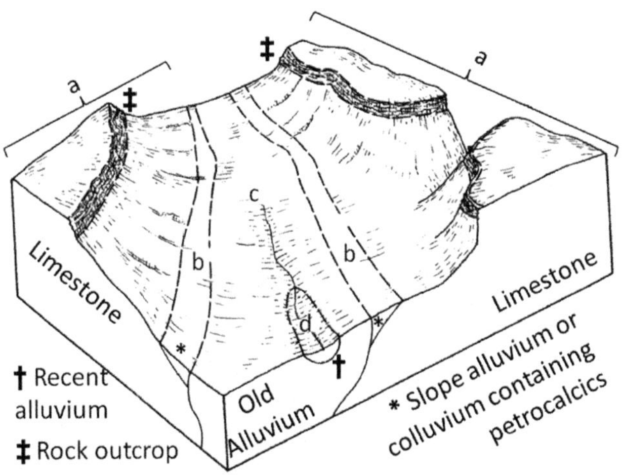

Fig. 9.28 Landscape distribution of great groups as related to landscape position in limestone bedrock-related landscapes in the Edwards Plateau. Southern Edwards Plateau (MLRA 81D): Haplocalcids and Torriorthents (*a*), Petrocalcids (*b*), Haplocambids and Haplocalcids (*c*), Haplocambids, Haplocalcids, and Haplustolls (*d*). Edwards Plateau, Western Part (MLRA 81A): Calciustolls (*a*), Petrocalcic Calciustolls (*b*), Haplustolls (*c*), Haplustolls (*d*). Modified from Rives (1980)

9.3.3.1 Northern Rio Grande Plain (MLRA 83A)

MLRA 83A is a large region with irregular boundaries in comparison with the Blackland Prairies and Southern Claypan, which bound its northeast border (Figs. 9.1, 9.2). Whereas the latter MLRAs contain high concentrations of Vertisols or Paleustalfs, respectively, the Northern Rio Grande Plain contains an abundance of Mollisols (Table 9.11). In the eastern portion of the Northern Rio Grande Plain, these Mollisols may be found on the same geologic formations (e.g., Oakville) of the Fayette Prairie of the Southern Blackland Prairie, but the strong vertic character of the Southern Blackland Prairie is diminished (Table 9.11). The soil differences at the frontier between this

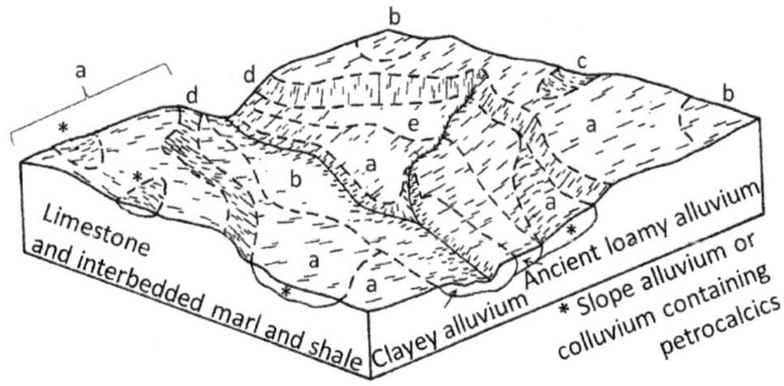

Fig. 9.29 Soil-landscape relationships of the Edwards Plateau, Central Part (MLRA 81B). Calciustolls (*a*), Haplustolls (*b*), Argiustolls (*c*), Calciustolls-rock outcrops and Haplustepts (*d*), Haplustepts (*e*). Modified from Blum (1982)

MLRA and those in LRR J to the east (Table 9.11) are not always distinct, with many series spanning this MLRA and those of LRR J to the east (Ressel and Brown 2004; Taylor 1977). For the most part, Argiustolls are restricted to the eastern two-third of the Northern Rio Grande Plain. Calciustolls, commonly with petrocalcic horizons, become more common to the west. Paleustalfs may be found throughout.

The northwest border of the MLRA occurs roughly at the delineation) between lower and upper Cretaceous materials along the system of Balcones faults; however, the majority of the parent materials in the MLRA are Tertiary fluvial/deltaic sediments of the coastal plain. The eastern portion receives more rainfall and is where dryland cropping occurs. Irrigation is common in the western portion, known as the 'winter garden'. Range and pasture are also major land uses, but what distinguishes the Northern Rio Grande Plain from its neighbors in the Edwards Plateau and South Texas is its suitability for row-cropped agriculture.

9.3.3.2 Western Rio Grande Plain (MLRA 83B)

In the Western Rio Grande Plain, plant available water is limiting due to salinity and the lack of rainfall (Table 9.6). One-third of the soils present are sufficiently affected by salinity to merit 'sal' 'natr', 'sodic', or 'hal' prefixes in their classification (Table 9.10). This underestimates salt-affected soils, however, since salts can trigger detrimental effects in plants at lower electrical conductivity (EC ≈ 4 dS m^{-1}) than those used as limits in soil taxonomy (at for example 30 dS m^{-1} for 'salic' horizons; Soil Survey Staff 2014). The other South Texas MLRAs are not comparably affected by salt (Table 9.10).

Approximately 52 % of soils in the Western Rio Grande Plain are sufficiently dry to belong to aridic or torric subgroups, indicating that these soils are on the dry side of the ustic moisture regime. An additional nine percent are sufficiently dry to be Aridisols (soils with moisture patterns typical of deserts). Other MLRAs of South Texas either lack Aridisols altogether or contain <1 % of such soils. Due to

Table 9.10 Percentage of soils in South Texas MLRAs with petrocalcic features, significant presence of salts or sodium, moisture conditions characteristic of desert soils, or sandy character

MLRA	Soils with 'Petro' in the subgroup[a] (%)	Soils with 'Sodic', 'Natr', 'Sal' or 'Hal' in the subgroup[b] (%)	Soils with 'arid' or 'torr' in the subgroup including Aridisols[c] (%)	Soils with 'psamm', or 'aren' in the great group or 'sandy' in the family[d] (%)
83A	6	3	23	8
83B	10	33	61	tr[e]
83C	39	1	29	1
83D	5	11	32	1
83E	1	16	24	57

[a]Petrocalcic horizon often within 100 cm of the surface
[b]Significantly salty horizon within the soil
[c]Limited plant available water over the course of the year
[d]High sand content
[e]'tr' indicates a percentage which does not round up to one or more

Table 9.11 Differences in prevalence of selected taxa and vertic character (excluding Vertisols) between the Northern Rio Grande Plain (MLRA 83A) and the MLRAs on its northeast border

LRR	MLRA	Mollisols (%)	Alfisols (%)	Vertisols (%)	Vertic character[a] (%)
I	83A	52	35	7	10
J	86A	28	25	43	36
	87A	3	85	2	17

[a]'ertic' as part of first term in the subgroup; Vertisols are not included

restricted rainfall and salt challenges, cropping in this MLRA is limited, and its primary agricultural use is as rangeland. Haplusterts occupy one quarter of the MLRA as do Inceptisols. Both occur on alluvial and residual materials, though Haplusterts tend to occur on alluvium and Inceptisols on Tertiary mudstones and sandstones.

9.3.3.3 Central Rio Grande Plain (MLRA 83C)

About 40 % of MLRA 83C is underlain by petrocalcic horizons (Table 9.10), and for the most part if shallow soils are encountered, it is because of an underlying petrocalcic horizon. However, thin soils (Durustepts) also occur on sandstone near the ridges of cuestas (Gary Harris, NRCS, pers. comm.), which occur in the northwest portion of the MLRA (Sackett 2011). Duripans are believed to have formed in these sandstone ridges due to the presence of soluble silica associated with tuffaceous (volcanic) materials (Sackett 2011). The northern boundaries of the MLRA closely adhere to the surficial Catahoula, Oakville, and Goliad formations, but the delineations of the MLRA and these geologic formations diverge in the lower portion of the MLRA. The Goliad Formation comprises the majority of the MLRA. When intact, the Goliad is often observed in well logs to have a caliche cap (Jacobs 1981), and hence, the upper part of the Goliad is itself a Paleosol. However, soils developing over the Goliad often contain no carbonates (Argiustolls) and when they do the petrocalcic appears to be neo-formed and not relict. This has resulted in the interpretation that much of the surface of the Goliad has been reworked (Sackett 2011). Petrocalcic-bearing Calciustolls, Calciustepts, and Paleustalfs tend to be on the summits of ridges and interfluves. Argiustolls and Calciustolls occupy sideslopes. Argiustolls, often with A horizons thickened by sedimentation, typically occupy drains (Fig. 9.30).

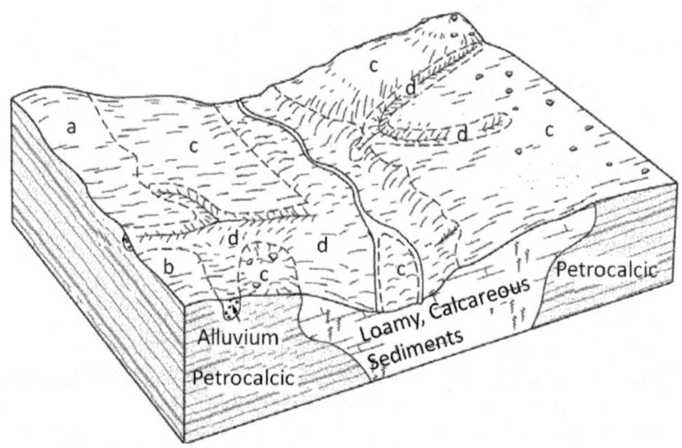

Fig. 9.30 Soil-landscape relationships in the Central Rio Grande Plain (MLRA 83C). Calciustepts and Haplustepts (*a*), Paleustalfs (*b*), Calciustolls (*c*), Argiustolls (*d*). Modified from Sackett (2011)

9.3.3.4 Lower Rio Grande Plain (MLRA 83D)

This MLRA is and has been a source of vegetable, sugar cane, and citrus production, and was delineated in terms of its potential for reliable citrus production (Gary Harris, NRCS, pers. comm.), due to the ease of access to reliable water (via the Rio Grande river) and accommodating winter

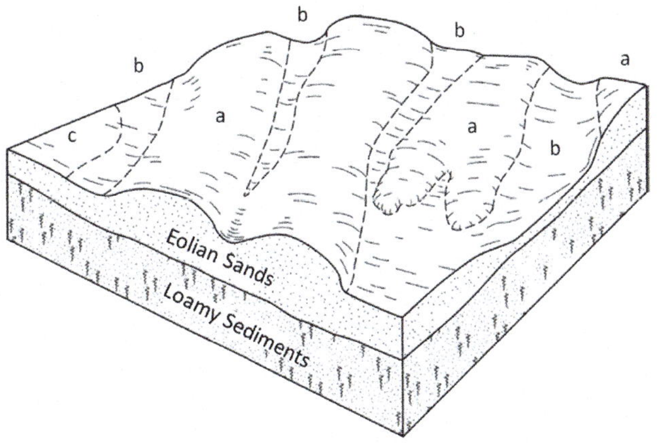

Fig. 9.31 Soil-landscape relationships in the Sandsheet Prairie (MLRA 83E). Ustipsamments (*a*), Paleustalfs (*b*), Haplustalfs (*c*). Modified from Haile and Brezina (2012)

temperatures. Much of the agriculture is irrigated. The region is undergoing extensive development. MLRA 83D is similar to 83A in the sense of not containing an overabundance of petrocalcic horizons or other limiting soil characteristics (Table 9.10). The predominating soils are Mollisols, Vertisols, and Inceptisols, soil orders that do not necessarily imply significant pedogenic development. This MLRA occupies much of the Rio Grande delta in the USA.

9.3.3.5 Sand Sheet Prairie (MLRA 83E)

In the Sand Sheet Prairie, Holocene, eolian materials have covered loamier Pleistocene or Tertiary landscapes (Fig. 9.31). The easternmost portion of the MLRA is the sandiest and most active, consisting of dunes and catenas of (well drained) Ustipsamments to (poorly drained) Psammaquents. To the west, in the interior of the Sand Sheet, the dunes have become stabilized. Here Paleustalfs, often associated with Ustipsamments (Fig. 9.31), have argillic horizons which commonly start at the contact between the overlying Holocene eolian sands and the underlying loamy materials (Gary Harris, NRCS, pers. comm.). These argillic horizons may in part be relict features of soils buried by the sand cap (James Akin, NRCS, pers. comm.). At the margins, the MLRA becomes loamier and the eolian cap thinner. Except in the active east, Alfisols are common throughout. Many finely textured Alfisols are affected by sodium (Natrustalfs).

Fig. 9.32 Dune overrunning an oak motte in Kenedy County, Texas. Trees are approximately twenty feet high. Modified from Haile and Brezina (2012)

Fig. 9.33 Future of the Eastern Cross Timbers? Post oaks in a shopping mall tree box. Euless, Texas, October 2014

In contrast to the rest of the Rio Grande Plain MLRAs in which few perpetually flowing streams may be found, in the Sand Sheet Prairie drainages are often not observable, due to the sand cap. In some locations, wind has removed Holocene sands, exposing the underlying Pleistocene sediments. These loamy materials can also be wind transported—in the form of sand-sized aggregates of mixed texture. These loamy aggregates do not travel long distances, but form 'clay dune' parent materials adjacent to the deflationary depressions from which they came. While most of the MLRA consist of former prairie becoming scrubland, mottes of live oaks occur on stabilized, sandier portions of the landscape until or unless these locations are overrun by shifting sands (Fig. 9.32).

Mostly occupied by very large ranches, this MLRA is used almost exclusively for grazing or wildlife hunts including naturalized exotic species. Row cropping does occur in the northern margin of the MLRA as it transitions to the clayier coastal prairie of MLRA 150A (LRR P). Cotton (*Gossypium hirsutum*) and sorghum (*Sorghum bicolor*) production occurs on soils with loamier surfaces whereas black-eyed peas (*Vigna unguiculata*) are produced where the sand cap is more pronounced (Fig. 9.33).

9.4 Conclusions

To the casual observer, rivers and ocean do not come to mind when viewing the landscapes of Central, South, and East Central Texas but, they should to the soil-minded observer. Residual parent materials are derived from either marine (Cretaceous) or alluvial (Tertiary) sediments and rocks.

Those not versed in soils or geology might also be tempted to assume incorrectly that the landscape is as it always has been. Belts of high folded mountains, volcanoes, and impossibly wide rivers have occupied these areas (Spearing 1991; Sellards et al. 1932) and might be expected to do so again (Swanson 1995). Climate change is expected to result in longer periods of drought and more intense rainstorms in at least part of this region (Norwine et al. 2007), which would affect soil development (formation or weathering of petrocalcic horizons, e.g.) and likely increase erosion rates. Erosion itself, biting into the underlying geological formations, will send the Cross Timbers and Prairies eastward as underlying rocks are exposed (Figs. 9.3, 9.21; Hayward and Allen 1987). Lastly, the vegetation to be observed across these landscapes is already very different than the original vegetation (Schmidly 2002) with many introduced species, the lack of regeneration of original species (e.g., Gabbard and Fowler 2007; Schmidt et al. 2008) or heightened densities of previously marginal species (Archer 1989). Anyone who has enjoyed a sunset among the silhouettes of gnarled post oaks can attest to the beauty of Texas landscapes. May these soil landscapes be preserved and not marginalized (Fig. 9.33). Individual landowners must take an active role since only a relatively small portion of the state is under public ownership.

References

Alizai HU, Hulbert LC (1970) Effects of soil texture on evaporative loss and available water in semi-arid climates. Soil Sci 110:328–332

Allison JE (1991) Soil survey of Lampasas County, Oklahoma. USDA-NRCS, Washington, DC. This and all county-level soil surveys. Available http://www.nrcs.usda.gov/wps/portal/nrcs/surveylist/soils/survey/state/

Archer S (1989) Have Southern Texas savannas been converted to woodlands in recent history? Am Nat 134:545–561

Baylor Geological Society (1979) The geomorphic evolution of the Grand Prairie. Tour guidebook. Baylor University, Waco

Baylor Geological Society (1983) Southeastern Llano County. Tour guidebook. Baylor University, Waco

Baylor Geological Society (1994) Stillhouse hollow reservoir. Tour guidebook. Baylor University, Waco

Blum EL (1982) Soil survey of Kimble County, Texas. USDA-NRCS, Washington

Bogard VA (1973) Soil survey of Pontotoc County, Oklahoma. USDA-NRCS, Washington

Bourlier BG, Sample GA, Swofford BG, Douthit G (1987) Soil survey of Cleveland County, Oklahoma. USDA-NRCS, Washington

Brooks CA (1978) Soil survey of Hill County, Texas. USDA-NRCS, Washington

Brooks CA, Rogers CA, Mayberry JH, McSpadden JO Jr, Mitchell WD, Huntsinger JW (1964) Soil survey of Ellis County, Texas. USDA-NRCS, Washington

Burgess DL (1977) Soil survey of Johnston County, Oklahoma. USDA-NRCS, Washington

Bryan F (1951) The grand prairie of Texas. In: Lozo FE (ed) The Woodbine and adjacent strata of the Waco area of Central Texas, Fondren Science Series 4. Southern Methodist University Press, Dallas, pp 1–11

Carson CD, Dixon JB (1983) Mineralogy and acidity of an inland acid sulfate soil of Texas. Soil Sci Soc Am J 47:828–833

Coburn WC (2007) Soil survey of Crockett County, Texas. USDA-NRCS, Washington

Cochran RA, Rives JL (1985) Soil survey of Big Bend National Park, Part of Brewster County, Texas. USDA-NRCS, Washington

Duttmann R, Petersen JF (2009) Eolian contributions to soil development in vernal pools: enchanted rock, Texas. Paper presented at the Association of American geographers annual meeting, Las Vegas, Nevada, 22–27 March 2009. https://www.lgi.geographie.uni-kiel.de/de/media/ss_12/120418_poster_enchanted_rock. Accessed 01 Mar 2015

Easterling D (1972) The Arbuckle group. In: Griffiths CE (ed) S.A.S.G.S. Spring field trip: paleozoic geology of the Arbuckle Mountains, Oklahoma. Tour guidebook. East Texas State Geological Society, Commerce

Fanning DS, Fanning MCB (1993) Soil: morphology, genesis, and classification. Wiley, Hoboken

Gabbard BL, Fowler NL (2007) Wide ecological amplitude of a diversity-reducing invasive grass. Biol Invasions 9:149–160

Gray F, Roozitalab MH (1976) Benchmark and key soils of Oklahoma: a modern classification system. Oklahoma State University Agricultural Experiment Station, Stillwater

Greenberg WA (1994) Hydrology and morphology of a toposequence of three claypan soils in Central Texas. Thesis, Texas A&M University

Greenwade JM (1996) Soil survey of Grimes County, Texas. USDA-NRCS, Washington

Haile NI, Brezina DN (2012) Soil survey of Kenedy and Kleberg Counties, Texas. USDA-NRCS, Washington

Hallmark CT (1993) The nature and origin of the Blackland Soils. In: Sharpless R, Yelderman JC (eds) The Texas Blackland Prairie: land, history and culture. In: Proceedings of the symposium of the natural regions of Texas April, 1990. Baylor University Program for Regional Studies, Baylor University, Waco, pp 41–47

Hawkins GP, Dodge CF, Butler JC (1974) Clay mineralogy of the Lewisville Member of the Cretaceous Woodbine Formation in the Arlington Area, Tarrant County, Texas. Gulf Coast Assoc Geol Soc Trans 24:283–290

Hayward OT, Allen PM (1987) Lampasas cut plain: form-process-evolution. Tour guidebook. Baylor Geological Society, Baylor University, Waco, p 88

Hill RT (1901) Twenty-first annual report of the United States Geological Survey, part VII-Texas, geography and geology of the Black and Grand Prairies, Texas with detailed descriptions of the Cretaceous formations and special reference to artesian waters. Government Printing Office, Washington, DC, pp 905

Hyde HW, Bruns RHL, Wilhelm MF, Porter AC (1992) Soil survey of Lavaca County, Texas. USDA-NRCS, Washington

Jacobs JL (1981) Soil survey of Hidalgo County, Texas. USDA-NRCS, Washington

Johnson KS (2008) Geologic history of Oklahoma. Oklahoma Geological Survey, Education publication 9, pp 8

Jurena MR (2005) Soil survey of Burleson County, Texas. USDA-NRCS, Washington

Jurena MR (2007) Soil survey of Lee County, Texas. USDA-NRCS, Washington

Klich I, Wilding LP, Pfordrsher AA (1990) Close-interval spatial variability of Udertic Paleustalfs in East Central Texas. Soil Sci Soc Am J 54:489–494

Loomis LE, Gabriel WJ (1991) Soils on late Holocene to mid Tertiary erosional and depositional geomorphic surfaces in the Central Edwards Plateau. Tour Guidebook for the Professional Soil Scientists Association of Texas Annual Meeting, Uvalde, Texas, February 14–15, pp 40

Macroberts MH, Macroberts BR (2003) The east-west transition of the flora of Texas: a biogeographical analysis. SIDA 20:1693–1700

Main DJ, Noto CR (2012) Paleoecology and paleoenvironments of the Woodbine Formation of North Texas: the Arlington Archosaur Site. Earthwatch Field Trip Guide, Summer 2012. www.academia.edu Accessed 08 Apr 2015

Marbut CF (1928) A scheme for soil classification. Proc First Int Cong Soil Sci 4:1–31

Marcy LE (1982) Habitat types of the Eastern Cross Timbers of Texas. Thesis, Texas A&M University

Mobley HL, Brinlee RC (1967) Soil survey of Comanche County, Oklahoma. USDA-NRCS, Washington

Moebius GE, Maxwell AJ (1979) Soil survey of Carter County, Oklahoma. USDA-NRCS, Washington

Monger HC, Gile LH, Hawley JW, Grossman RB (2009) The desert project—an analysis of aridland soil-geomorphic processes. Bulletin 798, New Mexico State University, Las Cruces, New Mexico, p 76

Montgomery JA (1993) The nature and origins of the Blackland Prairies of Texas In: Sharpless R, Yelderman JC (eds) The Texas Blackland Prairie: land, history and culture. Proceedings of the symposium of the Natural Regions of Texas April, 1990. Baylor University Program for Regional Studies, Baylor University, Waco, Texas, pp 24–40

Mutis-Duplat E, Gray MW (2011) Surface Geology. In: McCormick J (ed) Soil survey of Mason County, Texas. USDA-NRCS, Washington

Neitsch CL, Castille JJ, Jurena MR (1989) Soil survey of Leon County, Texas. USDA-NRCS, Washington

Norwine J, Harriss R, Yu J-H, Tebaldi C, Bingham R (2007) The problematic climate of South Texas1900–2100 In: Norwine J, Kuruvilla J (eds) The changing climate of South Texas 1900–2100: Problems and prospects impacts and implications. CREST-RESSACA, Texas A&M-Kingsville, Kingsville, Texas, pp 15–55

Noy-Meir I (1973) Desert ecosystems: environment and producers. Annu Rev Ecol Syst 4:25–51

Petersen JF, Duttmann R (2009) Depositional processes, stratigraphy, and soil development in vernal pools, Enchanted Rock, Texas. Paper presented German Geographical Society, Vienna, Austria. https://www.lgi.geographie.uni-kiel.de/de/media/ws_10_11/poster_enchanted_rock. Accessed 01 Mar 2015

Rabenhorst MC, Wilding LP (1986a) Pedogenesis on the Edwards Plateau, Texas: I. Nature and continuity of parent material. Soil Sci Soc Am J 50:678–687

Rabenhorst MC, Wilding LP (1986b) Pedogenesis on the Edwards Plateau, Texas: II. Formation and occurrence of diagnostic subsurface horizons in a climosequence. Soil Sci Soc Am J 50:687–692

Rabenhorst MC, Wilding LP (1986c) Pedogenesis on the Edwards Plateau, Texas: III. New model for the formation of petrocalcic horizons. Soil Sci Soc Am J 50:693–699

Rabenhorst MC, Wilding LP, Girdner CL (1984) Airborne dusts in the Edwards Plateau Region of Texas. Soil Sci Soc Am J 48:621–627

Rehage JA (1985) Hydrology and genesis of claypan soils in east central Texas. Dissertation, Texas A&M University

Ressel DD (1981) Soil survey of Tarrant County, Texas. USDA-NRCS, Washington

Ressel DD, Brown SE (2004) Soil survey of Fayette County, Texas. USDA-NRCS, Washington

Rives JL (1980) Soil survey of Pecos County, Texas. USDA-NRCS, Washington

Sackett JL III (2011) Soil survey of Duval County, Texas. USDA-NRCS, Washington

Sala OE, Parton WJ, Joyce LA, Lauenroth WK (1988) Primary production of the Central Grassland Region of the United States. Ecology 69:40–45

Schaetzl RJ, Anderson S (2005) Soils: Genesis and geomorphology. Cambridge University Press, Cambridge

Schmidly DJ (2002) Texas natural history: a century of change. Texas Tech University Press, Lubbock

Schmidt CD, Hickman KR, Channell R, Harmoney K, Stark W (2008) Competitive abilities of native grasses and non-native (*Bothriochloa* spp.) grasses. Plant Ecol 197:69–80

Sellards EH, Adkins WS, Plummer FB (1932) The geology of Texas, volume 1: stratigraphy. University of Texas Bulletin 3232. University of Texas, Austin, Texas

Smeins FE, Diamond DD (1983) Remnant Grasslands of the Fayette Prairie, Texas. Am Midl Nat 110:1–13

Soil Survey Staff (2014) Keys to soil taxonomy, 12th edn. USDA-NRCS, Washington

Soil Survey Staff (2015) Official soil series descriptions. USDA-NRCS. https://soilseries.sc.egov.usda.gov/osdname.asp. Accessed 09 June 2015

Spearing D (1991) Roadside geology of Texas. Mountain Press, Missoula

Stenzel HB (1935) Salient features of the geology of the Llano Uplift. In: Southwestern Geological Society (eds) Field trip through Llano Uplift. Tour guidebook, University of Texas Department of Geology, Bureau of Economic Geology, Austin, pp 1–5

Swanson ER (1995) Geo Texas: A guide to the earth sciences. Texas A&M University Press, College Station

Taylor FB (1977) Soil survey of Wilson County, Texas. USDA-NRCS, Washington

Texas County Estimates (2015) USDA-NASS, Washington, DC. http://www.nass.usda.gov/Statistics_by_State/Texas/Publications/County_Estimates/ce_pdf/. Accessed 18 June 2015

Tharp BC (1939) The vegetation of Texas. The Texas Academy of Science, Edinburg

USDA NRCS (2006) Land resource regions and major land resource areas of the United States, the Caribbean and Pacific Basin. USDA handbook, 296

Watterson A Jr, Boggard VA, Moebius GE (1984) Soil survey of Murray County, Oklahoma. USDA-NRCS, Washington

West LT, Wilding LP, Hallmark CT (1988) Calciustolls in Central Texas: II. Genesis of calcic and petrocalcic horizons. Soil Sci Soc Am J 52:1731–1740

Yelderman J (1998) Note number 9: true love on the Crosstimbers Blackland margin In: A field guide to the geography and geology of the Grand and Black Prairies of Texas. Baylor Geological Society, Waco, Texas, p 94-A

Soils of the Northern Lake States Forest and Forage Region: LRR K

10

Randall J. Schaetzl

10.1 Introduction and Overview

Within the Northern Lake States Forest and Forage Region (NLSFFR) are 17 Major Land Resource Areas (MLRAs), each of which is fairly internally homogeneous with respect to soils, landforms and land use (Fig. 10.1). The focus of this chapter is on the characteristics, distribution, classification, and genesis of soils within the NLSFFR, as well as on issues related to management and land use.

In this chapter, soils of MLRA 95B (Southern Wisconsin and Northern Illinois Drift Plain) in southeastern Wisconsin (Fig. 10.1) are not specifically discussed; discussion of the soils there will be in Chap. 11. Additionally, MLRA 96 (Western Michigan Fruit Belt) of northwestern Lower Michigan is included in this chapter even though it is technically not part of the NLSFFR, because its soils are similar to many of the other soils in northern Michigan, especially those within MLRA 94A. Thus, coverage of the NLSFFR, as slightly modified in extent in this chapter, spans northern Lower Michigan, all of Michigan's Upper Peninsula (UP), northern Wisconsin, and northeastern Minnesota and includes 17 MLRAs (Fig. 10.1).

The recently glaciated landscapes of the NLSFFR generally have gentle to moderately sloping topography (USDA-NRCS 2006). Most of the landscape has its origins in the last major glaciation. Nonetheless, two MLRAs in the NLSFFR occur south of the Late Wisconsin glacial border—one (94B) on areas of older drift in central Wisconsin (Attig et al. 2011a) and one that is dominated by the sandy, wet, former lake bed of Glacial Lake Wisconsin (Clayton and Attig 1989; Fig. 10.1). All other areas bear the conspicuous marks of recent (Late Pleistocene) glaciation, and most of the soils therein are formed in glacigenic sediments.

This LRR is dominated by Histosols, Alfisols, Spodosols, Inceptisols, and Entisols (Fig. 10.2). Histosols are common in lowland areas, such as in glacial kettles and on former glacial lake basins. Thus, with the exception of a few lake plains, most areas of Histosols are small in areal extent. Alfisols dominate large parts of the landscape where the parent materials are loamy or finer-textured, typical of tills or lacustrine sediment. Spodosols are more prevalent on sandier parent materials, and in the north. Young, sandy landscapes, e.g., sand dunes and recently exposed beaches, steeply sloping areas, and regions of shallow bedrock, are dominated by Entisols. Inceptisols are also found, commonly on wet, loamy lake plains and in areas of moderate slope where slope processes facilitate erosion and, thus, maintain soils in a minimally developed state. Much of northern Minnesota has a cover of Inceptisols and Entisols. A few areas of Vertisols and upland Mollisols are also mapped in parts of Minnesota, near the western and southwestern margins of the region. Small areas of lowland Mollisols are found in isolated areas across the region.

10.2 Geomorphology, Physiography, and Relief

The geologic and geomorphic history of this region involves repeated coverage by continental glaciers during the Pleistocene Epoch (Mickelson et al. 1983). Hence, most landscapes and soils are geologically young, i.e., less than about 18,000 years old, and many areas have deranged and poorly integrated drainage patterns. Kettle lakes and wetlands—many of them quite large—are common, having originated as buried ice blocks subsequently melted. Small, isolated hills and uplands, and hummocky moraines, are also common. Soils formed in bedrock residuum are found in only a few regions; most bedrock is either deeply buried by glacial drift or was scraped clean by the ice, leaving behind only hard bedrock that has not had enough time to weather into thick residuum. Therefore, most landforms have direct glacial origins, e.g., moraines, till plains, outwash plains,

R.J. Schaetzl (✉)
Michigan State University, East Lansing, MI 48823, USA
e-mail: soils@msu.edu

L.T. West et al. (eds.), *The Soils of the USA*, World Soils Book Series,
DOI 10.1007/978-3-319-41870-4_10

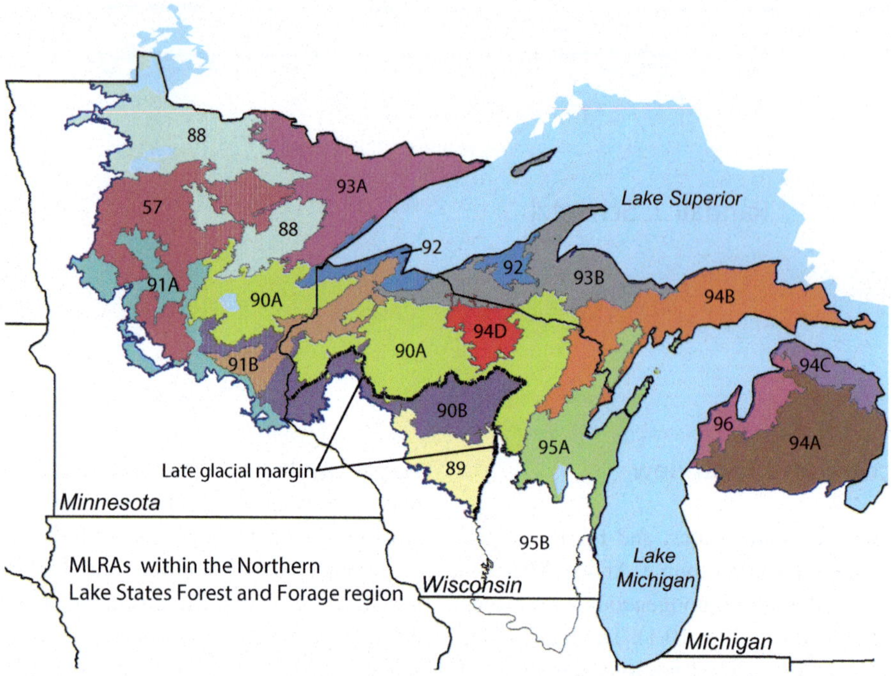

Fig. 10.1 Map showing the Major Land Resource Areas within the Northern Lake States Forest and Forage Region. Note that the soils of MLRA 95B, which is within the Northern Lake States Forest and Forage Region, is discussed in Chap. 11. Similarly, MLRA 96, which is included in a different LRR (Lake States Fruit, Truck Crop, and Dairy Region) IS included in this chapter, because its soils are fairly similar to the other soils in northern Michigan. Also shown in this figure is the limit of the last glaciation (Wisconsin stage) in the state of Wisconsin (*bold black line*). MLRA numbers and names: *57* Northern Minnesota Gray Drift, *88* Northern Minnesota Glacial Lake Basins, *89* Central Wisconsin Sands, *90A* Wisconsin and Minnesota Thin Loess and Till, Northern Part, *90B* Wisconsin and Minnesota Thin Loess and Till, Southern Part, *91A* Central Minnesota Sandy Outwash, *91B* Wisconsin and Minnesota Sandy Outwash, *92* Lake Superior Plain, *93A* Superior Stony and Rocky Loamy Plains and Hills, Western Part, *93B* Superior Stony and Rocky Loamy Plains and Hills, Eastern Part, *94A* Northern Michigan and Wisconsin Sandy Drift, *94B* Michigan Eastern Upper Peninsula Sandy Drift, *94C* Michigan Northern Lower Peninsula Sandy Drift, *94D* Northern Highland Sandy Drift, *95A* Northeastern Wisconsin Drift Plain, *95B* Southern Wisconsin and Northern Illinois Drift Plain, and *96* Western Michigan Fruit Belt

lacustrine plains, and including such common glacial landforms as eskers, kames and drumlins. Most soils are formed in recent and minimally weathered glacigenic sediments, whether the sediments are meters thick or exist as only a few cm of overburden above hard and glacially scoured bedrock. Areas of late Pleistocene- and Holocene-aged sand dunes also dominate some areas, especially near the current shores of the Great Lakes, and on inland, sandy surfaces that were formerly associated with proglacial lakes or outwash plains (Rawling et al. 2008; Arbogast 2009; Blumer et al. 2012; Loope et al. 2012).

Surface topography is, in many places, rolling, hummocky and irregular, due to its glacial origins and geologic youth. Most landscapes are low or moderate in slope. Most of the lowest relief areas are former lake plains and outwash plains. The highest relief areas are the hummocky end moraines, particularly in central Wisconsin and the western Upper Peninsula of Michigan, and the bedrock terrain of MLRA 93B, also in the western Upper Peninsula. Hummocky areas associated with formerly buried ice are, in some areas, impressive and extensive (Attig and Clayton 1993).

10.3 Climate

Upland soils in the NLSFFR have udic soil moisture regimes, typical of humid climates. Aquic soils with wet, reducing regimes, are mainly found in lowlands and on flat uplands with low permeability and minimal runoff. Large areas of wetlands, and hence, aquic soil moisture regimes, occur in the eastern UP of Michigan and in northern Minnesota. Although most of the region is in the frigid soil temperature regime, the southern parts of the more southerly MLRAs, e.g., 89, 91A, and 95A, are mesic. In northwestern Lower Michigan, MLRA 96, where the lake effect snowpacks insulate the soil and minimize wintertime

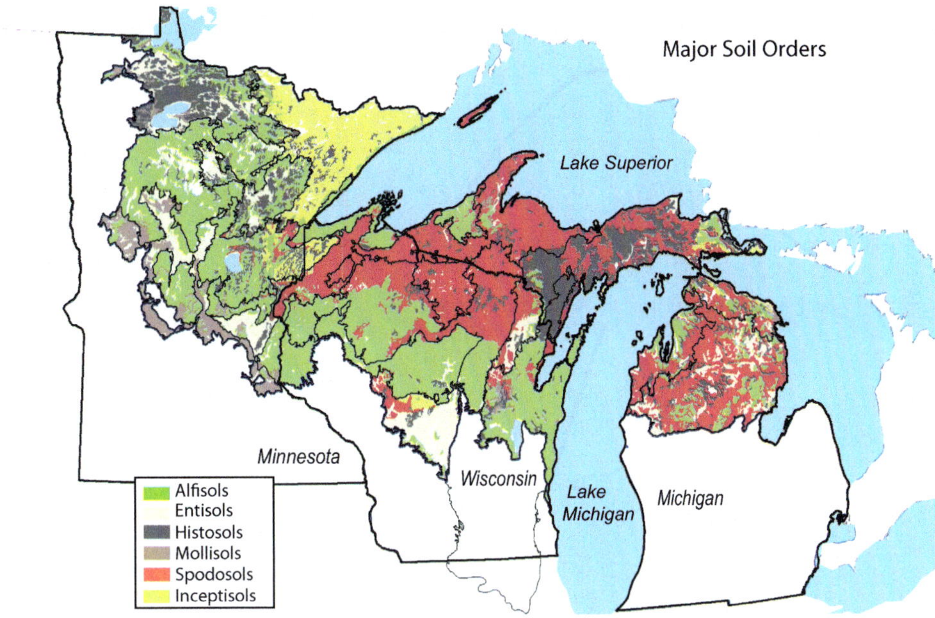

Fig. 10.2 Map of the major soil orders (excluding Vertisols, which are minor in extent) in the NLSFFR

freezing, is also within the mesic soil temperature regime (Schaetzl et al. 2005).

The NLSFFR has a strongly seasonal climate, with cold, often snowy, winters and warm, humid summers. Annual precipitation values are relatively uniform across the region, averaging between ≈750 mm in the eastern and northeastern sections of Michigan, to ≈850 mm in parts of central Wisconsin (Fig. 10.3). Precipitation totals fall off markedly in northwestern Minnesota; the extreme northwestern parts of the NLSFFR receive an average of only 500–550 mm of precipitation annually. These western areas mark the transition to the drier, prairie parklands of the Great Plains. Precipitation across the NLSFFR is fairly evenly partitioned throughout the growing season; there is no pronounced dry season, although February is commonly the driest month. Soils usually reach their driest point in late August and early September, before the start of predictable fall rains and then, winter snows (Schaetzl et al. 2015).

The frost-free period ranges from <90 days in the uplands of northern Lower Michigan, the western Upper Peninsula of Michigan, and the arrowhead region of Minnesota, to 170 days in central Wisconsin. Longer growing seasons also occur nearer the Great Lakes; these areas also have noticeably cooler summers and warmer winters, because of the slower response of the large water bodies to changes in air temperature and inputs of solar energy (Schaetzl and Isard 2002). As a result, many vegetation communities and soil types common to northerly sections of the NLSFFR have extended ranges to the south along the Great Lakes' shores. Additionally, soils typical of cooler regions, e.g., Spodosols and Histosols, tend to be better developed near the shores of Lakes Superior and Michigan. Land use patterns follow the lakes, as fruit (and some vegetable) production is typically better expressed on uplands near the Great Lakes, other things being equal (Hull and Hanson 2009).

Snowfall is common across the region in winter. Most sites in the region develop and maintain a persistent snowpack throughout at least part of the winter. Snowfall is particularly heavy, and snowpacks are thick and long-lasting, in areas immediately east and south of the Great Lakes (Muller 1966; Norton and Bolsenga 1993; Burnett et al. 2003). The importance of heavy lake effect snow to pedogenesis is discussed below. Climate data suggest that, in recent decades, snowfall is decreasing and snowpacks are thinning (Isard et al. 2007). In general, this trend will lead to colder soils in winter, because the snowpack tends to insulate the soil from frost and freezing (Isard and Schaetzl 1995).

In general, the udic soils of the NLSFFR receive adequate precipitation and snowmelt to allow for deep percolation (through the profile) at least once per year, and usually much more often. Commonly, this period of deep percolation is associated with spring snowmelt, and secondarily, with heavy, extended, fall or winter rains (Schaetzl et al. 2015). The latter types of events vary greatly both temporally and spatially. Most soils receive enough precipitation to flush salts and many types of soluble materials, e.g., carbonates, from the upper profile, in the few thousand years that have elapsed since $time_{zero}$. Most soils in the region, therefore, are leached of carbonates to a meter or more, depending on local hydrology and carbonate content of the parent material, i.e., if carbonates even exist in the parent material—see below.

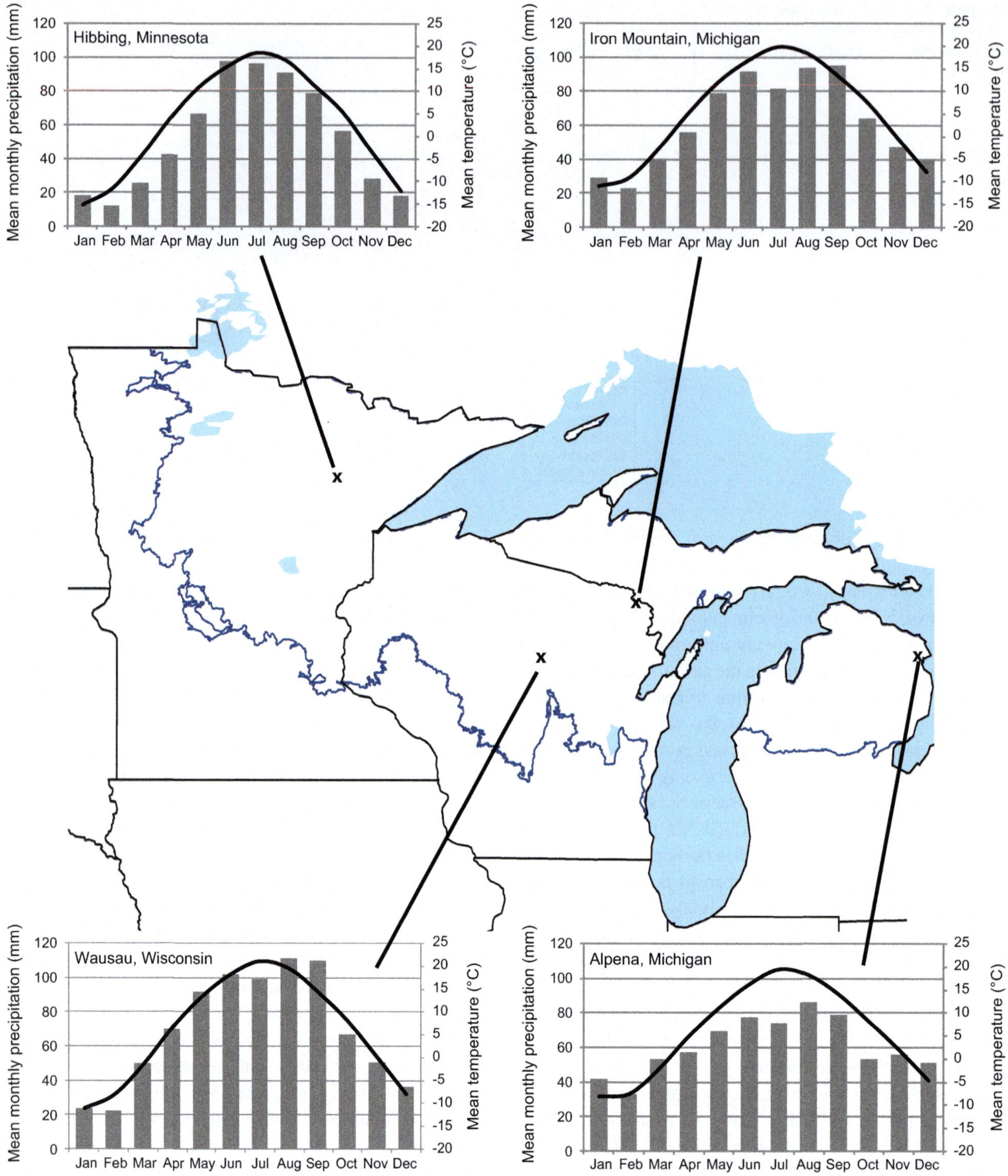

Fig. 10.3 Climographs for four representative sites within the NLSFFR. Most have been constructed using 1961–1990 data from the US National Weather Service

10.4 Vegetation and Fauna

Because of the humid, seasonal climate, and relatively fertile soils, most of the landscape was forested prior to European settlement. And much of it remains forested today (McNab and Avers 1994). Most of the forest was a mixture of coniferous species such as pine (*Pinus* spp.), hemlock (*Tsuga canadensis*), spruce (*Picea* spp.), and fir (*Abies* spp.), with broadleaf deciduous species such as maple (*Acer* spp.), basswood (*Tilia americana*), birch (*Betula* spp.), oak

(*Quercus* spp.), aspen (*Populus* spp.), and beech (*Fagus grandifolia*) (Nichols 1935; Curtis 1959). Many refer to this forest assemblage as the Laurentian Mixed Forest type, or simply as northern hardwoods (Nichols 1935; Stearns 1949). Wet soils in lowland areas tended to support more conifers, as did the driest upland sites, where pine species dominated on deep, infertile sands. In general, conifers tended to out-compete broadleaf trees on the end-members of the wetness and fertility spectrum—on the wettest and on the driest (sandy or shallow to bedrock) sites. In the far north, especially the arrowhead region of Minnesota, relatively pure boreal forest stands occur, but even here they are best developed in wet sites (McNab and Avers 1994). In northwestern Minnesota, drier climatic conditions led to a more discontinuous, parkland-type of forest community, as the region transitions to the western grasslands (Davis 1977).

Although fire is a dominant, natural disturbance factor in the forests of this region, it only indirectly affects the soils. Therefore, because of the near-continuous forest cover, tree uprooting is a common disturbance agent. Trees tear up soil as they uproot, often forming a pit at the former location of the roots, and an adjacent mound, located where the soil slumps off the roots (Fig. 10.4). Soil materials within the mound can be thoroughly mixed as they slump off the root plate, although in some cases, they are simply overturned in a more-or-less intact manner (Schaetzl 1986). Recent work on the longevity of the pit-and-mound topography formed by uprooting suggests that it can persist for millennia (Schaetzl and Follmer 1990; Šamonil et al. 2013), implying that the results of uprooting, i.e., the microtopography, may in many areas be a semipermanent part of the soil landscape. Pits and mounds affect pedogenesis by diverting water into the depressions and off the mounds, leading to enhanced leaching and soil formation in the pits and slower pedogenesis in the comparatively drier mounds (Schaetzl 1990). This type of disturbance leads to high spatial variability in the soil landscapes, especially at small scales (Meyers and McSweeney 1995; Kabrick et al. 1997; Šamonil et al. 2015).

Unforested areas in the NLSFFR generally are of three types: (1) bedrock outcrops, (2) recent and active sand dunes, and (3) very poorly drained sites with a water table at or above the surface. The wetter sites are typically covered in marsh or bog vegetation. Areas of open water—of all sizes—are extremely common in this region, and indeed, help define its character. And although it is not a dominant land use, many areas of agriculture do exist in the region, and as such, create unforested patches on the landscape.

Soil fauna play important roles in the soil system, particularly with respect to nutrient cycling and bioturbation (Salem and Hole 1968; Green et al. 1998). Invasive species are always a threat, and with respect to soils a particularly notable invasive is the common earthworm. Resner et al. (2011) reported on invasive earthworms in the forests of northern Minnesota. Here, bioturbation by earthworms has changed the soils by blurring horizon boundaries in the upper profile and dramatically affecting P and C distributions with depth.

10.5 Parent Materials

In this complex glacial landscape, determining the spatial variation in parent materials is perhaps the best way to understand and explain the composition and pattern of the soil landscape. Parent materials set the stage for soil genesis. And perhaps nowhere is parent material a more important soil forming factor than in young landscapes like this. Because of repeated Pleistocene glaciations, most soils in the NLSFFR have formed in sediments derived directly from the ice (till), indirectly from meltwater (outwash or lacustrine sediment), or due to later, secondary transport of glacial sediment by wind (dune sand or loess) or running water (alluvium) (Fig. 10.5). In many places, especially in northern Lower Michigan, the glacial sediments are many tens of meters thick (Rieck and Winters 1993).

An important, major division in parent material types across the region involves contents of $CaCO_3$ (carbonates)

Fig. 10.4 Photos showing the effects of uprooting on soils and soil landscapes. **a** An uprooted red pine tree, showing the root plate and the adjacent pit. Photo by RJS. **b** Pit-and-mound topography in northern Wisconsin, formed by millennia of tree uprooting. Photo by RJS. **c** Sequence of horizons across a 3600-year-old pit–mound pair in a northern Michigan Spodosol, showing the intact but buried horizons below the mound (*left*), and the spatially focused soil development in the pit (*far right*), as exemplified by deep E and B horizon tongues. Photo by Šamonil et al. (2013)

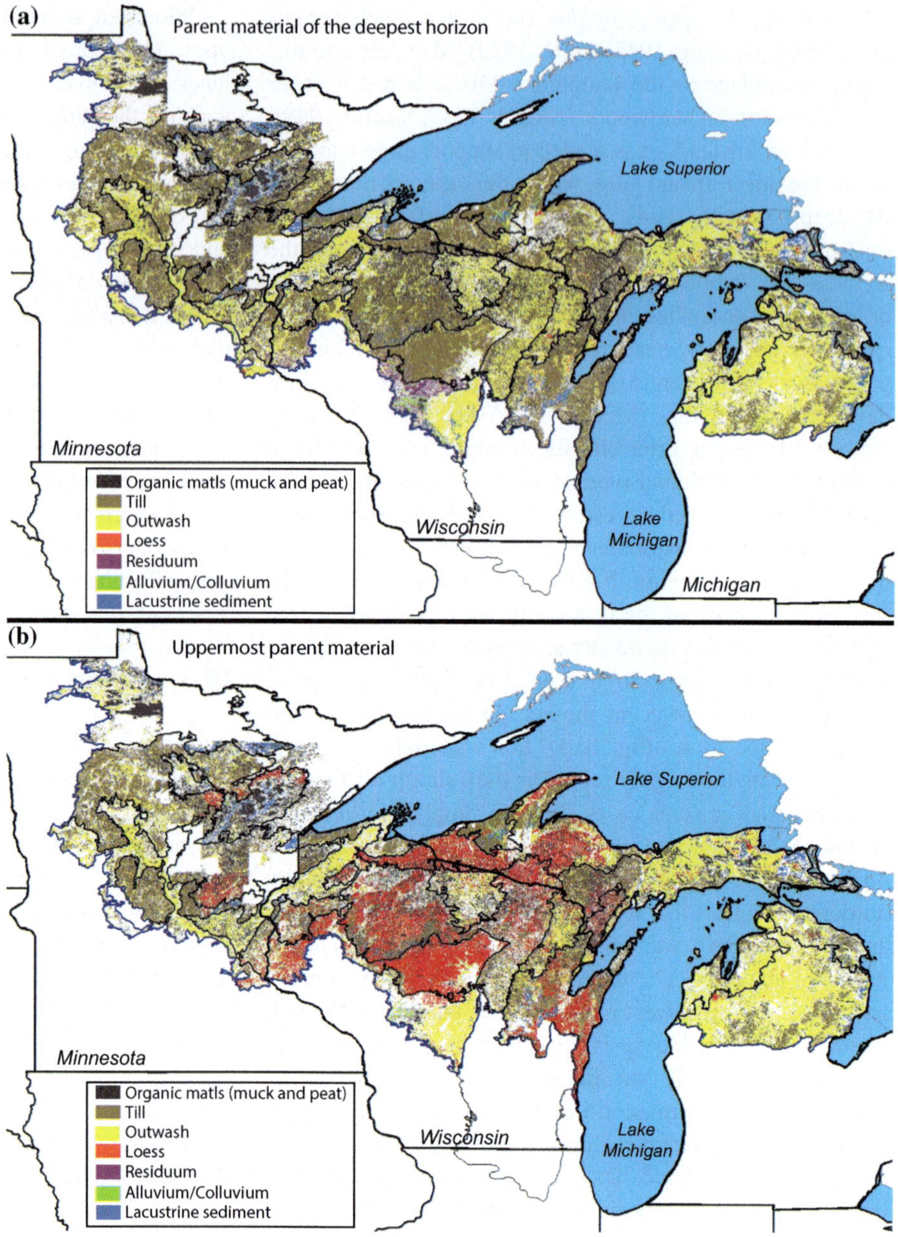

Fig. 10.5 Map of the major soil parent materials in the NLSFFR, as indicated by the Soil Survey Geographic Database (SSURGO) and the official soil series descriptions (OSDs). Areas of organic parent materials are very common, but because they normally occur in isolated, small pockets, they are not readily depicted on a map of this scale. **a** The parent material described at depth, i.e., in the lowermost horizon, in the OSD. **b** Shown here is the same map, but for the uppermost parent material. Note the common occurrence of loess (*red colors*) in this landscape, especially in Wisconsin and the western Upper Peninsula of Michigan. Because the loess here is seldom thicker than 100 cm, only rarely does it occur as the lowermost parent material in the OSD

(Fig. 10.6). In soils formed on calcareous parent materials, the bottom of the solum is usually determined by the depth of carbonate leaching. Soils that form in calcareous parent materials tend to be more fertile, but they also must be acidified and leached of carbonates before important processes like clay translocation and podzolization can begin, before fragipans can form, and before intense weathering of primary minerals can begin. Thus, soils forming in acidic (or at least noncalcareous) parent materials tend to have thicker sola; they have had a "pedogenic head start" of sorts. Parent materials across the north-central parts of the region have been derived from the acidic, igneous and metamorphic rocks of the Canadian Shield, which is centered in south-central Canada and extends into the western Upper Peninsula of Michigan, and into northern Wisconsin and Minnesota. As the glaciers moved across this landscape, they eroded mainly crystalline rocks, producing coarse-textured, acidic drift. In contrast, as the ice traversed the eastern Upper Peninsula of Michigan and parts of western Wisconsin and eastern Minnesota, it also eroded the overlying carbonate rocks like limestone and dolomite, along with shales (some of which are also calcareous) and acidic sandstones. Thus, parent materials in these regions tend to be strongly calcareous and (usually) finer-textured. An area of limestone

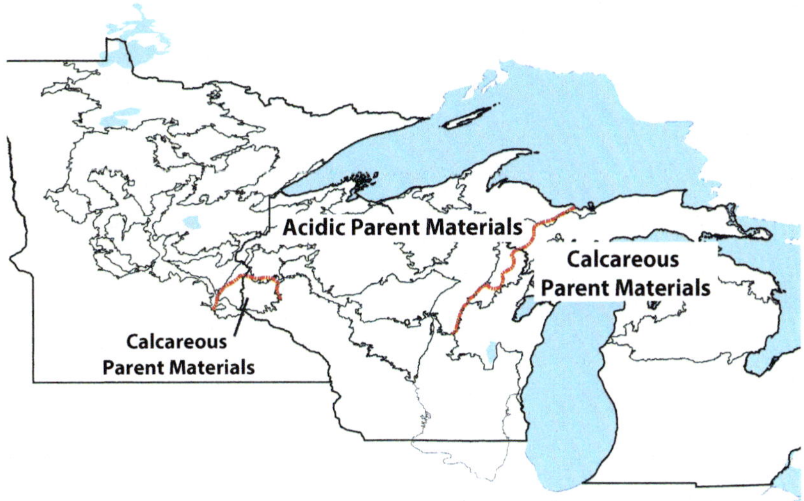

Fig. 10.6 Generalized boundary between acidic and calcareous parent materials in the region

Fig. 10.7 Examples of soil parent materials that are commonly found in the NLSFFR. **a** Sandy glacial till—a particularly rocky example. **b** Stratified sandy/gravelly outwash. **c** Lacustrine sediment—with strata of fine sand and silt. **d** Silty/flaggy colluvium at the base of a slope. Rocks from upslope have been incorporated into the colluvium. Photos by RJS

bedrock in western Wisconsin and eastern Minnesota has also led to calcareous parent materials there (Fig. 10.6).

Till parent materials are widespread; most are sandy-to-loamy in texture (Fig. 10.7). Most tills contain at least some coarse fragments, and some are extremely rocky. Upland soils in the region, particularly those in areas of morainic topography, are usually formed in till parent materials, which can vary markedly in texture and coarse fragment content across even short distances. Thus, soil landscapes formed in till often exhibit great spatial variability in texture, slope, and depth to water table, largely because of their direct glacial origins. These characteristics, combined with complex vertical stratification and frequent lithologic discontinuities (Schaetzl 1998), make mapping such soil landscapes extremely difficult. As a result, the soil maps in this region often have great complexity, with many small mapping units and untold numbers of map unit inclusions (Asady and Whiteside 1982; Miller and Schaetzl 2015). Indeed, this region may contain some of the most difficult and complex terrain in the US, from a soil mapping perspective (Khakural et al. 1993; Fig. 10.8).

As shown in Fig. 10.5, wide expanses of the region are comprised of sandy materials on outwash plains or lacustrine plains. These areas may range in wetness, depending on local depth to the water table, but the spatial variation in soils across them is generally lower than on other landscapes in the region. Many of the sandy lake plains, e.g., the Glacial Lake Algonquin plain in MLRA 94B or the Glacial Lake Wisconsin plain in MLRA 89, are wet. Conversely, many outwash plains, e.g., the Vilas outwash plains in MLRA 94D, are dominated by excessively drained, sandy soils. Some of these landscapes are spotted with small sand dunes which postdate the deposition of the outwash or lacustrine sediment by millennia (Loope et al. 2012).

Soils formed in residuum are not extensive in the NLSFFR, because most bedrock is covered by thick glacial

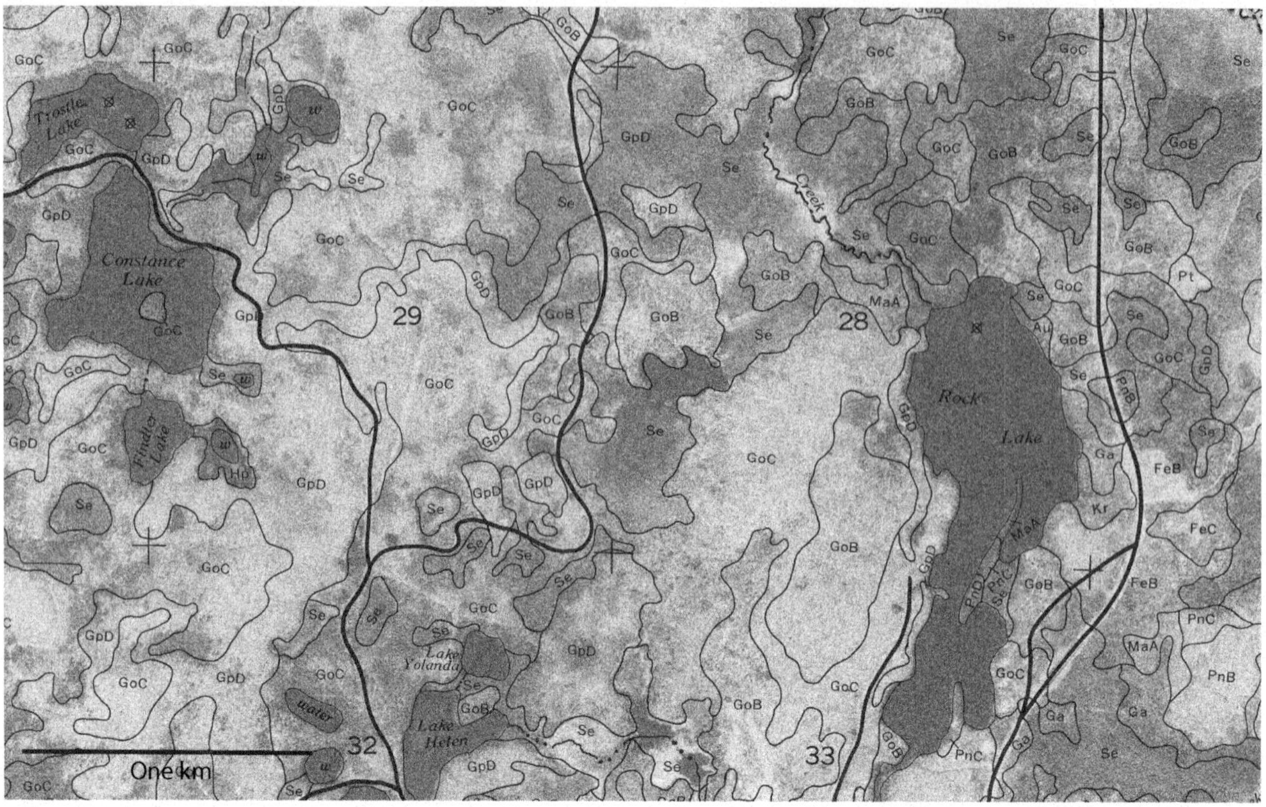

Fig. 10.8 Portion of the NRCS soil map for Vilas County, Wisconsin (Natzke and Hvizdak 1988), within the hummocky Winegar Moraine. Note the many lakes, the contorted outlines of many of the mapping units, and the many small, isolated mapping units within depressions. These characteristics are typical of glacial landscapes formed in till

deposits or has been scraped clean by the ice and has not had ample time to develop a weathered mantle. Similarly, soils formed in colluvium are not widespread, because in most areas slopes are not steep enough to initiate mass movement. Nonetheless, an area with considerable amounts of residual and colluvial soils does occur in the western part of MLRA 89 (but not on the lake plain proper), which lies outside of the last glacial border. The residual soils occur on sandy bedrock uplands, where soft Cambrian sandstones have weathered into sandy Entisols and Inceptisols. Colluvial parent materials here mainly occupy footslopes and lowland areas; often the colluvium is mixed with loess that had formerly accumulated on hilltops.

Other soils in the NLSFFR have developed in materials that postdate the recession of the ice, e.g., dune sand, alluvium, and loess. After ice recession, perhaps for a millennium or more, the landscape was unstable, winds were strong, and sediment was readily available for eolian transport (Schaetzl and Attig 2013; Schaetzl et al. 2014). Thus, many soils in the region are capped with several tens of cm of loess, which is usually silty but in places is loamy (Kabrick et al. 1997; Schaetzl 2008; Scull and Schaetzl 2011; Schaetzl et al. 2014; Fig. 10.5b). Most of the loess was locally derived from outwash plains, meltwater valleys, and abandoned lacustrine surfaces (Schaetzl and Loope 2008; Luehmann et al. 2013). The loess cover is particularly thick in parts of MLRA 90A, especially in Wisconsin. Only recently has the extent and the widespread nature of the loess "mantle" on soils in the NLSFFR been understood and appreciated. Many soils with a thin cap of silty loess exhibit loamy, rather than silty, textures in their upper profiles, sometimes because the silty loess has been mixed into the (normally) coarse-textured sediment below, resulting in a loamy mantle that was (previously) not recognized as loess. These types of loamy soils are particularly widespread in the western Upper Peninsula of Michigan (Luehmann et al. 2013). The loamy cap, even if thin and especially if it overlies sandy sediment, is important for sustaining productive forests. Without this cap, forest productivity on these landscapes would be considerably lower.

While the glaciers were present in the region, permafrost covered many parts of the landscape, south of the ice margin (Clayton et al. 2001). As the ice melted and the climate warmed, this permafrost thawed, leading to long periods of slope instability and erosion. This episode exposed many new parent materials, as soils above were stripped from slopes by solifluction (Attig and Muldoon 1989). Then, before vegetation colonized the landscape, erosion (assisted by the thawing permafrost) by running water contributed large amounts of sediment to local rivers, filling their valleys with alluvium. Other rivers were filled with alluvium derived directly from the melting ice, i.e., by sandy outwash. Today, wide expanses of alluvial soils occur in the major river valleys, many of which have one or more alluvial terraces. For these reasons, outwash is one of the main parent materials in the region, attesting to the vast amounts of meltwater that traversed the post-glacial landscape.

A second, more pronounced and temporally focused wave of erosion and valley alluviation occurred as settlers opened up the landscape to farming in the 1800s. Erosion of upland soils at this time led to valley filling and burial of preexisting soils in valley bottoms; this new sediment is called post-settlement alluvium (Faulkner 1998; Knox 2006).

In perhaps, no other Land Resource Region are Histosols such a dominant component of the soil landscape. Histosols are organic soils that have formed in decaying organic materials. They occur mainly in wet sites where organic matter decomposition is slowed by cold and anaerobic conditions (Kolka et al. 2011). This situation is typical of swamps, marshes, bogs, and fens, which are very common in the isolated depressions (kettles) and broad lake plains of these glacial landscapes (Fig. 10.9).

10.6 Time or Soil Age

All areas within the glacial limit are younger than approximately 18,000 years old, representing the approximate time when the ice started its retreat from its maximum position (Attig et al. 2011b; Schaetzl et al. 2014; Fig. 10.1). The ice was largely gone from the region by 11,000 years ago, implying that most soils in the NLSFFR are between 18,000 and 11,000 years old—quite young by worldwide pedogenic standards.

Soils south of the Late Wisconsin border in central Wisconsin are much older, having been formed in pre-Wisconsin aged deposits that may be a million years old or older. A few of these soils have weathered to such an extent that they classify as Ultisols (Mason et al. 1994), although most have been geologically rejuvenated by additions of loess generated during the late Wisconsin glaciation,

Fig. 10.9 The wet, very poorly drained plain of Glacial Lake Algonquin in the central Upper Peninsula of Michigan, in MLRA 94B. Histosols are widespread here. The wet conditions facilitate the continued formation of thick accumulations of peat. The upland in the background is a small sand dune. Photo by RJS

ca. 18,000–14,000 years ago (Stanley and Schaetzl 2011). Still others exist on old landscapes but have formed in less weathered parent materials, brought to the surface by slope instability and mass movements initiated as late Wisconsin permafrost thawed and solifluction was active (Mason and Knox 1997; Clayton et al. 2001). Indeed, most of the soils in the NLSFFR that are beyond the glacial border retain few, if any, characteristics of older, highly weathered soils that they probably exhibited prior to the last glacial advance.

Soils formed in sand dunes and alluvium can significantly postdate the last glaciations; many of these soils are Holocene in age. Additionally, along the shores of the Great Lakes are several terraces associated with former high water levels. Soils on these terraces and exposed lake beds generally date to 11,000 years ago or earlier (Barrett 2001; Schaetzl et al. 2002; Drzyzga et al. 2012). Lastly, along many bays and inlets of the Great Lakes are beach ridges formed by small-scale oscillations in lake level; soils here can be very young (Lichter 1998). Characteristics of soils on these beach ridges (and dunes that cap them) have been very useful in identifying the extent and ages of lake level fluctuations in the late Holocene (Thompson and Baedke 1997).

10.7 Pedogenesis

Pedogenesis can occur only after the landscape and its parent materials have stabilized. The immediate post-glacial landscape of the NLSFFR was probably a wet, unstable mess, with many areas of localized instability caused primarily by processes of thawing, slumping, washing, and blowing. Although there is at present no way to be certain, it may have taken hundreds to thousands of years before the glacial landscape stabilized and soils could begin forming, largely because of thawing permafrost and the melting of millions of buried (partially or wholly) ice blocks. The legacy of the latter is told today in the thousands of kettle lakes, swamps, and isolated depressions that so aptly characterize this landscape. Next, vegetation colonized the landscape, stabilizing the surface, trapping any loess that may have still been blowing around, and adding organic matter to the soil surface, forming O and eventually A horizons. Slowly, soils began to form.

On the acidic parent materials of the region (Fig. 10.6), pedogenic processes such as clay translocation (lessivage), weathering of primary minerals, and podzolization could proceed almost immediately. On the high-pH ($\approx$8.0–8.3), carbonate-rich materials, these processes were delayed until the carbonates could be leached from the upper profile, and the pH values lowered to at least 6.0. The decarbonation-leaching process probably took thousands of years. Carbonates leached more rapidly on coarser-textured parent materials like sands, while on many of the fine-textured parent materials, carbonates are today only leached to about a meter or less. Any clay that does get translocated in the latter types of soils stops at the upper carbonate boundary, forming a Bt (clay-enriched) horizon. In soils that are more deeply leached of carbonates, the depth of clay translocation is mainly due to depth of wetting fronts and the location of any textural discontinuities.

Soil formation across the region has led to different kinds of soil profiles, which the Soil Survey Staff (1999) has classified into several different orders. The discussion below focuses on soils by order. Occasionally, taxonomic suborders are also singled out.

10.7.1 Entisols

Entisols have minimal horizonation and/or thin profiles (Fig. 10.10). Many Entisols lack a B horizon, or have developed only a very weak B horizon. Many different types of soil landscapes contain small numbers of Entisols, for the reasons discussed above, even though the landscape itself may be dominated by soils of other orders. Several different reasons exist to explain why soils of this kind occur in the NLSFFR, not the least of which is that most soils here are less than a few thousand years old.

Some Entisols (Orthents) occur on clayey parent materials, where carbonates have only been leached to less than ca. 50 cm due to very low permeabilities and high runoff potentials. Water that runs off does not fully function in pedogenesis. Other Entisols (Orthents) occur on steep slopes, where high proportions of runoff again leave little water to percolate vertically and promote horizonation. Additionally, any water that runs off may erode the soil, keeping the profile thin and bringing unaltered parent material ever-closer to the surface. Many Entisols of this kind can be found in the clayey landscapes of MLRA 92. Some Entisols (Lithic Orthents) form in thin regolith (in this case, glacial sediments) over hard bedrock. Here, the overlying sediment, i.e., the soil's parent material, is so thin that a thick, well-developed soil profile cannot form in the accommodation space that remains. These types of soils are common in MLRA 93B, in the western UP of Michigan, and on the rolling bedrock hills of the Arrowhead region of Minnesota (MLRA 93A). Other Entisols (Psamments) are found on sand dunes and on dry, sandy outwash plains, where water tables are very deep. Many of these soils, especially on the dunes, are young and have not had ample time to form a well-horizonated profile. However, many areas of sandy Entisols occur on outwash plains and moraines that are >16,000 years old, e.g., those in MLRAs 94A and 91B. These soils are weakly developed because the sands are so coarse-textured that the soils remain dry throughout much of the summer. Vegetation is poor, often scrubby jack pine (*Pinus banksiana*) or pin oak (*Quercus palustris*) forest that burns frequently (Simard and Blank 1982). This type of

Fig. 10.10 Collage of photos of the typical soils found in the Northern Lake States Forest and Forage Region. Tape increments in cm. Photos by RJS

vegetation produces little litter to enhance pedogenesis (podzolization mainly, see below), and frequent fires burn up what litter is present (Mokma and Vance 1989; Schaetzl 2002). Soils formed in the recent alluvium of river valleys, especially those valley floors that flood commonly, are in the Fluvent suborder. These soils are prevented from developing thick profiles because of frequent additions of sediment from floods, or from high water tables (Aquents). On floodplains and elsewhere, Aquents have weakly developed profiles because of high water tables, which inhibit weathering and translocation. Aquents are common on the plain of Glacial Lake Wisconsin, for example, in MLRA 89.

Entisols are used for a variety of low-intensity purposes in the region. Many, if not most, remain in forest. Some, however, are cropped to hay or forage, and others are left in permanent pasture. The natural fertility of these soils is largely a function of the parent material, but in most cases the parent materials are too sandy, dry, wet, or clayey to be managed for agriculture, unless intensively fertilized, drained, or irrigated.

10.7.2 Inceptisols

Inceptisols are similar to Entisols in many ways, but have a slightly better developed profile (Fig. 10.10). Inceptisols are widely distributed in the NLSFFR and occur across a wide range of ecological settings, except for the sandiest landscapes. Soil Taxonomy (Soil Survey Staff 1999) does not allow most extremely sandy soils to be classified as Inceptisols. Two main Inceptisol suborders are recognized in the NLSFFR—Aquepts (Inceptisols with high water tables) and Udepts (all others). Upland Udepts in the region fall into two great groups—Eutrudepts and Dystrudepts. The former have higher contents of base cations (K, Ca, Mg, and Na) and nutrients, and tend to be found on carbonate-rich parent materials. Dystrudepts are more acidic and tend to be infertile; they are common on acidic parent materials (Figs. 10.2, 10.6). Udepts are especially common in the Arrowhead region of Minnesota (MLRA 93A). Many of the loamy soils on moraines and other, steeply sloping landscapes are also Inceptisols.

Considerable expanses of wet Inceptisols with high water tables (Aquepts) occur on the clayey lake plains of the eastern Upper Peninsula of Michigan (Fig. 10.11). Here, the high water table has inhibited pedogenesis, and thus, these soils have gleyed (gray, reduced, and waterlogged) Bg and Cg horizons.

Inceptisols, like Entisols, are often not intensively managed for agriculture, and most remain in forest. Aquepts on the clayey lake plains of the eastern UP are, however,

Fig. 10.11 The Chippewa Clay Plains region, part of MLRA 94B—the former floor of Glacial Lake Algonquin. The clay-rich Aquepts on this landscape are wet and calcareous at depths of less than a meter. Too wet and muddy to cultivate and re-plant each spring for grain crops, the soils are left in forage and cut annually for hay. The long summer days and cool temperatures combine to make hay production profitable in this otherwise wet, clayey landscape. Photo by RJS

successfully managed for hay production (Fig. 10.11). In other areas of Aquepts, subsurface drains are used to lower the water table and make agriculture possible. Many Udept landscapes are steeply sloping or erosion-prone, and hence, remain in forest or are used for pasture.

10.7.3 Alfisols

Alfisols have a strongly developed Bt horizon, which by definition is enriched in illuvial clay (Fig. 10.10). The clay is translocated from the upper profile by percolating water, but only if the pH is acidic (Schaetzl and Thompson 2015). Thus, soils on calcareous parent materials must first be leached of carbonates before clay translocation (termed *lessivage*) can begin. In most soils, ample decarbonation would have occurred after about 3000 to 5000 years. As the boundary between the (lower) calcareous zone and the (upper) leached zone deepens with time, the Bt horizon grows downward too, and often gets thicker. In many soils of this region, the boundary between the upper (eluvial) zone and the Bt horizon is not smooth, but undulating and wavy. Soil scientists refer to this type of morphology as interfingering or tonguing. Horizons that contain tongues or fingers of the E (eluvial) horizon in the Bt are called glossic horizons. Many soils in the NLSFFR are Glossudalfs—Alfisols with glossic features (Ranney and Beatty 1969; Bullock et al. 1974).

Alfisols can develop on a wide variety of parent materials, as long as the material contains some clay, which most parent materials in the NLSFFR do. Even sands can show signs of clay translocation, manifested as thin, often wavy, clay bands called lamellae (Schaetzl 1992; Rawling 2000). Bt horizons are very important ecologically, as they provide natural filters for percolating water, but more importantly, they enhance the soil's ability to retain nutrients and water. On many sandy landscapes, for example, forest type and productivity change when the sands have lamellae; even a few lamellae greatly enhance the productivity of the forest ecosystem (Host et al. 1988).

Everywhere across the NLSFFR ample precipitation exists to drive the process of lessivage, i.e., to produce at least some deep percolation events each year. Free drainage is also necessary for efficient clay translocation. For this reason, Bt horizons are often only weakly developed in some settings with high water tables; wet Inceptisols (Aquepts) then develop. If Bt horizons are weakly developed because the parent material is fine-textured and has low permeabilities, soil development will again be slowed, and Udepts (Inceptisols) will form.

Wide expanses of the NLSFFR have a mantle of Alfisols. In these landscapes, Histosols dot the lowlands and Entisols or Inceptisols may occur on areas of steep slope. But the matrix is an Alfisol landscape dominated by till parent materials that have undergone clay translocation—enough to produce (in places) what the locals may call a "clay pan." Alfisols have not only this "clay pan" but, as a consequence, they also have an upper profile that has lost clay, and some can even be quite sandy. Farming Alfisols for corn, hay, potatoes, and small grains is a common occurrence across the region. The upper part of the profile—the tilled part—can be loose and easy to till, whereas the subsoil is more clay-rich and thus, is effective at retaining water for deeply rooted crops (Fig. 10.12). This situation works well for the local farmers; agricultural production on Alfisols here is hampered more by climate than by soil. Indeed, many areas of Alfisols in Michigan's western Upper Peninsula and northern Minnesota could be brought into production if the growing season were only a little longer, or warmer. As it is, these areas support very productive forest ecosystems, especially valued for quality hardwood lumber and veneer products.

Wet Alfisols with high water tables classify as Aqualfs, whereas upland Alfisols are in the suborder Udalfs. Several different great groups of Udalfs also occur here, e.g., Hapludalfs, the "simple" ones, Glossudalfs with a glossic horizon, Fragiudalfs with a fragipan, and Fraglossudalfs with the combination. A fragipan is a hard, brittle, and dense pan that forms in the lower profile. Many Alfisols and Spodosols in the NLSFFR have fragipans. Fragipans are almost entirely found in the acidic parent material parts of the region (Fig. 10.6). Although fragipans inhibit deep rooting, many areas with these types of pans are still successfully managed for forest products and even crops.

10.7.4 Spodosols

On coarse-textured parent materials, and especially in the northern parts of the NLSFFR, Spodosols are the dominant soil order (Fig. 10.2). Spodosols form via a process called *podzolization*, which involves the translocation of organic materials, Fe and Al compounds in percolating water (Schaetzl and Harris 2011). Unlike clay translocation in Alfisols, the Fe and Al move in solution, as dissolved compounds. Soluble organic compounds also move in such soils, as a key part of the podzolization process. The result is a B horizon enriched in Fe, humus (decomposed and soluble compounds of organic matter) and Al. These compounds give the B horizon—the spodic horizon—its characteristic red, reddish brown, or black colors (Fig. 10.10). The more of these "spodic materials" in the B horizon, the darker and redder it becomes (Schaetzl and Mokma 1988). Eventually, the content of the spodic materials reaches a point where

Fig. 10.12 Irrigated potato production on the Antigo Flats—an outwash plain in northeastern Wisconsin that is covered by 60–100 cm of loess. The soils here are Alfisols and Inceptisols. The Antigo area is one of the region's main potato producing areas. Photo by RJS

they may cement the horizon into a substance called ortstein. Spodosols with ortstein represent the pinnacle of development for soils undergoing podzolization.

Upland Spodosols form best under a cool-cold, humid climate, where ample precipitation exists to drive soluble materials into the lower profile. (By necessity, this section does not discuss the wet Spodosols that are common in warm climates such as Florida.) Also necessary is a forest cover that produces acidic litter; the best trees for this purpose are many of the conifers, as well as beech (*Fagus* spp.), oak (*Quercus* spp.), hemlock (*T. canadensis*), and even some maples (*Acer* spp.). The litter produced by these trees accumulates on the soil surface, decomposing only slowly in the cool climate. As it decomposes, the soluble organic materials released are washed into the soil, complexing with Fe and Al compounds released by weathering of primary minerals. Once complexed, such compounds are readily translocated to the lower profile in percolating water. Thus, in locations farther south in the LRR that lack this type of forest cover, and areas to the west where grasslands are more common, Spodosols are absent.

Within the "Spodosol province" of the NLSFFR, Spodosols of varying degrees of development can be found. Strongly developed Spodosols also occur here; nowhere in the continental US are there better developed Spodosols (Schaetzl et al. 2015). On the margins of the Spodosol province, weakly developed Spodosols transition into sandy Entisols (Psamments).

The POD Index, developed by Schaetzl and Mokma (1988) to quantify the degree of development of spodic-like soils, shows the distribution of Spodosol development across the region—reaching its maximal expression in the eastern Upper Peninsula of Michigan (Fig. 10.13). Spodic development and the intensity of podzolization across the region appears to vary mainly as a function of climatic factors (Schaetzl and Isard 1991, 1996), or of climatic factors as they have (directly or indirectly) influenced vegetative cover and forest type (Mokma and Vance 1989; Schaetzl 2002). In particular, strong Spodosols usually coincide with areas of heavy snowfall and thick snowpacks. These snowbelt areas tend to occur about 20 to 70 km inland, west and/or south of the Great Lakes (Fig. 10.14). In these areas, thick, early snowpacks inhibit soil freezing and keep the soil and litter layers relatively warmer than at areas farther inland, which receive less snow (Isard and Schaetzl 1995, 1998). As a result, soils in snowbelt areas stay unfrozen and permeable throughout the winter. Then, in spring, large pulses of snowmelt water can freely infiltrate into and percolate through these soils (Schaetzl et al. 2015). The deep, reliable, and continuous snowpacks in snowbelt areas also insulate the fresh litter in the O horizon, facilitating its steady breakdown throughout the winter, thereby promoting the production of soluble organic acids. Wetting events that occur during snowmelt (and those associated with prolonged fall rain events) can readily translocate these soluble organic acids from the litter layer, deep into the mineral soil below. In sandy areas where podzolization is weak, less water is available for springtime percolation because of thinner snowpacks and the more frequently frozen soil, or the fall season is much drier. Much of the little snowmelt that exists in such areas runs off the still frozen surface and does not participate in pedogenesis.

Spodosols are mainly used for forest production and grazing. Some Spodosols that have formed in sandy materials that have 5–25 % silt + clay contents can develop a clay-rich horizon at depth, and these soils are more likely to be cropped because of their higher water- and nutrient-holding capacities. Potatoes and hay are common crops on such soils.

10.7.5 Mollisols

Mollisols have a thick, dark A horizon, rich in organic matter (Fig. 10.10). They are typical of drier climates and grassland ecosystems. Nonetheless, they can and do form in the NLSFFR, primarily where conditions permit the long-term accumulation of soil organic matter within the mineral soil (not on *top* of it—that is Histosols).

Soils accumulate organic matter when its production rate exceeds its rate of decay. This situation can occur in wet sites, where a high water table inhibits decomposition. These soils are called Aquolls, and their distribution is spotty and localized, often occurring immediately upslope from Histosols. Soils can even accumulate organic matter beneath forest, where large amounts of carbonate rocks are intermixed in the parent material (Schaetzl 1991). The carbonate minerals form strong bonds with the organic matter, slowing its decay. Lastly and most commonly, soils accumulate organic matter under grassland ecosystems, because grass roots form and die so rapidly that the processes of decay cannot keep pace. Eventually, an equilibrium state develops, where inputs of soil organic matter (mainly from root decay) nearly balance losses from decomposition. This "break-even point" typically occurs at about the depth of most grass roots, and in many Mollisols the soil organic carbon content at this depth is about 1.0–2.0 %. Because, the roots of grasses in these environments grow deep, the A horizon is much thicker than that of Alfisols formed under forests of more humid environments. Grasslands, or grassland-forest

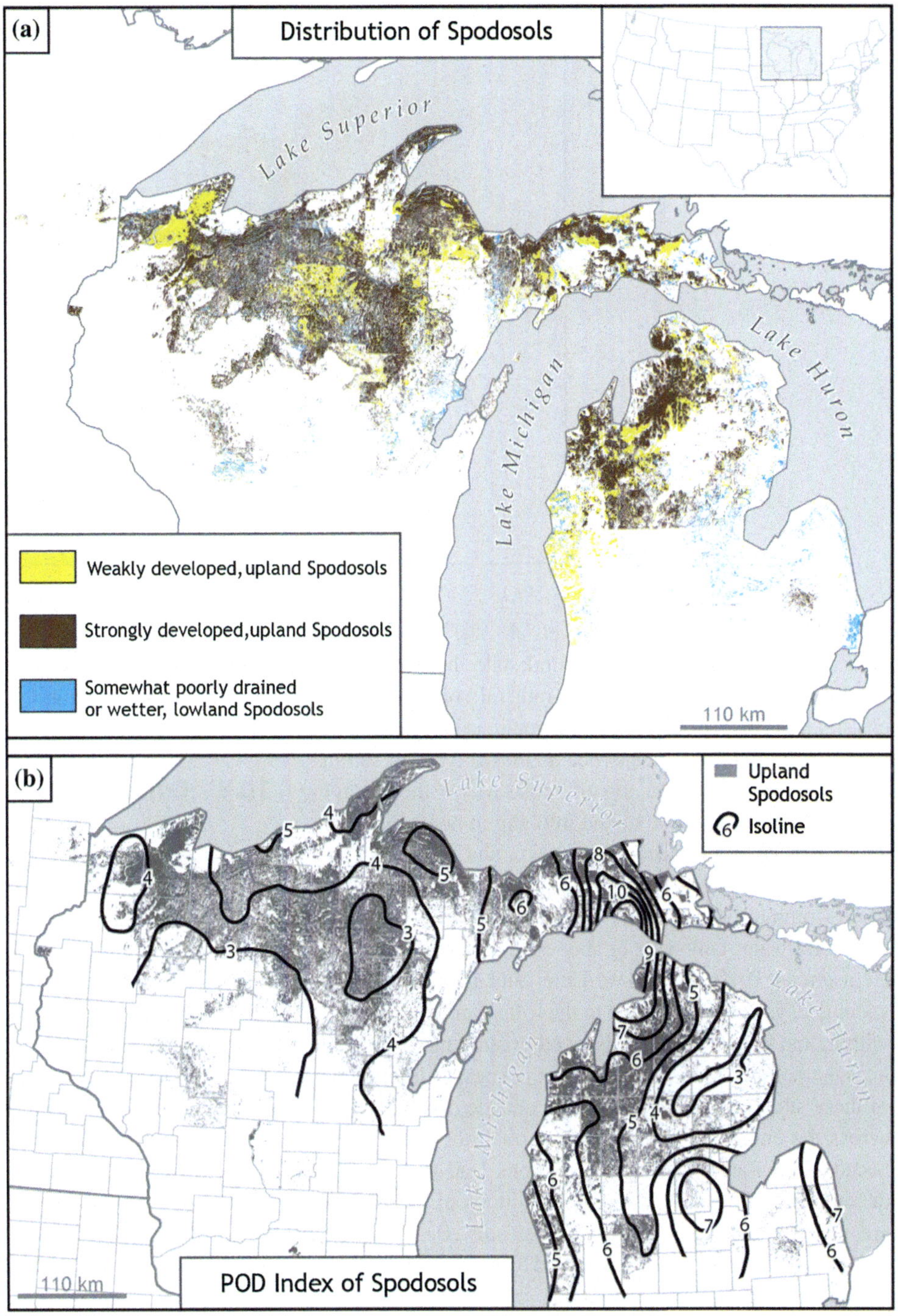

Fig. 10.13 Distribution of Spodosols in the Great Lakes region. **a** Spodosols, as grouped into three general categories based on taxonomic subgroup. Data source: NRCS SSURGO. **b** Isolines of interpolated POD Indices for upland Spodosols. Higher values indicate stronger soil development. So few upland Spodosols occur in Minnesota that the isolines were not interpolated for that area. After Schaetzl et al. (2015)

savanna ecosystems, occur in the far western parts of the NLSFFR, specifically in MLRA 91A, explaining why Mollisols occur here (Fig. 10.2).

Because of their high organic matter contents, Mollisols are highly productive soils. Aquolls are often drained and brought into agricultural production. If drainage is not required, as in Udolls, they are readily cultivated for many types of crops, with cereals and legumes being the most common types in this region. In most instances, Mollisols are the most productive soils of not only this region, but of most regions within the US (Schaetzl et al. 2012).

10.7.6 Histosols

Histosols are organic soils whose parent material is predominantly organic matter, e.g., leaves, grasses, roots, and wood, rather than mineral material, as in the other soil orders

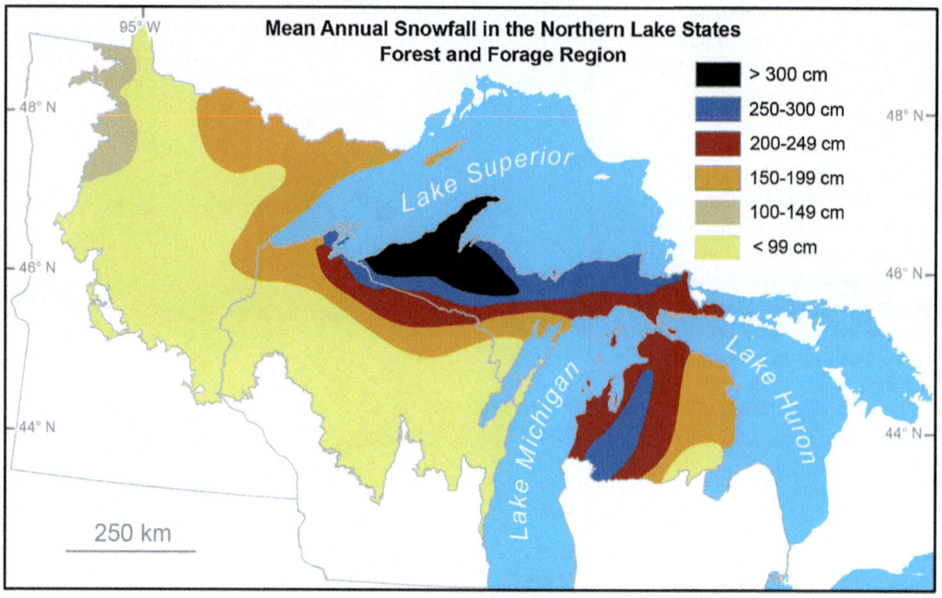

Fig. 10.14 Mean annual snowfall in the NLSFFR. Compiled from various sources

(Kolka et al. 2011; Kroetsch et al. 2011; Fig. 10.10). Although most soils, in their natural state, have an organic (O) horizon at the surface of the mineral soil, for a soil to classify as a Histosol, this organic layer must be at least 40 cm thick. This situation typically occurs in two types of settings. First, they can form on top of rock, where the decaying organic matter cannot be mixed into any mineral soil below, and hence, it simply accumulates (Fox and Tarnocai 2011). These soils are called Folists, and they are fairly rare. Much more commonly, Histosols form in lowlands where the long-term water table is at or above the surface (Gates 1942; Heinselman 1970; Fig. 10.9). Litter that falls into the cool water decays very slowly due to the (often) anoxic conditions within, and hence, partially decayed organic matter accumulates, sometimes to many meters in thickness. Although both of these situations occur commonly outside of the NLSFFR, across the entire US one normally would not find Histosols because the warmer climatic conditions lead to more rapid decomposition. That is, the cool conditions of the NLSFFR are vital for the development of Histosols. Because of the many isolated depressions left behind by the melting of buried ice in this young, glacial landscape, small-to-large bodies of Histosols are a part of almost every soil landscape. The consistent presence of small-to-medium pockets of Histosols adds great ecological diversity to the landscape. Areas of large expanses of Histosols also occur in the NLSFFR, mainly on Pleistocene lake plains.

Histosols are important as wetland habitat and play a key role in the Earth's carbon cycle. Obviously, they are also vital to the local and regional hydrologic cycles. Some Histosols have been drained for production of turfgrass or cranberries, and some are mined for peat and muck. Nonetheless, the vast majority of these soils remain essentially untouched. Most Histosols in the NLSFFR are naturally forested, unless too wet, where they transition to bog or marsh vegetation, and then (often) to open water.

10.8 Conclusions

The soil landscapes of the Northern Lakes States Forest and Forage Region are diverse and pedologically–geologically young. Many soils are just beginning to develop. Poorly integrated drainage networks and myriad of isolated depressions have led to an abundance of wet soils, and some of the greatest densities of Histosols anywhere in the US. Upland soils are mainly a product of their parent materials, as many soils have not had adequate time to overprint pedogenic characteristics onto the inherited sediments. Thus, the landscape's complexity is more geologic than pedologic. And it IS complex; perhaps some of the most complex and variable soil landscapes in the US are within this region. On top of this pedogenic diversity are a plethora of lakes and marshes; this is the land of far more than 10,000 lakes!

Most soils here have formed due to pedogenic processes associated with a cool, humid climate, under the influence of forest vegetation. Podzolization dominates on coarse-textured parent materials, leading to some of the strongest and best developed Spodosols in the US. On loamy and finer-textured materials, Bt horizons form readily, and if well-developed, these soils classify as Alfisols. Histosols are common in wetlands and lowlands, providing pedologic diversity to an already diverse soil landscape.

Most soils are minimally weathered, rich in nutrients and fertile, although many are too wet, or too dry and sandy to economically cultivate for agriculture. Still others are too far north and hence, too cold, for the production of most cash grains. For this reason, most of this landscape remains forested even today.

References

Arbogast AF (2009) Sand Dunes. In: Schaetzl RJ, Darden JT, Brandt D (eds) Michigan geography and geology. Pearson Custom Publishing, Boston, pp 274–287

Asady GH, Whiteside EP (1982) Composition of a Conover–Brookston map unit in southeastern Michigan. Soil Sci Soc Am J 46:1043–1047

Attig JW, Bricknell M, Carson EC, Clayton L, Johnson MD, Mickelson DM, Syverson KM (2011a) Glaciation of Wisconsin, 4th edn. Wisconsin Geological and Natural History Survey, Education Series Publication 36

Attig JW, Muldoon MA (1989) Pleistocene geology of Marathon County, Wisconsin. Wisconsin Geological and Natural History Survey, Information Circle 65

Attig JW, Clayton L (1993) Stratigraphy and origin of an area of hummocky glacial topography, northern Wisconsin, USA. Quat Int 18:61–67

Attig JW, Hanson PR, Rawling JE, Young AR, Carson EC (2011b) Optical ages indicate the southwestern margin of the Green Bay Lobe in Wisconsin, USA, was at its maximum extent until about 18,500 years ago. Geomorphology 130:384–390

Barrett LR (2001) A strand plain soil development sequence in Northern Michigan, USA. Catena 44:163–186

Blumer BE, Arbogast AF, Forman SL (2012) The OSL chronology of eolian sand deposition in a perched dune field along the northwestern shore of Lower Michigan. Quat Res 77:445–455

Bullock P, Milford MH, Cline MG (1974) Degradation of argillic horizons in Udalf soils in New York State. Soil Sci Soc Am Proc 38:621–628

Burnett AW, Kirby ME, Mullins HT, Patterson WP (2003) Increasing Great Lake-effect snowfall during the twentieth century: a regional response to global warming? J Clim 16:3535–3542

Clayton L, Attig JW (1989) Glacial Lake Wisconsin. Geol Soc Am Mem 173, p 80

Clayton L, Attig JW, Mickelson DM (2001) Effects of late Pleistocene permafrost on the landscape of Wisconsin, USA. Boreas 30:173–188

Curtis JT (1959) The vegetation of Wisconsin. University of Wis Press, Madison

Davis AM (1977) The prairie-deciduous forest ecotone in the upper Middle West. Ann Assoc Am Geogr 67:204–213

Drzyzga SA, Shortridge AM, Schaetzl RJ (2012) Mapping the stages of Glacial Lake Algonquin in Northern Michigan, USA, and nearby Ontario, Canada, using an isostatic rebound model. J Paleolimnol 47:357–371

Faulkner DJ (1998) Spatially variable historical alluviation and channel incision in west-central Wisconsin. Ann Assoc Am Geogr 88:666–685

Fox CA, Tarnocai C (2011) Organic soils of Canada: part 2. Upland organic soils. Can J Soil Sci 91:823–842

Gates FC (1942) The bogs of northern lower Michigan. Ecol Monogr 12:216–254

Green WP, Pettry DE, Switzer RE (1998) Formicarious pedons, the initial effect of mound-building ants on soils. Soil Surv Horiz 39:33–44

Heinselman ML (1970) Landscape evolution, peatland types, and the environment in the Lake Agassiz Peatlands Natural Area, Minnesota. Ecol Monogr 40:235–261

Host GE, Pregitzer KS, Ramm CW, Lusch DP, Cleland DT (1988) Variation in overstory biomass among glacial landforms and ecological land units in northwestern Lower Michigan. Can J For Res 18:659–668

Hull J, Hanson E (2009) The Fruit Industry. In: Schaetzl RJ, Darden JT, Brandt D (eds) Michigan geography and geology. Pearson Custom Publishing, Boston, pp 584–601

Isard SA, Schaetzl RJ (1995) Estimating soil temperatures and frost in the lake effect snowbelt region, Michigan, USA. Cold Reg Sci Technol 23:317–332

Isard SA, Schaetzl RJ (1998) Effects of winter weather conditions on soil freezing in southern Michigan. Phys Geogr 19:71–94

Isard SA, Schaetzl RJ, Andresen JA (2007) Soils cool as climate warms in Great Lakes region, USA. Ann Assoc Am Geogr 97:467–476

Kabrick JM, Clayton MK, McBratney AB, McSweeney K (1997) Cradle-knoll patterns and characteristics on drumlins in northeastern Wisconsin. Soil Sci Soc Am J 61:595–603

Khakural BR, Lemme GD, Mokma DL (1993) Till thickness and argillic horizon development in some Michigan Hapludalfs. Soil Surv Horiz 34:6–13

Knox JC (2006) Floodplain sedimentation in the Upper Mississippi Valley: natural versus human accelerated. Geomorphology 79:286–310

Kolka RK, Rabenhorst MC, Swanson D (2011) Histosols. In: Huang PM, Li Y, Sumner ME (eds) Handbook of soil sciences, 2nd edn. CRC Press, New York, pp 33-8–33-29

Kroetsch DJ, Geng X, Chang SX, Saurette DD (2011) Organic soils of Canada: part 1. Wetland organic soils. Can J Soil Sci 91:807–822

Lichter J (1998) Rates of weathering and chemical depletion in soils across a chronosequence of Lake Michigan sand dunes. Geoderma 85:255–282

Loope WL, Loope HM, Goble RJ, Fisher TG, Lytle DE, Legg RJ, Wysocki DA, Hanson PR, Young AR (2012) Drought drove forest decline and dune building in eastern USA, as the upper Great Lakes became closed basins. Geology 40:315–318

Luehmann MD, Schaetzl RJ, Miller BA, Bigsby M (2013) Thin, pedoturbated and locally sourced loess in the western Upper Peninsula of Michigan. Aeolian Res 8:85–100

Mason JA, Knox JC (1997) Age of colluvium indicates accelerated late Wisconsinan hillslope erosion in the Upper Mississippi Valley. Geology 25:267–270

Mason JA, Milfred CJ, Nater EA (1994) Distinguishing soil age and parent material effects on an Ultisol of north-central Wisconsin, USA. Geoderma 61:165–189

McNab WH, Avers PE (1994) Ecological subregions of the United States. U.S. For Serv Administrative Publ WO-WSA-5. Section descriptions and maps

Meyers NL, McSweeney K (1995) Influence of treethrow on soil properties in Northern Wisconsin. Soil Sci Soc Am J 59:871–876

Mickelson DM, Clayton L, Fullerton DS, Borns HW (1983) The Late Wisconsin glacial record of the Laurentide Ice Sheet in the United States. In: Wright HE Jr (ed) Late quaternary environments of the United States. Vol. 1: the late Pleistocene. University of Minn Press, Minneapolis, pp 3–37

Miller BA, Schaetzl RJ (2015) Digital classification of hillslope position. Soil Sci Soc Am J 79:132–145

Mokma DL, Vance GF (1989) Forest vegetation and origin of some spodic horizons, Michigan. Geoderma 43:311–324

Muller RA (1966) Snowbelts of the Great Lakes. Weatherwise 14:248–255

Natzke LL, Hvizdak DJ (1988) Soil survey of Vilas County, Wisconsin. USDA Soil Conservation Service, US Govt. Printing Office, Washington

Nichols GE (1935) The hemlock-white pine-northern hardwood region of eastern North America. Ecology 16:403–422

Norton DC, Bolsenga SJ (1993) Spatiotemporal trends in lake effect and continental snowfall in the Laurentian Great Lakes, 1951–1980. J Clim 6:1943–1956

Ranney RW, Beatty MT (1969) Clay translocation and albic tongue formation in two Glossoboralfs in west-central Wisconsin. Soil Sci Soc Am Proc 33:768–775

Rawling JE III (2000) A review of lamellae. Geomorphology 35:1–9

Rawling JE III, Hanson PR, Young AR, Attig JW (2008) Late Pleistocene dune construction in the central sand plain of Wisconsin, USA. Geomorphology 100:494–505

Resner K, Yoo K, Hale C, Aufdenkampe A, Blum A, Sebestyen S (2011) Elemental and mineralogical changes in soils due to bioturbation along an earthworm invasion chronosequence in Northern Minnesota. Appl Geochem 26:S127–S131

Rieck RL, Winters HA (1993) Drift volume in the southern peninsula of Michigan—a prodigious Pleistocene endowment. Phys Geogr 14:478–493

Salem MZ, Hole FD (1968) Ant (*Formica exsectoides*) pedoturbation in a forest soil. Soil Sci Soc Am Proc 32:563–567

Šamonil P, Daněk P, Schaetzl RJ, Vašíčková I, Valtera M (2015) Soil mixing and genesis as affected by tree uprooting in three temperate forests. Eur J Soil Sci 66:589–603

Šamonil P, Schaetzl RJ, Valtera M, Goliáš V, Baldrian P, Vašíčková I, Adam D, Janík D, Hort L (2013) Crossdating of disturbances by tree uprooting: can treethrow microtopography persist for 6,000 years? For Ecol Manag 307:123–135

Schaetzl RJ (1986) Complete soil profile inversion by tree uprooting. Phys Geogr 7:181–189

Schaetzl RJ (1990) Effects of treethrow microtopography on the characteristics and genesis of Spodosols, Michigan, USA. Catena 17:111–126

Schaetzl RJ (1991) Factors affecting the formation of dark, thick epipedons beneath forest vegetation, Michigan, USA. J Soil Sci 42:501–512

Schaetzl RJ (1992) Texture, mineralogy and lamellae development in sandy soils in Michigan. Soil Sci Soc Am J 56:1538–1545

Schaetzl RJ (1998) Lithologic discontinuities in some soils on drumlins: theory, detection, and application. Soil Sci 163:570–590

Schaetzl RJ (2002) A Spodosol–Entisol transition in northern Michigan. Soil Sci Soc Am J 66:1272–1284

Schaetzl RJ (2008) The distribution of silty soils in the Grayling Fingers region of Michigan: evidence for loess deposition onto frozen ground. Geomorphology 102:287–296

Schaetzl RJ, Attig JW (2013) The loess cover of northeastern Wisconsin. Quat Res 79:199–214

Schaetzl RJ, Drzyzga SA, Weisenborn BN, Kincare KA, Lepczyk XC, Shein KA, Dowd CM, Linker J (2002) Measurement, correlation, and mapping of Glacial Lake Algonquin shorelines in northern Michigan. Ann Assoc Am Geogr 92:399–415

Schaetzl RJ, Follmer LR (1990) Longevity of treethrow microtopography: implications for mass wasting. Geomorphology 3:113–123

Schaetzl RJ, Forman SL, Attig JW (2014) Optical ages on loess derived from outwash surfaces constrain the advance of the Laurentide ice from the Lake Superior Basin, Wisconsin, USA. Quat Res 81:318–329

Schaetzl RJ, Harris W (2011) Spodosols. In: Huang PM, Li Y, Sumner ME (eds) Handbook of soil sciences, 2nd edn. CRC Press, New York, pp 33-113–33-127

Schaetzl RJ, Isard SA (1991) The distribution of Spodosol soils in southern Michigan: a climatic interpretation. Ann Assoc Am Geogr 81:425–442

Schaetzl RJ, Isard SA (1996) Regional-scale relationships between climate and strength of podzolization in the Great Lakes region, North America. Catena 28:47–69

Schaetzl RJ, Isard SA (2002) The Great Lakes Region. In: Goudie A, Orme A (eds) The physical geography of North America. Oxford University Press, Oxford, pp 307–334

Schaetzl RJ, Knapp BD, Isard SA (2005) Modeling soil temperatures and the mesic-frigid boundary in the Great Lakes region, 1951–2000. Soil Sci Soc Am J 69:2033–2040

Schaetzl RJ, Krist FJ Jr, Miller BA (2012) A taxonomically based, ordinal estimate of soil productivity for landscape-scale analyses. Soil Sci 177:288–299

Schaetzl RJ, Loope WL (2008) Evidence for an eolian origin for the silt-enriched soil mantles on the glaciated uplands of eastern Upper Michigan, USA. Geomorphology 100:285–295

Schaetzl RJ, Luehmann MD, Rothstein D (2015) Pulses of podzolization: the importance of spring snowmelt, summer storms, and fall rains on Spodosol development. Soil Sci Soc Am J 79:117–131

Schaetzl RJ, Mokma DL (1988) A numerical index of Podzol and Podzolic soil development. Phys Geogr 9:232–246

Schaetzl RJ, Thompson ML (2015) Soils genesis and geomorphology, 2nd edn. Cambridge University Press, Cambridge

Scull P, Schaetzl RJ (2011) Using PCA to characterize and differentiate the character of loess deposits in Wisconsin and Upper Michigan, USA. Geomorphology 127:143–155

Simard AJ, Blank RW (1982) Fire history of a Michigan jack pine forest. Mich Acad 14:59–71

Soil Survey Staff (1999) Soil taxonomy. USDA-NRCS Agric. Handbook 436. US Govt. Printing Office, Washington

Stanley KE, Schaetzl RJ (2011) Characteristics and paleoenvironmental significance of a thin, dual-sourced loess sheet, North-Central Wisconsin. Aeolian Res 2:241–251

Stearns FW (1949) Ninety years of change in a northern hardwood forest in Wisconsin. Ecology 30:350–358

Thompson TA, Baedke SJ (1997) Strand-plain evidence for late Holocene lake-level variations in Lake Michigan. Geol Soc Am Bull 109:666–682

USDA-NRCS (2006) Land resource regions and major land resource areas of the United States, the Caribbean, and the Pacific Basin. US Department of Agriculture Handbook 296. U.S. Govt. Printing Office, Washington

11 Soils of the Central Feed Grains and Livestock Region and Lake State Fruit, Truck Crop, and Dairy Region USA: LRRs M and L

Zamir Libohova, Philip Schoeneberger, Doug Wysocki, and Phillip R. Owens

11.1 General Characteristics of the Regions

This chapter explores the soils and landscapes of the Central Feed Grains and Livestock Region and the Lake State Fruit, Truck Crop, and Dairy Region (Fig. 11.1). The Central Feed Grains and Livestock Region area is approximately 731,905 km^2. This region has been shaped by several glaciations, including the most recent Wisconsinan (10,500–30,000 years ago) in the northern part, the Illinoian (130,000–300,000 years ago) in the southcentral part, and the pre-Illinoian (500,000–2,500,000 years ago) in the southwest part (Fig. 11.2) (Dutch 2015a, b). As a result, the landscape is mostly a dissected glaciated plain with nearly level to gently rolling hills. The geology of this region is dominated by the Taylor Group (9.3 %) in the northwest, the Des Moinesian Series (9.1 %) from central Iowa to northwest Missouri and northeast Oklahoma, and the Ordovician (9.0 %) from southeast Ohio to northwest Illinois and from southwest of Wisconsin to northwest Minnesota (USGS 2015). To a lesser extent, it includes Devonian (6.1 %) and Silurian (4.5 %) rocks, which extend in the same direction as the Ordovician. Other underlying rocks are sedimentary/metasedimentary (3.4 %), granitic (3.3 %), and volcanic (1.4 %). Many other geologic formations are also present but to a very limited extend. The major soil orders for this region are Mollisols (47 %) and Alfisols (31 %), followed by Entisols (6.5 %), Inceptisols (4 %), and Histosols and Ultisols (both of which occur in approximately 1 % of the area) (SSURGO 2016). The soil temperature regimes of this region are predominantly mesic (90 %) (mean annual temperature between 8 and 15 °C) with limited extent of frigid (6.6 %) (mean annual temperature between 0 and 8 °C) occurring mostly in Minnesota (USDA-NRCS-NSSC 1994; Winzeler et al. 2012). The dominant soil moisture regimes are udic (59 %), followed by aquic (24 %). The mean annual precipitation is typically between 815 and 990 mm, but it ranges from as little as 485 to as much as 1220 mm. It increases from north to south and west to east and occurs mostly between spring and fall. The mean average annual temperature ranges from 4 to 17 °C, increasing from north to south. Prior to the settlement, this region was dominated by forest to the east and prairie to the west. This led to the development of Alfisols and Mollisols, respectively. Because of the favorable soils and climate, the region has become agricultural and today produces most of the corn, soybeans, and feed grains in the USA. The conservation measures applied in the region have had a limited success; thus, erosion and organic matter depletion remain the major soil concerns (Fig. 11.3). There are 27 MLRAs in the Central Feed Grains and Livestock Region.

The total surface area of the Lake State Fruit, Truck Crop, and Dairy Region, LRR L, is about 25,815 km^2. This region formed during the last glacial period (Wisconsin), around 10,500–30,000 years ago. As the glaciers melted and retreated, water filled in the basin of what once was at the heart of one of the largest glaciers in North America (Cordell et al. 2008; Schmus and Hinze 1985). The glaciers have retreated and advanced many times (Sugden and John 1976; Tranhaile 1998) and have carved and shifted the landscape. Due to these processes, unsorted and sorted materials have been transported and deposited many times over multiple spatial and temporal scales, leading to one of the most complex landscapes and soil distributions in the USA. The mixed geology of this region is as complex as the landscape

Z. Libohova (✉) · P. Schoeneberger · D. Wysocki
USDA-NRCS, Lincoln, NE, USA
e-mail: zamir.libohova@lin.usda.gov

P. Schoeneberger
e-mail: Philip.Schoeneberger@LIN.usda.gov

D. Wysocki
e-mail: doug.wysocki@lin.usda.gov

P.R. Owens
USDA-Agricultural Research Service, Booneville, AR, USA
e-mail: phillip.owens@ars.usda.gov

L.T. West et al. (eds.), *The Soils of the USA*, World Soils Book Series,
DOI 10.1007/978-3-319-41870-4_11

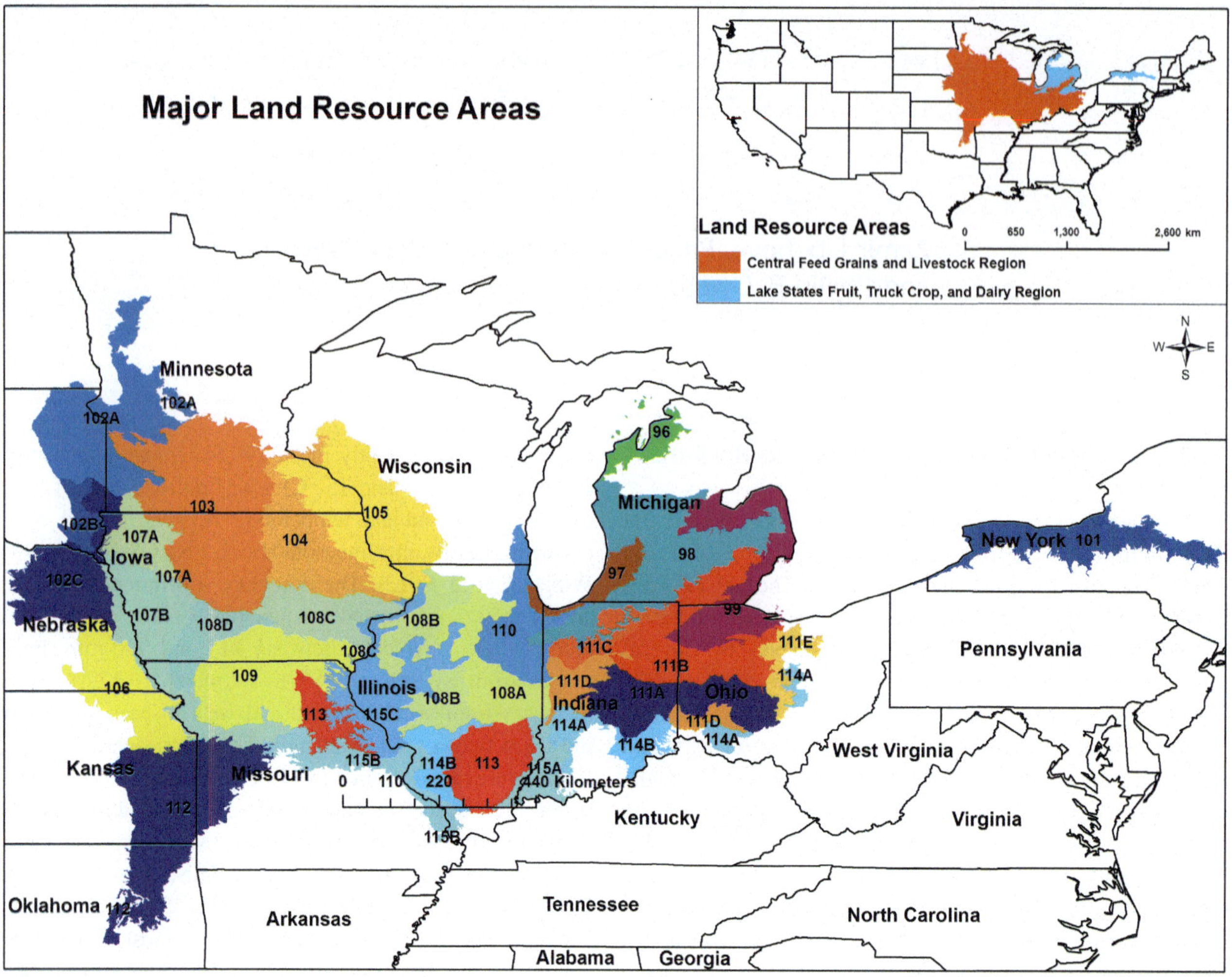

Fig. 11.1 Major land resource areas for the Great Lakes Area Land Resource Region and Central Feed Grains and Livestock Region

due to tectonic splits, which led to the creation of fault lines (Wayne 2007), combined with several glaciations. It is dominated by Devonian/Ordovician rocks (27 %), followed by sedimentary (13 %), granitic (10 %), Cambrian (9 %), and volcanic (8 %) rocks (USGS 2015). The major soil orders of this region are Alfisols (35 %) and Spodosols (24 %), followed by Entisols (13 %), Histosols (10 %), Inceptisols (10 %), and Mollisols (7 %) (USDA-NRCS-KSSL 2014). The soil temperature regimes of this region are mostly mesic (USDA-NRCS-NSSC 1994; Winzeler et al. 2012), and the soil moisture regime is mostly udic. Apart from cycles of glacier retreat and advance over an extended period of time, one of the most dramatic changes in the region has recently occurred, specifically the conversion of forest to farmland and the conversion of presettlement forest to other forest types (mostly aspen–birch forest due to logging) (Cole et al. 1999). The total amount of forested areas has declined by over 40 %, mostly due to the conversion of northern mesic and oak forests to farmland (Cole et al. 1999). Today, only 31 % of the forests remain. The average size of the contiguous forest units has declined by twice; however, aspen–birch forest units and non-forested units have more than doubled in unit size (Cole et al. 1999). Approximately 26 % of the area is under crop production, comprised mostly of truck crops, wheat, and dry beans with limited orchards near Lake Ontario and vineyards common in the Finger Lakes region. A significant area, almost 15 %, is urban, concentrated mostly around Buffalo, Rochester, and Syracuse. There are five MLRAs in the Lake State Fruit, Truck Crop, and Dairy Region (Table 11.1).

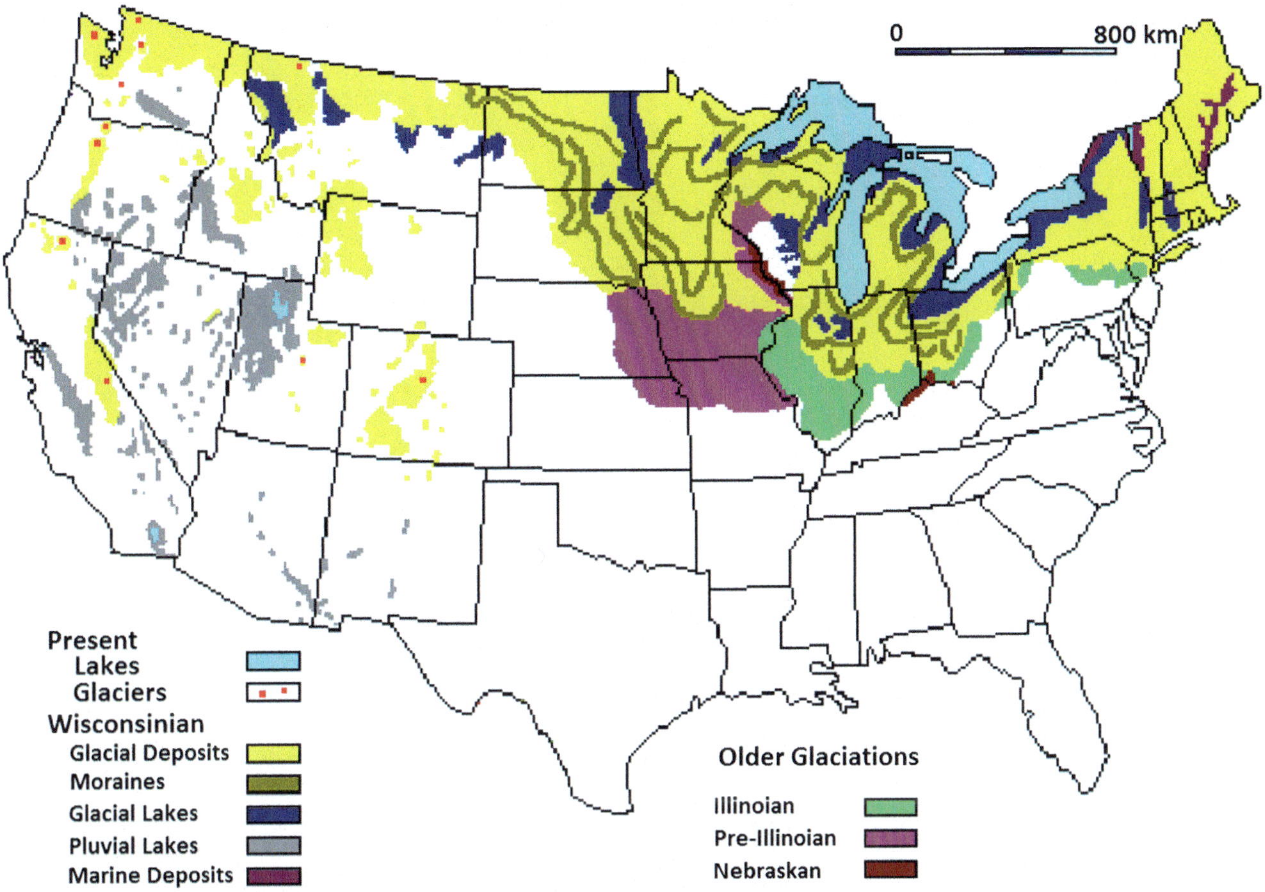

Fig. 11.2 Glaciation events and deposits in the Continental USA during the last 18,000 years. *Source* Steven Dutch, Natural and Applied Sciences, University of Wisconsin Green Bay

11.2 Central Feed Grains and Livestock Region, LRR M (MLRAs 102A/B/C, 103, 104, 105, 106, 107A/B, 108A/B/C/D, 109, 110, 111A/B/C/D/E, 112, 113, 114A/B, and 115A/B/C)

11.2.1 Rolling Till Prairie (MLRA 102A), Till Plains (MLRA 102B), and Loess Uplands (MLRA 102C)

11.2.1.1 Geology and Geomorphology

MLRAs 102A/B/C occur mostly on Cretaceous Pierre Shale, Precambrian rocks (MLRAs 102A/B), and the Dakota Sandstone Formation (MLRA 102C) overlain by glacial till of varying thickness (up to 60 m) and loess varying from 2 to 20 m in thickness. This area is part of the Western Lake Section of the Central Lowland Province of the Interior Plains. The landscape has been shaped by a series of glaciations and the subsequent retreats and disintegrations of the ice sheets over a period of 150,000 years (Sugden and John 1976). One of the most interesting landforms is Prairie Coteau, at an elevation of 610 m, which split the last continental ice sheet into the James and Des Moines Lobes. Located mostly on glacial margins, MLRAs 102A/B consist of stagnating moraines, end moraines, glacial outwashes, terraces, and flood plains. There are many small depressions, lakes, and ponds, commonly known as "prairie potholes" (Fig. 11.4).

11.2.1.2 Soil Properties

Mollisols is the dominant soil order in these MLRAs. The dominant parent materials are silty drift, glacial till, glacial outwash, and alluvium. The presence of various deposits, ranging from loess to glacial outwash, combined with landscape characteristics and slope position determines soil properties and their distribution (Fig. 11.5). Soils on uplands and in valleys are generally deep to very deep (Hapludolls, Argiaquolls, and Haplustolls) and well drained (Calciustolls and Calciudolls) to very poorly drained (Calciaquolls and Fluvaquents). Calcium carbonate ($CaCO_3$) is common in soils in the area and originates from glacial till parent materials rich in Ca or as a result of evaporative discharge caused by upward water movement (Richardson et al. 1992; Richardson 2010).

Fig. 11.3 The conversion of prairie land to cropland has led to decrease of soil organic matter thickness in addition to the depletion

The predominant soil textures are sandy loam and clay loam for surface horizons (A, Ap), silty clay loam for Bt and Btk horizons (30–50 % clay), and loam, sandy loam (~60 % sand), and silt loam (25–60 % silt) for BCk and Ck horizons. Soil reaction varies from 6.1 to 8.4 and commonly increases with depth due to the presence of carbonates. The carbonates originate from the till or are transported downslope in soils on glacial outwash and lakebeds. The organic matter (OM) content varies considerably with depth and slope position. For example, Hapludolls have a high OM content (up to 8 %) in the surface layer, but the content decreases to 4 % in the subsurface horizons and to less than 0.5 % in soil horizons at the contact with till or glacial outwash and sandy materials. Generally, soils are fertile due to their relatively young age, predominantly loess parent material, and temperature and moisture regimes. However, salinity and high concentrations of $CaCO_3$ can be limiting, especially for some Endoaquolls in alluvial depressions with low water permeability.

11.2.1.3 Soil Resource Management

Tall-grass prairie vegetation dominated these MLRAs prior to the settlement and led to the development of Mollisols.

Table 11.1 Dominant soil orders and some of their main climate characteristics in the Lake State Fruit, Truck Crop, and Dairy Region and Central Feed Grains and Livestock Region by major land resource areas

MLRA	Area (km^2)	Elev. (m)	Annual		Soil	Soil	Dominating
			Temp. (°C)	Precip. (mm)	Moist.	Temp.	Soil
					Regime	Regime	Orders
Central Feed Grains and Livestock Region							
102A	42,870	305–610	4–7	485–735	Aquic/Udic	Frigid	Mollisols
102B	5,735	350–575	6–9	585–660	Ustic	Mesic	Mollisols
102C	29,655	335–610	6–11	585–760	Ustic	Mesic	Mollisols
103	27,640	300–400	6–10	585–890	Aquic/Udic	Mesic	Mollisols/Alfisols
104	25,040	300–400	7–10	735–940	Aquic/Udic	Mesic	Mollisols/Alfisols
105	46,515	200–400	6–10	760–965	Udic	Mesic	Alfisols/Entisols/Mollisols
106	28,295	300–505	10–13	710–1,015	Udic	Mesic	Mollisols/Alfisols/Entisols
107A	11,590	340–520	7–9	660–790	Aquic/Udic	Mesic	Mollisols
107B	37,335	185–475	8–13	660–1,040	Aquic/Udic	Mesic	Mollisols/Alfisols/Entisols
108A	28,875	200–300	8–12	890–1,090	Aquic/Udic	Mesic	Mollisols/Alfisols
108B	19,300	200–300	8–12	840–990	Aquic/Udic	Mesic	Alfisols/Entisols/Inceptisols/Mollisols
108C	25,405	155–340	8–11	840–965	Aquic/Udic	Mesic	Mollisols/Alfisols/Entisols/Inceptisols
108D	14,195	210–460	9–11	840–940	Aquic/Udic	Mesic	Mollisols/Alfisols/Entisols
109	41,185	200–300	9–12	865–1,040	Aquic/Udic	Mesic	Alfisol/sMollisols
110	19,525	200	7–11	785–1,015	Udic/Aquic	Mesic	Alfisols/Histosols/Inceptisols/Mollisols
111A	28,445	205–380	9–12	915–1,090	Aquic/Udic	Mesic	Alfisols/Inceptisols/Mollisols/Histosols
111B	34,880	190–470	8–11	766–990	Aquic/Udic	Mesic	Alfisols/Inceptisols/Mollisols
111C	9,065	190–285	9–11	890–990	Aquic/Udic	Mesic	Alfisols/Mollisols/Entisols/Inceptisols
111D	13,880	160–320	10–12	915–1,090	Aquic/Udic	Mesic	Alfisols/Inceptisols/Mollisols
111E	7,720	175–425	9–11	890–1,040	Aquic/Udic	Mesic	Alfisols/Inceptisols/Mollisols
112	61,775	100–400	11–17	865–1,1145	Aquic/Udic	Thermic	Mollisols/Alfisols
113	33,150	200–300	11–14	915–1,170	Aquic/Udic	Mesic	Alfisols
114A	11,795	100–380	9–14	940–1,170	Aquic/Udic	Mesic	Alfisols/Inceptisols
114B	18,150	105–365	11–14	940–1,170	Aquic/Udic	Mesic	Alfisols/Inceptisols
115A	9,565	100–310	11–14	1,015–1,195	Udic/Aquic	Mesic	Alfisols/Entisols/Inceptisols/Mollisols
115B	20,955	100–310	12–14	965–1,220	Udic/Aquic	Mesic	Alfisols/Entisols/Inceptisols/Mollisols
115C	35,375	130–270	9–10	865–1,015	Udic/Aquic	Mesic	Alfisols/Entisols/Inceptisols/Mollisols
Lake State Fruit, Truck Crop, and Dairy Region							
96	7,230	175–295	5–9	760–915	Aquic/Udic	Mesic	Spodosols/Entisols/Alfisols/Histosols
97	7,995	200–305	8–11	890–1,015	Aquic/Udic	Mesic	Spodosols/Entisols/Alfisols/Histosols
98	49,050	175–335	7–10	735–1,015	Aquic/Udic	Mesic	Alfisols/Histosols/Mollisols
99	28,370	200–250	7–10	760–915	Aquic	Mesic	Alfisols/Inceptisols/Mollisols/Spodosols
101	25,815	100–400	5–10	735–1,145	Udic	Mesic	Alfisols/Inceptisols

Almost 70 % of the area under cropland is used for corn, soybeans, alfalfa, spring wheat, and oats. Forested areas occur as narrow strips along streams and rivers or as shelterbelts and windbreaks around farmsteads. The major resource management concerns are poor grazing in managed grassland and water and wind erosion. The estimated annual soil erosion is 5–6 ton ha^{-1}. In some areas, however, especially on steep slopes along drainageways and/or streams, it could exceed 10 ton ha^{-1}. Wind erosion is a concern especially on overgrazed or poorly managed grasslands. Irrigation is limited to valleys and flat areas where water supplies are available. Salinity and high $CaCO_3$ are concerns especially on footslopes where evaporative discharge occurs (Richardson et al. 1992; Richardson 2010).

Fig. 11.4 A view of prairie "pothole" landscape in Des Moines Lobe in central Iowa showing patterns soil wetness in close depressions *Source* Lynn Betts, USDA-NRCS Photo Gallery, http://photogallery.nrcs.usda.gov/res/sites/photogallery/Using.htm (Accessed on March 04, 2016)

11.2.2 Till Prairies (MLRAs 103 and 104) and Loess Hills (MLRAs 105, 106, 107A/B)

11.2.2.1 Geology and Geomorphology

The Till Prairies and Loess Hills are part of the Central Lowland Province of the Interior Plains. MLRA 103, known also as the Des Moines Lobe, is a gently rolling glaciated till plain with moraines and small natural lakes, marshes, and potholes occurring throughout. This MLRA is bordered to the east by MLRA 104, which is part of the dissected till plain section and has similar geomorphic characteristics. However, further east of MLRA 104 the landscape changes from gently sloping areas to larger summits and steeper and deeper V-shaped valleys with rock outcrops, high bluffs, caves, crevices, and sinkholes. This landscape, also known as the "Driftless Area," has a distinct abrupt, sharp-angled drainage network controlled by bedrock and has not been influenced by glacial ice, especially the Wisconsian. West to southwest of the Des Moines Lobe (MLRA 103), MLRAs 107A/B and 106 occur on the dissected till plain (Fig. 11.6). These MLRAs are dominated by an undulating to rolling glaciated till plain with broad ridgetops and limited steep slopes along the major stream valleys, such as that of the Missouri River. The underlying geology of the area is composed mostly of Paleozoic bedrock sediments (shale and limestone) underlying the glacial and loess deposits of various ages (Fig. 11.6) (Boellstorff 1978a, b; Mason et al. 2007). The major river valleys cut through various stratified bedrocks, and the glacial and loess deposits consist of alluvium composed of clay, silt, sand, and gravel (USDA-NRCS 2004, 2006).

11.2.2.2 Soil Properties

The soil orders in this region are dominantly Mollisols and, to a lesser extent, Alfisols and Entisols. The soils in the area have a mesic soil temperature regime, an aquic or udic soil

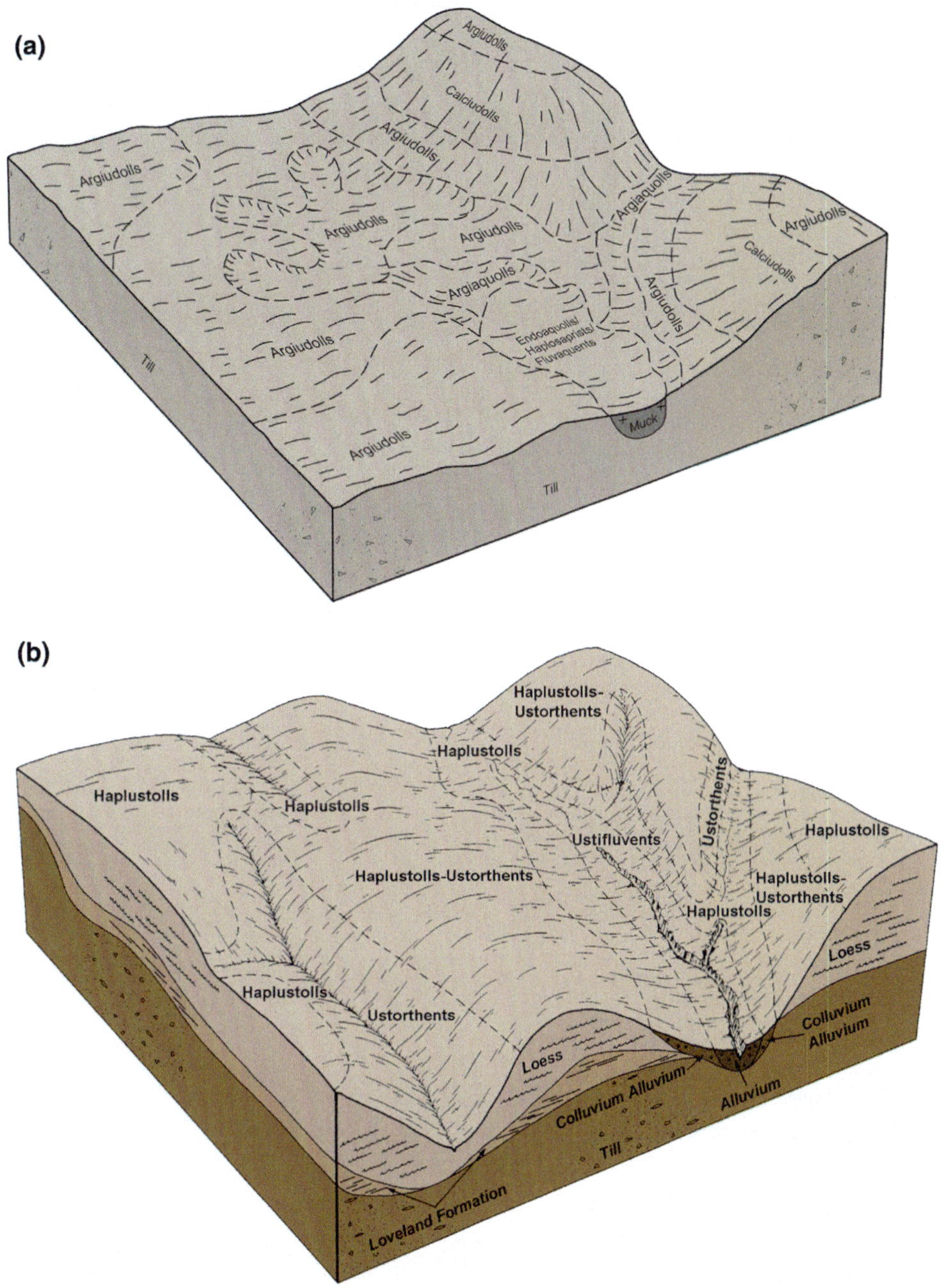

Fig. 11.5 **a** Typical pattern of soils and parent material in the Hapludolls–Calciudolls association on **a** till and **b** colluvium–alluvium. Haplustolls on loess uplands are associated with Ustorthents in loess in steep areas followed by Haplustolls on alluvium and colluvium

moisture regime, and mixed mineralogy. They vary from deep to very deep and are predominantly well drained (on dissected till plains and rolling hills) and poorly drained (in swales, valleys, and depressions). Hapludolls are associated mostly with loess-mantled areas on till plains and moraines and, to a limited extent, with alluvium on flood plains and fans and colluvium on footslopes (Fig. 11.7). The Alfisols occur mostly on steeper slopes bordering Udifluvents and Fluvaquents, which occur on alluvium and colluvium on bottomland flood plains and alluvial fans. While the majority of soils formed on loess over till, a limited number of soils, especially in MLRA 105 (Driftless Area), formed on loess over residuum on upland benches and, as a result, are redder compared to the more yellowish brown soils that formed in loess over till (USDA-SCS 1962) (Fig. 11.8). Soils that formed on loess are predominantly silt loam and silty clay loam throughout. Soil reaction varies from slightly acid (pH 6.1–6.5) to moderately alkaline (pH 7.9–8.4) depending on the underlying parent material and slope position. Thus, Hapludolls on uplands and summits are neutral in the surface

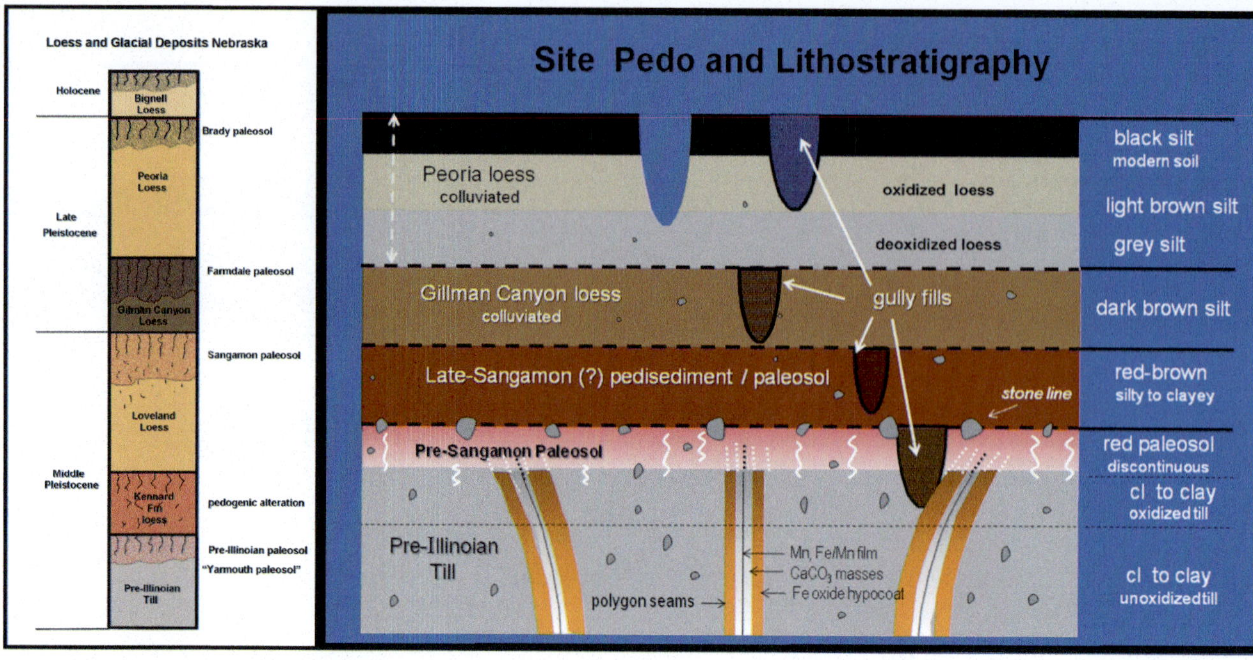

Fig. 11.6 Stratigraphy and chronosequence of till and loess deposits occurring in MLRAs 106 and 107. (Modified from Boellstorff 1978a, b; Mason et al. 2007)

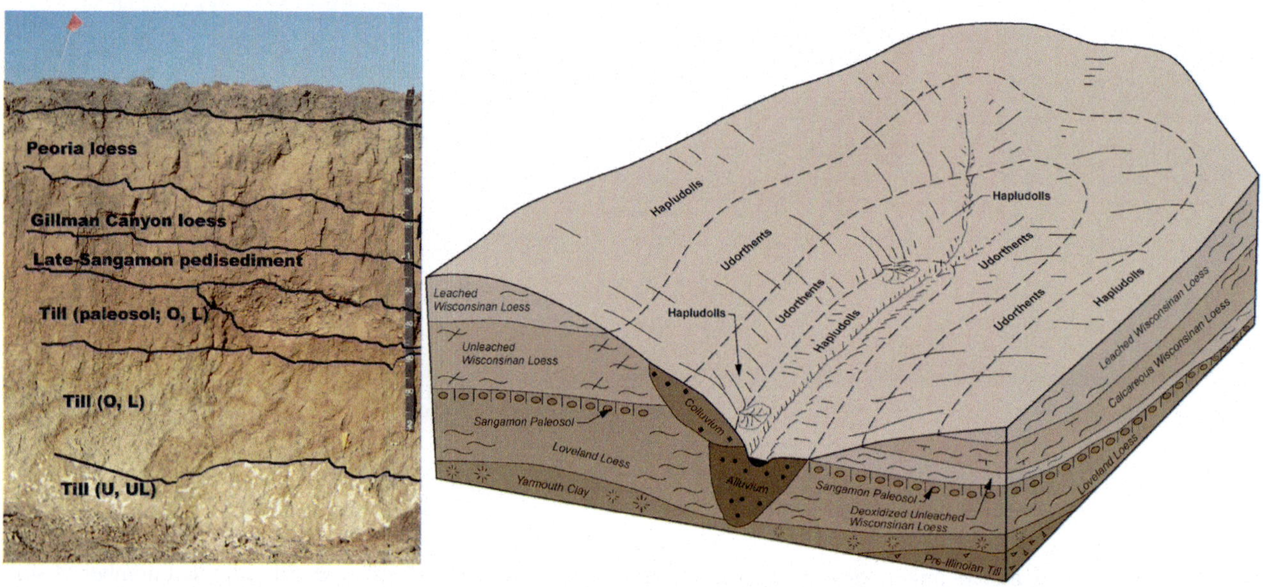

Fig. 11.7 Diagram showing the relationships of the upland soils (predominantly Hapludolls) in the eastern part of Shelby County, Iowa, to parent material and landforms (Wigginton et al. 2002)

layer but may become slightly acid if the underlying parent materials are residuum or sandstone or moderately alkaline if calcareous till underlies the loess. In addition, the soil reaction for soils on footslopes commonly becomes progressively alkaline with depth due to $CaCO_3$ moving downslope and accumulating in the soil by upward wicking at water discharge points. The content of organic matter (OM) varies considerably with depth and slope position. For example, Hapludolls have an OM content that is high (up to 4 %) in the surface layer but that decreases to 1.5 % in the subsurface horizons and to less than 0.5 % at the contact with till or sandstone. Due to the combination of parent material (mostly loess), climate, and historical vegetation, soils in the area are inherently very fertile.

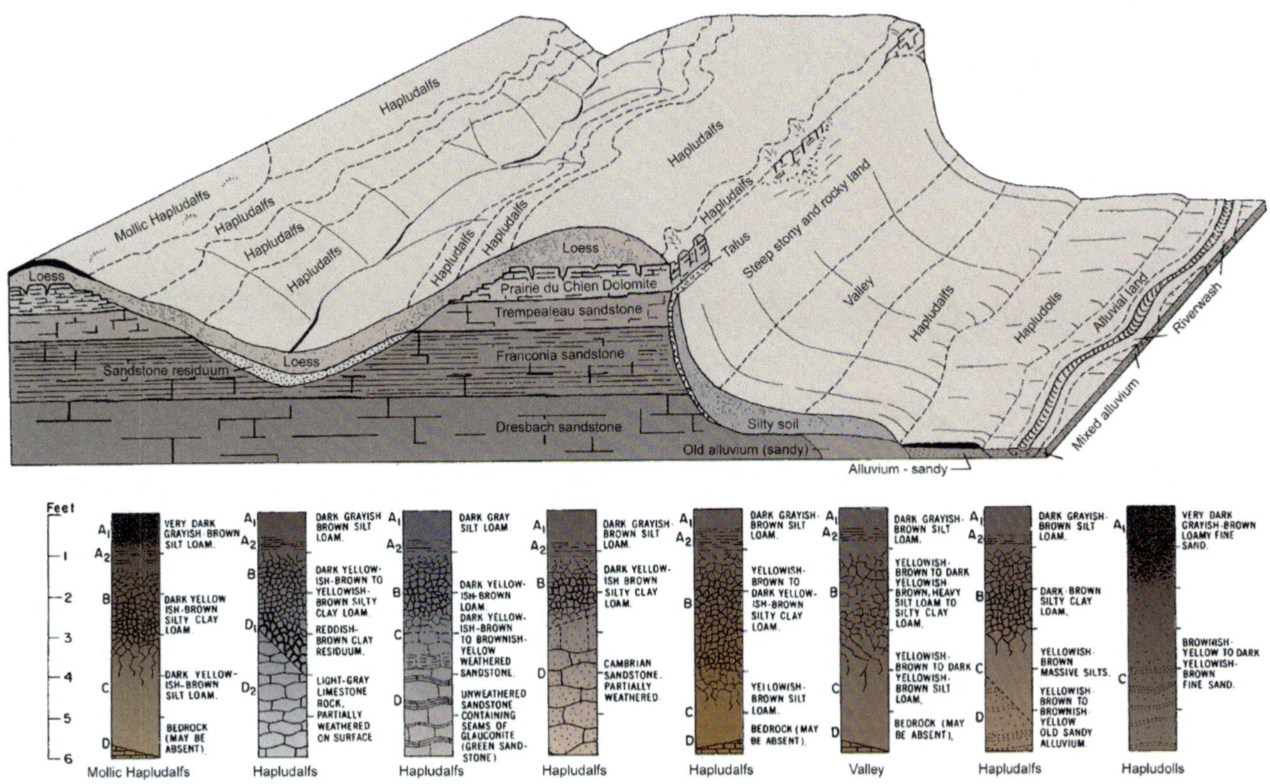

Fig. 11.8 Soil landscape of Buffalo County, Wisconsin, showing relationships among major soil series (Thomas et al. 1962)

11.2.2.3 Soil Resource Management

Tall- and short-grass prairie vegetation dominated this area, which is also known as the Prairie Peninsula Biome (NEON 2015). Various tree species still occur, mostly along drainageways, in major river valleys and depressions and on uplands. Almost the entire area is farmland used for the production of corn, soybeans, other feed grains, and hay. In addition to major vegetation changes, one of the most dramatic changes in the area has been to local hydrology as a result of the introduction of tile drains. In fact, the Central Feed Grains and Livestock Region has been identified as one of the major contributors to the Gulf of Mexico hypoxia phenomenon due to the unhindered delivery of chemically loaded drainage waters to local streams and the Mississippi River that has been enhanced by tile drains (EPA 2007). The major resource concerns for the area are organic matter depletion, water and wind erosion, and poor water quality (USDA-NRCS 2006). The estimated annual soil erosion is between 4 and 5 ton ha^{-1}. However, in some areas, especially on steep slopes along drainageways and/or streams, it could exceed 10 ton ha^{-1}. Highly eroded soil phases are common and, in some cases, comprise between 30 and 40 % of the soil map units (USDA-NRCS 2004; McCormick et al. 2004). Wind erosion is a concern especially on exposed soil surfaces.

11.2.3 Deep Loess and Drift (MLRAs 108A/B/C/D) and Iowa and Missouri Heavy Till Plain (MLRA 109)

11.2.3.1 Geology and Geomorphology

The majority of these MLRAs occur in the Dissected Till Plains Section of the Central Lowland Province of the Interior Plains. The soils and landscapes are relatively young and moderately dissected by local streams and major river networks arranged in a dendritic pattern. The resulting landscape is a combination of rolling plains and stream terraces adjacent to broad flood plains along major streams and rivers. Slopes are generally less than 15 %, except for areas along the major stream network where they are steeper. The geology of the area is comprised of rock formations overlain by till and loess deposits of various ages. The dominant geological deposits are Pennsylvanian shale, siltstone, and limestone (MLRAs 108A/B/C/D) and Mississippian shale and limestone (MLRA 109), which are commonly exposed along major streams (USGS 2015; USDA-NRCS 2006). The Illinoian, pre-Illinoian, and Wisconsin glacial till extends throughout the area and is covered, for the most part, by Wisconsin and/or Peoria loess deposits.

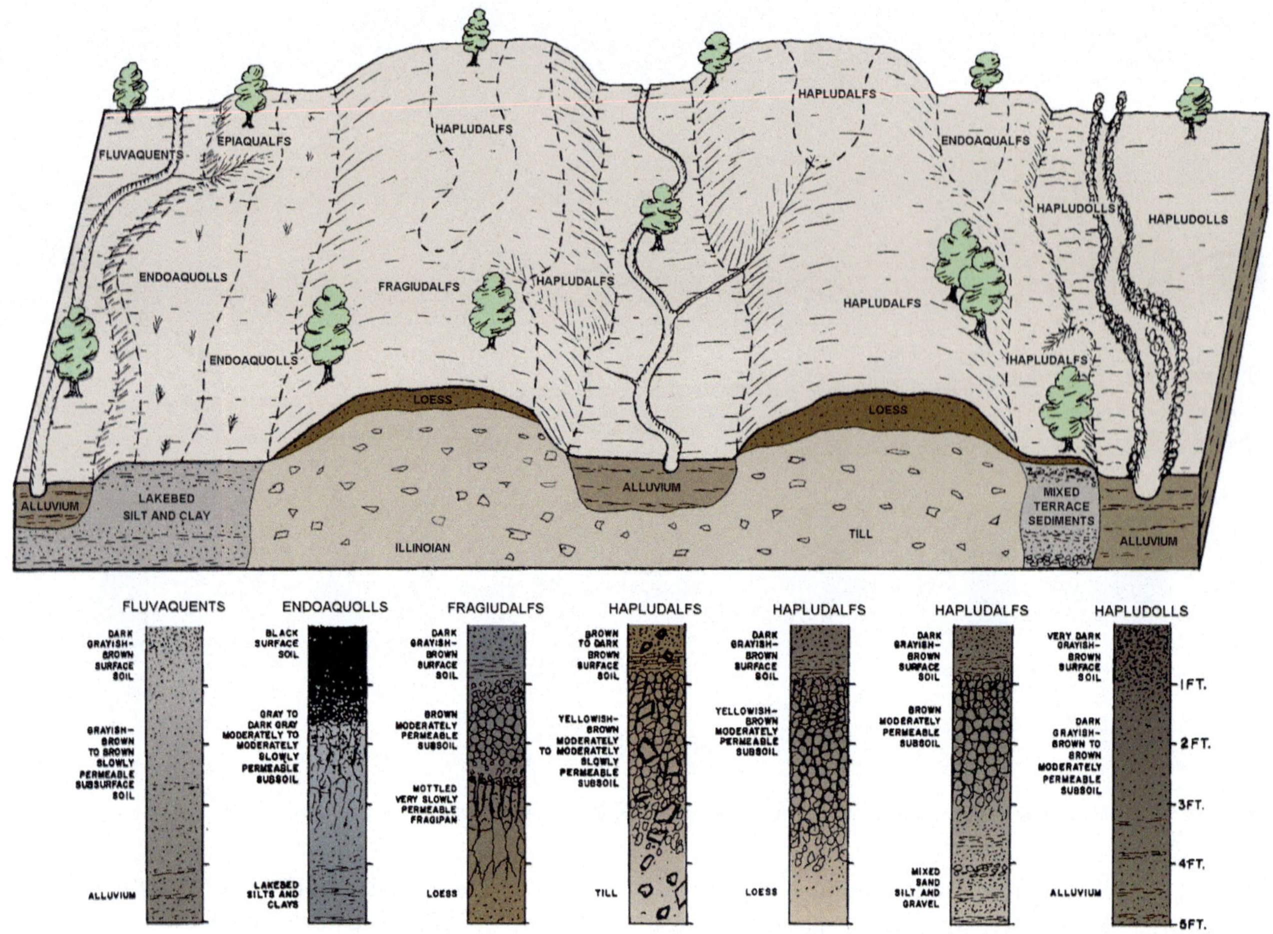

Fig. 11.9 Soil landscape of Wabash County, Illinois showing relationships among major soil series (Walker et al. 1964)

11.2.3.2 Soil Properties

The dominant soil orders are Mollisols and Alfisols, which occur mostly on loess-covered hills and uplands. Entisols and Inceptisols occur to a lesser extent in sloping areas adjacent to major streams and on broad interfluves (USDA-SCS 1964) (Fig. 11.9). The soils in the area are characterized by a mesic soil temperature regime, an aquic or udic soil moisture regime, and mixed mineralogy. They vary from deep to very deep and are predominantly well drained (deep loess on hills) and poorly drained (swales, valleys, and depressions). Argiudolls are associated mostly with loess over till and occur mainly on gently sloping to moderately steep uplands. Argiaquolls occur in similar upland areas but are poorly drained. Alfisols occur mostly in major stream valleys on stream terraces and in dissected upland drainageways. The majority of soils are Hapludalfs due to their landscape position, occurring mostly on sloping areas. Udifluvents occur to a limited extent on silty alluvium on flood plains or along much smaller upland drainage ways with Fluvaquents.

The majority of the soils formed from loess and are mostly silt loam and silty clay loam textures throughout. The clay content varies from 27 to 35 %, and few stones and pebbles commonly occur throughout the profile. Free carbonates occur generally at a depth of 100–200 cm. Soil reaction varies from slightly acid to moderately alkaline. Soils generally have a moderate permeability (0.5–1.5 cm h^{-1}) and a high available water-holding capacity (0.4–0.46 cm cm^{-1}) throughout due to the silt content. The content of organic matter varies from 4 % in the surface layer to less than 1 % below a depth of 30 cm. The soils in the region are some of the most productive soils in the world due to the climate and parent material (mostly loess).

11.2.3.3 Soil Resource Management

In the mid-1800s, the entire region was under tall- and short-grass prairie vegetation with a few woody species, mostly along drainageways and major river valleys, in depressions, and on uplands. More than 90 % of the area is farmland used for the production of corn, soybeans, other feed grains, and hay (USDA-NRCS 2006). Beef cattle and swine are also important sources of income. The major resource concerns are organic matter depletion, water and wind erosion, and poor water quality (USDA-NRCS 2006). The estimated annual soil erosion is between 5 and 5.5 ton ha^{-1}. However, in some areas, especially on steep slopes along drainage ways and/or streams, it could exceed 10 ton ha^{-1}. Highly eroded soil phases are common and, in some cases, comprise almost 35 % of the soil map units (USDA-NRCS 2007). Wind erosion is a concern especially on exposed soil surfaces. There are a few areas surface-mined for coal, especially in the northern part of MLRA 108B on the east side of the Illinois River. These areas are partially reclaimed.

11.2.4 Till Plain (MLRAs 111A/B/C/D/E)

11.2.4.1 Geology and Geomorphology

Most of these MLRAs are in the Till Plains Section of the Central Lowland Province of the Interior Plains. MLRA 111C is in the Eastern Lake Section. The entire area has been glaciated and is dominated by ground moraines from the Lake Erie ice lobe. The moraines are broken in many places by kames, lake plains, outwash plains, terraces, and stream valleys (Fig. 11.10). Some lake plains, such as the Kankakee, occur in front of the recessional moraines and are flat. Many stream valleys occur at the edges of the moraines, which represent different episodes of glacial advancement and retreat. The area is covered by a mix of materials, including glacial deposits of till, outwash, and lacustrine sediments dating to the Wisconsin or older glacial periods (USDA-NRCS 2002). Loess deposits of various thicknesses occur throughout the area. The underlying bedrock is dominantly Silurian and Devonian limestone and dolostone but includes Late to Middle Devonian shale, Early Mississippian black shale, and Middle to Early Mississippian siltstone (USGS 2015).

11.2.4.2 Soil Properties

The soil orders in these MLRAs are dominantly Alfisols, Inceptisols, and Mollisols with small isolated areas of Histosols. Hapludalfs occur on summits on till plains and moraines. Epiaqualfs and Endoaqualfs occur in broad flat depressions on till plains and moraines or outwash plains. Argiaquolls occur in depressions on moraines and till plains. Endoaquolls occur in depressions on till plains and moraines

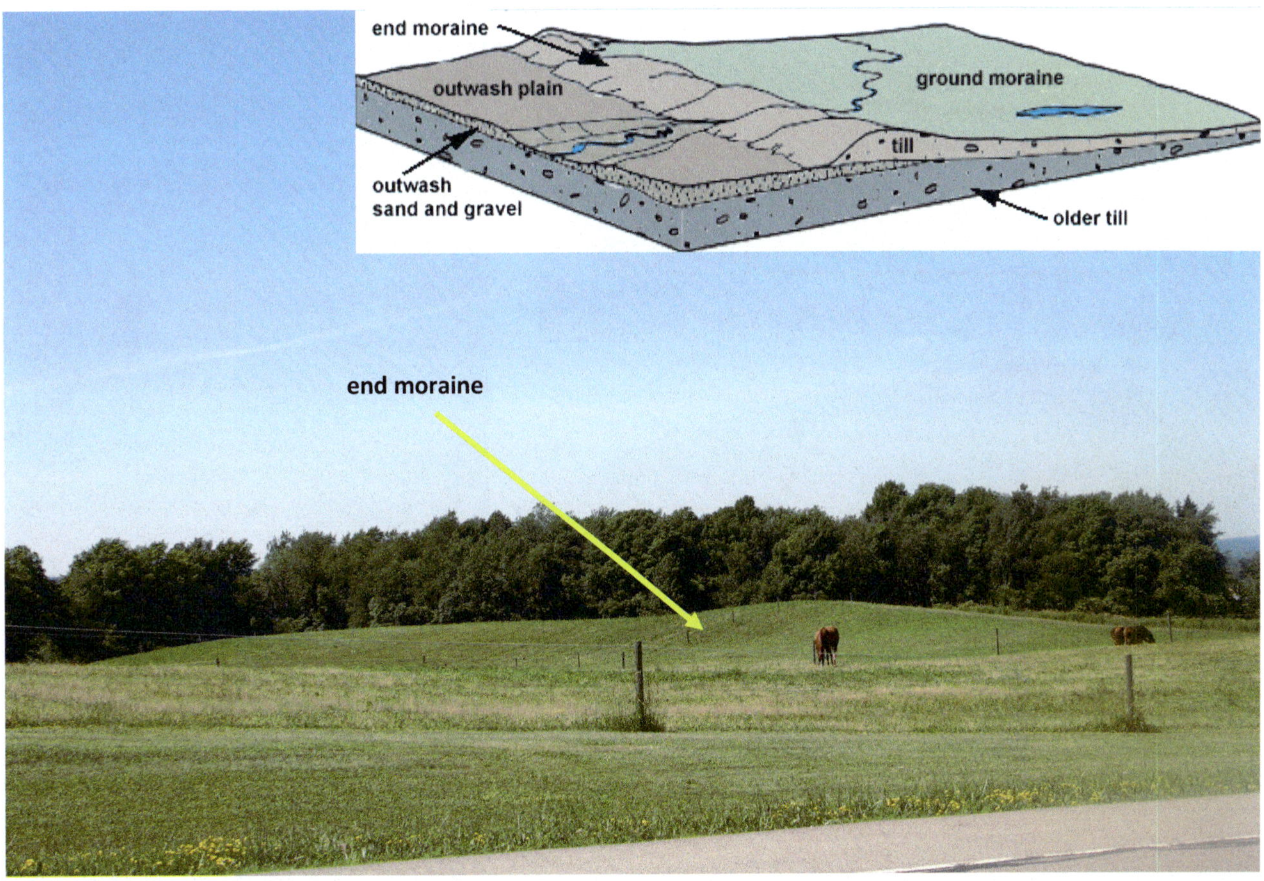

Fig. 11.10 Conceptual diagram of an end moraine [Illinois State Geological Survey, https://www.isgs.illinois.edu/outreach/geology-resources/end-moraines-end-glacial-ride (Accessed on March 04, 2016) and a ground moraine in Gainesville, New York, https://en.wikipedia.org/wiki/File:Ground_moraine_9004.jpg (Accessed on 2016)]

and lake and outwash plains (Fig. 11.11). Other Hapludolls, Endoaquepts, and Endoaquolls occur on flood plains. Haplosaprists occur in deep depressions and potholes. Soils in this region originated from loess in the uplands and are mostly silt loam and silty clay loam throughout. Soils on outwash plains have mixed parent materials that consist of loess, glacial fluvial, and sand. The clay content varies from 15 to 45 %, and few stones and pebbles commonly occur throughout the profile. Free carbonates occur generally at a depth of 100–200 cm but can occur at shallower depths in eroded soils on the lower parts of the landscape. Soil reaction varies from slightly acid to moderately alkaline. Soils generally have a moderate permeability (1.5–3.3 cm h^{-1}) and a highly variable available water-holding capacity, ranging from 0.05 to 0.25 cm cm^{-1} depending on whether the parent material is loess or sandy outwash (USDA-NRCS Soil Survey Staff 2015). The organic matter for most soils varies from 4 % in the surface layer to less than 1 % in the subsurface horizons (USDA-NRCS 1997). Haplosaprists, however, have between 12 and 18 % organic matter throughout the profile (Fig. 11.11). The soils in the region are very productive due to the climate and loess parent material.

11.2.4.3 Soil Resource Management

Prior to settlement, the majority of the eastern portion of the region was under hardwood forest, mostly pin oak (*Quercus palustris*), swamp white oak (*Quercus bicolor*), blackgum (*Nyssa sylvatica*), American sycamore (*Platanus occidentalis*), green ash (*Fraxinus pennsylvanica*), silver maple (*Acer saccharinum*), and cottonwood (*Populus sect. Aigeiros*) on the wetter soils. The hardwood forest gradually was replaced to the west with tall- and short-grass prairie vegetation and a few woody species, which occurred mostly along drainageways, in major river valleys, and in depressions on uplands. Most of MLRA 111D and the western portion of MLRA

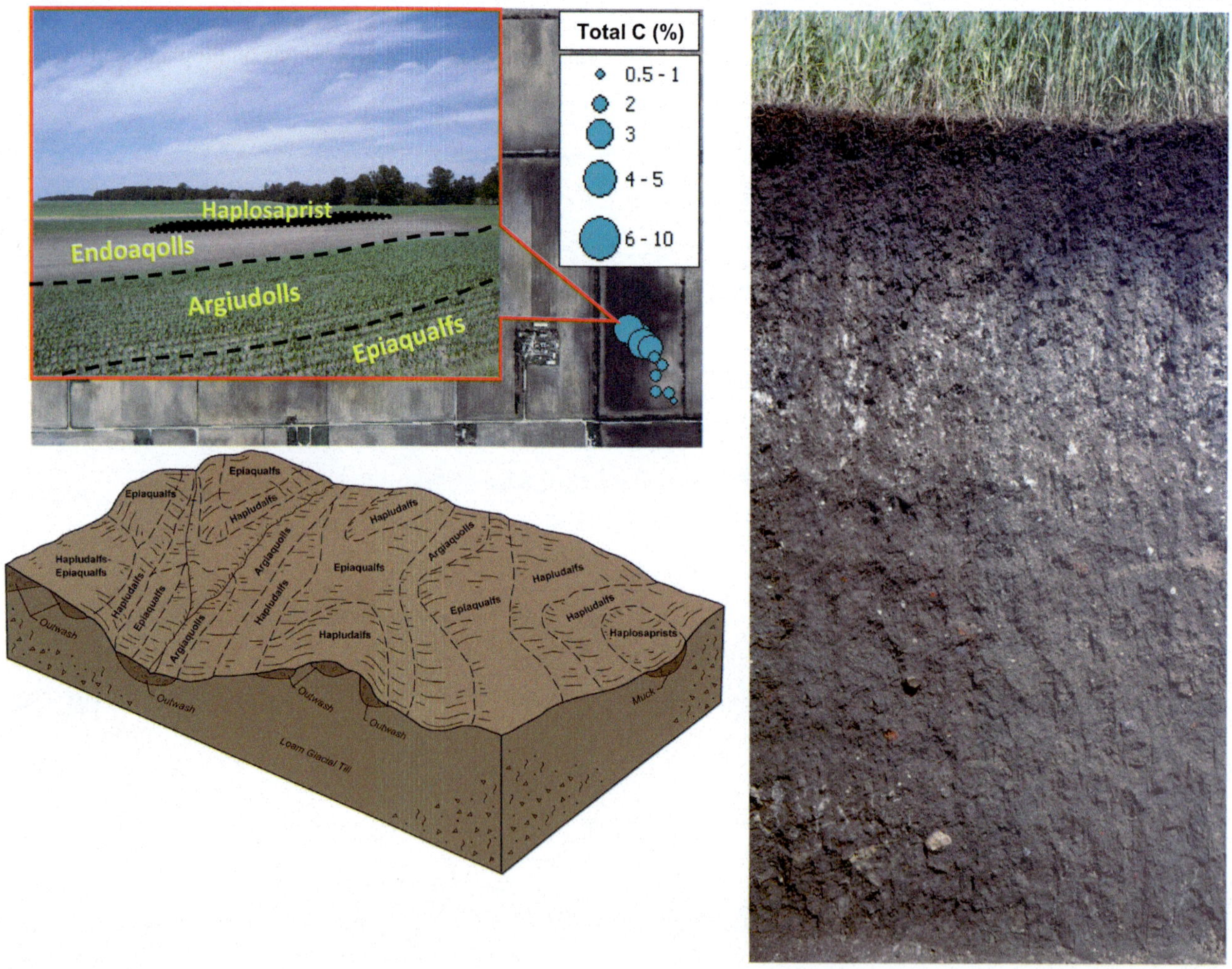

Fig. 11.11 Total soil carbon distribution for the surface horizon (0–25 cm) of an Epiaqualfs–Agriaquolls toposequence in Howard County, Indiana; a typical soil landscape of Elkhart County, Indiana, showing relationships among major soils (Brock et al. 1997); and a typical profile of a Mollisol

111C were part of the so-called Prairie Peninsula where mixed hardwood–prairie grass vegetation occurred. Today, more than 75 % of the region is farmland used for the production of corn, soybeans, other feed grains, and hay. Because the area includes or is near major urban centers such as Chicago, Indianapolis, and Cleveland, dairy farms and truck and canning crops are widespread (USDA-NRCS 2006). The major resource concerns for the region are organic matter depletion, water and wind erosion, and poor water quality (due to excessive sediments, nutrients, and pesticides in surface water and nutrients and pesticides in ground water) as well as loss of wildlife habitat and productive soil due to urban expansion (USDA-NRCS 2006). The estimated annual soil erosion is between 2 and 5 ton ha^{-1}. However, in some areas along drainageways and/or streams, the erosion rates could be higher. Highly eroded soil phases are common (USDA-NRCS 2007; Lensch et al. 2007) and have resulted in some cases to major shifts in soil, from Mollisols with thick surface layers enriched with organic matter to Alfisols with thinner surface layers depleted of organic matter.

11.2.5 Thin Loess and till Plains (MLRAs 114A/B), Central Mississippi Valley Wooded Slopes (MLRAs 115A/B/C), Central Claypan Areas (MLRA 113), Cherokee Prairies (MLRA 112)

11.2.5.1 Geology and Geomorphology

Most of these MLRAs occur in the Till Plains Section of the Central Lowland Province of the Interior Plains. MLRA 112 occurs in the Osage Plain Section. However, because the area represents mainly the southernmost extent of the older Illinoian glaciation and, to a limited extent, areas on the fringes of the latest glaciations (MLRA 112) (Fig. 11.12), the physiography, geology, and parent materials are very complex. The major tributaries, from the Mississippi River to the southwest and the Ohio River to the southeast, extended north through uplands, creating dissected till plains with well-defined broad flood plains and summits (USDA-NRCS 1995). Illinoian loess covers MLRA 113 and most of MLRAs 114A/B. It varies in thickness from 1 to 2 m on stable broad summits to less than 0.1–0.3 m on the steeper slopes along the major tributaries (USDA-NRCS 2006). MLRAs 115A/B/C are almost entirely covered by Wisconsin loess, which is more than 2 m deep in some areas, especially on broad summits. Only a thin layer of loess covers the northern portion of MLRA 112, while most of the southern part is covered by relatively older soils that formed from residuum weathered from sandstone and contain more clay (USDA-NRCS 1992; Tegeler et al. 1992). In glaciated areas, glacial drift deposits underlie loess, both of which are underlain by predominantly Pennsylvanian limestone and shale bedrock (MLRA 113) and Late Mississippian rocks that outcrop along the Mississippi and Ohio rivers and their tributaries. Meltwater outwash and lacustrine and alluvial deposits occur on stream terraces along major tributaries.

11.2.5.2 Soil Properties

The major soil orders in the region are Alfisols, Entisols, Inceptisols, and Mollisols (USDA-NRCS 2006). Hapludalfs occur on side slopes along drainage ways on upland till plains over loess, bedrock residuum, outwash, kames, moraines, stream terraces, and lacustrine sediment lake plains (Fig. 11.12). Paleudalfs occur on the broader gently sloping to steep side slopes over bedrock residuum and outwash, kames, moraines, and stream terraces. Fragiudalfs occur in positions similar to Paludalfs on till plains, and/or over bedrock residuum and outwash, kames, moraines, and stream terraces (USDA-NRCS 1995). Endoaquolls, Hapludolls, and Argiudolls occur on broad upland summits and in gently sloping areas as well as in major stream valleys. Fluvaquents and Udifluvents formed in alluvium on nearly level flats, on broad flood plains, and along the smaller upland drainage ways. The dominant soil textures are silt loam and silty clay loam mostly for upper horizons composed of loess but that transitions to sandy clay loam, sandy loam, or the gravelly analogs of these textures for subsurface horizons closer to the bedrock or glacial till or soils occurring on glacial fluvial and sand in outwash plains. The clay content varies from 24 to 35 %, and the content of rock fragments varies from 5 to 20 % for the horizons underlying the loess. However, for soils that formed in residuum weathered from shale and sandstone, especially in MLRA 113, the clay content is commonly as high as 60 % and smectitic. Free carbonates, where present, occur generally below a depth of 100 cm but can be closer to the surface in eroded areas on the lower parts of the landscape. Soil reaction varies from slightly acid to moderately alkaline. However, soils in MLRA 113, especially in the southern part, that formed from residuum weathered from shale and sandstone are moderately acid in the subsurface layer, with soil pH often as low as 5 (USDA-NRCS 1995). Soils generally have a moderate or high permeability (1.5–5.1 cm h^{-1}) but can have a moderately low permeability, especially in the southern part of MLRA 113 that formed from residuum weathered from shale and sandstone. Soils generally have a highly variable available water-holding capacity (0.08–0.25 cm cm^{-1}), depending on whether the parent material is loess, clay, or sandy outwash. For the most part, the organic matter content varies from 3 % in the surface layer to less than 1 % in the subsurface horizons. Some Endoaquolls, however, can have 5 % organic matter in the

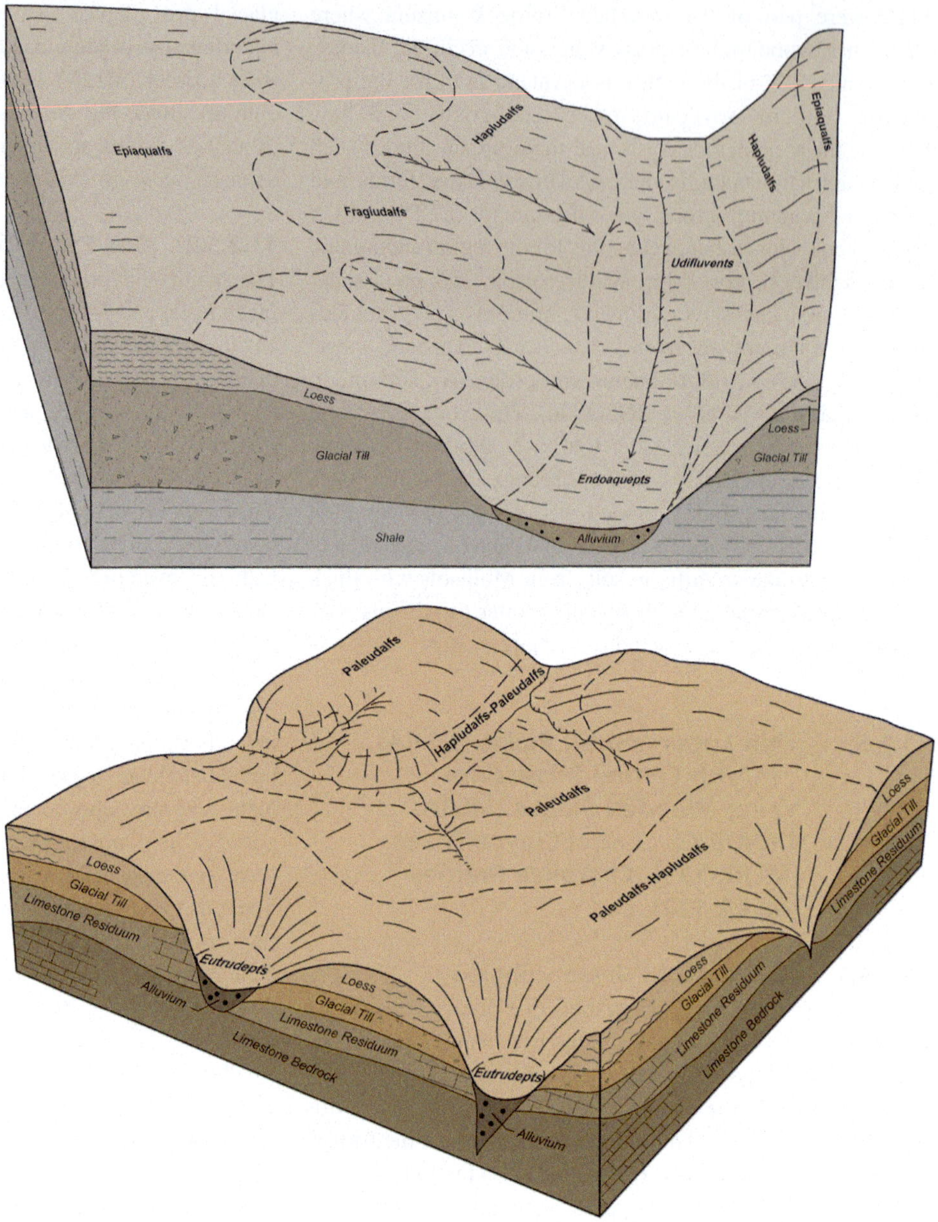

Fig. 11.12 Typical soil landscape of Paleudalfs–Fragiudalfs–Hapludalfs–Udifluvents in Cumberland County, Illinois, and a typical soil landscape of Owen County (McCarter et al. 1995), Indiana, showing relationships among Fluvaquents–Hapldalfs–Paleudalfs

surface horizons that deceases to less than 0.5 % in the subsurface horizons. The soils in the region are very productive due to the climate and loess parent material.

11.2.5.3 Soil Resource Management

Most of the MLRAs support natural hardwood forest, mostly oak (*Quercus*), hickory (*Carya*), beech (*Fagus*), and sugar maple (*Acer Saccharum*). MLRA 113 supports tall prairie grasses, mainly big bluestem (*Andropogon gerardii*), Indiangrass (*Sorghastrum nutans*), prairie dropseed (*Sporobolus heterolepis*), and switchgrass (*Panicum virgatum*). Most of the MLRAs are in cropland, varying in composition from 35 % (MLRA 115B) to 70 % (MLRA 113). The dominant crops are corn, soybeans, other feed grains, and hay for cattle and other livestock. Other grains such as winter wheat, oats, and sorghum are also grown. The production of tobacco, melons, potatoes, and apple and peach orchards occurs to a limited extent. Cotton and legumes grow in some parts of MLRA 113. The major soil resource concerns are wind and water erosion, flooding, and a limited available water-holding capacity which affects the majority of the dry-farmed crops. The maintenance of soil organic matter is of particular concern due to a climate that is warmer and wetter compared to that of the northern part of LRR M as well as to management-induced disturbances. Surface compaction and low soil pH are of concern especially in the southern part of MLRA 113 for soils that formed under residuum weathered from shale and sandstone.

11.3 Lake State Fruit, Truck Crop, and Dairy Region, LRR L (MLRAs 96, 97, 98, 99, and 101)

11.3.1 Western Michigan Fruit Belt (MLRA 96), Southwestern Michigan Fruit and Truck Crop Belt (MLRA 97), Southern Michigan and Northern Indiana Drift Plain (MLRA 98), Erie-Huron Lake Plain (MLRA 99), and Ontario-Erie Plain and Finger Lakes Region (MLRA 101)

11.3.1.1 Geology and Geomorphology

This region is in the Eastern Lake Section of the Central Lowland Province of the Interior Plains. For the most part, it is situated between three major lakes: Lake Michigan to the west, Lake Huron to the east, and Lake Erie to the southeast. The region is dominated by outwash and till plains and moraines. Mississippian–Silurian–Devonian rocks mostly composed of sandstone, limestone, and dolomite with some interbedded shale dominate the region. They are exposed only along the shores of Lake Michigan and Lake Huron. Significant deposits of glacial drift and unconsolidated sand and gravel outwash (commonly as thick as 180 m) overlay the bedrock, especially in MLRA 98, due to the curvature of the Michigan Basin.

11.3.1.2 Soil Properties

The major soil orders in the region are Alfisols, Inceptisols, Mollisols, Histosols, and Spodosols. Endoaqualfs and Epiaqualfs formed mostly on till, while Hapludalfs formed mostly on outwash plains, kames, terraces, and deltas over outwash or glacial drift parent materials and/or on till plains either exposed or covered by loess (Fig. 11.13). The majority of Epiaquepts and Endoaquepts formed in lacustrine deposits on lake plains and till. Endoaquolls and

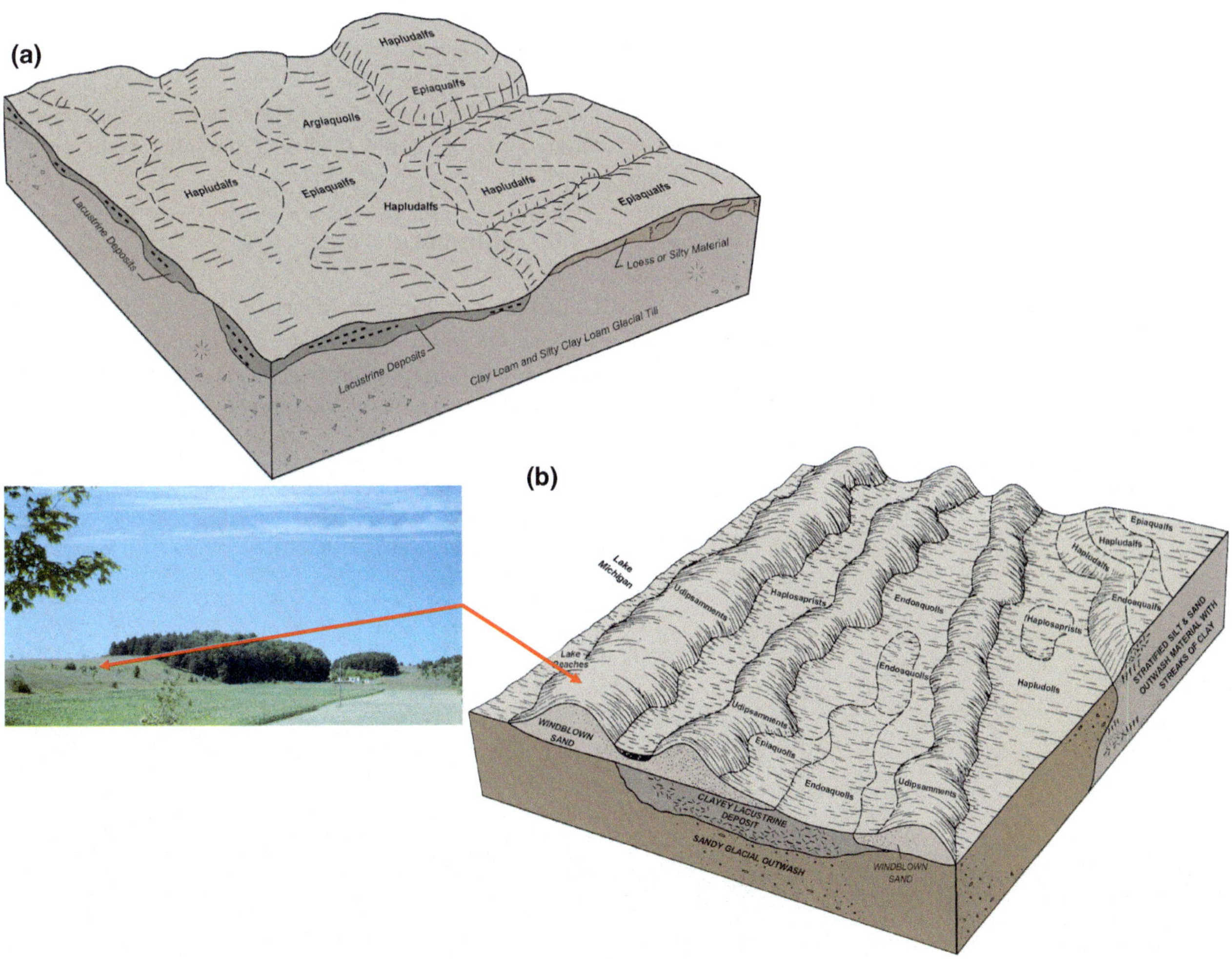

Fig. 11.13 **a** Typical soil landscape of Hapludalfs–Epiaqualfs–Argiudolls and **b** typical soil landscape showing relationships among Udipsaments–Epiaqualfs–Fluvaquents–Endoaquolls–Haplosaprists

Argiaquolls formed mostly in outwash or glacial drift over outwash on outwash plains, kames, terraces, and deltas, till, and loess over till. Udipsamments formed in sandy glaciofluvial deposits, while Haplosaprists, which occur in small depressions, formed from organic deposits on almost all types of parent material, from sandy glaciofluvial to till or loess over till. Spodosols, unique to the region, occur due to a combination of forest vegetation, high water tables, and sandy parent materials, which led to the formation of a thick organic layer over sandy materials (Fig. 11.14). The soil texture varies greatly due to the differences in parent material. The soils that formed on sandy glacial fluvial and outwash glacial drift over outwash on outwash plains, kames, terraces, and deltas are predominantly sand, fine sandy, loamy sand, loamy fine sand, coarse sandy loam, or sandy loam. Gravel occurs throughout the soil profile and varies from 5 % in the surface horizons to as much as 20 % in the subsurface horizons. The soils that formed in till or loess over till are mostly silty clay loam, clay loam, clay, or silty clay with a clay content between 15 and 48 %. Free carbonates occur generally below a depth of 100 cm. Soil reaction varies from strongly acid (pH 5–5.5) to moderately alkaline. Some soils, especially those that formed on sandy glacial fluvial and outwash glacial drift on outwash plains, can have a moderate or high permeability (1.5–50.1 cm h^{-1}). Soils that formed from loess or till can have a moderately low or low permeability (0.5 cm h^{-1}). The available water-holding capacity is highly variable, typically ranging from 0.08 to 0.25 cm cm^{-1}, depending on the parent material. For some Haplosaprists, it can be as high as 0.45 cm cm^{-1}. The content of organic matter mainly varies from 3 % in the surface layer to less than 1 % in the subsurface horizons. Some Haplosaprists, however, can have as much as 55–75 % organic matter throughout. The soils in the region are very productive due to the climate and loess parent material.

11.3.1.3 Soil Resource Management

Historically, the entire region was under a broadleaf deciduous forest comprised of a wide range of species, including white oak (*Quercus alba*), black oak (*Quercus velutina*), red maple (*Acer rubrum*), sugar maple (*Acer saccharum*), and beech (*Fagus*) (Cole et al. 1999). The majority of MLRA 96 is still covered by forest, but the extent decreases further south, where more than 75 % of the land is farmed (USDA-NRCS 2006). The orchards, vegetables, crops, and dairy farms that dominate the northern part of the region are gradually replaced with corn, soybean, winter wheat, canning crops (such as sugar beets), and fruit and truck crops in areas of sandy soils. Because the region is under diverse land uses, the soil resource concerns and management are very complex. Soil erosion by wind and water are major concerns. Water quality is threatened by point source pollution, such as from dairy farms, and non-point source pollution, such as from runoff containing pesticides and chemicals. Major wetlands in the region have been drained, which reduces the capacity of the soil to absorb and process runoff from agricultural activities. Because wide areas are dominated by coarse sandy materials, the water percolates quickly through the soil profile, leading to droughty conditions.

11.4 Summary

Land Resource Regions M and L located in northcentral USA are one of the areas in the USA with the youngest soils due to a cycle of glacial advancements and retreats. For more than 12,000 years, hardwood forest from east transitioned to

Fig. 11.14 Typical soil profile of a Spodosol and its landscape. *Source* Stan Buol, North Carolina University at https://www.flickr.com/photos/soilscience/5140042749/in/album-72157625094412337/

prairie vegetation to the west and both dominated the region. A combination of climate, vegetation, and predominantly loess materials leads to the establishment of fertile soils rich in nutrients and organic matter. Since its settlement in the mid-nineteenth century, the region underwent a deep transformation with agriculture development that overhauled the prairie vegetation and replaced it mostly with corn, soybean, and other cash crops. Today, LRR M and L are at the heart of the so-called bread basket of the USA and support high levels of food and livestock production. The use of fertilizers, pesticides, tillage, and other agricultural inputs from intensive agricultural practices has led to soil fertility and organic matter depletions in addition to physical loss of soil due to erosion. This combined with extensive drainage ditches and tile drains has accelerated water quality deterioration that has affected ground water, streams, and rivers all the way to the Gulf of Mexico.

References

Boellstorff JD (1978a) Chronology of some late Cenozoic deposits from the central United States and the ice ages. Neb Acad Sci Trans 6:35–49

Boellstorff JD (1978b) North American Pleistocene stages reconsidered in light of probable Pliocene–Pleistocene continental glaciation. Science 202:305–307

Brock RA, Chowdhery AA, Furr FG Jr, McBurnett SL, Patterson NT, Ross SS, Shively JL, Ziegler TR (1997) Soil Survey of Elkhart County, Indiana. USDA-Natural Resources Conservation Service, Washington

Cole KL, Davis MB, Stearns F, Guntenspergen G, Walker K (1999) Historical land cover changes in the Great Lakes Region. In: Sisk TD (ed) Perspectives on the land use history of North America: a context for understanding our changing environment. U.S. Geological Survey, Biological Resources Division, Biological Science Report USGS/BRD/BSR-1998-0003, Chapter 6

Cordell LS, Lightfoot K, McManamon F, Milner G (2008) Archaeology in America: an encyclopedia: ABC-CLIO, p 1. ISBN 978-0-313-02189-3

Dutch S (2015a) Natural and applied sciences. University of Wisconsin-Green Bay. https://www.uwgb.edu/dutchs/sitemap.htm

Dutch S (2015b) Glaciation events and deposits in the Continental USA during the last 18,000 years. https://www.uwgb.edu/dutchs/EarthSC202Notes/GLACgeog.HTM. Accessed 18 Aug 2015

Environmental Protection Agency Science Advisory Board (2007) Hypoxia in the Northern Gulf of Mexico: an update by the EPA Science Advisory Board. EPA-SAB-08-004, Washington. http://yosemite.epa.gov/sab/sabproduct.nsf/C3D2F27094E03F90852573B800601D93/$File/EPA-SAB-08-003complete.unsigned.pdf

Lensch RA, Bednarek RJ, Gundlach DM, McCormick JA, Oelmann DB, Younk AW (2007) Soil survey of Adams County, Iowa. USDA-Natural Resources Conservation Service, Washington

Mason JA, Joeckel RM, Bettis AE (2007) Middle to Late Pleistocene loess record in eastern Nebraska, USA, and implications for the unique nature of Oxygen Isotope Stage 2. Quat Sci Rev 26:773–792

McCarter P Jr, Nagel BG, Sanders FW, Brownfield S, Lehman S, Fields R, Shively JL, Patterson NT (1995) Soil Survey of Owen County, Indiana. USDA-Natural Resources Conservation Service, Washington

McCormick JA, Bednarek RJ Jr, Lensch RA, Pulido DJ (2004) Soil Survey of Shelby County, Iowa. USDA-Natural Resources Conservation Service, Washington

National Ecological Observatory Network (2015) http://www.neoninc.org/. Accessed 18 Dec 2015

Richardson J (2010) Interpreting hydrology from soil morphology. Soil Geomorphology Institute (SGI) lecture series. USDA-NRCS-NSSC, Lincoln

Richardson JL, Wilding LP, Daniels RB (1992) Recharge and discharge of groundwater in aquic conditions illustrated with flownet analysis. Geoderma 53:65–78

Schmus VWR, Hinze WJ (1985) The midcontinent rift system. Annu Rev Earth Planet Sci 13(1):345-83. doi:10.1146/annurev.ea.13.050185.002021

Soil Survey Staff. Soil Survey Geographic (SSURGO) database, United States Department of Agriculture, Natural Resources Conservation Service. http://websoilsurvey.nrcs.usda.gov/. Accessed 1 Jan 2016

Sugden DE, John BS (1976) Glaciers and landscapes. Edward Arnold, a division of Hodder and Stoughton Limited, Mill Road, Dunton Green, Sevenoaks, Kent TN 13 2YA by Butler and Tanner Ltd, Frome and London. ISBN 0 7131 5840 9

Tegeler RA, Engel EJ, Staley LR (1992) Soil survey of Cumberland County, Illinois. USDA-Soil Conservation Service, Washington

Tranhaile AS (1998) Geomorphology: a Canadian perspective. Oxford University Press, Don Mills. ISBN 0-19-541277-X

Thomas DD, Carrol PH, Wing GN, Haszel OE (1962) Soil survey of Buffalo County, Wisconsin. USDA-Soil Conservation Service, Washington

United States Department of Agriculture, Natural Resources Conservation Service (2006) Land Resource Regions and Major Land Resource Areas of the United States, the Caribbean, and the Pacific Basin. U.S. Department of Agriculture Handbook, Washington

United States Department of Agriculture, Natural Resources Conservation Service, Kellogg Soil Survey Laboratory (2014) Pedon database. http://ncsslabdatamart.sc.egov.usda.gov/

United States Department of Agriculture, Natural Resources Conservation Service, Soil Survey Adams County (2007), Iowa

United States Department of Agriculture, Natural Resources Conservation Service, Soil Survey Schuyler County (2004), Illinois

United States Department of Agriculture, Natural Resources Conservation Service, Soil Survey McDonough County (2002), Illinois

United States Department of Agriculture, Natural Resources Conservation Service, Soil Survey Delaware County (1997), Indiana

United States Department of Agriculture, Natural Resources Conservation Service, Soil Survey Owens County (1995), Indiana

United States Department of Agriculture, Natural Resources Conservation Service, Soil Survey Cumberland County (1992), Illinois

United States Department of Agriculture, Soil Conservation Service, Soil Survey Wabash County (1964), Illinois, 1957(17)

United States Department of Agriculture, Soil Conservation Service, Soil Survey Buffalo County (1962) New York, 1957(3)

United States Department of Agriculture, Natural Resources Conservation Service, National Soil Survey Center (1994) Soil Climate Regimes of the United States

United States Department of Agriculture, Natural Resources Conservation Service, Soil Survey Staff. Official Soil Series Descriptions. Accessed 16 Aug 2015

United States Geologic Survey (2015) Bedrock Geology Map of the United States of America. http://mrdata.usgs.gov/geology/state/. Accessed 18 Sept 2015

Walker GO, Fehrenbacher JB, Jorgensen IH, Buoy C, Johnson JR, Karraker E, Killough W, Steely EN (1964) Soil survey of Wabash County, Illinois. USDA-Soil Conservation Service, Washington

Wayne G (2007) The Great Lakes. Greystone Books and David Suzuki Foundation, Vancouver, pp 13, 21–26, 42–43. ISBN 978-1-55365-197-0

Wigginton M, Haley S, Shively J, Stephens N (2002) Soil survey of Boone County, Indiana. USDA-Natural Resources Conservation Service, Washington

Winzeler HE, Waltman SW, Waltman WJ, Libohova Z, Baudette D (2012) A methodology for examining changes in soil climate geography through time: U.S. soil moisture regimes for the period 1971–2000. Soil Sci Soc Am J 77:213–225. doi:10.2136/sssaj2012.0123

East and Central Farming and Forest Region and Atlantic Basin Diversified Farming Region: LRRs N and S

12

Brad D. Lee and John M. Kabrick

12.1 Introduction

The central, unglaciated US east of the Great Plains to the Atlantic coast corresponds to the area covered by LRR N (East and Central Farming and Forest Region) and S (Atlantic Basin Diversified Farming Region). These regions roughly correspond to the Interior Highlands, Interior Plains, Appalachian Highlands, and the Northern Coastal Plains.

12.2 The Interior Highlands

The Interior Highlands occur within the western portion of LRR N and includes seven MLRAs including the Ozark Highlands (116A), the Springfield Plateau (116B), the St. Francois Knobs and Basins (116C), the Boston Mountains (117), Arkansas Valley and Ridges (118A and 118B), and the Ouachita Mountains (119). This region comprises 176,000 km^2 in southern Missouri, northern and western Arkansas, and eastern Oklahoma (Fig. 12.1).

12.2.1 Climate

The mean annual precipitation ranges from 965 to 1675 mm increasing to the south. The average air temperature range from north to south is 12–17 °C. The average growing season ranges from 200 to 240 days decreasing to the north and with elevation. Most of the precipitation falls during the freeze-free period and occurs as high-intensity, convective thunderstorms in summer. An average of 305–355 mm of snowfall occurs annually in the Ozark Highlands, the Springfield Plateau, and the St. Francois Knobs and Basins MLRAs. In the southern half of the region, snowfall is uncommon.

12.2.2 Physiography

The topography of this region ranges from broad, gently rolling plains to steep mountains. In the northern portion of this region, much of the Springfield Plateau and the Ozark Highlands is a dissected plateau that includes gently rolling plains to steeply sloping hills with narrow valleys. Karst topography is common and the region has numerous sinkholes, caves, dry stream valleys, and springs. The region also includes many scenic spring-fed rivers and streams containing clear, cold water (Fig. 12.2). The elevation ranges from 90 m in the southeastern side of the region and rises to over 520 m on the Springfield Plateau in the western portion of the region. The relief ranges from 45 m on the gently rolling plains to more than 245 m in the more dissected parts of the region. The St. Francois Knobs and Basins occurs in the northeastern portion of the Ozark Highlands and is centered on an uplifted "dome" underlain by Precambrian igneous rocks. The topography includes deeply dissected "knobs" or hills, and wide basins and valleys. Elevations range from 135 m along the rivers to 540 m on the highest peak, Taum Sauk Mountain. Relief ranges from 30 m in the basins up to 305 m among the igneous knobs. Greater elevations and relief occur in the rugged Boston Mountains, which lies to the south of the Ozark Highlands MLRA. This dissected plateau features narrow ridges, steep slopes, and narrow valleys that provide more than 45–270 m of relief and elevations ranging from 60 to 850 m (Woods et al. 2004). In the Ouachita Mountains MLRA, valleys are narrow and the slopes are steep. Elevations range from 100 to 800 m and relief ranges from 15 to 480 m (Woods et al. 2004). The Arkansas Valley and Ridges (MLRAs 118A and 118B) occur between the Ozark Highlands and Ouachita

B.D. Lee (✉)
University of Kentucky, Lexington, KY, USA
e-mail: brad.lee@uky.edu

J.M. Kabrick
USDA Forest Service, Northern Research Station, Columbia, MO, USA
e-mail: jkabrick@fs.fed.us

L.T. West et al. (eds.), *The Soils of the USA*, World Soils Book Series,
DOI 10.1007/978-3-319-41870-4_12

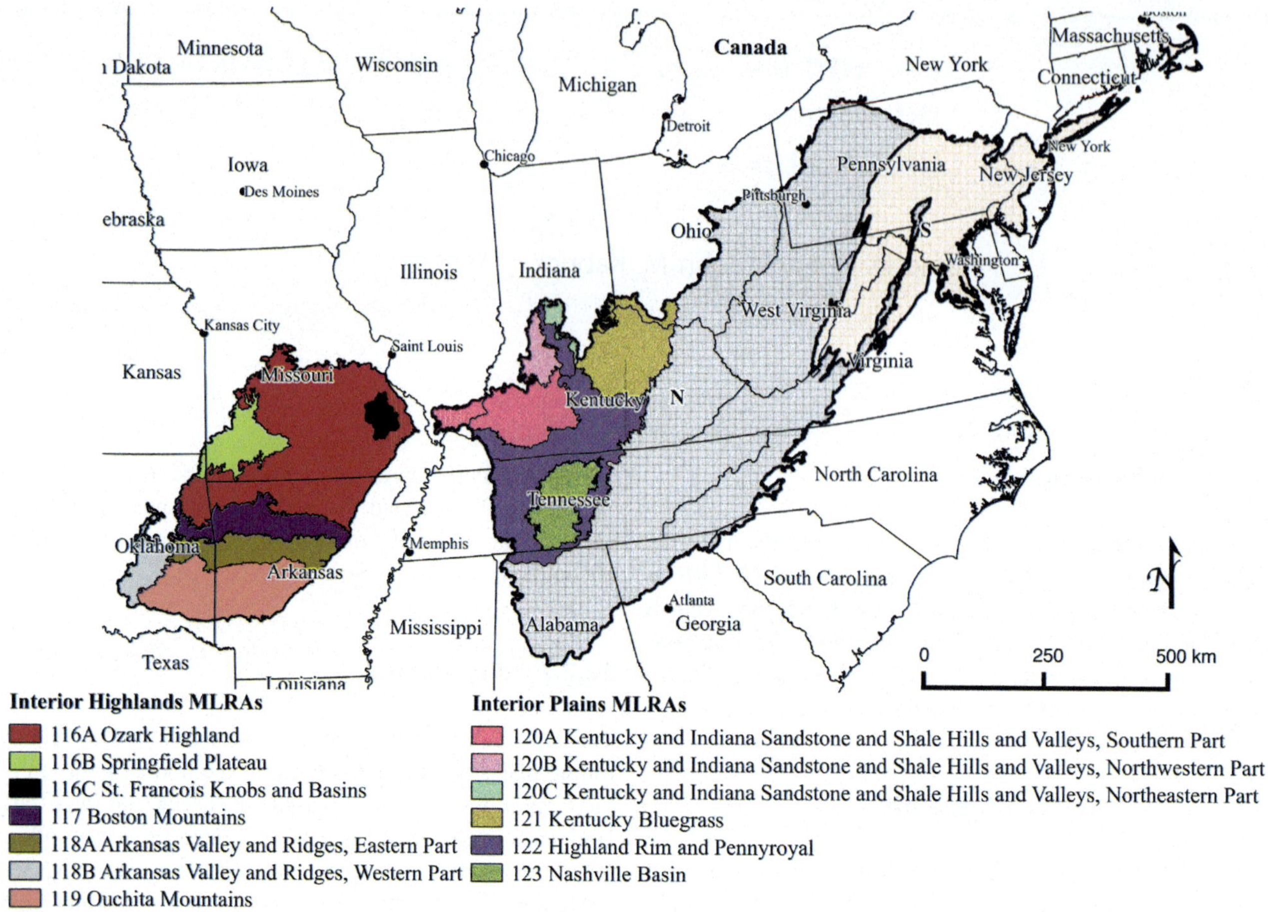

Fig. 12.1 Extent and distribution of MLRAs in the Interior Highlands (MLRAs 116A, 116B, 116C, 117, 118A, 118B, and 119) and Interior Plains (MLRAs 120A, 120B, 120C, 121, 122, and 123.) within LRR N

Mountains and consist of long broad ridges oriented east to west. Elevations range from 90 to 840 m and relief ranges from 15 to 330 m.

12.2.3 Geology

Most of the area is underlain by sedimentary rocks including dolostone, limestone, sandstone, siltstone, and shale although metamorphic rocks are also found in the Ouachita Mountains because of folding and faulting. The oldest rocks in the Interior Highlands occur in the St. Francois Knobs and Basins where Precambrian igneous rocks of volcanic origin are exposed in the center of the "Ozark Dome." These igneous rocks are flanked by Cambrian dolostone and sandstone that dip slightly away from the top of the dome. Ordovician dolostone and sandstone occur throughout the Ozark Highlands. Younger Mississippian limestone and dolostone, and Pennsylvanian sandstone and shale are found in the Springfield Plateau. Throughout the Ozark Highlands and Springfield Plateau, the limestone and dolostone contain horizontal chert beds that are numerous but are generally less than 30 cm thick. Thin, remnant loess deposits also can be found on broad ridges, particularly in the eastern portion of the Ozark Highlands nearest the Mississippi River.

The lithology of the Boston Mountains is more varied. Here nearly level layers of Pennsylvanian and Mississippian shale, sandstone, and siltstone occur. Sandstone, siltstone, and shale also occur in the Arkansas Valley and Ridges, but these are primarily Pennsylvanian in age. The folding and faulting that occurred in the Ouachita Mountains has exposed Ordovician shale and sandstone, Mississippian shale and sandstone, and Pennsylvanian shale, slate, quartzite, and sandstone.

12.2.4 Vegetation

The Interior Highlands occur along an ecotone between the mesophytic forests to the east and the tallgrass prairies to the west. Prior to European settlement, the vegetation comprised a mosaic of oak (*Quercus*) savannas, oak woodlands, pine

Fig. 12.2 The Ozark Highlands include many clear streams. Photograph credit Jaymi LeBrun

(*Pinus*) and oak-pine woodlands, and oak-hickory (Quercus-*Carya*) forests. Common tree species in savannas and woodlands included post oak (*Quercus stellata*), black oak (*Quercus velutina*), shortleaf pine (*Pinus echinata*), and blackjack oak (*Quercus marilandica*). Common tree species in oak-hickory forests included white oak (*Quercus alba*), northern red oak (*Quercus rubra*), and hickories (*Carya* spp.). Prairies and glades were dominated by warm-season grasses such as Indiangrass (*Sorghastrum nutans*), big bluestem (*Andropogon gerardii*), little bluestem (*Schizachyrium scoparium*), and switchgrass (*Panicum virgatum*), particularly in the western portion of the region. Anthropogenic fire along with topography, site conditions, and human population density were important determinants of the vegetation (Guyette et al. 2002). In general, savannas and oak woodlands occurred on the broad to gently rolling plains of the Springfield Plateau and the Ozark Highlands (Hanberry et al. 2014) and in the western portion of the Arkansas Valley and Ridges MLRA (Woods et al. 2004). Oak woodlands, sometimes mixed with shortleaf pine, and oak-hickory forests occurred on the steeper slopes of the more dissected regions of the Ozark Highlands, the St. Francois Knobs, and Boston Mountains (Hanberry et al. 2014, Woods et al. 2004). Oak-pine and pine woodlands occurred in large areas of the eastern Ozark Highlands and in the Boston Mountains.

12.2.5 Soils

The soils of the northern half of the Interior Highlands generally have a mesic soil temperature regime and those of the southern half of the region have a thermic soil temperature regime. Most of the soils are deep or very deep except along steep side slopes where geologic erosion has occurred. The soils are generally well drained unless they have a fragipan or a slowly permeable clay layer that inhibits vertical water movement. The permeability of most of the soils is slow to moderate. Most of the soils have udic or aquic soil moisture regimes and mixed or siliceous mineralogy. About 53 % of the soils are Ultisols, 34 % are Alfisols, and the remaining soils are Inceptisols (6 %), Mollisols (5 %), and Entisols (2 %). The soils vary considerably in depth, texture, and nutrient supply depending on their native vegetation, parent materials, and slope position.

Most of the soils formed under the influence of forest vegetation. Consequently, they have an ochric epipedon comprising a thin A horizon and one or more eluvial horizons overlying an argillic horizon (Alfisols or Ultisols) comprising a series of Bt horizons. Mollisols occur on level plains, particularly along the western boarder of the Interior Highlands where prairie or savanna was the native vegetation. Mollisols also occur adjacent to glades and rock outcrops where few trees are able grow well because of rooting limitations and the native vegetation was dominated by grasses, sedges, and forbs.

Many of the upland soils formed in slope alluvium (also called hillslope sediments or colluvium), residuum, or in a thin layer of slope alluvium overlying residuum. Over time, the downslope movement of the fine-textured sediments has concentrated coarse fragments in upslope positions. The slope alluvium in this region is either highly weathered and/or was derived from sandstone where quartz was the dominant mineral. Consequently, the soils that have slope alluvium as a parent material are gravelly or cobbly with siliceous mineralogy, low cation exchange capacity (CEC), low base saturation, and include Hapludults and Paleudults (Fig. 12.3). Conversely, soils that developed in residuum derived from limestone and dolostone have B horizons with mixed mineralogy, higher clay, and higher CEC than those in the slope alluvium. Part of the calcium and magnesium released during the weathering of the limestone and dolostone was retained in the soil resulting in these soils having a moderate to high base saturation and classifying as Hapludalfs and Paleudalfs (Fig. 12.4).

Where soils are shallow to moderately deep to bedrock, their properties greatly reflect the nature of the underlying lithology. They are clayey, have mixed mineralogy, and a moderate to high base saturation where derived from limestone or shale. Soils are loamy or sandy with siliceous mineralogy and low base saturation where derived from sandstone. For example, in the Boston Mountains, the Arkansas Valley and Ridges, and the Ouachita Mountains MLRAs, Dystrudepts occur on steep side slopes where the depth to the underlying sandstone is less than 50 cm and soil development is minimal. In many locations, sandstone is interbedded with limestone, dolostone, or shale and the mineralogy and soil textures of overlying soils have been influenced by multiple lithologies.

Fig. 12.3 Paleudults in the Interior Highlands formed in weathered slope alluvium. Photograph credit Satchel Gaddie

In upland waterways and in floodplains of this region, the soils reflect the properties of the source material from the surrounding uplands. However, because of their lower landscape position and the deposition of sediments during flood events, the soils on these landforms are enriched with organic carbon and nutrients. The sediments were sorted by the flood water during deposition, resulting in considerable lateral and vertical textural variation. Consequently, the soils of upland waterways and floodplains are variable in texture and organic carbon content and have moderate CEC, mixed mineralogy, and moderate to high base saturation. They include Hapludalfs, Hapludolls, Paleudalfs, Eutrudepts, and Udifluvents.

Fragipans (Fig. 12.5) are common throughout the region and occur along or above lithologic discontinuities in the soil, particularly on gentle slopes where the soils are more stable and soil movement is less likely to disrupt fragipan development. Although their genesis is poorly understood,

Fig. 12.4 Hapludalfs with gravelly slope alluvium overlying clayey residuum from dolostone. Photograph credit Satchel Gaddie

fragipans are thought to have formed where close packing by deposition and desiccation by trees has occurred (James et al. 1995). Many of these soils are Fragiudults or Fragiudalfs.

12.2.6 Land Use

The majority (50 %) of the land in this region is forested (Figs. 12.6 and 12.7), and forest cover ranges from 22 % in the Springfield Plateau to 72 % in the Ouachita Mountains. Harvesting of forest products is an important economic driver throughout the Interior Highlands on both private and public land, and forest products include lumber, cooperage, flooring, poles, posts, ties, pallets, blocking material, charcoal, and bedding material. About one-third of the land is used for forages for beef and dairy industry while the remaining land is split between 3 % water bodies (man-made lakes, rivers, and streams) and 5 % urban development, mostly along the interstate corridors, major highways, and large lakes. There is a small amount of cropland (6 %) primarily growing soybeans and winter wheat. This area contains one of the largest poultry production regions of the USA in northwestern Arkansas, northeastern Oklahoma, and southwestern Missouri.

Fig. 12.5 Fragipans like this one occurring from 75 to 120 cm below the soil surface are common in soils of the Interior Highlands. The scale is shown in *feet*

Fig. 12.6 Forests are an important resource in the Interior Highlands. Photograph credit Stephen Shifley

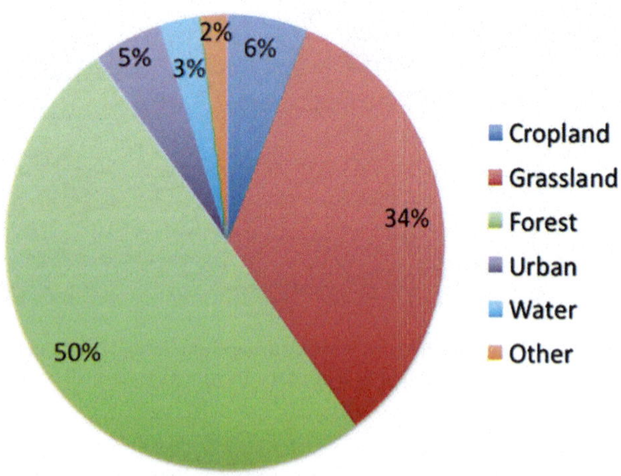

Fig. 12.7 Land use distribution in the Interior Highlands

12.3 Interior Plains

The Interior Plains occur within the central portion of LRR N and includes six MLRAs including the Kentucky and Indiana Sandstone and Shale Hills and Valleys (120A, B, C), Kentucky Bluegrass (121), Highland Rim and Pennyroyal Plateau (122), and Nashville Basin (123). This region comprises 131,720 km^2 in south central Indiana, central and western Kentucky, and central Tennessee (Fig. 12.1).

12.3.1 Climate

The mean annual precipitation ranges from 1040 to 1450 mm increasing to the south. Temperature averages 10–16 °C. The growing season lasts 180 days in the northern part of Kentucky and Indiana Sandstone and Shale Hills and Valleys and about 205 days in the Nashville Basin. Most of the precipitation falls during the freeze-free period and occurs as high-intensity, convective thunderstorms in summer. Snowfall is common in winter but does not remain on the surface for a long period of time.

12.3.2 Physiography

The topography of this region is mostly rolling plains with smaller areas of fairly rugged hills, swampy alluvial valleys (Fig. 12.8), and deeply entrenched rivers and streams in broad karst plains. Several large rivers traverse the region including the Ohio and several of its tributaries including the Tennessee, Cumberland, Kentucky, and Licking rivers. In the western part of the region (MLRAs 120 A, B, and C; Shawnee Hills), sandstone bluffs, steep-sided ridges, gentle hills, broad valleys, karst terrain, and gentle rolling lowland plains and bottomlands along major rivers and their tributaries are common. Elevation ranges from 105 m along the Ohio River to 325 m on high ridges.

In the Highland Rim and Pennyroyal Plateau to the south, elevation generally ranges from 200 to 300 m, and the uplands consist of broad, rolling to gently sloping interstream areas. This karst area is has numerous limestone sinks and the longest cave system (>640 km) in the world, Mammoth Caves. Many of the permanent streams are steeply incised and are few due to the internal drainage of the karst terrain.

The Highland Rim and Pennyroyal Plateau surrounds the Nashville Basin, and at the transition between these two MLRAs, the terrain is deeply dissected and consists of steep slopes between narrow, rolling ridgetops and narrow valleys. The inner part of the Nashville Basin is undulating and rolling with limestone sinks and abundant outcrops of limestone. Elevation is generally about 200 m ranging from 135 m in deep stream channels to 405 m on isolated hills. The terrain in the Kentucky Bluegrass is similar to the Nashville Basin and is gently rolling with isolated hills and

Fig. 12.8 Seasonally inundated cypress swamp. Photograph credit Jerry McIntosh

ridges, limestone sinks, and occasional limestone outcrops. Local relief is about 50–100 m on dissected hills and 25 m on the broad upland plains.

12.3.3 Geology

Most of the region is characterized by erosional landforms on slightly dipping Paleozoic carbonate rocks resulting in a series of dissected plateaus and rolling plains separated by scarps. Karst topography and residuum of variable thickness from the weathering of carbonate rocks is typical of most of the region (Newell 2001). The borders between the MLRAs commonly have abrupt slope changes due to changes in erosion rates between parent rocks. Sedimentary rocks underlying the Shawnee Hills include Early and Middle Pennsylvanian and Late Mississippian-age flat-lying, interbedded sandstone, shale, coal, and siltstone. The Highland Rim and Pennyroyal Plateau, Kentucky Bluegrass, and Nashville Basin are underlain by Ordovician- to Mississippian-age limestone, part of the Cincinnati Arch. Quaternary alluvium is found on valley floors and terraces along rivers and streams.

12.3.4 Vegetation

Oak-hickory (*Quercus-Carya*) forest is found in the uplands of Illinois and Kentucky and oak-gum-cypress (*Quercus-Liquidambar-Taxodium*) forest occupies the bottom lands throughout the section. Uplands are dominated by the white oak, black oak, shagbark hickory (*Carya ovata*) community; the blackjack oak, scarlet oak (*Quercus coccinea*), pignut hickory (*Carya glabra*) community occupies drier sites; and the beech (*Fagus grandifolia*), yellow-poplar (*Liriodendron tulipifera*), bitternut hickory (*Carya cordiformis*), sugar maple (*Acer saccharum*), white ash (*Fraxinus americana*) community occupies deep ravines. The southern flood plains along the Ohio and Wabash rivers are dominated by the sycamore (*Platanus occidentalis*), Kentucky coffeetree (*Gymnocladus dioicus*), sugarberry (*Celtis laevigata*), and honeylocust (*Gleditsia triacanthos*) community, with tupelo (*Nyssa* spp.) and baldcypress (*Taxodium distichum*) swamp communities occupying areas with more poorly drained soils. On undisturbed flood plains sites, Shumard oak (*Quercus shumardii*), sweetgum (*Liquidambar styraciflua*), swamp white oak (*Quercus bicolor*), cherrybark oak (*Quercus pagoda*), pin oak (*Quercus palustris*) as well as sedges and grass meadows are common. In the northern section (MLRA 120C), the protected areas and north-facing slopes support mixed forests of beech, sugar maple, and yellow-poplar.

12.3.5 Soils

The Interior Plains soils have a mesic soil temperature regime, udic or aquic soil moisture regime, and dominantly mixed mineralogy. Over half (58 %) of the soils are Alfisols while the remaining soils are Ultisols (20 %), Inceptisols (15 %), Mollisols (6 %), and Entisols (1 %). The majority of the less eroded uplands and ridges in the Shawnee Hills are covered with a veneer of loess (windblown silts). In the Shawnee Hills, the thickest loess veneer (several meters thick) is south of the Ohio River and it thins the southeast. Due to the abundance of loess in this area, most of the soils have a silt or silt loam surface texture. Underlying the loess is residuum derived from interbedded sandstone, shale, or siltstone. In uplands with thin loess over residuum, the soils range from moderately deep to very deep and are typically Hapludalfs. Most soils are well drained unless the soil has a fragipan that restricts vertical movement of water resulting in moderately well or somewhat poorly drained soils (Fragiudalfs). The fragipan horizons often occur at the lithologic discontinuity between loess and residuum or between two different loess depositions.

The surface loess deposit in this region is the Peoria Silt which was deposited between 27,000 and 17,000 years ago (Muhs et al. 2013) and is associated with the most recent glacial advance in the lower Midwest. Often underlying the Peoria Silt is the Roxana Silt, which was deposited between about 60,000 and 30,000 years ago (summarized in Curry et al. 2011) and is associated with the next-to-last glacial advance. In these loess-covered landscapes, Hapludults and Fragiudults are common on sideslopes and often have a limited plant available water holding capacity due to their shallow depth to a fragipan or horizons with fragic properties.

Within the Interior Plateau, fragipan formation has been described as primarily occurring in loess deposits 0.75–2.5 m thick that overlie residuum (interbedded sandstone and shale), or another loess layer. Soluble silica moves downward with percolating water and accumulates at the lithologic discontinuity. Removal of water from the zone immediately above the lithologic discontinuity by tree roots results in a preferential concentration of amorphous silica. As a result, silica precipitates between grain contacts and in the fine pores thus bonding the soil minerals resulting in horizons with brittle consistence and dense prismatic structure (Franzmeier et al. 1989; Karathanasis 1989).

On younger sediments in alluvial flood plains, soils range from well to poorly drained (Dystrudepts and Endoaquepts). Some of these poorly drained areas with sandy and loamy soil materials can be drained artificially through ditching or tile drainage. The latter is more common in this region and is accomplished by installation of 10–15 cm diameter perforated pipe approximately 0.5–1 m deep in the soil profile to remove free (zero tension) water by gravity. The perforated pipes are connected to a main pipe, often of a larger diameter, that discharges the collected water to a lower elevation surface outlet, often a ditch or creek. These drainage systems have been used in the corn belt of the Midwest for decades but in recent years have become more common in the poorly and somewhat poorly drained alluvial soils of the Interior Plains.

Endoaquepts, Epiaqualfs, and Hapludalfs formed in lacustrine sediments occur on lacustrine terraces or lake plains. Lacustrine sediments and the soils that form from these parent materials are fine textured (e.g., clay, clay loam, and silty clay loam) and are not suitable for drainage due to their high clay content and low saturated hydraulic conductivity, making them difficult to cultivate in wet growing seasons.

In the Highland Rim and Pennyroyal Plateau, the soils are generally well drained due to the lack of a water restrictive horizon, moderate to strong structure, and deep fractures in the underlying limestone making them well suited for row crop production. In the upland areas, soils have generally developed in deep clayey limestone residuum, old alluvium, or thin loess deposits over residuum or alluvium and are deep Paleudalfs and Paleudults (Smalley 1983) (Fig. 12.9). On younger alluvial surfaces near streams and on flood plains, Eutrudepts and Fluvaquents are common.

Soils in the Nashville Basin and Kentucky Bluegrass are similar as they share the same parent material derived from Ordovician-age limestone deposits associated with the Cincinnati Arch. The soils in these regions are moderately well to well drained, loamy or clayey in texture, and formed from residuum or thin loess over residuum. Upland soils are predominantly Hapludalfs, Paleudalfs, or Fragiudalfs while Eutrudepts are found in floodplains. Although most of the soils are Alfisols, there are large areas of Hapludolls in the uplands, and Aquolls are common on floodplains where native prairies once covered the landscape.

Fig. 12.9 Paleudalf (Crider series, State soil of Kentucky) is a well-drained soil formed from limestone residuum found in upland landscape positions of the Pennyroyal Plateau. Photograph credit Jerry McIntosh

12.3.6 Land Use

Much of the Interior Plains is in row crop production (27 %) of corn, soybeans, wheat, or tobacco on the plateaus and other gently sloping terrains as well as on recently drained floodplains and bottomlands (Fig. 12.10). Tobacco (burly and dark tobacco) is a very important crop for this region as Kentucky is the 2nd largest tobacco producer in the USA, and it makes up almost $400 million of Kentucky's Gross State Product (Brown 2007). Forests occupy 36 % of the area and commonly occur in fragmented woodlots on slopes too steep to farm or soils too shallow for row crops. Managed grasslands occupy 23 % of the landscape and

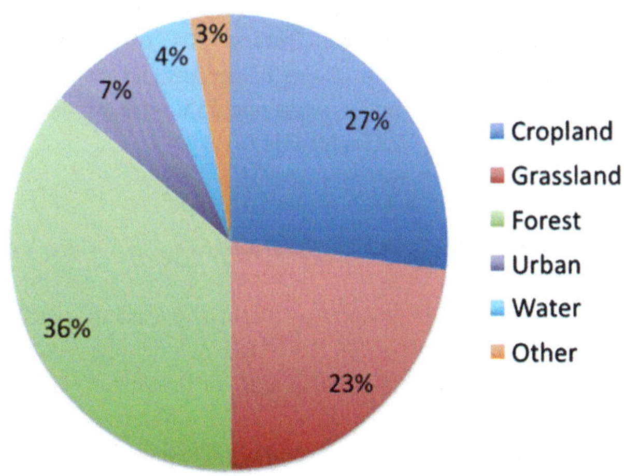

Fig. 12.10 Land use distribution in the Interior Plains

commonly occur on rolling terrain where soils are shallow. These areas are used for hay production or animal grazing operations. Of particular interest are the equine production facilities in the Kentucky Bluegrass, which has the largest

Fig. 12.11 Horses grazing on the Bluegrass, Interior Plains. Photograph credit University of Kentucky

equine sales in the nation (USDA NASS 2012). The success of the equine industry in this region has been attributed, in part, to the phosphorus-enriched limestone associated with the Cincinnati Arch (Harrison and Klotter 1997; McGrain 2001; Kentucky Geological Survey 2012) (Fig. 12.11). Urban areas occupy 7 % of the landscape and are mostly concentrated around major cities including Nashville TN, Louisville KY, and Cincinnati OH.

12.3.7 Appalachian Highlands

The Appalachian Highlands occur in the eastern part of LRR N and western part of LRR S and include nine MLRAs across 4 provinces: Appalachian Plateau Province (MLRAs 124 Western Allegheny Plateau, 125 Cumberland Plateau and Mountains, 126 Central Allegheny Plateau, 127 Eastern Allegheny Plateau, 129 Sand Mountain), Blue Ridge Province (MLRAs 130A and B Northern and Southern Blue Ridge Mountains), Valley and Ridge Province (MLRAs 128 Southern and 147 Northern Appalachian Ridges and Valleys), and Piedmont Province (MLRA 148 Northern Piedmont). The region comprises 358,115 km^2 in eastern Kentucky, eastern Ohio, southern Pennsylvania, West Virginia, western Maryland, western North Carolina, Eastern Tennessee, northern Alabama and Georgia, southeastern Pennsylvania, and west-central New Jersey (Fig. 12.12).

12.3.8 Climate

The average annual precipitation ranges from 915 mm to as high as 3025 mm at the highest elevations in the Blue Ridge Mountains. Orographic precipitation occurs in the higher elevations of the Appalachian Highlands due to moist air from the Gulf of Mexico being too heavy to cross the mountain range from west to east resulting in areas of high precipitation (>1900 mm) in some areas. Precipitation is lowest in autumn and generally increases with elevation. Snow is common in the high elevations of the northern latitudes; however, it melts quickly. In general, the fall is the driest season. The average annual temperature ranges from 8 to 17 °C and decreases with increased latitude and elevation.

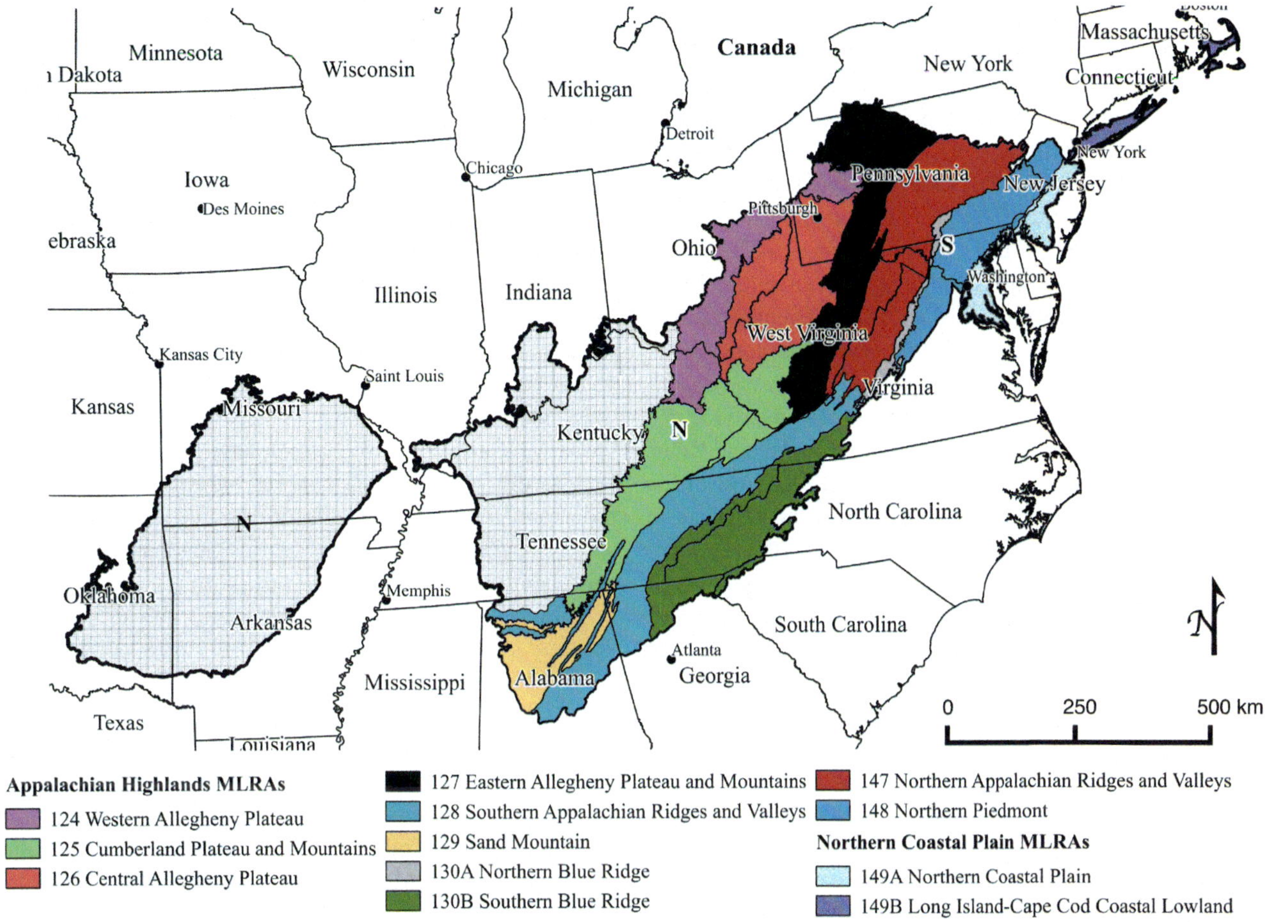

Fig. 12.12 Extent and distribution of MLRAs in the Appalachian Highlands (MLRAs 124, 125, 127, 128 130A, 130B, 128, 147, 148) and the Northern Coastal Plains (MLRAs 149A and 149B) within LRRs N and S

12.3.9 Physiography

Within the western part of this region, the Appalachian Plateau has narrow, level valley floors, rolling ridgetops, and hilly to steep ridge slopes. Elevations range from 200 m on the Ohio River flood plain to 1200 m in the mountains. Local relief is about 50–100 m. To the east, the Allegheny Front marks the termination of the plateau on the edge of the Appalachian Ridge and Valley Province. The Valley and Ridge Province is an area of folded and faulted parallel ridges and valleys that are carved in anticlines, synclines, and thrust blocks. Parallel sandstone and shale ridges are separated by narrow to moderately broad limestone and shale valleys. The ridges are strongly sloping to extremely steep and have narrow, rolling crests. The valleys are mainly level to strongly sloping. Elevations range from 100 m in the valleys to 800 m on the ridges and mountains. The Blue Ridge Mountains are rugged with steep slopes, sharp crests and narrow valleys. Streams have deeply dissected the landscape, flowing through gorges and gaps in the mountains. Broad valleys, basins, and rolling hills are extensive throughout the area. Elevations range from 250 m along the Potomoc River to 2010 m on Mt. Mitchell at the crest of the Great Smoky and Black Mountain Ranges (USDA 2006).

12.3.10 Geology

The Appalachian Highlands resulted from tectonic activity through the collision of tectonic plates from 480 to 220 million years ago. At about 480 million years ago, the oceanic plate began sinking underneath the North American continental plate below the current location of the Appalachian Mountains. Mountain building collisions continued including collisions of continental masses ancestral to North America and Africa about 270 million years ago. As a result, there is considerable metamorphism of the rocks, and in some places, igneous plutons formed and are scattered throughout the region. These igneous plutons are more resistant to weathering than surrounding rocks and are elevated relative to the surrounding less resistant metamorphic or sedimentary rocks at the surface (e.g., Looking Glass Rock).

The bedrock geology in the Blue Ridge Mountains consists mostly of Precambrian metamorphic rock formations with a few small bodies and of igneous and sedimentary rocks. The degree of metamorphism varies but generally decreases westward from gneiss, schist, and amphibolite to interbedded metasandstone, slate, phyllite, metasiltstone, and metaconglomerate. To the west of the Blue Ridge Mountains, the Valley and Ridge Province formed from tilted, Cambrian to Pennsylvanian aged sedimentary strata with varying resistance to erosion. The differential erosion resistance of the parent rocks is reflected in the topography with resistant sandstone and conglomerate bedrock on ridges, and less resistant shale and limestone underlying the valleys. The topographic orientation of the Valley and Ridge Province is dominantly northeast to southwest. West of the Valley and Ridge Province lies the Cumberland Plateau to the south and Allegheny Plateaus to the north. Both are underlain by relatively planar cyclic beds of sandstone, siltstone, clay, shale, and coal of Pennsylvanian age. The major river valleys are filled with unconsolidated deposits of clay, silt, sand, gravel, with outwash and glaciofluvial deposits in the northwestern part of the area in Pennsylvania. The lower slope positions of most hills are covered with a layer of colluvium.

The northern Piedmont geology includes an ancestral Atlantic Ocean basin with Triassic sandstone, shale, and conglomerates similar to those found in the western British Isles. The Piedmont geology is the result of the mountain building processes associated with a subduction zone where the oceanic crust slid under the continental crust. The resulting rocks in the Piedmont are metamorphic and are less resistant to erosion than some of the igneous rocks found in the Blue Ridge Mountains to the west. The northern boundary of the northern Piedmont marks the terminus of the Wisconsinan glaciers, but earlier periods of glaciation extend further south into New Jersey and eastern Pennsylvania.

12.3.11 Vegetation

This vast area supports deciduous forest vegetation. White oak, red oak, black oak, hickories (*Carya* spp.), and associated upland hardwoods are the major species. Scarlet oak, chestnut oak (*Quercus montana*), hickories along with scattered Virginia pine (*Pinus virginiana*), shortleaf pine, and white pine (*Pinus strobus*) grow on dry ridges and on the shallower soils. Yellow-poplar, black walnut (*Juglans nigra*), northern red oak, red maple (*Acer rubrum*), and other species requiring abundant water grow in sheltered coves, on footslopes and on north-facing slopes. Shagbark hickory, bitternut hickory, pignut hickory, and mockernut hickory (*Carya tomentosa*) also occur on these landforms. Oaks, blackgum (*Nyssa sylvatica*), flowering dogwood (*Cornus florida*), sassafras (*Sassafras albidum*), Virginia pine, pitch pine (*Pinus rigida*), and shortleaf pine grow mostly on ridgetops. Willows (*Salix* spp.), sycamore, sweetgum, and river birch (*Betula nigra*) grow on flood plains. In the Northern Piedmont, black walnut and black cherry (*Prunus serotina*) grow on the well-drained floodplain soils.

At higher elevations in the Blue Ridge Mountains, above 1525 m, red spruce and Fraser fir are the dominant tree species. In some areas at high elevation, grassy and heath "balds" are evident. These large meadows are dominated by grass species and are home to rare shade-intolerant plant species. Heath balds which often have organic, low pH or excessively drained soils support many shrubs including mountain laurel (*Kalmia latifolia*), rhododendron (*Ericaceae* spp.), hawthorn (*Crataegus* spp.), blueberry (*Vaccinium* spp.), and sandmyrtle (*Kalmia buxifolia*).

12.3.12 Soils

Most of the region's soils have a mesic temperature regime. However, the soils in the Blue Ridge Mountains above 1280 m have a frigid soil temperature regime. The soil moisture regime is udic, and the soils have mixed or siliceous mineralogy. Most of the soils are shallow in the uplands and deep to very deep in the valleys. Many are excessively drained due to coarse textures and fractured parent materials while others are somewhat poorly drained with fine textures and are skeletal. Erosion is a concern throughout the region on steep slopes. Most of the soils in the region are Ultisols (50 %), Inceptisols (28 %), and Alfisols (18 %). Entisols, Mollisols, Spodosols, Vertisols, and Histosols make up the remaining 4 % of the soils in this region. Hapludults are most common (33 %) and found in a variety of environments ranging from rolling and steep terrain on the Cumberland Plateau to valley floors in the Valley and Ridge Province. More deeply developed Paleudults (7 %) are common throughout the region, particularly in the Appalachian Ridges and Valleys where they form on areas underlain by limestone, on the lower footslopes, and in valleys. Fragiudults (6 %) are found throughout this region, particularly in the Eastern Allegheny Plateau and Northern Appalachian Ridges and Valleys.

Dystrudepts (26 %) formed from residuum or colluvium are common on hills, mountains, and ridges and can also be found on alluvium along major streams. Hapludalfs (15 %) formed on hillsides from parent materials with a high base saturation including limestone, calcareous shales, and colluvium.

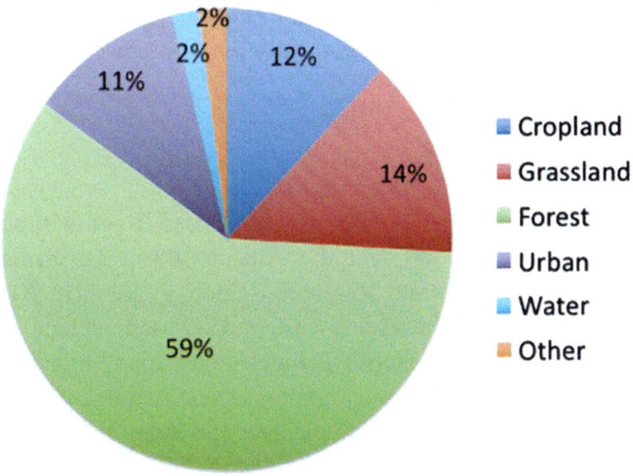

Fig. 12.13 Land use in the Appalachian Highlands

12.3.13 Land Use

Land use in the Appalachian Highlands is predominantly forest (59 %) (Fig. 12.13). Cropland and managed grasslands make up 26 % of the land area. A small amount of land is being used for coal extraction via subsurface and surface mining operations. The northern part of this region includes three of the top five coal producing states that produce 25 % of the coal produced in the USA (US Energy Information Administration 2013). With recent advances in horizontal well drilling coupled with hydraulic fracking, gas production from shale deposits in the northeastern part of this region is projected to be 115,000 m^3 day^{-1} in 2020 (Considine et al. 2009).

12.4 Northern Coastal Plain

The Northern Coastal Plains occur in the eastern portion of LRR S and include two MLRAs: Northern Coastal Plains (MLRA 149A) and Long Island-Cape Cod Coastal Lowland (MLRA 149B). Much of this area is in eastern Maryland, southwest and central New Jersey, Long Island, New York, and Cape Cod, Massachusetts (Fig. 12.12).

12.4.1 Climate

Average annual precipitation ranges from 1015 to 1220 mm. Most of the precipitation falls as high-intensity thunderstorms during summer. In the coastal plain, the seasonal snowfall ranges from little to none in the southern part of the region to as much as 750 mm in the northern part. The average temp is 10–14 °C. Due to the low elevation, areas adjacent to the coast and in river valleys are subject to storm surges associated with hurricanes.

12.4.2 Physiology

The Coastal Plain is low and partially submerged with many islands and marshes along the coast. The province consists of a series of south easterly dipping layers of unconsolidated sands and clays. Eroding streams have dissected the area, leaving a series of terraces across the landscape, which are most prevalent away from the coast where the terrain is more rolling. Nearest the coast the area is nearly level. Elevation ranges from sea level to 100 m, but local relief is 2–10 m. The north and west boundary between the Coastal Plain and the Piedmont Plateau has been identified as the fall line. This is an ill-defined boundary between the crystalline rocks of the Piedmont and the unconsolidated sediments of the Northern Coastal Plain. At the fall line, streams descend into the easily eroded unconsolidated sand and clay sediments of the Northern Coastal plains (Maryland Department of State Planning 1973).

12.4.3 Geology

Unconsolidated sediments deposited in the near-shore environment of late Cretaceous seas as they transgressed and regressed underlie most of this area. These materials overlay the rocks of the eastern Piedmont and extend submerged for another 125 km across the Atlantic Continental Shelf. In the northern part of this region, the southern extent of the Wisconsinan glacial advance is marked by remnants of moraines in the central lowlands of New Jersey, Long Island, and Cape Cod and sandy outwash plains that extend into the Atlantic Ocean. Sand dunes and tidal marshes are extensive along the coastline.

12.4.4 Vegetation

This area has a mixture of oak-hickory (*Quercus-Carya*) and other hardwoods, white pine-red pine (*Pinus strobus-Pinus resinosa*) forest, and pine-oak woodlands. Of particular interest in the region is the "pine barrens," which is an expansive area that is dominated by pitch pine. These forests are found in Long Island on sandy outwash and on undeveloped stretches of the bay side of barrier islands on the Atlantic shore. Upland forests in this region are characterized by a continuous gradient from pure pitch pine stands where the soils are sandy and nutrient poor at one end that grade into pure oak stands at the opposite end where soils are loamy and have a greater nutrient supply (McCormick 1979). Under the forest overstory, well-developed understories are dominated by scrub oak (various species of *Quercus*), mountain laurel, huckleberry (*Gaylussacia* spp.), and blueberry.

12.4.5 Soils

Over half of the soils in this region are Ultisols (58 %) while the majority of the remaining soils are Entisols (19 %), Inceptisols (15 %), and Histosols (5 %) with minor amounts of Spodosols (2 %) and Alfisols (1 %). The soils have a mesic soil temperature regime, and aquic or udic soil moisture regime, and mixed, siliceous, or glauconitic mineralogy. The soils are very deep, excessively to very poorly drained, and primarily loamy or sandy in texture. Hapludults are the most abundant Great Group found in the region and have formed in fluviomarine deposits on terraces, flats, near-shore marine deposits, sandy eolian deposits and silty loess deposits. Quartzipsamments occur on landscapes with thick eolian or marine sand deposits on dunes and flats along streams on sandy outwash plains. Dystrudepts, very deep and well-drained soils, are common in the glacial, coarse-textured outwash deposits. Haplosaprists formed in freshwater bogs and along stream corridors while Sulfihemists formed in organic deposits in estuarine and tidal marshes.

12.4.6 Land Use

The largest land use in this area is urban development (40 %, Fig. 12.14) in and surrounding large cities including Washington D.C., Baltimore, and New York City. Outside of these urban areas, much of the cropland (9 % of region) supports these urban areas by growing vegetables, corn, soybeans, small grains, and fruits. Turf farms and nurseries also serve the expanding urban sector. Forests occupy 23 % of the area. In the larger forested areas, pine pulpwood and hardwood lumber are the principle forest products. The coastal plain sediments are a source of groundwater for the most urbanized region of the USA. Erosion and nutrient runoff related to urban development and agriculture have degraded streams and subsequently the Chesapeake Bay.

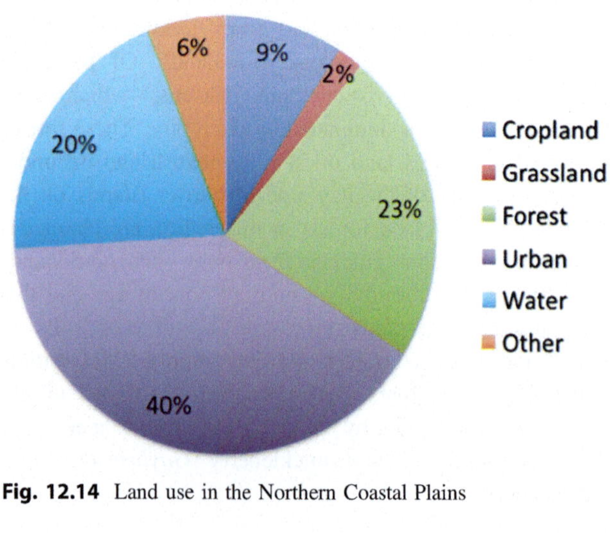

Fig. 12.14 Land use in the Northern Coastal Plains

12.5 Summary

The East and Central Farming and Forest Region (Land Resource Region N) and the Atlantic Basin Diversified Farming Region (Land Resource Region S) comprise 718, 550 km^2 in the central and eastern USA and include the Interior Highlands, Interior Plains, Appalachian Highlands, and the Northern Coastal Plains. These regions include nearly level to gently rolling plains in the Interior Plains and the Northern Coastal Plains and rugged hills and mountains in the Interior Highlands and Appalachian Highlands. The climate is mild with mean annual temperature ranging from 8 to 17 °C and rainfall is ample ranging from 915 to 3015 mm, with the greatest amount of precipitation occurring in the mountains in the Appalachian Highlands. The underlying bedrock consists of sedimentary rocks including limestone, dolostone, sandstone, siltstone, and shale in the Interior Highlands and in the Interior Plains, metamorphic rocks including gneiss, schist, amphibolite, and slate in the Appalachian Highlands, and unconsolidated coastal plain sediments and lesser amounts of glacial till and outwash in the Northern Coastal Plains. Diverse hardwood forests comprising several species of oaks and hickories occur throughout these regions. Yellow-poplar, black walnut, and northern red oak occur in sheltered coves and north-facing slopes of the Interior Plains and Appalachian Highlands. Pines occur on the drier slope positions or on coarse-textured soils except in the Interior Plains. Ultisols are the single most extensive soils of Land Resource Regions N and S comprising more than 40 % of the land area. Alfisols are the second most extensive soils (29 %) in Land Resource Region N and Inceptisols are the second most extensive soils (34 %) of Land Resource Region S. Other important soil orders include Entisols and Mollisols. Forests are the single most important land use followed by cropland and grassland. Urban land is more extensive near major metropolitan areas of Washington D.C., Baltimore, and New York City in the Northern Coastal Plains.

References

Brown LE (2007) Kentucky tobacco facts. USDA NASS Fact Sheet. http://www.nass.usda.gov/Statistics_by_State/Kentucky/Publications/Pamphlets/tobacco06.pdf. Accessed 16 October 2014

Considine T, Watson R, Entler R, Sparks J (2009) An emerging giant: prospects and economic impacts of developing the Marcellus Shale natural gas play. The Pennsylvania State University. http://marcelluscoalition.org/wp-content/uploads/2010/05/

EconomicImpactsofDevelopingMarcellus.pdf. Accessed 20 October 2014

Curry BB, Grimley DA, McKay ED (2011) Quaternary glaciation in Illinois. In: Ehlers J, Gibbard PL, Hughes PD (eds) Quaternary glaciations-extent and chronology: a closer look. Elsevier, Amsterdam, pp 467–485

Franzmeier DP, Norton LD, Steinhardt GC (1989) Fragipan formation in loess of the midwestern United States. In: Smeck NE, Ciolkosz EJ (eds) Fragipans: their occurrence, classification and genesis. SSSA Special Publication, vol 24. Soil Science Society of America, Madison, pp 69–97

Guyette RP, Muzika R, Dey DC (2002) Dynamics of an anthropogenic fire regime. Ecosystems 5:472–486

Hanberry BB, Kabrick JM, He HS (2014) Densification and state transition across the Missouri Ozarks landscape. Ecosystems 17(1): 66–81

Harrison LH, Klotter JC (1997) A new history of Kentucky. University Press of Kentucky, Lexington

James HR, Ransom MD, Miles RJ (1995) Fragipan genesis in polygenetic soils on the Springfield Plateau of Missouri. Soil Sci Soc Am J 59:151–160

Karathanasis AD (1989) Solution chemistry of fragipans—thermodynamic approach to understading fragipan formation. In: Smeck NE, Ciolkosz EJ (eds) Fragipans: their occurrence, classification and genesis. SSSA Special Publication, vol 24. Soil Science Society of America, Madison, pp 113–139

Kentucky Geological Survey (2012) Phosphate mining. https://www.uky.edu/KGS/im/phosphate.htm. Accessed 22 May 2015

Maryland Department of State Planning (1973) Natural soil groups and the Maryland generalized land use plan. Maryland Dept. of State Plan, HUD Proj. Num. MD-P-1008-100

McCormick J (1979) The vegetation of the New Jersey Pine Barrens. In: Forman RTT (ed) Pine Barrens: ecosystem and landscape. Academic Press, New York, pp 229–243

McGrain P (2001) Contributions to the economic geology of Kentucky. In: McDowell RC (ed) The geology of Kentucky—a text to accompany the geologic map of Kentucky. USGS Prof. Paper 1151-H. http://pubs.usgs.gov/prof/p1151h/economic.html. Accessed 22 May 2015

Muhs DR, Bettis EA, Roberts HM, Harlan SS, Paces JB, Reynolds RL (2013) Chronology and provenance of last-glacial (Peoria) loess in western Iowa and paleoclimatic implications. Quat Res 80:468–481

Newell WL (2001) Contributions to the geology of Kentucky: physiography. USGS Publication. http://pubs.usgs.gov/pp/p1151h/physiography.html. Accessed 15 Aug 2014

Smalley GW (1983) Classification and evaluation of forest sites on the Eastern Highland Rim and Pennyroyal. USDA Forest Service General Tech. Rep. SO-43, New Orleans, LA

United States Department of Agriculture (2006) Land resource regions and major land resource areas of the United States, the Caribbean, and the Pacific Basin. United States Department of Agriculture Handbook 296. http://www.nrcs.usda.gov/wps/portal/nrcs/detail/soils/survey/?cid=nrcs142p2_053624. Accessed 6 June 2015

United States Department of Agriculture National Agricultural Statistics Service (2012) Census of agriculture. http://www.agcensus.usda.gov/Publications/2012/. Accessed 22 May 2015

United States Energy Information Administration (2013) Latest coal production by state (2013). www.eia.gov. Accessed 20 October 2014

Woods AJ, Foti TL, Chapman SS, Omernik JM, Wise JA, Murray EO, Prior WL, Pagan J, Comstock JA, and Radford M (2004) Ecoregions of Arkansas (2 sided color poster with map, descriptive text, summary tables, and photographs) U.S. Geological Survey, Reston, VA. Scale 1:1,000,000

13 Soils of the Southeastern USA: LRRs O, P, and T

L.T. West, J.N. Shaw, and E.P. Mersiovsky

13.1 Introduction

The region addressed in this chapter is the southeastern USA, excluding peninsular Florida, which includes LRR O, Mississippi Delta Cotton and Feed Grains Region; LRR P, South Atlantic and Gulf Slope Cash Crops, Forest, and Livestock Region; and LRR T, Atlantic and Gulf Coast Lowland Forest and Crop Region (Figs. 13.1 and 13.2). The dominant soil order across LRRs P and T is Ultisols, which reflects the abundance of acid parent materials and relatively old landscapes that comprise these LRRs. Although Ultisols are dominant, soils across these LRRs have a great diversity of properties both within the Ultisols and among soils in other orders including Entisols, Inceptisols, Alfisols, Spodosols, and Vertisols. In contrast to LRRs P and T, LRR O is comprised of broad alluvial valleys with soils developed in alluvium deposited by the Mississippi, Arkansas, and Red Rivers. These alluvial valleys are dominated by soils in the Entisols, Inceptisols, Alfisols, and Vertisols orders with properties that reflect the composition, age, and type of the alluvial landforms on which they have developed.

The objective of this chapter is to provide a general overview of the genesis and properties of soils in the southeastern USA as related to the state factors of soil formation; climate, biota, parent material, relief, and age. Soil degradation, vulnerability to degradation, and other changes related to human manipulation are also discussed for selected regions. These discussions are not comprehensive, however, and the reader is referred to the references for additional details and discussion.

L.T. West retired from USDA-NRCS

L.T. West (✉)
USDA-NRCS, Fayetteville, AR, USA
e-mail: larrywestar@gmail.com

J.N. Shaw
Auburn University, Auburn, AL, USA
e-mail: shawjo1@auburn.edu

E.P. Mersiovsky
USDA-NRCS, Little Rock, AR, USA
e-mail: edgar.mersiovsky@ar.usda.gov

L.T. West et al. (eds.), *The Soils of the USA*, World Soils Book Series,
DOI 10.1007/978-3-319-41870-4_13

13.2 General Characteristics of the Region

The climate over this region is hot and humid and is classified as temperate to subtropical. It is characterized by long, hot summers and short, mild winters. Mean annual precipitation ranges from about 1100 to 1600 mm and falls almost entirely as rain. Precipitation is equally distributed throughout the year although fall months are driest. Precipitation mostly comes with frontal storms in late fall, winter, and early spring; as high-intensity, short-duration convective storms during the growing season; and as intense rainfall produced by tropical systems. Soil moisture regimes are udic and aquic. Mean annual air temperature across the region ranges from 12 to 22 °C. Thus, most of the region is in a thermic soil temperature regime. Northern Virginia and more northern areas are in a mesic soil temperature regime. The extreme southern parts of the region in Texas and Louisiana are in a hyperthermic temperature regime.

The region has a wide diversity of soil parent materials ranging from Holocene alluvium in alluvial valleys and near-shore coastal environments to saprolites formed from Precambrian metamorphic and igneous rocks in the Southern Piedmont. None of the region was glaciated during the Pleistocene. Thus, many landscapes in the region are relatively old, and most upland soils are in advanced stages of development. Superimpose on this a warm, humid environment, and many soils of this region exhibit morphological, chemical, and mineralogical characteristics commonly associated with highly weathered soils. As a whole, the region is best described as gently sloping to rolling with moderate local relief. Coastal areas and alluvial valleys are exceptions to this generality with low-relief landscapes that are commonly described as flat. Native vegetation over most of the region was mixed hardwood and coniferous forests, although there were extensive areas of savannah-like Longleaf pine (*Pinus palustris* Miller)—wiregrass (*Aristida stricta* Michx) ecosystems and scattered areas of native grasslands.

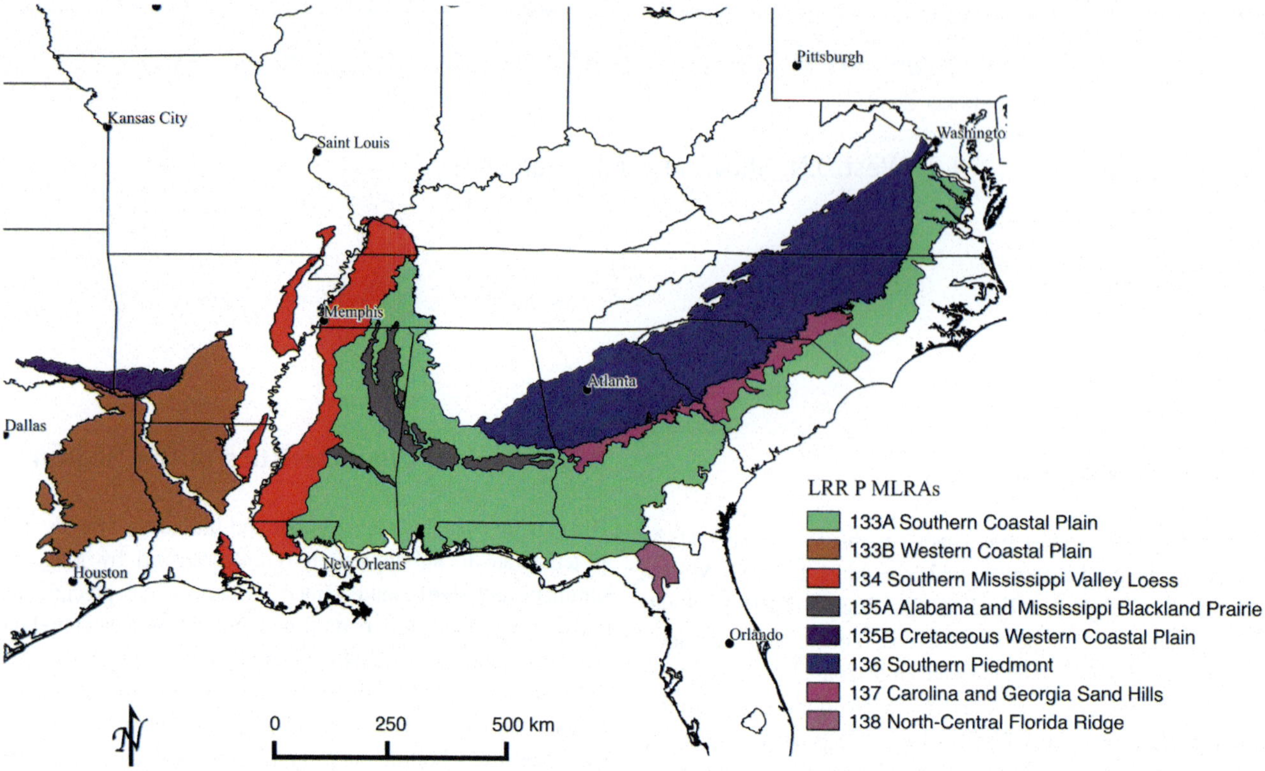

Fig. 13.1 Distribution of MLRAs in LRR P. MLRAs are Southern Coastal Plain, 133A; Western Coastal Plain, 133B; Southern Mississippi Valley Loess, 134; Alabama and Mississippi Blackland Prairie, 135A; Cretaceous Western Coastal Plain, 135B; Southern Piedmont, 136; Carolina and Georgia Sand Hills, 137; and North-Central Florida Ridge, 138

13.3 South Atlantic and Gulf Slope Cash Crops, Forest, and Livestock Region, Land Resource Region P (MLRAs 133A, 133B, 134, 135A, 135B, 136, 137, 138)

13.3.1 Southern Piedmont (MLRA 136)

The Southern Piedmont extends from Virginia southward to central Alabama (Fig. 13.1). It is bounded to the west and north by the Blue Ridge Mountains and similar landscapes with higher relief and to the east and south by the Southern Coastal Plain. The Piedmont is well dissected with rolling to hilly uplands and a well-defined drainage pattern. Stream valleys are generally narrow and are a minor component of the landscape. Maximum relief in the region is generally <350 m though higher isolated peaks occur (Murray 1961). The region slopes to the east and south with general surface slope being about 4 m km^{-1}. Uplands have commonly accordant summits constituting a broad plateau-like surface. The lack of topographic expression on the interfluves has been interpreted to suggest that the Piedmont is a peneplain (Thornbury 1965). Alternately, the accordant summits have been attributed to long-term weathering and uplift in the region (Pavich 1985, 1986).

The majority of the Southern Piedmont is underlain by metamorphic and igneous rocks ranging in age from Precambrian to late Paleozoic (Hack 1989). The parent rocks are dominantly felsic, but areas of intermediate and mafic rocks occur, often in intimate mixtures with the felsic country rock. In addition, mafic dykes intrude into more acidic rocks altering parent material composition and complicating soil patterns. Over most of the region, soils are underlain, on average, by 15–20 m of saprolite weathered from subjacent hard, crystalline rock (Hack 1989).

There are two areas of contrasting rock types within the Piedmont with soils that are considerably different from those in the rest of the region. The first is a series of scattered down-faulted basins that occur in South Carolina, North Carolina, and Virginia. These basins have less relief than most of the region and are underlain by sedimentary sandstone, siltstone, mudstone, and shale of Triassic and Jurassic age (Hack 1989; Daniels et al. 1999). The second is generally referred to as the Carolina Slate Belt, which is an area of low relief containing Cambrian and Precambrian slate, tuff,

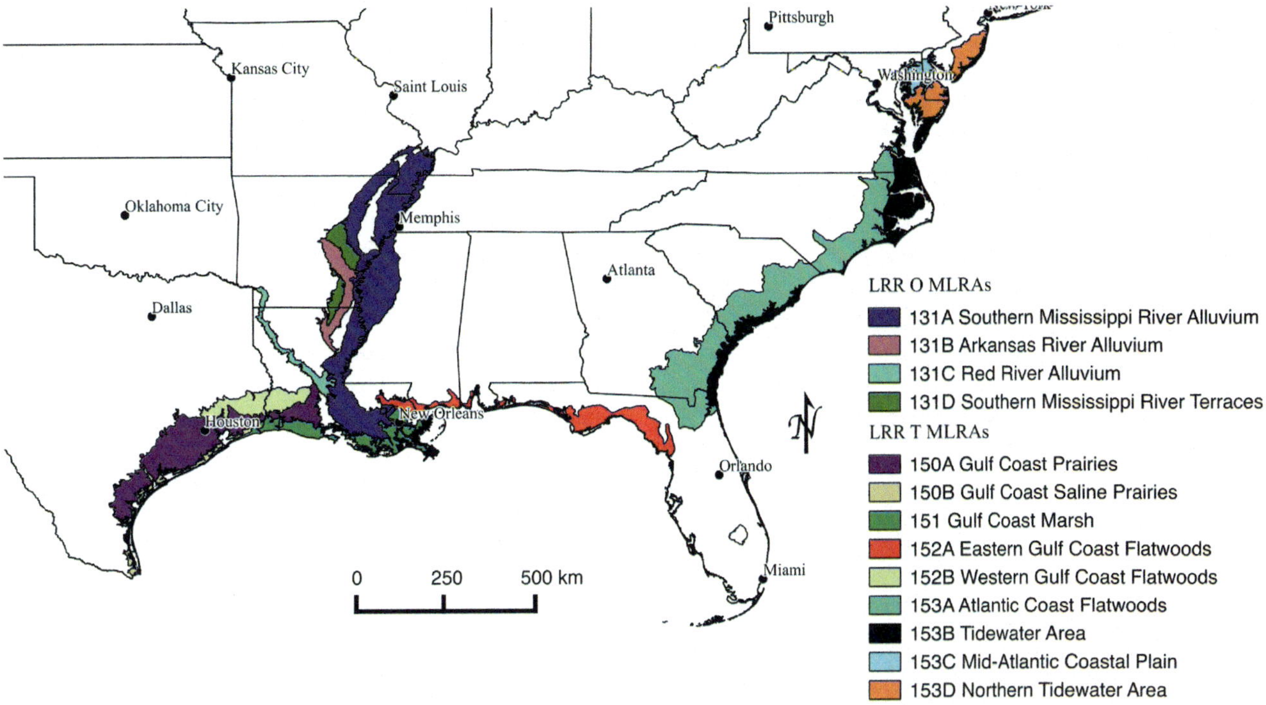

Fig. 13.2 Distribution of MLRAs in LRRs O and T. MLRAs in LRR O are Southern Mississippi River Alluvium, 131A; Arkansas River Alluvium, 131B; Red River Alluvium, 131C; and Southern Mississippi River Terraces, 131D. MLRAs in LRR T are Gulf Coast Prairies, 150A; Gulf Coast Saline Prairies, 150B; Gulf Coast Marsh, 151; Eastern Gulf Coast Flatwoods, 152A; Western Gulf Coast Flatwoods, 152B; Atlantic Coast Flatwoods, 153A; Tidewater Area, 153B; Mid-Atlantic Coastal Plain, 153C; and Northern Tidewater Area, 153D

and other low-grade metamorphic rocks of volcanic origin (Hack 1989).

13.3.1.1 Soil properties

Most upland soils are deep and well drained, and most are at least moderately permeable. Surface horizons are typically <20 cm thick and brown and have loamy textures and low organic matter content (Fig. 13.3). Moderately to severely eroded areas have thinner, redder, more clayey surface horizons due to tillage mixing with subsoil horizons. Upper subsoil horizons typically have clayey textures, and clay content decreases with depth into loamy BC and saprolite C horizons (Fig. 13.4). Subsoil color of well-drained upland soils is commonly red to yellowish red (Fig. 13.3), reflecting an abundance of Fe oxide and oxyhydroxide minerals, hematite and goethite, formed as Fe is released from biotite and other Fe bearing primary minerals as soils develop (Schwetmann 1993; Bigham et al. 2002).

Soils developed over felsic saprolite are typically acidic, have low base saturation and kaolinitic mineralogy. Clay content increases from A and E to Bt horizons (Fig. 13.4; argillic horizon), and most soils in the region are Ultisols. Solum thickness varies and along with subsoil color is a common basis for separation among series. Clay mineralogy is dominated by kaolinite. Hydroxy-interlayered vermiculite and similar clay minerals are common in upper subsoil horizons, and gibbsite is often present in subsoils (Shaw et al. 2010; Calvert et al. 1980). The abundance of kaolinite and other low-activity clay minerals results in subsoils having low cation exchange capacity (CEC), characteristics definitive of kandic horizons (CEC:clay and ECEC:clay ratio ≤ 16 and 12, respectively) and low shrink–swell potential. The dominant great group is Kanhapludults, which covers about 60 % of the MLRA.

Soils developed from intermediate and mafic saprolite typically have higher pH, CEC, base saturation, and lower Al saturation than the felsic-derived soils due to the greater content of basic cations in parent materials. The clay fraction of these soils often has more abundant smectite and other active 2:1 clay minerals (mixed mineralogy) resulting in lower K_s, and higher shrink–swell than soils formed in felsic saprolite. Many of the mafic-derived soils are Alfisols (Hapludalfs) with mixed or smectitic mineralogy although Ultisols (Hapludults) are also common (Fig. 13.5).

Within the Carolina Slate Belt, interfluves are irregular, and sharp topographic breaks are common. Deep soils generally occupy more gently sloping parts of the region, and soils <1 m to hard bedrock occur on convex parts of the landscape (Daniels et al. 1999). The fine grain size of the rocks in this region results in soils with higher silt and very

Fig. 13.3 Typic Kanhapludult from Southern Piedmont, MLRA 136. Sandy loam surface horizon overlying red clay Bt horizon at about 20 cm. Photo credit: John A Kelley, Soil Science@NC State

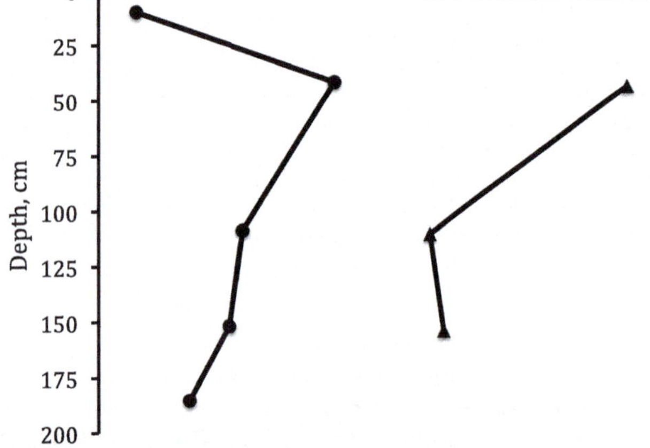

Fig. 13.4 Clay and saturated hydraulic conductivity (K_s) distribution with depth for a Typic Kanhapludult in the Southern Piedmont. Note that the horizon with highest clay is the horizon with maximum hydraulic conductivity

fine sand contents than the rest of the Piedmont. Surface textures are generally silty and subsoil textures range from silty clay loam to clay. As in other Piedmont soils, the soils are acidic and have low base saturation, and the dominant clay mineral is kaolinite. Many of the soils on these landscapes, however, have >10 % smectite and other 2:1 clays and are dominantly Hapludults with mixed mineralogy.

The Triassic basins are topographically lower and have less relief than the adjacent Piedmont landscapes. The Triassic rocks include sedimentary shale, sandstone, mudstone, siltstone, and conglomerate. Igneous diabase intrusions are common. These rock types are more erodible than the surrounding crystalline rocks, which contributes to the topographic low nature of the basins (Daniels et al. 1999). Most soils in this region are moderately permeable and well drained and have dominantly kaolinitic clays, although soils developed from diabase have more smectite and higher base saturation. The dominant great group in the region is Hapludults.

Soils in typically narrow floodplains of the Piedmont have variable texture depending on the depositional environment/stream valley geomorphic component on which they occur. Chemical properties of the alluvial soils reflect the properties of the soils from which the sediment originated. Thus, most are acid and have low organic matter content and low base saturation. Well-drained and moderately well-drained soils are common in floodplains although poorly drained soils often occur at intersections of backwater fluvial environments and toeslopes of uplands.

Soils in the Piedmont are often referred to as "highly weathered" because of their red clayey subsoils, low pH and base saturation, dominance of kaolinitic and hydroxyl-interlayered clays, and common occurrence of gibbsite and iron oxides (sesquioxides) in subsoil clays; properties often associated with advanced soil development (Shaw et al. 2010; Buol et al. 2011). While most upland landscapes in the Piedmont are relatively old compared with those in glaciated regions, they are much younger than the underlying saprolite (Pavich 1985, 1986), and stable landscape surfaces have been reported to have an age of 100,000 years or less (Schroeder et al. 2001). The color, texture, and mineralogical composition of felsic-derived soils in the region reflect the abundance of K-feldspar, biotite mica, and quartz in the acidic saprolite. Kaolinite and gibbsite are early weathering products of K-feldspar and biotite in the parent rocks (Calvert et al. 1980) and have been reported to be abundant in weakly developed soils over similar parent materials in the Blue Ridge Mountains (Norfleet and Smith 1989; Graham and Boul 1990).

Piedmont soils developed from felsic parent materials typically have the highest clay content in upper Bt horizons, and clay decreases with depth through lower solum Bt and transitional BC horizons to a minimum in underlying

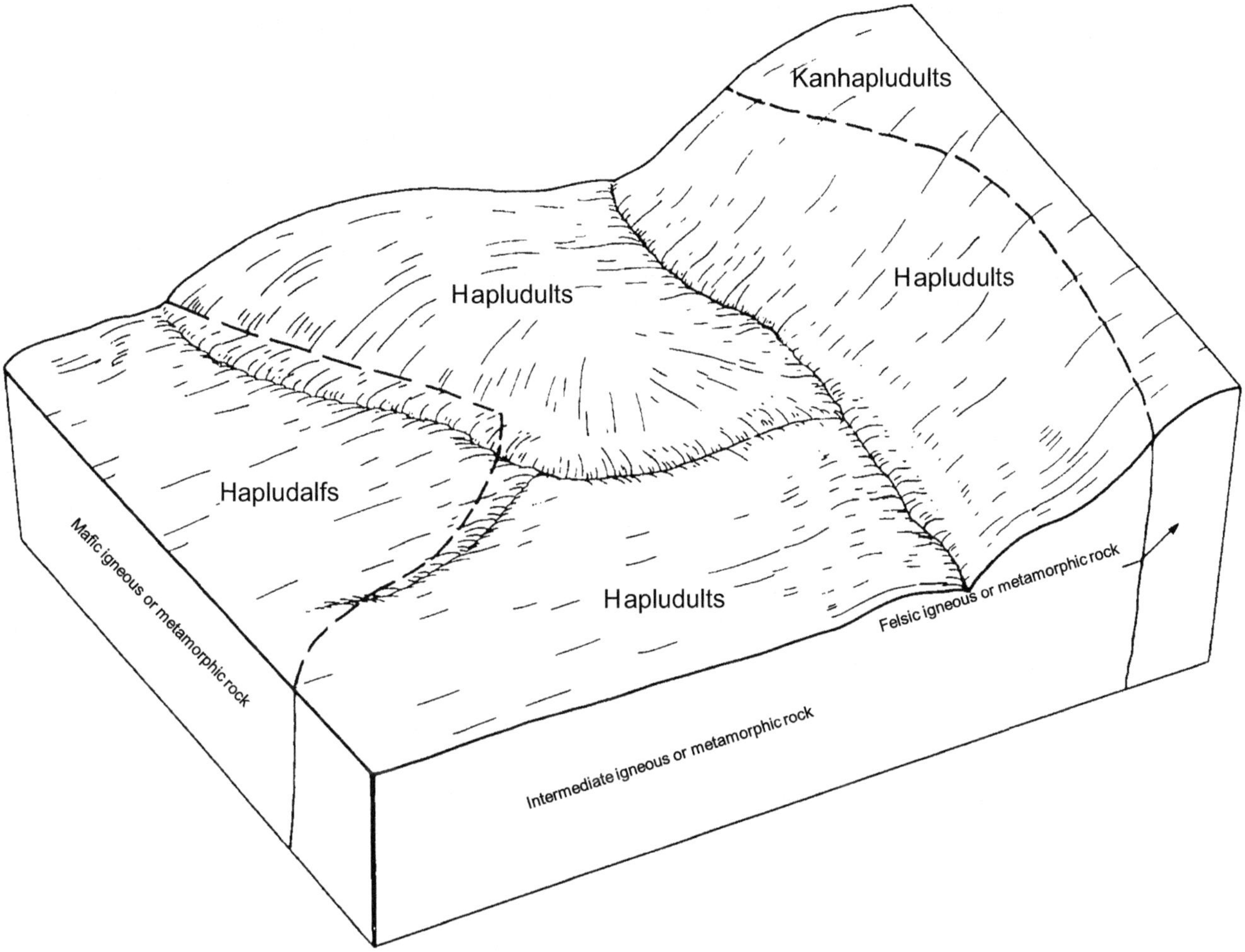

Fig. 13.5 Landscape distribution of great groups in the Southern Piedmont (MLRA 136) as related to type of parent material. (modified from Woody 1989)

saprolitic C horizons (Fig. 13.4). Upper Bt horizons have well-developed stable blocky structure that creates a network of macropores between peds (structural units). Thus, although clay content is high, the structural macropores readily transmit water, and upper Bt horizons typically have the highest K_s within the profile (Fig. 13.4). Deeper BC horizons have lower clay, but because soil structure is less well developed and pores may be plugged by clay and Fe oxide minerals, K_s is lower than in overlying horizons (O'Brien and Buol 1984; Schoeneberger et al. 1995; Vepraskas et al. 1996; West et al. 2008).

13.3.1.2 Historic Erosion

Upland soils in the Southern Piedmont are among the most eroded soils in the USA. At the time of European settlement in the early to mid-1700s, land was plentiful, and the climate was suitable for production of tobacco and cotton as cash crops and corn as animal feed. Intensive tillage coupled with moderate-to-steep slopes and high-intensity rainfall during the growing season resulted in extensive erosion including appreciable gully formation (Trimble 1974). One study (Trimble 1974) estimated the average loss of surface soil over the region, based on estimates of sediment volume in stream valleys, to be about 18 cm. Soil loss was not uniform across the region, however, and a comparison of a rare uncultivated area with nearby fields intensively cultivated since settlement suggested thickness of A and E horizons was similar in the two areas (McCracken et al. 1989). The area subjected to erosive land use peaked in about 1920 and has declined in the decades since. Forests, pastures, and urban lands have largely replaced cotton, corn, and tobacco fields, and erosion rates across the region have substantially declined.

The extensive erosion appreciably altered alluvial valleys in the region. Floodplains commonly have >1-m-thick deposits of cultural sediment (Trimble 1974; Ruhlman and Nutter 1999; Jackson et al. 2005). This recent sediment has altered floodplain hydrology and soils including quantity and distribution of hydric soils and wetlands, disruption of the connection between stream channels and floodplains, and high stream sediment loads from bank erosion (Ruhlman and Nutter 1999; Mukundan et al. 2010).

13.3.2 Southern and Western Coastal Plains (MLRAs 133A, 133B)

The Southern Coastal Plain is the largest geomorphic region in the south and occurs in a belt along the Atlantic and Gulf of Mexico coasts (Fig. 13.1; MLRA 133A). The Atlantic segment is about 250 km wide and broadens to about 650 km through parts of the Gulf segment. Elevations range from sea level along the Atlantic and Gulf coasts and increase gradually to about 200 m along the inner margin. The region is strongly dissected with nearly level to gently undulating valleys and gently sloping to steep uplands. Stream valleys are narrow in their upper reaches but broaden as they approach the coast where they have widely meandering channels. Local relief is mainly 3–6 m but may be as much as 25–50 m. The region is underlain by sediments ranging in age from Cretaceous at the interior margin to Holocene in coastal areas and alluvial valleys. Landscape-scale reworking of sediments during the Pleistocene was common. Topography varies from rolling hills in the interior to broad swampy flats at low elevations.

The Atlantic segment from Virginia to the Neuse River in North Carolina is composed of a series of Miocene to Holocene terraces that descend from the fall line (the boundary between the piedmont and coastal plain) to the coast, and rivers terminate in large estuaries (Thornbury 1965; Daniels et al. 1973). The coastwise terraces continue to the Florida peninsula south of the Neuse River, but barrier islands replace the estuaries along the coast (Daniels et al. 1973). Along the inner margin of the Coastal Plain south of the Neuse River, sandy Eocene and Cretaceous sediments extend southwestward across South Carolina and Georgia and comprise the Carolina and Georgia Sand Hills (Fig. 13.1; MLRA 137).

A short distance west of the Georgia–South Carolina boundary, the Coastal Plain widens due to an increase in the number and thickness of Eocene and Cretaceous beds. Lithologic variability and erodibility of the sediments also increase, and the Coastal Plain from central Georgia to the Mississippi River valley has a belted topography comprised of lowlands formed in erodible marls and clays separated by more resistant beds that form cuestas with in-facing escarpments (Thornbury 1965). The coastwise stepped terraces that comprise much of the Atlantic Coastal Plain become topographically important near the Gulf Coast, but these terraces are only about 30–80 km wide (Daniels et al. 1973).

The Western Coastal Plain (Fig. 13.1; MLRA 133B) is composed of topographic and lithologic belts similar to those found in the Southern Coastal Plain. In general, the Western Coastal Plain is wider than the Gulf portion of the Southern Coastal Plain, and consequently, major rivers have larger drainage basins and more extensive deltas. Low fluvial-deltaic terraces occur along the Gulf coast west of the Mississippi River (Thornbury 1965).

Landscapes across MLRAs 133A and 133B are generally well dissected with nearly level to gently undulating valleys and gently sloping to steep uplands. Soil parent materials are unconsolidated sands, silts, and clays, and textural characteristics of the soils generally reflect those of the parent material. Most upland soils are deep and acidic and have low base saturation. A horizons in upland soils are thin, and organic matter content is relatively low. The clay fraction is dominated by kaolinite although upper Bt horizons may have appreciable hydroxyl-interlayered vermiculite (Shaw et al. 2010; Carlisle and Zelazny 1974, 1976; Fiskell and Perkins 1970). Sand and silt separates are commonly >90 % quartz and other resistant minerals (Harris et al. 1989).

Thickness of sand or loamy sand A and E horizons overlying subjacent Bt horizons ranges from a few centimeters to more than 2 m (Fig. 13.6). The texture of Bt horizons is commonly sandy clay loam, but subsoil textures range from sandy loam to clay. Most upland soils on convex landscapes are well or moderately well drained. Valley bottoms near rivers and depressional ponds have shallow seasonal water tables and are poorly drained. Areas distant from stream valleys on broad low-relief interfluves also commonly have shallow seasonal water tables and poorly drained soils (Daniels et al. 1999), while soils near the margins of these interfluves have deeper water tables and well-drained soils (Daniels and Gamble 1967) (Fig. 13.7).

Upland soils in the Coastal Plain dominantly have a clay increase between A and E and subjacent Bt horizons with evidence of clay translocation (argillic horizon). In the Atlantic and eastern part of the Gulf sections (east of about Montgomery, AL), subsoil clays are dominated by kaolinite and other low-activity clays and often have kandic horizons. Kaolinite is also the dominant clay mineral in soils in the western part of the Southern Coastal Plain and the Western Coastal Plain, but the soils have a higher CEC due to greater abundance of 2:1 clay minerals and higher silt content, which contributes to the CEC. Thus, soils in western part of the Coastal Plain lack kandic horizons, and dominant great groups for upland soils are Paleudults or Hapludults depending on subsoil clay distribution (Soil Survey Staff 2014). In the eastern Gulf and Atlantic sections, the analogs for soils with kandic horizons, Kandiudults and Kanhapludults, dominate.

Plinthite Plinthite is a common feature in subsoils throughout the Coastal Plain (Fig. 13.8). More than 3,000,000 ha of soils with plinthite have been mapped, and more than 50 series in Plinthic subgroups of Paleudults and

Fig. 13.6 Variable sandy A and E horizon thickness in soils in the Southern Coastal Plain (MLRA 133A). **a** Plinthic Kandiudult, A + E horizons about 40 cm thick overlying Bt horizons, Btv horizons with plinthite begin at about 105 cm; **b** Arenic Kandiudult, A + E horizons about 65 cm thick overlying Bt horizons; **c** Grossarenic Kandiudult, A + E horizons about 160 cm thick overlying Bt horizons. Photo credits: John A Kelley, Soil Science@NC State

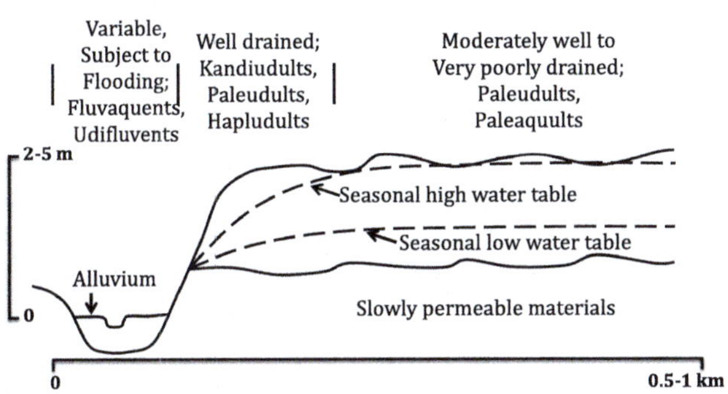

Fig. 13.7 Relationship of seasonal water table depths to distance from streams in Coastal Plains and Flatwoods landscapes (MLRAs 133A and 153A) (after Daniels and Gamble 1967)

Kandiudults have been recognized. Plinthite is an iron-rich (typically 5–10 % iron), firm to very firm (moist) humus-poor mixture of clay with quartz and other minerals (Soil Survey Staff 2014). It commonly occurs as dark red redox concentrations that occur as nodules or as more continuous forms that form platy, polygonal, or reticulate patterns. Plinthite changes irreversibly to ironstone (strongly or very strongly cemented) on exposure to repeated wetting and drying, especially if it is also exposed to heat from the sun. Laterally continuous plinthite common in tropical landscapes often has been referred to as laterite in the literature.

Plinthite forms by segregation of Fe associated with seasonal saturation and redox processes. Initially, the Fe segregations are soft brown, red, or dark red redox concentrations. Continued redox cycles results in additional concentration of Fe, and when Fe cementation has reached the point that the concentrations can be separated from the soil matrix but are less than strongly cemented, the concentrations are considered plinthite (Daniels et al. 1978). The Fe may be derived from within the horizon in which plinthite is found or may be concentrated from other horizons or from soils higher in the landscape. When plinthite is found in soils on higher positions, landscape inversion may have been active during landscape development.

Horizons containing plinthite have been reported to retard vertical water movement resulting in perched water tables and lateral subsurface flow across landscapes (Daniels et al. 1978; Carlan et al. 1985; Blume et al. 1987; Shaw et al. 1997). Hubbard and Sheridan (1983) reported that 25 % of the total rainfall over a 6-year period left a small watershed as shallow subsurface flow associated with plinthic soils. Although horizons with plinthite have lower K_s than overlying horizons with similar texture, subjacent BC horizons are commonly more water restrictive than the plinthic horizons (Carlan et al. 1985; Blume et al. 1987; Shaw et al. 1997). In addition to initiation of lateral subsurface flow,

Fig. 13.8 Plinthic Kandiudults from MLRA 133A. Btv horizons from 45 to 180 cm have nodular plinthite (red bodies). The plinthite becomes more continuous in horizons below 180 cm. Photo credit: John A Kelley, Soil Science@NC State

perching of water by restrictive BC horizons may be the cause of seasonal saturation and redox processes in overlying horizons that lead to plinthite formation (Shaw et al. 1997).

In the Atlantic and eastern part of the Gulf Coastal Plain, development of terraces associated with major streams was largely related to Pleistocene climate change (stream flow volumes as well as sea level fluctuation), and age of the terraces strongly influences soil development on the different aged surfaces. In general, base cation content decreases and low-activity clay minerals increase as terrace age increases (Shaw et al. 2003, 2010). Dystrudepts are common on Holocene terraces, Hapludults and Hapludalfs are found on late Pleistocene terraces, and Paleudults and Kandiudults occur on older Pleistocene surfaces (Shaw et al. 2010).

Pleistocene stream terraces are extensive in the Western Coastal Plain, and soils on the terraces commonly have silt loam A and E horizons overlying silt loam to clay Bt horizons. Seasonal water tables are often shallow, and many soils on terraces are somewhat poorly or poorly drained. Base saturation is generally higher than 35 % in lower subsoil horizons, and soils are mostly Alfisols.

13.3.3 Carolina and Georgia Sand Hills (MLRA 137)

The Carolina and Georgia Sand Hills (Fig. 13.1, MLRA 137) is a dissected, rolling to hilly upland underlain by Cretaceous and Tertiary deposits with abundant sand (Daniels et al. 1999). Soils are similar to those found in the Coastal Plain to the east and south and are deep, well drained, acidic, and have a dominance of kaolinite and other low-activity clays. Soils in the interior of interstream divides are often underlain by thick weathering zones that have sandy loam or sandy clay loam texture, firm to very firm consistence, retard vertical water movement, and perch water at their surface during wet seasons (Daniels et al. 1999). Soils with sand or loamy sand A and E horizons >50 cm thick overlying loamy Bt horizons (arenic and grossarenic subgroups) are common in this MLRA as are soils with sand textures to depths >2 m.

13.3.4 Alabama and Mississippi Blackland Prairie and Cretaceous Western Coastal Plain (MLRAs 135A and 135B)

Near the inland boundary of the Coastal Plain in Arkansas, Mississippi, and Alabama are outcroppings of Tertiary and Cretaceous marls, chalks, and calcareous clays (Fig. 13.1; MLRAs 135A and 135B). In Alabama and Mississippi, this region is commonly referred to as the "Black Belt" or "Blackland Prairie" because of the dark color of surface horizons of the soils. Calcareous and alkaline soils occur as islands interspersed among acid soils. Although this region was dominantly forested prior to European settlement, significant areas of relatively contiguous tallgrass prairie formed a coherent and significant ecosystem in this region, which was estimated to have covered about 140,000 ha of this region in 1830 (Barone 2005). Remnants of these ecosystems with little bluestem (*Schizachyrium scoparium*), indiangrass (*Sorghastrum nutans*), and other grasses, along with deciduous shrubs, red cedar (*Juniperus virginiana*), and mixed hardwoods exist today.

This region sits at a lower elevation than adjacent Coastal Plain landscapes due to the erodibility of the parent geology. Relief in these areas is generally low. A mosaic pattern of alkaline, acid, and acid over alkaline soils are found. Alkaline soils mostly form directly from Cretaceous chalk, while the more common acid soils form in Tertiary acid marine

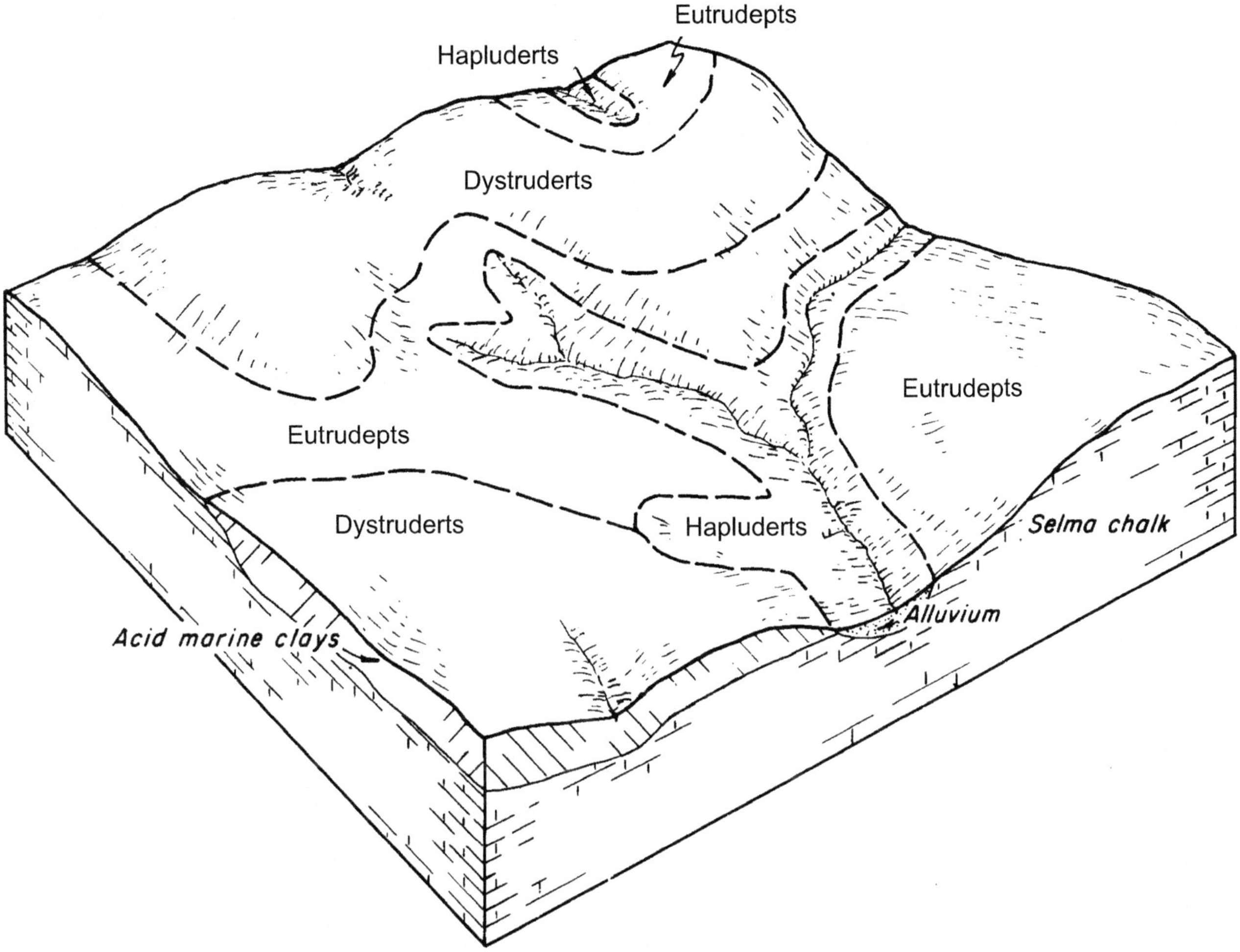

Fig. 13.9 Landscape distribution of great groups as related to landscape and type of parent material in the Alabama and Mississippi Blackland Prairies (MLRA 135A). (modified from Cotton 1971)

clays or high terrace deposits (Fig. 13.9), although weathering in the upper solum contributes to the acidification.

Upland soil surface horizons are commonly relatively thin, although mollic epipedons are found on lower landscapes. Deep soils overlying calcareous parent materials in these areas often have acidic A and upper B horizons, but pH and base saturation increase with depth. Many soils in the region are calcareous in lower horizons, and soils shallow to calcareous parent materials have high pH and carbonates throughout.

Soils are dominantly Alfisols, Vertisols, and Inceptisols. Eutrudepts and Udorthents underlain by chalk or other calcareous parent materials at depths <1 m are common on convex upland landscapes (Fig. 13.9). Dystruderts, Hapluderts, and Vertic Paleudalfs are common on uplands and are typically moderately well to somewhat poorly drained. Epiaquerts and Vertic Haplaquepts are common on floodplains of streams in the region. Soils developed from calcareous parent materials in the Blackland Prairies in Arkansas are dominantly well or moderately well-drained Paleudalfs with loamy or clayey Bt horizons. Dystruderts are also common. Most soils have thin light colored A horizons with relatively low organic matter.

Soils developed from calcareous parent materials have high native fertility and clay fractions dominated by smectite. Textures are commonly clayey, which when combined with abundant smectitic clay results in soils having high shrink–swell potential, high plasticity indices, very slow saturated hydraulic conductivity, and challenging soil physical properties. Relatively high erodibility and runoff make soils in the region easily eroded even on relatively gentle slopes, and many landscapes have experienced extensive erosion.

13.3.5 Southern Mississippi Valley Silty Uplands (MLRA 134)

Coastal Plain uplands east of the Lower Mississippi Valley are capped by thick loess deposits. On bluffs adjacent to the Mississippi River valley, the loess may be as thick as 20 m and thins eastward from the river valley (Wascher et al. 1947; Saucier 1974; Rutledge et al. 1996). Loess deposits are also common on the western side of the valley, but these deposits are generally thinner and less extensive. Crowley's Ridge, a prominent upland feature in the Mississippi River Valley, is capped by up to 20 m of loess (Thornbury 1965; West et al. 1980; Saucier 1974; 1994). Pleistocene braided streams terraces west of Crowley's Ridge are also capped by one or more loess deposits (Rutledge et al. 1985, 1996; Blum et al. 2000).

As many as five separate loess deposits, some with well-developed paleosols, have been identified in thick loess deposits at various locations along the Lower Mississippi Valley. The latest, most extensive surficial loess deposit in which the contemporary soil is formed is the late Wisconsin Peoria loess. For discussions of the properties, ages, and correlation of this and the older loesses, readers are referred to discussions by Saucier (1994), Autin (1996), Rutledge et al. (1996), and Blum et al. (2000).

In areas with thick loess on Coastal Plain deposits, valley sides are hilly to steep, and intervening ridges generally are narrow and rolling. Stream valleys are narrow in the upper reaches but broaden rapidly downstream and have wide, flat flood plains and meandering stream channels. Local relief is mainly 3–6 m but can be as much as 50 m. Relief is more subdued on the loess-covered Pleistocene terraces.

Properties of the loessial soils vary depending on loess thickness, physiography, landscape position, and natural drainage. Soils are dominantly very deep, but fragipans (water and root restrictive horizons) are common. Surface horizon textures are silt loam, and textures in the Bt (argillic) horizons are typically silt loam or silty clay loam with 18–35 % clay. Soil drainage class ranges from poorly to well drained depending on slope, landscape position, and presence of fragipan or other water restrictive horizon.

The chemical properties of loessial soils are well suited to production of agricultural crops. The pH of A and E horizons is strongly to moderately acid and increases in subsoil horizons. Organic C content in A horizons is generally <2 % and is often <1 % in areas of cropland. Mineralogy of the clay fraction is mixed with moderately high CEC. Sand and silt separates have sufficient weatherable minerals to supply nutrients for plant growth. However, these soils are highly erodible, occur on moderately steep slopes, and growing season rainfall often occurs as high-intensity convective storms. This combination has resulted in extensive erosion throughout these loess-covered landscapes.

About 2,500,000 ha of soils in these landscapes (about 40 %) have a fragipan that restricts water movement through the soil. The soils with fragipans are abundant both in hilly areas with thick loess and in areas of more subdued relief (Fig. 13.10). On low-relief landscapes, alternating reduction and oxidation associated with seasonal perching of water above the fragipan has resulted in clay dissolution through ferrolysis and development of silt coatings on ped faces (glossic features; Glossaqualfs and Fraglossudalfs) (Brinkman 1970; Ransom and Smeck 1986). Though not abundant, there are also scattered areas of loessial soils and soils developed in alluvium derived from loess that have high Na in the argillic horizon (natric horizon; exchangeable sodium percentage (ESP) >15 in the upper part of the argillic horizon), which results in clay dispersion, structure degradation, and low hydraulic conductivity in subsoils.

In areas of thick loess, soils are commonly Fragiudalfs, Hapludalfs, and Fraglossudalfs (Fig. 13.10). In thin loess east of the Mississippi valley, Coastal Plain materials with low base saturation occur within the solum, and upland soils are dominantly Hapludults. In this region, alluvium in floodplains and on low terraces is derived from the loessial uplands, and soils are commonly silty Fluvaquents and Udifluvents with <18 % clay.

13.4 Atlantic and Gulf Coast Flatwoods, Swamps, and Marshes, Land Resource Region T (MLRAs 150A, 150B, 151, 152A, 152B, 153A, 153B, 153C, 153D)

As elevations in the Coastal Plain decrease toward the sea, the landscape has less local relief, and upland interfluves become wider with large areas of poorly and very poorly drained soils. These landscapes are commonly termed "Flatwoods". These broad nearly level uplands are interspersed with swamps, estuaries, and lagoons. The drainage network is poorly integrated, and local relief is minimal. Water tables are commonly at or near the soil surface during part of the year except for better-drained soils on sandy ridges and near incised streams where gradients for water movement are greater (localized "red edge effect;" Daniels and Gamble 1967) (Fig. 13.7).

Flatwoods soil parent materials are unconsolidated Tertiary- to Quaternary-aged sediments with variable texture that reflects the environment in which they were deposited. Chemical and mineralogical properties of the sediments reflect properties of soils and rocks of the areas from which they were eroded.

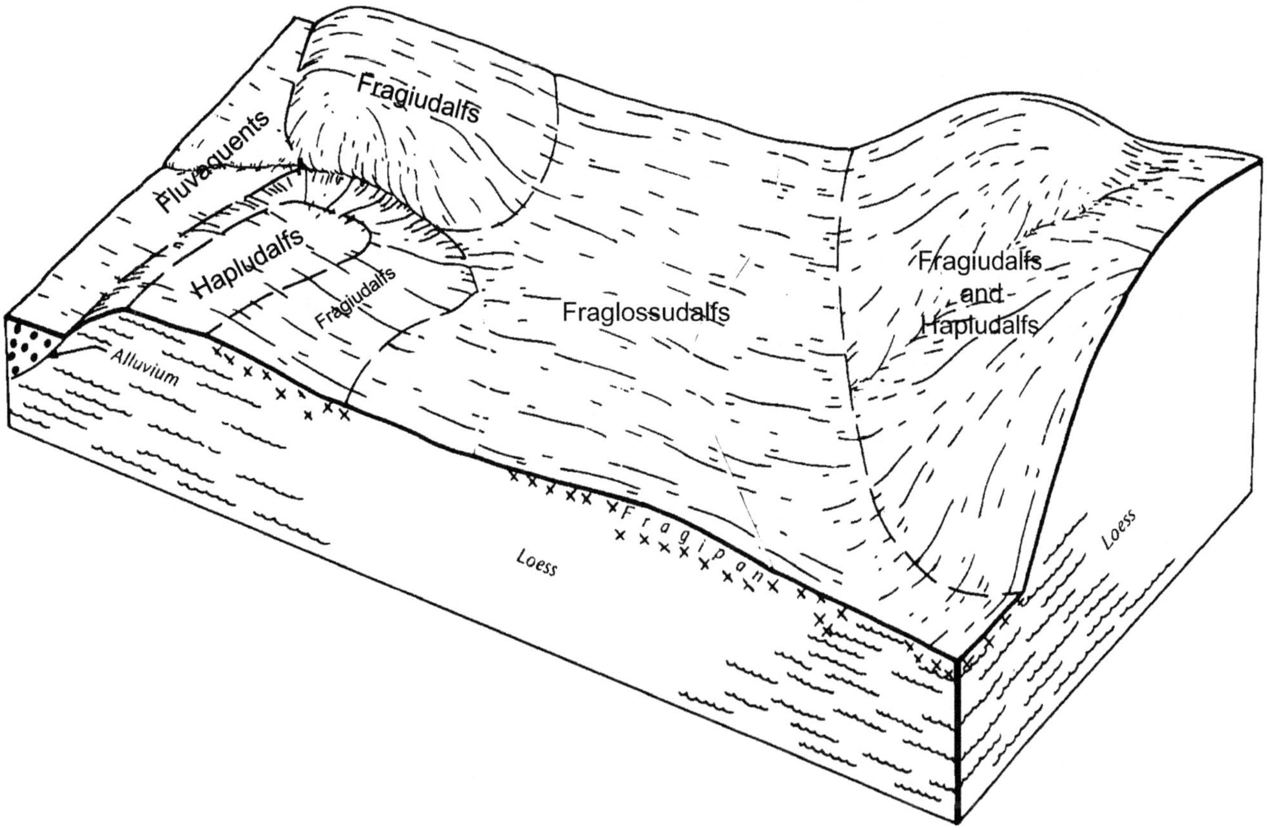

Fig. 13.10 Landscape distribution of great groups as related to loess thickness and landscape in the Loessial Hills (MLRA 134). (modified from Monteith 1990)

13.4.1 Atlantic Coast Flatwoods, Tidewater Area, Mid-Atlantic Coastal Plain, and Northern Tidewater Area (MLRAs 153A, 153B, 153C, 153D)

Soils along the Atlantic Coast Flatwoods (Fig. 13.2; MLRA 153A) are dominantly acid with low base saturation and have low native fertility reflecting origin of parent material sediments from upstream in the Piedmont or Coastal Plain. Somewhat poorly and poorly drained soils are abundant. A horizons are thin with low organic matter except in depressions and other very poorly drained landscapes. Subsoil textures vary widely and reflect the texture of the parent material. Most soils in these landscapes have A and E horizons with sand or loamy sand textures with subjacent loamy or clayey Bt horizons at variable depths up to 2 m (Paleaquults, Paleudults, and Endoaquults), but Bh (spodic) horizons are common in soils with sandy parent materials and soils with thick, sandy A and E horizons. Multiple spodic horizons within a profile are sometimes found.

Sandy parent materials and acid leachate from coniferous trees have resulted in an abundance of Spodosols (Alaquods) in the Flatwoods (Brasfield et al. 1973). These are classic ground water podzols and paucity of Fe in the sandy parent materials results in the spodic horizon being dark colored (Bh horizon) due to Al-organic complexes with little or no Fe (Fig. 13.11) (Holzhey et al. 1975; Sodek et al. 1990; Harris et al. 1995). These Spodosols occur on surfaces of variable age, some of which are older than is typical for northern, glacially influenced landscapes where Spodosols are found (Holzhey et al. 1975; Harris and Carlisle 1987). Spodic horizons of Flatwoods Spodosols can be thicker than is typical for Spodosols of cooler climates (Daniels et al. 1975), and as much as one-half of the organic C in Aquods may occur in Bh and B′h horizons (Stone et al. 1993).

Depth to the seasonal water table is an important factor influencing the distribution of Spodosols across Flatwoods landscapes (Daniels et al. 1999; Harris et al. 1995) (Figs. 13.12 and 13.13). Soils developed in sandy parent materials on relatively high landscape positions without

Fig. 13.11 Typic Alaquod from MLRA 153A. The Bh horizons (spodic horizons) underlie an albic E horizon at 30 cm and extend to about 45 cm. Underlying the Bh horizons are E′ horizons that extend to about 125 cm. A second spodic horizon (B′h horizon) underlies the E′ horizons. Photo credit: LT West

seasonal saturation lack a spodic horizon and are commonly Paleudults or Quartzipsamments. At the other end of the drainage catena, poorly and very poorly drained soils in broad drainageways and depressions also lack a spodic horizon (dominantly Endoaquults and Humaquepts). Spodic horizons and Spodosols are found on the parts of the landscape between these two extremes (Fig. 13.12).

A similar distribution of soils is found on the mainland margin and barrier islands along the Atlantic coast (Fig. 13.2; Tidewater Area, MLRA 153B). Quartzipsamments occur on stable dunes, and Alaquods, Endoaquults, and Humaquepts are found on lower positions on the islands. Soils in the back barrier salt marsh are dominantly Haplosaprists and Sulfaquents (Fig. 13.14).

Northern coastal areas in LRR T (Fig. 13.2; Mid-Atlantic Coastal Plain, MLRA 153C and Northern Tidewater Area, MLRA 153D) have soils with similar properties and landscape relationships. Soils are sandy and acid although they are not as deeply developed as their more southern counterparts. Histosols (Sulfihemists) are common in salt marshes and estuaries.

13.4.2 Gulf Coast Marsh, Eastern Gulf Coast Flatwoods, and Western Gulf Coast Flatwoods (MLRAs 151, 152A, 152B)

Soils along the Gulf coast from the Florida panhandle to Louisiana (Fig. 13.2; Eastern Gulf Coast Flatwoods, MLRA 152A) are similar to those on the Atlantic coast. Soils are dominantly poorly or very poorly drained. Common great groups are Alaquods, Paleaquults, Quartzipsamments, and Haplosaprists.

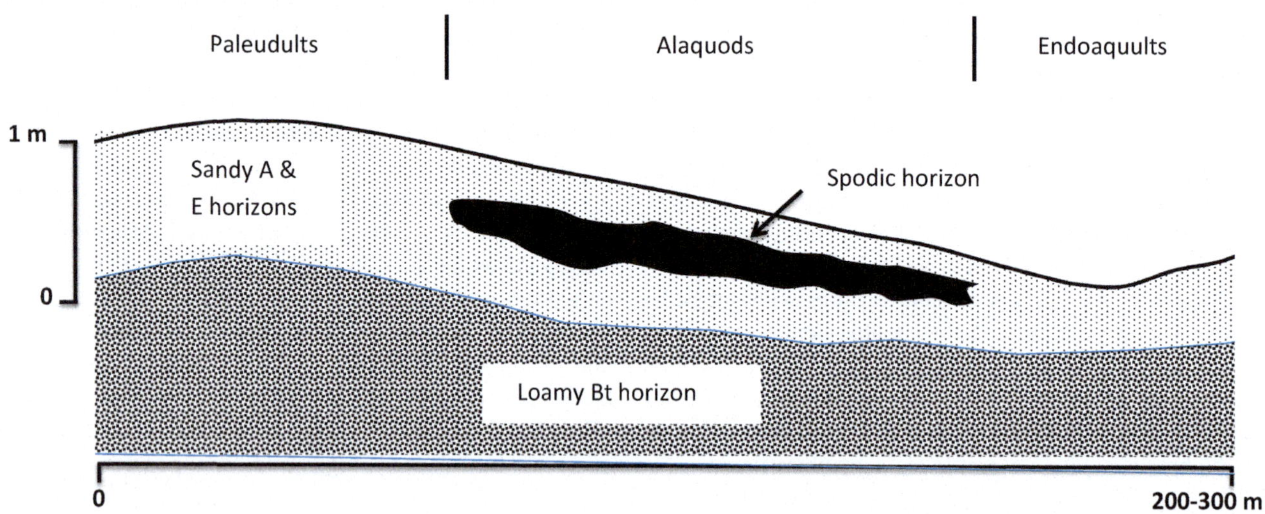

Fig. 13.12 Relation of spodic horizons to landscape position. Note the absence of spodic horizon in well-drained and poorly drained landscape components

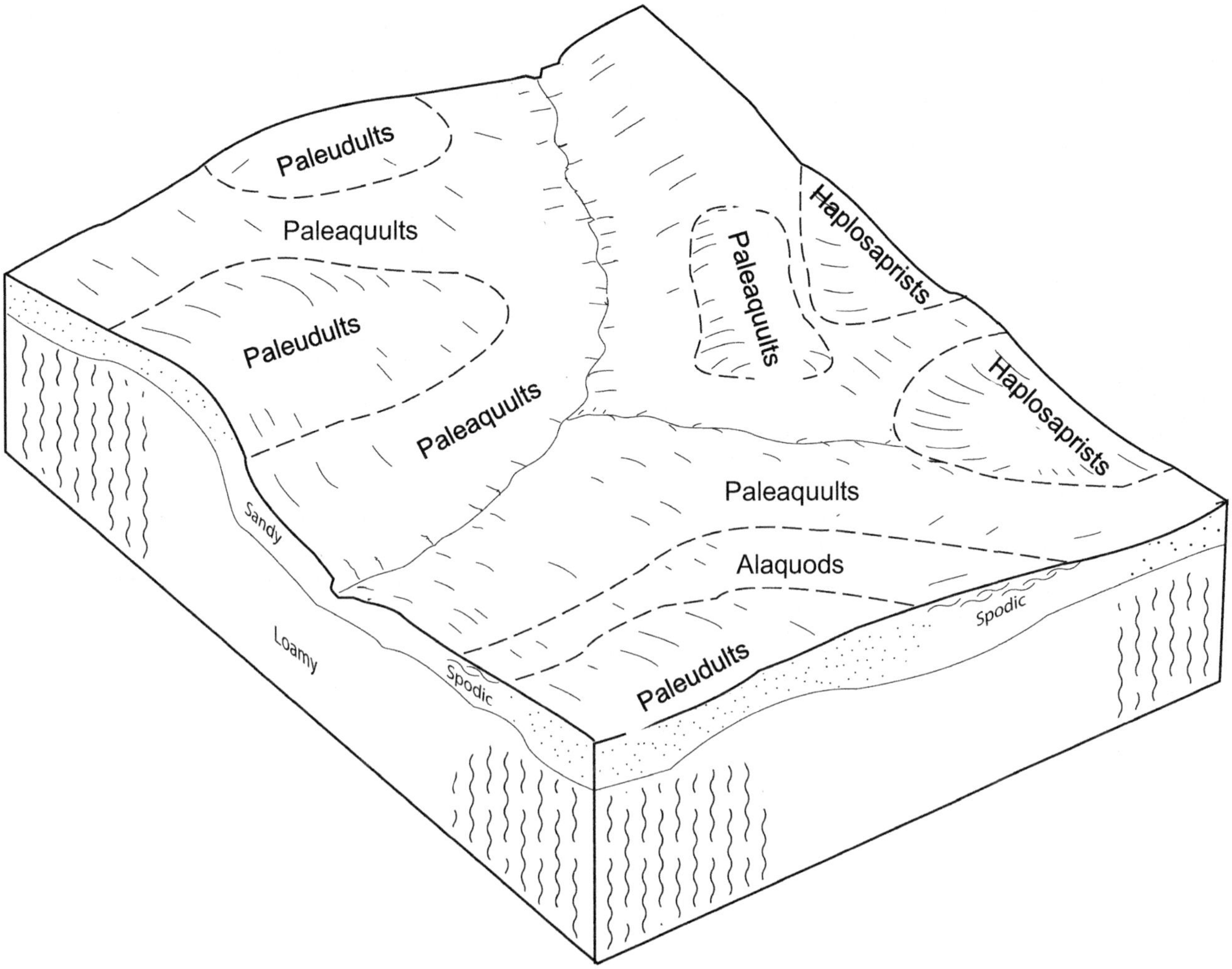

Fig. 13.13 Landscape distribution of great groups as related to landscape and A and E horizon thickness in the Atlantic and Gulf Coast Flatwoods (MLRAs 153A and 153B). (modified from Watts 1996)

The landscape along the Gulf coast in Louisiana (Fig. 13.2; Gulf Coast Marsh, MLRA 151) is dominantly marsh with low, narrow sandy ridges. Elevation ranges from sea level to about 2 m and the area has many rivers, lakes, bayous, oxbows, tidal channels, and manmade canals. Sediment is primarily fluvial-deltaic with near-shore deposits of fine sand, silt, and clay deposited by the Mississippi River during the Pleistocene. Because of organic matter accumulation in the marsh environment, Histosols are common although most have mineral materials within 2 m. Common great groups are Haplosaprists and Hydraquents.

Inland lies the Western Gulf Coast Flatwoods (Fig. 13.2; MLRA 152B) in western Louisiana and east Texas. The area is nearly level to gently sloping and has low local relief. Parent materials are Tertiary to Holocene age alluvial sediments with variable texture. Surface horizons are thin with low organic matter except in wetter areas on low landscapes. Subsoil texture varies and is strongly influenced by texture of the parent material. Most of the soils have acid subsoils and argillic horizons. Dominant soils are Alfisols (Glossaqualfs and Paleudalfs) and Ultisols (Paleudults), although Vertisols are common.

13.4.3 Gulf Coast Prairies and Gulf Coast Saline Prairies (MLRAs 150A, 150B)

The Gulf Coast Prairies and Gulf Coast Saline Prairies (Fig. 13.2; MLRAs 150A and 150B) occur as a strip about 30–50 km wide that extends from western Louisiana south along the Texas Gulf coast. The landscape is a level plain with low local relief. Parent materials are Pleistocene-age deltaic and lagoonal clays and loams derived from older rocks to the west. The alluvial materials are capped by loess in the eastern part of the region. In the Saline Prairies, most of the surface is covered by Pleistocene-age sand that has

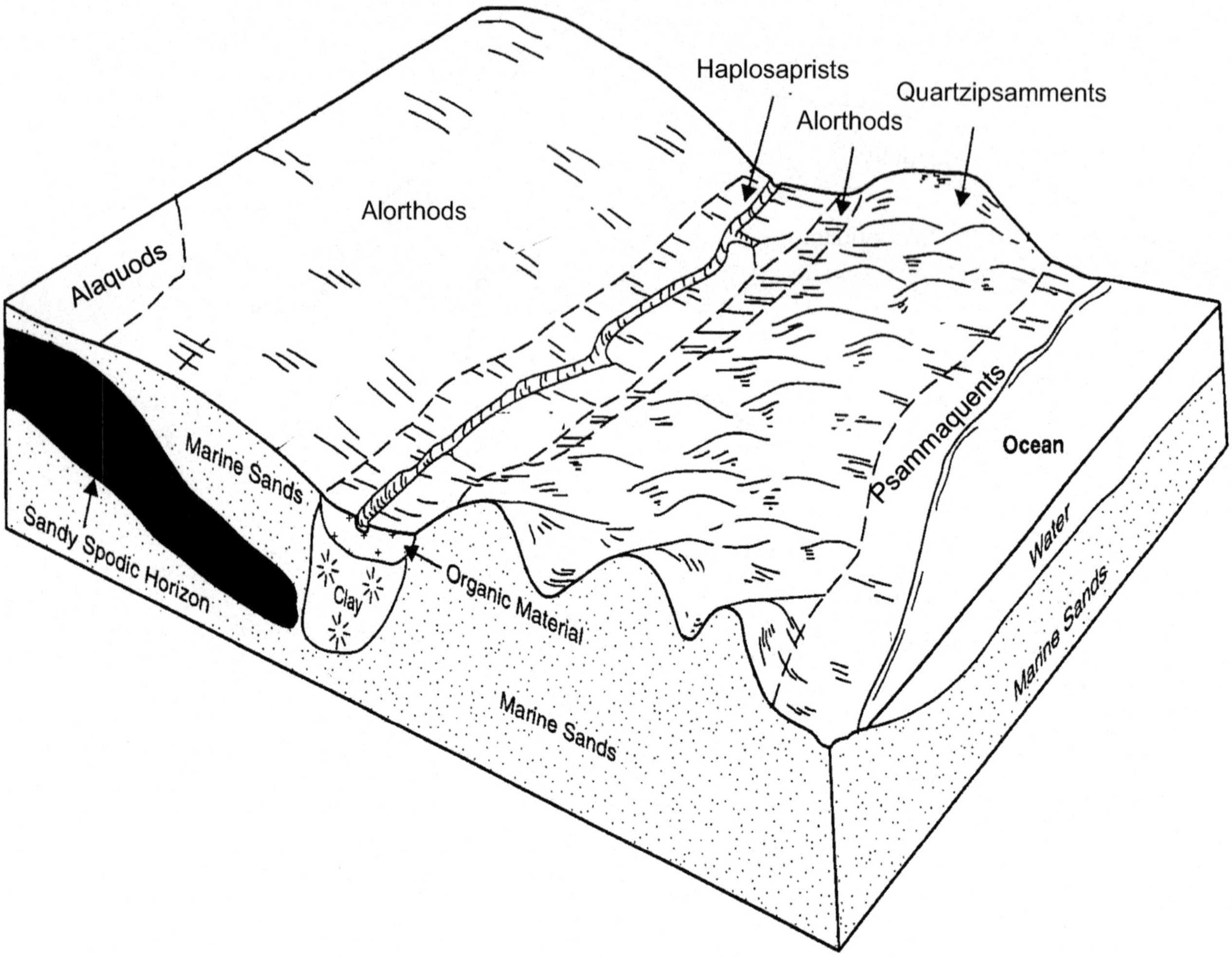

Fig. 13.14 Landscape distribution of great groups for coastal areas in the Tidewater Area (MLRA 153B). (modified from Watts 1998)

been reworked by the wind into mounds and dunes. The northern two-thirds of the region receives about 1200–1600 mm of rainfall annually. Precipitation decreases to about 700 mm at the southern end, and soils in this part of the region have an ustic moisture regime.

Surface horizons of soils in the Coast Prairie are slightly to moderately acid, and pH generally increases with depth. Many subsoils are calcareous, especially in the central and southern parts of the region. Subsoil textures are dominantly clayey, and many surface horizons have clay texture. Clay minerals are mostly smectite and other active 2:1 minerals, which combined with high clay contents impart high shrink–swell and low K_s to many soils in the region. Soils are well drained on convex ridges, but somewhat poorly and poorly drained soils dominate planar and concave landscapes that comprise most of the region. A horizons have moderate-to-high organic matter contents, and many have sufficient thickness to be a mollic epipedon. Vertisols (Hapluderts and Haplusterts), Mollisols (Argiaquolls), and Alfisols (Hapludalfs) are dominant soil Orders.

In the Gulf Coast Saline Prairie (Fig. 13.2; MLRA 150B), soils are sandier, often saline from tidal inundation and sea spray, and many have high contents of Na in subsoils. Surface horizons have sandy or loamy texture, alkaline pH, and relatively low organic matter. Subsoils have variable texture, but many are sandy (Psammaquents, Udipsamments, and Ustipsamments), especially on dunes on barrier islands and on back barrier bays and estuaries. In addition to Entisols, Alfisols (Natraqualfs) and Inceptisols (Halaquepts) are common in the region.

13.5 Mississippi Delta Cotton and Feed Grains Region, Land Resource Region O (MLRAs 131A, 131B, 131C, 131D)

The Southern Mississippi River alluvial valley is from 40 to 200 km wide and extends from the confluence of the Mississippi and Ohio Rivers near Cairo, IL to the Gulf of Mexico (Fig. 13.2). Much of the valley is occupied by

Holocene alluvium, but upland ridges and Pleistocene terraces break the continuity of the valley floor. These include Crowley's Ridge, about 70 m above the valley floor, that extends for about 300 km from southeast Missouri to east central Arkansas and Macon Ridge in extreme southeast Arkansas and northeast Louisiana. Macon Ridge is about 7–13 m higher than the valley floor and is composed of Pleistocene alluvium capped by loess (Daniels et al. 1973; Autin 1996). Also interrupting the alluvial valley is the Grand Prairie in Arkansas, which is a broad low-relief Pleistocene terrace interfluve that stands about 8–14 m higher than surrounding lowlands. Immediately west of Crowley's Ridge are a series of Pleistocene-age braided stream terraces formed by the ancestral Mississippi River before its diversion to the lowlands east of Crowley's Ridge (Saucier 1974, 1994; Blum et al. 2000). The oldest and highest of these terraces are capped by loess (Rutledge et al. 1985; Saucier 1994; Blum et al. 2000).

13.5.1 Southern Mississippi Valley Alluvium (MLRA 131A)

This region is comprised of terraces and flood plains along the Mississippi River and its major tributaries. Average elevations start at sea level in the southern part of the area and gradually rise to about 100 m in the northwestern part. Local relief can be as much as 5 m, but is considerably lower over most of the area. The region has fertile soils, smooth topography, abundant moisture, and a long growing season, which favor agricultural production. Levees have been constructed along the Mississippi River and major tributaries to protect cropland and communities from flooding. The region has also been extensively ditched to remove surface ponding to local streams.

Sediments in the valley are very thick deposits of sandy to clayey alluvium of Pleistocene to Holocene age. Landforms are level, depressional, or very gently undulating alluvial plains, backswamps, oxbows, natural levees, and terraces. Natural levees in the Mississippi River floodplain are up to 2 km wide with correspondingly wide backswamps and abandoned meander loops. Landform shapes range from convex on natural levees and undulating terraces to concave in oxbows. These shapes differentiate water-shedding positions from water-receiving positions, which greatly affect soil formation and hydrology. The relatively flat topography gives rise to a preponderance of somewhat poorly and poorly drained Vertisols (Epiaquerts), Inceptisols (Endoaquepts), and Alfisols (Endoaqualfs). Only on the highest convex part of natural levees and terraces are moderately well- and well-drained Hapludalfs found (Fig. 13.15).

Soils on natural levees and other convex surfaces have physical and chemical properties favorable for crop production and many other uses. Surface textures are silt loam to sandy loam, and most soils on natural levees have an argillic horizon and associated subsoil clay increase. Some of the soils are sandy throughout. On upper positions on natural levees and on convex surfaces of low terraces, soils are moderately well or well drained. A horizons are thin and have low contents of organic matter. Native fertility is relatively high, and soil pH throughout the profile is generally slightly acid to moderately alkaline. The soils typically have a high proportion of smectitic clays and high CEC. Saturated hydraulic conductivity is moderate or high. Typical great groups for soils on these landscapes are Udifluvents, Udipsamments, and Hapludalfs.

On lower parts of natural levees and planar surfaces of low terraces, soils are somewhat poorly or poorly drained. Textures are loamy and chemical properties are similar to those for soils on higher positions. Typical great groups are Endoaquepts, Endoaqualfs, and Epiaqualfs.

Soils in backswamps and oxbows are clayey and smectitic, have high shrink–swell potential, high plasticity indices, large cracks when dry, are difficult to till, and classify as Vertisols or in vertic subgroups of other great groups. These soils have very low K_s, which results in ponding of water at the soil surface after rainfall events unless the soils have been ditched for surface drainage. Organic C content of these clayey poorly drained soils is higher than that found in better-drained soils, and Histosols may occur in abandoned meander loops, especially in the southern part of the valley.

Although rainfall during the growing season is generally sufficient to support crop production, its distribution often results in periods of drought, and supplemental irrigation with ground water from the shallow alluvial aquifer is a common. As a result, groundwater withdrawal from this aquifer has increased substantially over the last 40 years, and elevation of the alluvial aquifer surface has declined (Schrader 2006, 2010). Lowering of the aquifer surface combined with levees and ditches to control flooding has reduced frequency and duration of seasonal saturation of soil horizons, and limited data suggest that redoximorphic features in the soils are not indicative of current soil hydrology.

13.5.2 Arkansas and Red River Alluvium (MLRAs 131B and 131C)

The valleys associated with lower part of the Arkansas River and the Red and Ouachita Rivers (Fig. 13.2; MLRAs 131B and 131C) have similar physiography and landforms to that in the Mississippi River valley. Maximum local relief is about 3 m, but most relief is considerably lower. Soil parent materials are alluvial. The sediments deposited in the Arkansas and Red River valleys were wholly or partially derived from the Permian Red Beds of north Texas and southwest Oklahoma

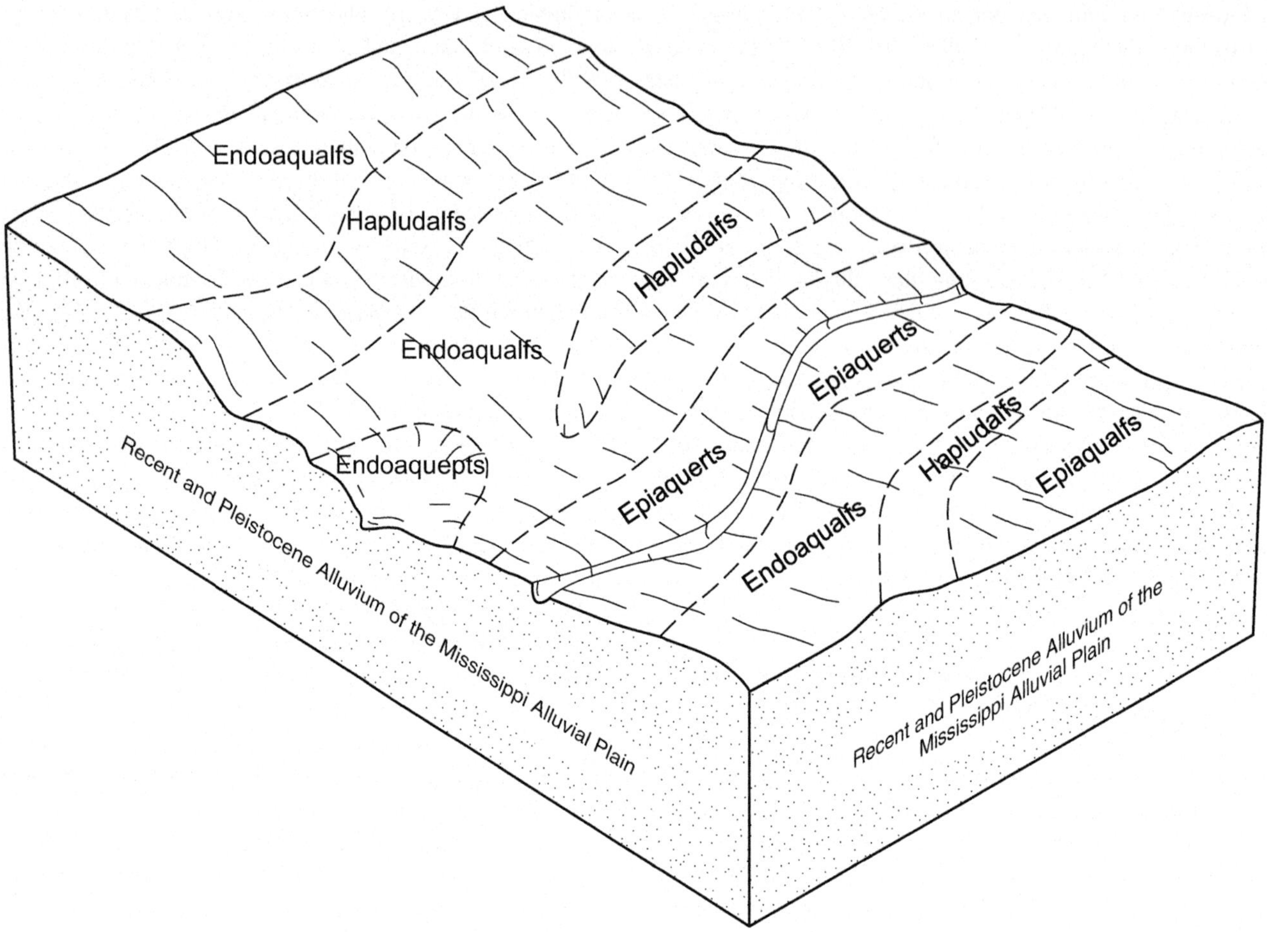

Fig. 13.15 Landscape distribution of great groups as related to landscape in the Mississippi River Alluvial Plain (MLRA 131A). (modified from Huddleston 1990)

and have a red color that often changes little during soil development. These soils tend to remain red even if the soils are subject to periodic saturation and reduction that typically form redoximorphic features (Brown et al. 1972).

Soils in these alluvial valleys have similar properties to soils in the lower Mississippi River valley. They are very deep, dominantly poorly and somewhat poorly drained, and have chemical properties favorable for crop production. Native fertility is relatively high, soil pH throughout the profile is generally slightly acid to moderately alkaline, carbonates are common in deeper horizons, and smectite is abundant in the clay separate.

Soils on natural levees are loamy and generally have favorable physical properties. Most have an argillic horizon. Dominant great groups include Epiaqualfs, Hapludalfs, Eutrudepts, and Udifluvents. Clayey soils in backswamps and oxbows have high shrink–swell potential, are difficult to till, and have very low K_s. Dominant great groups are Epiaquerts and Epiaquepts in the Arkansas valley and Hapluderts in the Red River valley.

13.5.3 Southern Mississippi River Terraces (MLRA 131D)

The Pleistocene-age terraces within the Mississippi River valley (Fig. 13.2; MLRA 131D) are level to gently sloping. Soil parent materials are alluvial deposits from the Mississippi and Arkansas rivers of varying age and depositional environments. Many of the terraces, especially Macon Ridge and the higher terraces west of Crowley's Ridge in Arkansas, are capped by loess. Dominant great groups include Fragiudalfs, Albaqualfs, Hapludalfs, and Fraglossudalfs.

The Grand Prairie region in Arkansas is a broad flat plain with relief commonly less than 1 m. This terrace is part of the Pleistocene-age Prairie Complex described by Saucier (1994). Parent materials are loess-derived silty alluvium overlying alluvium from the Arkansas River. Before European settlement, the region was a tallgrass prairie with interlaced hardwoods. Although most the region was grassland, soils are dominantly Alfisols instead of Mollisols. Soils in the region typically have silt loam A and E horizons that abruptly overly clayey Bt horizons with smectitic clays and very low saturated hydraulic conductivity. The combination of the silt loam surface and very slowly permeable subsoil makes these soils well suited to flooded rice production, and this region is a major rice production area. Its proximity to the Mississippi migratory bird flyway and preponderance of surface ponding during the winter also make it an excellent habitat for ducks and other water fowl and a major duck hunting destination.

13.6 Summary

Land Resource Regions (LRRs) O, P, and T encompass the southeastern USA with the exception of the Florida peninsula (LRR U). Climate across the region is uniformly warm and humid. The region, however, has a wide variety of soil parent materials and topographies that give rise to a broad range of soils. Soils in LRR O have primarily developed in Holocene and Pleistocene alluvium associated with floodplains, and terraces of the Mississippi, Arkansas, and Red Rivers and are dominantly Alfisols, Vertisols, and Inceptisols. The region is agriculturally productive due to its fertile soils, smooth topography, abundant moisture, and a long growing season. Parent materials in the Southern Piedmont (LRR P) are dominantly saprolite from Precambrian to late Paleozoic acid igneous and metamorphic rocks, and topography is gently rolling. Most upland soils in the region are Ultisols that are deep, well drained, and moderately permeable. Loamy surface horizons are thin and overly red clayey subsoils dominated by kaolinite and other low-activity clays. Although the area was historically used for crop production and was subject to considerable erosion, the area is currently dominantly forest and pasture. The Southern and Western Coastal Plains are underlain by sediments ranging in age from Cretaceous at the interior margin to Holocene, and topography varies from rolling hills to broad swampy flats at low elevations. Most upland soils are Ultisols that are deep and acidic and have low base saturation although appreciable areas of Alfisols, Inceptisols, and Vertisols are found whose properties are related to parent material characteristics. Crops, forests, and pasture are dominant land covers across the region. LRR T is the lowest elevation part of the Southern Coastal Plain. Poorly and very poorly drained Ultisols, Spodosols, and Histosols are common in this LRR east of the Mississippi River. The region west of the Mississippi River (Louisiana and Texas) has similar landscapes, but soils have thicker surface horizons, more fertile subsoils with active clays and are dominantly Mollisols, Alfisols, and Vertisols. Ultisols and Spodosols are dominantly used for forest production and pasture. In the western part of the region, Alfisols and Vertisols commonly are used for production of row crops.

References

Autin WJ (1996) Pleistocene stratigraphy in the southern Lower Mississippi Valley. Eng Geol 45:87–112

Barone JA (2005) Historical presence and distribution of prairies in the black belt of Mississippi and Alabama. Castanea 70:170–183

Bigham JM, Fitzpatrick RW, Schulze DG (2002) Iron oxides. In: Dixon JB, Schulze DG (eds) Soil mineralogy with environmental applications. Soil Science Society of America, Madison, pp 323–366

Blume LJ, Perkins HF, Hubbard RK (1987) Subsurface water movement in an upland coastal plain soil as influenced by plinthite. Soil Sci Soc Am J 51:774–779

Blum MD, Guccione MJ, Wysocki DA, Robnett PC, Rutledge EM (2000) Late Pleistocene evolution of the lower Mississippi River valley, southern Missouri to Arkansas. Geol Soc Am Bull 112:221–235

Brasfield JF, Carlisle VW, Johnson RW (1973) Spodosols-soils with a spodic horizon. In: Buol SW (ed) Soils of the southern states and Puerto Rico. Southern cooperative series Bull 174, North Carolina State University, Raleigh, pp 57–60

Brinkman R (1970) Ferrolysis, a hydromorphic soil forming process. Geoderma 3:199–206

Brown DA, Nash VE, Caldwell AG, Bartelli LJ, Carter RC, Carter OR (1972) A monograph of the soils of the southern Mississippi River valley alluvium. Southern Cooperative Series Bull 178, North Carolina State University, Raleigh

Buol SW, Southard RJ, Graham RC, McDaniel PA (2011) Soil genesis and classification. Wiley, Hoboken

Calvert CS, Buol SW, Weed SB (1980) Mineralogical characteristics and transformations of a vertical rock-saprolite soil sequence in the North Carolina piedmont: I. Profile morphology, chemical composition, and mineralogy. Soil Sci Soc Am J 44:1096–1103

Carlan WL, Perkins HF, Leonard RA (1985) Movement of water in a Plinthic Paleudult using a bromide tracer. Soil Sci 139:62–66

Carlisle VW, Zelazny LW (1974) Mineralogy of selected Florida Paleudults. Soil Crop Sci Soc Florida Proc 33:136–139

Carlisle VW, Zelazny LW (1976) Subjacent mineralogy of selected Florida Paleudults. Soil Crop Sci Soc Florida Proc 35:192–195

Cotton JA (1971) Soil survey of Greene County, Albama. USDA-Natural Resources Conservation Service, Washington

Daniels RB, Gamble EE (1967) The edge effect in some Ultisols in the North Carolina Coastal Plain. Geoderma 1:117–124

Daniels RB, Gamble EE, Holzhey CS (1975) Thick Bh horizons in the North Carolina Coastal Plain: I. Morphology and relation to texture and soil ground water. Soil Sci Soc Am J 39:1177–1181. doi:10.2136/sssaj1975.03615995003900060039x

Daniels RB, Allen BL, Bailey HH, Bienroth FH (1973) Physiography. In: Buol SW (ed) Soils of the Southern States and Puerto Rico. Southern Cooperative Series Bull 174. North Carolina State University, Raleigh, pp 3–16

Daniels RB, Boul SW, Kleiss HJ, Ditzler CA (1999) Soil systems in North Carolina. Bull 467, Soil Science Department, North Carolina State University, Raleigh

Daniels RB, Perkins HF, Hajek BF, Gamble EE (1978) Morphology of discontinuous phase plinthite and criteria for its field identification in the southeastern US. Soil Sci Soc Am J 42:944–949

Fiskell JGA, Perkins HG (1970) Selected coastal plain soil properties. Bull 148, Southern Coop Series

Graham RC, Boul SW (1990) Soil-geomorphic relations on the Blue Ridge Front: II. Soil characteristics and pedogenesis. Soil Sci Soc Am J 54:1367–1377

Hack JT (1989) Geomorphology of the Appalachian highlands. In: Hatcher RD Jr, Thomas WA, Viele GW (eds) The Appalachian–

Ouachita Orogen. vol F-2, The Geology of North America, Geological Society of America, Boulder, pp 459–470
Harris WG, Carlisle VW (1987) Clay mineralogical relationships in Florida Haplaquods. Soil Sci Soc Am J 51:481–484
Harris WG, Crownover SH, Comerford NB (1995) Experimental formation of Aquod-like features in sandy coastal plain soils. Soil Sci Soc Am J 59:877–886
Harris WG, Hollien KA, Carlisle VW (1989) Pedon distribution of minerals in coastal plain Paleudults. Soil Sci Soc Am J 53:1901–1906
Holzhey CS, Daniels RB, Gamble EE (1975) Thick Bh horizons in the North Carolina Coastal Plain: II Physical and chemical properties and rates of organic additions from surface sources. Soil Sci Soc Am J 39:1182–1187. doi:10.2136/sssaj1975.03615995003900060040x
Hubbard RK, Sheridan JM (1983) Water and nitrate–nitrogen losses from a small upland, coastal plain watershed. J Environ Qual 12:291–295
Huddleston JS (1990) Soil survey of Carroll County, Mississippi. USDA-Natural Resources Conservation Service, Washington
Jackson CR, Martin JK, Leigh DS, West LT (2005) A southeastern piedmont watershed sediment budget; evidence for a multi-millenial agricultural legacy. J Soil Water Cons 60:298–310
McCracken RJ, Daniels RB, Fulcher WE (1989) Undisturbed soils, landscapes, and vegetation in a North Carolina Piedmont virgin forest. Soil Sci Soc Am J 53:1146–1152. doi:10.2136/sssaj1989.03615995005300040026x
Monteith SE (1990) Soil survey of Lauderdale County, Tennessee. USDA-Natural Resources Conservation Service, Washington
Murray GE (1961) Geology of the Atlantic and Gulf Coastal province of North America. Harper and Brothers, New York
Mukundan R, Radcliffe DE, Ritchie JC, Risse LM, McKinley RA (2010) Sediment fingerprinting to determine the source of suspended sediment in a Southern Piedmont stream. J Environ Qual 39:1328–1337. doi:10.2134/jeq2009.0405
Norfleet ML, Smith BR (1989) Weathering and mineralogical classification of selected soils in the Blue Ridge Mountains of South Carolina. Soil Sci Soc Am J 53:1771–1778
O'Brien EL, Buol SW (1984) Physical transformations in a vertical soil-saprolite sequence. Soil Sci Soc Am J 48:354–357
Pavich MJ (1985) Appalachian Piedmont morphogenesis: weathering, erosion, and Cenozic uplift. In: Morisawa M, Hack JT (eds) Tetonic geomorphology, Proceedings on 15th annual Binghamton geomorphology symposium, Boston, pp 299–319
Pavich MJ (1986) Processes and rates of saprolite production and erosion on a foliated granitic rock of the Virginia Piedmont. In: Coleman SM, Dethier DP (eds) Rates of chemical weathering of rocks and minerals. Academic Press, New York, pp 551–590
Ransom MD, Smeck NE (1986) Water table characteristics and water chemistry of seasonally wet soils of southwestern Ohio. Soil Sci Soc Am J 50:1281–1290
Ruhlman MB, Nutter WL (1999) Channel morphology evolution and overbank flow in the Georgia piedmont. J Am Water Resour Assoc 35:277–290
Rutledge EM, West LT, Omakupt M (1985) Loess deposits on a Pleistocene age terrace in Eastern Arkansas. Soil Sci Soc Am J 49:1231–1238
Rutledge EM, Guccione MJ, Markewich HW, Wysocki DA, Ward LB (1996) Loess stratigraphy of the Lower Mississippi Valley. Eng Geol 45:167–183
Saucier RT (1974) Quaternary geology of the Lower Mississippi Valley. Research Series 6, Arkansas Archeological Survey, Fayetteville
Saucier RT (1994) Geomorphology and quaternary geologic history of the Lower Mississippi Valley, vol 1. US Corps of Engineers, Mississippi River Comm, Vicksburg
Schoeneberger PJ, Amoozegar A, Buol SW (1995) Physical property variation of soil and saprolite continuum at three geomorphic positions. Soil Sci Soc Am J 59:1389–1397
Schrader TP (2006) Status of water levels and selected water-quality conditions in the Sparta-Memphis Aquifer in Arkansas, Spring–Summer 2003. US Geological Survey Scientific Investigations Report 2006-5021
Schrader TP (2010) Water levels and selected water-quality conditions in the Mississippi River Valley alluvial aquifer in eastern Arkansas, 2008. US Geological Survey Scientific Investigations Report 2010-5140
Schroeder PA, Melear ND, Bierman P, Kashgarian M, Caffee M (2001) Gibbsite growth ages for the regolith in the Georgia Piedmont. Geochimca et Cosmochimica Acta 65:381–386
Schwetmann U (1993) Relations between iron oxides, soil color, and soil formation. In: Bigham JM, Ciolkosz EJ (eds) Soil color. Soil Science Society of America, Madison, pp 51–69
Shaw JN, Hajek BF, Beck JM (2010) Highly weathered mineralogy of select soils from Southeastern U.S. Coastal Plain and Piedmont landscapes. Geoderma 154:447–456
Shaw JN, Odom JW, Hajek BF (2003) Soils on quaternary terraces of the Tallapoosa River, Central Alabama, USA. Soil Sci 168:707–717
Shaw JN, West LT, Truman CC, Radcliffe DE (1997) Hydraulic properties of soils with water restrictive horizons in the Georgia Coastal Plain. Soil Sci 162:875–885
Sodek F III, Carlisle VW, Collins ME, Hammond LC, Harris WG (1990) Characterization for selected Florida soils. Res 81-1. University of Florida, Gainesville
Soil Survey Staff (2014) Keys to soil taxonomy, 12th edn. USDA-Natural Resources Conservation Service, Washington
Stone EL, Harris WG, Brown RB, Kuehl RJ (1993) Carbon storage in Florida spodosols. Soil Sci Soc Am J 57:179–182
Thornbury WD (1965) Regional geomorphology of the United States. Wiley, New York
Trimble SW (1974) Man-induced soil erosion on the Southern Piedmont. Soil and Water Conservation Society, Ankeny
Vepraskas MJ, Guertal WR, Kleiss HJ, Amoozegar A (1996) Porosity factors that control the hydraulic conductivity of soil-saprolite transition zones. Soil Sci Soc Am J 60:192–199
Wascher HL, Humbert RP, Cady JG (1947) Loess in the southern Mississippi Valley: identification and distribution of the loess sheets. Soil Sci Soc Am Proc 12:389–399
Watts FC (1996) Soil survey of Baker County, Florida. USDA-Natural Resources Conservation Service, Washington, DC
Watts FC (1998) Soil Survey of City of Jacksonville, Duval County, Florida. USDA-Natural Resources Conservation Service, Washington, DC
West LT, Abreu MA, Bishop JP (2008) Saturated hydraulic conductivity of soils in the Southern Piedmont of Georgia, USA: Field evaluation and relation to horizon and landscape properties. Catena 73:174–179
West LT, Rutledge EM, Barber DM (1980) Sources and properties of loess deposits on Crowley's Ridge in Arkansas. Soil Sci Soc Am J 44:353–358
Woody WE (1989) Soil survey of Gaston County, North Carolina. USDA-Natural Resources Conservation Service, Washington

14 Northeastern Forage and Forest Region: LRR R

M.A. Wilson and R.K. Shaw

14.1 Introduction

Land Resource Region R (Fig. 14.1) encompasses most of New England and stretches from northern Maine through northern New Jersey, Pennsylvania, and Ohio (USDA-NRCS 2006). It is composed of nine Major Land Resource Areas (MLRAs), several physiographic provinces, and a wide array of geology (Table 14.1). Most of the region was ice covered during the last glacial advance; thus, soils in this area are uniformly young, generally less than 18,000 years old.

14.1.1 Climate

The climate can be characterized as temperate, with long cold winters and short mild summers. Regional climate varies widely from north to south, and also by elevation, topography, and proximity to bodies of water (Atlantic Ocean and Great Lakes) (Ciolkosz et al. 1984). The Appalachian Mountains are the major factor controlling regional climate variability and create an orographic effect for weather fronts. In general, higher elevations receive greater rainfall and have lower temperatures. Also, aspect is important at higher mountainous elevations with northern aspects being cooler, moister, and generally having greater organic matter accumulation than the southern aspect.

Rainfall across LRR R varies from about 960 to 1270 mm yr^{-1} and generally increases from the western part of the region to the Atlantic seaboard (Fig. 14.2). Soils have primarily an udic soil moisture regime (SMR) (Soil Survey Staff, 2014). Each MRLA has soils with aquic SMRs, though generally these soils are concentrated along the Great Lakes and St Lawrence River area.

Elevations across LRR R range from sea level to 1917 m on Mt. Washington in New Hampshire. Mean annual air temperatures range from about 11 °C (from Connecticut through central Pennsylvania) to about 3 °C in northwestern Maine (Fig. 14.2). The soil temperature regime (STR) is mesic in the lower half of the LRR and frigid in higher elevations of north central Pennsylvania (Waltman et al. 1997), northern New York, Vermont, New Hampshire, and Maine. Soils have a cryic STR at higher elevations in Maine, New Hampshire, and Vermont (e.g., at elevations greater than 700 m in Maine) (Hughes et al. 1993). In Maine, forested soils with an O horizon were reported to have a frigid STR as compared to a cryic STR for agricultural soils lacking an O horizon (Roberts 2008). The moderating effect of the Great Lakes on the climate ranges from increased snowfall from lake effect snows (up to 3500 mm of snow near Lake Ontario) to increased temperature and length of the growing season in the southwestern part of LRR R along Lake Erie (Fig. 14.2).

14.1.2 Vegetation

LRR R has native forest vegetation that was regionally variable. In general, central hardwood forests [white oak (*Quercus alba* L.)-red oak (*Quercus rubra* L.)-hickory (*Carya ovata* (Mill.) K. Koch)-white pine (*Pinus strobus L.*)] covered much of western New York and the eastern seaboard from Connecticut and Rhode Island to southern Maine. Northern Pennsylvania, central New York, and mountainous regions of Vermont and New Hampshire are in the northern deciduous hardwood forests [American beech (*Fagus grandifolia* Ehrh.)-birch (*Betula aleghaniensis*)-maple (*Acer rubrum* and *Acer saccharum*)]. The Adirondack Province of the Northeastern Mountains MLRA and most of central and northern Maine are in the boreal conifer forest region [spruce (*Picea rubens and Picea mariana)*-fir (*Abies*

M.A. Wilson (✉)
USDA-NRCS, Lincoln, NE, USA
e-mail: mike.wilson@lin.usda.gov

R.K. Shaw
USDA-NRCS, Somerset, NJ, USA
e-mail: richard.shaw@nj.usda.gov

L.T. West et al. (eds.), *The Soils of the USA*, World Soils Book Series,
DOI 10.1007/978-3-319-41870-4_14

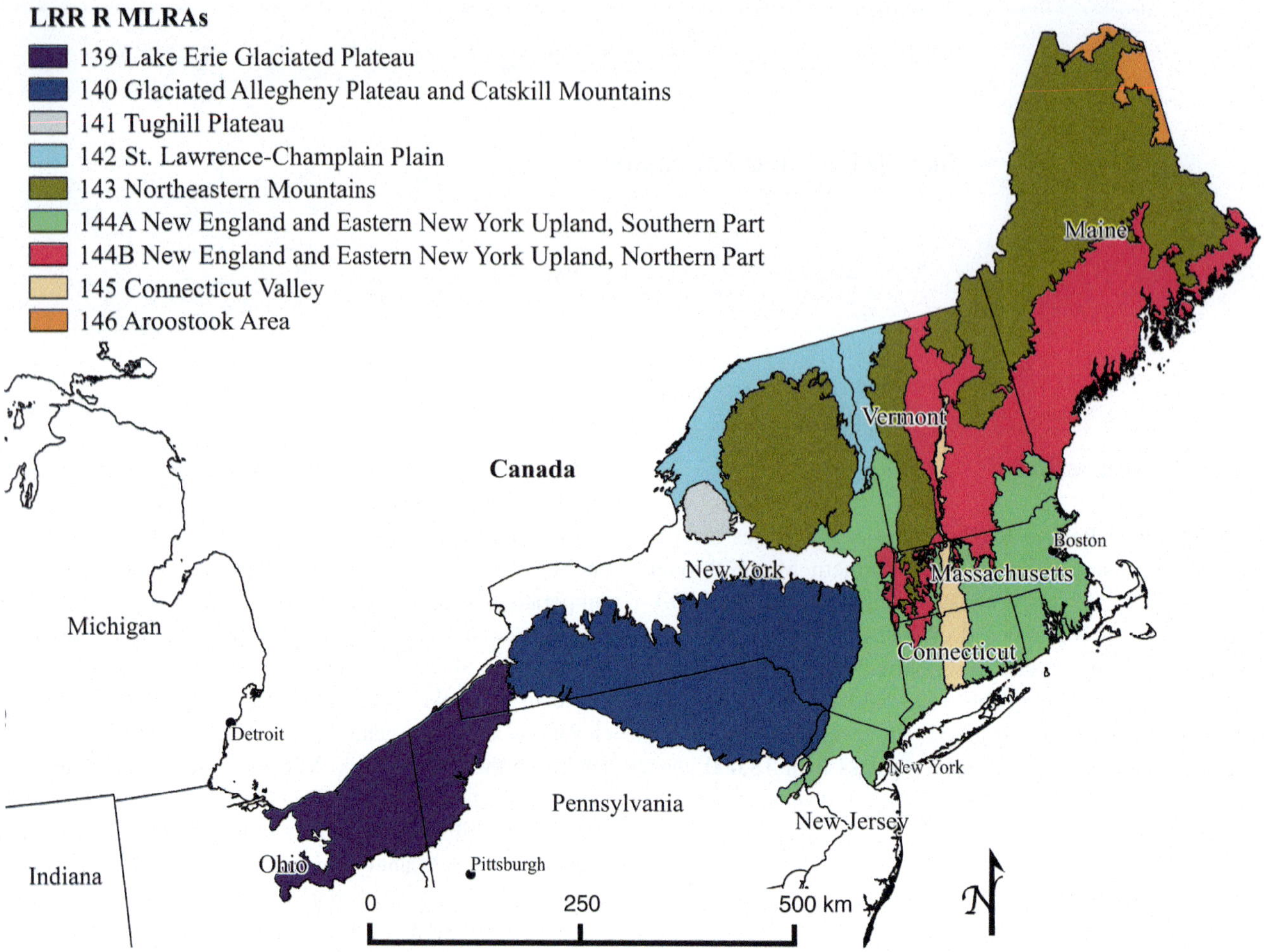

Fig. 14.1 Major land resource areas within Land Resource Region R

balsamea)] (Ciolkosz et al. 1984). Forest types vary with elevation; e.g., boreal conifer forests occupy higher elevation, while northern deciduous hardwoods occur in lower elevations in the northern part of the LRR.

Immediately following the retreat of the final Wisconsin glaciation, the northern New England region was generally a treeless tundra. Tree species were introduced from the south and gradually spread northward. The early-to-middle Holocene era is believed to have been about 2 °C warmer than currently, and white pine was a dominant species. During the past 1000 years (during the Little Ice Age), boreal forest species expanded into the LRR R (Jacobson 2000). Red Spruce is a codominant species in the climax vegetation of the boreal conifer forest and is currently abundant in higher elevation ecosystems with acid Spodosols and Inceptisols (Blum 1990). This tree species was once more widespread in the central and northern Appalachian Mountains, but its abundance has been reduced by logging, and its range reduced due to acid deposition and possibly climate change (Adams and Eagar 1992).

Forested regions of New England were heavily logged since the time of European settlement, which began in the 1600s. A large percentage of the land area was subsequently developed for agricultural use, and composition and regional patterns of forests were altered (Fuller et al. 1998; Morin et al. 2001). Since the late 1800s, pastures and fields have been abandoned and these areas have been undergoing reforestation by a gradual succession of different forest communities. For example, in Middlesex County, Massachusetts, 75–80 % of the land had been cleared for agricultural use by the late 1700s, compared to only 6 % agricultural land in the county in 2002 (Peragallo 2009). Currently, there are more areas of forest in this region than any time in the past 200 years (Foster et al. 2010). Modern threats to forests include invasive diseases and insects, climate change, and urban sprawl. Due to climate change, ecosystems in this region are undergoing increases in temperature that will bring about transitions to other plant/tree communities across the LRR and will introduce new pests and weed species in the region (Morin et al. 2001; Park et al. 2014; Horton et al. 2014).

Table 14.1 Physiographic provinces and sections for each MLRA in LRR R

Province	Section	MLRA								
		139	140	141	142	143	144A	144B	145	146
		Lake Erie Glaciated Plateau	Glaciated Allegheny Plateau and Catskill Mountains	Tughill Plateau	St. Lawrence-Champlain Plain	Northeastern Mountains	New England and Eastern New York Upland, Southern Part	New England and Eastern New York Upland, Northern Part	Connecticut Valley	Aroostook Area
Appalachian Highlands Region										
Valley and Ridge	Middle		X							
Valley and Ridge	Hudson Valley				X		X			
St. Lawrence Valley	Champlain				X					
Appalachian Plateau	Mohawk			X	X	X				
Appalachian Plateau	Catskills		X							
Appalachian Plateau	Southern New York	X	X							
Appalachian Plateau	Allegheny Plateau		X							
Appalachian Plateau	Kanawha	X	X							
New England	Seaboard Lowlands						X	X		
New England	New England Upland					X	X	X	X	X
New England	White Mountain					X				
New England	Green Mountain									
New England	Taconic				X			X		
Adirondack					X	X				
Interior Plains Region										
Central Lowland	Eastern Lake	X		X	X					
Central Lowland	Till Plains	X								

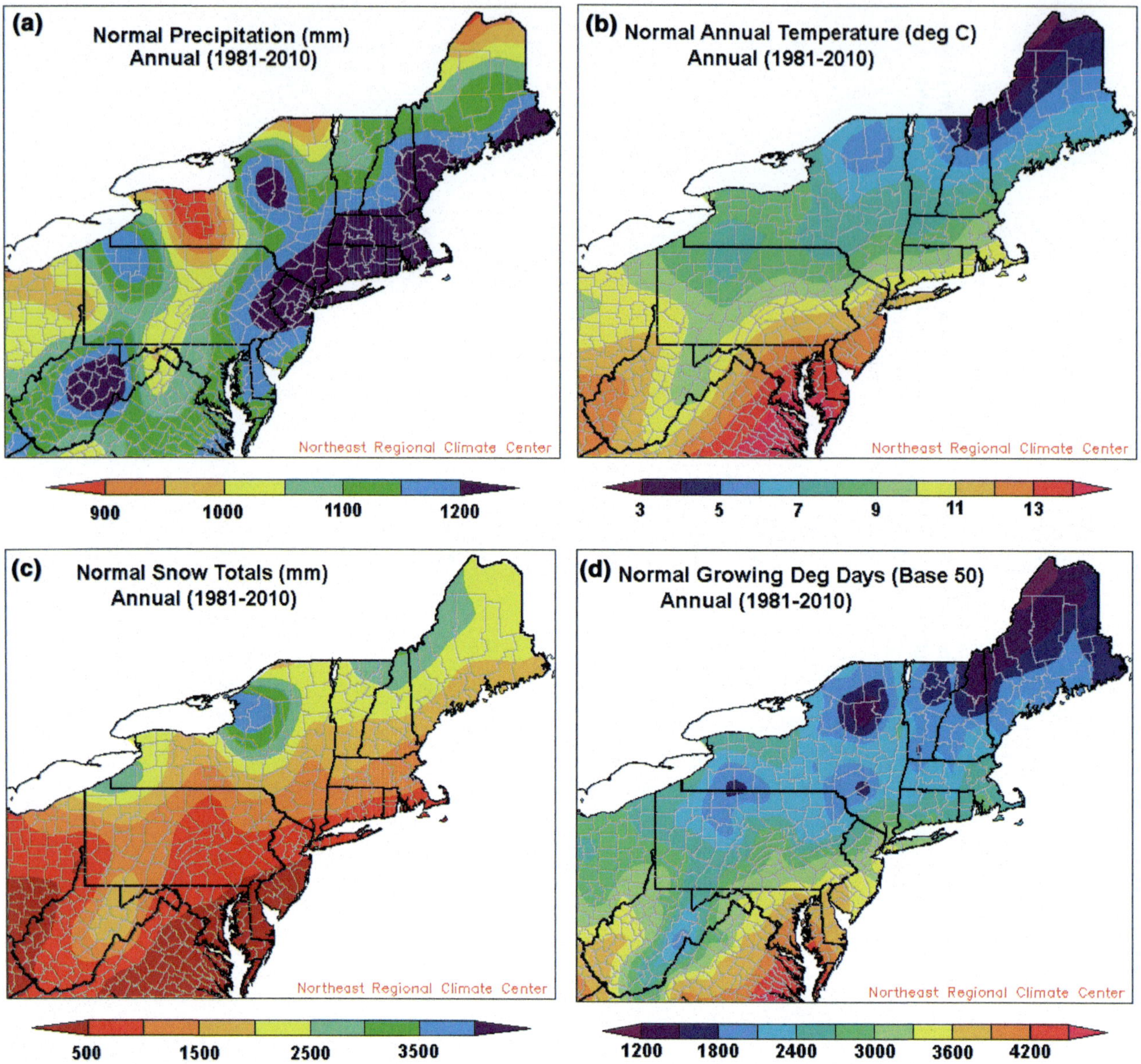

Fig. 14.2 Climate of New England: **a** mean annual precipitation, **b** mean annual air temperature, **c** normal annual snow totals, and **d** normal growing degree-days. *Source* NOAA, Northeast Regional Climate Center at Cornell University. http://www.nrcc.cornell.edu/page_northeast.html

The large pool of soil organic C in forested regions in the northeast USA is recognized as increasingly important with the current interest in carbon sequestration (greenhouse gas mitigation) and climate change (Hoover et al. 2012; Zhu and Reed 2014). Size of this C pool in forested ecosystems is governed by drainage and moisture availability, pedogenesis, and plant productivity. Carbon accumulates and is better preserved in poorly drained soils due to lack of oxidation during periods when water tables are high. On uplands sites in Maine, Raymond et al. (2012) report that greater amounts of C accumulate in somewhat poorly drained rather than poorly drained landscape positions, likely due to greater vegetative production.

14.1.3 Parent Materials

The northeast part of the USA has a very complex geologic history and is geologically quite variable (Ciolkosz et al. 1984; Eusden et al. 2013). The oldest surficial strata are igneous and metamorphic rocks of the Precambrian Grenville Formation, exposed in the Adirondack Mountains of northern New York as well as the Green Mountains of Vermont. Various types of sedimentary rocks were formed in subsequent geologic eras. Also, folding, faulting, and lifting of the strata occurred and these events during the Permian era resulted in the Appalachian relief with metamorphosis of existing materials. Erosion occurred during the late Mesozoic and Cenozoic times, but subsequent differential uplift has created the mountainous elevations that are present today. For example, Mount Washington in New Hampshire was calculated to have about 13 km of rock material overlying the current summit that has been eroded (Eusden et al. 2013).

The entire area has been glaciated multiple times, most recently about 24–28,000 years ago during the Pleistocene epoch when the glacial maximum occurred (Braun 2011). During this advance, the Laurentide ice sheet covered eastern Canada and was up to 3500 m thick. It profoundly influenced soils and landscapes of New England and LRR R; landscapes were eroded, morainal till deposits were left on glacial boundaries, rivers and drainage systems altered, proglacial lakes formed, outwash sediments were deposited in valleys, and silt-sized eolian loess was derived from glacial sediments and redeposited by the wind (Porter 1983).

Thus, upland soils have been forming since the retreat of the last glacier (between 18,000 ybp in New Jersey and New York and 13,500 ybp in northern Vermont and Maine), while alluvial soils on flood plains and tidal marsh soils are much younger (Ridge et al. 2012). This is the only glacial episode that is generally visible, though there is evidence of multiple glaciations in certain locations (e.g., coastal New England and southern Maine) (Drake 1971). Multiple till layers on a landscape do not necessarily signify separate glacial advances (Goldthwait 1971). These layers could be a changing pattern of deposition of a single episode. For example, a compacted till could be deposited by the glacial advance, while loamy, friable till may develop from sediments left by the melting glacier.

The geologic strata in any particular area impacts the nature of the till deposit as most sediments composing soil parent materials were derived from local bedrocks; i.e., glacial deposits were generally moved only a short distance from their source. Soils developed from till derived from granitic and gneissic rocks commonly are sandy with rounded or subrounded rock fragments. These soils often have a quartz-rich sand fraction, with moderate amounts of feldspars, and biotite and muscovite mica. In the clay fraction, mica alteration products include hydroxy-interlayered vermiculite and smectite, as well as small amounts of kaolinite.

Till-derived soils from schistose and phyllitic rocks are higher in silt, with flat rock fragments. In northwestern Pennsylvania, the low relief and easily eroded shaly bedrock resulted in thick till deposits, while in northeastern Pennsylvania, higher relief and more resistant sandstone bedrock resulted in ridge crests of bare rock, thin till deposits on uplands, and thicker glacial deposits in valleys (Braun 2011).

Till occurs in two general forms, based on location of the till at deposition: (a) subglacial or lodgment till (formerly referred to as basal till) and (b) supraglacial till or ablation till (melt-out) (Schoeneberger et al. 2012; Goldthwait and Matsch 1989). Subglacial till was deposited by moving glacial ice (generally on landforms such as drumlins and moraines), is finer grained than ablation till, and is highly compacted due to weight of the overlying glacier. Drumlins, smooth elongated hills of subglacial till, are a common glacial landform developed from the laminar flow of glaciers and are generally oriented north to south in LRR R. Subsoil horizons developed in compacted subglacial till generally meet requirements of a densic horizon (Soil Survey Staff 2014) and have firm to very firm moist consistence, are root restrictive, but are not cemented as they will slake in water. Supraglacial till is coarser textured, uncompacted (friable), and permeable. Soils derived from subglacial materials typically are more poorly drained and have greater land use restriction compared to soil developed in supraglacial till. The bulk density of the substratum (Cd) horizons of soils formed in subglacial till is commonly higher (up to 1.9 Mg m^{-3}) compared to the substratum of those formed in the friable till. In general, the saturated hydraulic conductivity is moderately high to high throughout the profile of soils formed from supraglacial till, while the substratum conductivity is reduced in those pedons developed from dense till. The Cd horizons tend to seasonally perch water (seasonal high water table), creating temporary seasonal reduction. This feature is a limitation for onsite septic systems and dwellings with basements.

Composition of till reflects changes in bedrock over which glaciers moved from north to south. Depth of till deposits is governed by filling of preglacial valleys, variable lithology and deposition of till, and erosion during the post-glacial periods (Milliron et al. 2007). Also, till thins from the edge of the ice movement to the source, as well as from the valley floor to the upper parts of the landscape in mountainous areas (Goldthwait 1971). Because of the heterogeneity in lithology and particle size of till deposits, rock fragments of mixed lithology are common in these soils and rock fragment content is a major land use limitation in most New England soils.

Glaciofluvial landforms are the result of water flowing from the melting glacier, creating deposition of stratified materials filling stream valleys that form terraces, outwash plains, kames, eskers, and deltas. Outwash materials, deposited when glacial stream velocity decreased due to slope, are generally sorted and stratified by the moving water. Kettles (basins) may also be present, created when blocks of ice were detached from the melting glacier and subsequently covered with outwash deposits. A kettle is formed when the ice melted. Kettles may be filled with water (ponds), organic materials (bogs), or remain dry (Fig. 14.3).

Glacial lakes were common throughout LRR R following the recession of the Laurentide ice sheet. Many of these lakes eventually filled with fine-grained sediments or drained

Fig. 14.3 Acidic bog formed in a kettle in Maine

when their natural dams breached. Glacial lacustrine and marine deposits are present in several MLRAs and are principal components of the St. Lawrence-Champlain Plain MLRA. Lacustrine deposits in former glacial lakebeds are generally fine textured with appreciable amounts of silt and clay.

Thin mantles of loess from glacial lake beds and river floodplains created by melting glaciers have been identified across the LRR. These silty-textured deposits are generally thickest in areas near the Great Lakes and gradually thin eastward, though loess deposits have been identified in the Northeastern Mountains of central Maine and in subaqueous soils on the coast of Rhode Island.

14.1.4 Soils and Pedogenesis

Differentiation of soils in the glaciated landscapes of LRR R is generally due to the type of glacial deposit (till versus glaciofluvial materials, density of till (subglacial versus supraglacial till), depth of till over bedrock, degree of soil development, subsoil color, lithology/mineralogy of parent material, rock fragment content, and presence/absence of a fragipan. Due to forest vegetative cover, most soils have (or had if cleared for agricultural or urban use) organic horizons at the surface that overlie the uppermost mineral A or E horizon.

There are seven soil orders represented in this LRR (Table 14.2). Inceptisols are the dominant order overall. There is a wide range in Inceptisol composition among MLRAs, ranging from 20 % in MLRA 143–93 % in MLRA 140. Dystrudepts are the most prevalent great group composing nearly one half of the total area of Inceptisols (Table 14.3). Other important great groups of Inceptisols in the region are Endoaquepts, Fragiudepts, and Fragiaquepts. Conceptually, Inceptisols have limited pedogenic development and in LRR R, the limited development is related to time since glacial retreat, but may also be due to the coarse texture and siliceous mineralogy of common acid sandstone and shale parent materials. In addition, many soils are located on steeply sloping landscapes, where erosion and colluvial processes are active.

Fragipans are common in this region and impact properties such as hydraulic conductivity and root penetration. Fragipans perch water during the spring wet-season, creating a seasonally high water table. In the late summer, moisture perched above the fragipan has evaporated and water is limiting for crop growth as rooting depth into the subsoil is limited. Fragipans were recognized in many soils in early soil surveys, but it was later concluded that fragipans in many soils were actually horizons developed in dense basal till, especially in the Adirondacks and northern New England (Lindbo and Veneman 1989). Many soils in the LRR, however, have features including bleached prism faces, high bulk density, and lack of roots that are indicative of fragipans (Lindbo and Veneman 1993). Currently, fragipans are regarded as more common in better developed Alfisols of southern New York and northern Pennsylvania than in Inceptisols in this region (Table 14.3).

Spodosols are the second most common soil order in the LRR (Table 14.2), with Haplorthods composing 90 % of Spodosols (Table 14.3; Fig. 14.4). Areal extent of Spodosols generally increases from south to north across the region, and

Table 14.2 Percentage of soil orders in each MLRA for Land Resource Region R

	MLRA								
	139	140	141	142	143	144A	144B	145	146
	%								
Alfisols	65.4	5.8	4.0	21.1	0.3	4.8	0.0	0.0	0.0
Entisols	1.9	0.2	5.0	8.1	1.0	10.0	3.1	19.6	0.0
Inceptisols	26.9	93.2	43.3	46.7	20.4	75.4	44.9	72.6	33.2
Histosols	0.9	0.5	4.0	2.9	5.4	5.3	3.1	1.9	0.0
Spodosols	0.0	0.0	43.1	19.9	72.8	2.4	48.9	5.0	66.8
Mollisols	0.7	0.0	0.6	1.2	0.0	0.1	0.0	0.8	0.0
Ultisols	4.3	0.2	0.0	0.0	0.0	2.1	0.0	0.2	0.0

Table 14.3 Great groups in Land Resource Region R, with percent of LRR for each

Great group	Percent of order	Great group	Percent of order
Inceptisols		*Entisols*	
Dystrudepts	44.4	Udorthents	43.0
Endoaquepts	15.8	Udipsamments	39.5
Fragiudepts	13.2	Fluvaquents	4.3
Fragiaquepts	9.1	Psammaquents	3.8
Eutrudepts	6.6	Endoaquents	3.1
Humaquepts	5.9	Udifluvents	2.0
Epiaquepts	4.2	Quartzipsamments	1.8
Humudepts	0.7	Cryorthents	0.8
Fragiochrepts	0.1	Epiaquents	0.6
Cryaquepts	<0.1	Sulfaquents	0.4
		Sulfiwassents	0.3
Spodosols		Psammowassents	0.2
Haplorthods	89.6	Frasiwassents	0.1
Haplohumods	4.3		
Endoaquods	2.6	*Histosols*	
Humicryods	1.4	Haplosaprists	69.5
Fragiorthods	1.1	Cryofolists	11.3
		Haplohemists	9.4
Alfisols		Udifolists	5.3
Hapludalfs	27.1	Sulfihemists	2.6
Fragiaqualfs	20.9	Haplofibrists	1.3
Fragiudalfs	19.9		
Endoaqualfs	18.9	*Mollisols*	
Epiaqualfs	12.3	Endoaquolls	59.4
		Argiaquolls	26.4
Ultisols		Hapludolls	12.6
Fragiudults	63.3	Epiaquolls	1.6
Fragiaquults	25.0		
Hapludults	11.6		

they are more common in soils with frigid or cryic STRs. The cool, humid climatic conditions in northern parts of LRR R, combined with the sandy, quartz-rich parent materials and acid-producing spruce-fir forest vegetation create conditions favorable for the development of albic and spodic horizons (Jongmans et al. 1997; Lundström et al. 2000).

Fig. 14.4 Sandy, acidic Haplorthod in Northeastern Mountains of Maine

The process of spodic horizon formation involves the translocation (vertical and/or lateral) of Fe and Al complexed with organic acids derived from surface conifer litter. The gray or white albic (E) horizons principally reflect the residual quartz grains following the loss of clays, base cations, Fe, and Al during the eluviation process. Dark reddish brown to black illuvial horizons (Bs, Bhs, Bhm) are created by the accumulation of spodic materials composed of organic matter, Fe, and Al in the subsoil (Wilson and Righi 2010). Darker, organic-rich spodic materials (Bh horizons) typically accumulate above the redder, Fe/Al rich materials (Bs horizons) (McKeague et al. 1971; De Coninck 1980).

Shoji and Yamada (1991) reported an abundance of allophane, imogolite, and Al/Fe humus complexes in Bs horizons of New England Spodosols. Ferrihydrite was identified as the principal iron oxide, while clay minerals were expansible 2:1 minerals in E horizons and chloritized 2:1 minerals in Bs horizons. Spodic horizons typically have chemical and physical characteristics which result in their classification in the isotic soil mineralogy family class (Soil Survey Staff 2014). This class is characterized by a high NaF pH ($\geq$8.4) and a 1500 kPa water retention to clay ratio $\geq$0.6. The abundance of poorly crystalline (short-range order) components and substantial variable charge result in significant P fixation (Wilson et al. 2002).

In many areas of northern New England, Inceptisols and Spodosols are found on similar landscape positions. When soils with spodic horizons are used for agriculture and tilled, the albic/spodic sequence is often destroyed. Without the morphological evidence of these horizons, these soils are likely mapped as Inceptisols.

Alfisols are another important Order in LRR R. Alfisols are more pedogenically developed than Inceptisols and are common soils in glaciated areas of central New York. These soils have commonly formed over calcareous till that has enhanced translocation of clays (Ciolkosz et al. 1989). Some Alfisols (and Ultisols) can also be found at the southern end of the Wisconsin glaciation in New Jersey, where soils have had a slightly longer time for development.

Mollisols and Ultisols compose a very minor area in the region. Entisols and Histosols are also of only minor extent but are important components of landscapes. Organic soils form in bogs and other low positions within landscapes (depressions and drainageways) where water tables are elevated. Entisols are common in alluvial valleys along rivers and streams, and for soils formed in human-altered and human-transported materials.

14.2 MLRA 139: Lake Erie Glaciated Plateau

The Lake Erie Glaciated Plateau MLRA (27,770 km^2) is part of two quite different physiographic provinces: the Appalachian Plateau and Central Lowlands (Table 14.1). Much of the southern part of the MLRA is in the glaciated Appalachian Plateau province. Rocks in this part of the Appalachian Plateau are predominantly sandstone, siltstone, and shale of upper Devonian, Mississippian, and Pennsylvanian age. Bedrock is commonly overlain by mantles of till, stratified outwash deposits (kames and eskers), and lacustrine sediments from former glacial lakes. The Central Lowlands province occurs along Lake Erie and is nearly level. This area was a Wisconsin-aged glacial lake and is commonly referred to as the Lake Plain (Milliron et al. 2007). Some areas of the Lake Plain were depositional sites for iron ores (bog iron) and were mined for several years in the mid-1800s.

The native vegetation in the region was northern deciduous hardwood. The climate varies somewhat between the two physiographic provinces that compose the MLRA. Along Lake Erie in the Central Lowlands province, the climate is relatively more moderate (more frost free days and a longer growing season), and suitable for specialty crops such as grapes, fruit trees, and vegetables (Milliron et al. 2007).

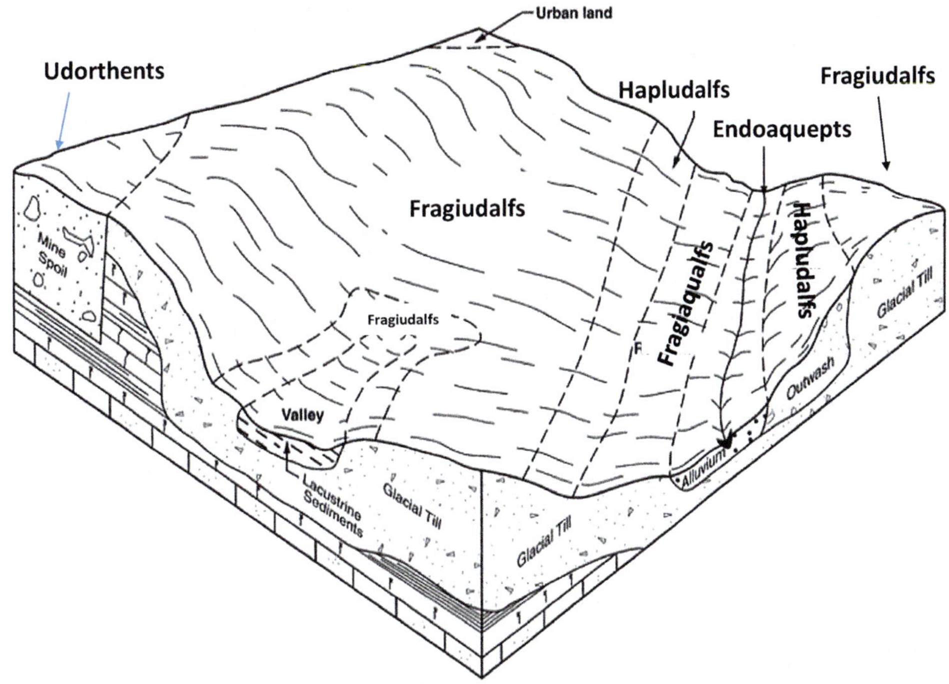

Fig. 14.5 Soil association composed of Fragiudalfs and Fragiaqualfs, located in the Appalachian Plateau Section of MLRA 139 (Milliron and Buzard 2007)

Lack of adequate drainage can limit soil productivity unless artificially drained (e.g., tiles, ditches). The growing season is shorter in the glaciated Appalachian Plateau region of the MLRA. Dairy farming is important in this region as well as row-crop production (corn, oats, and wheat) and forestry.

This MLRA is unique in LRR R as Alfisols make up 65 % of the soils in this area, followed by Inceptisols (27 %), and Ultisols (4 %) (Table 14.2). Fragiaqualfs and Fragiudalfs are the dominant great groups, and these soils are common on gently to steeply rolling landscapes in the dissected glacial Appalachian plateau area of the MLRA (Fig. 14.5). These soils are very deep, moderately well to somewhat poorly drained, on till plains that vary from gently sloping to steep. They have fragipans in the subsoil, resulting in seasonally high water tables in winter and spring. These soils have poor fertility, and their root zone is strongly to very strongly acidic.

The association of Epiaquepts and Udipsamments (Fig. 14.6) from Ashtabula County is an example of soils of the Lake Plain region. The very deep and somewhat poorly drained Epiaquepts are formed in glaciolacustrine sediments with a thin mantle of loess, while the moderately well drained Udipsamments are formed on a beach ridge deposit. Land use limitations are generally seasonal wetness, compaction, and potential for groundwater contamination.

14.3 MLRA 140: Glaciated Allegheny Plateau and Catskill Mountains

MLRA 140 is principally in the Southern New York Section of the Appalachian Plateau Province and is about 57,975 km^2 (Table 14.1). The Allegheny plateau, part of the Appalachian Mountains, extends from west and central New York, south to Kentucky and West Virginia. It is composed of sedimentary rocks (principally acid shales, sandstones, and siltstones) of Devonian age that have been subsequently eroded. The Catskill Mountains, on the east side of the MLRA, are a dissected plateau formed from sediments deposited during the Devonian and Mississippian eras (principally sandstones). MLRA 140 is composed of the part of these plateaus that was glaciated, most recently during the late Wisconsin glacial period. Upland landscapes are mantled by till, and outwash covers valley floors. Some glacial lake sediments are also present in valleys.

Vegetation in MLRA 140 is generally northern deciduous hardwood forest. This area was first settled by the Europeans in about 1796, and most of the landscapes were cleared for agriculture (pasture and limited row crops such as corn or small grains) and forest-based industries (Morin et al. 2001). A large portion of the upland areas has soils and slopes that have made it less favorable for agricultural production and

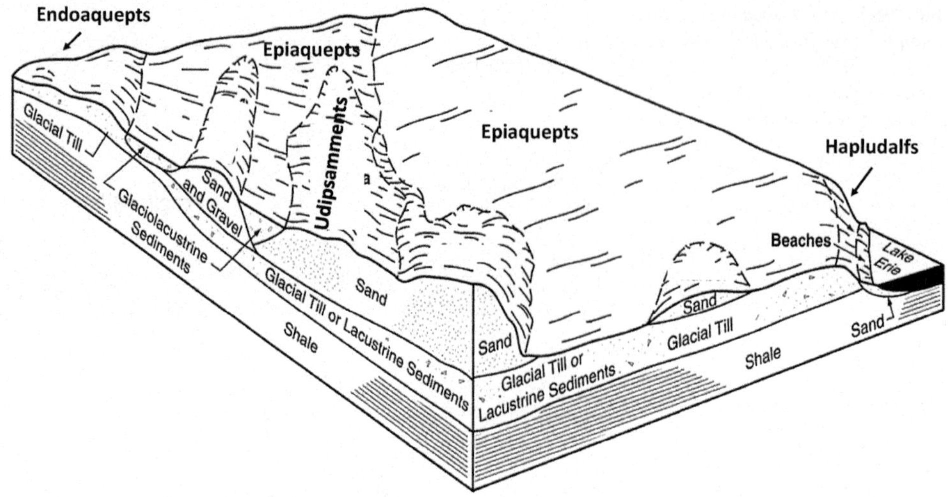

Fig. 14.6 Soil association located in the Lake Plain section of MLRA 139. Ashtabula County, Ohio Soil Survey (Milliron et al. 2007)

have been abandoned in recent years. Thus, many areas are idle or have reverted to woodlands. Despite the steep landscapes, the valley floors can be quite broad. Some less sloping areas and valleys are still managed for pasture, hay, corn, small grains, and other crops generally supporting the local dairy industry. Small vegetable farms are also common. Soils that lack fragipans (such as on outwash plains) are more productive for row crops, pastures and hay, and locally grown vegetables. Dairy farming, beef and livestock production, maple syrup production, and forestry enterprises are economically important industries as well.

This MLRA contains both mesic and frigid STRs. Elevation of much of the Glaciated Plateau lies between 366 and 610 m, with the highest elevations in the southern part of the area (Fenneman 1938). Precipitation is well distributed throughout the year, with frequent winter snows. Inceptisols compose 93 % of the soils in MLRA 140 (Table 14.2). Common upland soils in the MLRA are Fragiaquepts, derived from Wisconsin-aged dense basal till. These deep, somewhat poorly drained soils lie in concave to linear glaciated uplands and have a dense fragipan in the subsoil (Fig. 14.7). When located on broad ridge tops or footslopes of hills, the soils can be nearly level to sloping (0–15 % slope), but slopes can range up to 35 % on steeper side slopes. Thus, erosion and wetness are major resource concerns in these soils. These acid soils are deep with loamy surface textures and have a high percentage of rock fragments. They have moderate amounts of organic C in surface horizons, which varies with land use and management. The bulk density of the fragipan can range up to 1.9 Mg m^3. Associated soils occupying higher adjoining landscape positions are Fragiudepts and very deep, poorly to very poorly drained Fragiaquepts occur in upland depressions.

Dystrudepts are common in areas with thin mantles of till over bedrock (Fig. 14.7). These soils are typically moderately deep (depth to bedrock ranging from 50 to 100 cm) and are well drained. They also range from moderately to extremely acid and have abundant rock fragments. Dystrudepts developed from glaciofluvial materials on outwash plains, kames, terraces, and alluvial fans are common soils in valleys. They are typically deep, well to excessively well drained. These soils formed on more level landscape positions and are well suited for row-crop production, though water-holding capacity and rock fragment content are major limitations.

14.4 MLRA 141: Tughill Plateau

The Tughill Plateau (MLRA 141) is 3,045 $km^{2,}$ and most of this area is in the Appalachian Plateaus Province of the Appalachian Highlands and the Central Lowland Province of the Interior Plains (Table 14.1). On the plateau, surficial geologic strata are principally Ordovician-aged sandstone, with some shale and siltstone. In the eastern and northern edge of the MLRA along the Black River valley, Precambrian-aged granites and gneiss are also found. All of the uplands have a mantle of till, much of it subglacial till. Some glacial lake sediments and moraines occur in the southwest part of the area and glaciofluvial deposits cover terraces, kames, eskers, and outwash plains.

MRLA 141 is technically a cuesta, a series of hills composed of multiple geologic strata that have topography ranging from rolling to steep; with wide, generally level ridges on top of plateaus and hilly to steep-side slopes. Thus, much of the extensive drainage network follows this change in geology and flow directions are principally northeast to southwest. These glaciated landforms can be steeply dissected, with slopes ranging up to 90 %. Due to these extreme slopes, the potential for surface runoff and erosion can be high if the soils are not vegetated.

Soils in this region have a frigid STR. In the wintertime, there is abundant snowfall (lake effect snow) due to the

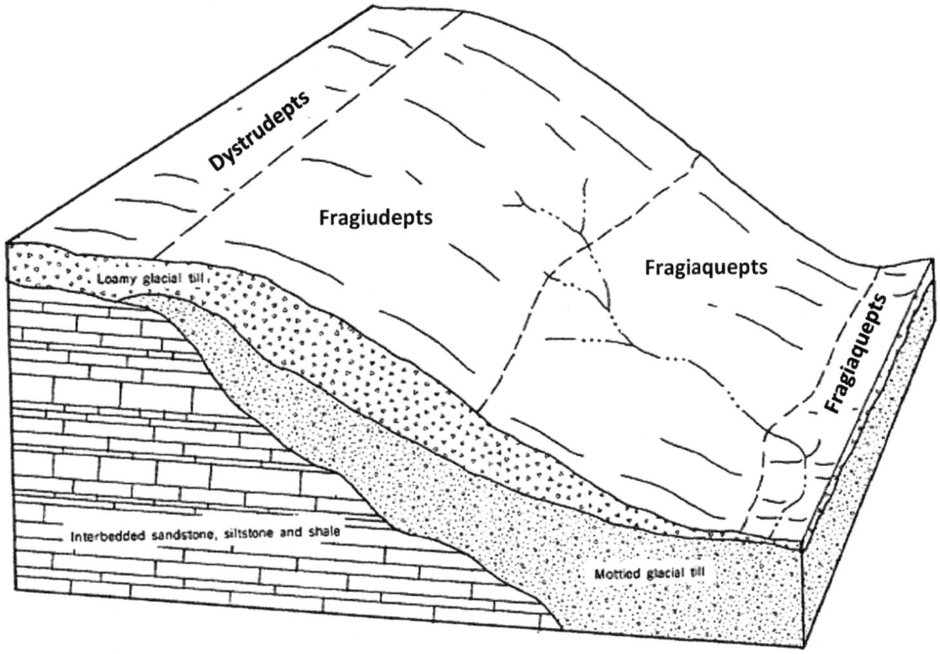

Fig. 14.7 Common soil drainage sequence on a landscape in MLRA 140 in Chenango County, NY (Crandall 1985). Dystrudepts, well drained; Fragiudepts, moderately well drained; Fragiaquepts, somewhat poorly drained and poorly to very poorly drained

proximity to Lake Ontario. The cooler, moister conditions of the Tughill Plateau area compared to MLRA 140 result in a higher percentage of Spodosols (principally Fragiorthods, Haplorthods, and Fragiaquods), composing 43 % of the soils in the MLRA. Inceptisols make up 47 % of the MLRA (principally Fragiaquepts, Dystrudepts, and Endoaquepts). There are minor amounts of Alfisols, Histosols, and Entisols. Much of the uplands are covered with dense till and soils with fragipans. Fragiorthods occupy higher knolls and ridges on the landscape and typically have a higher sand content compared to Fragiudepts and Fragiaquepts.

Soils on glaciofluvial landforms lack a fragipan and are typically Dystrudepts, Haplorthods, Udorthents, and Udipsamments. Very poorly drained Endoaquepts are extensive on flats and depressional areas of moraines and outwash plains. Larger wet areas commonly have very deep, very poorly drained Histosols vegetated by species such as tamarack (*Larix laricina* (Du Roi) K. Koch), black spruce (*Picea mariana* (Mill.) Britton, Sterns and Poggenb.), and paper birch (*Betula papyrifera* Marshall).

Land use is principally forests, with principal products of pulpwood, Christmas trees, and maple syrup. Limited acreage of pasture and cropland producing forage and small grains for dairy cattle are also present. Nearly all soils are acidic with low base saturation, though more alkaline soils are possible in areas where limestone is incorporated in the till. Soils generally have a high to very high saturated hydraulic conductivity due to the coarse textures, though hydraulic conductivity is reduced in fragipans. If the land use is pasture, overgrazing (loss of vegetation) or grazing while wet (compaction) are major land use concerns.

14.5 MLRA 142: St. Lawrence-Champlain Plain

The St Lawrence-Champlain Plain MLRA (18,240 km^2) is a horseshoe shaped region that lies along the St Lawrence River to the north, extending along three sides of the Tughill Plateau (MRLA 141). This MLRA is located in several physiographic provinces, including the Hudson Valley Section of the Valley and Ridge Province, Champlain Section of the St. Lawrence Valley, and the Mohawk Section of the Appalachian Plateau Province (Table 14.1). On the east side of the Adirondacks, this MLRA encompasses soils along the valley of Lake Champlain and lies on the border of New York and Vermont.

This MLRA is principally a glacial lake plain. The area has a mantle of till, outwash, lacustrine, and marine deposits forming the parent materials for soils (Trevail 2006). A thin mantle of loess, deposited during glacial recession, is also present. Inceptisols compose 47 % of the soils across the MLRA, with great groups such as Eutrudepts, Dystrudepts, Endoaquepts, and Epiaquepts (Table 14.2). Spodosols (Haplororthods, Endoaquods, Fragiorthods, Fragiaquods) are 20 % of the area, while Alfisols (21 %), Entisols (8.1 %), and Histosols (3 %) are also found. Soils have either mesic or frigid STRs, and native vegetation on the glacial lake plains is central hardwood forest. This area has a greater population density and more agricultural activity than adjacent MLRAs. Forest production is the major land use in the MLRA, with lumber and pulpwood as the main products. But many soils are cleared and used for hay and pasture, with corn, wheat, oats, and apples produced. Dairy and beef cattle operations are common.

The predominant landform of the MLRA consists of nearly level areas of sandy deltas and lacustrine basins or plains. Lake Vermont, a proglacial lake formed during the recession of the Laurentide Ice Sheet about 14,000 years ago, encompassed the present-day Lake Champlain. When the ice sheet melted, marine water inundated the isostatically depressed St. Lawrence and Champlain lowlands. Isostatic rebound of these lowlands about 10,000 ybp resulted in formation of the current Lake Champlain. Thus, marine deposits are present on the perimeter of Lake Champlain and lacustrine deposits occur at a further distance from the current lake shore. Figure 14.8 illustrates an association of Hapludalfs and Endoaqualfs formed in calcareous estuarine and glaciolacustrine clays. The pH is moderately acidic in the solum, increases with depth, and carbonates commonly occur from 100 to 150 cm. The saturated hydraulic conductivity can be moderately high in the surface, but low to very low in the subsoil. These soils remain wet until late in the spring and are wet early in the fall. Another common association of soils are Epiaqualfs, Endoaquepts and Epiaquepts developed in glacial lacustrine, and marine deposits on lake plains and uplands mantled with lake sediments (Trevail 2006) (Fig. 14.9). They are level to nearly level, and are somewhat poorly to very poorly drained, especially when found in depressional areas. These soils can have a seasonably high water table that impacts crop production.

In other parts of the MLRA, soils are derived from glacial sediments overlying either limestone/dolomite or more acidic granites, gneisses, and schists. Soils developed in till derived from schist, granite, and gneiss are acidic and are commonly shallow and moderately deep Haplorthods, while soils forming from limestone and dolomites are Eutrudepts. These soils with calcareous parent materials vary in depth to bedrock and have a densic contact in the lower substratum when developed from dense till deposits. The better drained soils are on convex knolls and ridges, and the poorer drained catena members are on foot slopes or nearly flat areas. They have a high base status and often have carbonates in the lower depths. Runoff varies based on slope, and saturated hydraulic conductivity is moderately high to high in the solum, but decreases in the substratum of soils with Cd horizons. Many of these soils are cleared and used for hay, pasture, corn, or small grains due to their more fertile nature.

14.6 Northern Mountains Region: MLRA 143: Northeastern Mountains and MLRA 144B: New England and Eastern New York Upland, Northern Part

Northeastern Mountains MLRA 143 (95,465 km^2) lies in several physiographic provinces of the Appalachian Highlands Region: Adirondack, Appalachian Plateaus, and New England (Table 14.1). It consists of four distinct areas: the Adirondack Mountains in New York, White Mountains of New Hampshire and northern Maine, Green Mountains in Vermont, and Berkshires in Massachusetts. The Northeastern Mountains MLRA is home of the highest peaks in New England, most occurring in New Hampshire, with 17 peaks over 1400 m in elevation. Principal bedrock in MLRA 143 is metamorphic rock such as gneiss, schist, slate, marble, and quartzite. Igneous rocks (principally granite and granodiorite) are also present, intruded into the metamorphic rocks. These different geologic units have been folded and faulted during the uplift of mountainous areas. While this entire area was glaciated by the Laurentide glacier, there is evidence of

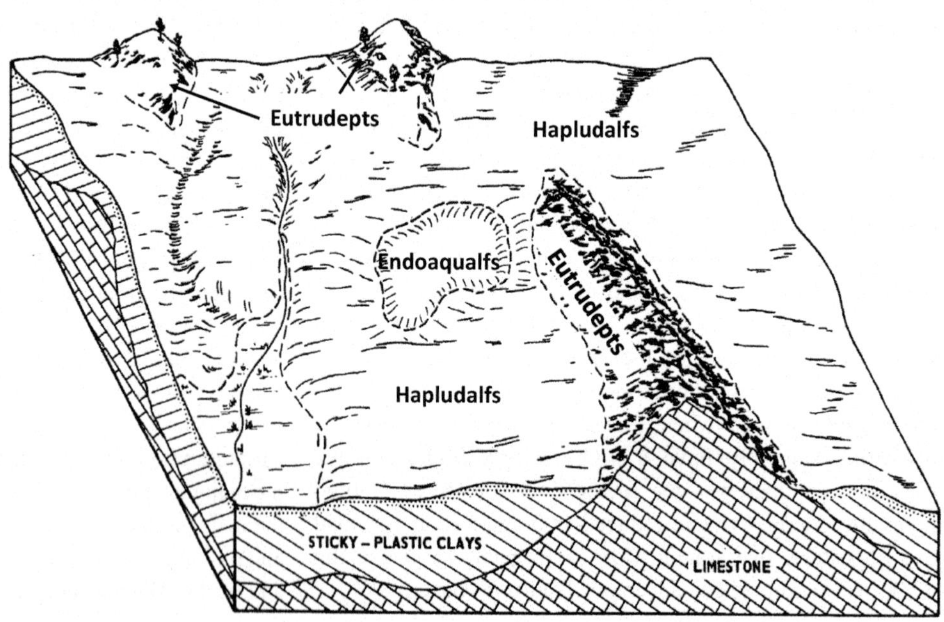

Fig. 14.8 Soil association in MRLA 142 Chittenden County, Vermont (Allen 1974)

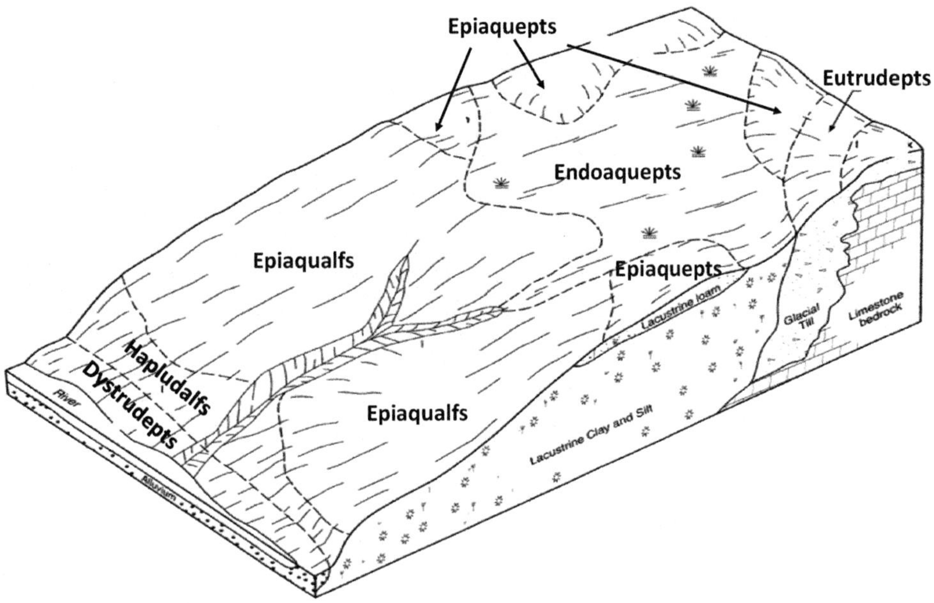

Fig. 14.9 Soil association in MLRA 142 in Clinton Co., NY. Lake plain landscapes have a high water table that impacts land use (Trevail 2006)

alpine glacial events at higher elevations of the White Mountains (Eusden et al. 2013) and in Maine (Caldwell and Davis 1983). Soils in the Northeastern Mountains MLRA generally have a frigid STR, though cryic STRs occur at high elevations in the Adirondack Mountains. The MLRA is composed of 73 % Spodosols, 20 % Inceptisols, and 5 % Histosols (Table 14.2). Spodosols are principally Haplorthods (65 % of total area), with minor amounts of Humods, Cryods, and Aquods. Inceptisols are principally Endo- and Humaquepts (15 %), with lesser amounts of Udepts.

MLRA 144B (New England and Eastern New York Upland, Northern Part) occupies 53,125 km^2 and lies principally in the New England Uplands Section of the New England Physiographic Province. A part of the western area is in the Taconic Section and another area of the MLRA in southeastern Maine is in the Seaboard Lowlands, both in the same New England Province (Table 14.1). Bedrock geology is generally gneiss, granite, and schist, but some limestone and dolomite units are exposed in the southern and western sections of the MLRA. Soils developed from limestone and dolomite have higher pH and base saturation (Eutrudepts) than soils developed from more acidic rocks and are important for agricultural production. Soils in MLRA 144B have a frigid STR and are composed of 49 % Spodosols and 45 % Inceptisols (Table 14.2). Haplorthods are the dominant great group in the MLRA, composing 47 % of the land area, while Udepts and Aquepts suborders compose the Inceptisol. Landscapes in MLRA 144B generally consist of glaciated hills and mountains (Fig. 14.10). Lakes, bogs, and other wetlands are numerous. The relief is prominent, and mountains have steep sideslopes ranging from 3 to 80 %.

The large percentage of Spodosols that occur in these two MLRA regions contrasts to MLRA 144A, directly to the south, that has only 2 % Spodosols. MLRA 143 and 144B have a colder climate (frigid soil temperature regime) than 144A (mesic soil temperature regime) contributing to a greater amount of spruce-fir vegetation and conditions that influence the development of spodic features.

Both MLRAs (143 and 144B) have boreal conifer forests that occupy higher elevations on mountain summits, while a combination of northern deciduous hardwoods and spruce-fir forests occupy middle slopes. Also, white pine (*Pinus strobus L.)* and eastern hemlock (*Tsuga canadensis* (L.) Carrière) are common species. Forestry, comprising over 75 % of the land use in both MLRAs, is the main economic enterprise. Also, there are some dairy and beef farms and corn, potatoes, apples, and beans are main cash crops. In general, this is a sparsely populated area, though the majority of the population in Maine lives in MLRA 144B.

Moderately deep (50–100 cm), well drained Halplorthods are common in these two MLRAs (Fig. 14.11) on convex upper slopes and backslopes. Some shallow (25–50 cm deep to bedrock), excessively drained Haplorthods are also found on upper backslopes (Fig. 14.12). These soils commonly overlie areas of loamy, friable till over schist, gneiss, phyllite, or granite. Slopes may range from 0 to 50 %. These soils are moderately to extremely acidic in the solum and have very low base saturation. Soils have a moderately high to high saturated hydraulic conductivity, but erosion is a hazard due to the sloping landscapes. Available water-holding capacity is moderate and restricted by depth of soil. Most areas are wooded due to the steep slope. Conservation tillage, cover crops, residue management, and rotational grazing are important if used for agriculture.

Fig. 14.10 Typical glacial landscape in MLRA 144B (Caledonia County, VT)

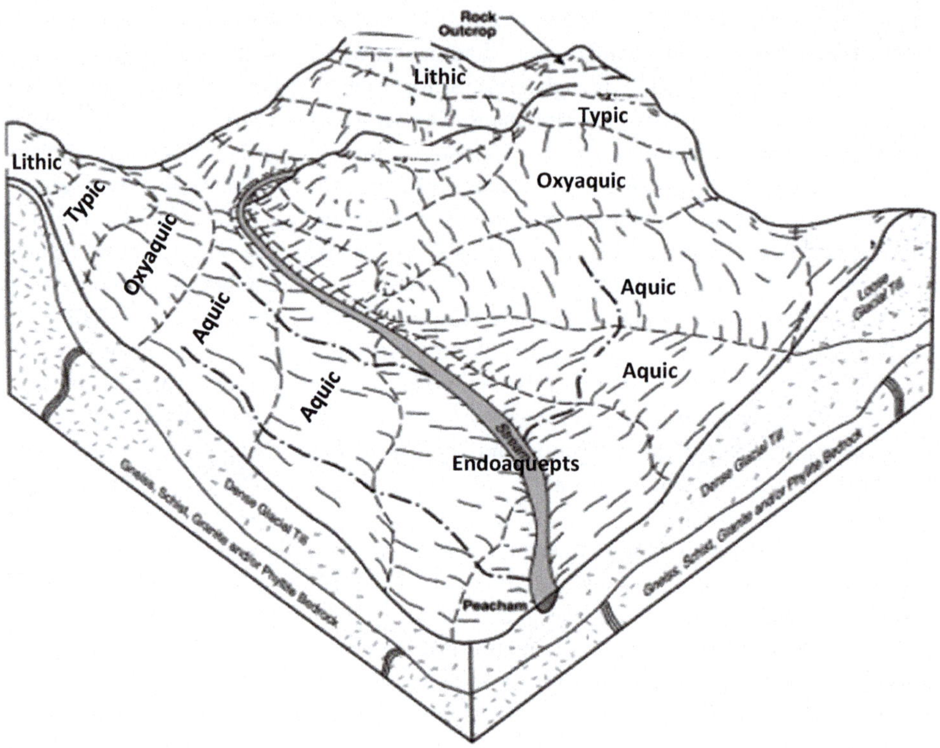

Fig. 14.11 Soil association from MLRA 144B illustrating the subgroups of Haplorthods across the landscape (Hedstrom 2003)

14.7 MLRA 144A: New England and Eastern New York Upland, Southern Part

MLRA 144A (48,180 km^2) is composed of two provinces in the Appalachian Highlands Division. The northern part of the area in Vermont and New York occurs in the Valley and Ridge Province, while the remainder lies in the New England Province (Table 14.1). The most common rocks are granite, gneiss, and schist. Sandstones, shales, and limestones are common in the Pennsylvania and southeast portion of the province in New York, while dolomite and limestone dominate the rock types in the valleys of northwestern Connecticut. An area of Triassic red beds and associated igneous rocks is also found in eastern New Jersey

Fig. 14.12 Profile of a Lithic Haplorthod in MLRA 144B. The profile has 18 cm of O horizon (Oi and Oa), an albic E horizon, Bhs horizon, and Bs horizon. The contact with the R horizon is at 62 cm

and New York. The entire area has been glaciated, and the southern boundary of the MLRA is the furthest extent of glaciation on the eastern coastal area.

This MLRA has a mesic STR. The area is extensively wooded, principally being central hardwood forest, with maple (Acer rubrum and Acer saccharum), birch (*Betula aleghaniensis*), and eastern hemlock (*Tsuga canadensis* (L.) Carrière) as associated forest components. The principal land use is forestry used for wood products, hunting, and recreation. Large areas of urban development are also present. Many rural areas have been cleared and used for cultivated crops, hay, or pasture, and, especially along the Atlantic coast. Dairies, truck crops, apple, and nursery stock production are also common.

Landscapes are generally rolling hills and valleys that are not as mountainous or hilly as compared to MLRA 144B. Soils are acidic and rocky. Inceptisols compose 74 % of the soils in MLRA 144A, and Dystrudepts are the dominant great group, occupying 58 % of the total area (Table 14.2). Other soil orders are much less extensive, e.g., Entisols (10 % of the MLRA), Histosols (5 %), and Alfisols (5 %). Spodosols compose only 2 % of the MLRA.

Till is the dominant parent material in the area and ranges in composition and depth. Most soils formed in acid till derived from schists, granites, gneiss, or phyllite. Generally, soils from granites, schists, and gneiss have loam and fine sandy loam textures with moderate amounts of rock fragments. They are moderately to strongly acid, have a high water-holding capacity, and are suitable for agriculture (corn, silage, apples) and urban use on less sloping landscapes. Dystrudepts are mapped on both subglacial till with dense substratum (Cd horizon) and on landscape components with friable, loamy (supraglacial) till (Fig. 14.13). They commonly have a moderate amount of rock fragments.

Till in certain areas may also be derived from glacial materials high in iron sulfide bearing micaceous schist.

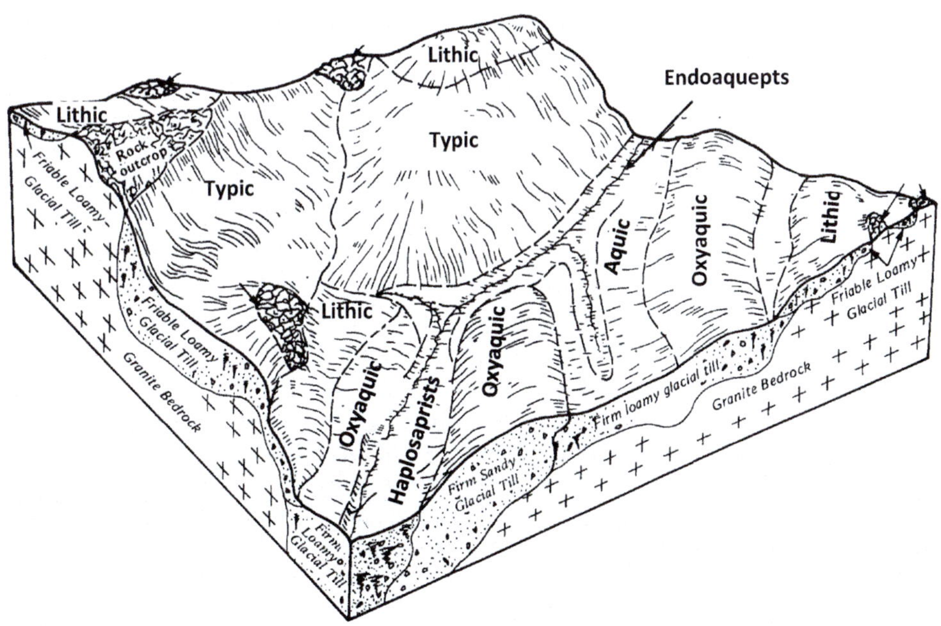

Fig. 14.13 Soil association composed principally of various subgroups of Dystrudepts in MLRA 144A (Peragallo 1989)

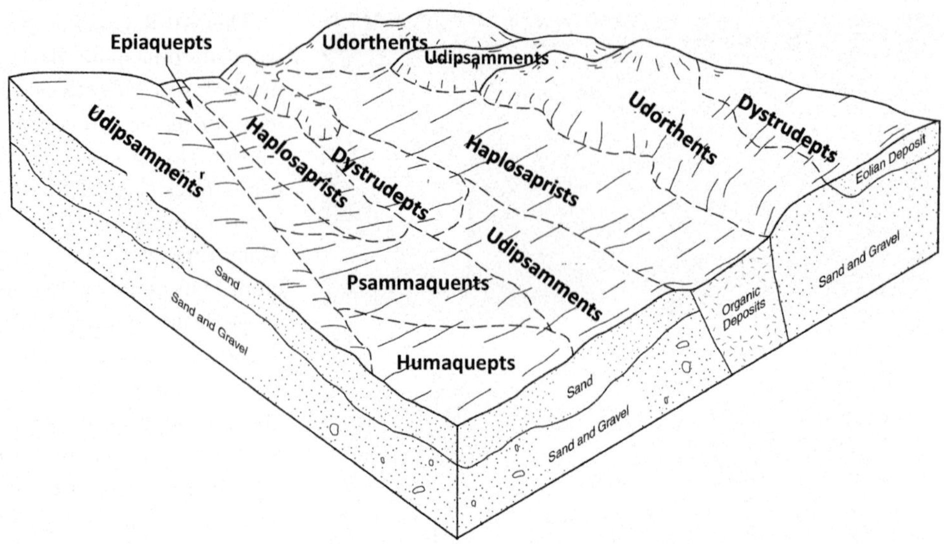

Fig. 14.14 A typical pattern of soils and underlying material of sandy-textured soils in MLRA 144A (Peragallo 2009)

These Dystrudepts are highly weathered, post-active sulfate soils and are extremely to moderately acid (pH 3.5–6.0). Also, these acid sulfate soils have a high content of citrate dithionite-extractable Fe (3–7 %) and moderate amounts of gibbsite, resulting in a parasesquic family mineral class (McVey 2006).

Soils on terraces, outwash plains, kames, and eskers are typically sandy and poorly developed. However, some areas may contain surface mantles of locally derived finer textured windblown deposits with better developed soil profiles. Figure 14.14 illustrates this type of landscape with sandy-textured Udorthents, important soils in this MLRA region and across New England. This figure also illustrates the occurrence of organic-rich Haplosaprists in the lowest part of the landscape. These organic deposits are common in bogs or depressions in level uplands and alluvial plains and can range up to several hundred hectares in size. These soils are acidic, very poorly drained and generally ponded, with over 120 cm of highly decomposed (sapric) organic materials. Organic soils are poorly suited for agriculture or woodlands, but native vegetation can be cleared for cranberry production.

Soils formed in human-altered and human-transported materials are common in the many urbanized areas of the MLRA. These include loamy Eutrudepts and Dystrudepts, as well as skeletal Udorthents high in artifacts (Fig. 14.15), and Udipsamments formed in dredged materials. Soils on coastal margins are also important in this area. These Sulfihemists tidal marsh soils are deep and very poorly drained. They occur on level landscapes that are influenced by tides and inundated daily with salt water. They range from strongly acid to mildly alkaline. Due to the wetness and flooding, these soils are not suitable for any land use or development, but perform important filtering and buffering functions, and serve as wildlife habitat. Also, subaqueous soils are mapped on the estuarine areas along the coasts of Rhode Island and Connecticut (Demas and Rabenhorst 1999; Payne and Turenne 2009; Bradley and Stolt 2003). In addition, subaqueous soils are mapped in inland areas of LRR R, such as Lake Champlain and freshwater ponds in Rhode Island.

14.8 MLRA 145: Connecticut Valley

This MLRA (5,520 km^2) is in the New England Upland Section of the New England Province and occupies a rift valley (Table 14.1). The soils in this area formed since the last glacial retreat 10,000–12,000 years ago, from a variety of parent materials including alluvium, glaciolacustrine, glaciofluvial, till, and organic deposits along with lesser amounts of human-altered and human-transported materials. Lacustrine sediments from glacial Lake Hitchcock extend from west central New Hampshire to central Connecticut. Bedrock in the region includes the Triassic–Jurassic sedimentary red beds and igneous rocks of the Hartford Basin. The native vegetation in the area is oak (northern red (*Quercus rubra* L.), black (*Quercus velutina* Lam.), white (*Quercus alba* L.), eastern white (*Pinus strobus L.*) and pitch pine (*Pinus rigida* Mill.), eastern hemlock (*Tsuga canadensis* (L.) Carrière), and birch (*Betula aleghaniensis*). Land use today is principally woodland, with some agriculture. There is a considerable amount of urban land use as well.

The MLRA is composed of 72 % Inceptisols and 20 % Entisols, with lessor areas of Histosols (2 %) and Spodosols (5 %) (Table 14.2). Soils generally are in the mesic STR, but

Fig. 14.15 An Udorthent in MLRA 144A that is derived from human-altered, human-transported materials

some frigid areas occur in the northern part of the MLRA. Dystrudepts are by far the most common great group, with Udipsamments, Udorthents, and Endoaquepts also important. The Connecticut River flood plain is a distinct geomorphic landform in the valley (Fig. 14.16). Soils (mainly Udifluvents, Udisamments, and Fluvaquents) range from sandy to silty in texture and include most drainage classes. These alluvial soils are used extensively for agriculture, wildlife habitat, and recreation. A common group of soils in this area is Udorthents, Haplosaprists, and Udipsamments derived from glaciofluvial materials on outwash plains and have loamy sand and sand textures, are very deep, excessively drained, with high or very high saturated hydraulic conductivity. Landscapes are generally cleared and used for hay, pasture, vegetable, tobacco, or silage production, or urban development.

Glaciolaustrine deposits occupy a large segment of the valley extending from central Connecticut to southern Vermont in the area once flooded by glacial Lake Hitchcock. Fine-textured Endoaquepts, Eutrudepts, and Humaquepts are the dominate soils. These slowly permeable soils often have a seasonal high water table and severe limitation for onsite septic systems. Some areas are used for pasture, hay land, or vegetable production; other uses include wildlife habitat and a limited amount of urban development. The clayey deposits are mined for brick production in a few locations.

Fig. 14.16 Soils along the Connecticut River Valley in MLRA 145 (unpublished, developed by D Parizek, NRCS, CT)

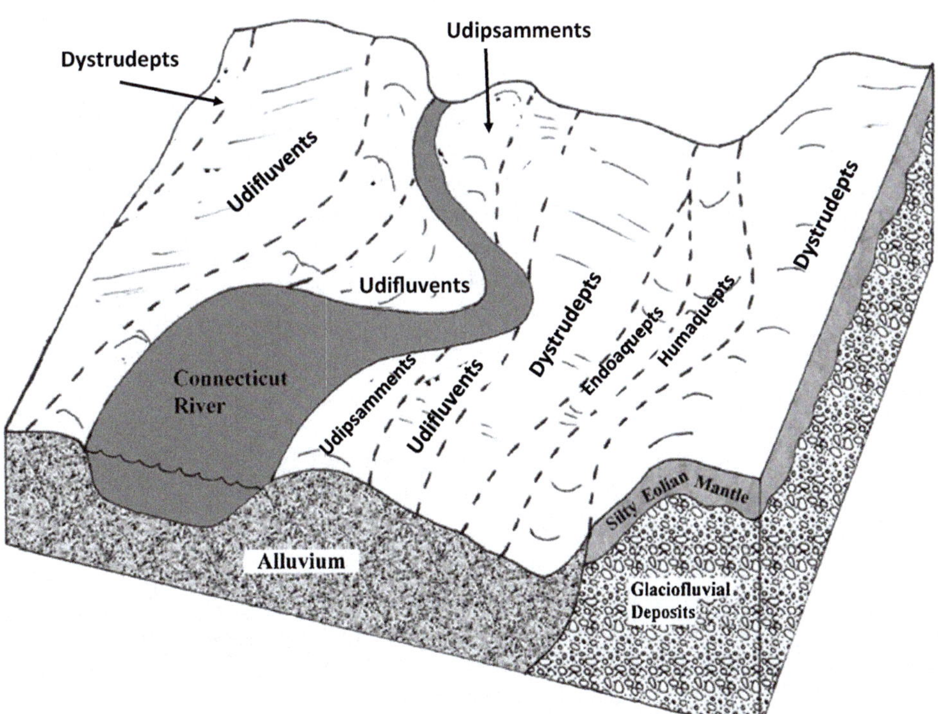

14.9 MLRA 146: Aroostook Area

This MLRA (3,305 km^2) lies in the northernmost part of Maine in the New England Uplands Section of the New England Province (Table 14.1). The climate in the area is cool and humid, with a frigid STR. The Cary Mills Formation, composed of limestone mixed with shale and siltstone, forms the bedrock in the eastern part of the area. Soils formed from this parent material are important for potato production in the state. The remainder of the MLRA has a variety of igneous and metamorphic rocks such as granite, gabbro, diorite, schist, gneiss, and argillite. In addition, dikes composed of basalt and diabase cross the area. Higher elevations are covered with a thin mantle of till, while deeper deposits of subglacial till occupy lower landscape positions. Outwash and glacial lake deposits are also common.

Land use is principally forest (saw lumber and pulpwood industry), and native vegetation is a mixture of coniferous and deciduous trees (sugar maple (Acer saccharum), American beech (*Fagus grandifolia* Ehrh.), birch (*Betula aleghaniensis*, quaking aspen (*Populus tremuloides* Michx.), red spruce (*Picea rubens)*, balsam fir (*Abies balsamea* (L.) Mill.), and white pine (*Pinus strobus L*)). The principal agricultural crop is potatoes, as well as some broccoli, oats, canola, and barley.

Spodosols compose 67 % of the area (Table 14.2). Typic Haplorthods are commonly mapped in the MLRA and are deep, well-drained soils on till plains and ridges. The till is generally a mix of limestone, calcareous sandstone, and shale; thus, soils have a range in pH from 5.5 to 8.0, depending on the till composition. These soils generally have less than 15 % clay, with an equal mix of sand and silt. Commonly, the spodic horizon is absent due to land use or tree throw. For this same reason, the O horizon is commonly absent in soils that have been cultivated. Inceptisols (Endoaquepts, Eutrudepts, Humaquepts) are the other dominant order (32 %), with Entisols as a minor component in alluvial-filled valleys. Broad flats or depressional areas of organic soils are also present.

14.10 Conclusion

Soils in LRR R are principally suited for forest land use, though agriculture (row crops, dairies, truck farms) is present in all MLRAs of the region. Upland soils are generally composed of till and use is limited by the cool climate as well as both chemical and physical properties; principally acidity, low water-holding capacity, and slope. The cool, humid climate and increasing boreal conifer forest vegetation is responsible for the increasing abundance of Spodosols from south to north in the region. Over all, the climate and abundance of forest land cover result in a high carbon content throughout this region. Forest management will be increasingly important in the future as soils in this area are major carbon sinks. With the advent of climate change, maximizing forest productivity will be important in sequestering carbon and mitigating greenhouse gases produced from burning of fossil fuels by our industrial, modern society.

References

Adams MB, Eagar C (1992) Impacts of acidic deposition on high-elevation spruce-fir forests—results from the Spruce-Fir Research Cooperative. For Ecol Manag 51:195–205. doi:10.1016/0378-1127(92)90485-r

Allen GW (1974) Soil survey of Chittenden County, VT. USDA-NRCS, Washington, DC

Blum BM (1990) Picea rubens Sarg. red spruce. In: Burns RM, Honkala BH (tech coord) Silvics of North America. Conifers Agric Handb 654, vol 1. US Department of Agriculture, Forest Service, Washington, DC, pp 250–259

Bradley MP, Stolt MH (2003) Subaqueous soil-landscape relationships in a Rhode Island estuary. Soil Sci Soc Am J 67:1487–1495

Braun DD (2011) The glaciation of Pennsylvania, USA. Dev Quat Sci 15:521–529

Caldwell DW, Davis PT (1983) The timing of alpine glaciation of Mt. Katahdin. In: Caldwell DW, Hanson LS (eds) A guidebook for the Greenville-Millinocket regions, north central Maine: New England intercollegiate geological conference, 75th annual meeting, Boston University, Boston, MA, pp 79–86

Crandall L (1985) Soil survey of Chenango County, NY. USDA-NRCS, Washington, DC

Ciolkosz EJ, Gardner TW, Sencindiver JC (1984) Chapter 2: Geology, physiography, vegetation, and climate. In: Cunningham RL, Ciolkosz EJ (eds) Soils of the Northeast, U.S. Pennsylvania Agricultural Experiment Station Bulletin 848, pp 2–47

Ciolkosz EJ, Waltman WJ, Simpson TW, Dobos RB (1989) Distribution and genesis of soils in the Northeastern United States. Geomorphology 2:285–302

De Coninck F (1980) Major mechanisms in formation of spodic horizons. Geoderma 24:101–128

Demas GP, Rabenhorst MC (1999) Subaqueous soils: pedogenesis in a submerged environment. Soil Sci Soc Am J 63:1250–1257

Drake LD (1971) Evidence for ablation and basal till in east-central New Hampshire. In: Goldthwait RP (ed) Till, a symposium. Ohio State University Press, Columbus, pp 73–91

Eusden JD, Thompson WB, Fowler BK, Davis PT, Bothner WA, Boisvert RA, Creasey JW (2013) The geology of New Hampshire's White Mountains. Durand Press, Lyme, NH

Fenneman NM (1938) Physiography of eastern United States. McGraw-Hill, New York, NY

Foster D, Kittredge D, Donahue B, Motzkin G, Orwig D, Ellison A, Hall B, Colburn B, D'Amato A (2010) Wildlands and woodlands: a vision for the forests of Massachusetts. Harvard Forest, Harvard University, Petersham, MA. http://harvardforest.fas.harvard.edu/sites/harvardforest.fas.harvard.edu/files/publications/pdfs/HF_WandW.pdf

Fuller JL, Foster DR, McLachlan JS, Drake N (1998) Impact of human activity on regional forest composition and dynamics in central New England. Ecosystems 1:76–95

Goldthwait RP (1971) Introduction to till, today. In: Goldthwait RP (ed) Till, a symposium. Ohio State University Press, Columbus, pp 3–26

Goldthwait RP, Matsch CL (1989) Genetic classification of glacigenic deposits (Final report of the commission on genesis and lithology of glacial quaternary deposits of the International Union for Quaternary Research (INQUA). A.A. Balkema Publ, Rotterdam

Hedstrom GT (2003) Soil survey of Franklin County area and part of Somerset County, Maine. USDA-NRCS, Washington, DC

Hoover CM, Leak WB, Keel BG (2012) Benchmark carbon stocks from old-growth forests in northern New England, USA. For Ecol Manag 266:108–114

Horton R, Yohe G, Easterling W, Kates R, Ruth M, Sussman E, Whelchel A, Wolfe D, Lipschultz F (2014) Chapter 16: Northeast. Climate change impacts in the United States: the third national climate assessment. In: Melillo JM, Terese TC, Richmond XX, Yohe GW (eds) U.S. Global Change Research Program, pp 371–395. doi:10.7930/J0SF2T3P

Hughes PA, Lesniewicz KA, Rourke RV (1993) Water tables, soil temperatures, and morphological characteristics in selected Maine soils. Maine Agric Exp Stn Tech Bull 150. University of Maine, Orono, Maine

Jacobson GL Jr (2000) Post-glacial changes in vegetation and climate in Northern New England. Rhodora 102(911):246–247

Jongmans AG, van Breemen N, Lundstrom U, van Hees PAW, Finlay RD, Srinivasan M, Unestam T, Giesler R, Melkerud PA, Olsson M (1997) Rock-eating fungi. Nature 389:682–683

Lindbo DL, Veneman PLM (1989) Fragipans in the northeastern United States. In: Smeck NE, Ciolkosz EJ (eds) Fragipans: their occurrence, classification, and genesis. SSSA Spec Publ 24. SSSA, Madison, WI, pp 11–32

Lindbo DL, Veneman PLM (1993) Morphological and physical properties of selected fragipan soils in Massachusetts. Soil Sci Soc Am J 57:429–436

Lundström US, van Breemen N, Bain D (2000) The podzolization process: a review. Geoderma 94:91–107

McKeague JA, Brydon JE, Miles NM (1971) Differentiation of forms of extractable iron and aluminium in soils. Soil Sci Soc Am Proc 35:33–38

McVey SJ (2006) Separating soil series using extractable iron and aluminum and the use of extractable iron, aluminum, and pH in the separation of red soils with differing geneses in southern New England, USA. MS Thesis, University of Connecticut

Milliron EL, Buzard RW (2007) Soil survey of Columbiana County, Ohio. USDA-NRCS, Washington, DC

Milliron EL, Prebonick ST, Svoboda JR (2007) Soil survey of Ashtabula County, Ohio. USDA-NRCS, Washington, DC

Morin RS, Liebhold AM, Gottschalk KW, Twardus DB, Acciavatti RE, White RL, Horsley SB, Smith WD, Luzader ER (2001) Forest health conditions on the Allegheny National Forest (1989–1999): analysis of forest health monitoring surveys. NA-TP-04-01. U.S. Department of Agriculture, Forest Service, Northeastern Area, State and Private Forestry, Newtown Square, PA

Park A, Puettmann K, Wilson E, Messier C, Kames S, Dhar A (2014) Can boreal and temperate forest management be adapted to the uncertainties of 21st century climate change? Crit Rev Plant Sci 33 (4):251–285. doi:10.1080/07352689.2014.858956

Payne MK, Turenne J (2009) Mapping the "New frontier" of soil survey: Rhode Island's MapCoast partnership. Soil Surv Horiz 50:86–89

Peragallo TA (1989) Soil survey of Norfolk and Suffolk Counties, Massachusetts. USDA-NRCS, Washington, DC

Peragallo TA (2009) Soil Survey of Middlesex County, Massachusetts. USDA-NRCS, Washington, DC

Porter SC (1983) Introduction. In: Porter SC (ed) Late-quaternary environments of the United States. vol 1: The Late Pleistocene, pp xi–xiv

Raymond JE, Fernandez IJ, Ohno T, Simon K (2012) Soil drainage class influences on soil carbon in a New England forested watershed. Soil Sci Soc Am J 77:307–317

Ridge JC, Balco G, Bayless RL, Beck CC, Carter LB, Dean JL, Voytek EB, Wei JH (2012) The new North American Varve chronology: a precise record of Southeastern Laurentide Ice Sheet deglaciation and climate, 18.2–12.5 kyr BP, and correlations with Greenland ice core records. Am J Sci 312:685–722

Roberts WJ (2008) Frigid forests vs. cryic fields: soil temperature regimes of two different cover types in Aroostook County, Maine. Soil Surv Horiz 49:42–44

Schoeneberger PJ, Wysocki DA, Benham EC, Soil Survey Staff (2012) Field book for describing and sampling soils (ver 3.0). USDA-NRCS, National Soil Survey Center, Lincoln, NE

Shoji S, Yamada H (1991) Comparisons of mineralogical properties between tephra-derived Spodosols from Alaska and nontephra-derived Spodosols from New England. Soil Sci 152 (3):162–183

Soil Survey Staff (2014) Keys to Soil Taxonomy, 12th edn. Soil Survey Staff, Washington, DC

Trevail TD (2006) Soil survey of Clinton County, New York. USDA-NRCS, Washington, DC

USDA-NRCS (2006) Land resource regions and major land resource areas of the United States, the Caribbean, and the Pacific Basin. U. S. Department of Agriculture Handbook 296

Waltman WJ, Ciolkosz EJ, Mausbach MJ, Svoboda MD, Miller DA, Kolb PJ (1997) Soil Climate Regimes of Pennsylvania. Penn State Agr Exp Station Bull 873

Wilson MA, Burt R, Ottersberg RJ, Lammers DA, Thorson TD, Langridge RW, Kreger AE (2002) Isotic mineralogy: criteria review and application in Blue Mountains, Oregon. Soil Sci 167:465–477

Wilson MA, Righi D (2010) Chapter 13: Spodic materials. In: Stoops G, Marcelino V, Mees F (eds) Interpretations of micromorphological features of soils and regoliths. Elsevier, Amsterdam, pp 251–273

Zhu Z, Reed B (eds) (2014) Baseline and projected future carbon storage and greenhouse-gas fluxes in ecosystems of the eastern United States: U.S. Geological Survey Professional Paper 1804. doi:10.3133/pp1804

15 Soils of Florida and the Caribbean: LRRs U and Z

Mary E. Collins

15.1 Land Resource Region U

15.1.1 Subtropical Fruit, Truck Crop, and Range Region

This region of the USA exists completely in the state of Florida (Fig. 15.1). The area of Florida south of a diagonal line from directly south of the city of Jacksonville in the extreme northeast corner of the state (Atlantic Ocean side) to the city of Cedar Key (Gulf of Mexico side), including the Florida Keys, comprise Land Resource Region U (LRR U). Thus, the major cities in Florida (Orlando, Gainesville, Tampa, St. Petersburg, Miami, Ft. Meyers, Naples, and Key West) are located within this region. Because of the continuing population growth in this region, there are limited resources, especially quantity and quality of freshwater, to support the influx of people. For example, it is estimated that 26×10^9 l d^{-1} are withdrawn from surface (about 40 %) and subsurface (about 60 %) water sources (USDA-NRCS 2006).

The land surface varies from rolling hills in the central part of Florida (Fig. 15.2) to level flatwoods (Fig. 15.3) to freshwater swamps and marshes along the coasts (Fig. 15.4). The elevation in this region goes from sea level to about 50 m on the central ridge (also known as the "Backbone of Florida"). The landscape positions of major soils in the region are shown in Figs. 15.5, 15.6, and 15.7.

Native vegetation includes live oak (*Quercus virginiana*), turkey oak (*Quercus laevis*), slash pine (*Pinus elliottii*), longleaf pine (Pinus palustris), Sable palmetto maidencane (*Panicum hemitomon*), and various types of mangroves and grasses. Agricultural crops include winter vegetables (Fig. 15.8), truck crops, citrus (Fig. 15.9), forestland, and grazing lands for livestock (Fig. 15.10). The parent material is dominantly sandy and/or loamy marine sediments, but also includes alluvial deposits, aeolian sands, limestone bedrock, and organic matter.

Even though Florida was never glaciated, the physiography of the state was influenced by the melting (interglacial periods) and the forming (glacial periods) of the glaciers (Watts and Collins 2008). Waxing and waning of the glaciers resulted in formation of eight marine terraces, and the Florida coastline has been stable for the last 5000 years. Therefore, ages of the parent material of the soils are as young as recent alluvial deposits to millions of years on the Central Florida Ridge (Watts and Collins 2008). Precipitation across the region ranges from 1115 to 1575 mm all as rainfall. Snow is extremely rare. More than 50 % of the precipitation comes in the months of June, July, and August as very intense tropical storms (Fig. 15.11). Fall and winter are somewhat dry. Air temperatures average 20–25 °C (Fig. 15.12). Summers are hot and humid, and winters are mild and relatively dry (USDA-NRCS 2006). White frost may occur during the winter months.

The soils in LRR U are excessively to very poorly drained. The soil moisture regime (SMR) reflects this drainage with soils having an udic or aquic SMR. All of the soils have a hyperthermic soil temperature regime (STR) (mean annual soil temperature at 50 cm ≥ 22 °C) except for those in the Florida Keys in extreme south Florida. The Florida Keys have an isohyperthermic STR (summer and winter soil temperatures differ by <6 °C). The soils are dominantly sandy in texture (e.g., Quartzipsamments) although loamy subsurface horizons are common (e.g., Paleudults). In the Everglades Agricultural Area (EAA) and other local areas through the region, the soils are organic (e.g., Haplosaprists). General characterizes of the soils in LRR U are presented in Table 15.1. Information on location, size, and land use concerns for the MLRAs is shown in Table 15.2.

M.E. Collins (✉)
University of Florida, Gainesville, FL, USA
e-mail: mec@ufl.edu

L.T. West et al. (eds.), *The Soils of the USA*, World Soils Book Series,
DOI 10.1007/978-3-319-41870-4_15

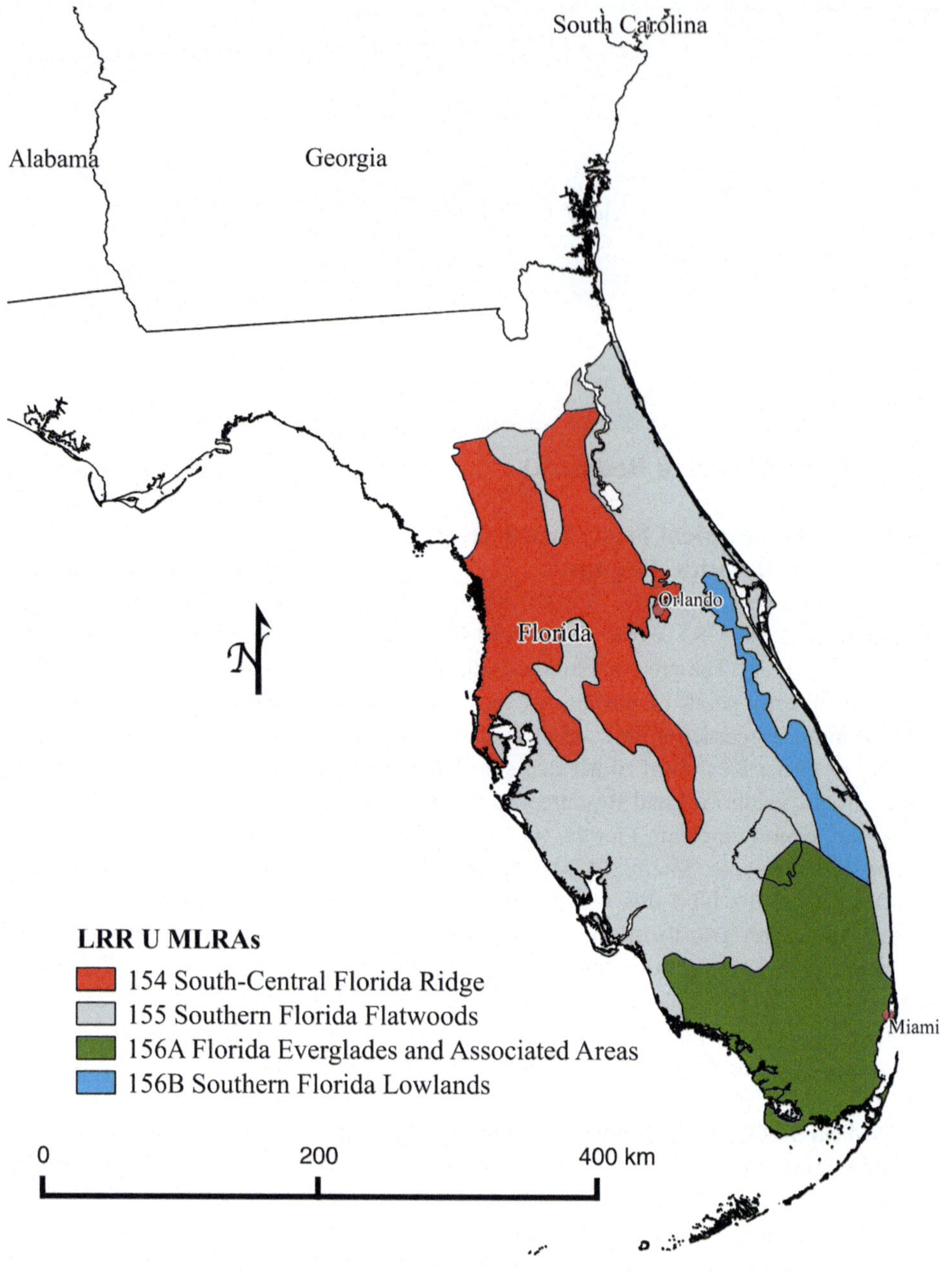

Fig. 15.1 Extent of and MLRAs comprising LRR U

15.1.2 General Characteristics—South-Central Florida Ridge (MLRA 154)

The South-Central Florida Ridge (MLRA 154) is located along the central ridge in Florida and continues to the Gulf of Mexico in the west. The MLRA extends west from Gainesville in the north to Lake Placid in the south and includes the cities of Clearwater and St. Petersburg (Fig. 15.1). It covers approximately 21,470 km^2. The topography varies from nearly level near the Gulf of Mexico to rolling hills in the central part of the state. The elevation ranges from 25 to 50 m in the majority of the area but can be as low as sea level to as high as 100 m in some locations. Agriculture in the MLRA includes citrus, pasture, livestock, and truck crops (Table 15.1). Citrus, especially oranges and grapefruit, were the leading agricultural crops until the region experienced destructive freezes in the mid-1980s. Much of the land which once was orange groves now has either sand pine (*Pinus clausa*) grown for Christmas trees or urban development.

The climate in this area is hot and humid. Precipitation ranges from 1170 to 1420 mm, all as rain. Average temperatures are from 20 to 23 °C. Winters are warm and dry, and summers are hot and rainy. The native vegetation consists of turkey oak, bluejack oak (*Quercus incana*), and longleaf pine.

Fig. 15.2 Rolling hills of central Florida. Much of this area was in orange groves. Now the area has been converted to residential use

Fig. 15.3 Typical level flatwoods landscape in Florida

Ultisols and Entisols are the dominant soil Orders in this MLRA. They are excessively to somewhat poorly drained formed in sandy to loamy marine sediments, limestone, and sandy aeolian materials (Table 15.1). The depth to limestone is very shallow in some areas; thus, sinkholes are very common.

15.1.2.1 Soils of the MLRA 154

The soils in MLRA 154 are mostly loamy or sandy Paleudults and Quartizpsamments with udic SMR and hyperthermic STR. The soils representing this area are the Millhopper (loamy, siliceous, semiactive, hyperthermic Grossarenic Paleudults) and the St. Lucie (hyperthermic, uncoated Typic Quartizpsamments) series.

Millhopper soils (Fig. 15.13) are located on rolling uplands (slopes range from 0 to 8 %) and are very deep (>200 cm thick), slightly to strongly acid, moderately well drained with slow runoff. They formed in deep sandy over loamy marine sediments. These soils are used for improved pasture or for various cultivated crops (e.g., peanuts, watermelon, corn, and hay grasses). Native vegetation includes live oak, laurel oak (*Quercus laurifolia*), and post oak (*Quercus stellata*).

St. Lucie soils (Fig. 15.14) are very deep and excessively drained with very rapid permeability. They occur on dune-like ridges, rises, and knolls (0–20 % slopes). They formed in very deep aeolian or marine sandy sediments. The

Fig. 15.4 Mangroves in near coastal environments

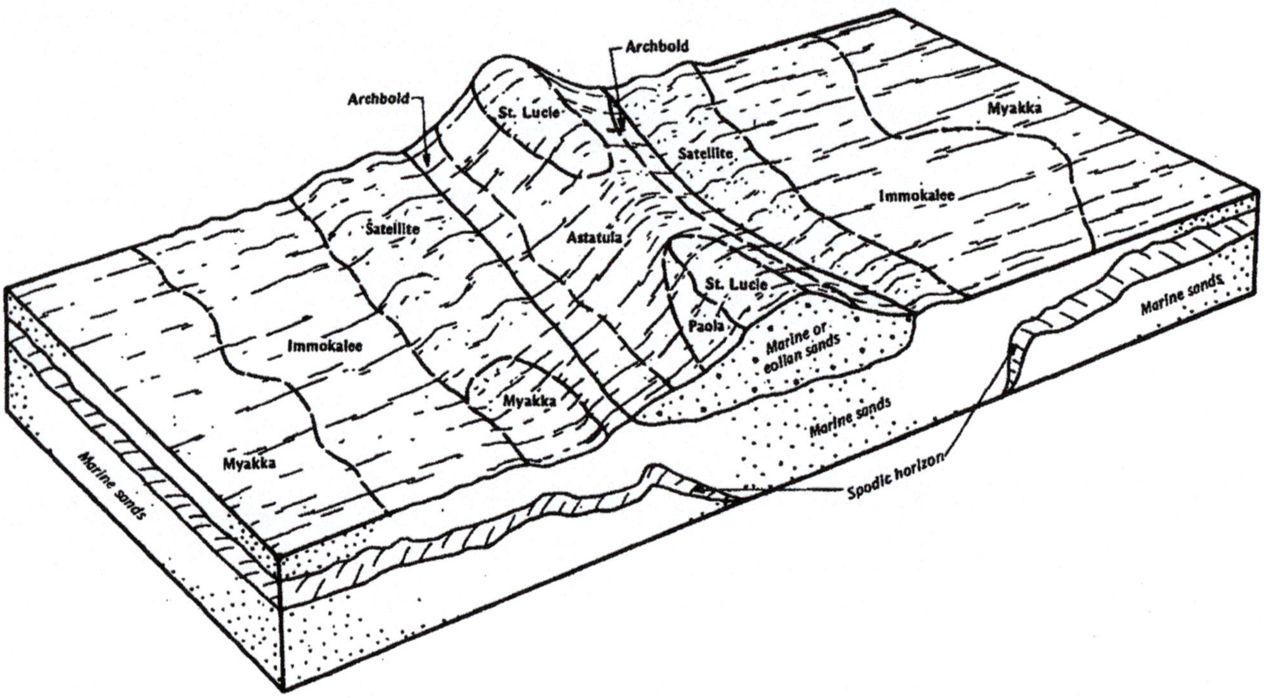

Fig. 15.5 Landscape position of the St. Lucie and Myakka soils. From Soil Survey Staff (1987)

soil consists of a thin A horizon with sand texture and very low organic C that overlies a white (10YR 8/1) sand C horizon. Because they are composed of dominantly quartz sand with very few nutrients and very low available water, the vegetation is scrub forest, e.g., sand live oak (*Quercus geminate*), sand pine, and pricklypear cactus (*Opuntia humifusa*). The genesis of the St. Lucie soils is still a point of debate. A common theory is that the white sand comprising the C horizon was the original parent material. Others consider the white sand to be a very thick albic E horizon and fine material originally present in the parent material has been eluviated to more than 200 cm.

15.1.3 General Characteristics—Southern Florida Flatwoods (MLRA 155)

The MLRA extends from Gainesville in the north to Immokalee in the south and from the Atlantic Ocean on the east to the Gulf of Mexico in the west (Fig. 15.1). This area

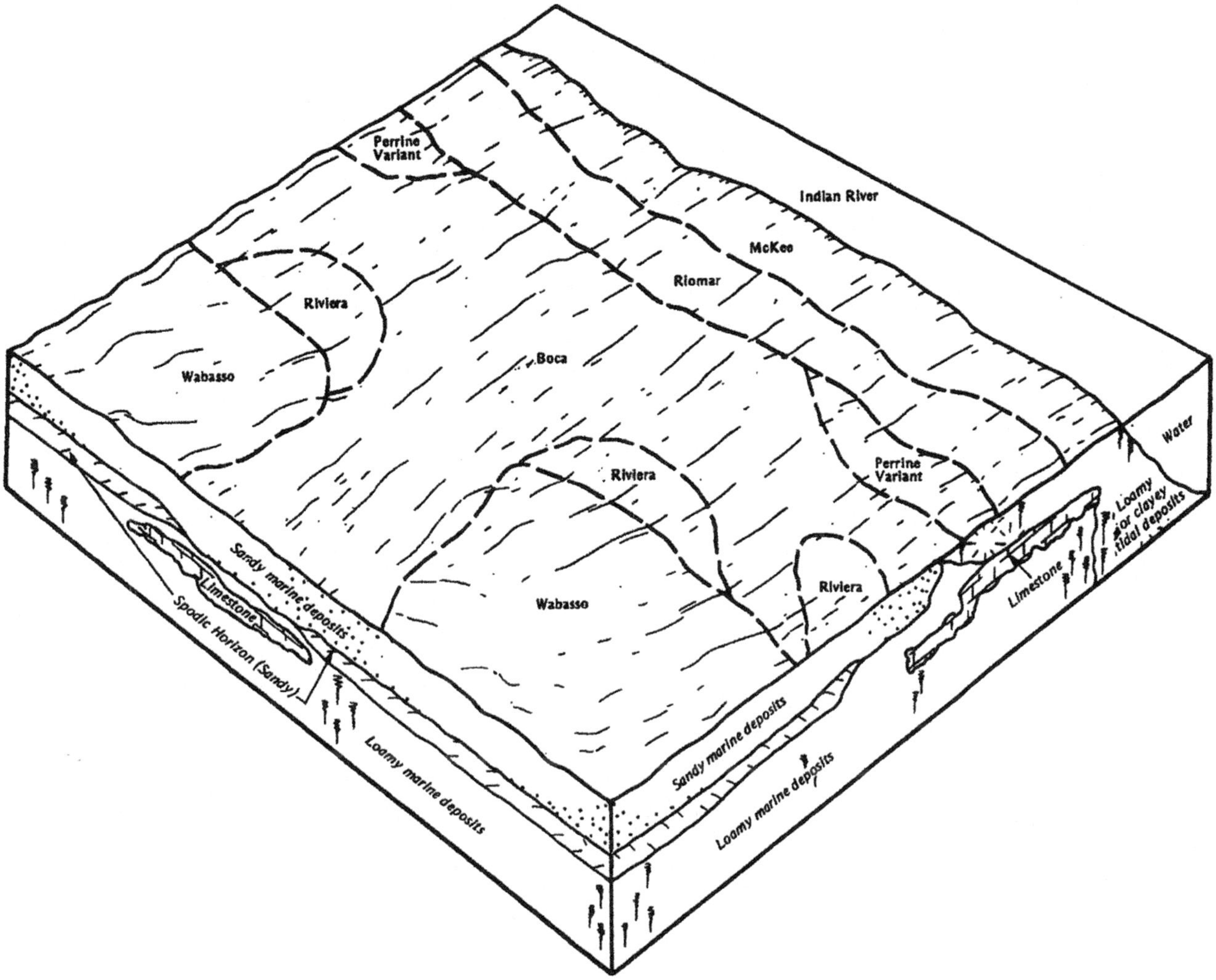

Fig. 15.6 Landscape position of the Riviera soil. From Soil Survey Staff (1987)

encompasses several sizeable cities including Tampa, Daytona Beach, Fort Meyers, part of Orlando, and Miami. The MLRA extends over 48,135 km^2, and the landscape varies from nearly level coastal plains to rolling uplands. The elevation rises from sea level at the coastline to 25 m in the interior of the state. The climate is typical of this area of the state with an average of 1120–1525 mm of annual rainfall, and annual temperatures averaging 20–24 °C. Common agricultural production includes improved pasture and native range used for cattle grazing, many types of winter vegetables, and citrus and fruits crops in the southern part of the MLRA. Native vegetation is slash pine, cabbage palm, and live oak.

Soils in the MLRA are Alfisols, Entisols, and Spodosols. Thus, they have a wide range in soil properties (Table 15.1).

15.1.3.1 Soils of the MLRA 155

The soils in this MLRA are loamy or sandy, deep to very deep, poorly to very poorly drained with a hyperthermic STR and aquic SMR. They formed in sandy marine sediments as well as alluvial materials. Soils representing this MLRA are the Malabar (loamy, siliceous, active, hyperthermic Grossarenic Endoaqualfs), Pineda (loamy, siliceous, active, hyperthermic Arenic Glossaqualfs), Basinger (siliceous, hyperthermic Spodic Psammaquents), and Myakka (sandy, siliceous, hyperthermic Aeric Alaquods) series.

Malabar soils (Fig. 15.15) are very deep, poorly drained, neutral to moderately acid, slowly permeable soils located in sloughs, shallow depressions, and floodplains. The parent material is thick deposits of sandy and loamy marine material. The Malabar soils are used primarily for rangeland as well as for citrus and truck crops. Native vegetation includes slash pine, saw palmetto (Serenoa repens), and cypress wax myrtle. In very poorly drained depressions, the vegetation is largely St. Johnswort (*Hypericum perforatum*) or maidencane.

Pineda soils (Fig. 15.16) are very deep, nearly level, poorly drained, neutral to strongly acid soils with very slow permeability and are located on broad low flats, hammocks,

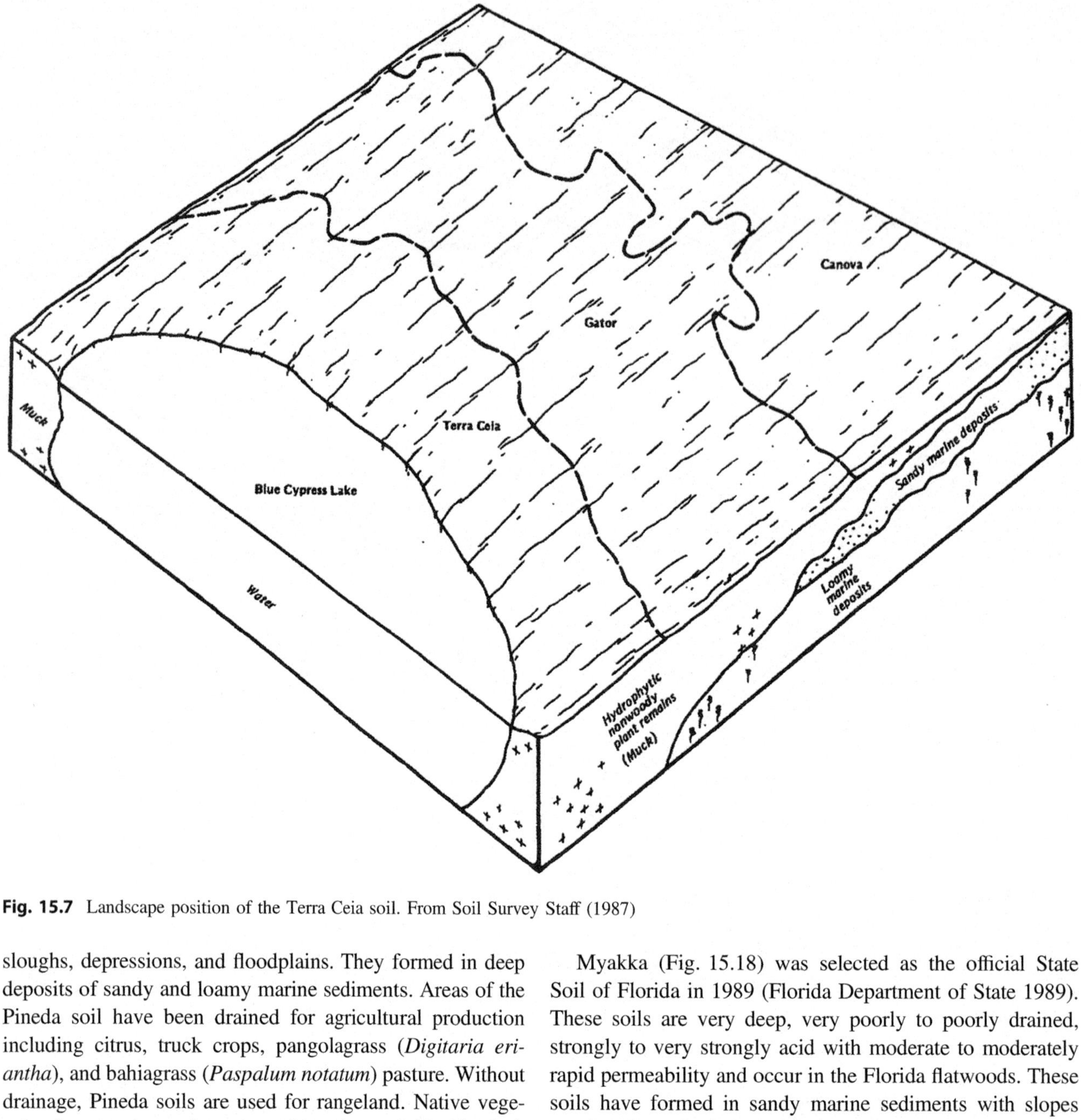

Fig. 15.7 Landscape position of the Terra Ceia soil. From Soil Survey Staff (1987)

sloughs, depressions, and floodplains. They formed in deep deposits of sandy and loamy marine sediments. Areas of the Pineda soil have been drained for agricultural production including citrus, truck crops, pangolagrass (*Digitaria eriantha*), and bahiagrass (*Paspalum notatum*) pasture. Without drainage, Pineda soils are used for rangeland. Native vegetation is extensive and consists of mostly south Florida slash pine, bald cypress (*Taxodium distichum*), wax myrtle (*Myrica cerifera*), saw palmetto, and blue maidencane.

Basinger soils (Fig. 15.17) are very deep, very poorly to poorly drained, strongly to very strongly acid soils with rapid permeability and are located in sloughs, low flats, depressions, and ill-defined drainageways. Their parent material is sandy marine sediment with slopes ranging from 0 to 2 %. Most of the Basinger soils are used for improved pasture and rangeland. If drained, they can be used to produce truck crops. Natural vegetation includes slash pine, long leaf pine, southern slash pine with an understory of gallberry (*Ilex glabra*), sabal palmetto (sabal palmetto), and saw palmetto.

Myakka (Fig. 15.18) was selected as the official State Soil of Florida in 1989 (Florida Department of State 1989). These soils are very deep, very poorly to poorly drained, strongly to very strongly acid with moderate to moderately rapid permeability and occur in the Florida flatwoods. These soils have formed in sandy marine sediments with slopes ranging from 0 to 8 %. Myakka soils are used for commercial forest production or rangeland. Some areas have been converted to truck crops, improved pasture, or citrus. Native vegetation is slash pine and longleaf pine. The understory includes saw palmetto and wax myrtle.

15.1.4 General Characteristics—Florida Everglades and Associated Areas (MLRA 156A)

The area of MLRA 156A extends south from the city of Miami on the east coast and Naples on the Gulf coast

Fig. 15.8 Tomatoes growing on very shallow and very gravelly soils located in MLRA 156A—Florida Everglades and Associated Areas

Fig. 15.9 Citrus groves in MLRA 155—South Florida Flatwoods

(Fig. 15.1). It includes Big Cypress National Preserve, the Everglades National Park, and the Florida Keys. It occupies an area of approximately 17,920 km^2. The physiography is level, low coastal plain with large areas of swamps and marshes. The elevation is mostly below 5 m on beach ridges and dunes, but can be as high as 25 m. The dominant bedrock is limestone, which is commonly overlain by sandy deposits of Pleistocene age. The average annual air temperature is 23–25 °C with almost all year frost free. The average rainfall is 1015–1575 mm with more than half coming in the months of June thru September, primarily as intense tropical rainstorms. Most of the native vegetation is wet grasses [sawgrass (*Cladium mariscus*), pickleweed (*Salicornia L*), buttonbush (*Cephalanthus occidentalis*), and maidencane]. Bald cypress is common in swamps and mangrove trees are common in saltwater areas. Approximately 35 % of this MLRA is owned

Fig. 15.10 Improved pasture in MLRA 155—South Florida Flatwoods

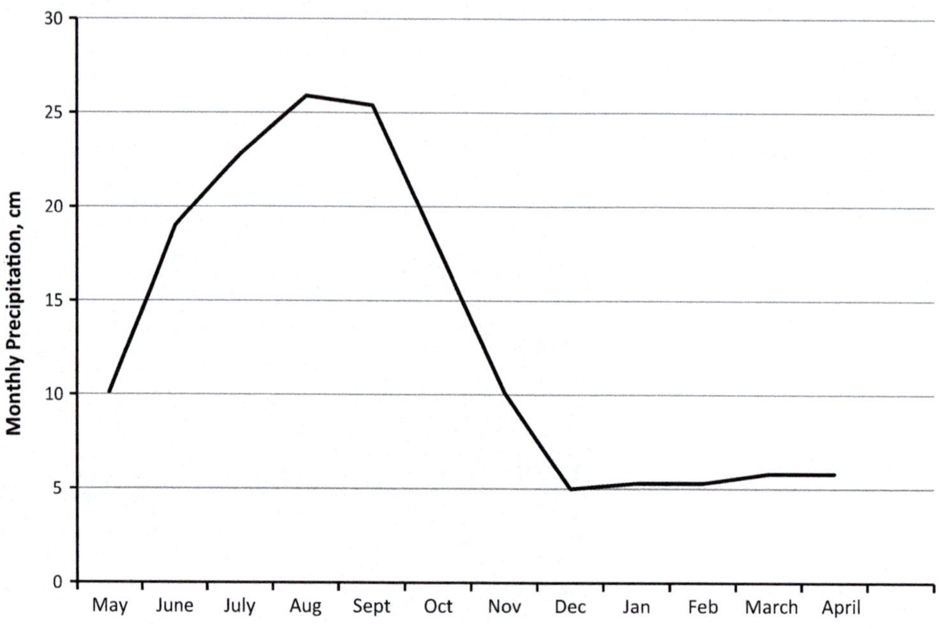

Fig. 15.11 Average monthly precipitation in Florida. Also shown are the "wet season" and "dry season" months (National Weather Service 2015)

by the US government (e.g., Everglades National Park and Big Cypress National Preserve). A limited area of cropland is used to grow winter vegetables and citrus. Sugarcane is grown on the Histosols south of Lake Okeechobee. A large portion of the area is used for hunting, fishing, and other leisure activities.

Soils representing this MLRA are the Krome (loamy-skeletal, carbonatic, hyperthermic Lithic Udorthents), Perrine (coarse-silty, carbonatic, hyperthermic Typic Fluvaquents), Hallandale (siliceous, hyperthermic Lithic Psammaquents), and Terra Ceia (euic, hyperthermic Typic Haplosaprists) series.

15.1.4.1 Soils of the MLRA 156A

Krome soils are very shallow (in some areas <25 cm thick to bedrock). They are moderately well drained, mildly alkaline, and moderately permeable soils overlying limestone. They formed in outcrops of Pleistocene-age Miami Oolitic Limestone and loamy residual material covering the "pitted" oolitic limestone. These soils are located in very shallow to deep pockets in the limestone that are infilled by the loamy sediments. Slopes are from 0 to 2 %, and Krome soils have higher native fertility than many other soils in the LRR. Tomatoes, beans, avocadoes, and limes are the traditional crops grown on Krome soils.

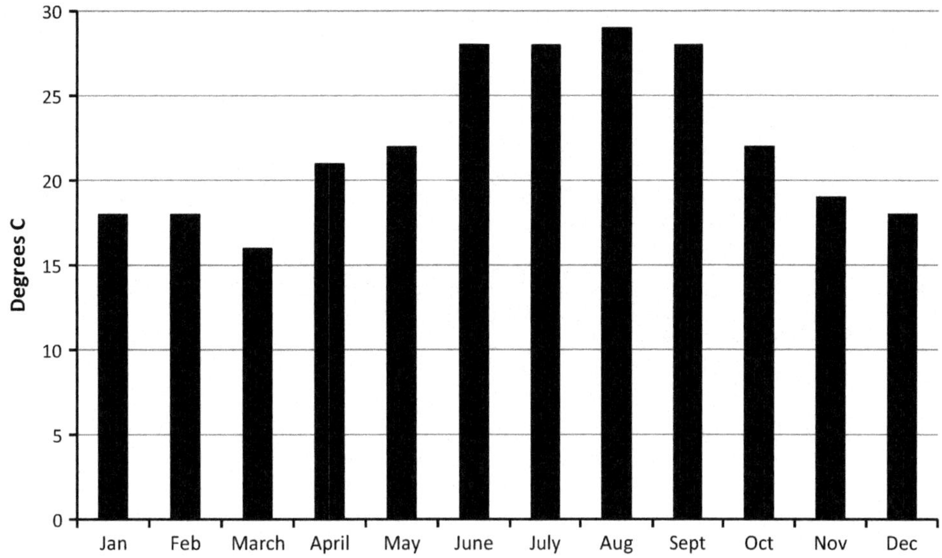

Fig. 15.12 Average monthly temperatures (2013) for Florida (Florida Climate Center 2015)

Table 15.1 General characteristics of the soils in LRR U

MLRA	Classification of major soils	Drainage	Climate	Vegetation	Agriculture	PM/relief/elevation	SMR/STR
154	Loamy or sandy Paleudults and Quartzipsamments	Excessively to Somewhat Poorly	Precipitation; 1170–1420 mm. Temperature; 20–23 °C	Turkey oak, Bluejack oak, and Longleaf pine	Livestock, citrus, and truck crops	Sandy aeolian and marine sediments/Nearly level to rolling/25–50 m	Udic, aquic/hyperthermic
155	Loamy or Sandy Endoaqualfs, Glossaqualfs, Alaquods, and Psammaquents	Poorly to Very poorly	Precipitation; 1120–1525 mm. Temperature; 20–24 °C	Slash pine, cabbage palm, and live oak		Sandy marine sediments and floodplains/Nearly level coastal plain/0–25 m	Aquic/hyperthermic
156A	Sandy Udorthents, Fluvaquents, Psammaquents, and Haplosaprists	Moderately well drained to Very poorly drained	Precipitation; 1115–1575 mm. Temperature; 23–25 °C	Sawgrass, pickleweed, willow, buttonbush, maidencane, mangrove	Winter vegetables, citrus, sugarcane	Marine sediments, limestone/Level/0–25 m	Aquic, Udic/hyperthermic, isohyperthermic
156B	Loamy or Sandy Glossaqualfs, Endoaqualfs, Argiaquolls, Psammaquents, and Haplosaprists	Poorly to Very poorly drained	Precipitation; 1170–1525 mm. Temperature; 22–24 °C	Native grasses, forbs, sedges, mixed pine, cabbage palm	Citrus, rangeland, forestland	Young marine sediments/Level/0–25 m	Aquic/hyperthermic

Perrine soils (Fig. 15.19) are moderately deep, poorly drained, moderately alkaline, moderately slowly to moderately permeable soils. They are located in lowlands (slopes are 0–1 %) along the Atlantic Coast and formed from marine or freshwater deposits of calcareous silty and loamy sediments that cover the Miami Oolite Formation. Near the coastline, various mangroves including white mangrove (*Laguncularia racemosa*) trees live. Perrine soils can be used for production of vegetables and ornamentals if properly drained. In natural environments, the vegetation includes sawgrass, reeds, and palm trees.

Hallandale soils (Fig. 15.20) are shallow, poorly to very poorly drained, strongly acid to neutral, and rapidly permeable soils on slopes of 0–2 %. They occur on broad low flats and in sloughs, shallow depressions, and tidal areas. The parent material is sandy sediment of marine origin over limestone. Depth to the seasonal water table in these soils depends on physiography and time of the year. As an example, in depressional areas, these soils are typically covered with water for 6–9 months during the wet season. In tidal areas, the soils are covered with water every day. Hallandale soils are commonly used for rangeland and

Table 15.2 General location, size, and major land concerns of the MLRAs in LRR U

MLRA	Location	Size of MLRA (km^2)	Major land concerns
154	Area is entirely in Florida located along the central ridge in Florida and continues to the Gulf of Mexico in the west. The MLRA is from west Gainesville in the north to Lake Placid in the south and along the Gulf Coast in the west and includes the cities of Clearwater and St. Petersburg	21,470	Wind erosion, soil organic matter content, soil productivity, water quality/quantity, and urbanization
155	Area is entirely in Florida located from east Gainesville in the north to Immokalee in the south and from the Atlantic Ocean on the east to the Gulf of Mexico in the west. Because of its extensive area, there are sizeable cities within its boundary such as Tampa, Daytona Beach, Fort Meyers, part of Orlando, and Miami	48,135	Wind erosion, soil organic matter content, soil productivity, water quality/quantity, and urbanization
156A	The area of MLRA 156A goes from the city of Miami south on the east coast and Naples on the Gulf coast. It includes Big Cypress National Preserve, the Everglades National Park, and the Florida Keys	17,920	Wind erosion, soil organic matter content, soil productivity, soil subsidence, water quality/quantity, and urbanization
156B	This MLRA is unique in that it has no major cities located within its boundaries. It is a long narrow strip (north to south) on the near the east coast of Florida	4750	Wind erosion, soil organic matter content, soil productivity, and water quality/quantity

pasture. In non-tidal areas, the vegetation includes pineland threeawn (*Aristida stricta*), bluejoint panicum (*Panicum tenerum*), cypress, slash pine, and saw palmetto. Because the tidal areas are covered with water daily, the vegetation consists of seashore saltgrass (*Distichlis spicata*), needlegrass (*Nassella E. Desv.*), rushes (*Juncus* sp.), and saltwort (*Batis maritime*).

Terra Ceia soils (Fig. 15.21) are very deep (>127 cm of sapric organic materials), very poorly drained, slightly to moderately acid, rapidly permeable Histosols (organic soils). Because they are organic soils, the parent material is herbaceous plant material. Slopes are 0–1 %, and the soils occur in nearly level freshwater marshes or in depressions in flatwoods. If Terra Ceia soils have been drained, they are very productive for agricultural commodities, e.g., sugarcane and sweet corn. If left in their natural state, the vegetation is cypress, blackgum (Nyssa sylvatica), sabal palmetto (*Sabal palmetto*), sawgrass, sedges (*Cyperaceae*), reeds, and maidencane.

15.1.5 General Characteristics—Southern Florida Lowlands (MLRA 156B)

MLRA 156B is unique in this LRR because it has no major cities located within its boundaries. It is a long narrow strip (north to south) on the east coast of Florida (Fig. 15.1). It is the smallest of the MLRAs in LRR U, about 4750 km^2. The topography is nearly level with a few hummocks that are 1–2 m in height. The parent material is sandy marine deposits of Pleistocene age. Rainfall is 1170–1525 mm with most of the rain coming from June to September. The temperature varies slightly from 22° to 24°. Native vegetation is composed of slash pine and sabal palmetto (Sabal palmetto) with saw palmetto, cordgrass (*Spartina* sp.), and bluestem understory. Private farms and ranches make up most of the land use in the region, and citrus is the chief agricultural crop.

Soils representing this MLRA are the Riviera (loamy, siliceous, active, hyperthermic Arenic Glossaqualfs), Felda (loamy, siliceous, superactive, hyperthermic Arenic Endoaqualfs), Floridana (loamy, siliceous, superactive, hyperthermic Arenic Argiaquolls), Basinger (siliceous, hyperthermic Spodic Psammaquents), and Terra Ceia (euic, hyperthermic Typic Haplosaprists) series.

15.1.5.1 Soils of the MLRA 156B

Riviera soils (Fig. 15.22) are very deep, moderately acid to moderately alkaline, poorly drained, very slowly permeable soils on broad low flats and depressions. Slopes are 0–2 %. They formed in stratified sandy and loamy marine deposits. These soils are used for citrus, winter truck crops and improved pasture when drained. The vegetation in undrained areas is slash pine, saw palmetto, and cypress trees.

Felda soils (Fig. 15.23) are very deep, slightly acid to moderately alkaline, very poorly to poorly drained, moderately permeable soils on flat (0–1 %) landscapes such as drainageways, sloughs, and depressions. This soil formed in stratified marine deposits of sands and clays. For Felda soils in cleared areas to be agriculturally productive, they must be drained, and water table depths need to be managed. In drained areas, the soils are used for citrus, truck crops, and pastureland. In its native state, the vegetation is cypress, pond pine (*Pinus serotina*), slash pine, and sabal palmetto.

Fig. 15.13 Millhopper series in MLRA 154—South-Central Florida Ridge. These moderately well drained soils formed in deep sandy material over loamy marine sediments and occur on rolling uplands

Fig. 15.14 St. Lucie series in MLRA 154—South-Central Florida Ridge. These excessively drained soils formed in very deep aeolian or marine sandy sediments and occur on dune-like ridges, rises, and knolls

Floridana soils (Fig. 15.24) are very deep, slightly acid to slightly alkaline, very poorly drained and have slow to very slow permeability. They occur on broad flats, flood plains, and depressions with 0–1 % slopes. The parent material for this soil is thick beds of sandy and loamy marine sediments. As with the Felda soil, if areas of Floridana soils have been cleared and depth to water tables are managed, truck crops and citrus are grown. Native vegetation varies depending on landscape position. In depressions, the native vegetation is cypress. On flood plains, the vegetation is commonly sweetgum (*Liquidambar styraciflua*), blackgum, and cypress.

Basinger and Terra Ceia soils in this MLRA have similar properties to Basinger and Terra Ceia soils in MLRA 156A.

15.1.6 Resource Concerns

Many of the resource concerns in LRR U are related to population increases, rapid urbanization, and growth in tourism; all of which create stresses on water quality and quantity and degradation of soils. In the 1980s, Florida's population grew by an astounding 32.7 %; in the 1990s by 23.5 %; and in the 2000s by 17.6 % to its current level of approximately 19.3 million (State of Florida 2014). Florida is ranked as the top travel destination in the world (State of Florida 2015), and the state had about 87.3 million visitors in 2011 (a record number). For almost all of Florida, the primary source of drinking water is groundwater from the Floridan aquifer including the cities of Tallahassee, Jacksonville, Gainesville, Orlando, Daytona Beach, Tampa, and St. Petersburg as well as thousands of wells (domestic, industrial, and irrigation) throughout the state (Borisova and Carriker 2014). High amounts of water withdrawal from the aquifer are depleting its water levels more rapidly than they are being recharged. The abundance of sandy soils across the region also creates a potential contamination problem with improper applications of nutrients and pesticides. Soil health is emerging as a concern in many agricultural areas in the

Fig. 15.15 Malabar series in MLRA 155—Southern Florida Flatwoods. These very poorly and poorly drained soils formed in very deep sandy and loamy marine sediments and occur in sloughs, depressions, and along flood plains

Fig. 15.16 Pineda series in MLRA 155—Southern Florida Flatwoods. These poorly drained soils formed in thick deposits of sandy and loamy marine sediments and occur on broad low flats, hammocks, sloughs, depressions, and poorly defined drainageways and flood plains

state where intensive production of vegetable crops has degraded and potentially will continue to degrade the soil resource without careful management (Grabowski 2015).

15.2 Land Resource Region Z—Caribbean Region

The Caribbean Region (LRR Z) (Fig. 15.25) includes Puerto Rico (94 %), the US Virgin Islands (4 %), and other smaller nearby islands (2 %). Soils on these small islands will not be discussed. The land area of the region is approximately 9310 km^2. Within the region are three main physiographies: humid and semiarid mountains, valleys, and coastal plains. Annual rainfall varies from as little as 250 mm in the semiarid coastal plain to more than 2200 mm in the humid high elevations. Average annual air temperature only varies from 21 °C in the humid uplands to 26° on the semiarid coastal plain. Parent materials of the soils are marine sediments, alluvium, limestone, and volcanic bedrock/residuum. The topography of the region varies greatly from the relatively level coastal plain to the very steep, very strongly sloping mountains. The native vegetation also has a very wide range depending on the climate. There are numerous native and improved grasses and tree species that grow on the different islands. The land is approximately 96 % privately owned with food crops such as plantains, bananas, yams, vegetables, and some citrus fruits as the main crops. Much of the better agricultural lands have been converted to urban advances, highways, and recreational areas.

Because of the wide range in the soil forming factors on these islands, there are seven soil Orders in the region: Entisols, Inceptisols, Alfisols, Ultisols, Oxisols, Mollisols, and Histosols. The soil moisture regimes are aquic in the depressional areas (e.g., swamp and marshes), ustic on the relatively drier areas of the islands (e.g., semiarid coastal plains), and udic in more humid areas (e.g., high elevations). The vast majority of the soils have an isohyperthermic soil temperature regime.

Fig. 15.17 Basinger series in MLRA 155—Southern Florida Flatwoods. These very poorly and poorly drained very deep soils formed in sandy marine sediments and occur in flats, sloughs, depressions, and poorly defined drainageways

Fig. 15.18 Myakka series, the official State Soil of Florida, in MLRA 155—Southern Florida Flatwoods. These very poorly to poorly drained very deep soils formed in sandy marine deposits and occur in mesic flatwoods

General characterizes of the soils in LRR Z are presented in Table 15.3. Within the LRR are MLRAs 270—Humid Mountains and Valleys, 271—Semi-arid Mountains and Valleys, 272—Humid Coastal Plains, 273—Semi-arid Coastal Plains. Information on location, size, and land use concerns for the MLRAs is in Table 15.4. The locations of MLRAs 270–273 are presented in Fig. 15.25.

15.2.1 General Characteristics—Humid Mountains and Valleys (MLRA 270)

The Humid Mountains and Valleys MLRA is located entirely in central Puerto Rico (Fig. 15.25). It is approximately 4660 km^2 in area (the largest MLRA in the LRR) and contains several state parks and forest preserves.

As the name implies, the resource area is very steep and mountainous (elevations range from 50 to more than 1300 m) with associated narrow valleys. Landslides are common (Fig. 15.26). There are three mountain ranges: the Cordillera Central, Sierra de Luquillo, and the Sierra de Cayey. El Yunque, the famous rainforest on the island, is at an elevation of about 1065 m and is located southeast of San Juan.

Soils representing this MLRA are the Caguabo (loamy, mixed, active, isohyperthermic, shallow Typic Eutrudepts), Humatas (very-fine, parasesquic, isohyperthermic Typic Haplohumults), Los Guineos (very-fine, kaolinitic, isothermic Humic Hapludoxs), and Nipe (very-fine, ferruginous, isohyperthermic Typic Acrudoxs) series.

15.2.1.1 Soils of the MLRA 270

Caguabo soils are shallow (bedrock within 50 cm), slightly acid, well drained, moderately permeable soils on side slopes of highly dissected uplands (12–60 % slopes). Commonly, they occur on lower positions of dissected volcanic uplands with elevation up to about 550 m (Fig. 15.27). These soils developed from weathered basalt bedrock. Pastureland is the common agricultural use for Caguabo soils. Various native grasses, shrubs, and trees are located in native areas of these soils.

Fig. 15.19 Perrine series in MLRA 156A—Florida Everglades and Associated Areas. These poorly drained moderately deep soils formed in calcareous silty and loamy freshwater deposits or marine residues and occur on broad low flats near the Atlantic Ocean (*scale* is in feet)

Humatas soils are very deep, very strongly acid, well drained, moderately slowly permeable soils located on side slopes and ridges of highly dissected uplands (5–60 % slopes) (Fig. 15.27). These soils formed in fine-textured (loamy and clayey) residuum, which weathered from basic igneous rock. Agricultural use includes pasture, food crops, and coffee production. In its native state, vegetation includes grasses, shrubs, and trees.

Los Guineos soils are very deep, very strongly to extremely acid, moderately permeable soils on side slopes of mountains or deeply dissected plateaus of uplands (Fig. 15.28). Slopes range from 5 to 60 %. Los Guineos soils formed in residuum from sandstone. Most of the areas where Los Guineos soils occur are forested.

Nipe soils (Fig. 15.29) are very deep, moderately to strongly acid, moderately permeable soils on ridge tops of hills and mountains (2–20 % slopes). Parent material for these soils is iron-rich residuum, which weathered from serpentine bedrock. High amounts of Fe oxide and oxyhydroxide minerals typically found in Nipe soils results in their having a net positive charge (Acrudox great group) in contrast to a net negative charge typical for most soils. This soil supports urban development with limited agricultural use. Native vegetation

Fig. 15.20 Hallandale series in MLRA 156A—Florida Everglades and Associate Areas. These very poorly and poorly drained shallow soils formed in thin sandy marine deposits over limestone and are situated on broad flats, sloughs, tidal areas, and shallow depressions

Fig. 15.21 Terra Ceia series in MLRA 156A—Florida Everglades and Associate Areas. These very poorly drained very deep soils formed in sapric herbaceous plant materials and occur in freshwater marshes and depressions

is Brazil beauty-leaf (*Calophyllum brasiliense*), Monkey-apple (*Annona glabra*), and *Clusia clusioides*.

15.2.2 General Characteristics—Semi-arid Mountains and Valleys (MLRA 271)

This MLRA is about 70 % in Puerto Rico, 20 % in the US Virgin Islands, and 10 % on the outlying small islands

Fig. 15.22 Riveria series in MLRA 156B—Southern Florida Lowlands. These poorly drained very deep soils formed in stratified sandy and loamy marine sediments occur on broad flats and in depressions (*scale* is in feet)

Fig. 15.23 Felda series in MLRA 156B—Southern Florida Lowlands. These very poorly to poorly drained very deep soils formed in stratified, unconsolidated sands, and clays of marine origin and are found in drainageways, sloughs, depressions, and flood plains

(Fig. 15.25). The total area is only 1365 km^2. In Puerto Rico, there are several major towns and cities (e.g., Cabo Rojo, Yauco, and Arroyo) within this MLRA. In the Virgin Islands, all of St. John and St. Thomas as well as the cities of Christiansted and Federiksted on St. Croix are within the MLRA.

The part of this MLRA in Puerto Rico is comprised of semi-arid mountains (moderately steep to very steep) that extend from east to west across the island. Elevations are from 50 to 395 m. The area of the MLRA in the Virgin Islands is also mountainous. The highest mountain peaks in the MLRA are above 335 m. The geology of the MLRA is diverse with rocks ranging in age from the Jurassic to Tertiary. Soil parent materials consist of a mixture of limestone and volcanic rock. Annual rainfall varies from 890 to 1140 mm near the sea and increases to more than 1500 mm at high elevations in the interior of the island. Moisture is lost thru evapotranspiration, and the air temperatures are high (annual average is 26 °C with <8° difference between mean summer and winter temperatures).

Vegetation is comprised of grasses (hurricanegrass (*Fimbristylis cymosa*), guineagrass (*Megathyrsus maximus*), buffelgrass (*Cenchrus ciliaris*), and Egyptian grass (*Dactyloctenium aegyptium*) and trees (black olive (*Olea europaea*), turpentine (*Syncarpia glomulifera*), Christmas tree and guayacan (*Tabebuia chrysantha*)). Almost 40 % of the MLRA is in grassland while private forested areas are about 25 %.

Soils representing this MLRA in Puerto Rico are the Descalabrado (clayey, mixed, superactive, isohyperthermic, shallow Typic Haplustolls), Callao (clayey, mixed, superactive, isohyperthermic, shallow Typic Dystrustepts), Jacana (fine, mixed, superactive, isohyperthermic Vertic Haplustolls), and Guayama (clayey, mixed, isohyperthermic shallow Typic Haplustalfs) series.

In the US Virgin Islands, soils representing this MLRA are the Annaberg (loamy-skeletal, mixed, superactive, isohyperthermic Lithic Haplustolls), Fredriksdale (clayey-skeletal, vermiculitic, isohyperthermic, Lithic Haplustolls), and Victory (loamy-skeletal, mixed, superactive, isohyperthermic Typic Haplustepts) series (Soil Survey Staff 2000).

Fig. 15.24 Floridana series in MLRA 156B—Southern Florida Lowlands. These very poorly drained very deep soils formed in thick beds of sandy and loamy marine sediments and occur on broad flats, flood plains, and depressions

15.2.2.1 Soils of the MLRA 271

Puerto Rico Descalabrado soils are shallow, well drained, neutral, with moderate permeability and are located on slopes from 2 to 60 %. They occupy side slopes and hilltops and are in mountainous areas. The mollic epipedon (Mollisols order) is thin because of the shallow depth to the limestone parent material (<50 cm). Most of the areas where this soil occurs are used for pastureland. If cropped, main crops grown are corn, pigeon peas, avocado, and mangoes.

Callabo soils are moderately deep, well drained, slightly acid to neutral, and are moderately permeable on strongly dissected uplands (12–60 % slopes). The parent material for Callabo soil is residuum weathered from volcanic bedrock, which can be as shallow as 70 cm. As with the Descalabrado soils, the land use for these soils is mainly pastureland with some cultivated areas.

Jacana soils are moderately deep, well drained, moderately acid to slightly alkaline, and moderately slowly permeable. These soils have a dark colored A horizon >25 cm thick with moderate amounts of organic C. They are located on fans, foot slopes, and lower regions of small hills with slopes of 2–25 %. Parent material is weathered volcanic rock. Pasture grasses include Guineagrass (*Megathyrsus maximus*) and buffelgrass (*Cenchrus ciliaris*). Cultivated areas are used for tomatoes, peppers, pigeon peas, and mangoes.

Guayama soils are well drained, slightly acid to moderately alkaline soils that are shallow to bedrock. Slopes are typically 12–60 %. These soils occur on side slopes and summits of hills and mountains. They formed in a complex parent material of residuum and colluvium weathered from basalt, chert, and rhyolite. This soil is dominantly used for pasture.

Fig. 15.25 Extent of and MLRAs comprising LRR Z

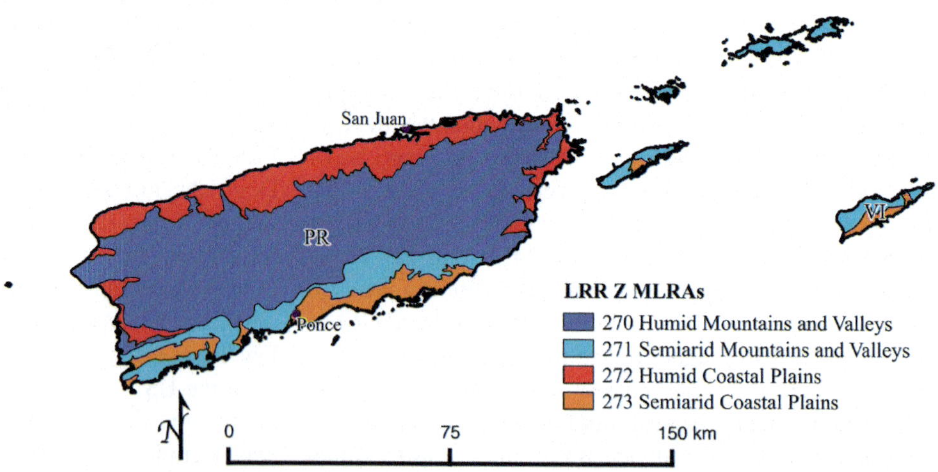

Table 15.3 General characteristics of the soils in LRR Z

MLRA	Classification of major soils	Drainage	Climate	Vegetation	Agriculture	PM/Relief/Elevation	SMR/STR
270	Clayey or Loamy Eutrudepts, Haplohumults, Hapludox, Acrudox	Well drained	Precipitation: 2030–2160 mm Temperature: <21 °C at high elevations and <23 °C at low elevations	Carpetgrass, whorled dropseed, pendejuelo, knotroot bristlegrass, creeping wheatgrass, St. Augustine grass, woodland grass, foxtail grass	Coffee, plantains, bananas, taniers, yams, and pigeon peas	Volcanic rocks that formed below sea level during the Cretaceous/steep to very steep/50–1340 m	Udic, perudic/isohyperthermic or isothermic
271	Clayey Haplustolls, Dystrustepts, Haplustolls Haplustepts	Well drained	Precipitation: Puerto Rico; 890–1525 mm St. Croix; 1015–1270 mm St. Thomas; 890–1065 mm St. John; 1015–1395 mm Temperature; ~26 °C all year	Hurricanegrass, guineagrass, Mexican bluegrass, buffelgrass, southern sandbur, Egyptian grass, Kleberg's bluestem, flame tree, white oak, thorny bushes including cati in semi-arid areas	Pasture with native grasses, natural forests and hardwoods such as mahogany, teak, and eucalyptus	Volcanic rock/Mostly mountainous/50–395 m	Ustic/Isohyperthermic
272	Udults, Udox, Aquolls, Aquepts, Udolls, Saprists, Aquents, Rendolls, Udalfs	Poorly to well drained	Precipitation: 1525–1650 mm increasing with elevation Temperature; ~25 °C all year	Caribgrass, streambank millet, para grass, beach sedge, Durban crowfoot grass, Jamaica fingergrass, lovegrass, flame tree, white oak, beachgrass, St. Augustine grass	Pineapple, hayland, sod grasses, citrus	Alluvium, marine sediments (coastal plains, floodplains), limestone karst/0–700 m	Mostly ustic, some areas of udic/Isohyperthermic
273	Aquolls, Ustolls, Usterts, Aquerts	Excessively well to somewhat poorly drained	Precipitation: 760–1145 mm, lower (255–760 mm) in some areas. Temperature: ~26 °C all year	The dominant plant species in this area are beachgrass, southern sandbur, saltwort, bermudagrass, Mexican bluegrass, Egyptian grass, matojo de piramide, whorled dropseed	Hay crop, native and improved pasture, plantains, bananas, avocados, mangos, citrus	Alluvium, marine sediments/level to gently rolling/2–75 m	Ustic/Isohyperthermic

Table 15.4 General location, size, and major land concerns of the MLRAs in LRR Z

MLRA	Location	Size of MLRA (km^2)	Major land concerns
270	Central Puerto Rico	4660	Water erosion, mass movement of soil, soiltilth, fertility, water quality and urbanization
271	Puerto Rico (71 %), the Virgin Islands (20 %), and the outlying islands of Vieques and Culebra (9 %)	1365	Water erosion, soil organic matter, soil tilth, water quality and urbanization
272	Northern coast of Puerto Rico, but it includes small areas on the east and west coasts	2500	Water erosion, soil organic matter, soil tilth, fertility, soil crusting, water quality and urbanization
273	Southern coast of Puerto Rico (81 %) and mostly on the southern side of St. Croix in the Virgin Islands (10 %). The outlying islands of Vieques, Desecheo, and Mona make up the 9 %	785	Water erosion, soil organic matter surface compaction, soil tilth, fertility, water quality, and urbanization

Virgin Islands Annaberg soils are shallow (volcanic bedrock within 50 cm), well drained, neutral, and have moderately permeability. Slopes range from 2 to 60 % and are associated with summits and backslopes on volcanic hills and mountains. Mollic epipedons in these soils are relatively thin because of the shallow bedrock. Because slopes are typically steep, Annaberg soils are used mostly for pasture, but some areas have been converted to urban development (residential and commercial). Rangeland vegetation is dominantly native grasses and shrubs.

Fig. 15.26 Colluvial soils are common in the high elevation on the US Virgin Islands. Because of the heavy rainfalls and volcanic parent material, there many landslides on the islands

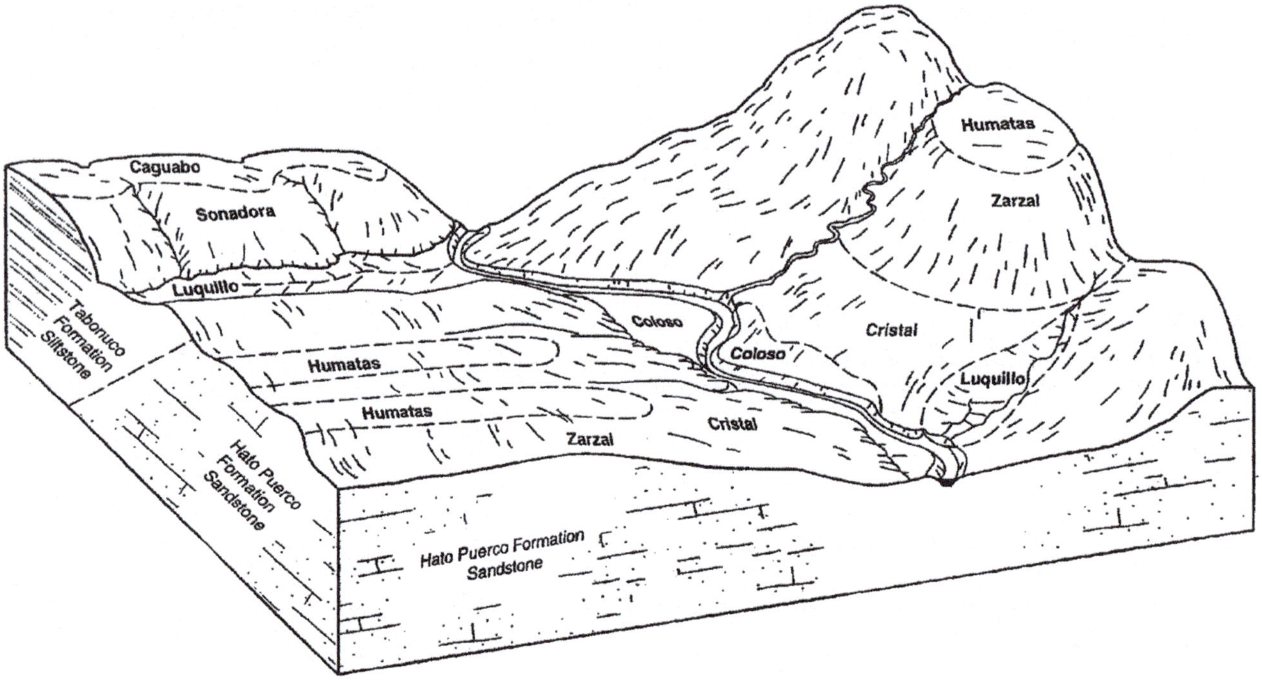

Fig. 15.27 Landscape positions of the Humatas and Caguabo soils in Puerto Rico (Soil Survey Staff 2002)

Fredriksdal soils are also shallow, well drained, neutral, and slowly permeable. They are typically located on summits and side slopes of volcanic hills and mountains. Slopes range from 12 to 90 %. As with the Annaberg soils, these soils developed from igneous bedrock. Most of these soils support small-scale residential development.

Victory soils are moderately deep, well drained, neutral to slightly acid, moderately permeable soils with slopes ranging from 2 to 70 %. They differ from the Annaberg and Fredriksdal soils in that these soils formed in fine-textured residuum from weathered volcanic bedrock and lack mollic epipedons (Inceptisols). Victory soils are on summits and

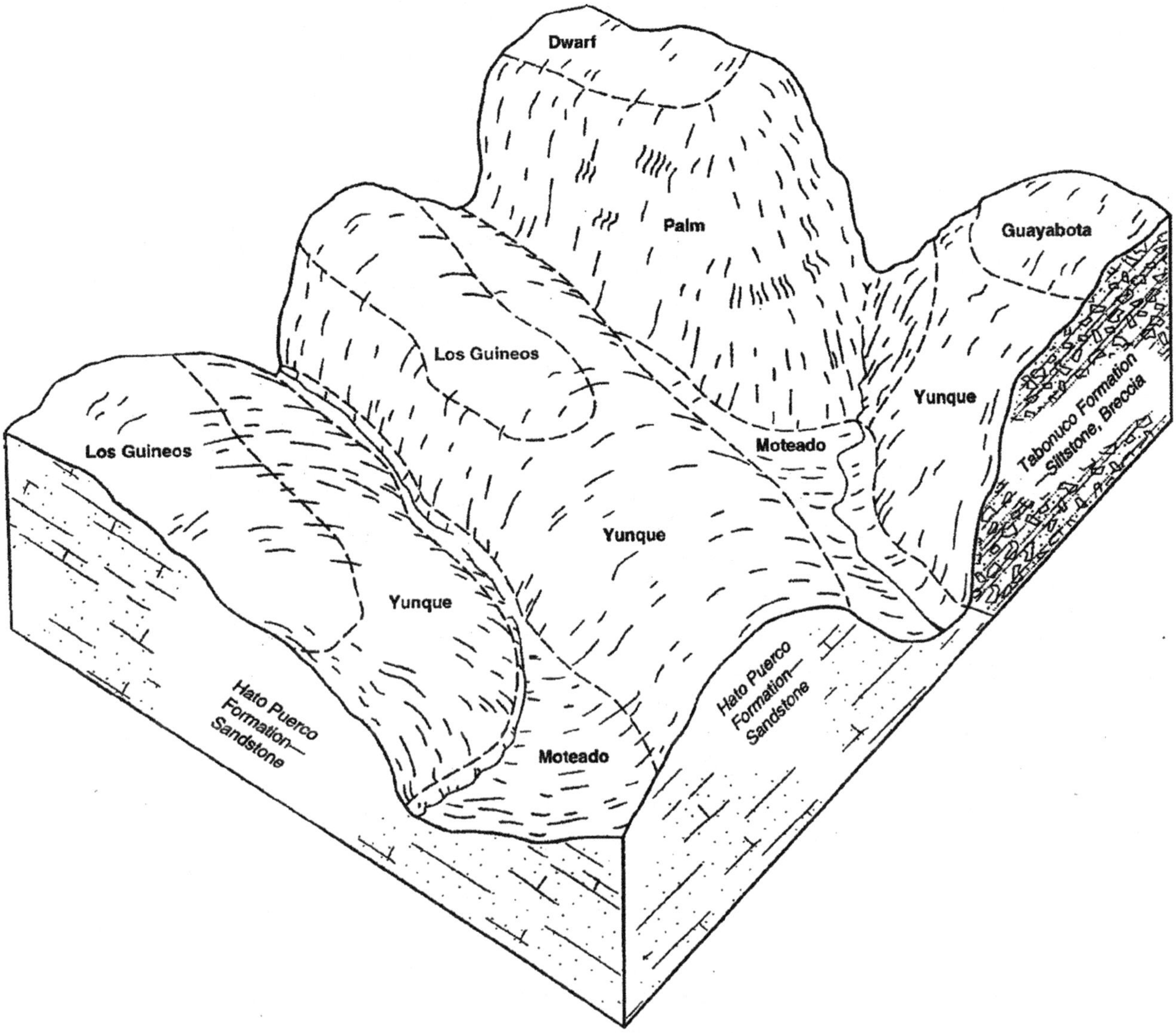

Fig. 15.28 Landscape positions of the Los Guineos soil in Puerto Rico (Soil Survey Staff 2002)

side slopes of volcanic hills and mountains. Pasture and housing developments are common land uses for this soil.

Other soils of significance are the Isaac (fine, mixed, active, isohyperthermic Typic Argiustoll), Magens (fine, oxidic, isohyperthermic Typic Haplustult), Cramer (clayey, mixed, active, isohyperthermic, shallow Typic Haplustoll), and Southgate (loamy-skeletal, mixed, active, isohyperthermic Lithic Haplustept) series.

Isaacs soils (Fig. 15.30) are moderately deep to very deep, well drained, medium acid to neutral, moderately permeable soils on slopes ranging from 5 to 40 %. These soils are commonly located on backslopes and footslopes of dissected uplands. The parent material is weathered extrusive volcanic rocks. Sola are commonly moderately deep (46–84 cm) deep and bedrock occurs at 51–183 cm. Areas of this soil have been converted to housing developments (Fig. 15.31).

Magens soils (Fig. 15.32) are very deep, well drained, medium to strongly acid, and moderately permeable. They occur on moderately steep backslopes (30–50 % slope) of dissected volcanoes. The parent material is similar to the Isaacs and Cramer soils; weathered volcanic material. A few areas of the Magens soils are in cultivated crops or pasture. Areas of Magens soils also have been converted to non-agricultural uses.

Cramer soils (Fig. 15.33) are shallow, well drained, slightly acid to neutral, and are moderately permeable. These soils occur on summits and side slopes of volcanic mountains, and slopes range from 2 to 90 %. The Isaacs and Cramer soils are located on similar landscape positions, and Cramer soils, like Isaacs, have developed from weathered volcanic materials. Cramer soils are shallower to the volcanic rock (25–50 cm) than Isaacs soils. Most of the areas mapped as Cramer soils are devoted to rangeland or

Fig. 15.29 Nipe soil series in MLRA 270—Humid Mountains and Valleys (Puerto Rico). These soils are very deep, moderately to strongly acid, moderately permeable on ridge tops of hills and mountains (2–20 % slopes). The parent material is iron-rich residuum which weathered from serpentine bedrock

Fig. 15.30 Isaacs soil series in MLRA 271—Semi-arid Mountains and Valleys (Virgin Islands) These soils are moderately deep to very deep, well drained, medium acid to neutral, moderately permeable on slopes ranging from 5 to 40 %

small-scale tropical fruit. Other soil areas have been converted to residential or commercial uses.

Southgate soils (Fig. 15.34) are shallow, well drained, very strongly to slightly acid, and moderately permeable. Slopes range from 2 (summits) to 90 % (backslopes) on volcanic hills and mountains. Southgate soils are similar to the Cramer series except that Cramer soils have a thicker, darker A horizon (mollic epipedon) than Southgate soils. As with the Isaacs, Magens, and Cramer soils, the parent material is weathered volcanic bedrock. Rangeland is the common land use along with urban development.

15.2.3 General Characteristics—Humid Coastal Plains (MLRA 272)

The Humid Coastal Plains, an area of approximately 2500 km^2, are found mainly on the north coast of Puerto Rico with a few small areas on the east and west coasts (Fig. 15.25). More than half of the population of Puerto Rico lives in several large cities (e.g., San Juan) within this MLRA.

There are two diverse landscapes in this region; relatively flat alluvial plains and terraces near the coast and, irregular karst landscapes associated with limestone further from the coast (Fig. 15.35). The flat alluvial areas are relatively young (Quaternary age). The low, rolling hills in the karst region are the result of eroded, dissected limestone. The Tertiary-aged limestone can also produce steep cliffs along the coast. Thus, the elevations can range from 0 to 700 m. Depressional areas in the flatter landscapes may have swamps and lagoons. The climate is rather predictable (except for hurricanes) in that the average precipitation is 240–260 mm and average air temperature is 25 °C. The vegetation in dry areas is a diverse suite of grasses and trees. In areas of poorly or very poorly drained soils, the vegetation consists mainly of red (*Rhizophora mangle*), white, black (*Avicennia germinans*), and button mangroves (*Conocarpus erectus*), as well as southern cattail (*Typha domingensis*), leatherfern (*Acrostichum aureum*), and para grass (*Urochloa mutica*). Approximately 33 % of the land area of this region is urban, and the region supports a large population.

15.2.3.1 Soils of the MLRA 272

There is a very wide range of soils that are located in the Humid Coastal Plains, and landscape position has a strong influence on their properties. All of the soils have an isohyperthermic soil temperature regime, and most of the better drained soils on terraces and uplands have an ustic soil moisture regime. Soils vary across the MLRA. Udults and Udox are common on the coastal plains, Aquolls and Aquepts occur on floodplains, and Saprists and Aquents

Fig. 15.31 Areas of Isaacs soils have been converted to housing developments on the US Virgin Islands

Fig. 15.32 Magens soil series in MLRA 271—Semi-arid Mountains and Valleys (Virgin Islands). The soils are very deep, well drained, medium to strongly acid, moderately permeable

Fig. 15.33 Cramer soil series in MLRA 271—Semi-arid Mountains and Valleys (Virgin Islands). They are shallow, well drained, slightly acid to neutral, moderately permeable on summits, and side slopes of volcanic areas

occupy swamps and marshes in small depressions. Mollisols (Rendolls, Udolls) and Alfisols (Udalfs) are common in extensive areas of weathered limestone across the region.

15.2.4 General Characteristics—Semi-arid Coastal Plains (MLRA 273)

The Semi-arid Coastal Plain (Fig. 15.25) is principally on the southern coast of Puerto Rico with very small areas on the south side of St. Croix. This is the smallest (only 785 km^2) MLRA in this Land Resource Region. Physiography of the MLRA is complex. In Puerto Rico, about 50 % of the coastal region slopes from sea level to 2–15 m. The other half of the coastal region rises to an elevation of approximately 75 m. On St. Croix, elevations in the MLRA range from 3 to 45 m. Near the coast, slopes are nearly level but increase to a more rolling topography further from the sea.

The geology within the MLRA is also complex, but is similar to that of the lower elevations in MLRA 271. The

Fig. 15.34 Southgate soil series in MLRA 271—Semi-arid Mountains and Valleys (Virgin Islands). These soils are shallow, well drained, very strongly to slightly acid, moderately permeable on slopes ranging from 2 (summit) to 90 % (backslopes of volcanic hills and mountains)

parent materials are mostly Quaternary alluvial gravel, sand, silt, and clay near the coast. At the higher elevations where slopes are steeper, parent materials are limestone and volcanic rock. The vegetation across the region is diverse. In drier areas, the plant species includes beachgrass (*Ammophila*), southern sandbur (*Cenchrus echinatus*), saltwort (*Batis maritime*), bermudagrass (*Cynodon dactylon*), flame tree, white oak (*Quercus alba*), leucaena, and black olive (*Olea europaea*) to name a few. In wetter areas, the vegetation includes red, white, black, and button mangroves, southern cattail, leatherfern, water panicum, and para grass.

The climate is very uniform with rainfall ranging from 750 mm near the coast to about 1100 mm on the inland hills. There are isolated areas where the rainfall is as low as 250 mm. Rainfall comes primarily as afternoon showers. The amount of rain depends on the time of the year (e.g., rainy in April and May and drier in December to March). Average air temperature is 26 °C with very little deviation.

Approximately 50 % of the land use is grassland and urban development. Cropland comprises only 20 % of the area.

15.2.4.1 Soils of the MLRA 273

Mollisols and Vertisols are the dominant soils in this resource area. In Puerto Rico and Ct. Croix, the soil temperature regime is isohyperthermic with mostly an ustic soil moisture regime. In a small area in Puerto Rico, the soils have an aridic soil moisture regime. On floodplains in Puerto Rico, the soils are Aquolls (somewhat poorly drained) and Usterts (excessively to well drained). Vertisols (Usterts and Aquerts) are located on the high terraces.

On St. Croix, the soils are Mollisols (Ustolls) and Vertisols (Usterts) on alluvial fans and terraces. Ustolls are located in valleys associated with limestone hills and mountains.

Fig. 15.35 Irregular topography associated with karst landscapes in MLRA 272—Puerto Rico

15.3 Summary and Conclusions

The soils in peninsular Florida (LRR U) and the Caribbean islands (LRR Z) reflect the environmental factors; climate, topography, parent materials, vegetation, and time; that strongly influence properties of soils across the landscape. In LRR U, the majority of the soils have formed in sandy marine sediments although alluvial deposits, aeolian sands, limestone bedrock, and organic matter are also important soil parent materials in the region. Elevations range from sea level along the Atlantic and Gulf coast up to about 50 m on the central Florida ridge. Low relief, especially near the coasts, and numerous depressions has resulted in appreciable areas of poorly and very poorly drained soils with aquic moisture regimes. The region is warm and humid. Annual air temperatures across the region range from about 20 to 25 °C and do not vary greatly among the seasons which results in the soils having an isohyperthermic soil temperature regime. Precipitation ranges from about 1100 to 1500 mm annually and often falls as short duration high intensity storms associated with tropical depressions. Soils across the region are commonly acid, moderately permeable, and have thin A horizons with relatively low contents of organic C. They are dominantly sandy Entisols (Psamments and Psammaquents) and Spodosols (Alaquods) although in areas with underlain by loamy sediments Ultisols are common (Paleudults, Hapludults, Paleaquults). Soils in low relief and depressional areas are commonly Histosols (Saprists). Because of their properties, commonly sandy with low inherit fertility, soils across the region are fragile and are subject to degradation without careful management.

Soils in Puerto Rico and the US Virgin Islands (LRR Z) are also products of their environment for development although conditions in LRR Z are considerably different from those in LRR U. Elevation ranges from sea level to more than 1220 m on Puerto Rico and local relief often results in steep terrain with soils subject to landslides and high rates of erosion. Soils across the region have developed in a very wide range of parent materials including volcanic materials such as ash and basalt, limestone, and coastal beach and dunal sediments. Rainfall varies widely across the islands ranging from about 890 to more than 2100 mm. The range in topography, precipitation, and parent materials has resulted in soils with widely varying properties that include Ultisols, Oxisols, Inceptisols, Mollisols, Alfisols, Entisols, and Histosols. As is the case in LRR U, the soils are fragile and subject to degradation if mismanaged. Steep slopes in mountainous areas make the soils extremely vulnerable to erosion if cleared for crop production. The soils can also be easily degraded as the result of urbanization which is rapidly occurring in the region.

References

Borisova T, Carriker RR (2014) Florida's water resources. EDIS Publication No. fe757. https://edis.ifas.ufl.edu/fe757

Florida Climate Center (2015) Monthly normals 1981–2010. http://climatecenter.fsu.edu/products-services/data

Florida Department of State (1989) State soil of Florida. http://dos.myflorida.com/florida-facts/florida-state-symbols/state-soil/

Grabowski J (2015) Study helps Florida ranches, growers build soil health. USDA-NRCS News Release. http://www.nrcs.usda.gov/wps/portal/nrcs/detail/fl/newsroom/releases/?cid=NRCSEPRD335284

National Weather Service (2015) Normals by month. http://www.srh.noaa.gov

Soil Survey Staff (1987) Soil survey of Indian river county, Florida. http://www.nrcs.usda.gov/Internet/FSE_MANUSCRIPTS/florida/FL061/0/indian.pdf

Soil Survey Staff (2000) Soil survey of the US Virgin islands. http://www.nrcs.usda.gov/Internet/FSE_MANUSCRIPTS/puerto_rico/PR690/0/VI.pdf

Soil Survey Staff (2002) Soil survey of Caribbean National Forest and Luquillo Experimental Forest, Commonwealth of Puerto Rico. http://www.nrcs.usda.gov/Internet/FSE_MANUSCRIPTS/puerto_rico/caribbeanNF_PR2002/CNF.pdf

State of Florida (2014) Office of Economic and Demographic Research. Tallahassee, FL. http://edr.state.fl.us/Content/population-demographics/reports/econographicnews-2014v1.pdf

State of Florida (2015) Florida quick facts. http://www.stateofflorida.com/facts.aspx

USDA-NRCS (2006) Land resource regions and major land resource areas of the United States, the Caribbean, and the Pacific Basin. U. S. Department of Agriculture Handbook 296

USDA-NRCS (2015) Official soil series descriptions. https://soilseries.sc.egov.usda.gov/osdname.asp

Watts C, Collins ME (2008) Soils of Florida. Soil Science Society of America, Madison, Wisconsin

Soils of Hawaii and the US Territories in the Pacific: LRRs Q and V

16

Michael P. Robotham, Robert T. Gavenda, and Christopher W. Smith

16.1 Introduction

The islands of the State of Hawaii, the US territories of Guam and American Samoa, and the Commonwealth of the Northern Mariana Islands (CNMI) make up the area that is sometimes referred to as the "American Pacific". This chapter is a discussion of soils and landscapes in the region, which is covered by the Pacific Basin Region (LRR Q) and the Hawaii Region (LRR V). Within LRR Q, the chapter only focuses on those island groups that are explicitly under the political jurisdiction of the US government (American Samoa, Guam, and the Commonwealth of the Northern Mariana Islands—CNMI) (Fig. 1.3). It does not include the US Affiliated Nations: the Federated States of Micronesia, the Republic of Palau, and the Republic of the Marshall Islands (all of which have NCSS Soil Surveys from when these areas were Trust Territories), nor does it discuss smaller US possessions where soil information is unavailable including the Northern Hawaiian Islands, Johnston Island, Wake Island and Palmyra Atoll.

16.2 General Characteristics of the Region

The Hawaii and Pacific Basin LRRs span an area that is larger in size than the contiguous 48 states. However, despite the large expanse, the actual land area is <20,000 km^2 (75 % of that total in the Hawaiian Islands).

The region sits within the "true tropics" and, with the exception of high elevation areas on the Big Island of Hawaii and on Maui, is characterized by a tropical maritime climate that is generally warm to hot and dry to humid. Hawaii and the Mariana Islands (Territory of Guam and the Commonwealth of the Northern Mariana Islands) are located in the Northern Hemisphere while the Territory of American Samoa is in the Southern Hemisphere. Hawaii and American Samoa are within the trade wind belt that blows from the northeast in Hawaii and the southeast in American Samoa for much of the year. These prevailing winds create a rainfall gradient from the eastern to the western sides of the islands. The eastern, or windward, areas are characterized by high rainfall (as much as 4000–5000 mm year^{-1}) that generally increases with elevation while the western and southwestern, or leeward, sides of the islands experience significantly less rainfall that rapidly decreases with elevation to as low as 250 mm year^{-1} near the coast in leeward areas (Fig. 16.1). Rainfall is uniformly high (over 3000 mm year^{-1}) throughout American Samoa. Annual rainfall amounts in the Mariana Islands are similar to much of Hawaii, but the climate is characterized by much more pronounced wet and dry seasons. Annual rainfall decreases slightly from Guam (2400–3000 mm year^{-1}) north to the CNMI (2200–2500 mm year^{-1}). The wettest part of the year is from November to March in American Samoa and most of Hawaii and from July to November in the Mariana Islands. The driest part of the year is from June to September in Hawaii and American Samoa and from February to April in the Mariana Islands (Fig. 16.2). Humidity is high year-round throughout this region. Tropical storms, including the intense storms referred to as hurricanes in the Eastern Pacific and typhoons in the Western Pacific, are a regular occurrence (major storms occur every 5–20 years on average). Severe storms can occur throughout the year but most commonly occur in the late summer (between July and November in the Northern Hemisphere and between January and May in the Southern Hemisphere).

The entire region consists of islands ranging in size from very small offshore islets (<1 km^2) to Hawaii Island (often referred to as the "Big Island") with an area of over

M.P. Robotham (✉)
USDA-NRCS, Lincoln, NE, USA
e-mail: Michael.robotham@lin.usda.gov

R.T. Gavenda
USDA-NRCS, Merizo, Guam
e-mail: Gb96921@yahoo.com

C.W. Smith
USDA-NRCS, Washington, DC, USA
e-mail: Christopherw.smith@wdc.usda.gov

L.T. West et al. (eds.), *The Soils of the USA*, World Soils Book Series,
DOI 10.1007/978-3-319-41870-4_16

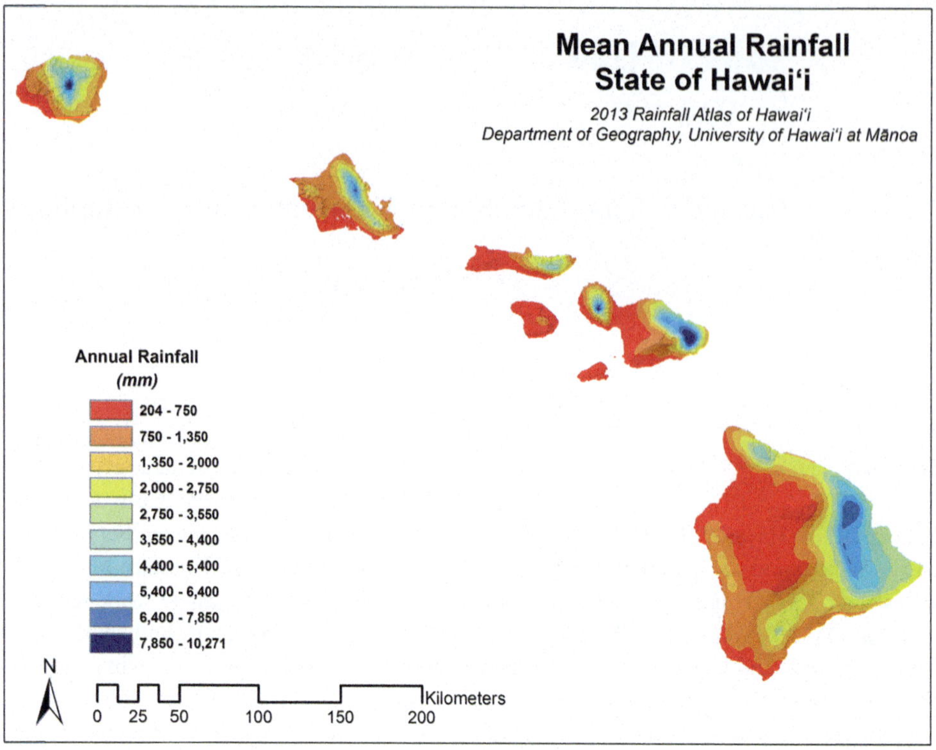

Fig. 16.1 Mean annual rainfall for the State of Hawaii (Giambelluca et al. 2014)

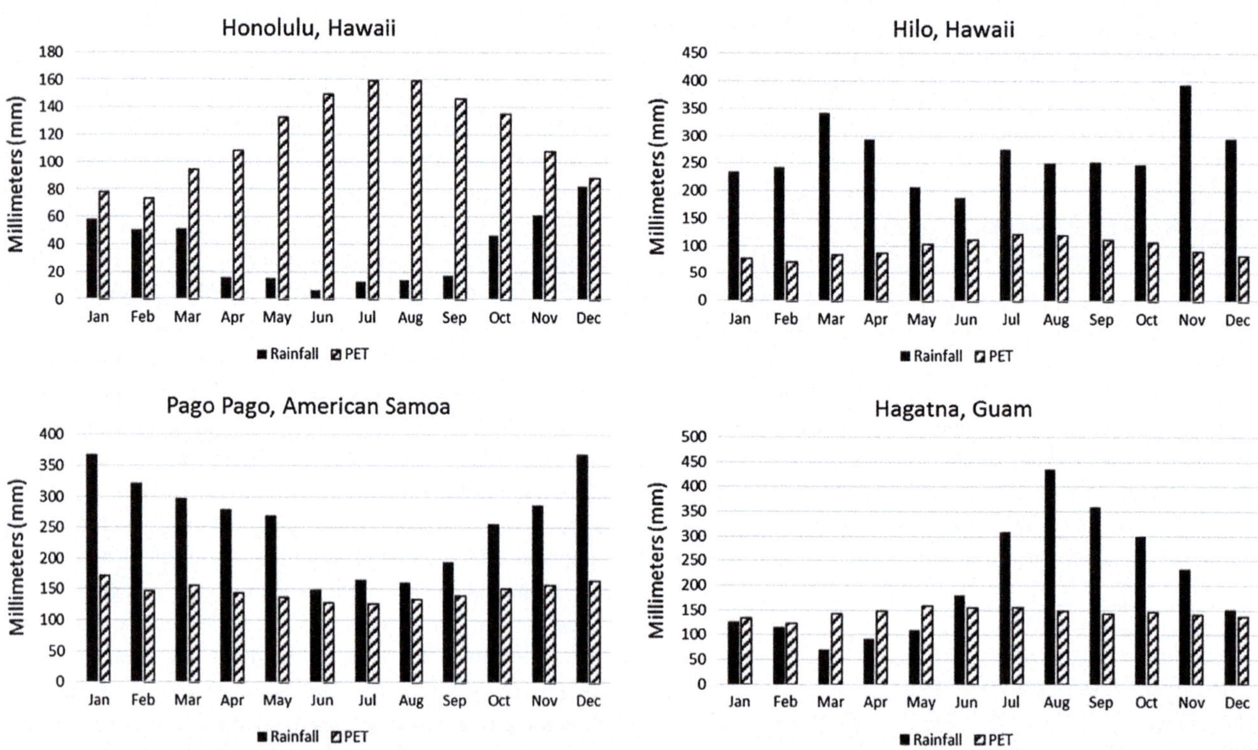

Fig. 16.2 Monthly rainfall and potential evapotranspiration (PET) for Honolulu (leeward Hawaii), Hilo (windward Hawaii), Pago Pago (American Samoa) and Hagatna (Guam). Adapted from NOAA (2015a) (rainfall) and USDA-NRCS (2015a) (PET)

10,000 km^2. Elevation ranges from sea level to over 4200 m at the top of Mauna Kea on the Big Island. All of the islands in the region are of volcanic origin, but the ages of the islands vary widely. The oldest islands are the southern islands in the Mariana Island archipelago (Guam, Rota, Tinian and Saipan), which consist of severely eroded volcanic peaks that date from the Eocene (Tracey et al. 1964) surrounded by coralline limestone plateaus. In contrast, the main Hawaiian Islands, the islands of American Samoa, and the northern parts of the Mariana Island chain have formed from more recent volcanic activity (ongoing on the Big Island and the northern Mariana Islands). Niihau and Kauai, the oldest of the main Hawaiian Islands, were initially formed approximately 5.8 million years before present and the age of the Hawaiian Islands decreases to the south and east along the island chain. The islands of American Samoa are of similar age to Oahu (approximately 1–2 million years). Small areas of more recent volcanic activity (also called post-erosional volcanic eruptions) occur on Kauai, Oahu, Kahoolawe and American Samoa. Soils formed on coralline limestone plateaus are found on Maui, Oahu, and Kauai. A majority of the soils throughout the region are formed from volcanic parent materials (ash, cinders, weathered volcanic rock) but soils formed on coralline limestone plateaus are common in the Marianas Islands. Soils formed from alluvium and colluvium are found in depositional landscapes throughout the region. With the exception of the high mountain areas of the Big Island and Maui, most soils in these regions fall into the isohyperthermic or isothermic temperature regimes. Soil moisture regimes follow the local climate and range from perudic in the high rainfall areas on the windward sides of the main Hawaiian Islands and in American Samoa, to ustic in parts of Hawaii and throughout the Mariana Islands, to aridic in the driest parts of Hawaii. Nearly all of the soil orders included in Soil Taxonomy, except Gelisols and Spodosols, can be found in this region and the dominant soil orders vary across the islands. Taken at a regional scale, Oxisols, Andisols, and Inceptisols are most common.

Due to the unique geographic scope of LRRs Q and V and their associated MLRAs, discussion of the characteristics and genesis of soils in these LRRs is organized into three broad groups which are used to frame the remainder of this chapter: the younger Hawaiian Islands consisting of the entire Big Island and eastern part of Maui which is dominated by the Haleakala volcano (Fig. 1.3; MLRAs 157, 159A, 159B, 160, 161A, 161B, and 162); the older Hawaiian Islands and American Samoa consisting of West Maui, Kahoolawe, Lanai, Molokai, Oahu, Kauai and Niihau as well as the islands of American Samoa (Fig. 1.3; MLRAs 158, 163, 164, 165, 166, 167, and 197); and the Mariana Islands consisting of Guam, Rota, Tinian, Saipan and the smaller islands in the northern part of the archipelago (Fig. 1.3; MLRAs 190, 191 and 192).

16.3 The Younger Hawaiian Islands

The younger islands include the Big Island and the eastern portion of Maui. The oldest subaerial rocks are estimated to have formed approximately 0.7 million years before present (BP) and 1.3 million years BP for Hawaii and Maui respectively (Tilling et al. 2010). The islands are shield volcanoes made from innumerable aa and pahoehoe lava flows. Pahoehoe lava is less viscous than aa and has a smoother, ropey surface. Being more viscous, aa lava does not spread as much as pahoehoe lava and the surface of flows typically has several meters of relief above the surrounding landscape. Organic matter and volcanic ash accumulate on the surface of pahoehoe lava. The surface of an aa lava flow is comprised of clinkers with extensive void space that eventually fills with organic matter and volcanic ash. The overlapping and overriding layers of these two kinds of lava flows have built landscapes with numerous stepped surfaces, which are constructional, not erosional, and that have a series of ages. Adjacent landscapes may differ in age by more than 10,000 years, with the older lavas supporting ash-derived soils more than 1 m thick and the youngest lavas with organic soils only a few centimeters thick.

16.3.1 Parent Materials

Basaltic and olivine basalt lavas (aa, pahoehoe), cinders, and ash are the dominant soil parent materials throughout the Pacific Islands. Cinder cones on East Maui and the Big Island have more potassium (K) feldspars relative to older ash deposits and flow materials. Late stage alkali basalts, hawaiite, mugearite, and trachyte, occur on Mauna Kea and Haleakala (MacDonald et al. 1983). Volcanic ash has been deposited throughout the Big Island and East Maui through multiple events. Although geologists consider the Hawaiian volcanoes to produce limited amounts of ash, Andisols formed in ash, found mostly on the Big Island and Maui, comprise the majority of soils in Hawaii. There are small areas of calcareous sand inshore of beaches and alluvium in older valleys.

16.3.2 Climate

Windward versus leeward differences in precipitation in all of Hawaii are most pronounced on the Big Island and East Maui due to the elevation and mass of these volcanoes. All three major volcanoes (Mauna Kea, Mauna Loa and Haleakala) exceed 3000 m. The summit of Mauna Kea is the highest point in the islands with an elevation of 4205 m and was glaciated during the Pleistocene (Porter 1979). On the windward side of the mountains, rainfall increases with elevation until the inversion elevation (the "top" of the trade winds) at approximately 1700 m at which point it begins to decrease. This leads to arid conditions at high elevations on all three major peaks (annual precipitation is about 380 mm per year) (Juvik and Juvik 1999) (Fig. 16.1). Temperatures decrease with elevation following the adiabatic lapse rate with a decreased cooling rate above the inversion elevation. Blocking of trade winds by the massive volcano allows a sea breeze to flow inland and upslope creating a rain belt on the leeward side of Mauna Loa.

16.3.3 Soil Formation

As a lava flow cools, it is immediately populated, mainly in fractures and between clinkers, by microorganisms. These are followed by mosses and ferns and later by shrub and tree vegetation (mainly Metrosideros polymorpha or commonly, Ohia Lehua). The litter and other organic matter creates Folists (Fig. 16.3) (unsaturated Histosols) as the first stage of soil formation on lava flows in all environments. However, productivity and rate of formation differ between windward and leeward sides. A Folist can persist until volcanic ash accumulates to sufficient thickness that a shallow Andisol develops.

Among Andisols on the island of Hawaii, there is a correlation between ash thickness and age of lava surface except in areas close to ash-producing cinder cones. On the leeward side of the Big Island, there is the general lack of buried soils, which are common in ash-derived soils and are found on windward Big Island and on leeward Mauna Kea. The lack of buried soils suggests that ash accumulates at a slow rate and that topsoil formation occurs at a similar rate, which masks evidence of thin ash deposits. Although ash accumulation is slow, an abrupt ash-lava contact in soils that are about 10,000 years and younger indicates that ash deposition occurs much faster than weathering of the underlying basaltic materials. In some instances, one can still observe the characteristic ropey surface of original pahoehoe lava flow (Fig. 16.4).

Mineralogy varies among the Andisols and depends on the extent of weathering. Andisols with hydrous (water-loving) properties are formed under continuous moist conditions in high rainfall (perudic) areas (Fig. 16.5). These soils are

Fig. 16.3 Profile of the Kekake series (Lithic Ustifolists) a very shallow, moderately well drained, organic soil over pahoehoe lava on the Big Island of Hawaii

Fig. 16.4 Profile of the Kainaliu series (Andic Haplustolls), a moderately deep soil formed from volcanic ash over aa lava on the Big Island of Hawaii

dominated by amorphous minerals and have high surface area. Vitric (glassy) Andisols have low surface area due to large particle size, young age, and slow weathering due to relatively low precipitation. These soils are typically found in close proximity to cinder cones and often contain visible cinders and small fragments of pumice. Medial (intermediate) Andisols are formed in conditions between these two extremes and are most commonly found in areas where seasonal drying occurs (drier udic and ustic environments). The surface area of these soils is intermediate between the hydrous and vitric Andisols.

Fig. 16.5 Profile of the Kealakekua series (Typic Hydrudands), a moderately deep, hydrous Andisol found on the Big Island of Hawaii

16.3.4 Soil Management

In some circumstances, Folists are cleared and graded for crop production, particularly for perennial crops like coffee and avocados in the Kona coffee belt and papaya and other fruit crops in windward areas. Clearing and grading of Folists generates sand size fragments from the ripped clinkers and pahoehoe. When coupled with reduced organic inputs and oxidation of soil organic matter (SOM) due to exposure and increased oxygen, the increased mineral content of the soil results in the soils being classified as Entisols instead of the original Histosols. While oxidation of SOM does not constitute erosion, it is a common form of soil loss for all Histosols, including Folists (Fig. 16.6).

Fig. 16.6 Papaya orchard on ripped very shallow Folists over pahoehoe lava on the Big Island of Hawaii (photo courtesy of Tony Rolfes, USDA-NRCS)

In the Andisols, the soil erodibility has been evaluated by rain simulator experiments and is quite low (Dangler et al. 1976). However, former sugarcane fields have been observed to be as much as 0.75 m lower in elevation than the soil in the adjacent forest (Amy Koch, personal communication). This drastic landscape change is believed to have been caused by a combination of over 100 years of cropping systems on these steep slopes that included intensive tillage and historic sugar cane harvest practices that removed significant soil material with harvested roots, and increases in soil bulk density and to soil desiccation (termed irreversible shrinkage) caused by these management practices.

The formation of tillage pans has also contributed to increased erosion rates in areas where Andisols have been used intensively for agriculture. Bulk densities in Andisols typically range from 0.15 to 0.9 Mg m^{-3}. Because the bulk density of these soils is so much lower than that of other "common" mineral soils, it was often asserted that tillage pans could not form in Andisols. However, field observations have shown that tillage pans are often commonly found in areas where heavy equipment and intensive tillage was used—particularly in former sugar cane plantations. Even though the bulk density may still be <1.0 Mg m^{-3}, tillage pans capable of retarding root penetration and water movement on these sloping fields cause water to saturate the surface plow layer, generating runoff and associated rills and gullies (Fig. 16.7).

Many Andisols on the Big Island and Maui can be and have been degraded through inappropriate management, and the observed difference in soil organic matter content of soils in native forest as compared to those in tilled fields suggests great potential for carbon sequestration in soils through management (Saunders and Robotham 2010). With the closure of the sugar plantations, former sugar cane lands have been returned to fallow grasslands and plantation forests. Additional small areas have been planted to specialty crops such as papaya or ginger, and other areas have been reforested with high value native and introduced species.

Fig. 16.7 Severe water erosion on Udands in a Big Island sweet potato field exposing the underlying tillage pan (photo courtesy of Tony Rolfes, USDA-NRCS)

Continued improvements in the management of these soils present a tremendous opportunity for carbon sequestration in this landscape. However, further research is needed to identify appropriate management systems and to document the impacts of management changes.

16.3.5 Water and Nutrient Management

Soil texture (specifically clay content) is one of the primary diagnostic soil properties for soil classification and is also widely and successfully used by soil scientists to predict soil behavior. However, due to strong aggregation of clay-sized particles into silt and sand-sized water-stable aggregates, textural analysis has proven to be an inappropriate means to characterize and predict behavior in most Andisols. As a consequence, soil mineralogy has proven to be a more effective way to group these soils and to identify issues and challenges for management.

The upland Hydrudands, due to the very large surface area of the poorly crystalline components that is accentuated by SOM and a very high porosity and proportion of fine pore space, retain in excess of 100 % water (w/w) at 33 and 1500 kPa tension (Sollins et al. 1988). The high water retention makes these soils desirable for agricultural production although high soil moisture limits when tillage can be safely and properly performed. In addition, soil fertility management can be difficult in these soils. Due to the high surface area and high Fe content of most Andisols, particularly those found in Hawaii, phosphorous is "fixed" to the point where a previously unfertilized soil may require over 1000 kg ha^{-1} to provide enough P for crop growth (Fox 1980). It was formerly believed that P fixation could never be overcome in Andisols due to the continual formation of new P sorption sites in poorly crystalline components. However, for Hydruands with a long history of P fertilization, soil test P can be high or very high. Additional application of P may not be needed for production of many crops on these soils. Where soils with high P concentrations erode to water bodies, reduction reactions can solubilize Fe components binding the P which would release disproportionately large amounts of P to the water body. Phosphorus loading in coastal waters has not been shown to create significant issues for the marine environment at this time, probably due to waves and currents. However, delivery of large amounts of P could potentially become a problem in enclosed bodies of water such as coastal wetlands or smaller bays.

Exchangeable Al often creates problems for crop production in tropical systems. Fortunately, measured KCl extractable Al values are generally low in most Andisols on the Big Island and East Maui. However, values of over 6 cmol kg^{-1} soil have been measured in soils at high elevations. Liming rates required to bring these soils to a pH of 6.5 as is commonly recommended for most mainland soils can be as high as 15 T ha^{-1}. A liming philosophy that addresses Ca levels and Al saturation instead of achieving a certain pH has been successfully used on these soils. K levels are inherently low in these Andisols, but K is generally sufficient for low intensity pasture production (contributions of K from salt spray cannot be discounted). Supplemental potassium fertilization is generally required for agronomic crop production.

Folists on pahoehoe lava are typically <20 cm thick, and <1 m thick on aa lava. Even though water and nutrient holding capacity is high on a per cm thickness basis, the thin Folists have low available water holding capacity (AWC) and (CEC). These Histosols, however, form organo-P complexes and maintain P in a form more readily available for plants as compared to Andisols. However, this relatively high P availability, coupled with thin soils, allows P to leach more readily than nearly any other soil except for sandy soils (Yost, personal communication). The shallow thickness also puts the soils at high risk for nitrate leaching especially if rapid-release fertilizers are applied. To compound the leaching risk, water typically flows rapidly through these landscapes to the ocean via sub-surface pathways in the underlying lava flows. Fortunately, the potential for nitrate and phosphorus-induced algal blooms is minimized in coastal waters by active currents and wave action. Since water and nutrients may move rapidly, a key to managing these soils is using good soil health practices that include mulching and cover crops, and small frequent fertilizer applications.

16.4 The Older Hawaiian Islands and American Samoa

West Maui, Kahoolawe, Lanai, Molokai, Oahu, and Kauai, plus American Samoa are considered the "older" islands with ages ranging from 1 to 5.3 million years (Tilling et al. 2010). There has been considerable stream erosion dissecting landscapes on these islands. Parts of some islands have been over-steepened by giant landslides, which have likely caused large local tsumanis. After the original shield building phase of island construction, which was followed by extensive erosion, there have been phases of renewed volcanic activity that created many cinder cones and deposits of volcanic ash. The deposits related to this renewed volcanism are referred to as post-erosional volcanics. Diamond Head Crater on Oahu is the most famous of these deposits.

16.4.1 Parent Materials

The ages above refer to the lavas. Most soils, however, formed on moderately to gently sloping flanks and plateaus that are typically considerably younger than the age of the island. Generally, Andisols (from cinder cones) and Inceptisols (on steep mountain sideslopes) can be said to occur on the youngest upland surfaces and Oxisols/Ultisols on the oldest surfaces. Kahoolawe was formed at about the same time as West Maui. As Haleakala volcano (East Maui) was growing, Haleakala ash buried Kahoolawe ash as the cinder cones on Kahoolawe ceased eruptions and Haleakala remained active. Late stage alkali basalts, hawaiite, mugearite, and trachyte, are also present on Haleakala volcano (East Maui) and in West Maui. Ash from post-erosional volcanic eruptions is present on Niihau, Kauai, Oahu, and Kalaupapa on Molokai (MacDonald et al. 1983). Most of the sodium and potassium, as well as the small amounts of silica, present in the volcanic parent materials have been weathered out in high rainfall areas.

Geologists mostly assume that the soils on the older Hawaiian Islands developed from the underlying lava. If the parent material for most soils on the lava flow landscapes of the older islands was generated similarly to that of the Big Island (Hawaii), then it is reasonable to say that the Oxisols and Ultisols of these "older" islands formed in volcanic ash from either the original volcanoes that formed the islands or from eruptions on other islands. The upper solum of Oxisols in Central Oahu have been shown through stratigraphy and mineralogy to have developed in volcanic ash although these soils no longer have andic properties (Gavenda 1989). In addition to this "older ash," younger volcanic ash deposits from late stage and post-erosional cinder cones are the parent material for small areas of Andisols and soils with andic influence on Maui, Oahu and Kauai.

In addition to soils formed in multiple ages of ash and volcanic parent materials, steep gradient streams have provided alluvial materials in areas of central Maui, Oahu, and Kauai as well as stream outwash areas and floodplains on all islands in Hawaii and American Samoa. The steep topography and high rainfall, particularly in windward areas, has also resulted in frequent landslides, which have led to soils developed in colluvium at the base of steeper slopes. Although not common, there are also small areas including coastal plains on Oahu (Ewa Plain, North Shore) and Kauai (Mana Plain) where soils have formed from a combination of sediments eroded from upstream volcanic materials, weathering of the limestone bedrock, and deposition of volcanic ash from late stage eruptions.

16.4.2 Climate

As noted under the discussion of the "younger" islands, Hawaii and American Samoa have a trade wind climate with a strong partitioning of precipitation as the mountains intercept the rain on the windward side. The mountains on all the higher islands not in a rain shadow are important in "catching" the trade winds and increasing precipitation as moist air rises and cools. However, none of the mountains on any of the older islands extend above the inversion layer. This leads to a situation where rainfall often "spills over" the mountains creating a very strong precipitation gradient on the leeward side of the islands. An example of this is the strong orographic effect of the Koolau mountains on Oahu. Average annual precipitation ranges from 4000 mm at the top of the mountains to 600 mm approximately 10 km downslope in Waikiki (Fig. 16.1). Areas such as West Molokai, which lack sufficient elevation to capture orographic rainfall have a dry ustic to aridic soil moisture regime.

The geologic evolution of the islands has also resulted in changes in climate regimes for certain islands over time. The presence of highly weathered soils on Kahoolawe and Lanai suggests that these islands experienced a more typical trade-wind driven rainfall pattern in the past. However, the growth of Haleakala served to decrease the precipitation on Kahoolawe and Lanai by creating a rain shadow that now impacts both islands. Haleakala also creates a venturi effect of trade winds wrapping around and converging on Kahoolawe creating generally stronger winds and accompanying evapotranspiration rates. Wind erosion potential is greatest on Kahoolawe of all the islands. Limited information is available about the soils on Niihau, the smallest of the major Hawaiian Islands, but, given its location in the rain shadow of Kauai, a similar sequence of early soil development and subsequent climate change is believed to have occurred. Glacial and interglacial sea level changes have also affected rainfall distribution as modeled by Ruhe (1975).

16.4.3 Soil Formation

From observations on the Big Island of lava flows overlapping prior flows or stopping short of prior flows, the landscape surface on the older islands evolved in a stair step fashion creating treads and risers. Ruhe (1975) observed these features on Oahu and suggested these were erosional surfaces. Further observations since publication of these observations have not been able to link treads and risers across dissections suggesting that these features are depositional rather than erosional. Features in soil profiles suggesting that the soils are derived from ash blanketing the flows include presence of saprolite from flow material abruptly overlain by red and brown, highly weathered clay materials that are rich in iron and aluminum oxides (oxidic materials) and a lithologic discontinuity deep to moderately deep in the profile. Additionally, there are no erosional features at tread and riser contacts suggesting that volcanic ash falls covered existing landscapes and the soils on the stepped surfaces are formed in these volcanic ash deposits. For reasons unknown, the volcanic ash retained andic properties (low bulk density, non-crystalline mineralogy) in some areas but weathered to oxidic mineralogy dominated by 1:1 clays (kaolinite) and iron and aluminum oxides in most of the area. The andic to oxidic transition occurs over short distances and can be seen in individual soil profiles. The geomorphology and mineralogy strongly suggest that these are not two separate deposits but derive from the same original volcanic ash parent material (Gavenda 1989) (Fig. 16.8).

Where the discontinuity is shallow enough in the profile, these soils have been classified as Ultisols even though the soil formation processes do not follow those typically associated with Ultisols in other regions. In Hawaii, it appears that the discontinuity is an erosional surface that truncated soil profiles. Volcanic ash was subsequently deposited on that surface. The ash has weathered resulting in illuviation of clay that has produced clay coatings (argillans) on soil aggregates in the underlying material. Many of these soils also have kandic horizons and meet the chemical criteria for Oxisols (Unpublished letter from R. Stoops to R. Gavenda).

The interaction between parent materials, climate conditions, and vegetative growth on the windward sides of the older Hawaiian islands including East Molokai, Oahu and Kauai as well as the islands of American Samoa have resulted in the development of a number of soil series that classify as Mollisols, particularly in American Samoa. Many of these soils are formed in oxidic parent materials, but warm temperatures and abundant year-round rainfall have created conditions for lush vegetative growth and the associated accumulation of organic matter leading to development of a deep, dark surface horizon with sufficient CEC to meet the criteria for a mollic epipedon and for Mollisols.

In areas of central Maui, the coastal plains of Oahu, and Kauai, steep gradient streams have deposited alluvial materials partially composed of abraded basalt fragments that

Fig. 16.8 Profile of the Kolekole series (Oxic Dystrudepts), a deep kaolinitic soil found on Oahu, Hawaii. The topsoil formed in dust blown in from Asian deserts. Color changes lower in the profile are the result of different degrees of weathering of the volcanic ash parent material. The volcanic ash overlies a truncated soil formed in brown stream deposits

weather to smectite and other clay minerals. Thus, soils have mixed or smectitic mineralogy classes depending on the moisture regime at the deposition site. This mineralogy in aridic and dry ustic soil moisture regimes may result in development of Vertisols and Vertic Mollisols. Vertic soils also occur on toeslope positions in more arid areas. In more moist areas, Mollisols have interlayered 2:1 clay minerals, kaolinite, and Fe oxihydroxides with high base saturation and moderate amounts of soil organic matter.

Small areas of younger ash-derived soils (Andisols) are found within the Oxisol/Ultisol landscapes on Oahu, Kauai and Tutuila (American Samoa) in the vicinity of late stage and post-erosional volcanic cinder cones. Shallow soils formed from volcanic ash and/or alluvial parent materials are found over calcareous coastal plain substratum in coastal areas such as the Ewa plain on the south shore of Oahu, areas on the north shore of Oahu, and the Mana plain on Kauai.

Older soils on stable surfaces have accumulated deposits of tropospheric dust. The dust has been identified via isotope analysis as having come largely from the Asian deserts. The dust contains quartz and muscovite that are not present in Hawaiian ash or lava flow material. As evaluated from research ships, dust presently falls into the open ocean in the area of the Hawaiian Islands at a rate of approximately 1 mm 1000 $year^{-1}$ and is generally correlated with precipitation. Topsoils comprised of dust are commonly about 20 cm thick on Oahu and Kauai (Jackson et al. 1971). Approximately 2 m of dust accumulation has been observed in the Alakai Swamp, a high elevation, high rainfall area on Kauai, and on Mt. Kaala on Oahu (Smith, personal communication).

Information is limited for the steep mountainous soils of Oahu, Kauai, Molokai and West Maui. A preliminary soil survey of the Hanalei watershed on Kauai indicated that most of the soils of the steeplands are deep, acid, clayey Inceptisols over saprolite. Inceptisols formed in colluvium located in concave toeslope positions are typically skeletal with the <2 mm material having similar characteristics to those in non-skeletal counterparts. Extremely steep areas that are subject to landslides typically contain very shallow organic soils (Folists) from the fern vegetation growing in the slightly weathered saprolite cliff faces. Consequently, most landslides in these regions do not produce significant amounts of sediment. Stream bank failures appear to be more significant contributors to stream sedimentation.

Sea cliffs on Kahoolawe, Molokai, Lanai, Kauai, Oahu (Moore et al. 1989), and Tutuila are the result of catastrophic landslides severing up to half of the original volcanoes and depositing them into the ocean. An abrupt lithologic discontinuity has been identified in many of the older soils of Central Oahu. The cause of this discontinuity is debatable, but one hypothesis is that the giant landslide that created the cliffs of Windward Oahu generated a local tsunami that stripped off the existing soil creating an erosion surface. The landslide may have caused large phreatic eruptions of volcanic ash as seawater came in contact with hot magma. This ash was then deposited on the erosion surface. Erosion surfaces that appear to have been covered by volcanic ash have also been observed on West Molokai and Lanai.

16.4.4 Soil Management

Plantation era agricultural production, particularly sugarcane, caused severe "legacy" erosion in the agricultural landscapes on all of the older islands, particularly in areas of steeper slopes including West Maui and Central Oahu. Intensive agriculture is still common in many of these areas with crops like corn and a wide variety of vegetables now grown in areas formerly in sugarcane. The shift to crops other than sugarcane has resulted in smaller field sizes and reduced use of heavy equipment. However, many of the specialty crops that replaced the plantation crops require more extensive and more frequent tillage, which can increase erosion risk. The Oxisols and Ultisols common in these areas are relatively resistant to sheet and rill erosion, but erosion can and does occur in these landscapes.

In some cases, extensive use of intensive tillage methods have broken down the normally stable sand and coarser sized aggregates of clay in surface layers of Oxisols into smaller silt and sand sized micro-aggregates. Because of associated increase in ease of detachment and transport of these aggregates, these areas are at a much higher risk for both wind and water erosion as compared to native soils. In addition, tillage pan formation is common, particularly under the influence of heavy equipment such as was used in the sugarcane and pineapple industry. With the disappearance of these cropping systems, abandoned land has been left with tillage pans, and smaller farming operations now using these soils do not have the equipment to break up the pans. The presence of these tillage pans reduces the root zone available for crops to extract both water and nutrients, increases runoff, and exacerbates the risk of erosion.

Organic matter management is very important in these soils, especially those in the oxidic mineralogy class where the CEC is typically low. Increasing organic matter can facilitate better nutrient management by increasing the soils CEC and can also increase the soil's water holding capacity. Soils found in the windward areas of these islands where vegetative cover has been maintained can have high levels of organic matter. These levels, relative to other soils in non-tropical environments, are due to long term accumulation from high levels of vegetative biomass. In contrast, soils on the drier leeward sides of the islands generally tend to be lower in organic matter that those on the windward side. In both regions, historical agricultural practices decreased soil organic matter levels. Organic matter levels are increasing in soils that have been left in fallow, especially in windward areas, but the low inherent productivity of the drier areas will slow the accumulation of organic matter. Soils that remain in agriculture are likely to remain low in organic matter as typical crops being grown do not produce the root mass or residue required to increase soil organic matter concentration. Use of cover crops and other soil improvement technologies would increase rates of organic matter accumulation, but these are hampered by high costs for seed and the need for irrigation water if cover crops are to be grown in drier areas during dry months.

16.4.5 Water and Nutrient Management

Because of the wide range of properties of soils on the older Hawaiian Islands and in American Samoa, these soils exhibit a wide range of water and nutrient retention properties. The relatively small areas of soils with smectitic mineralogy (Vertisols and Vertic Mollisols) found primarily in ustic and arid leeward areas have relatively high inherent water holding capacity and CEC. Although these soils can be difficult to manage, especially when wet, they are favored by many small growers due to their higher inherent fertility and water holding capacity (Fig. 16.9). Water runoff and associated nutrient or pesticide loss can occur in these soils, but most are found in flatter landscapes where these losses are typically not an issue. Leaching is also not typically seen as an environmental risk in these soils due to their low saturated hydraulic conductivity.

Shallow oxidic soils on all islands that formed in volcanic ash or alluvium deposited over calcareous bedrock are generally desirable for agriculture due to available cations, especially Ca and Mg, from the underlying bedrock. However, their generally shallow depth causes challenges for water and nutrient management including relatively low water holding capacity. In addition, appropriate nutrient and pesticide management practices coupled with good irrigation management are required to reduce the risk of leaching nutrients and pesticides through the soil, into the permeable bedrock and subsequently into the underlying aquifer.

Fig. 16.9 Small acreage irrigated vegetable production on a Vertisol (shrink-swell clay soil) on Oahu, Hawaii

The inherent fertility of the Oxisols and Ultisols varies by rainfall and original ash composition. More arid areas on Maui and Oahu can have moderate amounts of potassium on the exchange complex. The pH and base saturation can be relatively high in soils derived from base rich ash (such as Eutrotorrox). Acid soils formed from highly weathered ash, such as the Dystrudepts on Kauai, may exhibit the lowest CEC of all volcanically derived soils in Hawaii and may have weathered to the point that they are deficient in silica and require calcium silicate application for certain crops. P fixation potential is high in most of the Oxisol/Ultisols, but many areas have had the P fixation satisfied by decades of fertilization.

A unique property of some Oxisols and Ultisols as well as acid Andisols is the presence of anion exchange capacity (AEC) (Uehara and Gillman 1981). Although generally less than 2 cmols(+)kg^{-1}, the AEC in these soils is sufficient to retard the movement of nitrate. This is particularly true in former pineapple production lands where acidification of the soil was encouraged to retard nematode growth. Deep coring has shown a nitrate pulse above the water table (Deenik and Uehara 1997), and there is concern that the shift to other types of crops and accompanying liming will increase the pH, decrease the AEC of the soils, and allow adsorbed nitrates to be released and leach to the water table.

16.4.6 Other Management Considerations

On Kahoolawe, the fragile soil condition coupled with especially strong winds, grazing by goats and cattle, and lastly, wildfires as a result of the island being used for a bombing range have led to approximately 1.5 m thickness of soil lost to wind and water erosion. The erosion has exposed a paleo-Oxisol (Smith and Nakamura, personal observations). The exposed paleo surface is wind and saltation burnished, which has clogged pores, reduced water infiltration and seed germination, and exacerbated runoff and gully formation. Reseeding these "scoured" areas has been attempted by windrowing with hay bales to trap sediment on the upslope side of the bales. The accumulated material provides a shallow seedbed for germination of seed falling from the bales. Some success has been achieved and efforts continue to restore this highly degraded environment (Fig. 16.10). Similar erosion has occurred on the northern slopes of Molokai and Lanai.

Fig. 16.10 Native plant re-establishment on highly degraded soils on the Island of Kahoolawe, Hawaii (photo courtesy of Gregory Koob, USDA-NRCS)

The steep upland areas on all islands in Hawaii (Fig. 16.11) provide important ecosystem services such as native plant and animal species habitat, recreation, water catchment and storage, and hunting. These areas are home to the last remnant native plant communities but only on elevations and topographies where non-native plant communities and feral pigs have not encroached. Areas accessible to feral pigs have been devastated by the creation of trails and wallows resulting in loss of native plants. Pig trails and wallows also change forest soils that have high infiltration rates into highly compacted areas with low infiltration and high runoff and erosion. Standing water in these areas can harbor mosquitoes and contaminated runoff flowing from these areas into streams can spread fecal borne diseases. Approximately 50 % of low to mid mountain landscapes have been affected.

In addition to serving as important habitat for native species and source of ground and stream water recharge, the steeply sloping uplands of American Samoa also are used for agricultural production. The indigenous management system in these areas was characterized by a multi-species, multi-story cropping system that maintained surface and canopy cover throughout the year and thereby greatly reduced soil and organic matter losses. The shift away from a subsistence economy to a market economy has led to an increase in these steep landscapes being used for monocrop agriculture creating the conditions for extensive amounts of erosion and organic matter loss. However, in recent years, awareness of these problems has greatly increased and many growers have adopted management systems such as hillside ditches with vegetative barriers, strip cropping, and reduced tillage that allow them to produce cash crops while greatly reducing soil and organic matter losses (Fig. 16.12).

Many hydric soils in the coastal lowlands of Hawaii and Samoa are well suited for taro production and have been traditionally used for taro production and fishponds by the

Fig. 16.11 Steep cliffs in the upper reaches of the Hanalei Valley on the Island of Kauai, Hawaii

Fig. 16.12 Multi-story, minimum-tillage cropping system including banana, taro and root crops on steep slope in American Samoa

Native Hawaiians and Samoans since soon after the islands were initially settled. Taro (*Colocasia esculenta*) is the staple starch in Polynesian culture and was the primary cultivated crop for the native peoples of both Hawaii and Samoa (Fig. 16.13). These wetland soils were managed for hundreds of years as low input systems. However, with the rise of commercial agriculture and the availability of nitrogen fertilizer, management practices were put in place that resulted in poor nitrogen fertilizer efficiency due to volatilization of residual N upon draining, thus oxidation to nitrate then subsequent flooding producing NO_x. Leaching of N also occurs throughout the season due to the flooded conditions. In many areas, especially on Oahu, these coastal wetlands were converted to duck ponds and rice paddies. The converted wetlands subsequently were filled by sediment from eroding uplands and by urban development. As much as 1.25 m of fill materials over the original wetland surface have been documented in urban Honolulu (Bergmanis 2004). The combination of both deliberate conversion and degradation through erosion from upland areas have led to significant reductions in coastal wetlands on all of the Hawaiian islands since non-Hawaiian settlement (van Rees and Reed 2014).

16.5 The Mariana Islands

The Mariana Island chain is located just west of the major subduction zone where the Pacific plate is moving to the northwest under the Philippine plate. The deepest point in the world's oceans is found in the Mariana Trench associated with this tectonic activity (NOAA 2015b). The island chain extends generally south to north for nearly 2500 km from the US territory of Guam to Farallon de Pajaros, the northernmost of the 4 larger and 10 smaller islands that make up the Commonwealth of the Northern Mariana Islands (CNMI). All of the islands are volcanic in origin and range in age from 40 million years (Guam) to the current time in areas of ongoing volcanic activity in the northern islands.

Fig. 16.13 Lowland (flooded) taro cultivation on hydric soils on the Big Island of Hawaii (photo courtesy of Amy Koch, USDA-NRCS)

16.5.1 Parent Materials

The core of the southern islands, particularly Guam but also Saipan, Rota, and Tinian, is volcanic material dating from the original volcanism that created the islands approximately 20–40 million years ago. Because this volcanism is associated with the Mariana subduction zone, these lavas have andesitic mineralogy including much higher aluminum content than those found in Hawaii and American Samoa. Another common parent material is the coralline limestone that makes up the plateaus found on the larger, southern islands including major portions of Guam, Saipan, Rota, Tinian and Aguijan. The soils on the island of Rota are primarily derived from volcanic ash over limestone, and ash-derived soils are also found in small areas on Northern Saipan and on Tinian. The source of the ash was not volcanic activity on these islands as eruptive activity ceased before the volcanoes were covered by limestone and emerged above sea level. The origin of this ash has not been established, but may have come from volcanic activity in the Philippines. The volcanoes in the Northern Islands of the CNMI are too young (Quaternary) to be the source of the parent material for the highly weathered ash-derived soils to the south. There is no evidence that soils on the other southern Mariana Islands are derived from volcanic ash, but, the presence of ash-derived soils on Rota and Saipan as well as convincing evidence of ash-derived soils in Hawaii and American Samoa suggests that other, highly weathered soils may have been developed, at least in part, from volcanic ash. The parent materials for the soils of the Northern Islands are characterized, almost exclusively, by relatively recently deposited volcanic materials including ash, cinders and pumice. The remaining parent materials are volcanic alluvium and coralline sand coastal deposits.

16.5.2 Climate

The climate in the Mariana Islands differs somewhat from that of Hawaii and American Samoa. Although annual rainfall amounts are in excess of 2000 mm throughout the Marinas and are similar to those in much of windward Hawaii and American Samoa, the islands experience a distinct dry season ranging from 4 months (January–April) on Guam (Young 1988) to 6 months (December–May) in the CNMI (Young 1989). Leeward areas in Hawaii also have strong dry seasons but have much lower annual rainfall amounts. Nearly all soils in the Marianas have an ustic moisture regime. This distinct dry season also has significant implications for management.

16.5.3 Soil Formation

Given the relatively consistent moisture and temperature distribution over the Mariana Islands, parent material is the primary factor that can be used to differentiate the soils found in these areas and which drives the soil properties that affect soil use and management.

16.5.4 Soil Management

16.5.4.1 Soils Formed in Older Volcanic Materials (MLRA 192)

The oldest soils in the Pacific Islands and in many ways the most challenging soils to manage are the highly weathered soils found in Southern Guam and to a much lesser extent in Rota, Tinian and Saipan. In addition to being highly weathered due to their age and the climate conditions (high temperatures and rainfall), the original volcanic materials are significantly different than the basaltic lavas and ash that characterize Hawaii and American Samoa. Because it is located on the plate margin, volcanic parent materials in the Marianas are generally more acidic (more potassium and sodium relative to magnesium and calcium) and have higher concentrations of aluminum (Schmidt 1957). Weathering has removed most of the more mobile cations (Na, K) and much of the silica leaving soils dominated by Fe and Al oxides.

The landscape is highly eroded and consists of low residual hills intersected by broad to narrow valleys. The upland soils formed from the original volcanic parent materials, likely mixed with more recent ash, typically have very low CEC and are especially vulnerable to degradation with loss of organic matter. Soils formed from alluvial materials in the valleys are generally more fertile and higher in organic matter. Scattered areas of wetlands are also present in valleys and back-beach environments.

Paleo-botanical evidence (Athens and Ward 2004) indicates that southern Guam was dominated by forest vegetation but that the area was rapidly converted to a fire dominated grassland landscape after permanent human settlement approximately 4300 years ago. When populations were low, the land could be successfully managed through the use of shifting cultivation and long periods of natural fallow. However, more intensive land use, poor grazing management, and frequent fire has resulted in significant areas of highly degraded soils where natural regeneration of vegetation no longer occurs due to the exposure of dense saprolite and infertile subsoils that support sparse vegetative growth (Fig. 16.14).

Fig. 16.14 Profile of a highly eroded Akina series (Inceptic Haplustox) developed in volcanic parent materials in Southern Guam. The *pinkish colored* material to the *left of the shovel* is dense saprolite. The species of fern behind the shovel is an indicator of low fertility and/or high (toxic to most plants) levels of soluble aluminum

At the present time, agricultural activities in these areas are primarily confined to small areas of alluvial soils in the valley bottoms. Some sloping areas are mechanically cultivated for high value vegetable crops. However, these systems require significant inputs of commercial fertilizer and lime and have proven not to be sustainable over the long term. The adoption of reduced tillage, vegetative barriers and cover crops have shown promise as a way to build long-term soil fertility and reduce soil degradation in these landscapes, particularly on Saipan, but these practices are not widely used.

Given that most of these areas are not intensively managed, fire (primarily anthropogenic) has become the primary driving force in these landscapes, especially on Guam. By removing vegetative cover, fires, especially those followed by high rainfall events, can create the conditions for significant soil erosion. This leads to the exposure of highly infertile subsoil horizons in upland areas and to significant sediment deposition in streams, coastal wetlands and near shore ocean environments. Sediment pollution has been identified as the major contributor to coral reef degradation in nearly all of the coastal areas on the larger Mariana Islands, especially on Guam and Saipan. Historic and ongoing sediment deposition has also created significant expenses for the US Navy and for commercial shipping interests where regular dredging has been required to maintain ship channels.

16.5.4.2 Soils Formed on Coralline Limestone Plateaus (MLRA 191)

The second major group of soils found in the Mariana Islands is formed on coralline limestone plateaus. These plateaus, found in Northern Guam, Saipan, Rota, Tinian and Aguijan, consist of very porous limestone that formed as tectonic activity raised the islands above sea level. (Tracey et al. 1964). These areas are generally nearly flat and bounded by steep escarpments. Because of the high porosity of the underlying limestone, few streams are present. Soils in these areas are generally shallow, but areas of deeper soils are present where "pockets" occur in the underlying limestone. Although the mineralogy and chemical properties of both soils are similar, the shallow soils classify as Entisols whereas the deep soils are Oxisols (Fig. 16.15). Moderately deep soils classify as Alfisols and Mollisols. These soils are characterized by primarily oxidic materials, which were originally postulated to have been deposited through the accumulation of impurities through the dissolution of the soluble limestone (Tracey et al. 1964). Geologists, however, have not seen evidence for the 30–100 m of dissolved limestone that would be needed to produce even 30 cm of soil from limestone impurities. Recent understanding of soil forming processes throughout the Mariana Islands has led us to hypothesize that these soils are primarily formed in situ from volcanic ash that has been deposited on these landscapes over millennia with inputs from Asian aerosolic dust (Birkeland 1999). The origin of this volcanic ash remains unclear with volcanism in the Philippines as the most likely source. The presence of Andisols on Rota, Tinian and Saipan further supports this ash deposition hypothesis.

Because of the presence of soluble coralline materials at a relatively shallow depth, these soils are generally more fertile than the soils derived from the older volcanic materials. There is abundant calcium, and the neutral pH results in Al compounds that do not interfere with plant growth. In addition, because of their relatively gentle slopes, these areas are easier to access and cultivate. Consequently, most of the agricultural activity in the Marianas is concentrated in these landscapes. The primary limitation for agriculture in these areas is water. Since the soils are often <20 cm thick over limestone, they have limited water holding capacity. This property, coupled with the seasonal rainfall in the Marianas, limits non-irrigated production to the rainy season. Nitrogen and phosphorus are also in short supply in these ecosystems so commercial fertilizer use is common and necessary for crop production.

Since these areas are generally flat, soil erosion is typically not a concern. Unfortunately, the combination of shallow soils and porous limestone bedrock makes these systems highly vulnerable to groundwater pollution, especially leaching of nitrates. Elevated nitrate levels have been found in drinking water sourced from groundwater on Guam (Guam Waterworks Authority 2013) and on Saipan, Tinian and Rota (Commonwealth Utility Corporation 2013). Although agriculture is only one potential source of this nitrate contamination, management of nitrogen sources, both inorganic fertilizers and manures, and irrigation scheduling are extremely important to reduce the risk of nitrate leaching in these systems. Nutrient management has been extensively promoted as a way to increase nitrogen use efficiency through split applications at rates that correspond to crop needs and thereby reduce nitrogen fertilizer applications. In addition, the use of leguminous cover crops as part of the crop rotation has been encouraged as a way to reduce production costs as well as reducing nitrate leaching risk. Improvements in irrigation management, including soil moisture monitoring to improve irrigation timing and the increased use of micro-irrigation technologies (micro-sprinklers and drip irrigation), are also being promoted as having the dual benefit of reducing production costs and reducing pollution risks. Although pesticides and herbicides are not currently perceived as a major risk in these systems, increased intensification and commercialization in the agricultural sector that have increased use of agricultural chemicals that may potentially present significant risk to ground water supplies.

Urban development also presents a significant risk to soils and the overall environment in the Mariana Islands, especially on Guam and Saipan where the population is concentrated. The risk is particularly acute on the limestone plateaus, which are generally flat and conducive to urban development. Commercial and residential development in these landscapes, especially in the last 20–30 years, has resulted in significant increases in impervious surface, which increases water runoff and the associated risks of sediment and nutrient transport to coastal environments. In addition, there has been a significant increase in areas devoted to golf courses, lawns, and other urban vegetated landscapes throughout the islands, which can be major contributors of nutrients and agricultural chemicals to ground and surface waters without proper management.

16.5.4.3 Soils Formed on Young Volcanic Parent Materials (MLRA 190)

In contrast to the Southern islands, the soils of the ten Northern Islands in the CNMI are primarily formed from more recently deposited volcanic materials. Volcanic activity in this region has been ongoing for as long as five million years, and numerous volcanoes remain active including the volcano on Pagan, which last erupted in 1981 (Smithsonian Institution 2015). The island of Anatahan also experienced a major eruption in 2005. A reconnaissance soil survey of the Northern CNMI (USDA-NRCS 2015b) identified the major soils as Andisols derived primarily from ash, cinders, pumice and other ejected materials.

Fig. 16.15 Profile depth variation of the mostly deep and very deep Yigo series (Eutric Acrustox) over limestone in Guam. Depth variation shown in the top photo is not common. The bottom photo shows the more typical situation of a generally consistent soil thickness over the underlying limestone

Although there are areas on the larger islands that have been used for small-scale agriculture in the past, the Northern Islands are officially uninhabited. The islands and the associated marine environment remain as natural areas and are used on an occasional basis for subsistence hunting and fishing. Much of this area is located within the Marians Trench Marine National Monument, which is jointly managed by the CNMI government, the United States Fish and Wildlife Service (USFWS) and that National Oceanic and Atmospheric Administration (NOAA) (USFWS 2015). The US military currently uses small areas in the islands for training purposes and a significant expansion of these activities has been proposed (Cave 2015).

16.6 Conclusions

The US Pacific Islands of Hawaii, Guam, the Commonwealth of the Northern Mariana Islands, and American Samoa, are some of the most unique, diverse and ecologically important areas of the United States. These attributes extend to the soils found in the islands and the opportunities and constraints that the soils place on management of these environments. As described in detail in this chapter, the islands themselves and the soils found there range in age from hours and days in areas of ongoing volcanic activity to 40 million years in Guam. Other soil forming factors vary widely including elevation which ranges from sea level on the coasts of all islands to over 4200 m at the top of Mauna Kea on the Big Island of Hawaii, and rainfall which ranges from less than 2500 mm $year^{-1}$ in leeward areas of the Hawaiian Islands to over 50,000 mm $year^{-1}$ in windward areas in Hawaii and in American Samoa. Soil parent materials are dominated by volcanic ash and lava, but also include Aeolian dust from Asian deserts and calcareous materials from ancient coral reefs.

Given this diversity of soil forming factors, it is no surprise that the islands also feature a wide range of soils ranging from shallow organic soils (Folists) over aa and pahoehoe lava flows, to young soils that developed in volcanic ash (Andisols), to very deep, highly weathered soils (Oxisols and Ultisols) on stable landscapes of the older islands. All of these soils support unique natural ecosystems and all present different and unique management challenges. This chapter has attempted to provide a brief overview of both the variety of soils found in the islands and the complex issues associated with their management for various uses with a focus on agriculture. Readers who are interested in additional details and information are encouraged to consult the references listed below as well as the wealth of materials ranging from formal research papers to anecdotal and historical accounts that are available in libraries, bookstores and online.

References

Athens JS, Ward JV (2004) Holocene vegetation, Savanna origins, and human settlement of Guam. In: Attenbrow V, Fullagar R (eds) A Pacific odyssey: archeology and anthropology in the Western Pacific. Papers in honor of Jim Specht. Records of the Australian Museum, Supplement 29. Australian Museum, Sydney, pp 15–30

Bergmanis CE (2004) Evaluation and characterization of limiting factors to grass growth in a tropical urban park. M.S. thesis, University of Hawaii at Manoa

Birkeland PW (1999) Soils and geomorphology. Oxford University Press, New York

Cave J (2015) The Pentagon wants to bomb the hell out of this tiny Pacific island. The Huffington Post (online) 29 May 2015. http://www.huffingtonpost.com/2015/05/29/pagan-island-marines-military_n_7342168.html. Accessed 4 June 2015

Commonwealth Utilities Corporation (2013) Summary of primary drinking water quality results for 2012. Commonwealth Utility News, 1 July 2013

Dangler EW, El-Swaify SA, Ahuja LR, Barnett AP (1976) Erodibility of selected Hawaii soils by rainfall simulation. ARS Publication W-35. United States Department of Agriculture, Agricultural Research Service (ARS) Western Region, Berkeley, CA

Deenik JL, Uehara G (1997) Nitrate retention in the vadose zone. Poster presented at the ASA-CSSA-SSSA international annual meeting, Anaheim, CA, 27–31 October 1997

Fox RL (1980) Soils with variable charge: agronomic and soil fertility aspects. In: Theng BKG (ed) Soils with variable charge, New Zealand Society of Soil Science, Offset Press, Palmerston North, NZ, pp 195–224

Gavenda, RT (1989) Soil genesis and landscape evolution in Central Oahu, Hawaii. Dissertation, University of Hawaii at Manoa

Giambelluca TW, Shuai X, Barnes ML, Alliss RJ, Longman RJ, Miura T, Chen Q, Frazier AG, Mudd RG, Cuo L, Businger AD (2014) Evapotranspiration of Hawai'i. Final report submitted to the U.S. Army Corps of Engineers—Honolulu District, and the Commission on Water Resource Management, State of Hawai'i

Guam Waterworks Authority (2013) 2012 Annual water quality report

Jackson ML, Levelt TWM, Syers JK, Rex RW, Clayton RN, Sherman GD, Uehara G (1971) Geomorphological relationships of tropospherically derived quartz in the soils of the Hawaiian Islands. Soil Sci Soc Am Proc 35:515–525

Juvik JO, Juvik SP (eds) (1999) Atlas of Hawaii. University of Hawaii Press, Honolulu

Macdonald GAA, Abbott AT, Peterson FL (1983) Volcanoes in the sea, 2nd edn. University of Hawaii Press, Honolulu

Moore JG, Clague DA, Holcomb RT, Lipman PW, Normark WR, Torresan ME (1989) Prodigious submarine landslides on the Hawaiian ridge. J Geophys Res 94(B12):17465–17484

National Oceanic and Atmospheric Administration (NOAA) (2015a) NOAA National Centers for Environmental Information: Climate Data Online. http://www.ncdc.noaa.gov/cdo-web/. Accessed 5 June 2015

National Oceanic and Atmospheric Administration (NOAA) (2015b) National Ocean Service Ocean Facts: How deep is the ocean? http://oceanservice.noaa.gov/facts/oceandepth.html. Accessed 4 June 2015

Porter SC (1979) Hawaiian glacial ages. Quatern Res 12:161–187

Ruhe RV (1975) Geomorphology: geomorphic processes and surficial geology. Houghton Mifflin Company, Boston

Saunders A, Robotham M (2010) Effects of land use on soil properties in a Hawaiian volcanic ash soil. Poster presented at the ASA-CSSA-SSSA International Annual Meeting, Long Beach, CA, 31 Oct–4 Nov 2010

Schmidt RG (1957) Petrology of the volcanic rocks. In: Geology of Saipan: Mariana Islands, US Geological Survey professional paper 280. US Government Printing Office, Washington, DC, pp 127–176

Smithsonian Institution (2015) National Museum of Natural History Global Volcanism Program. http://www.volcano.si.edu/volcano.cfm?vn=284170. Accessed 4 June 2015

Sollins PG, Robertson GP, Uehara G (1988) Nutrient mobility in variable and permanent-charge soils. Biogeochemistry 6:181–199

Tilling RI, Heliker C, Swanson DA (2010) Eruptions of Hawaiian volcanoes—past, present, and future. U.S. Geological Survey General Information Product 117. US Geological Survey, Reston, VA

Tracey JI, Schlanger SO, Stark JT, Doan DB, May HG (1964) General geology of Guam. Geological survey professional paper 403-A. US Government Printing Office, Washington, DC

Uehara G, Gillman GP (1981) The mineralogy, chemistry, and physics of tropical soils with variable charge clays. Westview Press, Boulder

United States Department of Agriculture Natural Resources Conservation Service (USDA-NRCS) (2015a) Java Newhall Simulation Model (jNSM). http://www.nrcs.usda.gov/wps/portal/nrcs/detail/soils/survey/class/?cid=nrcs142p2_053559. Accessed 5 June 2015

United States Department of Agriculture Natural Resources Conservation Service (USDA-NRCS) (2015b) Soil Survey of Agrigan, Alamagan, Anatahan, Pagan, and Sarigan, Commonwealth of the Northern Marianas Islands. http://www.nrcs.usda.gov/wps/portal/nrcs/surveylist/soils/survey/state/?stateId=PB. Accessed 4 June 2015

United States Fish and Wildlife Service (USFWS) (2015) Mariana Trench Marine National Monument, Northern Mariana Islands. http://www.fws.gov/refuge/mariana_trench_marine_national_monument/. Accessed 4 June 2015

van Rees CB, Reed JM (2014) Wetland loss in Hawai'i since human settlement. Wetlands 34(2):335–350

Young, FJ (1988) Soil Survey of the Territory of Guam. United States Department of Agriculture (USDA) Soil Conservation Service

Young, FJ (1989) Soil Survey of the Territory of the Islands of Aguijan, Rota, Saipan and Tinian, Commonwealth of the Northern Mariana Islands. United States Department of Agriculture (USDA) Soil Conservation Service

17 Soils of Alaska: LRRs W1, W2, X1, X2, and Y

C.L. Ping, M.H. Clark, D. D'Amore, G.J. Michaelson, and D.K. Swanson

17.1 Introduction

Alaska is a vast area of 1,508,120 km^2, an area about 25 % of the contiguous USA. It borders northwestern Canada to the east and is bounded by the Arctic Ocean to the north and the Pacific Ocean to the south. Alaska has the northernmost point of the USA at Barrow (Lat. 71.4°N) and extends over 19° in latitude from north to south. The state has 5 Land Resource Regions (LRRs) (USDA-NRCS 2006): Southern Alaska (W1), the Aleutian Alaska (W2), Interior Alaska (X1), Western Alaska (X2), and Northern Alaska (Y). In this discussion, LRRs are organized into three regions: Northern Alaska (Y), Interior and Western Alaska (X1, X2), and Southern and Aleutian Alaska (W1, W2) (Fig. 17.1).

17.2 Northern Alaska (LRR Y)

This northern region of Alaska (Fig. 17.1) includes six Major Land Resources Areas (MLRAs): the Northern Seward Peninsula, Selawik Lowlands, Western Brooks Range Mountain Foothills and Valleys, Northern Brooks Range Mountains, Arctic Foothills, and the Arctic Coastal Plains. The region covers 325,345 km^2, about 22 % of the landmass of Alaska.

C.L. Ping (✉) · G.J. Michaelson
University of Alaska Fairbanks, Palmer, AK, USA
e-mail: cping@alaska.edu

G.J. Michaelson
e-mail: gjmichaelson@alaska.edu

M.H. Clark
USDA-NRCS, Palmer, AK, USA
e-mail: mclark@matonline.net

D. D'Amore
USDA Forest Service, Juneau, AK, USA
e-mail: ddamore@fs.fed.us

D.K. Swanson
USDI-National Park Service, Fairbanks, AK, USA
e-mail: David_K_Swanson@nps.gov

L.T. West et al. (eds.), *The Soils of the USA*, World Soils Book Series,
DOI 10.1007/978-3-319-41870-4_17

17.2.1 Physiography

This region consists of mountains, hills, floodplains, and extensive coastal plains and deltas. Elevation ranges from sea level to 2749 m at the summit of Mt. Chamberlain in the Brooks Range. The northern Brooks Range consists of folded and faulted strata uplifted during the Cretaceous period. The mountains and hills were extensively glaciated during the Pleistocene. To the north, the rolling hills, ridges, and plateaus extend to the gently rolling to level, fluvial deposits of the Arctic Coastal Plain. Periglacial features, such as patterned ground, thaw-lake basins, pingos, beaded drainages, and gelifluction lobes, are common throughout the region. The southwest part of the region, extending into the Seward Peninsula, includes floodplains, rolling lowlands, and mountains. Surface water drains to the north and west into the Arctic Ocean, but the western and southern Seward Peninsula drains into the northern Bering Sea (Wahrhaftig 1965).

17.2.2 Climate

The Arctic climate is characterized by short, cool summers and long, cold winters, with low precipitation and low evapotranspiration. Freezing temperatures can occur in any month. The climate varies with distance from the Arctic Ocean or Bering Sea and elevation. The mean annual precipitation (MAP) ranges from 10 to 25 cm at lower elevations in the northern and western parts of the region and from 75 to 100 cm at higher elevations in the Brooks Range and Seward Peninsula. The mean annual air temperature (MAAT) ranges from −13 to −5.6 °C. The mean annual soil temperature (MAST) at 50 cm in soils ranges from −5 to −9 °C (Osterkamp and Romanovsky 1996), which qualifies the soils for the pergelic soil temperature class (Soil Survey Staff 2014). Growing degree days based on a 4.4 °C threshold range from 531 at Prudhoe Bay to 1171 at Kotzebue (Western Regional Climate Center 2014).

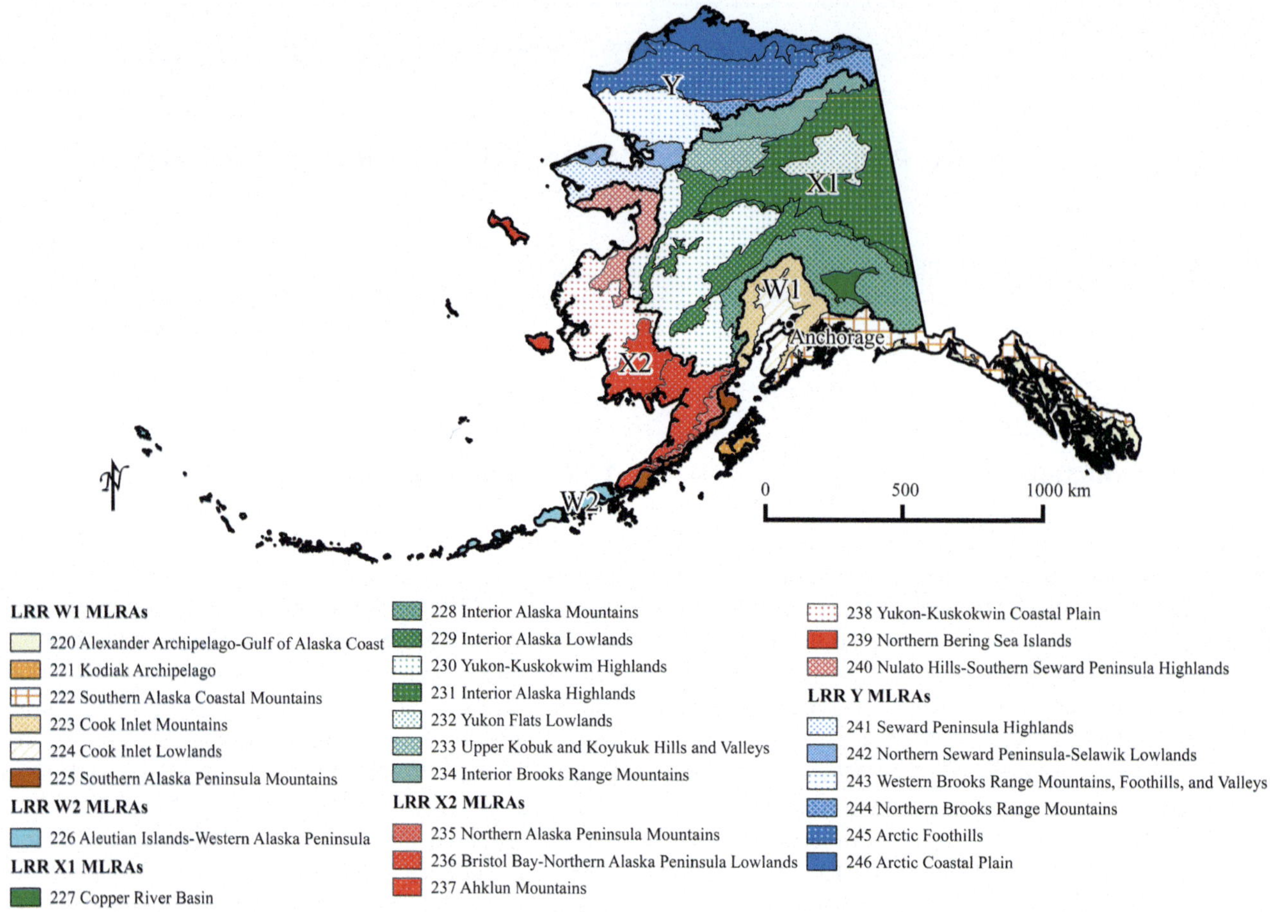

Fig. 17.1 STATSGO map showing the land resource regions (*LRR*) and major land resources areas (*MLRA*) in each LRR (*W1* Southern Alaska, *W2* the Aleutian Alaska, *X1* Interior Alaska, *X2* Western Alaska, *Y* Northern Alaska)

17.2.3 Permafrost Distribution

Permafrost is continuous across this LRR (Péwé 1975; Ferrians 1994) (Fig. 17.2). Permafrost temperature ranges from −2 to −10 °C and reaches more than 650 m deep at Prudhoe Bay (Davis 2001). Permafrost often consists of massive ground ice forming patterned ground surface features, extensive throughout the Arctic Coastal Plain and low-relief areas throughout the LRR.

17.2.4 Geology and Parent Material

The bedrock of the Brooks Range is primarily siliceous in the middle to west-central, carbonaceous in the east, and a mixture of silicious and metamorphic rocks in the west regions. The mountains were extensively glaciated during the Pleistocene. Mountain glaciers were confined to deep valleys flowing out as tongues or lobes southward into Interior Alaska and northward shaping much of the Arctic Foothills (Hamilton 2003). Glacial drift in foothills north and south of the Brooks Range becomes progressively younger approaching the flank of the range, from late Tertiary to late Holocene. Mixed colluvium and alluvium are common on mountain footslopes. Loess deposits are widespread on the plains and hills and eolian sands underlay the loess to the southwest and central plain, with fluvial sands and gravel in the central coastal plain (Péwé 1975; Carter 1981, 1988). Fine-textured marine deposits underlying the loess deposit are concentrated at the northwest part of the coastal plain (Péwé 1975). Volcanic ash deposits are common on the northern Seward Peninsula. Extremely ice-rich loess deposits known as "yedoma" are locally present (Kanevskiy et al. 2011, 2013).

17.2.5 Vegetation

The tundra vegetation cover of Northern America is divided into 5 bioclimate subzones based on mean July temperature (MJT) and summer warmth index (SWI, sum of mean monthly temperatures greater than 0 °C) (Walker et al. 2005). Subzones A and B are not present in Alaska, while subzone C is on

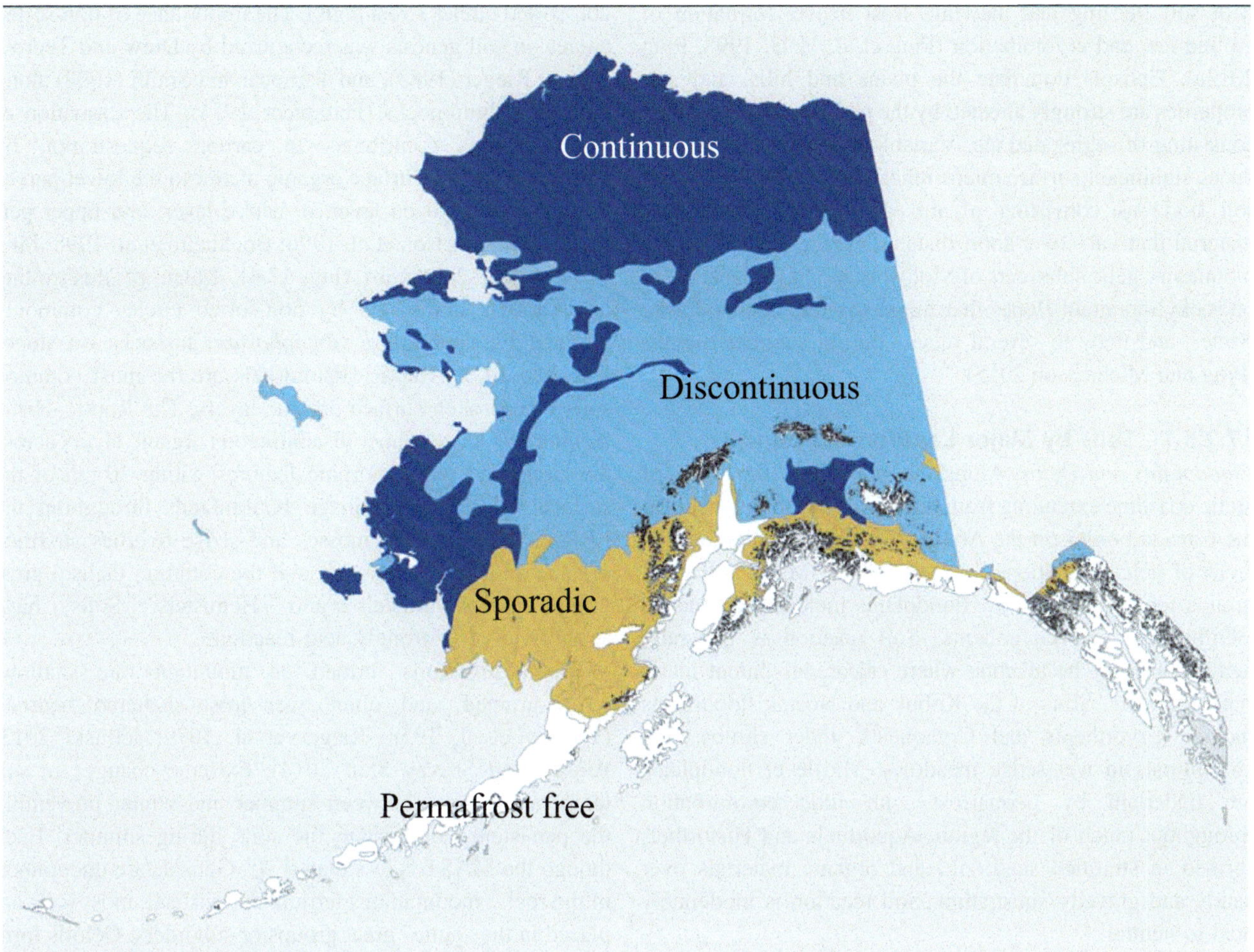

Fig. 17.2 Permafrost distribution in Alaska

the northern tip of the Arctic Coastal Plain at latitudes mostly at 70° N or higher. It has MJT of 6–7 °C and SWI of 9–12 °C mo. Subzone D occupies most of the Arctic Coastal plain and the northern tip of Seward Peninsula with MJT of 8–9 °C and SWI of 12–20 °C mo. Subzones C and D have vegetation dominated by sedges (*Carex, Eriophorum*), dwarf shrubs (*Salix, Dryas*), mosses, and lichens. Subzone E occupies most of the Arctic Foothills. It has MJT of 10–12 °C and SWI of 20–35 °C mo. with vegetation dominated by low shrubs (*Vaccinium, Empetrum, Salix, Betula, Ledum*), tussock sedge (*Eriophorum, vaginatum*), mosses, and lichens. Walker and Everett (1991) classified vegetation land cover types on North Slope of Alaska based on soil substrate pH and drainage. On the Arctic Coastal Plain and Arctic Foothills (subzones C and D), the wet and moist acid tundra (WAT and MAT) dominate areas west of the Colville River, and wet and moist non-acid tundra (WNT and MNT) dominate areas east of the river. On the Arctic Foothills (subzone E), the MNT dominates the northern fringe of the foothills and also some of the southern foothills that are covered by most recent calcareous glacial drift (Walker and Everett 1991; Walker et al. 2005). The MAT dominates the rest of the foothills. Narrow valleys with deep organic deposits amid the foothills are either WAT or WNT. On exposed ridge tops and mountain slopes, the land cover type is classified as "barren" with sparse sedge, forbs, and lichens (CAVM Team 2003). The mountains, foothills, and valleys of the southeastern Seward Peninsula and western Brooks Range area support both white spruce (*Picea glauca*) and black spruce (*Picea mariana*) forests along high floodplains and glaciated plains and on lower mountain slopes. These forests represent the northernmost extent of significant boreal forest in Alaska suggesting a warmer microclimate compared with most northern land resource areas.

17.2.6 Major Soils and Their Properties

The main soil forming processes in this region include accumulation of organic matter, hydromorphism due to water saturation above the permafrost, and processes associated with soil freezing and thawing: frost heave, formation of ground ice, and cryoturbation (Ping et al. 1993, 1998, Ping 2013a). Gelisols dominate the plains and hills, and soil properties are strongly affected by the presence of permafrost consisting of segregated ice. Variable moisture content produces significant surface micro-relief with warped or broken soil horizons consisting of mixed mineral and organic material that vary over short distances (Ping 2013b). In the mountains, gelic suborders of Mollisols and Inceptisols occur on rocky mountain slopes that freeze to great depths during winter, and thaw to several meters during summer months (Ping and Michaelson 2015).

17.2.6.1 Soils by Major Landform Groups

Floodplains and Deltas Along the more than 10,600 km of arctic coastline extending from northern Seward Peninsula to the Canada border on the Arctic Ocean, there are extensive areas of delta formations and floodplains. Major soil great groups represented on low floodplains include Gelaquents, Gelifluvents, and Gelorthents. Soil reaction is generally neutral but may be alkaline where calcareous parent materials dominate. Soils of the Kobuk and Noatak floodplains include Cryorthents and Cryaquents under shrubs, and Cryofibrists in wet sedge meadows. Mid-level floodplains are underlain by permafrost with little cryoturbation throughout much of the region. Aquorthels and Historthels formed in stratified sand, silt, and organic materials over sandy and gravelly substratum. Soil reaction is moderately acid to neutral.

Arctic Coastal Plains The coastal plains along the perimeter of the Beaufort, Bering, and Chukchi Seas are characterized by a myriad of pattern ground, thaw lakes, and features formed by collapse of ground ice. The most abundant are ice wedge polygons that form due to vertical ice lenses growing in reoccurring seasonal ground cracks, and appear as honeycomb pattern over the landscape (Drew and Tedrow 1962; Washburn 1973). Each micro-feature of the polygon has unique soil components such as Glacistels in the trough, Histoturbels along the rim due to deformation of soil horizons caused by ice wedge expansion, and Historthels or Aquiturbels in the polygon center (Ping et al. 2008a; Fig. 17.3). Glacistels have a glacic layer (ice wedge or ground ice) within 100 cm of the surface. In addition to alluvial parent materials on the coastal plain, there is an extensive Late Pleistocene eolian sand belt stretching from the central to the western region (Carter 1981). The dominant soils are Psammorthels and Psammoturbels (Michaelson et al. 2013a).

Arctic Foothills Extensive across the Northern LRR, this rolling landscape includes both deglaciated and non-glaciated areas mantled by loess deposits over glacial drift with variable rock fragment content. Loess parent materials provide favorable conditions for ice segregation and the formation of surface micro-features such as non-sorted circles (frost boils). The importance of non-sorted circles on soil genesis was recognized by Drew and Tedrow (1962), Rieger (1983), and Tarnocai and Smith (1992) along with earth hummocks (Pettapiece 1974). The formation of these features contributes to carbon sequestration by frost-churning the surface organic matter to the lower part of the seasonally frozen layer, or active layer, and upper permafrost (Michaelson et al. 1996; Bockheim et al. 1998; Ping et al. 1998, 2008a, b) (Fig. 17.4). Much of this rolling landscape is dominated by non-sorted circles commonly outlined by cottongrass (*Eriophorum*) tussocks on slopes less than 14 %. Ruptic Histoturbels are the most common soils with broken surface organic layers. The Ruptic–Histic Aquiturbels have thinner discontinuous organic layers across the circle and redoximorphic features within 50 cm of the surface. Along the drainage bottomlands throughout the LRR, a thick layer of mosses and sedge overlies stratified organic and fluvial deposits, and the common Gelisol great groups are Fibristels and Hemistels. Soils have strongly-to-very strongly acid reaction.

Mountains Soils formed on mountains are shallow, coarse-grained, and often over frost shattered bedrock (Tedrow et al. 1958; Rieger et al. 1979; Jelinski 2013; Alaska Soil Survey Staff 2014). Extreme changes in soil temperature occur between summer and winter preventing the persistence of frost in the soils during summer. Even though the MAST is less than 0 °C, Gelisols are uncommon in the rocky mountainous terrain, and instead, most soils are placed in the "gelic" great groups or suborders. Gelolls form on calcareous and dark sedimentary rocky slopes where humus-rich A horizons with high base saturation form under herbaceous or shrub vegetation (Ping and Michaelson 2015, Fig. 17.5). Gelepts are more typical of mixed sedimentary, igneous, and metamorphic rock types. Base status is strongly controlled by rock type with Dystrogelepts common on acid igneous and metamorphic rocks, and Haplogelepts often associated with mixed sedimentary rocks.

17.2.6.2 Soil Physical Properties

Soil Horizonation Gelisols comprise over half of the soils within this Region. Only in Orthels are the soil horizons relatively smooth and parallel to the ground surface. Most Gelisols in Arctic Alaska are Turbels. The surface organic horizons are often broken due to frost cracking and heave. The underlying mineral horizons are either involuted or mixed with organic materials churned from the surface. In the lower active layer and upper permafrost, there is a zone of organic material mixed in a mineral matrix with sharp boundaries due to frost-churning (Ping 2013a) (Fig. 17.4b).

Ground Ice The most striking feature of soils in this region is the presence of ice. Ice forms as ice lenses with thickness of <1 to 20 mm or ice belts thicker than 20 cm, or vertical ice wedges from a few cm wide when newly formed

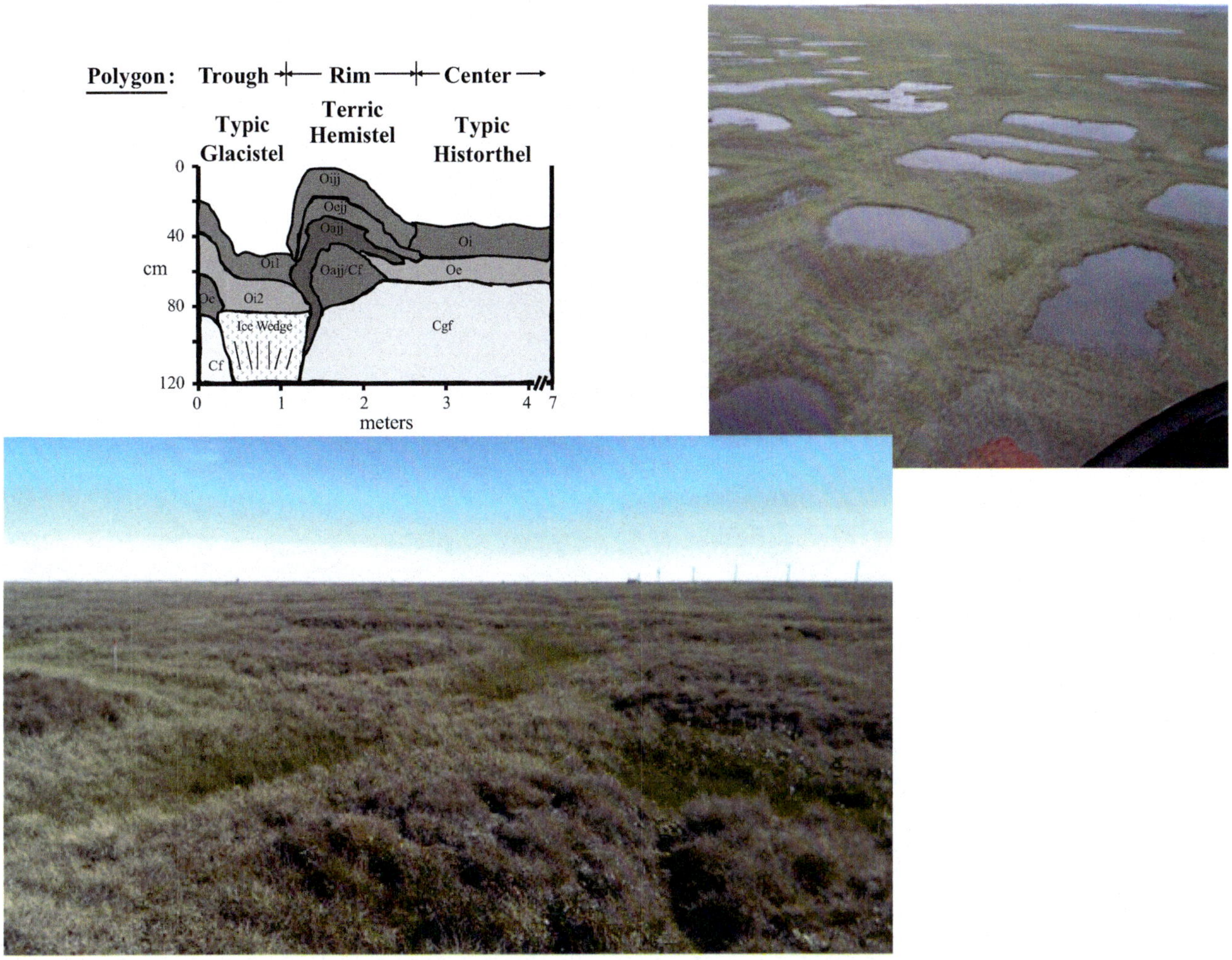

Fig. 17.3 Ice wedge polygons on the Arctic Coastal Plains and soil components across the microtopography formed by ice wedge polygons (*upper left*)

up to several meters wide. The total average volumetric ice content, including ice wedge, segregated ice, and pore ice, ranges from 30 to 80 % on the foothills (Ping et al. 1998) up to 43–89 % on the coastal plains (Kanevskiy et al. 2013, Fig. 17.6). Hence, thawing of permafrost due to climate warming or surface disturbance will result in thermokarsting and ground subsidence.

Soil Texture and Structure On the Arctic Coastal Plain, most soils have loamy textures and variable rock fragment content (Michaelson et al. 2013a). On the Arctic Foothills, soils formed in thick, eolian deposits have silt loam texture, and those formed in gravelly glacial deposits have loamy textures (Hamilton 2003; Ping et al. 1998, 2008a). The clay content is low in this region but generally increases with the age of moraines in the Arctic Foothills (Ping et al. 1998). In this Region soil structure is influenced by cryogenic processes. Granular or crumb structures form on soil surfaces due to combined effects of freeze–thaw and root disturbance (Gubin 1993). Platy structures are common in active layers due to ice lens formation during seasonal freezing. The upper permafrost layers have lenticular or reticulate cryostructure (structure within the permafrost layer) formed by ice lenses and ice veins. This layer is called the transient layer and is formed by fluctuations of the permafrost table from perturbations in climate and vegetation (French and Shur 2010). Below the transient layer, there is an ice-rich intermediate layer with more than 60 % ice by volume due to permafrost table fluctuation on a century scale (Shur 1988). Commonly observed cryostructures are shown in Fig. 17.7.

17.2.6.3 Biological, Chemical, and Mineralogical Properties

Soils in the arctic tundra have large organic matter accumulations due to slow decomposition under the cold and wet climate. The total carbon contents in the recently formed Cryofluvents or Gelorthents are low, usually <15 kg C m^{-2} (Michaelson et al. 1996). Elsewhere on the Arctic Coastal Plain, most soils are cryoturbated with high carbon stores

Fig. 17.4 Examples of cryoturbated profiles of Gelisols in northern Alaska: **a** A Historthel formed under moist acidic tundra on an alluvial fan; **b** an Aquiturbel formed in association with non-sorted circles under non-acidic tundra, the bottom of the pit was scraped clean to show the cryoturbated organic matter mixing with reduced mineral matrix (*gray color*); **c** a Histoturbel formed in association with non-sorted circles under acid tundra, note the effect of organic horizon thickness on the depth of the active layer (*pit bottom*); and **d** In an exposed face, the thawed profile of a Histoturbel showing the cross section of a non-sorted circle with organic matter cryoturbated down from the edge of the circle

estimated at 40 to more than 100 kg C m^{-2} measured to 3 m depth (Ping et al. 2011). Aquiturbels formed under MNT have lower soil organic carbon (SOC), averaging 35–45 kg C m^{-2}, than those under Histoturbels in MAT, averaging 40–60 kg C m^{-2} (Michaelson et al. 2013b). Soil organic matter plays an important role in controlling soil chemical properties. In most Gelisols, there is a strong correlation between cation exchange capacity (CEC) and SOC but not clay content (Ping et al. 2005a). Most clay minerals are from parent materials instead of weathered in situ (Borden et al. 2010).

17.2.6.4 Soil Function, Ecosystem Services, and Land Use

The extensive wetland complex in this Region provides important wildlife habitat for migratory birds. The entire Region supports large mammal populations including caribou, moose, Dall sheep, and both grizzly and polar bears. The large native Alaskan population in this Region relies on renewable resources for their sustenance. Oil and associated infrastructure development is the single largest human impact in the Region. A major concern in land use is the integrity of permafrost and practices that minimize the impact of the thermo regime of permafrost. Soils of this Region store a large quantity of carbon, which is crucial to future climate change (Ping et al. 2008b). Climate warming could lead to a release of carbon currently stored in permafrost as carbon dioxide or methane (Euskirchen et al. 2009).

17.3 Interior (LRR X1) and Western Alaska (LRR X2)

The Interior Alaska LRR includes central Alaska, from the south slopes of the Brooks Range to the north slopes of the Alaska Range (Fig. 17.1). It also includes the Copper River Basin and its surrounding mountains. The region makes up

Fig. 17.5 A typical limestone landscape in the Brooks Range where Haplogelolls are common (*upper right*). Haplogeloll formed in limestone parent material in the Brooks Range (*left*). Stone circles form in depressions in the Brooks Range (*right*)

671,835 km^2. The western Alaska region begins at the Bering Sea and includes the Alaska Peninsula, Bristol Bay lowlands, and the southern Seward Peninsula, an area covering 236,585 km^2. The combined area of these two LRRs amounts to 60 % of the land surface of Alaska.

17.3.1 Physiography

Interior Alaska consists of several broad, nearly level lowlands with elevations mostly below 500 m and rounded mountains with elevations up to about 2000 m (Wahrhaftig 1965). It also includes the rugged high mountains of the Alaska Range, Wrangell Mountains, and southern Brooks Range. Western Alaska includes steep, rugged mountains and hills, broad valleys, and the northern Bering Sea islands. The rest of the region consists of coastal lowlands with about 80 % covered by lakes and interconnecting wetlands.

17.3.2 Climate

The subarctic continental climate in Interior Alaska has short warm summers and long cold winters. The MAP ranges from about 15 cm in the northeast lowlands to >150 cm in the Alaska Range. Summer afternoon thunderstorms are common in valleys and at the lower elevations in the mountains. Lightning-caused wildfires burn many thousands of hectares annually. The MAAT ranges from −13 to −2 °C. Frost-free days (FFD) range from 90 to 110 days in lowlands, and freezing temperatures may occur in any month in the mountains. In Western Alaska, the climate ranges from coastal maritime to subarctic continental. Summers are short and cool, and winters are long and cold. The MAP in the region ranges from 33 to 200 cm. The MAAT ranges from −4 to 2 °C. Frost may occur in any month. Strong winds are common, especially in winter. Snow covers the ground for approximately 7–9 months each year.

Fig. 17.6 Arctic Coastal Plain; thermokarsting caused by thawing of ground ice (*top*) and coastal erosion due to melting of ground ice and thawing of permafrost along the Beaufort Sea coast, northern Alaska (*bottom*)

17.3.3 Permafrost Distribution

Permafrost is discontinuous across both regions (Péwé 1975; Ferrians 1994) (Fig. 17.2), with permafrost temperatures from near freezing to about −2 °C. In the Tanana–Yukon–Kuskokwim Highlands and the Copper Basin, permafrost underlies most of the north slopes, gentle footslopes, and most toe slopes. In the lowlands, permafrost underlies most of the landscape except Holocene terraces, alluvial fans, and active floodplains. Ground ice occurs as finely segregated ice crystals, thin discontinuous ice lenses, and large masses of ice. In clayey and silty deposits, ice content averages 30–60 % of the dry weight of the soil (Nichols 1956). In the Interior, massive ground ice and thermal erosion are common but less prevalent in the Copper River Basin (Péwé 1982).

Fig. 17.7 Common cryostructures in Gelisols: **a** ice belt; **b** reticulate; **c** suspended (ataxitic); and **d** lenticular grading into reticulate structure at 59 inches

17.3.4 Geology and Parent Material

Bedrock Bedrock of the central Brooks Range consists entirely of stratified sedimentary rocks of Paleozoic and Precambrian age with the eastern range primarily Paleozoic and early Jurassic volcanic and igneous rocks. The highland terrains in the Interior and Western Alaska are mainly Permian to Lower Cretaceous stratified sedimentary, igneous, and volcanic rocks in the north, with Precambrian and Paleozoic metamorphic sedimentary rocks in the central and west part. Cretaceous, Tertiary, and Quaternary volcanic rocks dominate the southern Seward Highlands, the Alaska Peninsula, and the Bering Islands.

Parent Material Residual and colluvial parent material occur on gentle to steep slopes. Soils have loamy textures with a high percentage of rock fragments. Most of the region was not glaciated during the Pleistocene. Glacial moraines are present in mountain valleys and locally as piedmont on basin floors near the mountains. Coarse-grained glacial outwash deposits with very gravelly or cobbly sand texture are widespread in river valleys near the Alaska Range. Most upland landforms of these LRRs are blanketed with a layer of loess that consists mainly of silt and very fine sand transported from exposed braided riverbeds. During Illinoian and Wisconsin glaciations (550–400 and 80–10 kA, respectively), silt from the Yukon and Tanana Rivers and their tributaries was carried by southerly winds and deposited on the Yukon–Tanana uplands with thickest deposits on south-facing slopes and valley bottoms (Péwé 1975). Loess deposition continues today in areas near modern glacial-fed rivers. Large sand dunes transported by strong winds during the Late Pleistocene and early Holocene were deposited between the Yukon and Tanana Rivers (Péwé 1975; Dijkmans and Koster 1990). Volcanic ash deposits are widespread in the Northern Alaska Peninsula due to its proximity to the volcanoes on the Peninsula and the Aleutian Islands. Thin (<25 cm) volcanic ash layers blanket the uplands of the northeastern Interior region (Lerbekmo et al. 1975).

Alluvium is widespread along rivers and streams throughout these LRRs with particle size varying from silt to sand to large cobbles. On more gentle gradients, alluvium has a stratified cap of sand thru silt-size particles over the sandy and gravelly substrate. Glaciofluvial deposits are extensive in the lower mountain areas and alluvial fans in the glaciated area in southwest Alaska that includes the Bristol Bay area. Lacustrine deposits are common throughout these LRRs. A unique feature of the Copper River Basin is the extensive areas of clayey and calcareous lacustrine deposits in a large proglacial lake that occupied the basin during the Pleistocene, about 35,000–9000 years ago (Ferrians et al. 1989). Small areas of lacustrine deposits are identified in other parts of these LRRs.

17.3.5 Vegetation

Vegetation communities of Interior and Western Alaska vary with slope, aspect, drainage, soil reaction, flooding, and fire history. On the warm and drier south-facing upland slopes, white spruce often in mixed stand with paper birch (*Betula alaskana*) dominates the canopy with alder (*Alnus crispa*), the common understory. Quaking aspen (*Populus tremuloides*), a serial species, usually dominates drier sites on ridge tops or shallow southerly slopes. Endless expanses of stunted black spruce (*Picea mariana*) woodlands are broken only by cottongrass tussock and sedge–sphagnum bogs. A mosaic of shrub and hardwood communities common throughout the lowlands is indicative of the profound influence of fire. Areas of sedge and buckbean (*Menyanthes* L.) fens fringe the many intermontane basins throughout this region which are relatively warm with nutrient-rich groundwater. On floodplains, the vegetation succession follows the age of the flooded surface and frequency of flooding: from horsetail (*Equisetum* spp.) and sedge (*Carex* spp.) wet meadows in oxbows and cutoff meanders where soils are saturated and frequently flooded, to willow (*Salix* spp.) and alder (*Alnus* spp.) shrub communities on slightly higher and drier surfaces, to poplar (*Populus balsamifera*) forest, and eventually white spruce forest on the highest and rarely flooded positions. Stream terraces no longer disturbed by flooding support stunted black spruce woodlands and cottongrass tussocks.

17.3.6 Major Soils and Their Properties

17.3.6.1 Soil Genesis

In Interior Alaska, topography plays a key role in soil formation and variation in soil properties because it creates different microclimates due to differences in slope, aspect, permafrost distribution, drainage, and vegetation communities within broad landscape units (Furbush and Schoephorster 1977; van Cleve et al. 1993; Swanson 1996a; Ping et al. 2005b) (Fig. 17.8). The pedogenic processes in these regions include freeze–thaw, paludification, brunification, reduction–oxidation, and podzolization. Little soil leaching occurs in Interior Alaska, except in areas with either higher precipitation or highly permeable parent materials. Accumulation or deposition of secondary salts or carbonates is limited to floodplains, and regions of extreme continental climate. Soils with secondary carbonate deposition have been reported in the warmest and driest parts of northeastern interior Alaska (Yukon Flats) (unpublished NRCS data). Reports of pedogenic carbonates elsewhere in Interior Alaska represent exhumed horizons from buried Pleistocene

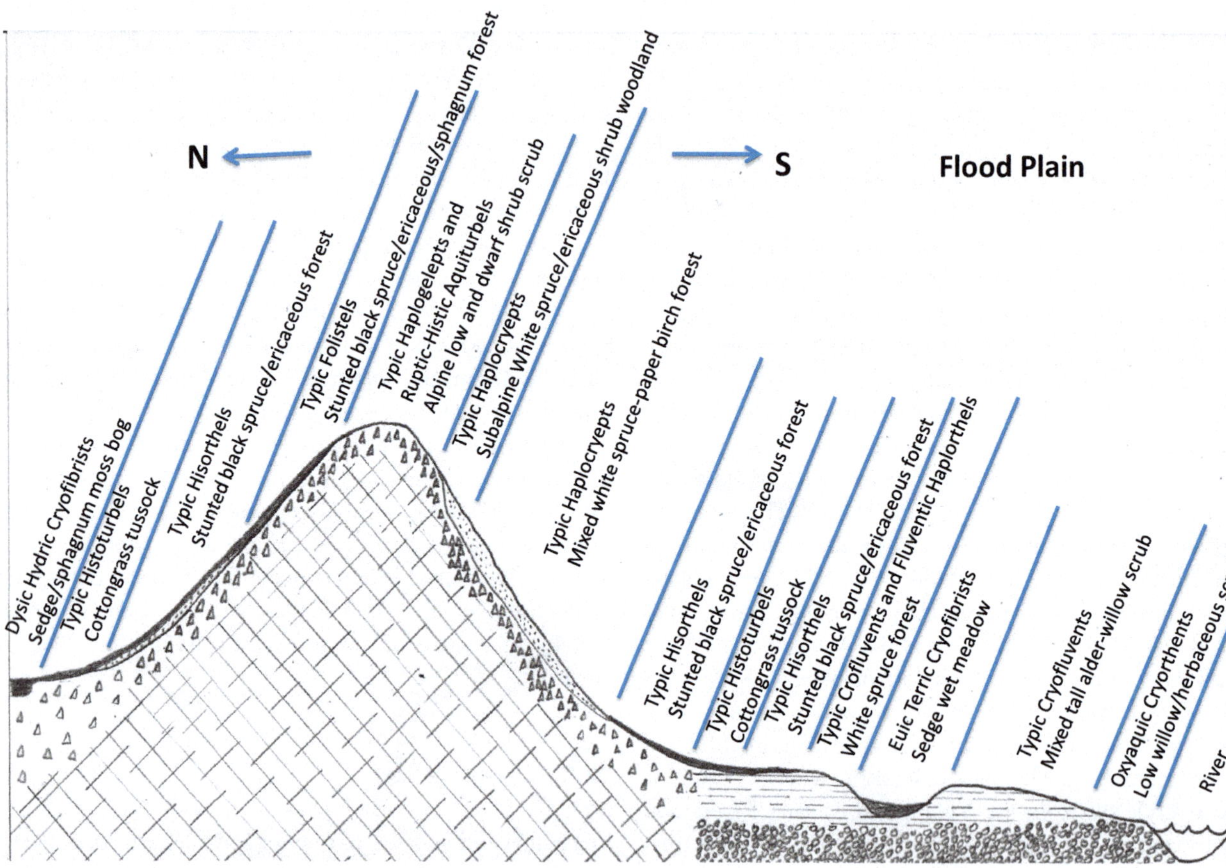

Fig. 17.8 Soil–landform relationships in the hilly terrain of Interior Alaska [modified from Furbush and Schoephorster 1977]

soils (Dijkmans et al. 1986) or transient evaporative crusts on river silt bars that do not persist (Van Cleve et al. 1993).

Role of Fire in Soil Genesis and Classification Fire is an integral part of the interior boreal forest ecosystem with a reoccurrence of 50–130 years (Dyrness et al. 1986). Fire destroys the forest and surface organic horizons and exposes and warms the mineral soils causing thawing of permafrost where present, changing soil drainage and resetting the vegetation succession (Viereck et al. 1983; Dyrness et al. 1986; Swanson 1996b). Thus, interior Alaska soils are classified based on three highly dynamic soil properties: organic layer thickness, permafrost, and saturation caused by perched water table over permafrost. Properties are subject to change as vegetative succession progresses following fire (Fig. 17.9). A common progression includes a soil sequence of Cryaquents and Haplocryepts, three to seven years following fire that change to Haplorthels with low ice content permafrost at approximately 50 years following fire, and eventually back to the pre-disturbance Historthels and Histoturbels at >90 years following fire.

Boreal Mountains and Hills Soils on southerly aspects on mountains and hills with slopes exceeding 16 % are warm and lack permafrost. They have lower organic matter content and thinner surface organic horizons than their counterparts on more northerly aspects. The most common soils are Haplocryepts that formed in loess over colluvium under mixed deciduous–conifer forests. Dystrocryepts are common on sites with increased moisture or where soils formed in acid igneous or metamorphic parent material. Water accumulates in swales and toe slopes with wetness indicators including redoximorphic features, and thick and organic-rich mineral soils thus qualify as Cryaquepts in the forest and Gelaquepts in the alpine zone.

Northerly aspects with slopes greater than 16 % are cooler and wetter than south-facing slopes of similar gradient because they receive less solar energy, evaporate less water, and exhibit delayed snowmelt. Thus, more available water enhancing soil acidification and cool summer temperatures retard organic matter decomposition. Slope movement enhanced by the annual freeze–thaw cycle promotes the mixing of soil horizons (Wu 1984), a process referred to as gelifluction. Cryoturbated soils are classified within the Turbel suborder. Folistels dominate slopes in excess of 40 % where unsaturated organic horizons more than 25 cm thick overlie thin mineral horizons over frozen shattered bedrock. Aquiturbels or Haploturbels are cryoturbated permafrost soils that occur on footslopes of all aspects (Swanson 1996a).

Floodplains and Lowlands The primary soil forming processes on floodplains include fluvial processes, hydromorphism, and alkalinization (Clark and Kautz 1999; Van Cleve et al. 1993). Cryofluvents and Cryaquents occupy the riparian zone and have stratified sand and silt with seasonally high water tables. During the dry and warm early summers, a brown-to-whitish surface soil crust is often observed with alkaline soil reaction. Soils on high floodplains are rarely flooded and have carbonates and soluble salts removed from the upper part of the soil (Clark and Duffy 2006). Haplorthels are common on some of these surfaces with thick organic horizons and permafrost. In lowlands and depressions, soil drainage is restricted by landscape position and presence of permafrost. The wet and cold conditions favor organic matter accumulation. Histels, organic soils with organic layers >40 cm, generally occur in bogs and fens (Fig. 17.10). The alluvial soils in the Yukon–Kuskokwim Delta area are subject to seasonal inundation, and soil formation follows the depositional sequence of alluvial bars, from Cryaquents in the youngest bars, to Cryofluvents, and Dystrocryepts on stabilized levees (Woodgate 2015). In enclosed oxbow or old channels, Cryofibrists form in very poorly drained sedge meadows and *Sphagnum* bogs. Peat plateaus or frost mounds form in these bogs due to paludification, and frozen ground eventually develops under the thick organic layers. The soils are either Histoturbels or Sapristels (Woodgate 2015). Histoturbels, cryoturbated permafrost soils with thick surface organic horizons, occur on footslopes of all aspects and some level terraces (Swanson 1996a), and Haploturbels occur where drainage is better. The summer depth of thaw increases after wildfire, sometimes to the point where the Gelisols change to Inceptisols (Moore and Ping 1989; Swanson 1996b, Fig. 17.9).

Alpine Zone The alpine zone occurs at high elevations throughout the Interior LRR generally above 1200 meters elevation and consists of low-to-dwarf shrub communities. Most soils consist of rocky parent materials. Deep annual freeze–thaw cycles of the soil are indicated by the unique pattern ground features such as earth hummocks, non-sorted circles, stone circles, and stone stripes, as observed in the Steese and White Mountains (Geisler and Ping 2013). Haplogelepts are common on mixed mineralogy parent materials with Haplogelolls on dark or calcareous sedimentary rocks like shale or limestone. Dystrogelepts are common on acidic rock types like schist and diorite. Aquiturbels dominate more gently sloping summits, shoulders, and backslopes where non-sorted circles underlain by permafrost are common (Geisler and Ping 2013). Gelorthents dominate colluvial material such as those under the stripes, and Histoturbels dominate the inter-stripe tundra (Fig. 17.11). In Interior Alaska, mainly the glaciated valleys of the Alaska Range, Spodosols, mainly Haplogelods, occur in loamy-textured glacial drift under alpine shrub tundra (Clark and Duffy 2006).

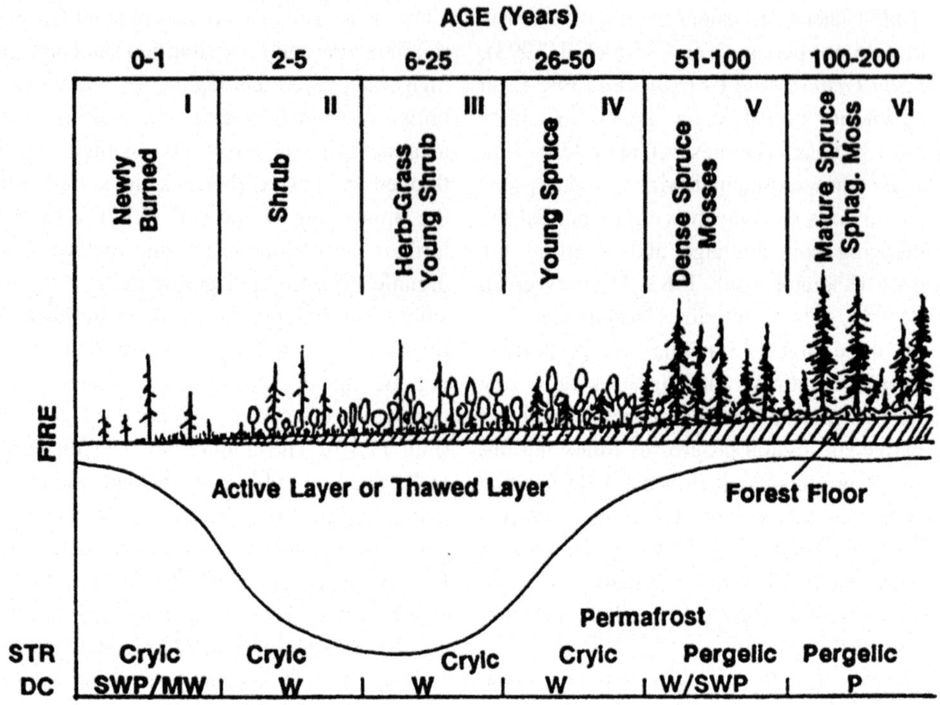

Fig. 17.9 Change of soil temperature and moisture regimes following vegetation succession, after fire in the boreal forest in Interior Alaska

17.3.6.2 Physical Properties

Most of the soil structure observed in soils of these LRRs is the product of frost processes, and platy soil structure is common to deep well-drained soils with loamy textures. Coarse-textured soils with high rock fragment content tend to conduct heat more effectively, and thus disfavor the formation of permafrost. Strong granular and subangular blocky structures formed in carbonate-rich clayey lacustrine materials in the Copper River Basin (Clark and Kautz 1998). With the exception of low floodplains and extremely steep mountain slopes, organic mats mantle most soils within these LRRs. Surface organic layers are crucial in maintaining permafrost in soils by insulating the soil from warm summer temperatures common in the Interior Alaska LRR.

17.3.6.3 Biological, Chemical, and Mineralogical Properties

Soil reaction in surface mineral horizons of most upland soils is slightly to moderately acid on the drier southern slopes and moderately to strongly acid on the wet northerly slopes due to increased acidification from thicker surface organic layers. Soil reaction in subsoils formed in deep loess is nearly neutral. Soil reaction in surface horizons changes abruptly from pH 4.6 before to pH 8.0 or more after fire. But once the moss layer is reestablished after the fire the soil pH drops rapidly. Organic carbon contents vary greatly with landform, vegetation community, and permafrost distribution. In well-drained Haplocryepts on south-facing slopes, organic mats are thin; the SOC stocks generally less than 12 kg C m^{-2}. Historthels and Histoturbels dominate the more northerly aspects where the organic layers are thick, SOC stocks increase to 20–40 kg C m^{-2}. Elsewhere in lowlands, Histels formed in thick organic deposit store over 100 kg C m^{-2} (Michaelson et al. 2013b). CEC in soils of this Region is tied closely to organic matter content since the clay content is typically low (Ping et al. 2005a) or the reactivity is low, such as the chlorite-dominated lacustrine sediments in the Copper River Basin (Clark and Kautz 1998). The mineralogy of most soils at the family level in Interior Alaska is mixed. Chemical weathering to form clay minerals occurs at a negligible rate in cold dry climates with most of the clays being inherited from the parent material (Dement 1962).

17.3.6.4 Soil Function, Ecosystem Services, and Ecological Sites

Water dynamics, crop, and other primary soil functions within the Interior and Western LRRs are numerous and complex. Extensive riverine wetlands along the Interior Alaska Lowland, Yukon Lowlands, and Yukon–Kuskokwim Coastal Plain provide a myriad of habitat types that house some of the richest diversity of plant species and wildlife in Alaska. Important wetland functions include surface water storage especially during spring floods, denitrification, carbon storage, and exchange between the terrestrial and

Fig. 17.10 Vegetation dominated by ericaceous shrubs and lichen on a peat plateau in Kanuti National Wildlife Refuge (**a**), A Folistel on a peat plateau in Kanuti National Wildlife Refuge in interior Alaska (**b**). A Cryosaprist with layers of tephra at 75–80 cm formed in a bog dominated by ericaceous shrubs (**c**) and lichen in Cook Inlet Lowland, southern Alaska (**d**)

Fig. 17.11 Stripes formed along the mountain slopes in the alpine zone of the Steese Mountains, interior Alaska (**a**); Histoturbels formed in the shrub tundra in the concave inter-stripe area (**b**); 0–60 cm on tape; and Gelorthents formed on the convex strip (**c**)

riparian systems. Comprehensive ecological site descriptions are available for Interior Alaska in soil and ecological site inventories completed for Denali National Park and Preserve (Clark and Duffy 2006), Yukon–Charley Rivers National Preserve (Parry 2013), Gulkana River (Clark and Kautz 1999), Delta River (Clark 2005), and Nushagak–Mulchatna Rivers (Schmit 2012).

17.3.6.5 Soil Management and Changes Due to Climate and Land Use

The major soil management issues are related to permafrost, especially ice-rich permafrost. Any activities that remove the vegetative cover and/or the surface organic horizon would alter the thermal regime of the soils and lead to thawing of the permafrost and subsequently land subsidence (Péwé 1954). Frost heave is a major concern to urban development and linear development such as pipelines, roads, and powerlines. The Interior Alaska LRR has the largest acreage of developed agriculture in Alaska including the Big Delta southeast of Fairbanks where hay and barley have historically grown. Small tracts of farmland are developed elsewhere in villages throughout the region. Cool temperature during the growing season limits the variety and yields of crops in the Western Alaska LRR. The Interior Alaska LRR experiences a significant moisture deficit from mid-May through mid-July, and irrigation systems have been installed to offset this water deficit.

17.4 Southern Alaska (LRR W1) and Aleutian Alaska (LRR W2)

The Southern Alaska Land Resource Region (LRR) includes Alexander Archipelago–Gulf of Alaska Coast, Kodiak Archipelago, Southern Alaska Coastal Mountains, Cook Inlet Mountains, Cook Inlet Lowlands, and Southern Alaska

Peninsula Mountains MLRAs (Fig. 17.1). The Aleutian Alaska LRR includes Aleutian Islands–Western Alaska Peninsula MLRA. These Regions makes up 246,710 km^2, about 18.2 % of the land surface of Alaska.

17.4.1 Physiography and Bedrock Geology

These Regions have rugged, rounded mountains with varied bedrock types and glaciated lowlands along a narrow arc of islands and coastal mountains. The regions are underlain with Paleozoic, Mesozoic, and Lower Tertiary stratified sedimentary rocks. Cretaceous and Tertiary intrusive rocks are common in southeast Alaska's accreted terrains. Elevation ranges from 760 m along the boundary with the Cook Inlet Lowlands to 6195 m at the summit of Mt. McKinley (known as Denali to Alaskans). These mountainous terrains consist of steep, rugged, terrain, massive glaciers, ice fields, and moraines. Floodplains and stream terraces on valley floors grade to alluvial fans. The Cook Inlet Lowlands are a broad expanse of gently sloping to rolling plains and low- or moderate-relief hills bordered by the lower slopes of the surrounding mountains. Depressions and shallow basins on plains are dotted with thousands of small and medium-size lakes and interconnecting wetlands. The Aleutian Islands–Western Alaska Peninsula and the South Alaska Peninsula Mountains are made of Quaternary and Tertiary volcanic rocks. The terrain consists of steep, low-to-moderately high, rounded mountains and isolated volcanic cones on the islands, and rugged, low-to-moderately high mountains dissected with narrow, high-gradient valleys. Glaciers and ice fields are common at the upper elevations on the highest peaks in the area.

17.4.2 Climate

The climate ranges from maritime in the Aleutian Islands–Western Alaska Peninsula, to transitional maritime–continental climate in the Cook Inlet Mountains, Cook Inlet Lowlands, and Southern Alaska Coastal Mountains, and hyper-maritime in the Alexander Archipelago–Gulf of Alaska Coast. The mean annual precipitation (MAP) ranges from 35 cm in Anchorage in the central Cook Inlet Lowlands to 200–300 cm in the Southern Alaska Peninsula Mountains to over 600 cm in the Alexander Archipelago–Gulf of Alaska Coast. The average annual snowfall ranges from 75 to 180 cm in the Cook Inlet Lowlands to 2000 cm in the Southern Alaska Coastal Mountains. The mean annual air temperature (MAAT) ranges from 2 to 8 °C with the exception of the Cook Inlet Lowlands and Cook Inlet Mountains with MAAT ranging from −3 to 2 °C. The freeze-free period ranges from less than 60 days to more than 140 days. The Aleutian Alaska, lower Cook Inlet, and southeast Alaska Regions experience strong winds associated with large pressure systems in the Gulf of Alaska.

17.4.3 Parent Materials

Glacial deposits Southern Alaska was heavily glaciated during the Pleistocene. Glaciation deposited glacial drift as compact till and ablation till moraines, glaciolacustrine, and glaciofluvial material on lower mountain slopes, hills, and coastal plains around the Gulf of Alaska. Most of the glacial till is poorly sorted, loamy-textured sediment with rock fragment content up to 70 %. Broad outwash plains are found in the lower Susitna Valley and western Kenai Peninsula. Soils formed in these materials are mantled with loess or volcanic ash.

Colluvium Soil materials on steep mountains consist of fine-to-coarse material originating as frost rived rock fragments or glacial drift that have moved downslope. The colluvium has rock fragments ranging from gravel to boulder. In the steep mountains around the Gulf of Alaska and southeast Alaska, colluvium accounts for 65–90 % of the local parent materials (USDA-NRCS 2006).

Alluvium Extensive glacially derived rivers are wide, braided, and sparsely vegetated with expanses of sandy and gravelly materials in a sinuous pattern of bars and channels. Texture ranges from silt to large cobbles depending on stream gradient and flooding regime. Gravel and cobble size alluvium is common as both bar and channel deposits on low floodplains. Stratified sand and silt deposits typical of overbank deposits on higher floodplains form a layer of variable thickness over a coarser textured substrate.

Eolian Deposits These include loess primarily from braided, barren, glacially influenced river systems and tephra deposits from volcanic eruptions. In the post-Wisconsin glaciation, loess deposits occur in the Cook Inlet Lowlands, Gulf of Alaska Coast, and the Southern Alaska Coastal Mountains. Sand dunes, deposited during the Holocene, are common in the coastal areas of the Cook Inlet Lowlands and the Aleutians. Holocene volcanic deposits from the Aleutian Alaska Region covered most of southern Alaska. Mt. Edgecumbe erupted during the late Quaternary to early Holocene, and stratified tephra covered much of the landscape around Sitka.

Glacio-Marine Late Pleistocene glacial retreat and ice melt coupled with attenuated rebound of the islands along the ice margin created extensive zones of marine silt deposition in southeast Alaska. The uplifted, slowly permeable marine sediment known as the Gastineau Formation (Miller 1973) resulted in extensive low-lying areas of Histosols adjacent to the present-day shoreline extending to the footslopes of many islands in southeast Alaska.

17.4.4 Vegetation

Southern Alaska illustrates a clear pattern of vegetation community following the altitudinal gradient. In the Southern Alaska Coastal and Cook Inlet Mountains true alpine vegetation is present at higher elevations below the permanent snow line or glaciers with dwarf scrubs (*Salix* spp., *Empetrum, Vaccinium, Luzula* spp.) and lichens. Subalpine vegetation occurs below the alpine zones. The vegetation varies from alder and bluejoint reedgrass (*Alnus-Calamagrostis)* to ericaceous shrubs (*Betula, Empetrum, Salix, Vaccinum*), bryophytes, and lichens. Conifers and mixed forest dominate the foothills and uplands. In southeast Alaska, conifer species include western and mountain Hemlock (*Tsuga*), Sitka spruce (*Picea*), redcedar (*Thuja*), and yellow-cedar (*Xanthocyparis*). Hardwoods including red alder (*Alnus*) and black cottonwood (*Populus*) occur on disturbed areas such as snowslides and floodplains. Shrubs are common in the understory and include blueberry and red huckleberry (*Vaccinum*), and rusty menziesia (*Menziesia*). In the Cook Inlet Lowlands, forests of white spruce, Kenai birch (Betula), and balsam poplar (*Populus*) dominate well or moderately well drained sites and black spruce with understory dominated by Labrador tea on poorly drained sites. Hybrid spruce and Sitka spruce occur on the lower Kenai Peninsula. The dominant understory includes alder, willow, devil's club (*Oplopanax*), rusty menziesia and various ferns. Often patches of bluejoint grass and fireweed occur in openings throughout the forest and subalpine. There are large areas of wet sedge meadows and bogs with various ericaceous shrubs and stunted black spruce. In the Aleutians, the vegetation at higher elevation is dominated by dwarf shrub scrub. Meadows of mid-to-tall grass and herbs are extensive at lower elevation. Low ericaceous shrub communities occupy wet depressions. Beach ryegrass (*Leymus*) grows on sand dunes along the coast. Aleutian shield-fern, the only endangered plant species currently listed for Alaska, is on Adak and Attu Islands.

17.4.5 Major Soils and Their Properties

17.4.5.1 Soil Genesis

The soils throughout this region have a cryic soil temperature regime with some frigid soil temperature regimes at lower elevations, and aquic or udic soil moisture regimes. Most of these soils are subject to moderate-to-strong leaching from inland to more coastal areas. Perudic soil moisture regimes exist at high elevations of southeast Alaska and South Alaska Coastal Mountains where annual precipitation exceeds 250 cm. Soils on steep slopes with convex shape or soils formed in coarse texture materials where drainage is unimpeded favor podzolization. Andisolization, the process associated with weathering of tephra, is codominant with podzolization throughout most MLRAs in this region. Hydromorphism, the process associated with saturated and anaerobic conditions, occurs on concave or planar slopes. Scattered areas of Gelisols exist in the northern Cook Inlet in low-lying, cold-air drainages where site and soil properties favor permafrost.

The Cook Inlet Lowlands and Cook Inlet Mountains MLRAs

Soil development in the Cook Inlet Lowlands or Highlands follows a climate gradient based on geographic proximity to the Gulf of Alaska. Periods of moisture deficit are common, especially during spring in areas distal from the Gulf. However, areas proximal to the Gulf of Alaska lack periods of moisture deficit, and soil development is influenced by topography and flow of water.

Spodosols Spodosols are the dominant soils on uplands under conifer forest, and their development is enhanced by accumulation of acidic coniferous organic matter in O horizons. Organic acids from decomposed organic matter strip Fe and Al from E horizons, forming organometallic compounds, which precipitate in subsurface Bs or Bhs horizons. In the Cook Inlet basin, Haplocryods are the dominant soils which have a weakly developed spodic horizon due to the relatively low MAP (35–75 cm). Humicryods developed in areas where MAP ranges from 75 to 150 cm, with increased leaching and more humus incorporated into the subsurface horizons as indicated by their very dark-reddish-brown-to-black colors (Fig. 17.12). Poorly drained Cryaquods form where water accumulates on concave surfaces such as toe slopes and footslopes (Clark and Kautz 2002).

Inceptisols Downward leaching is limited in the Inlet region due to either low MAP, steep slopes, or continuous surface additions of loess. At the southern end of the Cook Inlet Lowlands, Haplocryepts formed in deep loess deposits from the glacial floodplain of the Knik and Matanuska Rivers. On steep mountain slopes, Dystrocryepts form in acid igneous rocks such as diorite, and Haplocryepts form in basic metamorphic rocks with high base saturation. Cryaquepts are abundant throughout the Cook Inlet Mountains and Lowlands MLRAs due to hydromorphic processes occurring on concave surfaces on all landforms where water accumulates and anaerobic conditions prevail for a significant part of the year. The primary indicators of hydromorphism include the accumulation of organic matter, a gleyed or reduced mineral soil matrix, and/or the presence of significant redoximorphic features.

Andisols Andisols formed in volcanic ash deposits under grass or herbaceous vegetation on foothills and slopes. The

Fig. 17.12 Spodosol (Andic Humicryods) formed in volcanic ash under the shrub vegetation of the upper backslopes in the Alaska Range of South-Central Alaska (*top*). Note the yellowish brown volcanic ash layer at 30 cm. Andic Haplocryod formed in volcanic ash under spruce forest in the lowlands of South-Central Alaska (*bottom*)

dense roots of grasses and herbs form a humus-rich A horizon. Andisols with dark A horizons are Humicryands, and those with lighter ones are Haplocryands. On the plateau at the west side of the Inlet region, Melanocryands developed under alder shrubs. On the southwest Kenai Peninsula, lighter colored Fulvicryands developed under bluejoint grass. Cryaquands occur in poorly drained sites. Under the high MAP, the highly weathered volcanic ash developed into Hydrocryands in the northeastern part of the Kodiak Islands, Kruzof Island, and Baranof Island (Fig. 17.13).

Histosols Histosols prevail in the lowlands and valley bottoms throughout the entire southern land resource region. Saturated conditions favor accumulation of organic materials forming shrub/sphagnum bogs, sedge fens, and bogs, as well as wet forests. Nutrient status in these soils is often controlled by water movement, which may carry nutrients from adjacent uplands or underlying mineral substrates into the organic surface layers with significant influence on productivity (Julin and D'Amore 2003). Within this region, Cryofibrists are common to sedge fens and bogs with consistently high water tables. Cryohemists are more common in shrub/sphagnum bogs or sedge bogs where water table fluctuates during summer. Cryosaprists are typical along the perimeter of wetland complexes and raised bogs and are associated with shrub communities or stunted black spruce forest with highly fluctuating water tables. Many Histosol soil profiles within the Cook Inlet Lowlands have layers of tephra from Alaska Peninsula and Aleutian Island volcanoes (Fig. 17.10).

Gelisols Gelisols occur in limited areas at the northern fringe of the Inlet Lowlands and Mountains because of sporadic permafrost. Some Aquorthels and Historthels occur on the northerly slopes on the foothills and lowlands where black spruce dominates the canopy with ericaceous shrub understory and moss ground cover.

Entisols Entisols occur on floodplains, stream terraces as Cryofluvents (D'Amore et al. 2011), and colluvial slopes as Cryorthents throughout the whole LRR.

Aleutian Islands–Western Alaska Peninsula

Andisols and Histosols dominate this MLRA. The soils in the area have a cryic soil temperature regime and udic or aquic soil moisture regimes. Well-drained Haplocryands and Vitricryands formed in moderately thick or thick deposits of silty-to-sandy volcanic ash and coarse sandy-to-gravelly cinders over basalt bedrock on most landforms (Fig. 17.13). Dystrocryepts occur where coarse marine sediments underlie thin volcanic deposits. Along the margins of streams and lakeshores, poorly drained or very poorly drained Cryofibrists formed in thick deposits of organic material. Cryopsamments occur on sand dunes along the beach.

Fig. 17.13 Andisols: Haplocryand formed in volcanic ash deposits of southern Alaska (**a**) and formed under grassy slopes on Adak Island, Aleutian Islands, Alaska (**b**), Cryaquand (**c**) formed in a wet meadow on the Cook Inlet Lowlands (**d**), and Humicryand (**e**) formed under mixed forest (**f**) in the upper Cook Inlet Lowlands

Southeast and Coastal Gulf of Alaska

The perhumid coastal temperate rainforest zone has a cryic soil temperature regime and udic-to-perudic soil moisture regimes. Persistent precipitation creates a strong leaching component to soil development on upper slopes and retention of water on lower slopes. The strong relationship between soil development and topographic position leads to a typical soil catena where Spodosols dominate uplands under conifer forest and Histosols dominate lowlands with bogs and fens (D'Amore et al. 2015a). Humicryods accumulate organic matter in subsurface horizons, Aquods form under wet conditions on lower slopes, and Haplocryods have intermediate characteristics (Fig. 17.14). An association containing both an Entisol and Spodosol on floodplains and terraces is found along channels at the distal end of post-glacial valleys (D'Amore et al. 2011). Andisols are common on the outer coast near the zone of late Pleistocene and early Holocene volcanic eruptions. Histosols form on slopes up to 20 % and are common on lower slope landforms and uplifted marine terraces that fringe the present coastline (D'Amore and Lynn 2002) (Fig. 17.14). Cryofibrists form in bogs dominated by moss. Cryohemists occur on slopes with ericaceous vegetation while Cryosaprists occur in mineral-rich fens with high pH. Calcareous fens occur in areas with limestone bedrock (McClellan et al. 2003).

17.4.5.2 Physical Properties

Most mineral soils have loamy to loamy-skeletal textures. Coarse textures are typical of post-glacial landscapes with abundant glacial drift. Soils in aquic suborders formed in less dense till over compact till along footslopes and areas shallow to bedrock. High organic matter content lowers the bulk density of mineral soils (Alexander 1989). A unique weathering product in Andisols is the allophone-like clay minerals, which have thixotropic properties due to its high water holding capacity and irreversible drying properties (Ping et al. 1989). Most of these soils have bulk densities less than 0.9 Mg m^{-3}, and medial mineralogy classes indicating they can hold 40–100 % (w/w) of water at the wilting point.

Fig. 17.14 A Histosol formed in raised peat plateau (*left*) and a Haplocryod formed in glacial till in the temperate rainforest in southeast Alaska (*right*)

17.4.5.3 Biological, Chemical, and Mineralogical Properties

Most soils in these regions are acid due to high leaching conditions. The pH of Bs horizons of Cryods averaged 5.4–6.0 in MAP zone of 35–50 cm and 4.4–5.2 in MAP zone of >100 cm. Following the same moisture gradient, base saturation decreased from 22 to <5 % and soil carbon contents increased from 3–5 % to 5–20 %. Exchangeable Al increased from 0.5 to >2 cmol kg^{-1}. Andisols have high phosphate fixing capacity because of the Al- and or Fe-humus complexes in the A horizons and amorphous minerals like allophone in the Bw horizons. The diverse rock assemblages in the region lead to mixed soil mineralogy. The dominant clay minerals are poorly crystallized allophone-like Al-silicates, hydrolyzed iron oxides like goethite, ferrihydrite, and minor amounts of phyllosilicates including vermiculite, mica, kaolinite, and chlorite. These phyllosilicates are likely inherited from glacial parent materials (Heilman and Gass 1974; Ping et al. 1988). The sand fraction of the volcanic ash contains glass or glass-coated particles, feldspar, ferromagnesium minerals, and quartz.

17.4.5.4 Soil Function, Ecosystem Services, and Ecological Sites

The volcanic ash-derived soils provide carbon storage through stable compounds formed between the amorphous minerals and humus. The carbon stores range from 20 to 30 kg C m^{-2} in Haplocryods and Haplocryands to 50–70 kg C m^{-2} in Humicryods and Melanocryands and >100 kg C m^{-2} in Histosols (Michaelson et al. 2013b). These soils are the source of dissolved organic carbon, Fe and other nutrients for the watershed and aquatic systems (D'Amore et al. 2015b). The temperate coastal rainforest is productive in timber and has supported timber harvest on over 450,000 ha since 1950 (USDA Forest Service, 2014).

17.4.5.5 Land Use and Soil Management

The Haplocryepts in the south Cook Inlet region have been cleared of spruce forest for farm land since the 1940s. The Matanuska-Susitna Valley area of the Cook Inlet region is the fastest growing area in the state with urban expansion and more land clearing due to a doubling of population over 30 years. The soils are fertile but subject to severe wind erosion after the native vegetation is removed. Southeast Alaska hosts the largest national forest in the USA. It has a long history of logging, but recently the management has been shifted to watershed management while logging remains on some native lands. Slope stability is of great concern due to logging and road building. The farm lands cleared from Andisols and Spodosols experience phosphate deficiency because the strong P-fixing capacity of allophone–imogolite minerals and iron hydroxides.

17.5 Summary

The state of Alaska has a large geographic span and diverse soil environments. Soils in the state cover 7 of the 12 soil orders in Soil Taxonomy, including vast areas of permafrost and associated soils. These soils contain large amount of carbon, but also have ice-cemented permafrost, which presents special problems and challenges to land use and management. The increased rate of permafrost thawing not only changes soil hydrology, but also shifts the carbon balance from an atmospheric sink to source. The South-Central region of the state contains a large area of arable land once cleared of native forest vegetation. This area shows great potential for future food production with increased temperature and precipitation.

References

Alaska Soil Survey Staff (2014) USDA-Natural Resources Conservation Service. http://www.nrcs.usda.gov/Internet/FSE_DOCUMENTS/nrcs142p2_035817.zip. Accessed 09/18/2014

Alexander AE (1989) Bulk density equations for southern Alaska soils. Can J Soil Sci 69:177–180

Bockheim JG, Walker DA, Everett LR, Nelson FE, Shiklomanov NI (1998) Soils and cryoturbation in moist nonacidic and acidic tundra in the Kuparuk River Basin, arctic Alaska, USA. Arc Alp Res 30:166–174

Borden PW, Ping CL, McCarthy PJ, Naidu S (2010) Clay mineralogy in Arctic Tundra Gelisols, Northern Alaska. Soil Sci Soc Am J 74:580–592

Carter LD (1981) A Pleistocene sand sea on the Alaskan Arctic Coastal Plain. Science 211:381–383

Carter LD (1988) Loess and deep thermokarst basins in arctic Alaska. In: Proceedings of the fifth international conference on permafrost, Trondheim, Norway, vol 1. Tapir, Trondheim, pp 706–711

CAVM Team (2003) Circumpolar Arctic vegetation map (1:7,500,000 scale), Conservation of Arctic Flora and Fauna (CAFF) Map No. 1. U.S. Fish and Wildlife Service, Anchorage, Alaska

Clark MH (2005) Soil survey area of the Delta River Area, Alaska. United States Department of Agriculture. Web Soil Survey. http://websoilsurvey.nrcs.usda.gov/

Clark MH, Duffy MS (2006) Soil survey of the Denali National Park and Preserve. United States Department of Agriculture. Web Soil Survey. http://websoilsurvey.nrcs.usda.gov/

Clark MH, Kautz DR (1998) Soil survey of Copper River Area, Alaska. United States Department of Agriculture. Web Soil Survey. http://websoilsurvey.nrcs.usda.gov/

Clark MH, Kautz DR (1999) Soil survey area of the Gulkana River Area, Alaska. United States Department of Agriculture Web Soil Survey. http://websoilsurvey.nrcs.usda.gov/

Clark MH, Kautz DR (2002) Soil survey area of the Matanuska-Susitna Valley Area, Alaska. United States Department of Agriculture. Web Soil Survey. http://websoilsurvey.nrcs.usda.gov/

D'Amore DV, Bonzey NS, Berkowitz J, Rüegg J, Bridgham SD (2011) Holocene soil-geomorphic surfaces influence the role of salmon-derived nutrients in the coastal temperate rainforest of Southeast Alaska. Geomorphology 10:205–209

D'Amore DV, Lynn WC (2002) Classification of forested histosols in Southeast Alaska. Soil Sci Soc Am J 66:554–562

D'Amore DV, Ping CL, Herendeen PA (2015a) Hydromorphic soil development in the perhumid coastal temperate rainforest of Alaska. Soil Sci Soc Am J 79:698–709. doi:10.2136/sssaj2014.08.0322

D'Amore DV, Edwards RT, Herendeen PA, Hood E, Fellman JB (2015b) Dissolved organic carbon fluxes from hydropedologic units in Alaskan coastal temperate rainforest watersheds. Soil Sci Soc Am J 79:378–388. doi:10.2136/sssaj.09.0380

Davis N (2001) Permafrost: a guide to frozen ground in transition. University of Alaska Press, Fairbanks, p 351

Dement JA (1962) The morphology and genesis of the subarctic brown forest soils of central Alaska. Ph.D. Dissertation, Cornell University, Ithaca, NY

Dijkmans JWA, Koster EA, Galloway JP, Mook WG (1986) Characteristics and origin of calcretes in a subarctic environment, Great Kobuk Sand Dunes, northwestern Alaska, USA. Arct Alp Res 18:377–387

Dijkmans JWA, Koster EA (1990) Morphological development of dunes in a subarctic environment, central Kobuk Valley, northwestern Alaska. Geogr Ann 72A(1):93–109

Drew JV, Tedrow JCF (1962) Arctic soil classification and patterned ground. Arctic 15(2):109–116

Dyrness CT, Viereck LA, Van Cleve K (1986) Fire in Taiga communities of Interior Alaska. In: Van Cleve K, Chapin III FS, Flanagan PW, Viereck LA, Dyrness CT (eds) Forest ecosystems in the Alaskan Taiga: a synthesis of structure and function. Ecological Studies, vol 57. Springer, New York, pp 74–86

Euskirchen ES, McGuire AD, Chapin FS III, Yi S, Thompson CC (2009) Changes in vegetation in northern Alaska under scenarios of climate change, 2003–2100: implications for climate feedbacks. Ecol Appl 19:1022–1043. doi:10.1890/08-0806.1

Ferrians OJ (1994) Permafrost in Alaska. In: Plafker G, Berg HC (eds) The geology of North America V.G-1 Geological Society of America, Boulder, pp 845–854

Ferrians OJ, Nichols DR, Williams JR (1989) Copper River Basin. Quaternary geology and permafrost along the Richardson and Glen highways between Fairbanks and Anchorage, Alaska: Fairbanks to Anchorage, Alaska July 1–7, 1989, 1965, pp 39–44

French H, Shur Y (2010) The principles of cryostratigraphy. Earth Sci Rev 101:190–206. doi:10.1016/j.earscirev.2010.04.002

Furbush CE, Schoephorster DB (1977) Soil survey of Goldstream-Nenana Area, Alaska, with maps. USDA Soil Conservation Service, US Government Printing Office, Washington, DC

Geisler E, Ping CL (2013) Effects of periglacial processes on soil formation in Subarctic Alaska. In: Abstracts of the soil science society annual meetings, Tampa, FL 4–7 Nov 2013

Gubin SV (1993) Late Pleistocene soil formation on coastal lowlands of northern Yakutia. Pochvovedeniye 10:62–70 (in Russian)

Hamilton TD (2003) Glacial geology of the Toolik Lake and upper Kuparuk River regions. Biological Papers of the University of Alaska no. 26. Institute of Arctic Biology, University of Alaska Fairbanks

Heilman PE, Gass CR (1974) Parent materials and chemical properties of mineral soils in southeast Alaska. Soil Sci 117:21–27

Jelinski NA (2013) Cryoturbation in the central Brooks Range. Soil Horizons, Alaska. doi:10.2136/sh13-01-0006

Julin KR, D'Amore DV (2003) Tree growth on forested wetlands of southeastern Alaska following clearcutting. West J Appl For 18:30–34

Kanevskiy M, Shur S, Fortier D, Jorgenson MT, Stephani E (2011) Cryostratigraphy of late Pleistocene syngenetic permafrost (yedoma) in northern Alaska, Itkillik River exposure. Quat Res 75:584–596. doi:10.1016/j.yqres.2010.12.003

Kanevskiy M, Shur Y, Jorgenson MT, Ping CL, Michaelson GJ, Fortier D, Stephani E, Dillon T, Tumskoy V (2013) Ground ice of the upper permafrost of the Beaufort Sea coast of Alaska. Cold Reg Sci Technol 85:56–70. doi:10.1016/j.coldregions.2012.08.002

Lerbekmo, JF, Westgate JA, Smith DGW, Denton GH (1975) New data on the character and history of the White River volcanic eruption, Alaska. Quaternary Studies, vol 13. Royal Society of New Zealand, Wellington, pp 203–209

McClellan MH, Brock T, Baichtal JF (2003) Calcareous fens in southeast Alaska. USDA Forest Service, Pacific Northwest Research Station, Research Note PNW-RN-536

Michaelson GJ, Ping CL, Clark MH (2013a) Alaska soil carbon and nitrogen database. http://www.uaf.edu/files/snras/Alaska-Soil-C-and-N-Pedon-Data-Table-5-24-2013.xlsx

Michaelson GJ, Ping CL, Clark MH (2013b) Soil pedon carbon and nitrogen data for Alaska: an analysis and update. Open J Soil Sci 3:132–142. doi:10.4236/ojss.2013.32015

Michaelson GJ, Ping CL, Kimble JM (1996) Carbon storage and distribution in tundra soils of Arctic Alaska, USA. Arct Alp Res 28:414–424. doi:10.2307/1551852

Miller RD (1973) Gastineau channel formation: a composite glaciomarine deposit near Juneau, Alaska. Geological survey bulletin, contributions to stratigraphy, USGS Report #1394, Washington, DC

Moore JP, Ping CL (1989) Classification of permafrost soils. Soil Surv Horizons 30:98–104

Nichols DR (1956) Permafrost and ground-water conditions in the Glenallen area. U.S. Geological Survey Open-file Report 56–91, Alaska

Osterkamp TE, Romanovsky VE (1996) Characteristics of changing permafrost temperatures in the Alaskan Arctic, USA. Arct Alp Res 28:267–273

Parry N (2013) Soil survey of Yukon-Charley Rivers National Preserve. United States Department of Agriculture. Web Soil Survey. http://websoilsurvey.nrcs.usda.gov/

Pettapiece WW (1974) A hummocky permafrost soil from the subarctic of northwestern Canada and some influences of fire. Can J Soil Sci 54(4):343–355

Péwé T (1954) Effect of permafrost on cultivated fields, Fairbanks area, Alaska, U.S. Geol Surv Bull 989-F:315–351

Péwé T (1975) Quaternary Geology of Alaska. Geological Survey Professional Paper 835, U.S. Geological Survey. U.S. Govt. Printing Office, Washington, DC

Péwé TL (1982).Geologic hazards of the Fairbanks area, Alaska, vol 15. Alaska Division of Geological & Geophysical Surveys

Ping CL (2013a) Gelisols: part I. Cryogenesis and state factors of formation. Soil Horizons. doi:10.5410.2136/sh2013-54-3-gc

Ping CL (2013b) Gelisols: Part II. Classification and related issues. Soil Horizons. doi:10.2136/sh2013-54-4-gc

Ping CL, Bockheim JG, Kimble JM, Michaelson GJ, Swanson DK, Walker DA (1998) Characteristics of cryogenic soils along a latitudinal transect in arctic Alaska. J Geophy Res 103:28917–28928

Ping CL, Lynn W, Smith CAS (1993). Redoximorphic features in permafrost soils. In: Gilichinsky DA (ed) Post-seminar proceedings, joint Russian-American seminar on cryopedology and global change. Russian Academy of Sciences, Puschino, Russia, pp 233–244

Ping CL, Michaelson GJ (2015) Guidebook: Alaska soil geography field study on permafrost-affected soils. SNRE-AFES Bulletin 118. University of Alaska Fairbanks, AK

Ping CL, Michaelson GJ, Guo L, Jorgenson T, Kanevskiy M, Shur Y, Dou F, Liang J (2011) Soil carbon and material fluxes across the eroding Alaska Beaufort Sea coastline. J Geophys Res Biogeosci. doi:10.1029/2010JG001588

Ping CL, Michaelson GJ, Kimble JM, Romanovsky VE, Shur YL, Swanson DK, Walker DA (2008a) Cryogenesis and soil formation along a bioclimate gradient in Arctic North America. J Geophys Res 113:G03S12. doi:10.1029/2008JG000744

Ping CL, Michaelson GJ, Jorgenson T, Kimble JM, Epstein H, Romanovsky VE, Walker DA (2008b) High stocks of soil organic carbon in North American Arctic region. Nat Geosci 1(9):615–619. doi:10.1038/ngeo284

Ping CL, Michaelson GJ, Kimble JM, Walker DA (2005a) Soil acidity and exchange properties of cryogenic soils in arctic Alaska. Soil Sci Plant Nutri 51(5):649–653

Ping CL, Michaelson GJ, Packee EC, Stiles CA, Swanson DK, Yoshikawa K (2005b) Soil catena sequences and fire ecology in the boreal forest of Alaska. Soil Sci Soc Am J 69:1761–1772

Ping CL, Shoji S, Ito T (1988) The properties and classification of three volcanic ash derived pedons from Aleutian Islands and Alaska Peninsula, Alaska. Soil Sci Soc Am J 52:455–462

Ping CL, Shoji S, Ito T, Takahashi T, Moore JP (1989) Characteristics and classification of volcanic ash-derived soils in Alaska. Soil Sci 148:8–28

Rieger S (1983) The genesis and classification of cold soils. Academic Press, San Diego

Rieger S, Schoephorster DB, Furbush CE (1979) Exploratory soil survey of Alaska. US Department of Agriculture, Soil Conservation Service, Washington, DC

Schmit S (2012) Soil survey report of the Nushagak-Muchatna area, Alaska. USDA-NRCS, Anchorage

Shur YL (1988) The upper horizon of permafrost soils. In: Proceedings of the fifth international conference on permafrost, vol 1. Tapir Publishers, Trondheim, pp 867–871

Soil Survey Staff (2014) Key to soil taxonomy, 12th edn. USDA Natural Resources Conservation Service, Washington, DC

Swanson DK (1996a) Soil geomorphology on bedrock and colluvial terrain with permafrost in central Alaska, USA. Geoderma 71: 157–172

Swanson DK (1996b) Susceptibility of permafrost soils to deep thaw after forest fires in interior Alaska, USA, and some ecological implications. Arct Alp Res 28(2):217–227

Tarnocai C, Smith CA (1992) The formation and properties of soils in the permafrost regions of Canada. In: Gilichinsky DA (ed) Cryosols: the effects of cryogenesis on the processes and peculiarities of soils formation, proceedings of the 1st international conference on cryopedology, 10–14 Nov, 1992 Pushchino, Russia, pp 21–42

Tedrow JCF, Drew JV, Hill DE, Douglas LA (1958) Major genetic soils of the arctic slope of Alaska. J Soil Sci 9:35–45

USDA Forest Service (2014) Tongass young-growth management strategy. Tongass National Forest, Region 10, Juneau, AK

USDA-NRCS (2006) Land resource regions and major land resource areas of the United States, the Caribbean, and the Pacific Basin. U. S. Department of Agriculture Handbook 296

Van Cleve K, Dyrness T, Marion GM, Erickson R (1993) Control of soil development on the Tanana River floodplain, interior Alaska. Can J For Res 23:941–955

Viereck LA, Van Cleve K, Foote MJ (1983) Vegetation, soils, and forest productivity in selected forest types in interior Alaska. Can J For Res 13(5):703–720

Wahrhaftig C (1965). Physiographic division of Alaska. U.S. Geological Survey Professional Paper 482

Walker DA, Everett KR (1991) Loess ecosystems of northern Alaska: regional gradient and toposequence at Prudhoe Bay. Ecol Monogra 61(4):437–464

Walker DA, Raynolds MK, Daniëls FJA, Einarsson E, Elvebakk A, Gould WA, Katenin AE, Kholod SS, Markon CJ, Melnikov ES, Moskalenko NG, Talbot SS, Yurtsev BA, The other members of the CAVM Team (2005) The circumpolar Arctic vegetation map. J Veg Sci 16:267–282

Washburn AL (1973) Periglacial processes and environment. St. Martin's Press, New York

Western Regional Climate Center (2014) Cooperative climatological data summaries. http://www.wrcc.dri.edu/summary/climsmak.html

Woodgate M (2015) Vegetation succession and pedogenesis on the Yukon-Kuskokwim Delta floodplain near St. Mary's, Alaska. M.S. Thesis, School of Natural Resources and Extension, University of Alaska Fairbanks

Wu TH (1984) Soil movements on permafrost slopes near Fairbanks, Alaska. Can Geotech J 21:699–709

Human Land-Use and Soil Change

18

Skye A. Wills, Candiss O. Williams, Michael C. Duniway,
Jessica Veenstra, Cathy Seybold, and DeAnn Presley

18.1 Introduction

Soil change is the central, if under-recognized, component of land and ecosystem changes (Yaalon 2007). Soils change naturally over a long timescale (decades to millennia) in response to soil-forming factors (biota, climate, parent material, time, and topography). However, human land-use pressures are currently the driving force in maintaining, aggrading, and degrading soil properties across nearly all ecosystems. Traditionally, in order to simplify and standardize the relationships between soils and soil-forming factors, pedology and soil survey have often focused on "natural" or "virgin" soil (e.g., Hilgard 1860; Jenny 1980), but many argue that humans should be thought of as a part of soil genesis and formation (Amundson and Jenny 1991; Yaalon and Yaron 1966; Bidwell and Hole 1965).

Landscapes and soils have been altered by wide-scale conversion to agriculture, use of vegetative products, and development for direct human use. Land-use impacts can be gradual or abrupt, subtle, or catastrophic (Table 18.1). The interactions between environmental changes and geomorphic and biotic feedback loops vary across temporal and spatial scales depending on the setting (Monger and Bestelmeyer 2006). The effects of land use can linger for decades to centuries and beyond (Hall et al. 2013; Jangid et al. 2011; Sandor et al. 1986). While each land resource region has some specific soil–land use interactions, this chapter will focus on general uses and topical areas: croplands, wetlands, grazing lands (both pasture and rangelands), and forest lands with smaller sections devoted to special issues including acid sulfate soils, strip-mined lands, and cold soils.

18.1.1 Concepts of Soil Change

Soil change refers to the variation of soil properties in one location over time. The concept of "soil change" has been proposed as a framework for understanding and documenting the impact of human use and management on soil properties and function (Arnold et al. 1990; Palm et al. 2007; Richter and Markewitz 2001; Robinson et al. 2012; Tugel et al. 2005, 2008). Management of soil resources, directly or indirectly, can alter soil properties and soil functions both negatively and positively. While all management (even its absence) impacts soil conditions, some impacts are intensive and site specific while others are extensive (Grigal 2000). The direction of change depends on the nature of the management, the goals of the management action, and the framework for measuring outcomes. While management actions to increase soil fertility might lead to improved soil function (for instance, crop productivity) at a local scale, a broader watershed scale might assess decreased soil function (for instance, nutrient filtering and buffering that influence eutrophication).

Soil change in this chapter refers to disturbances caused directly or indirectly by human land use and management. Palm et al. (2007) refers to soil degradation as a change that causes a reduction in ecosystem function (as in Fig. 18.1).

S.A. Wills (✉) · C.O. Williams · C. Seybold
USDA-NRCS, Lincoln, NE, USA
e-mail: Skye.wills@lin.usda.gov

C.O. Williams
e-mail: Candiss.williams@lin.usda.gov

C. Seybold
e-mail: Cathy.seybold@lin.usda.gov

M.C. Duniway
US Geological Survey, Moab, UT, USA
e-mail: mduniway@usgs.gov

J. Veenstra
Flagler College, St. Augustine, FL, USA
e-mail: JVeenstra@flagler.edu

D. Presley
Kansas State University, Manhattan, KS, USA
e-mail: deann@ksu.edu

L.T. West et al. (eds.), *The Soils of the USA*, World Soils Book Series,
DOI 10.1007/978-3-319-41870-4_18

Table 18.1 Types of soil change ranging from landscape alteration to cyclical fluctuations

Landscape alteration	Taxonomic classification	Phase distinctions	Surface properties	Cyclical fluctuations
Drastically altered arrangement of materials • Removal or transport of material • Gully erosion • Landslides	Alteration or mixing of horizons • Extension erosion or deposition changes soil interacting with the environment • Large changes in nutrient or base status due to long-term management	Distinction soil features remain intact with relatively unaltered subsurface features (control section) • Low-to-moderate erosion • Drainage	Dynamic soil properties that change with land use or management • Nutrients • Aggregates • Organic matter	Regular, periodic, or cyclical fluctuations • Soil moisture • Temperature • Water table
Change in soil and ecosystem functions				
High, catastrophic change to functions relating to soil stability and productivity	Variable, depends on the taxonomic system used and classes crossed	Moderate, soil capacity, or potential may remain intact, but sensitive functions may be disrupted	Moderate to low, depends on type and magnitude of soil property change as well as the constraints upon the ecosystem	Variable. Typical fluctuations occur as part of highly functioning soil systems. However, alteration to the timing and amplitude of fluctuations can greatly modify soil functions

Adapted from Tugel et al. (2008)

Table 18.2 Summary of effects of tillage and residue management on soil physical properties in small grains cropping systems at the end of 20 years in interior Alaska

Tillage/residue	Effects on soil properties
Intensive[a]	Had the most detrimental effects on soil physical properties and was found to be the least effective in reducing soil loss during high wind events
Autumn chisel	Promoted greater roughness, aggregation, and residue cover compared with intensive tillage; random roughness was greater for autumn chisel compared to all other tillage treatments. Infiltration was greater compared to all other tillage treatments
Spring disk	Promoted greater roughness, aggregation, and residue cover compared with intensive tillage
No tillage	Had larger aggregates, greater soil strength (penetration resistance and shear stress at the surface), wetter soil, and greater residue cover compared to all other tillage treatments; resulted in an organic layer on the soil surface (after 20 years) which suppressed infiltration; had higher saturated hydraulic conductivity and retained more water against gravitational and matric forces versus intensive tillage
Crop residue management[b]	Influenced residue biomass and cover (the response was dependent on tillage treatment). No tillage resulted in 100 % residue cover, while residue cover ranged from 2 to 4 % on all other tillage treatments. Crop residue management did not influence soil properties

Summarized from Sharratt et al. (2006a, b)
[a]Intensive = autumn chisel and spring disk
[b]Crop residue management treatments were residue removed and residue retained after harvest

The resistance and resilience of a soil refer to the capacity of a particular soil to respond to a disturbance (Seybold et al. 1999). A soil is resistant if it does not change functionality after a disturbance or management impact occurs. Lal (1993) uses the term soil stability for the same concept. A soil is resilient if it recovers its initial qualities after a disturbance (Seybold et al. 1999). Resilience can be applied to landscapes, soil profiles, or individual properties and are important for predicting and assessing soil change.

18.1.2 Measuring Soil Change

An assessment of soil change depends upon the metrics of soil and ecosystem function evaluated. Richter and Markewitz (2001) illustrated an approach that used an in-depth evaluation of an experimental forest to represent regional soil change that has occurred over centuries. Richter et al. (2011) expanded on the limits of a soil management approach and emphasized the need to understand soil

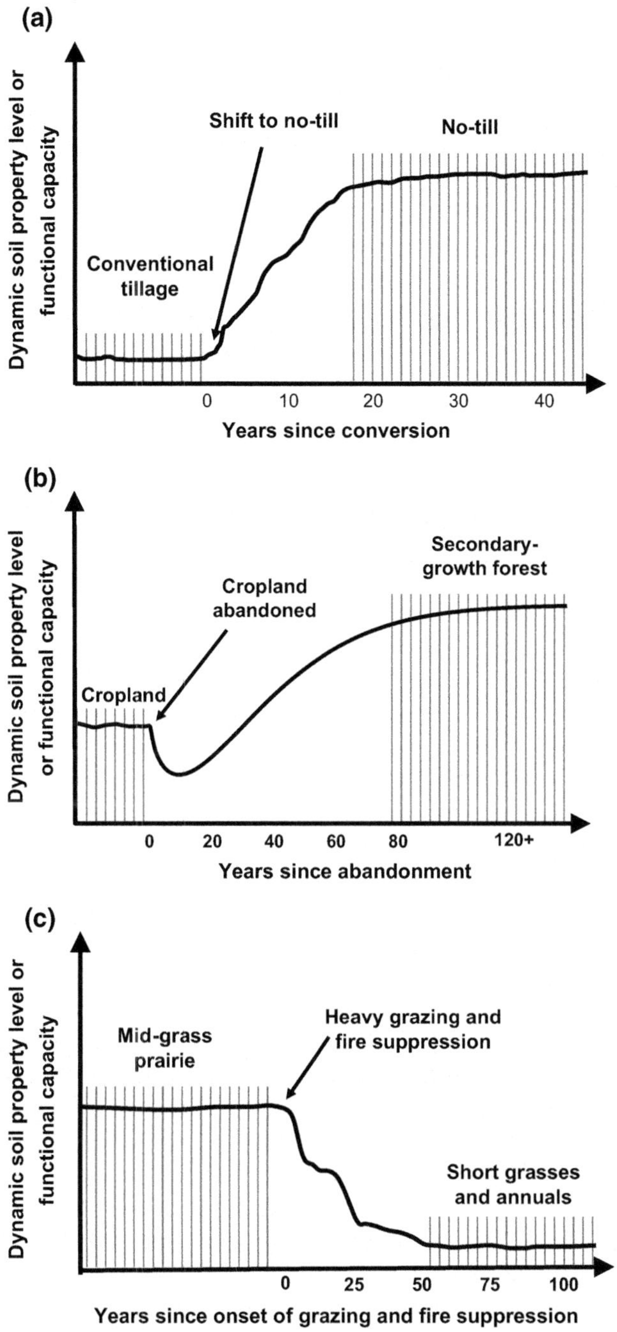

Fig. 18.1 Three examples of expected soil organic C change after management change. The *arrows* point to a disturbance that may have positive or negative impacts: **a** Conversion from conventional tillage to no till. West and Post (2002) found that maximum SOC accumulation (average increase of 0.6 kg m^{-2}) occurred within 15–20 years after conversion to no till. **b** Secondary forest succession after field abandonment. *Curve* from 1 to 80 years based on soil N level in a Minnesota chronosequence (Zak et al. 1990). SOC followed a similar trajectory and increased from 1.25 to 2.10 kg m^{-2} during the same period. **c** Change from reference state after grazing and absence of fire in a grassland system. Adapted from Tugel et al. (2005) and based on data from Archer et al. (2001). Archer et al. (2001) simulated a SOC reduction of 16 % in a sandy loam and 29 % in a clay loam over a period of about 50 years after the onset of heavy grazing and fire suppression

genesis through the concepts of the Anthropocene (the "age of humans"). Ecology and economic disciplines have attempted to document soil change in terms of the natural environment and ecosystem functions and to apply monetary metrics to ecosystem services (Costanza et al. 1997; Goudie 2013).

The spatial and temporal frameworks implemented in soil change studies depend upon the nature of the agents of change, system attributes, and types of soil change of most concern (Table 18.1; Fig. 18.1). However, since there are often no good records of historical soil properties, techniques such as "space-for-time substitution" have been used to compare soils that differ only in management or other disturbance (Pickett 1989). This technique requires that other soil-forming factors (parent material, climate, organisms, relief, and time) be held constant to the extent possible and that the conceptual model of soil change be explicitly considered and documented (Tugel et al. 2008). Soil survey and environmental covariates that represent soil-forming factors can be used to choose comparable locations. When possible, monitoring the same location over time improves understanding of the processes and mechanisms of soil change. Variability in soil with spatial extent, management applications, and weather phenomena means that measuring and interpreting soil change are not simple or straightforward. While humans have been active managers of soil and ecosystem resources for millennia (Nir 1983), the degree and amount of change have accelerated in recent centuries (Richter et al. 2011). In the United States, the time of European settlement is often used as a reference to assess soil change across ecosystems and regions (i.e., Richter and Markewitz 2001). The scale and type of soil change being investigated may require alternate reference states for relevant comparisons (Fig. 18.1).

When measurements of soil change are focused on management-relevant timelines (months to decades) and spatial scales (e.g., field or paddock), the concept of soil quality is useful. Soil quality and soil health are terms used to indicate the ability of soils to perform functions, such as support and stability, productivity, and cycling, filtering, and buffering (Doran and Parkin 1994; Karlen et al. 1997; Singer and Ewing 2000). Assessments of soil quality and health are often used to assess soil change and intermediate scales of human management (i.e., Tugel et al. 2005).

18.1.3 Monitoring and Modeling Soil Change

Soil as a component of agricultural and environmental ecosystems is well studied. However, there is growing emphasis on long-term studies that focus on change in soil properties (i.e., university- and government-led agricultural research). While soil degradation may occur quickly as a result of practices on the land, studies have shown that agricultural management improvements of soil properties are likely to be observed and quantified in long-term studies (Lal 1993; McVay et al. 2006, Wood and Edwards 1992). The importance of quantifying land management practices on soil properties is well identified. However, ongoing soil sampling and monitoring to quantify changes in soil properties as a result of management and/or environmental changes is relatively new and is greatly needed to develop guidelines and criteria for decision-making (Lal 1993; Lawrence et al. 2013). Long-term studies also allow for a wide range of environmental conditions in addition to land management. These conditions allow for the inclusion of episodic events (i.e., drought) or slower events such as changes in land use and climate. In addition, long-term studies provide necessary data for the calibration and validation of crop and environmental simulation models used to forecast changes (Hobbie et al. 2003; Robertson et al. 2008; Williams et al. 2008).

Guidelines and criteria for understanding how management practices influence changes in soil properties can only be developed if long-term experimental data are available (Lal 1993). Since the inclusion of repetitive soil sampling in long-term studies is relatively new and questions relating to long-term impacts of management on soil properties may be difficult to ascertain from data collected from networks, alternative approaches such as space-for-time (Pickett 1989; Tugel et al. 2005; Blois et al. 2013) and computer modeling (Kelly et al. 1997; Smith et al. 1997; Izaurralde et al. 2006) have been used.

Space-for-time studies encompass analyses in which current spatial phenomena are used to understand and model processes that are unobservable (Blois et al. 2013) such as the impact of management change on soil properties that were implemented in the past and their likely impact long term. Space-for-time measurements have limitations in soil management research due to changes in climate and management over time. Computer models are more commonly used in soil management research to estimate the effects of management on changes in soil properties and are increasingly important tools for short- and long-term assessments (Williams and Sharpley 1989; Lowrance et al. 2000; Izaurralde et al. 2006; Williams et al. 2008, 2013) especially when measured data are limited and forecasting is necessary. The disadvantage of using computer models, however, is the inherent uncertainty. Soils are part of a complex system, and models may not be able to simulate such complexities as simultaneous interaction between physical, chemical, biological, climatic conditions, and the calibration process.

18.2 Soil Change in Croplands

Soils in the USA have been altered to provide food and fiber for millennia. While Native Americans altered landscapes for hunting and farming, European settlement and clearing for agriculture resulted in a marked acceleration of soil change (Richter and Markewitz 2001). Eastern forests were first cleared for production agriculture in the early 1800s (Ramankutty et al. 2010), and the impact on soil properties can only be postulated by observing relatively undisturbed areas or inferring from modern-day disturbance studies (as in space-for-time). The immediate impacts of deforestation likely include increased soil temperature, altered moisture regime, and modified nutrient cycles (Vitousek et al. 1983).

As agricultural development progressed westward, grasslands were plowed and wetlands drained for cropland uses (Ramankutty and Foley 1999). Even where native vegetation regenerated, the effect of cropping has lingered in soil properties from decades (Bellemare et al. 2002; Guzman and Al-Kaisi 2010) to centuries (McLauchlan 2006; Sandor et al. 1986). While using soils for cropland can cause multiple disturbances to soil functions through the introduction of monocultures, the addition of inorganic fertilizers and pesticides, and irrigation, the most direct impact comes from physical disruption.

The Morrow plots at the University of Illinois were established in 1876 to answer questions about agricultural management (Odell et al. 1982). Initial results indicated that growing corn depleted soil organic matter and associated soil fertility. Introducing inorganic fertilizer and crop rotations improved yields but did not return soil conditions to their original status. Other long-term observation plots such as the Sanborn Field in Missouri (Brown 1993), the Magruder plots in Oklahoma (Webb et al. 1980), and long-term studies in Oregon (Rasmussen et al. 1998) and Nebraska (Peterson et al. 2012), all reported decreases in SOC over time with cropping. The world's oldest continuous cotton experiment, called "The Old Rotation" in Auburn, AL, has demonstrated that changes to management systems (tillage and rotation) can increase SOC concentrations (and improve other soil properties and functions) (Hubbs et al. 1998; Mitchell et al. 2008; Prieto et al. 2002).

Tillage changes the physical arrangement of soil particles, which has cascading effects on chemical and biological properties (Bronick and Lal 2005). The disruption of soil structural integrity with tillage leads to the breaking apart of soil aggregates and breakdown of soil organic matter. Cambardella and Elliott (1992) studied Mollisols in Iowa under native prairie or sod and intensively cropped conditions. They found that carbon and nitrogen dynamics were linked to particulate organic matter (POM) and soil structural units or aggregates. The soil structure was disturbed due to tillage, therefore leading to a breakdown of larger aggregates and decrease in SOC. Six et al. (1998) found similar results and showed that tillage intensity was related to POM and structural stability and later linked aggregate sizes directly to the SOC they contained (Six et al. 2000).

Tillage also leads to changes in mechanical impedance (Bennie 1990), which can be described in terms of soil bulk density and soil strength (often measured as penetration resistance). These physical properties impact plant growth, water infiltration, and water storage (Dao 1993; Ehlers et al. 1983; Lampurlanés and Cantero-Martínez 2003). Changes in physical properties may increase erosion on summits and sideslopes and deposition in footslopes and depressions (e.g., Daniels et al. 1985; Beach 1994; Konen 1999; Norton 1986), and therefore can alter watershed hydrology (Johnson et al. 1980). In some cases, erosion and deposition can alter the soil profile and therefore, the classification (taxa) and mapping of soil bodies (Indorante et al. 2014; Fenton 2012).

Limiting the impact of agricultural practices on erosion has long been a goal of conservation efforts, and recent conservation efforts are focusing on soil health with emphasis on soil biology (Morgan 2009; Lindbo et al. 2014). There are complex interactions between soil types, management systems, and the impact of tillage on soil ecology (Kladivko 2001). Microbial diversity and activity in the soil have been shown to be a signal of the overall function of soil chemical and physical systems (Torsvik and Øvreås 2002; Wander et al. 1995; Wardle et al. 1999).

New developments in agricultural technology (conservation or reduced and no-tillage cropping systems) have led to improvements in soil properties. For instance, Blanco-Canqui et al. (2009) reported that a no-till cropping system increased SOC and decreased bulk density. Rhoton et al. (2002) observed improvements with low-tillage systems in organic matter and physical structure, which lead to increased infiltration and decreased runoff. Furthermore, developments in crop rotations, cover crops, and reduced chemical disturbance (inputs of inorganic nutrients and pesticides) have led to improvements in soil biological function (Doran 1980; Parkinson and Coleman 1981; Spedding et al. 2004). Although the improvements in soil properties were observed in these studies, recovery of some soil properties may take longer than most management time frames (Fig. 18.2). See Sect. 18.2.3 for a discussion of erosion that cannot be reversed on a human timescale with even the best management.

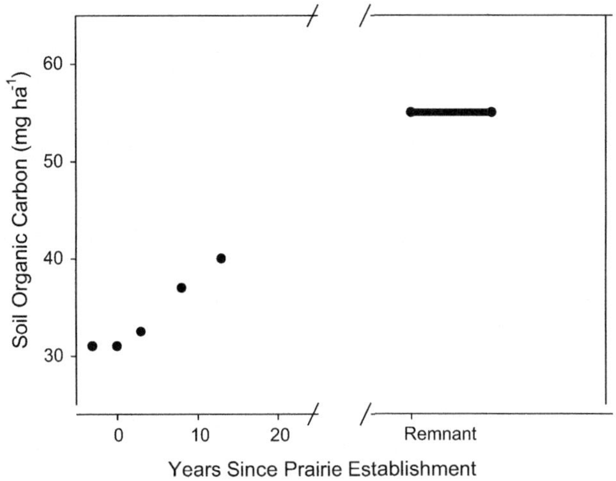

Fig. 18.2 Soil organic C after tillage and restoration in an Iowa Mollisol. Initial tillage was approximately 150 years ago. Each point represents one site. Adapted from Guzman and Al-Kaisi (2010)

18.2.1 Impacts of Cropping on Soils of the Great Plains of the USA

Mollisols are the predominant soils in the Great Plains (Chap. 8), and because of inherent soil fertility, the soils in this region are intensively cropped. Strong gradients in annual precipitation and temperature are present in this region, and lead to a range of management choices by agricultural producers, including crop rotations, cropping intensity, amendments, and tillage operations. Access to irrigation water from either aquifers (such as the Ogallala, High Plains, and Equus Beds) or surface water is also an important factor in crop production in the Great Plains. The Great Plains is an extremely important center of agricultural production. However, it is also a region that has numerous challenges including losses of soil from erosion, low precipitation in the west, intense thunderstorms, and declining aquifer levels for irrigation. Water is the most limiting resource to attaining maximum crop yield potential (Nielsen et al. 2005). During the 1930s, the farming of marginal lands in the Great Plains, combined with a prolonged drought, culminated in dust storms and soil destruction of disastrous proportions (Goudie and Middleton 1992). This period, known as the "Dust Bowl," inflicted great hardships on the people and the land, and has been called the greatest ecological disaster to have occurred in the USA (Cook et al. 2009). The erosion and deposition events of the dust bowl created shallow low-fertility soils in some areas and buried soils in others leading to reduced agricultural productivity (Hansen and Libecap 2003; Hornbeck 2009).

A traditional Great Plains dryland cropping system was wheat/fallow with conventional tillage during the fallow period for weed control. During the past ~20 years, primarily due to developments in chemical weed control, a shift has occurred toward reduced or no tillage, and has allowed for more intensive rotations, thus reducing the amount of fallow. Declines in soil organic matter have been documented with tillage and summer fallow, and changing to no-till or more intensive crop rotation has been found to increase soil organic matter levels, though not to the original levels (Hartman et al. 2011; Mikha et al. 2012; Peterson et al. 1998). For these reasons, and because of agronomic and environmental advantages including increased stored water, decreased wind and water erosion, increased C storage, and overall increased grain yields, no-till practices have increased (Reicosky and Saxton 2007). Decreasing the fallow period or rotating wheat with other crops has shown to sequester C at greater rates than either the wheat-fallow or continuous wheat (West and Post 2002; Sainju et al. 2006). McVay et al. (2006) summarized five long-term studies across a precipitation gradient in Kansas and observed that decreased tillage, increased fertilization, and crop rotations that included at least one cereal crop in the rotation increased SOC in the surface 5 cm, while no significant changes were observed at other depths.

Irrigation has been observed to affect the soil organic matter of some Great Plains soils. Bordovsky et al. (1999) found that irrigation led to increases in total C for semi-arid sandy soils in Texas, while Lueking and Schepers (1985) observed increased total C for sandy soils in Nebraska. Williams (2001) observed increased SOC with irrigation for drier upland soils, and no increase for wetter lowland soils in a study conducted on fine-textured soils of Konza Prairie in eastern Kansas. Presley et al. (2004) measured no changes in SOC for the upper 0.3 m of irrigated silt loams and silty clay loams in western Kansas, as compared with dryland soils sampled from the corners of a center-pivot irrigated field. However, it is critical to note that irrigation is not the only management practice that differs; dryland fields receive lower inputs of N and produce less biomass than the irrigated portion of the field and are also less productive. It seems that the effect of irrigation is dependent upon inherent soil properties that include soil texture, drainage class, and/or slope position, among others.

18.2.2 Impacts of Cropping in the Southeastern USA

In the southeastern USA, agriculture created some unique problems due to the predominance of low-fertility Ultisols. In the 1800 s, farmers in the southeastern USA managed these soils by shifting agriculture, clearing forests, and growing cotton and tobacco with little to no fertilizer additions until the land would no longer yield, and then moving their fields (Gray 1933; Trimble 1974).

In addition to depleted soil fertility, shifting to row-crop agriculture practices in this region accelerated erosion rates. Ultisols can be very stable under native forest cover, but once the forest cover is removed, in combination with the high-intensity rainfall that is common to the region, they can be subject to high degrees of soil loss (Trimble 1974). The years 1860–1920 were the period of the most erosive land use in the Southern Piedmont (Fig. 13.1). In 1934, 3100 km^2 of the Piedmont was reported to be destroyed by gully erosion with 85 % of the affected area being in South Carolina and Georgia (Trimble 1974). An extreme example of this type of soil loss from the nearby Coastal Plain region is Providence Canyon State Park in southwest Georgia, which is referred to as "Georgia's Little Grand Canyon" (Fig. 18.3). In 1820, the Creek Native Americans ceded the land area that included modern-day Providence Canyon; settlement and row-crop agriculture quickly followed. The removal of forest led to gully erosion and the development of a canyon 50 m deep and several hundred meters wide in places (Froede and Williams 2004). Paleolimnological studies confirm that the period of most extreme historical erosion in the canyon occurred between 1840 and 1880 (Hyatt and Gilbert 2000).

After the US Civil War (1865), continuous cultivation and fertilization became more common (Sheridan 1979), whereas soil conservation practices became more common after the 1930s (Trimble 1974). Studies from the Calhoun Experimental Forest in South Carolina showed that long-term cultivation and fertilization of the Ultisols in the southeastern USA induced deep and long-lasting changes in soil properties including changes in bulk density, pH, nutrients, and exchangeable cations (Richter and Markewitz 2001). Also in the southeastern USA, Levi et al. (2010) reported similar results and suggested that anthropogenic amendments have actually changed the soil classification from Ultisols to Alfisols.

An example of soil change caused by tillage for agricultural production resulted from the development of the rock plow in 1951 (US Patent 2573977[1]). The rock plow was designed to break up "solid formations of rock, coral, or slag" and prepare the soil for fruit trees and row-crop production. The rock plow was designed for the shallow, oolitic limestone soils of south Florida. Before the rock plow was developed, growers used dynamite to blast holes in the ground in order to plant fruit trees (Derr 1998). As of 2007, 339,960 ha. in south Florida have been transformed by the rock plow, while only 14,074 ha. of the native soil remain

[1]Any use of trade, product, or firm names is for descriptive purposes only and does not imply endorsement by the US Government.

Fig. 18.3 The walls of Providence Canyon, in southeastern Georgia. This massive gully formed when the land was cleared for agriculture in the early 1800s. A thick Ultisol is exposed at the surface. (Photograph credit: Aaron Daigh)

(Soil Survey Staff 2007). Rock plowing results in soils that are a several centimeters thicker than the native soils, and much gravellier (from 12 % in the native soil to 35–70 % in the rock plowed soil). Therefore, despite the fact that management accelerated physical weathering, the resulting soil textures are coarser, rather than finer than the native soils.

18.2.3 Impact of Cropping on Arid and Irrigated Soils

Dry soils, referred to as Aridisols in Soil Taxonomy (Soil Survey Staff 1999), pose unique management challenges and respond to cropping systems in unique ways. Dry soils may have taken long periods to develop and have limited inputs for restoring functionality; their resistance to change and resilience may be low (Homburg and Sandor 2011; Seybold et al. 1999). The risk of detrimental change can be high for arid systems under many climate change scenarios (Herrick and Beh 2015). For a more complete discussion of soil change dynamics in arid and semi-arid systems see the section on grazing lands (Sect. 18.5).

Irrigation has been an important part of agriculture and cropping for decades, but has become increasingly widespread in the modern era, since 1800 (Michael 2008). Irrigation alters the distribution of salts by leaching soluble salts downward while evapotranspiration moves salts toward the surface. The resulting distribution of salts in the soil profile is determined by the crops grown, weather conditions, water quality, and water quantity applied as well as the drainage of excess irrigation water. Application of poor-quality irrigation water can lead to accumulations of sodium and trace elements (Dregne 2011; Grattan 2002). Most crop production on dry soils requires irrigation, but much of the USA's irrigation water is applied in semi-humid and humid areas (Fig. 18.4). The effects of irrigation can be very localized or regional (Ferguson and Maxwell 2012). See the discussion of irrigation in the Great Plains (Sect. 18.2.3).

18.3 Soil Change in Wetlands

While wetland soils are recognized as environmentally and ecologically important and legally protected in the USA, many USA wetland soils have been drained for agricultural production. The agricultural use of these potentially highly productive soils was recognized, and, in fact, encouraged by the Swamp and Overflowed Lands Act of 1850 (Wright 1907). Drainage districts were established in these states, and wetlands across the nation were drained. This resulted in significant management-induced soil change on these lands. Drainage of wetlands for agriculture in the Midwest may have resulted in SOC losses of 30–50 % across the region (Baker et al. 2007). More recently, the value of wetland and hydric soils for ecosystem services have been recognized and attitudes about wetland drainage have changed (Gopal 2000). Wetlands are being managed for flood control, water quality, and wildlife habitat (Barbier et al. 2011; Brander et al. 2006; Zedler and Kercher 2005).

Wetland drainage is a direct alteration of the water table (e.g., James and Fenton 1993). As water tables were lowered through artificial drainage, peat or organic soils that

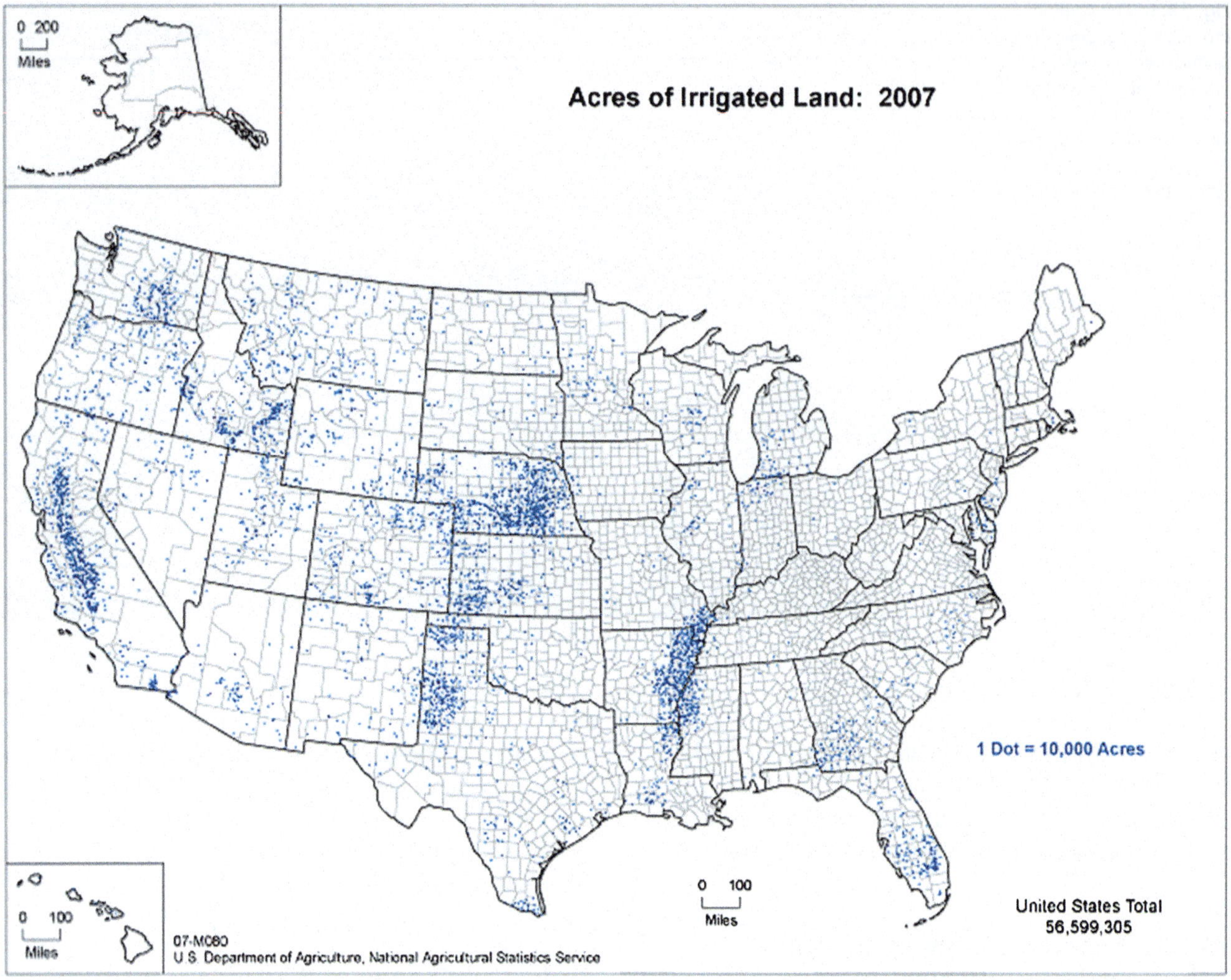

Fig. 18.4 Acres of irrigated land from the 2007 Census of Agriculture (USDA-National Agriculture Statistics Service (NASS) (2007)

had built up over time due to anaerobic conditions and slow decomposition were aerated and exposed to oxidation and wind erosion. In many places, this caused the soil to subside and become lower relative to set benchmarks (Parent et al.1982; Kohake et al. 2010). The majority of the subsidence is accounted for in gaseous as opposed to aqueous CO_2 fluxes (Deverel and Rojstaczer 1996), suggesting that the losses are largely related to increased rates of organic matter decomposition. In some areas, subsidence occurred rapidly after the soil was first drained for agriculture and development. For example, in the California Sacramento-San Joaquin Delta, an estimated 55–80 % of the original peat layer was lost due to subsidence, which equates to a loss of 2900–5700 metric tons of organic C ha^{-1} (Drexler et al. 2009). Other soil changes associated with the drainage and tillage of organic matter-rich wetland soils include increased bulk density, decreased labile organic matter contents, and increased mineral content (Drexler et al. 2009; Shih et al. 1998).

Rates of organic soil subsidence have decreased over time. In 1950, Weir estimated that the rates of subsidence were 2.8–11.7 cm $year^{-1}$ in the California Delta (Weir 1950), and in the same time period, Stephens and Johnson (1951) estimated that the subsidence rate in the Everglades Agricultural Area in Florida was 3 cm $year^{-1}$. By 1998, that rate had decreased to 1.4 cm $year^{-1}$ (Shih et al. 1998; Fig. 18.5). In the California Delta, the subsidence rate has decreased to 0.5–1.0 cm $year^{-1}$ (Deverel and Rojstaczer 1996). The reduction in subsidence rate is likely due to the maintenance of higher water tables in both regions. The Everglades Agricultural Area of Florida (EAA), during the first half of the twentieth century, was largely farmed for vegetables, and the perception was that water tables needed to be low. However, by the mid-twentieth century, more of the EAA was being farmed for sugarcane, a crop that is more tolerant of high water tables.

Not all wetlands have organic soils and wetland soils change for reasons other than drainage and subsidence.

Fig. 18.5 Soil change through subsidence. This image shows the amount of subsidence that has occurred at the University of Florida's Everglades Research and Education Center, in Belle Glade, FL, since 1924 when a concrete post was driven into the organic soil so that it was level with the soil surface at the time. This photograph was taken in 2013 and represents about 1.8 m of soil loss. (Photograph credit: Alan Wright)

Riparian wetlands occur adjacent to streams and thus receive sediment deposition from the upstream watershed. Depressional wetlands receive deposition from a more limited area. Wetlands can retain high levels of nutrients and pollutants from water and sediments (Johnston 1991; Gilliam 1994), and these additions may alter the biogeochemistry within the wetland soils themselves (Morris 1991; Rokosch et al. 2009). Craft and Casey (2000) found that deposition in Georgia wetlands was dependent on past and current anthropogenic disturbance within the surrounding watershed. Freeland et al. (1999) found that there were higher rates of deposition in wetlands adjacent to cultivated fields in North Dakota. Sediments and nutrients deposited in wetlands impact the vegetation and hydrology of the systems (Mahaney et al. 2005; Preston et al. 2013).

Tidal or salt marshes occur along coasts and estuaries and have flooding characteristics determined by the tides (Adam 2002). Soils in these coastal marshes may be mineral or organic, and the salt balance in the marsh is determined by local geomorphology. They have low redox potential and are often very biologically active (Rabenhorst 2001; Seybold et al. 2002). Coastal marshes have been managed for forage and grazing for centuries (Gedan et al. 2009), and they are increasingly under heavy pressure from human development including pollution, introduced species, altered hydrology (including nutrient and sediment loadings), and climate change (Adam 2002; Scavia et al. 2002; Kirwan and Megonigal 2013). Many have been degraded and even destroyed by direct and indirect human management on the Atlantic (Coverdale et al. 2014; Kearney et al. 2002), Gulf (Turner 2011), and Pacific coasts (Craft et al. 2003; Kennish 2001).

Kirwan et al. (2010) summarized the stability of coastal marshes as the balance between accretion and submergence where biological and physical factors interact to modify the marsh environment. Anthropogenic activity modifies marsh hydrology through changes in the quality, quantity, or distribution of water. On the gulf coast of Louisiana, DeLaune et al. (1983) found that the tidal marshes were being submerged due to decreases in sediment loads caused by changes in waterway management (such as dykes and levees for flood protection). Changes in water quality such as excess nutrient loading can also lead to coastal marsh degradation (Turner 2011; Deegan et al. 2012). Sea-level rise stresses the interaction between biophysical properties and can lead to soil submergence and weakening of marsh strength and integrity which eventually leads to erosion of the marsh into open water (Adam 2002; Kirwan and Megonigal 2013).

18.4 Soil Change in Forest Lands

While the most visible and dramatic changes in forested soils can occur when they are cleared for agriculture or plantations, forest management (silviculture) can induce remarkable levels of soil change (Binkley and Fisher 2012). Stand establishment (regeneration) and harvesting techniques can

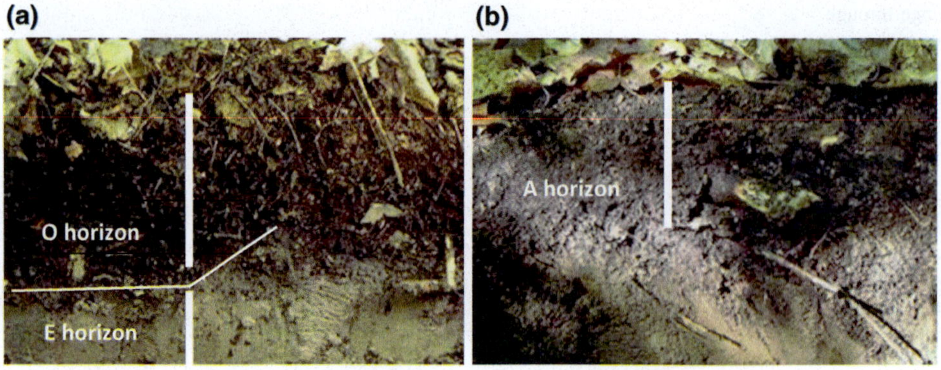

Fig. 18.6 Changes to O and A horizons caused by introduction of invasive earthworms. **a** Before earthworm invasion, a thick forest floor of organic materials (*dark* O horizon) on top of the mineral soil (*light-colored* E horizon) with a distinct boundary between the two. **b** After earthworm invasion, the forest floor has been replaced by a layer of thick black topsoil (A horizon) made mostly from earthworm casts. There are no longer distinct organic (O) and mineral horizons. (Photograph credit: Great Lakes Worm Watch http://www.nrri.umn.edu/worms/forest/soil.html)

disturb the soil surface and can thus increase soil erosion and negatively impact soil productivity (Swanson et al. 1989; Megahan 1990). Loss of forest floor (litter and woody debris), compaction (increase in bulk density), and access routes may contribute to increased erosion in managed forests (Elliot et al. 1998; Grigal 2000). Due to the slow growth of most forest species, recovery from soil disturbance can take extended lengths of time (decades to centuries).

The most apparent soil change in forestland at the pedon scale takes place at the interface between air, forest floor, and soil. Many mechanical actions from harvesting disturb the soil surface inadvertently with trails, skids, and other physical impacts from heavy machinery. The soil surface may be deliberately disturbed through management techniques such as prescribed burns and scarification (exposing mineral soil) to improve forest seedling regeneration (Page-Dumroese et al. 1997). In southeastern forests, bedding radically alters the forest floor by mechanically creating raised beds to plant plantation seedlings. Physical disturbance may remove organic material, therefore altering bulk density and disrupting the nitrogen cycle as well as leading to accelerated soil erosion (Fox et al. 1986; Gent et al. 1984; Grigal 2000; Jandl et al. 2007; Jurgensen et al. 1997; Mclaughlin et al. 2000). Surface disruption can also cause lingering changes to soil biochemistry and microbial communities (Amaranthus et al. 1989; Hartmann et al. 2012).

In the eastern USA, forest clearing has created massive erosion and deposition events, which can be observed as multiple buried soils along terraces (Bierman et al. 2005). Forests in the northern portion of the USA may experience higher soil temperature and impaired soil gas exchange due to harvesting and site preparation (Ballard 2000). In the western USA, changes to fire regimes (through suppression, prevention, or more recently, prescribed burns) can cause changes in forest floor that may vary in duration due to soil and landscape variables as well as management actions (Certini 2005). These types of soil change can have impacts at watershed and larger scales such as stream quality and flooding from increased runoff (DeBano et al. 1998).

Some of the most widespread impacts of human management on forested soils are indirect. Soil acidification through industrial output (deposited across forests in the northeastern USA as acid rain) has caused alteration of chemical and biological processes in forest ecosystems (Robarge and Johnson 1992). This has been shown to lead to changes in forest productivity and species composition related to changes in Al concentrations and pH of the soil (Drohan and Sharpe 1997). The spread of introduced earthworms has altered the biological and structural systems of forest floors in northern USA hardwood forests (Fig. 18.6) Invasive earthworms consume O horizon material and leave behind castes that mix decomposed organic material and mineral soil. A horizons in these systems can be thickened with increased bulk density and altered N and P cycling (Hale et al. 2005). The presence of invasive earthworms has been shown to be correlated with anthropogenic activities such as recreation (fishing and cabins) and transportation (roads and trails) (Holdsworth et al. 2007). In the northwestern USA, current fire regime and past fire history have interacted to produce forests that may not be well suited to thrive or survive under future climate change scenarios (Whitlock et al. 2003).

18.5 Soil Change in Grazing Lands

Large ungulate grazing shapes ecosystems through herbivory (consumption of plant matter by ungulates), physical impact (hoof impact, wallowing, etc.), and deposition (urine, dung, and carcasses). Grazing lands, defined here as lands where grazing by domestic livestock is currently or has historically been a major land-use activity, are typically divided into two general types: (1) pasture lands, characterized by intensive management (e.g., irrigation and application of fertilizer and seed) and replacement of a native vegetation community with seeded forage; and (2) rangelands, which are characterized by low management inputs and retention of native vegetation communities. The management impacts on soils can differ dramatically between range and pasture lands; thus, it is a useful distinction for discussions of soil change.

In pasture systems, direct effects of grazing are additive to those caused by the direct impacts of management on soils. For example, seedbed preparation, seeding, and fertilization can cause soil change similar to those seen in croplands. Similarly, irrigation can change the growing environment and soil solution composition leading to increased plant production and altered carbon and nutrient dynamics. Soil changes due to pasture management include direct impact on the soil surface and changes in nutrient cycling due to species composition, animal deposition, and fertilizer inputs. Grazing may increase bulk density (Greenwood and McKenzie 2001), specifically in areas of concentrated hoof action (Tate et al. 2004), although sometimes less so than conversion to tilled cropland (Franzluebbers et al. 2000). In North Dakota, Liebig et al. (2006) found that pasture management (including stocking rate, species mix, and fertilizer treatment) changed soil bulk density, pH, and nitrogen levels. In the southeastern USA, conversion of cropland to pasture has been shown to increase SOC content, but the depth distribution is unlikely to be equivalent to native systems (Franzluebbers et al. 2000). While soil degradation (loss of beneficial surface material and characteristics) in pasture lands is recognized as a potential result of poor pasture management, the exact mechanisms of these responses are not fully known (Bilotta et al. 2007). However, active management of pastureland can also provide opportunities to manage undesirable soil change (more so than in rangelands).

In rangelands, management options that effect soil change are more limited. Short-term (annual) options are typically only associated with animal management (timing, intensity, and duration of grazing). Although management of rangelands is usually not intensive, the impacts of domestic herbivores on these landscapes can be substantial. Cows (the most common grazer) have been proposed as an agent of geomorphic change by Trimble and Mendel (1995). Active degradation of western USA rangelands was particularly widespread and acute after the expansions of the railroads and prior to the Taylor Grazing Act of 1934. In a study using high alpine lakes of the southern Rocky Mountains, Neff et al. (2008) documented an increase in dust production by more than fourfold beginning in the mid- to late nineteenth century (Pacific Railway Act was signed in 1862) and subsequent decrease in the early to mid-1900s. This dust is sourced to the dry regions of northern Arizona and attributed to the removal of vegetation and disturbing of exposed soil surfaces by cattle (Neff et al. 2008). The impacts of grazing on soils are highly variable depending on the ecosystem, soil and management system in question, as well as the soil properties evaluated (Bilotta et al. 2007). In some instances (e.g., fertile soils in mesic environments), the effects of grazing on plant and soils can be gradual and readily reversible with minor adjustments to grazing strategies (Briske et al. 2003). However, in many areas, particularly drier regions (aridic moisture regimes) that are characterized by low and variable rainfall, sparse vegetation cover, and erosion-prone soils, poor grazing management can cause ecosystems to cross biophysical thresholds. In these instances, the changes to soils and vegetation persist when grazing pressure is reduced or eliminated and usually require active restoration (Briske et al. 2003).

Range improvement projects that use herbicide, fire, heavy equipment, or hand tools to manipulate rangeland vegetation communities are a common approach for improving wildlife habitat, increasing forage production, controlling fuels, or restoring post-fire landscapes in the western USA. The degree to which these activities affect soil change is primarily dependent on the amount that the soil is disturbed during implementation and project success in altering species composition (e.g., woody to herbaceous) or abundance. Woody species removal (in fuel treatments or for range improvements) can increase the amount of woody litter incorporated into soils (Ross et al. 2012). Treatments that increase herbaceous cover and productivity may be associated with a multitude of changes to soil quality (van Auken 2000). The use of heavy equipment for both removal of woody vegetation and application of seed disturbs the soil surface and can leave large areas susceptible to erosion by wind and water. For example, broad-scale soil disturbance associated with post-fire seeding treatments following the Milford Flat Fire in Central UT was associated with extreme wind erosion in the most sensitive landscape settings, resulting in rates of wind-driven sediment flux that rank among the highest ever recorded in North America (Miller et al. 2012). Furthermore, negative feedbacks between broad-scale soil destabilization and wind-driven saltating sands further hampered establishment of desired species in these aridic soils (Duniway et al. 2015). Indeed, direct manipulations of rangeland plant communities are more likely to be successful in more mesic soil systems than aridic (Knutson et al. 2014), suggesting desirable impacts on soil change and dynamic soil properties are much more likely in mesic than aridic soil systems.

In rangelands, soil change and changes in plant community composition and cover are often coupled. Grazing

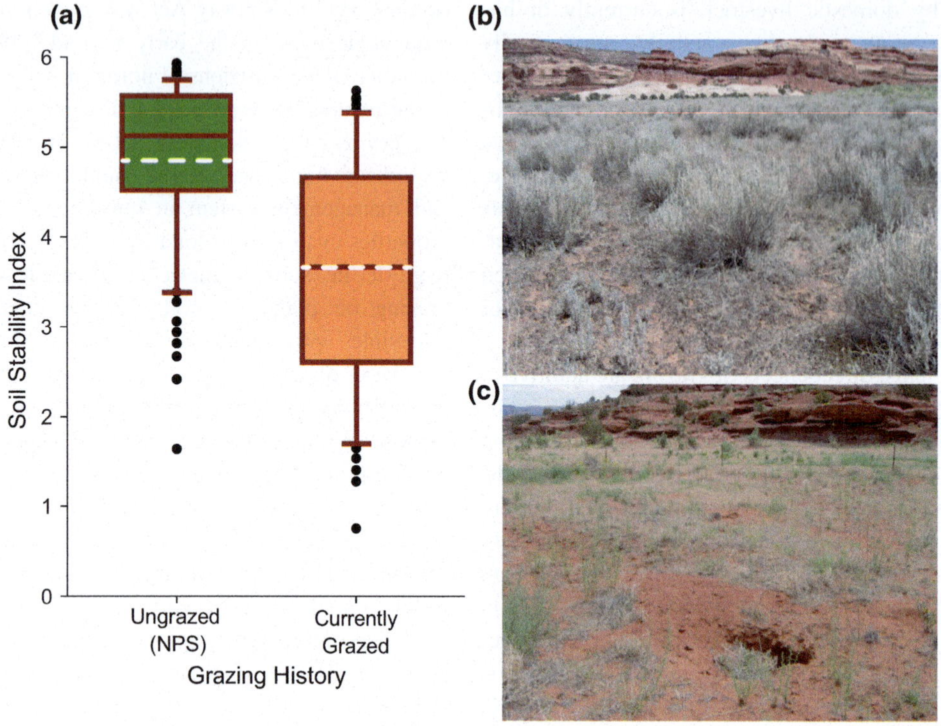

Fig. 18.7 Soil change caused by grazing: **a** Effects of grazing history on near surface soil aggregate stability from a rangeland setting in southeastern Utah, USA. Average aggregate stability is significantly greater in ungrazed than grazed pastures ($p < 0.001$) (aggregate stability collected following Herrick et al. 2001). Photographs showing **b** well-developed biological crust cover in a national park, compared to **c** heavily grazed area outside the park. *Box plots* show plot mean (*white dashed horizontal line*), median (*solid horizontal line*), 25th and 75th quantiles (*lower and upper bounds of box*), 10th and 90th percentile (*whiskers*), and outliers (*black dots*) aggregate stability values. Data are from 80 plots sampled within Canyonlands National Park (ungrazed for 40 years) and 65 plots in the surrounding private, state, and federal lands that are open to grazing (Miller et al. 2012). All sites were located on similar soils based on NRCS Ecological Site Classification

alters the structure of plant communities by selectively grazing or disrupting growth and reproduction of some species (Augustine and McNaughton 1998; Belsky 1992; Cingolani et al. 2005). Grazing can also interact with woody, weedy, and invasive species by altering competitive interactions, effects on fire frequency (via effects on fine fuels), and seedling germination (MacDougall and Turkington 2005; DiTomaso 2000). Grazing alters carbon and nutrient cycling directly through removal of plant biomass reducing the amount of biomass that is returned to the soil as litter (Holland et al. 1992) and indirectly through changes in vegetation structure, composition, and growth (Biondini et al. 1998; McIntyre and Lavorel 2007).

Dynamic soil properties in semi-arid and arid rangelands are characteristically patchy, and understanding how soil quality soil quality varies spatially has emerged as an important indicator of range land health (Herrick 2000; Pyke et al. 2002). Some of the most important indicators of soil change are not simply plot averages but measures of vascular plant distribution, soil attributes both under plant canopies and in patches between plants. For example, the development of biological and physical soil crusts is common in the patches between plant canopies. Biological crusts (and physical crusts to some extent) can increase the stability of these exposed surfaces, lowering erodability by wind and water (Belnap and Gillette 1998; Belnap 2006). Direct hoof action by grazers disturbs soil surfaces, breaking up fragile crusts, and thereby greatly increasing soil erodability in areas heavily trafficked by grazers (Belnap and Eldridge 2003). Similarly, the development of water-stable soil aggregates will covary with both plant and crust cover, typically greater in perennial plant or crust patches than in bare interspaces (Fig. 18.7; Herrick et al. 2001). The patchy nature of arid and semi-arid rangelands has led to the development of semi-quantitative approaches for evaluating soil and site stability that explicitly accounts for spatial variability (e.g., Interpreting Indicators of Rangeland Health (Pellant et al. 2005) and Pedoderm and Pattern Classes (Burkett et al. 2011)).

18.6 Soil Change in Extreme Conditions

Some of the most dramatic examples of soil change occur when sensitive soils or areas are mishandled or poorly managed. Acid sulfate soils undergo extreme (and negative) changes after they are exposed and oxidized. Cold soils are sensitive to heat and other human-induced environmental changes. Some land uses, such as strip-mining, alter the landscape entirely producing massive amounts of soil change. Extreme soil change is typically non-recoverable. Land-use options are limited by the impacts of past use and management.

18.6.1 Acid Sulfate Soils

An extreme example of management-induced soil change is the production of sulfuric acid in acid sulfate soils. In such soils, sulfuric acid is being produced (active), has been produced (post-active), or could be produced (potential). Potential acid sulfate soils are typically formed under anaerobic conditions in coastal or tidal sedimentary environments where iron sulfide minerals accumulate (Dent and Pons 1995). When these minerals oxidize, they produce sulfuric acid, which lowers soil pH to 3.5 or less, which is a level below which most plants will not grow. These minerals oxidize, and the soils become active acid sulfate soils through drainage, surface deposits of dredged material, or earth-moving associated with development, construction, and mining (Bradshaw 1997; Fanning 2006).

The Food and Agriculture Organization of the United Nations (FAO) suggests that there are 0.1 million ha of potential and active acid sulfate soils in North America, which is only a small proportion of the estimated world total of 12.6–18.1 million ha. (Andriesse and van Mensvoort 2006). Although they have not been formally mapped, there are several locations within the USA where the formation of sulfuric horizons (low pH, evidence of oxidation of sulfide minerals) has been observed, such as dredged materials from the San Francisco Bay of California, the Baltimore Harbor in Maryland, the tidal Pocomoke River in Maryland, reclaimed marshland soils in Florida, clay landfill caps in New York and Maryland and exposed soil as a result of construction activities in Virginia (Grass et al. 1962; Calvert and Ford 1973; Fanning and Burch 1997; Demas et al. 2004, Fanning 2004).

18.6.2 Management Impacts in Soils of Cold Climates

Cold climate soils are those that formed in high latitudes or at high altitudes that have a mean annual soil temperature (at 50 cm depth) of <8 °C (Ping 2005) and mean summer (June, July, August) soil temperature of <15 °C (Soil Survey Staff 1999). About 16–18 % of the USA consists of cold climate soils of which <1 % is farmland. Almost all of the cold soils (in the USA) are located in Alaska, most in the tundra and boreal forest ecoregions. Features of cold soils include permafrost (continuous and discontinuous), formation of ground ice, and cryoturbation from freeze–thaw cycles. Permafrost impedes soil drainage creating high moisture contents and can result in the accumulation of solutes in the active layer (Alekseev et al. 2003). Above the permafrost is the active layer (seasonally thawed layer), that can range from 0.2 to 5 m or more in thickness (Marchenko and Etzelmüller 2013). In a natural system, permafrost and ground ice are buffered or insulated by surface vegetation and surface organic layers (O horizons). Removal of this buffer results in recession of the permafrost table and an increase in depth of the active layer. Once the thermal equilibrium is disrupted, increased thawing can result (Brown 1997).

Near Fairbanks Alaska, clearing of an area of natural boreal forest vegetation and removal of the 10-cm-thick surface organic layer resulted in a drop in the permafrost table from 1 to 5 m (Kallio and Rieger 1969). When ground ice melts, the surface topography is changed and increased melting can lead to major ground disturbances known as thermokarst (Fig. 18.8). Anthropogenic disturbances such as construction of roads, trails, airfields, trampling, and agriculture can have profound effects on the landscape due to subsidence and thermokarst over relatively short timescales (Nelson et al. 2002). Melting permafrost on ice-rich slopes can cause soil erosion and gullying (Brown 1997). Farming in cold soils leads to a thickening of the active layer (deeper permafrost), which results in changes in leaching, oxidation, and reduction as well as cryogenic processes (Ping 2005). Overgrazing and excessive trampling can result in disturbed or destroyed surface organic layers or surface vegetation cover, leading to permafrost melting and erosion (Forbes 1999; Ping 2005). Increased dust load associated with gravel roads and pads can alter tundra nutrient cycling and nutrient regimes (Moorhead et al. 1996) and has several ecological consequences (Walker and Everett 1987).

The active layer and upper permafrost can contain large quantities of organic C compared to soils in temperate ecosystems as a result of frost churning (Bockheim et al. 1999; Ping et al. 1997; Ping 2013; Tarnocai 2009). In the subarctic, land-use change has been found to increase soil temperatures by 4–5 °C, lengthen the season of biological activity by 2–3 weeks, and enhance plant residue decomposition by 25 % (Grünzweig et al. 2003). Land-use change from forest to agricultural land results in increased CO_2 emissions, but can be minimized by selecting relatively C-poor soils for land-use change and by implementing C

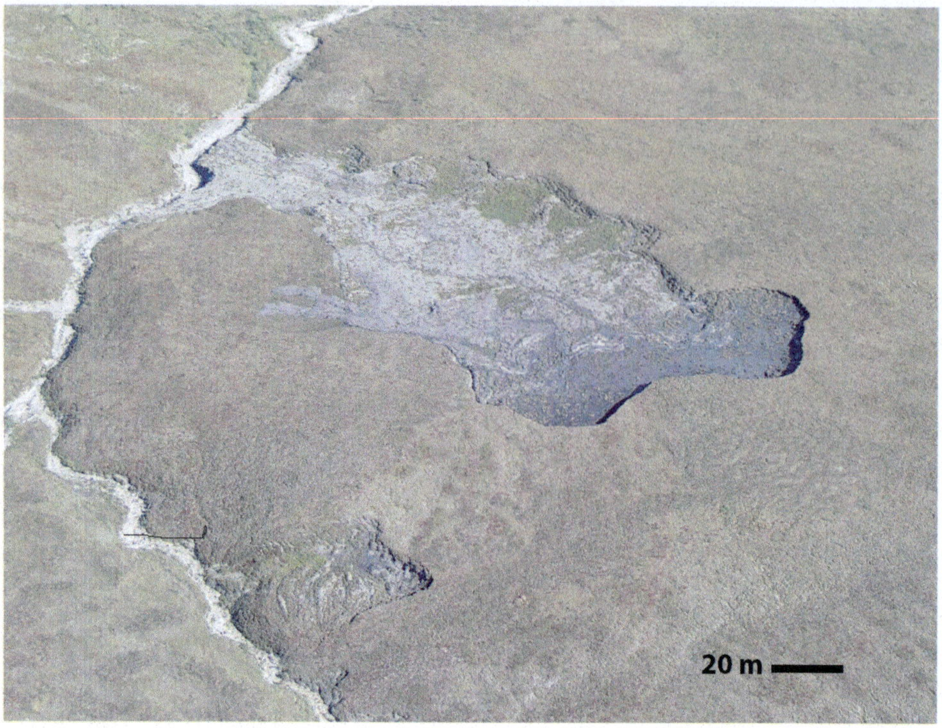

Fig. 18.8 Aerial photograph of thermokarst in tundra on the North Slope of Alaska. The upper ice-rich permafrost is thawing and water draining, resulting in the settling and collapse of the surface. (Photograph credit: Cathy Seybold)

preservation management strategies (e.g., perennial crops, low tillage) (Grünzweig et al. 2004). Increases in the number of freeze–thaw cycles can decrease aggregate size distribution and can lower infiltration rates (Fouli et al. 2013). Conversation tillage options have been shown to be viable in Alaska to promote infiltration and conserve organic matter (Sharratt et al. 2006a; Sparrow et al. 2006). Zhang et al. (2012) found net C gains and improved soil quality after 18 years in the conservation reserve program (CRP), over that of native forest in subarctic Alaska. The CRP is a US government program where environmentally sensitive land is removed from agriculture production for a period of time.

18.6.3 Impacts of Strip-Mining for Phosphates

An example of extreme soil change is that caused by management associated with phosphate strip-mining as is done in Florida, North and South Carolina, Tennessee, Utah, Idaho, Montana, and Wyoming (USGS 2014a). Phosphorus mining began in the late 1800s and expanded in the 1900s in North and South Carolina, Tennessee, and Florida. The mining industry grew largely as a response to the increasing demand for phosphorus fertilizer by farmers in the southeast USA (Richter and Markewitz 2001). Today 85 % of USA-mined phosphorus comes from mines in Florida and North Carolina, while the remainder comes from Idaho and Utah (USGS 2014b).

In the phosphorus mining regions in North Carolina and Florida, 3–15 m of overburden is stripped off to allow excavation of the phosphate-rich material below (Marion 1986). After excavation, the phosphate is separated from the soil and unconsolidated material leaving behind overburden fill, sand tailings, and clay-slime (Brown et al. 1992) each of which have different soil characteristics. Overburden fill tends to have characteristics most closely related to the native soils of the area, but they are mixed and spatially variable (Wallace and Best 1983). The native soils in the phosphate-mining region of Florida are Spodosols with a fine sand texture, low pH, low cation exchange capacity (CEC), and low organic matter. The native soils of the phosphate-mining region of North Carolina are Ultisols with fine sandy loam surface texture, low pH, and low CEC. The overburden fill sites are characterized by the mixing of many native soil horizons and tend to have finer surface textures, greater CEC, higher water-holding capacity (WHC), and higher pH than nearby native soils (Hawkins 1973; Wilson and Hanlon 2012). Also, overburden fill surface horizon bulk density is usually higher and subsoil bulk density is usually lower than the natural soils (Gee et al. 1978; Chambers et al. 1994). Overburden fill sites can be suitable for agriculture or development.

Sand tailings are the quartz sands that have been separated from the clay and phosphate during the mining process. Phosphorus and calcium concentrations tend to be three to five times greater in these soils than those found in undisturbed or native soils (Davis et al. 1991), but the soils have low CEC, WHC, organic matter, K, and micronutrients (Mislevy and Blue 1981) These sand tailings can be used for agriculture, but nutrients and WHC can be limiting. They are highly suitable for development as the sands are quite stable surfaces compared to the clay-slime settling areas (Wilson and Hanlon 2012). The final waste sediment from the phosphate-mining process is the clay-slime slurry, which is pumped to retention basins and allowed to dry. About 40 % of phosphate-mined reclamation areas are clay-settling basins (Wilson and Hanlon 2012). The resulting soil textures of these sites are largely clay that is high in smectites and has high shrink swell potential (Hawkins 1973). They are heavy, sticky, and slow to dry, which makes them poorly suited for development, but they have been productive for agricultural crops (Mislevy et al. 1990).

In Florida, phosphate mining has impacted 526,000 ha., disturbing 2500 ha. annually, and 25–30 % of the lands impacted by phosphate mining are isolated or hydrologically connected wetlands (FDEP 2010). Although, reclaimed phosphate-mined land soils differ from native soils, soil development in created wetlands on overburden fill, and sand tailings mimics that of natural wetland soil formation; that is, organic matter accumulates, C:N ratio decreases, and bulk density decreases with increasing wetland age (Nair et al. 2001).

18.7 Conclusions

Soil change is a general term that can apply to small localized disturbances and responses or to broad landscape scale disruptions. Human management impacts soil properties across all kinds of soils and land uses. Land-use conversion is often accompanied by soil change, but management decisions within individual land uses can have meaningful impacts on soil properties. Tillage has produced notable changes in soil physical structure and carbon cycles across the USA croplands. Wetland soils have been changed by drainage, sedimentation, and changes in sea level. Forest silviculture often results in changes to the forest floor and surface soil horizons. Grazing impacts are highly dependent on the specific conditions of a location and the management system in place, inputs, and disruptions. Soils interact with ecosystems through feedbacks with plants, animals, and water. Some systems are resistant to change, some are resilient and can recover after disruption, and still others will undergo catastrophic change when disturbed. Anthropogenic soil change should continue to be studied and assessed to allow for better quantification of the impacts on soil function and ecosystem services.

References

Adam P (2002) Saltmarshes in a time of change. Environ Conserv 29 (01):39–61

Alekseev A, Alekseeva T, Ostroumov V, Siegert C, Gradusov B (2003) Mineral transformations in permafrost-affected soils, North Kolyma Lowland, Russia. Soil Sci Soc Am J 67:596–605

Amaranthus MP, Trappe JM, Molina RJ (1989) Long-term forest productivity and the living soil. In: Perry DA, Meurisse R, Thomas B, Miller R, Boyle J, Means J, Perry CR, Powers RF (eds) Maintaining the long-term productivity of Pacific Northwest Forest ecosystems. Timber Press, Portland, pp 36–52

Amundson R, Jenny H (1991) The place of humans in the state factor theory of ecosystems and their soils. Soil Sci 151:99–109. doi:10.1097/00010694-199101000-00012

Andriesse W, van Mensvoort MEF (2006) Acid sulfate soils: distribution and extent. In: Lal R (ed) Encyclopedia of soil science, 2nd edn. Taylor and Francis Group, New York

Archer S, Boutton TW, Hibbard KA (2001) Trees in grasslands: biogeochemical consequences of woody plant expansion. In: Schultze ED, Harrison SP, Heimann M, Hollan EA, Lloyd J, Prentice IC, Schimel D (eds) Global biogeochemical cycles in the climate system. Academic Press, San Diego, pp 115–137

Arnold RW, Szabolcs I, Targulian VO (eds) (1990) Global soil change. Report of an IIASA-ISSS-UNEP task force on the role of soil in global change. International Institute for Applied System Analysis, Laxenburg, Austria

Augustine DJ, McNaughton SJ (1998) Ungulate effects on the functional species composition of plant communities: herbivore selectivity and plant tolerance. J Wildl Manag 1998:1165–1183

Baker JM, Ochsner TE, Venterea RT, Griffis TJ (2007) Tillage and soil carbon sequestration—what do we really know? Agric Ecosyst Environ 118(1):1–5

Ballard TM (2000) Impacts of forest management on northern forest soils. For Ecol Manag 133(1):37–42

Barbier EB, Hacker SD, Kennedy C, Koch EW, Stier AC, Silliman BR (2011) The value of estuarine and coastal ecosystem services. Ecol Monogr 81(2):169–193

Beach T (1994) The fate of eroded soil: sediment sinks and sediment budgets of agrarian landscapes in southern Minnesota, 1851–1988. Ann Assoc Am Geogr 84(1):5–28

Bellemare J, Motzkin G, Foster DR (2002) Legacies of the agricultural past in the forested present: an assessment of historical land-use effects on rich mesic forests. J Biogeogr 29:1401–1420

Belnap J (2006) The potential roles of biological soil crusts in dryland hydrologic cycles. Hydrol Process 20(15):3159–3178

Belnap J, Eldridge D (2003) Disturbance and recovery of biological soil crusts. Biological soil crusts: structure, function, and management. Springer, Berlin, pp 363–383

Belnap J, Gillette DA (1998) Vulnerability of desert biological soil crusts to wind erosion: the influences of crust development, soil texture, and disturbance. J Arid Environ 39(2):133–142

Belsky AJ (1992) Effects of grazing, competition, disturbance and fire on species composition and diversity in grassland communities. J Veg Sci 3(2):187–200

Bennie AT (1990) Growth and mechanical impedance. In: Waisel Y, Eshel A, Kafkafi U (eds) Plant roots. The hidden half. Marcel Dekker, New York, pp 393–414

Bidwell OW, Hole FD (1965) Man as a factor of soil formation. Soil Sci 99(1):65–72

Bierman PR, Howe J, Stanley-Mann E, Peabody M, Hilke J, Massey CA (2005) Old images record landscape change through time. GSA Today 15(4):1–6. doi:10.1130/1052-5173(2005)015

Bilotta GS, Brazier RE, Haygarth PM (2007) The impacts of grazing animals on the quality of soils, vegetation, and surface waters in intensively managed grasslands. Adv Agron 94:237–280

Binkley D, Fisher R (2012) Dynamics of forest soils. In: Binkley D, Fisher R (eds) Ecology and management of forest soils. Wiley, London, pp 189–288

Biondini ME, Patton BD, Nyren PE (1998) Grazing intensity and ecosystem processes in a northern mixed-grass prairie, USA. Ecol Appl 8:469–479

Blanco-Canqui H, Stone LR, Schlegel J, Lyon DL, Vigil MF, Mikha MM, Stahlman PW, Rice CW (2009) No-till induced increase in organic carbon reduces maximum bulk density of soils. Soil Sci Soc Am J 73(6):1871–1879

Blois JL, Williams JW, Fitzpatrick MC, Jackson ST, Ferrier S (2013) Space can substitute for time in predicting climate-change effects on biodiversity. Proc Natl Acad Sci 110(23):9374–9379

Bockheim JG, Everett LR, Hinkel KM, Nelson FE, Brown J (1999) Soil organic carbon storage and distribution in arctic tundra, Barrow, Alaska. Soil Sci Soc Am J 63:934–940

Bordovsky DG, Choudhary M, Gerard CJ (1999) Effect of tillage, cropping, and residue management on soil properties in the Texas rolling plains. Soil Sci 164:331–340

Bradshaw A (1997) Restoration of mined lands—using natural processes. Ecol Eng 8:255–269

Brander LM, Florax RJ, Vermaat JE (2006) The empirics of wetland valuation: a comprehensive summary and a meta-analysis of the literature. Environ Resource Econ 33(2):223–250

Briske DD, Fuhlendorf SD, Smeins FE (2003) Vegetation dynamics on rangelands: a critique of the current paradigms. J Appl Ecol 40 (4):601–614

Bronick CJ, Lal R (2005) Soil structure and management: a review. Geoderma 124:3–22

Brown MT, Tighe RE, McClanahan TR, Wolfe RW (1992) Landscape reclamation at a central Florida phosphate mine. Ecol Eng 1:323–354

Brown JR (1993) Sanborn field: a capsule of scientific agricultural history in central Missouri. Missouri Agric. Experiment Station, Columbia

Brown J (1997) Disturbance and recovery of permafrost terrain. In: Crawford RMM (ed) Disturbance and recovery in Arctic Lands. Kluwer Academic, Dordrecht, pp 167–178

Burkett LM, Bestelmeyer BT, Tugel AJ (2011) A field guide to pedoderm and pattern classes. Agriculture handbook, vol 1. http://jornada.nmsu.edu/files/FieldGuidePedodermPattern.pdf. Accessed 15 May 2015

Calvert DV, Ford HW (1973) Chemical properties of acid sulfate soils recently reclaimed from Florida marshland. Soil Sci Soc Am Proc 37:367–371

Cambardella CA, Elliott ET (1992) Particulate soil organic-matter changes across a grassland cultivation sequence. Soil Sci Soc Am J 56:777–783. doi:10.2136/sssaj1992.03615995005600030017x

Certini G (2005) Effects of fire on properties of forest soils: a review. Oecologia 143(1):1–10

Chambers JC, Brown RW, Williams BD (1994) Reclamation success on Idaho's phosphate mines. Restor Ecol 2(1):4–16

Cingolani AM, Noy-Meir I, Díaz S (2005) Grazing effects on rangeland diversity: a synthesis of contemporary models. Ecol Appl 15 (2):757–773

Cook BI, Miller RL, Seager R (2009) Amplification of the North American "Dust Bowl" drought through human-induced land degradation. Proc Natl Acad Sci 106(13):4997–5001

Costanza R, d'Arge R, de Groot R, Farber S, Grasso M, Hannon B, Limburg K, Naeem S, Paruelo J, O'Neill RV, Paruelo J, Raskin R, Sutton P, van den Belt M (1997) The value of the world's ecosystem services and natural capital. Nature 387:253–260

Coverdale TC, Brisson CP, Young W, Yin SF, Donnelly JP, Bertness MD (2014) Indirect human impacts reverse centuries of carbon sequestration and salt marsh accretion. PLoS One 9(3): e93296

Craft CB, Casey WP (2000) Sediment and nutrient accumulation in floodplain and depressional freshwater wetlands of Georgia, USA. Wetlands 20(2):323–332

Craft CB, Megonigal P, Broome S, Stevenson J, Freese R, Cornell J, Zheng J, Sacco J (2003) The pace of ecosystem development of constructed *Spartina alterniflora* marshes. Ecol Appl 13(5):1417–1432

Daniels RB, Gilliam JW, Cassel DK, Nelson LA (1985) Soil erosion class and landscape position in the North Carolina Piedmont. Soil Sci Soc Am J 49(4):991–995

Dao TH (1993) Tillage and winter wheat residue management effects on water infiltration and storage. Soil Sci Soc Am J 57(6):1586–1595

Davis M, Brown MT, Doherty S, Tighe RE (1991) Techniques and guidelines for reclamation of phosphate mined lands. In Brown MT, Tighe RE (eds) Florida Institute Phosphate Research publication no. 03-044-095

DeBano LF, Neary DG, Folliott PF (1998) Fire effects on ecosystems. Wiley, London

Deegan LA, Johnson DS, Warren RS, Peterson BJ, Fleeger JW, Fagherazzi S, Wollheim WM (2012) Coastal eutrophication as a driver of salt marsh loss. Nature 490(7420):388–392

DeLaune RD, Baumann RH, Gosselink JG (1983) Relationships among vertical accretion, coastal submergence, and erosion in a Louisiana Gulf Coast marsh. J Sediment Res 53:1

Demas SY, Halt AM, Fanning DC, Rabenhorst MC, Dzantor EK (2004) Acid sulfate soils in dredged materials from tidal Pocomoke Sound in Somerset County, MD, USA. Aust J Soil Res 42:537–544

Dent DL, Pons LJ (1995) A world perspective on acid sulphate soils. Geoderma 67(3–4):263–276

Derr M (1998) Some kind of paradise: a chronicle of man and the land in Florida. University Press of Florida, Gainesville

Deverel S, Rojstaczer S (1996) Subsidence of agricultural lands in the Sacramento-San Joaquin Delta, California: role of aqueous and gaseous carbon fluxes. Water Resour Res 32(8):2359–2367

DiTomaso JM (2000) Invasive weeds in rangelands: species, impacts, and management. Weed Sci 48(2):55–265

Doran JW (1980) Soil microbial and biochemical changes associated with reduced tillage. Soil Sci Soc Am J 44(4):765–771

Doran JW, Parkin TB (1994) Defining and assessing soil quality. SSSA Spec Publ 35:3–3

Dregne HE (2011) Soils of arid regions. Elsevier, Amsterdam

Drexler JZ, Fontaine C, Deveret S (2009) The legacy of wetland drainage on the remaining peat in the Sacramento-San Joaquin Delta, California, USA. Wetlands 29(1):372–386

Drohan JR, Sharpe WE (1997) Long-term changes in forest soil acidity in Pennsylvania, USA. Water Air Soil Pollut 95(1–4):299–311

Duniway MC, Palmquist E, Miller ME (2015) Evaluating rehabilitation efforts following the Milford Flat Fire: successes, failures, and controlling factors. Ecosphere 6:art80

Ehlers W, Köpke U, Hesse F, Böhm W (1983) Penetration resistance and root growth of oats in tilled and untilled loess soil. Soil Tillage Res 3(3):261–275

Elliot WJ, Page-Dumroese D, Robichaud PR (1998) The effects of forest management on erosion and soil productivity. In: Lal R (ed) Soil quality and soil erosion. Soil and Water Conservation Society, CRC Press LLC, Boca Raton, p 195

Fanning D (2004) Acid sulfate soils. In: Muckel GB (ed) Understanding soil risks and hazards: using soil to identify areas with risks and hazards to human life and property. United States Department of Agriculture, USA

Fanning D (2006) Acid sulfate soils. In: Lal R (ed) Encyclopedia of soil science, 2nd edn. Taylor and Francis Group, New York

Fanning D, Burch SN (1997) Acid sulfate soils and some associated environmental problems. Adv Geoecol 30:145–158

FDEP (2010) Rate of reclamation report. Florida Department of Environmental Protection. http://www.dep.state.fl.us/water/mines/docs/2010ror.pdf. Accessed 1 July 2014

Fenton TE (2012) The impact of erosion on the classification of Mollisols in Iowa. Can J Soil Sci 92(3):413–418

Ferguson IM, Maxwell RM (2012) Human impacts on terrestrial hydrology: climate change versus pumping and irrigation. Environ Res Lett 7(4):044022

Forbes BC (1999) Land use and climate change on the Yamal Peninsula of north-west Siberia: some ecological and socio-economic implications. Polar Res 18:367–373

Fouli Y, Cade-Menum BJ, Cutforth HW (2013) Freeze thaw cycles and soil water content effects on infiltration rate of three Saskatchewan soils. Can J Soil Sci 93:485–496

Fox TR, Burger JA, Kreh RW (1986) Effects of site preparation on nitrogen dynamics in the southern Piedmont. For Ecol Manag 15 (4):241–256

Franzluebbers AJ, Stuedemann JA, Schomberg HH, Wilkinson SR (2000) Soil organic C and N pools under long-term pasture management in the Southern Piedmont USA. Soil Biol Biochem 32 (4):469–478

Freeland JA, Richardson JL, Foss LA (1999) Soil indicators of agricultural impacts on northern prairie wetlands: Cottonwood Lake Research Area, North Dakota, USA. Wetlands 19(1):56–64

Froede CR Jr, Williams EL (2004) The origin, development, and eventual consolidation of the canyons comprising Providence Canyon State Park Stewart County, Georgia. Southeast Geol 43 (1):39–50

Gedan K, Silliman BR, Bertness MD (2009) Centuries of human-driven change in salt marsh ecosystems. Mar Sci 1:117–141

Gee GW, Bauer A, Decker RS (1978) Physical analyses of overburden materials and mine land soils. Reclamation of drastically disturbed lands. American Society of Agronomy, Madison, pp 665–686

Gent JA, Ballard R, Hassan AE, Cassel DK (1984) Impact of harvesting and site preparation on physical properties of Piedmont forest soils. Soil Sci Soc Am J 48(1):173–177

Gilliam JW (1994) Riparian wetlands and water quality. J Environ Qual 23(5):896–900

Gopal B (2000) Wetlands and agriculture: Are we heading for confrontation or conservation. http://www.srs.fs.usda.gov/pubs/20153/. Accessed 14 Jan 2015

Goudie AS (2013) The human impact on the natural environment: past, present, and future. Wiley, London

Goudie AS, Middleton NJ (1992) The changing frequency of dust storms through time. Clim Change 20(3):197–225

Grass LB, Aronovici VS, Mucket DN (1962) Some chemical characteristics of submerged and reclaimed sediments of the San Francisco Bay System. Proc Soil Sci Soc Am 26:453–455

Grattan S (ed) (2002) Irrigation water salinity and crop production. UCANR Publications

Gray LC (1933) History of agriculture in the southern United States to 1860. The Carnegie Institution, Washington

Greenwood KL, McKenzie BM (2001) Grazing effects on soil physical properties and the consequences for pastures: a review. Anim Prod Sci 41(8):1231–1250

Grigal DF (2000) Effects of extensive forest management on soil productivity. For Ecol Manag 138(1):167–185

Grünzweig JM, Stephen D, Sparrow S, Yakir D, Chapin FS III (2004) Impact of agricultural land-use change on carbon storage in boreal Alaska. Glob Change Biol 10:452–472

Grünzweig JM, Sparrow SD, Chapin FS III (2003) Impact of forest conversion to agriculture on carbon and nitrogen mineralization in subarctic Alaska. Biogeochemistry 64:271–296

Guzman J, Al-Kaisi M (2010) Soil carbon dynamics and carbon budget of newly reconstructed tall-grass prairies in south central Iowa. J Environ Qual 39:136–146

Hale CM, Frelich LE, Reich PR, Pastor J (2005) Effects of European earthworm invasion on soil characteristics in northern hardwood forests of Minnesota, USA. Ecosystems 8:911–927

Hall SJ, Trujillo J, Nakase D, Strawhacker C, Kruse-Peeples M, Schaafsma H, Briggs J (2013) Legacies of prehistoric agricultural practices within plant and soil properties across an arid ecosystem. Ecosystems 16(7):1273–1293

Hansen ZK, Libecap GD (2003) Small farms, externalities, and the Dust Bowl of the 1930s (No. w10055). National Bureau of Economic Research

Hartman MD, Merchant ER, Parton WJ, Gutmann MP, Lutz SM, Williams SA (2011) Impact of historical land-use changes on greenhouse gas exchange in the US Great Plains, 1883–2003. Ecol Appl 21(4):1105–1119

Hartmann M, Howes CG, Van Insberghe D, Yu H, Bachar D, Christen R, Nilsson R, Hallam SJ, Mohn WW (2012) Significant and persistent impact of timber harvesting on soil microbial communities in Northern coniferous forests. ISME J 6(12):2199–2218

Hawkins GJ (1973) Physical chemical, and mineralogical properties of phosphatic clay slimes from the bone valley formation. Master's thesis, University of Florida

Herrick JE (2000) Soil quality: an indicator of sustainable land management. Appl Soil Ecol 15(1):75–83

Herrick JE, Beh A (2015) A risk-based strategy for climate change adaptation in dryland systems based on an understanding of potential production, soil resistance and resilience, and social stability. Sustainable intensification to advance food security and enhance climate resilience in Africa. Springer, Berlin, pp 407–424

Herrick JE, Whitford WG, De Soyza AG, Van Zee JW, Havstad KM, Seybold CA, Walton M (2001) Field soil aggregate stability kit for soil quality and rangeland health evaluations. Catena 44(1):27–35

Hilgard EW (1860) Report on the geology and agriculture of the state of Mississippi. E. Barksdale, Jackson

Hobbie JE, Carpenter SR, Grimm NB, Gosz JR, Seastedt TR (2003) The US long term ecological research program. Bioscience 53:21–32

Holdsworth AR, Frelich LE, Reich PB (2007) Effects of earthworm invasion on plant species richness in northern hardwood forests. Conserv Biol 21:997–1008

Holland EA, Parton WJ, Detling JK, Coppock DL (1992) Physiological responses of plant populations to herbivory and their consequences for ecosystem nutrient flow. Am Nat 140(4):685–706

Homburg JA, Sandor JA (2011) Anthropogenic effects on soil quality of ancient agricultural systems of the American Southwest. Catena 85(2):144–154

Hornbeck R (2009) The enduring impact of the American Dust Bowl: short and long-run adjustments to environmental catastrophe (No. w15605). National Bureau of Economic Research

Hubbs MD, Reeves DW, Mitchell CC (1998) Measuring soil quality on the 'old rotation'. Special reports-University of Arkansas Agricultural Experiment Station, vol 186, pp 50–54

Hyatt JA, Gilbert R (2000) Lacustrine sedimentary record of human-induced gully erosion and land use change at Providence Canyon, southwest Georgia, USA. J Paleolimnol 23:421–438

Indorante SJ, Kabrick JM, Lee BD, Maatta JM (2014) Quantifying soil profile change caused by land use in Central Missouri Loess Hillslopes. Soil Sci Soc Am J 78(1):225–237

Izaurralde RC, Williams JR, McGill WB, Rosenberg NJ, Jakas MC (2006) Simulating soil C dynamics with EPIC: model description and testing against long-term data. Ecol Model 192:362–384

James HR, Fenton TE (1993) Water tables in paired artificially drained and undrained soil catenas in Iowa. Soil Sci Soc Am J 57(3):774–781

Jandl R, Lindner M, Vesterdal L, Bauwens B, Baritz R, Hagedorn F, Johnson DW, Minkkinen K, Byrne KA (2007) How strongly can forest management influence soil carbon sequestration? Geoderma 137(3):253–268

Jenny H (1980) The soil resource, origin and behavior. Springer, New York. ISBN 0-387-90453-X

Jangid K, Williams MA, Franzluebbers AJ, Schmidt TM, Coleman DC, Whitman WB (2011) Land-use history has a stronger impact on soil microbial community composition than above ground vegetation and soil properties. Soil Biol Biochem 43(10):2184–2193

Johnson WC, Cantwell HD, Dean TA (1980) The impact of agricultural settlement on Canadian Sandy Creek, Oklahoma. In: Proceedings of the Oklahoma Academy of Science, vol 60

Johnston CA (1991) Sediment and nutrient retention by freshwater wetlands: effects on surface water quality. Crit Rev Environ Sci Technol 21(5-6):491–565

Jurgensen MF, Harvey AE, Graham RT, Page-Dumroese DS, Tonn JR, Larsen MJ, Jain TB (1997) Review article: Impacts of timber harvesting on soil organic matter, nitrogen, productivity, and health of inland northwest forests. For Sci 43(2):234–251

Kallio A, Rieger S (1969) Recession of permafrost in a cultivated soil of interior Alaska. Soil Sci Soc Am J 33(3):430–432

Karlen DL, Mausbach MJ, Doran JW, Cline RG, Harris RF, Schuman GE (1997) Soil quality: a concept, definition and framework for evaluation. Soil Sci Soc Am J 61:4–10. doi:10.2136/sssaj1997.03615995006100010001x

Kearney MS, Rogers AS, Townshend JR, Rizzo E, Stutzer D, Stevenson J, Sundborg K (2002) Landsat imagery shows decline of coastal marshes in Chesapeake and Delaware Bays. Eos Trans Am Geophys Union 83(16):173–178

Kelly RH, Parto WJ, Crocker GJ, Grace PR, Klfre J, Körschens M, Poulton PR, Richter DD (1997) Simulating trends in soil organic carbon in long-term experiments using the century model. Geoderma 81:75–90

Kennish MJ (2001) Coastal salt marsh systems in the US: a review of anthropogenic impacts. J Coast Res 17:731–748

Kirwan ML, Guntenspergen D, Alpaos A, Morris JR, Mudd SM, Temmerman S (2010) Limits on the adaptability of coastal marshes to rising sea level. Geophys Res Lett 37:L23401

Kirwan ML, Megonigal JP (2013) Tidal wetland stability in the face of human impacts and sea-level rise. Nature 504:53–60

Kladivko E (2001) Tillage systems and soil ecology. Soil Tillage Res 61(1–2):61–76. doi:10.1016/S0167-1987(01)00179-9

Knutson KC, Pyke DA, Wirth TA, Arkle RS, Pilliod DS, Brooks ML, Chambers JC, Grace JB (2014) Long-term effects of seeding after wildfire on vegetation in Great Basin shrubland ecosystems. J Appl Ecol 51:1414–1424

Kohake DJ, Hagen LJ, Skidmore EL (2010) Wind erodibility of organic soils. Soil Sci Soc Am J 74:250–257

Konen ME (1999) Human impacts on soils and geomorphic processes on the Des Moines Lobe. Iowa State University, Iowa

Lal R (1993) Tillage effects on soil degradation, soil resilience, soil quality, and sustainability. Soil Tillage Res 27:1–8

Lampurlanés J, Cantero-Martínez C (2003) Soil bulk density and penetration resistance under different tillage and crop management systems and their relationship with barley root growth. Agron J 95 (3):526–536

Lawrence GB, Fernandez JJ, Richter DD, Ross DS, Hazlett PW, Bailey SW, Ouimet R, Warby RAF, Johnson AH, Lin HS, Kaste JM, Lapenis AG, Sullivan TJ (2013) Measuring environmental change in forest ecosystems by repeated soil sampling: a North American perspective. J Environ Qual 42:623–639

Levi MR, Shaw JN, Wood CW, Hermann SM, Carter EA, Feng Y (2010) Land management effects on near-surface soil properties of southeastern US coastal plain Kandiudults. Soil Sci Soc Am J 74 (1):258–271

Liebig MA, Gross JR, Kronberg SL, Hanson JD, Frank AB, Phillips RL (2006) Soil response to long-term grazing in the northern Great Plains of North America. Agric Ecosyst Environ 115 (1):270–276

Lindbo DL, Moebius-Clune B, Hatfield J, et al (2014) Creating common ground: activities of the soil health dialog workgroup. In: EGU general assembly conference abstracts, vol 16, p 14555. http://adsabs.harvard.edu/abs/2014EGUGA.1614555L. Accessed 30 July 2014

Lowrance R, Altier LS, Williams RG, Inamdar SP, Sheridan JM, Bosch DD, Hubbard RK, Thomas DL (2000) REMM: the riparian ecosystem management model. J Soil Water Conserv 55(1):27–34

Lueking MA, Schepers JS (1985) Changes in soil carbon and nitrogen due to irrigation development in Nebraska's Sandhills soils. Soil Sci Soc Am J 49:626–630

MacDougall AS, Turkington R (2005) Are invasive species the drivers or passengers of change in degraded ecosystems? Ecology 86 (1):42–55

Mahaney WM, Wardrop DH, Brooks RP (2005) Impacts of sedimentation and nitrogen enrichment on wetland plant community development. Plant Ecol 175(2):227–243

Marchenko S, Etzelmüller B (2013) Permafrost: formation and distribution, thermal and mechanical properties. In: Shroder J (ed) Giardini R, Harbor J (eds) Treatise on geomorphology. Glacial and periglacial geomorphology, vol 8. Academic Press, San Diego, pp 202–222

Marion WR (1986) Phosphate mining regulations, reclamation, and revegetation. Florida Institute of Phosphate Research, Bartow, FL. Publication no. 03-043-040

McIntyre S, Lavorel S (2007) A conceptual model of land use effects on the structure and function of herbaceous vegetation. Agric Ecosyst Environ 119(1):11–21

McLauchlan K (2006) The nature and longevity of agricultural impacts on soil carbon and nutrients: a review. Ecosystems 9(8):1364–1382

McLaughlin JW, Gale MR, Jurgensen MF, Trettin CC (2000) Soil organic matter and nitrogen cycling in response to harvesting, mechanical site preparation, and fertilization in a wetland with a mineral substrate. For Ecol Manag 129(1):7–23

McVay KA, Budde JA, Fabrizzi K, Mikha MM, Rice CW, Schlegel AJ, Peterson DE, Sweeney DW, Thompson C (2006) Management effects on soil physical properties in long-term tillage studies in Kansas. Soil Sci Soc Am J 70:434–438

Megahan WF (1990) Erosion and site productivity in western Montane forest ecosystems. In: Harvey AE, Neuenschwander LG (eds) Proceedings of the management and productivity of Western Montane forest soils, Boise, ID, April 10–12, 1990. USDA Forest Service General Technical Report INT-280, pp 146–150

Michael AM (2008) Irrigation: theory and practice. Vikas Publishing House Pvt Ltd, India

Mikha MM, Vigil MF, Benjamin JG (2012) Long-term tillage impacts on soil aggregation and carbon dynamics under wheat-fallow in the central Great Plains. Soil Sci Soc Am J 77:594–605

Miller ME, Bowker MA, Reynolds RL, Goldstein HL (2012) Post-fire land treatments and wind erosion—lessons from the Milford Flat Fire, UT, USA. Aeol Res 7:29–44

Mislevy P, Blue WG (1981) Reclamation of quartz sand-tailings from phosphate mining: I. Tropical forage grasses. J Environ Qual 10(4):449–453

Mislevy P, Blue WG, Roessler CE (1990) Productivity of clay tailings from phosphate mining: II. Forage crops. J Environ Qual 19:694–700

Mitchell CC, Delaney DP, Balkcom KS (2008) A historical summary of Alabama's old rotation (circa 1896): the world's oldest, continuous cotton experiment. Agron J 100(5):1493–1498

Monger HC, Bestelmeyer BT (2006) The soil-geomorphic template and biotic change in arid and semi-arid ecosystems. J Arid Environ 65:207–218

Moorhead DL, Linkins AE, Everett KR (1996) Road dust alters extracellular enzyme activities in tussock tundra soils, Alaska, USA. Arct Alp Res 28(3):346–351

Morgan RPC (2009) Soil erosion and conservation. Wiley, London

Morris JT (1991) Effects of nitrogen loading on wetland ecosystems with particular reference to atmospheric deposition. Annu Rev Ecol Syst 22:257–279

Nair VD, Graetz DA, Reddy KR, Olila OG (2001) Soil development in phosphate-mined created wetlands of Florida, USA. Wetlands 21 (2):232–239

Neff JC, Reynolds RL, Belnap J, Lamothe P (2008) Multi-decadal impacts of grazing on soil physical and biogeochemical properties in southeast Utah. Ecol Appl 15:87–95

Nelson FE, Anisimov OA, Shiklomanov NI (2002) Climate change and hazard zonation in the circum-arctic permafrost regions. Nat Hazards 26:203–225. doi:10.1023/A:1015612918401

Nielsen DC, Unger PW, Miller PR (2005) Efficient water use in dryland cropping systems in the Great Plains. Agron J 97(2):364–372

Nir D (1983) Man, a geomorphological agent: an introduction to anthropic geomorphology. Springer, Berlin

Norton LD (1986) Erosion-sedimentation in a closed drainage basin in northwest Indiana. Soil Sci Soc Am J 50(1):209–213

Odell RT, Walker WM, Boone LV, Oldham MG (1982) The morrow plots: a century of learning. Agricultural Experiment Station, College of Agriculture, University of Illinois Bulletin 775, Urbana-Champaign, IL

Page-Dumroese DS, Jurgensen MF, Harvey AE, Graham RT, Tonn JR (1997) Soil changes and tree seedling response associated with site preparation in northern Idaho. West J Appl For 12(3):81–88

Palm C, Sanchez P, Ahamed S, Awiti A (2007) Soils: a contemporary perspective. Annu Rev Environ Resour 32:99–129

Parent LE, Millette JA, Mehuys GR (1982) Subsidence and erosion of a Histosol. Soil Sci Soc Am J 46:404–408

Parkinson D, Coleman DC (1981) Microbial communities, activity and biomass. Agric Ecosyst Environ 34:3–33

Pellant M, Shaver P, Pyke D, Herrick JE (2005) Interpreting indicators of rangeland health, version 4. Technical reference 1734-6. US Department of the Interior, Bureau of Land Management, National Science and Technology Center, Denver, CO. BLM/WO/ST-00/001 +1734/REV05

Peterson GA, Halvorson AD, Havlin JL, Jones O, Lyon DJ, Tanaka DL (1998) Reduced tillage and increasing cropping intensity in the Great Plains conserves soil C. Soil Tillage Res 47(3):207–218

Peterson GA, Lyon DJ, Fenster CR (2012) Valuing long-term field experiments: quantifying the scientific contribution of a long-term tillage experiment. Soil Sci Soc Am J 76(3):757–765

Pickett STA (1989) Space-for-time substitution as an alternative to long-term studies. In: Likens GE (ed) Long-term studies in ecology: approaches and alternatives. Springer, New York, pp 110–135

Ping CL (2005) Cold region soils. In: Reference module in earth systems and environmental sciences, from encyclopedia of soils in the environment. Elsevier, London, pp 268–276

Ping CL (2013) Gelisols: part I. Cryogenesis and state factors of formation. Soil Horizons 54(3):1–5

Ping CL, Michaelson GJ, Kimble JM (1997) Carbon storage along a latitudinal transect in Alaska. Nutr Cycl Agroecosyst 49:235–242

Presley DR, Ransom MD, Kluitenberg GJ, Finnell PR (2004) Effects of thirty years of irrigation on the genesis and morphology of two semiarid soils in Kansas. Soil Sci Soc Am J 68:1916–1926

Preston TM, Sojda RS, Gleason RA (2013) Sediment accretion rates and sediment composition in Prairie Pothole wetlands under varying land use practices, Montana, United States. J Soil Water Conserv 68 (3):199–211

Prieto GS, Reeves DW, Shaw JN, Mitchell CC (2002) Impact of conservation tillage on soil carbon in the 'Old Rotation'. In: Proceedings of 25th annual conservation tillage conference, vol 25, pp 277–282

Pyke DA, Herrick JE, Shaver P, Pellant M (2002) Rangeland health attributes and indicators for qualitative assessment. J Range Manag 55(6):584–597

Rabenhorst MC (2001) Soils of tidal and fringing wetlands. In: Richardson JL, Vepraskas MJ (eds) Wetland soils: genesis, hydrology, landscapes, and classification. CRC Press, Boca Raton, pp 301–316

Rasmussen PE, Albrecht SL, Smiley RW (1998) Soil C and N changes under tillage and cropping systems in semi-arid Pacific Northwest agriculture. Soil Tillage Res 47(3):197–205

Ramankutty N, Foley JA (1999) Estimating historical changes in global land cover: Croplands from 1700 to 1992. Global Biogeochem Cycles 13:997–1027

Ramankutty N, Heller E, Rhemtulla J (2010) Prevailing myths about agricultural abandonment and forest regrowth in the United States. Ann Assoc Am Geogr 100(3):502–512

Reicosky DC, Saxton KE (2007) The benefits of no-tillage. In: Baker CJ, Saxton KE (eds) No-tillage seeding in conservation agriculture. Food and Agriculture Organization of the United Nations, Rome, pp 11–20

Rhoton FE, Shipitalo MJ, Lindbo DL (2002) Runoff and soil loss from midwestern and southeastern US silt loam soils as affected by tillage practice and soil organic matter content. Soil Tillage Res 66(1):1–11

Richter DdeB, Bacon AR, Megan LM, Richardson CJ, Andrews SS, West L, Wills S, Billings CA, Cambardella CA, Cavallaro N, DeMeester JE, Franzluebbers AJ, Grandy AS, Grunwald S, Gruver Hartshorn AS, Janzen H, Kramer MG, Ladha JK, Lajtha K, Liles GC, Markewitz D, Megonigal PJ, Mermut AR, Rasmussen C, Robinson DA, Smith P, Stiles CA, Tate RL, Thompson A, Tugel AJ, Van Es H, Yaalon D, Zobeck TM (2011) Human-soil relations are changing rapidly: proposals from SSSA's cross-divisional soil change working group. Soil Sci Soc Am J 75:2079–2084

Richter DdeB, Markewitz D (2001) Understanding soil change: soil sustainability over millennia, centuries, and decades. Cambridge University Press, Cambridge

Robarge WP, Johnson DW (1992) The effects of acidic deposition on forested soils. Adv Agron 47:1–83

Robertson GP, Allen VG, Boody G, Boose ER, Creamer NB, Drinkwater LE, Gosz JR, Lynch L, Havlin JL, Jackson LE, Pickett STA, Pitelka L, Randall A, Reed SA, Seastedt TR, Waide RB, Wall DH (2008) Long-term agricultural research: a research, education, and extension imperative. Bioscience 58:640–645. doi:10.1641/B580711

Robinson DA, Hockley N, Dominati E, Lebron I, Scow KM, Reynolds B, Emmett BA, Keith AM, de Jonge, LW, Schjonning P, Moldrup P, Jones SB, Tuller M (2012) Natural capital, ecosystem services, and soil change: Why soil science must embrace an ecosystems approach. Vadose Zone J 11(1): 0–0

Rokosch AE, Bouchard V, Fennessy S, Dick R (2009) The use of soil parameters as indicators of quality in forested depressional wetlands. Wetlands 29(2):666–677

Ross MR, Castle SC, Barger NN (2012) Effects of fuels reductions on plant communities and soils in a Piñon-juniper woodland. J Arid Environ 79:84–92

Sainju UM, Lenssen A, Caesar-Tonthat T, Waddell J (2006) Tillage and crop rotation effects on dryland soil and residue carbon and nitrogen. Soil Sci Soc Am J 70:668–678

Sandor JA, Gersper PL, Hawley JW (1986) Soils at prehistoric agricultural terracing sites in New Mexico: I. Site placement, soil morphology, and classification. Soil Sci Soc Am J 50:166–173. doi:10.2136/sssaj1986.03615995005000010031x

Scavia D, Field JC, Boesch DF, Buddemeier RW, Burkett V, Cayan DR, Fogarty M, Harwell MA, Howarth RW, Reed DJ, Royer TC, Sallenger AH, Titus JG (2002) Climate change impacts on US coastal and marine ecosystems. Estuaries 25(2):149–164

Seybold CA, Herrick JE, Brejda JJ (1999) Soil resilience: a fundamental component of soil quality. Soil Sci 164(4):224–234

Seybold CA, Mersie W, Huang J, McNamee C (2002) Soil redox, pH, temperature, and water-table patterns of a freshwater tidal wetland. Wetlands 22(1):149–158

Sharratt B, Zhang M, Sparrow S (2006a) Twenty years of tillage research in subarctic Alaska II. Impact on soil strength, aggregation, roughness, and residue cover. Soil Tillage Res 91:75–81

Sharratt B, Zhang M, Sparrow S (2006b) Twenty years of conservation tillage in subarctic Alaska II. Impact on soil hydraulic properties. Soil Tillage Res 91:82–88

Sheridan RC (1979) Chemical fertilizers in southern agriculture. Agric Hist 53:308–318

Shih SF, Glaz B, Barnes RE Jr (1998) Subsidence of organic soils in the Everglades Agricultural Area during the past 19 years. Soil Crop Soc Fl Proc 57:20–29

Singer MJ, Ewing S (2000) Soil quality. In: Sumner ME (ed) Handbook of soil science. CRC Press, Boca Raton

Six JA, Elliott ET, Paustian K, Doran JW (1998) Aggregation and soil organic matter accumulation in cultivated and native grassland soils. Soil Sci Soc Am J 62(5):1367–1377

Six JA, Elliott ET, Paustian K (2000) Soil macroaggregate turnover and microaggregate formation: a mechanism for C sequestration under no-tillage agriculture. Soil Biol Biochem 32(14):2099–2103

Smith P, Smith JU, Powlson DS, McGil WB, Arah JRM, Chertov OG, Coleman K, Franko U, Frolking S, Jenkinson DS, Jensen LS, Kelly RH, Klein-Gunnewiek H, Komarov AS, Li C, Molina JAE, Mueller T, Parton WJ, Thornley JHM, Whitmore AP (1997) A comparison of the performance of nine soil organic matter models using datasets from seven long-term experiments. Geoderma 81:153–225

Soil Survey Staff (2007) Soil extent mapping tool. Natural Resources Conservation Service and Pennsylvania State University. http://apps.cei.psu.edu/soiltool/. Accessed 16 Jan 2015

Soil Survey Staff (1999) Soil taxonomy: a basic system of soil classification for making and interpreting soil surveys, 2nd edn. Natural Resources Conservation Service. U.S. Department of Agriculture Handbook 436

Sparrow SD, Lewis CL, Knight CW (2006) Soil quality response to tillage and crop residue removal under subarctic conditions. Soil Tillage Res 91:15–21

Spedding TA, Hamel C, Mehuys GR, Madramootoo CA (2004) Soil microbial dynamics in maize-growing soil under different tillage and residue management systems. Soil Biol Biochem 36(3):499–512

Stephens JC, Johnson L (1951) Subsidence of organic soils in the upper Everglades region of Florida. Soil Sci Soc Fl Proc 11:191–237

Swanson FJ, Clayton JL, Megahan WF, Bush G (1989) Erosional processes and long-term site productivity. In: Perry DA, Meurisse R, Thomas B, Miller R, Boyle J, Means J, Perry CR, Powers RF (eds) Maintaining the long-term productivity of pacific northwest forest ecosystems. Timber Press, Portland, pp 67–81

Tarnocai C (2009) Arctic permafrost soils. In: Margesin R (ed) Permafrost soils. Soil biology series, vol 16. Springer, Berlin, pp 3–16

Tate KW, Dudley DW, McDougald NK, George MR (2004) Effect of canopy and grazing on soil bulk density. Rangel Ecol Manag 57 (4):411–417

Torsvik V, Øvreås L (2002) Microbial diversity and function in soil: from genes to ecosystems. Curr Opin Microbiol 5(3):240–245

Trimble SW (1974) Man-induced soil erosion on the southern Piedmont, 1700–1970. Soil Conservation Society of America, Madison

Trimble SW, Mendel AC (1995) The cow as a geomorphic agent—a critical review. Geomorphology 13(1):233–253

Tugel AJ, Herrick JE, Brown JR, Mausbach MJ, Puckett W, Hipple K (2005) Soil change, soil survey, and natural resources decision making. Soil Sci Soc Am J 69:738–747. doi:10.2136/sssaj2004.0163

Tugel AJ, Wills SA, Herrick JE (2008) Soil change guide. In: Procedures for soil survey and resource inventory. Version 1.1. National Soil Survey Center, Lincoln, NE

Turner RE (2011) Beneath the salt marsh canopy: loss of soil strength with increasing nutrient loads. Estuar Coasts 34(5):1084–1093

USDA-National Agriculture Statistics Service (NASS) (2007) Census of agriculture: volume 3 special studies part 1. http://www.nass.usda.gov/research/2007mapgallery/album/Farms/Land_in_Farms_and_Land_Use/slides/Acres%20of%20Irrigated%20Land.html. Accessed online 15 Jan 2015

USGS (2014a) Mineral resources online spatial data.US Geological Survey. http://mrdata.usgs.gov/mineral-resources/phosphate.html. Last modified 3 September 2014. Accessed 26 Nov 2014

USGS (2014b) Mineral commodity summaries: phosphate rock. US Geological Survey, Reston, pp 118–119

van Auken OW (2000) Shrub invasions of North American semiarid grasslands. Annu Rev Ecol Syst 31:197–215

Vitousek PM, Bolin B, Cook RB (1983) The effects of deforestation on air, soil, and water. In: Bolin B, Cook RB (eds) Major Biogeochem Cycles Interact. John Wiley & Sons, Chichester, pp 223–245

Walker DA, Everett KR (1987) Road dust and its environmental impact on Alaskan taiga and tundra. Artic Alpine Res 19(4):479–489

Wallace PM, Best GR (1983) Enhancing ecological succession: 6. Succession of endomycorrhizal fungi on phosphate strip mined lands. In: Proceedings of the 1983 symposium on surface mining, hydrology, sedimentology, and reclamation. University of Kentucky, Lexington

Wander MM, Hedrick DS, Kaufman D, Traina SJ, Stinner BR, Kehrmeyer SR, White DC (1995) The functional significance of the microbial biomass in organic and conventionally managed soils. Plant Soil 170:87–97

Wardle DA, Yeates GW, Nicholson KS, Bonner KI, Watson RN (1999) Response of soil microbial biomass dynamics, activity and plant litter decomposition to agricultural intensification over a seven year period. Soil Biol Biochem 31:1707–1720

Webb BB, Tucker BB, Westerman RL (1980) The Magruder Plots: taming the prairies through research. Agricultural Experimental Station, Oklahoma State University, Stillwater

Weir WW (1950) Subsidence of peat lands of the Sacramento-San Joaquin delta, California. Hilgardia 20:37–56

West TO, Post WM (2002) Soil organic carbon sequestration rates by tillage and crop rotation: a global data analysis. Soil Sci Soc Am J 66:1930–1946

Whitlock C, Shafer SL, Marlon J (2003) The role of climate and vegetation change in shaping past and future fire regimes in the northwestern US and the implications for ecosystem management. For Ecol Manag 178(1):5–21
Williams MA (2001) Influence of water on the carbon and nitrogen dynamics of annually-burned tallgrass prairie. Ph.D. dissertation, Kansas State University, Manhattan
Williams JR, Sharpley AN (eds) (1989) EPIC-erosion-productivity impact calculator: 1. Model documentation, USDA Tech. Bulletin No. (1768) USDA, Washington, DC
Williams JR, Izaurralde RC, Steglich EM (2008) Agricultural Policy/Environmental eXtender Model: Theoretical Documentation Version 0604. BREC Report No. 2008–17. Texas A&M University. Blackland Research and Extension Center, Temple, TX
Williams CO, Lowrance R, Bosch DD, Williams JR, Benham E, Dieppa A, Hubbard R, Mas E, Potter T, Sotomayor D, Steglich EM, Strickland T, Williams RG (2013) Hydrology and water quality of a field and riparian buffer adjacent to a mangrove wetland in Jobos Bay Watershed, Puerto Rico. Ecol Eng 56:60–68. doi:10.1016/j.ecoleng.2012.09.005
Wilson M, Hanlon EA (2012) Florida reclaimed phosphate mine soils: characteristics, potential uses, and management considerations. University of Florida Extension Publication #SL370
Wood CW, Edwards JH (1992) Agroecosystem management effects on soil carbon and nitrogen. Agric Ecosyst Environ 39:123–138
Wright JO (1907) Swamp and overflowed lands of the United States: ownership and reclamation, vol 76. United States Department of Agriculture, Government Printing Offices, USA
Yaalon DH (2007) Human-induced ecosystem and landscape processes always involve soil change. Bioscience 57:918–919
Yaalon DH, Yaron B (1966) Framework for man-made soil changes: an outline of metapedogenesis. Soil Sci 4(102):272–277
Zak DR, Grigal DF, Gleeson S, Tillman D (1990) Carbon and nitrogen cycling during old-field succession: constraints on plant and microbial biomass. Biogeochemistry 11:111–129
Zedler JB, Kercher S (2005) Wetland resources: status, trends, ecosystem services, and restorability. Annu Rev Environ Resour 30:39–74
Zhang M, Sparrow SE, Van Veldhuizen B, Masiak DT (2012) Soil quality under different land uses in a subarctic environment in Alaska. J Land Use Sci 7(1):109–121.

19 Future Challenges for Soil Science Research, Education, and Soil Survey in the USA

Patrick J. Drohan

19.1 Introduction

US soil scientists have succeeded tremendously in helping to lessen soil/ecosystem degradation and meet the commodity demands of an ever-growing world population. This is reflected in increased food production over time, improved management of many ecosystems, recent advances in human health due to access to food, and educational/discovery opportunities due to time freed from food production. However, while many of the nation's citizens are more educated and well-off as a result of these achievements, the majority of the US population does not have to manage their soil/land on a daily basis, and thus are disconnected from understanding the challenges of soil science and management.

Soil science is in a period of change. The popularity of soil science research, which could be argued to reflect state and federal priorities via funding support, appears strong. Baveye et al. (2006) and Baveye and Jacobson (2009) provide publishing trend data from several sources that show a steady growth in publishing of soil science-related topics, but not always by soil scientists. A check of the same databases used in these papers shows this trend is continuing (Fig. 19.1), which is good news because it reflects an ever-growing sphere of influence and interest in soil. However, the fact that a growing number of non-soil scientists are publishing this research has concerned some (Baveye et al. 2006). Increased interest in soil is also highlighted by US Senate Resolution 440, passed June 23, 2008 (Brown 2008), which recognized the importance of soil and need to continue studying soil. The Soil Science Society of America has been successful at increasing public awareness via the "Dig It" exhibit (Drohan et al. 2010; Megonigal et al. 2010) and popular press children's book "Know Soil, Know Life" (Lindbo et al. 2012). The Geological Society of America is also promoting soil science via a new Soils and Soil Processes Interdisciplinary Interest Group. In contrast, to this interest in soils though are frustrating trends in education reflecting fewer undergraduate and graduate students in soil science academic programs (Baveye et al. 2006; Collins 2008; Havlin et al. 2010) even though employers are seeking students to fill positions (Havlin et al. 2010). This is coupled with some US soil science programs losing academic positions, merging with other departments and losing the "soils" identity in the department name, and eliminating their undergraduate program (Baveye et al. 2006; Collins 2008; Havlin et al. 2010).

Soil scientists believe there is much we can still learn from studying soil, and are genuinely concerned as to whether we will be capable of maintaining the necessary depth of the discipline for scientific discovery and knowledge application. During the last 10 years, many soil scientists have addressed this concern and made their own suggestions for where soil science should place its efforts (see Hartemink 2006; Baveye 2006, 2015; Hopmans 2007; Baveye et al. 2011; Adewopo et al. 2014; and several of the above citations). This chapter presents several challenges that effect not just soil science but also our civilization's immediate future. They are listed in order of this author's perception of their importance. By addressing them in order, the problems most limiting the sustainability of our soil resource (and thus future of the USA) and the soil science profession can be more quickly and efficiently solved. Applicability of these points to other countries is not guaranteed given the state of soil science and the problems of those countries differ (Hartemink and McBratney 2008; Hartemink et al. 2008; Baveye et al. 2010).

P.J. Drohan (✉)
Pennsylvania State University, State College, PA, USA
e-mail: pjd@psu.edu

L.T. West et al. (eds.), *The Soils of the USA*, World Soils Book Series,
DOI 10.1007/978-3-319-41870-4_19

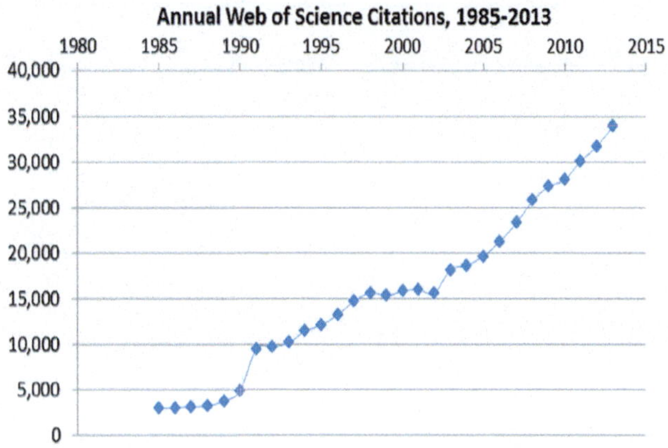

Fig. 19.1 Annual Web of Science citations involving the word "soil" between 1985 and 2013

19.2 Challenges to the Future of Soil Science

19.2.1 Challenge 1: Soil Erosion Is Still a Substantial Threat to US Food Production

The early efforts of the US Soil Survey, and broader SCS (now NRCS), to combat erosion and land degradation following the 1930s Dust Bowl (USDA-SCS 1992; Egan 2006) is an example of a highly successful effort to modify human behavior via science and education. This came about via the collaborative efforts of farmers, land managers, federal agencies, land grant universities, and local government.

The extensive 80+ years of conservation practices since the Dust Bowl have dramatically slowed, but not stopped, soil erosion. Soil scientists use a concept called "tolerable soil loss" (Johnson 1987) to examine erosion rates. In theory, some soil can be lost as long as more soil is made to replace it. Comparing calculations of the soil lost to soil produced allows a land manager to evaluate what land use practices need to be adjusted to slow erosion so that the soil loss is stabilized, or better yet, slowed so that soil building occurs. Today, many areas of the USA are still facing erosion in excess of soil building rates which can limit crop yields (Cruse et al. 2006; Verheijen et al. 2009), and simple physics dictates that at some point, the remaining effective soil depth will be too thin to maintain current crop yields.

Soil erosion also wastes energy. The effect of erosion on crops (lower yields) can be lessened by increasing fertilizer application. However, if improperly applied, nutrients carried in surface or subsurface water, or via erosion, can degrade stream water quality (Hatfield et al. 2013). Thus, maintaining yields on eroded lands results in more energy used in total via mining of materials to make more fertilizer. Efficiency improvements affecting agriculture, such as natural gas enabled Haber Bosch ammonia synthesis (Erisman et al. 2008), have in part facilitated this situation. Smil (2000) estimates 40–50 % of the Earth's population is alive because of the nitrogen produced from Haber Bosch ammonia synthesis. This is, in essence, the percent of the population living beyond what the land might produce without this technology. Cropping shifts to meet biofuel production goals, most notably the production of ethanol from corn (Donner and Kucharik 2008) (Fig. 19.2, Platteville, WI, USA), have further amplified energy waste and accelerated erosion. Corn (*Zea mays*) has been extensively replanted across the USA on many lands formerly enrolled in conservation protection programs. Corn is often planted at some point with tillage, and the erosion resulting from this tillage (even conservation tillage) may lead to more soil loss and nutrient-induced water pollution. Corn also typically uses only half of its annual nitrogen application. As energy becomes more expensive due to scarcity and rising demand, its products (fertilizer and food) become increasingly expensive. Given its relatively poor fertilizer usage efficiency, corn therefore appears to be quite a poor choice for biofuel production if "sustainability" is a goal, especially when the potential effect of erosion and resulting indirect energy waste are considered.

Erosion is not solely a "crop production issue." Rangeland erosion is still a substantial problem in the USA (Webb et al. 2014), especially during periods of drought. Urban and suburban development is also responsible for erosion of landscapes and degradation of water quality (Wolman 1967; Hogan et al. 2014). When urban and agricultural land management problems intersect, understanding the root causes of the resulting problem can be complex and solutions seem insurmountable. For example, summer of 2014 saw the buildup of microtoxins derived from blue-green algae in Lake Erie, which reached high enough levels to result in the shutdown of Toledo, Ohio's water supply (Lee 2014). Phosphorus surface runoff from agricultural fields was believed partly responsible, but other potential sources such as lawn fertilizers were also examined. Since the Lake Erie episode, phosphorus runoff from agricultural lands has been identified as a likely cause of the algae growth and millions of dollars are being invested (Jackson 2014) to track phosphorus runoff among other potential nutrient sources.

How can soil scientists begin to correct the problems of erosion? One solution is to begin managing the ecosystems receiving the agricultural pollutants rather than the producers (Donner and Kucharik 2008). This solution is still avoiding the direct cause of the problem, which is human behavior. Dr. William Mitsch suggests that geo-engineered riparian zones and wetlands can help reduce problems like agriculturally derived nitrogen (Mitsch et al. 2005, 2008). Across the Mississippi River Basin alone, implementation of these practices could result in 22,000 km^2 of created and restored wetlands. This greatly increases opportunities for soil

Fig. 19.2 Example of land disturbances affecting modern soils. *Clockwise from upper left* shale gas development in the Northern Appalachians; urban renewal in Cleveland, OH; Agricultural runoff in the Chesapeake Bay watershed; corn field along river near Platteville, WI

scientists to collaborate and help correct the problem. While geo-engineered riparian zones and wetlands will not stop erosion, they may solve water quality problems. Alternative solutions deal directly with adjusting human behavior causing erosion and encompass developing new models of soil science formal and public education (*Challenge* 2 below), and refocusing efforts on public education (*Challenge* 3). But ultimately, what will be required is an effort that reshapes agriculture to protect soil and its ecosystem services to society (*Challenge 4* below). This is especially true given the projected western droughts (Cook et al. 2015) and their ultimate effect on much of US agriculture.

19.2.2 Challenge 2: Sustaining Formal Soil Science Education and Training

In the past 100 years, the bulk of soil science and soil science education have been spearheaded by US land grant universities. The land grant institution has been, and still is to some degree, a key part of the success of soil science in the USA. Land grant universities, established under the Morrill Acts of 1862 and 1890, were originally put in place to further agricultural education and research, and today, they are still a model for agricultural extension. Land grant institutions have assisted greatly in carrying out the science and mapping of soil in conjunction with the USDA Soil Conservation Service (SCS) and later the USDA Natural Resources Conservation Service (NRCS) Soil Survey Program.

While excellent soil science is taught at other schools, fully staffed land grant university soil science departments have typically employed a soil mineralogist, physicist, chemist, pedologist, microbiologist, or ecologist, and often, multiples of these positions. They also have positions focused on applied aspects of the science that work collaboratively with crop scientists, agronomists, foresters, agricultural engineers, etc. Unfortunately, a combination of decreased soil science faculty, and the merger of soil science programs with other disciplines over the last 15 years, is resulting in a rapid decline in soil science educational expertise across much of the USA. Under the current model of US higher education, this will make it more difficult to produce well-rounded soil scientists with depth of expertise in soil science.

This decline in soil science faculty is due to multiple interlinked factors. Soil science faculty has been replaced by faculty in other academic fields. In some cases, these new hires are educated in a targeted field with more federal funding available than that available for soil science research (genomics for example), which not only helps fund research but also helps support the academic institution's infrastructure and improves ranking metrics for universities (Baveye and Jacobson 2008). Fewer students majoring in soil science may also have contributed to declining soil science faculty numbers (fewer faculties are needed to teach). In addition, there are fewer and smaller Land Grant Extension Programs today where soil scientists are staffed and fewer federal positions in which soil science graduates have been employed. Undergraduate student enrollment has shifted to new programs (Environmental Science or Management) (Collins 2008; Havlin et al. 2010) created in response to demand, and in part due to the boon in 1970s environmental legislation and shifting demographics from farm operations to manufacturing in urban–suburban areas. An example of this in the Penn State College of Agricultural Sciences is a student population of over 70 % from a non-farm background. Penn State is not a unique example.

Soil science, along with many other fields of the natural or physical sciences, may be at further risk to decline due to the cost of education far exceeding salaries an undergraduate soil science degree might secure. The cost of tuition at US colleges and universities rose 757 % between 1980 and 2009 (CCAP 2014). During the great recession, the cost of education rose 9.3 % for 4-year state schools, and tuition is expected to rise more as state support for state-funded institutions drops further (Barr and Turner 2013). A 4-year soil science or forestry degree at Penn State University (main campus) now costs an instate student ~$140,000 (out of state is ~$190,000) (Penn State University 2014). The anticipated starting salary for graduates from either of these two majors is between $35 and $50,000 per year. Higher education has made up for state shortfalls in funding by increasing enrollment of out-of-state and international students who pay higher tuition rates (Stephens 2013). The cost of education may further increase as investment companies remodel education (Denneen and Dretler 2012), and state funding remains low or continues to drop (Barr and Turner 2013). See Baveye and Jacobson (2008) for a discussion of broader research on the increasing investment in Academia. Finally, soil science may further decline simply because fewer students are forecasted to attend colleges and universities in some parts of the USA. This projected decrease in students attending college in many states is colloquially referred to as the "pupil cliff"; in some states, the drop will be >15 % (Kiley 2013).

Given these changes, it appears that soil science faces substantial challenges in remaining an independent discipline at many land grant universities. This is likely unavoidable given current momentum but will not mean the end of soil science training and education. The growth of soil science is possible, but will require new model(s) of soil science education very different from the current one.

19.2.2.1 Model 1: A Multi-institution Approach with Regional Specialization

This model would consist of a national curriculum offered online through the Soil Science Society of America or similar organization in consort with a few leading land grant universities located in different regions of the country. Introductory courses would be taught through the Soil Science Society of America (SSSA) by distance education methods with upper-level specialized courses taught by soil scientists in regional institutions. Perhaps a negative aspect of this model is that only a few schools would likely have a viable soil science program. Such a model could take advantage of distance education techniques to make up for this problem. Employing technology now commonly used in a variety of simulation computer programs and video games would dramatically enhance current soil science education. Frankly the possibilities for education and training are unlimited given the state of current technology.

19.2.2.2 Model 2: Industry, Land Grant School, Philanthropist, or the Federal Government Develops a One/Two-year Technical Program in Soil Science

Students entering such a program would be coming from a 4-year Environmental Management/Science or Geoscience backgrounds. A 1/2-year program would focus on the upper-level courses in soil science: mineralogy, pedology, chemistry, physics, biology. The student would take 15–20 credits in soil science-specific classes (classroom and laboratory) during the academic year with summer being an intensive field camp covering important field soil science skills. The end result of this process would be some type of certification as a soil scientist, perhaps through the SSSA Council of Soil Science Examiners (CSSE).

19.2.2.3 Model 3: Soil Science Education Evolves to Be Taught at Non-land Grant Public Institutions

This model would be difficult to implement due to problems associated with the current rising costs of college education. However, many non-land grant public universities are more affordable than larger land grant institutions, and can thus reach a broader segment of the population. The 4-year ratio of debt load to salary is more favorable to students at such schools. Examples of somewhat similar models exist now at the California Polytechnic State University in San Luis Obispo, University of Wisconsin, Platteville, and a few other institutions.

19.2.2.4 Model 4: Soil Science Taught at Vocational Schools

Several educational systems around the world limit the number of students who advance on to university education. This limitation may be due to limited government support of higher education and in other cases, may reflect the reality that not all citizens are qualified to continue their education at universities. Such systems use some testing at about 18 years of age to discern whether one goes on to university versus vocational/apprenticeship education. Vocational institutions teaching soil science could be regionally located and meet the demand for excellent soil scientists to fill many career niches. Soil scientists have prided themselves on being a scientist. Thus, this model may not be favorably looked upon due to potentially less science being taught in a vocational setting.

19.2.3 Challenge 3: Public Education Promoting Human Behavior that Protects Soil

Much is known about soil science, how to manage landscapes to minimize soil degradation, and how to educate people about soil. We also understand poor soil management has potentially weakened (Stevenson et al. 2015) and/or contributed to the collapse of prior civilizations (Diamond 2005). One of the greatest threats to soil, after erosion, is product waste.

Human behavior that results in the waste of products produced from soil is a substantial challenge for soil science and society and a reflection of its maturity. Food, fiber, and fuel are produced with valuable water, fertilizer, and energy (Foley et al. 2011). While it is important that people are taught that this waste results in greater costs to society, in environmental degradation and a decline in human health (and likely know this already to some degree), we must find a way to change behavior resulting from this education. Mass education of the population is a potential solution, but such a program must be omnipresent in the K-12 curriculum (Havlin et al. 2010), and carry through to popular advertising. For example, to see an advert supporting purchases that have protected soil from erosion and improved water quality should be as commonplace as ones for a 6000 cal fast-food meal that contributes to diabetes, heart disease, soil erosion, and water quality degradation. Adverts should also be placed where the general public lives: suburban and urban areas, on transportation infrastructure (busses, trains, billboards) and on radio, television, Internet advertising, etc. The linkage between the health and stability of a civilization and soil must be commonplace knowledge if we are to change food production and consumption habits, reduce waste of precious resources soil helps supply, and end water quality degradation due to erosion and nutrient runoff.

Funding a mass education effort will not be easy. Looking to the federal government to solve this problem is likely unrealistic given budget limitations and other issues demanding limited available funds. Local efforts are likely best and help invest people in community. Local efforts also can be more affordable with in-kind donations of advertising space and people's time. However, large-scale efforts should be attempted by state or federal governments. Funds raised could come from product "branding" for soil/watershed protection, a tax on food to offset conservation, crowd source funding, and philanthropic support. The latter is likely a very realistic solution and substantially underappreciated.

19.2.4 Challenge 4: Reshaping Agriculture to Protect Soil and Its Ecosystem Services to Society

Scientists recognize that an uncomfortable shift in our way of life may be rapidly approaching due to climate change (Karl et al. 2009) and associated long-term drought (Cook et al. 2015) and that transformative adaptations may be necessary for survival of the human species (Kates et al. 2012). Preparing ahead of time can help to lessen how uncomfortable this may be in terms of economics, food supply, and ecosystem stability. These factors have been recently highlighted as key aspects of "soil security" (McBratney et al. 2014), which is a useful rebranding of the ideas highlighted nearly 100 years ago by Charles Kellogg in the 1938 Yearbook of Agriculture "Soil and Men" (Kellogg 1939).

Correcting current land uses that may not be aligned with promoting economic, food supply, and ecosystem stability will be difficult (Meyfroidt 2013) and require economic incentives or new forms of legislation, but they will be needed. The 2013 National Agricultural Biotechnology Committee Report stated "A nationwide systems analysis is needed across agriculture to assess where the most water-efficient crops are being grown in each region. If that analysis concludes that certain crops are performing inefficiently, what infrastructure needs to be developed, and what social shifts need to occur for improvement? (NABC 2013)" Soil scientists must be a key element of this successful transition by helping producers adapt to the suitabilities and limitations of new local agricultural operations, helping to map and design food sheds (Peters et al. 2009) that could be improved by land shaping and identification of soils suitable for geo-engineering to remove limiting physical or chemical characteristics

Soil management can also be influenced by the consumer purchase and the business producing the commodity. Soil scientists must educate consumers that it is, in part, their responsibility, and within their power via purchasing, to help improve soil management and diminish unsustainable soil management practices. However, it must be clear to the purchaser that the added expense is directly linked to soil conservation (Wognum et al. 2011; Trienekens et al. 2012). While public education of soil management and food production may help change human behavior by, in part, reshaping buying habits, the agribusiness industry is an element of agriculture that has a tremendous effect on soil through practices to which contract farmers must adhere. These practices may be unsustainable and have contributed to severe nutrient imbalances associated with animal agriculture (Beegle 2013; Sims et al. 2005), complicated effects of agricultural management on invasive and native plants (Mortensen et al. 2012; Egan et al. 2014), and contributed to extensive erosion of Mollisols in the central USA. (Den Biggelaar et al. 2001).

Reshaping the agricultural landscape to promote better soil management is potentially becoming more difficult. The number of farmers and agricultural land per capita is decreasing, farms are getting larger, and the price of agricultural land is increasing (NASS 2014). Larger farms are more expensive to buy, especially by an agricultural undergraduate who is $140–190,000 in debt. These lands can be bought by developers for housing or retail centers, large agribusiness companies, or overseas investors. Land ownership has been highlighted throughout history as an important element of democracy (Kellogg 1939; Russett 1964), and we should pay attention to the current trends in US farmland, who grows our food, and how.

19.2.5 Challenge 5: Federal Funding for Soil Science Research and Soil Management

Millions of dollars have been spent on federal grants that examine some aspect of soil, yet we still have a substantial problem with soil management and use that transcends society due to poor human behavior. This problem exists, in part, because research funding has not effectively targeted the needs of society and funding that is too small to solve the problem. A root cause for this problem could be the intense specialization of fields that now study soil in some way, which has been noted by Baveye et al. (2006). Kellogg (1939) warned of the risk this could have in causing the scientific community to lose sight of the "big picture of soil." US soils are as diverse as its present population, and the problems US soils face demand diversified solutions that federal funding programs seem to be missing; perhaps due to the nationwide model for funding scientific research. Typically, one request for proposals is made to the entire country. Perhaps this should be changed to a model that solicits regional proposals, especially if regional adaptations to climate change and population pressures on resources are to be meaningful and purposeful.

How research is conducted at universities may be hindering societal changing results. Typically, a researcher, the Principal Investigator, applies for a grant (alone or with co-Principal Investigators) and the grant is awarded. While today's grants in the realm of agriculturally based soil research are often focused on large multidisciplinary projects, leadership of the team is usually one to three scientists. There are examples of USDA Agricultural Research Service laboratories with specific focuses in aspects of soil science, e.g., the US Salinity Laboratory and the National Soil Erosion Research Lab. Far more effective would be a whole department of soil *and* other scientists focused on solving a specific problem. For example, using Challenge 1 of soil erosion as an example, we see a problem driven by human behavior, often in response to economics, knowledge of how to farm, perhaps stubbornness to keep farming the same way, and a perceived lack of incentive to change. Some would argue that we need a research effort in soil science that transcends just agriculture and on par with the likes of the "Manhattan Project." In contrast, Jogelekar (2012) argues that we need to stop comparing every Big Science Project to the Manhattan Project.

The Manhattan Project was much more engineering and practical chemistry than pure science, and it was much more project management than pie-in-the-sky thinking. It had a well-defined and rather well-understood goal and was therefore very far from the kind of pure scientific endeavors to which it is compared. The Apollo Project to send humans to the Moon was another enterprise steeped in engineering and virtually removed from pure science, both in its methods and in its fruits.

Challenge 1 is exactly an example "Manhattan Project" for soil science for the very reason Jogelekar (2012) critiques many "Big Science" projects. To solve a problem as large as US soil erosion requires an extensive modification of human behavior across an array of demographics, and a reshaping of where and how agriculture is conducted. We already know how to do this; we just need to implement it.

19.3 Future Soil Science Research Needs

To help meet the needs of an evolving society with an unpredictable climate future, national research efforts in soil science should focus on issues that are critical for stable commodity production, that secure ecosystem services, and that protect human health and welfare as climate change evolves.

19.3.1 Urban/Suburban Soils

Given most of the US population lives in urban/suburban areas, there is a tremendous need for research in order to better understand the potential interactions and needs surrounding urban soils and human health including food supply creation, ecosystem restoration, urban ecosystem services, redevelopment strategies, and pollutant cleanup/remediation from past, chronic pollution, and/or natural disasters.

Soils in urban and suburban environments have become a more popular topic of research over the last 20 years, but never have been a primary focus of the National Cooperative Soil Survey (NCSS) and only studied intensively by a relatively small body of US soil scientists. The bulk of urban soil research has been conducted elsewhere in the world where nations and their cities are older; repurposing of land is a necessity, and environmental legislation is often stricter (which drives much soil pollutant research). The NCSS has explored ways to inventory urban soils and communicate soil information to users (Scheyer and Hipple 2005), and has coarsely mapped urban soils, e.g., NRCS (1998). Conventional NRCS soil mapping in urban areas is often not very useful, however, given the scale of variability and short temporal stability of urban soils. Site-specific mapping of key parameters (pollutant concentrations, density, hydrologic measures) is necessary but is needed at the parcel level at the time of a specific project. Use of pre-development Soil Survey information can be useful for understanding parent materials that might be comprised of urban soil.

Quantifiable standards for uses of urban soils that express limits or boundaries would be extremely useful and save much money misspent in poor urban design. For example, what is the appropriate bulk density that might restrict root growth for a range of urban plantings versus the bulk density that will restrict root growth, versus the bulk density that will not impede root growth. Coarse estimates exist (Scheyer and Hipple 2005), but more extensive data is necessary for the majority of plantings used in urban environments. The economic significance of such data is great, especially when one considers the number of vegetated coastal urban protection zones being developed around the country to protect against flood losses (Hallegatte et al. 2013).

Currently, there is much research investigating storm proofing coastal cities against flood surges. As climate change alters the intensity, frequency, and temporal patterns of precipitation (Pyke et al. 2011; Newcomer et al. 2014) and threatens coastal communities (Rosenzweig and Solecki 2014; Wagner et al. 2014), the management of stormwater has become a greater priority of many cities. Soil scientists must step forward and play a substantial role in such projects by vocalizing/demonstrating their skills. The profession of soil science should strive to produce a cohort of future soil scientists with broader environmental training, and additional degrees in civil and environmental engineering in order to better meet the needs of such projects.

Combined sewer overflow (CSO) systems (McLellan et al. 2007; Lund et al. 2014) across the USA can contribute substantial water pollution during storm events. Recognition of this pollution and new US Environmental Protection Agency policies (USEPA 2013) have resulted in efforts to examine disconnecting such systems and routing stormwater locally into green infrastructure (Shuster et al. 2014; Shuster and Garmestani 2014; Lucas and Sample 2015). In addition, spray irrigation of tertiary treated sewage continues to be examined as a technique to recharge local aquifers and reduce flows to water bodies. One of the fastest growth areas of soil science research will be that which examines the fate of the thousands of chemicals we put into our soil via these types of onsite waste disposal (Casey et al. 2005; Woodward et al. 2014), and that we potentially ingest via water from the aquifers receiving these waste products. In addition, there is much value in examining the legacy of soil pollutants that can persist or can be redistributed numerous times in the local urban environment (Laidlaw et al. 2012).

In and around cities, and even in the outskirts of rural areas, there is a vast amount of land in the USA that is already disturbed, but minimally managed or unmanaged. Examples include corridors along highways or municipality owned land in urban and suburban environments. We know little about the degree of disturbance of these soils and how readily these lands could be managed to some form of improved ecosystem or agricultural use. These lands provide key locations for easy access to, and quick development of, regional food supplies, biofuels, carbon sequestration management zones, onsite wastewater disposal for aquifer recharge, and pollinator/biodiversity refuges.

Support of such a land management effort can come from soil scientists who develop a nationwide data set specific to making easy calculations of maximum potential soil carbon sequestration given various land uses, carbon addition methods, types of carbon, and differing soils. Access to such information is critical for helping educate land managers on how to sequester soil carbon. We must make it easy for people to calculate for their own soil anywhere in the USA the theoretical maximum amount of soil carbon that can be stored given a number of management strategies. This is not to oversell the notion that climate change can be mitigated by soil carbon to the point climate change is not a problem, but rather to guide carbon accounting strategies and land planning projects using them. The NCSS should develop the online tools to make such calculations just as they have developed Web Soil Survey.

19.3.2 Human Health

One of the greatest untapped areas of research in soil science is the exploration of the many potential relationships between human health and soil (Brevik and Burgess 2012). Soil obviously plays a role in supporting human health by facilitating the production of clean water, food, shelter, clothing, etc. However, more must be learned especially about the subtle linkages between: soil chemistry, plant nutrition, and human health (Albrecht 1944; Albrecht 1957; Townsend et al. 2003); disease and dust exposure (Buck et al. 2013; Baumann et al. 2015); exposure risks to soils with different degrees of natural radioactivity (Myrick et al. 1983); endocrine disruptors (Woodward et al. 2014); and other contaminants (Ruby et al. 1999).

19.3.3 Subaqueous Soils

The growing study of subaqueous soils (Demas and Rabenhorst 1999; Erich et al. 2010; Rabenhorst and Stolt 2012) has critical implications for supporting sustainable coastal and inland waters, especially given the uncertain effect of climate change. Coastal ecosystems supply food products, can buffer against storm damage, and support great species diversity. Inland freshwater systems encompass a diverse array of water body types society is directly dependent upon including drinking water reservoirs, recreational impoundments, and inland open-water wetlands, streams, and rivers. The sheer economies of scale and potential influence of these soils on the people of the USA has not yet been recognized by soil scientists. For example, estuarine regions of the country are estimated to support ~43 % of the population and be worth half of the US economy's output, not just seafood production but all businesses in these areas (Kildow et al. 2009).

The field of subaqueous soil science has the potential to increase awareness of soil science in a way that has not been seen in decades. Areas of current study include carbon sequestration potential (Miilar et al. 2015), soil climate (Salisbury and Stolt 2011), coastal acidification (Still and Stolt 2015), sea-level rise (Hussein 2009), and coastal development (Donohue et al. 2009).

19.3.4 Soil Security

The notion of soil security should be at the forefront of all soil research activities. Whether soil security is recognized somehow more formerly (Drohan and Farnham 2006; McBratney et al. 2014) or via recognition of the externalities affected by soil management (health, economic, and welfare goals (Millennium Goals)) a broader, likely International, body should put the issue of soil security front and center with world leaders. The International Union of Soil Scientists is one entity that could help move such an idea forward, but they alone cannot achieve this goal. A collection of non-governmental organizations, political leaders, and scientific organizations must work together to bring soil security to the forefront of sustainability discussions. Research on soil security issues is ripe and open to many multidisciplinary avenues never before, or rarely attempted in soil science.

19.4 A New Vision for Soil Survey

Federal downsizing of the Soil Survey program has been extensive and quick resulting in far fewer employees than in the last 100 years. This has caused a substantial loss of institutional knowledge and weakened the capability of the program to respond to new needs and opportunities. Not at all clear is whether the Soil Survey will ever return to its former size and influence. How can the Soil Survey work within its new funding and support model?

Given the Soil Survey's mission, information delivery is paramount to success. Web Soil Survey has had a somewhat difficult time becoming a viable tool to deliver soil information, but has improved with time. More intuitive and better designed delivery tools, like the SoilWeb application for mobile phones and Google Earth (Beaudette and O'geen 2010), provide easy access to the broad suite of Soil Survey information wherever one can access an Internet connection. The success of SoilWeb demonstrates how valuable quick adoption of new technologies can be to soil education. More data delivery tools, like the prior proposed carbon sequestration tool or the Iowa soil erosion forecast tool (Cruse et al. 2006), could do much to promote soil science and conservation.

Via further adoption of new technology, the NCSS led by the USDA-NRCS could revolutionize soil science education, training, and outreach via game-based learning. Game-based learning (Prensky 2005) and the use of multi-agent systems (Bousquet et al. 2002) to create virtual societies can provide for deep learning experiences. Incorporating state-of-the-art immersive digital "sandbox" environments would allow users to explore their virtual world in a way very similar to one exploring the natural world. In theory, a student in such a game-based environment would interact with a soil issue in many ways including sociocultural aspects, profile and landscape description, design of data collection, laboratory analysis, data interpretation, and information delivery. Multi-agent systems help enhance decision-making skills and have been applied in learning exercises for ecosystem management (Bousquet et al. 2002). Game designs with technologies that visually and mechanically provide an example foundation for the depth possible with a soil science and management educational game include Giants Software Farming Simulator, SCS Software's Euro Truck Simulator, and the US Army Proving Grounds. For an example of the data interfaces that allows users to see graphs depicting relationships reflecting their decision making, see 2kGames Civilization or Railroad Tycoon. While gaming technology is not a perfect substitute for field experiences and may further separate humans from the natural environment (Kahn et al. 2009), such technology adoption can be a very effective educational experience (Robertson and Howells 2008) and bring soil science to the general population in a much quicker, affordable, and efficient way.

A well-known problem with the Soil Survey is that map units in areas not used for production of a crop commodity consist of series that are very broad in their range of properties. One example of such a series is the state soil of Pennsylvania, Hazelton, which can form from three different lithologies and on slopes from 0 to 80 % (Soil Survey Staff 2013). Such series and their extents do little to convey the important relationships between soils and the ecosystems in which they occur. While this problem is a result of the early focus of Soil Survey on supporting agriculture, it is well past the time for this to be rectified. Rapidly advancing analysis techniques in digital soil mapping (Behrens et al. 2010; Brus et al. 2011; Nauman and Thompson 2014; Odgers et al. 2014) are improving existing soil maps and their rate of development. Coupled with new, affordable field sensors or lower-cost laboratory techniques, the Soil Survey could rapidly increase the collection and use of soil data. Focus should be on techniques that are the most common in a variety of fields of soil science, soil properties that are desired to be sampled frequently but are expensive to measure, and properties that are affordable to measure but measurements are imprecise due to methodology [e.g., visible and near-infrared reflectance spectroscopy (Baveye and Laba 2014) to infer heavy metals]. This can be accomplished via intensely focused cooperative research projects funded partly by the Soil Survey.

The USDA-NRCS Soil Science Division (SSD) is currently funding small research projects, which helps bring new ideas into the soil survey and maintains relationships with academic units around the country. Many SSD staff have worked on these projects and received valuable training. The SSD will typically fund a suite of projects in different areas, but that fall within defined areas of focus. This model may not serve agency priorities as well as it could. A more productive model might be to focus the limited resources now available on just one area of research for 3–5 years, and rapidly advance knowledge and its dissemination. An example of a past project reflecting this model might be the "Desert Project" (Monger et al. 2007). The NCSS should strive for new, ground-breaking collaborations with companies at the cutting edge of technology development and perhaps outside the typical realm of soil and agricultural scientists.

Several examples in this chapter have highlighted the need for specific quantification of soil properties and their ranges for acceptable use. Examples were given for bulk density in urban soils (Scheyer and Hipple 2005) and carbon sequestration accounting. The physical and economic constraints of continued mapping of a country as large as the USA are prohibitive for the SSD. Refocusing the NCSS skill set on the development of site-specific soil assessment, diagnosis and treatment could result in much more effective models of data collection and soil knowledge application developed in conjunction with researchers. This could also lead to substantial revenue generation via new mapping contracts and studies. However, to be most effective, this information must also be highly accessible to a variety of levels of soil knowledge, and the success of projects must be made an example of via broad public outreach.

Federal downsizing of the NRCS SSD may result in full or semi-privatization of the NCSS. This could be a positive outcome for US soil science if properly managed. Models of such a potential US Soil Survey Program exist: the James Hutton Institute in Scotland and AgResearch in New Zealand. A private or semi-private soil survey might allow for more specific project development, without the burden of federal bureaucracy that now may hamper innovation.

19.5 Conclusion

If soil scientists are to be proactive about changing their science, changes must promote solutions very different from the current reality. This is especially true if the science is to survive and if we are to solve the world's soil problems. The problems facing soil science and society can be solved, but human behavior must be altered to achieve success. This is the commonality among many of the challenges discussed in this chapter, and nearly 100 years ago by Charles Kellogg. Soil scientists should look forward to the changes that are needed to enhance the future of soil science and Soil Survey because these changes will result in a civilization much more stable, proactive, and humane because of soil scientists.

References

Adewopo JB, VanZomeren C, Bhomia R et al (2014) Top-ranked priority research questions for soil science in the 21 century. Soil Sci Soc Am J 78(2):337–347

Albrecht WA (1944) Soil fertility and national nutrition. J Am Soc Farm Manag Rural Appraisers

Albrecht WA (1957) Soil fertility and biotic geography. Geogr Rev 47 (1):86–105

Barr A, Turner SE (2013) Expanding enrollments and contracting state budgets. The effect of the great recession on higher education. Ann Am Acad Polit Soc Sci 650(1):168–193

Baumann F, Buck BJ, Metcalf RV et al (2015) The presence of asbestos in the natural environment is likely related to mesothelioma in young individuals and women from Southern Nevada. J Thorac Oncol. doi:10.1097/JTO.0000000000000506

Baveye PC (2006) A future for soil science. J Soil Water Conserv 61 (5):148–151

Baveye PC (2015) Grand challenges in the research on soil processes. Front Environ Sci 3:10

Baveye PC, Jacobson AR (2008) Soil science education and the "age of money": reflections and concerns for the near future. Water Air Soil Pollut 187:1–4

Baveye PC, Jacobson AR (2009) Comment on "A soil science renaissance" by AE Hartemink and A McBratney. Geoderma 151(3):126–127

Baveye PC, Laba M (2014) Visible and near-infrared reflectance spectroscopy is of limited practical use to monitor soil contamination by heavy metals. J Hazard Mater 285:137–139

Baveye PC, Jacobson AR, Allaire SE et al (2006) Whither goes soil science in the United States and Canada? Soil Sci 171(7):501–518

Baveye PC, Camargo FA, Poss R (2010) The discipline of soil science is not doing too badly…under different skies. Soil Sci 175(6):313–314

Baveye PC, Rangel D, Jacobson AR, Laba M, Darnault C, Otten W, Radulovich R, Camargo FAO (2011) From dust bowl to dust bowl: soils are still very much a frontier of science. Soil Sci Soc Am J 75 (6):2037–2048

Beaudette DE, O'geen AT (2010) An iPhone application for on-demand access to digital soil survey information. Soil Sci Soc Am J 74(5):1682–1684

Beegle D (2013) Nutrient management and the Chesapeake Bay. J Contemp Water Res Educ 151(1):3–8

Behrens T, Zhu AX, Schmidt K et al (2010) Multi-scale digital terrain analysis and feature selection for digital soil mapping. Geoderma 155(3):175–185

Bousquet F, Barreteau O, D'Aquino P et al (2002) Multi-agent systems and role games: an approach for ecosystem co-management. In: Janssen M (ed) Complexity and ecosystem management: the theory and practice of multi-agent approaches. Edward Elgar, Cheltenham

Brevik EC, Burgess LC (eds) (2012) Soils and human health. CRC Press, Boca Raton

Brown S (2008) Senate resolution 440, 110th congress (2007–2008). https://www.congress.gov/bill/110th-congress/senate-resolution/440. Accessed 23 Feb 2015

Brus DJ, Kempen B, Heuvelink GBM (2011) Sampling for validation of digital soil maps. Eur J Soil Sci 62(3):394–407

Buck BJ, Goossens D, Metcalf RV et al (2013) Naturally occurring asbestos: potential for human exposure, Southern Nevada, USA. Soil Sci Soc Am J 77(6):2192–2204

Casey FX, Šimůnek J, Lee J et al (2005) Sorption, mobility, and transformation of estrogenic hormones in natural soil. J Environ Qual 34(4):1372–1379

Center for College Affordability and Productivity (CCAP) (2014) the truth about college costs. http://centerforcollegeaffordability.org/data/quick-facts/. Accessed 23 Feb 2015

Collins ME (2008) Where have all the soils students gone? J Nat Resour Life Sci Educ 37:117–124

Cook BI, Ault TR, Smerdon JE (2015) Unprecedented 21st century drought risk in the American Southwest and Central Plains. Sci Adv. doi:10.1126/sciadv.1400082

Cruse R, Flanagan D, Frankenberger J et al (2006) Daily estimates of rainfall, water runoff, and soil erosion in Iowa. J Soil Water Conserv 61(4):191–199

Demas GP, Rabenhorst MC (1999) Subaqueous soils pedogenesis in a submersed environment. Soil Sci Soc Am J 63(5):1250–1257

Den Biggelaar C, Lal R, Wiebe K et al (2001) Impact of soil erosion on crop yields in North America. Adv Agron 72:1–52

Denneen J, Dretler T (2012) The financially sustainable university. Bain and Company, Boston. http://www.augustana.edu/documents/president/Augustana2020/Bain2012.pdf. Accessed 23 Feb 2015

Diamond J (2005) Collapse: how societies choose to fail or succeed. Penguin Publishing Group, Westminster

Donner SD, Kucharik CJ (2008) Corn-based ethanol production compromises goal of reducing nitrogen export by the Mississippi River. PNAS 105(11):4513–4518

Donohue SW, Stolt MH, Gold A et al (2009) Human-transported material soils of urbanizing estuarine landscapes. Soil Sci Soc Am J 73(5):1587–1596

Drohan PJ, Farnham TJ (2006) Protecting life's foundation. Soil Sci Soc Am J 70(6):2086–2096

Drohan PJ, Havlin JL, Megonigal JP et al (2010) The "Dig It!" Smithsonian soils exhibition: lessons learned and goals for the future. Soil Sci Soc Am J 74(3):697–705

Egan T (2006) The worst hard time. Houghton Mifflin Harcourt, Boston

Egan JF, Graham IM, Mortensen DA (2014) A comparison of the herbicide tolerances of rare and common plants in an agricultural landscape. Environ Toxicol Chem 33:696–702

Erich E, Drohan PJ, Ellis LR et al (2010) Subaqueous soils: their genesis and importance in ecosystem management. Soil Use Manag 26(3):245–252

Erisman JW, Sutton MA, Galloway J et al (2008) How a century of ammonia synthesis changed the world. Nat Geosci 1:636–639

Foley JA, Ramankutty N, Brauman KA et al (2011) Solutions for a cultivated planet. Nature 478(7369):337–342

Hallegatte S, Green C, Nicholls RJ et al (2013) Future flood losses in major coastal cities. Nat Clim Change 3:802–806

Hartemink AE (2006) The future of soil science. International Union of Soil Science, Wageningen

Hartemink AE, McBratney A (2008) A soil science renaissance. Geoderma 148(2):123–129

Hartemink AE, McBratney A, Minasny AB (2008) Trends in soil science education: looking beyond the number of students. J Soil Water Conserv 63(3):76–83

Hatfield JL, Cruse RM, Tomer MD (2013) Convergence of agricultural intensification and climate change in the midwestern United States: implications for soil and water conservation. Mar Freshw Res 64 (5):423–435

Havlin J, Balster N, Chapman S et al (2010) Trends in soil science education and employment. Soil Sci Soc Am J 74(5):1429–1432

Hogan DM, Jarnagin ST, Loperfido JV et al (2014) Mitigating the effects of landscape development on streams in urbanizing watersheds. J Am Water Resour Assoc 50(1):163–178

Hopmans JW (2007) A plea to reform soil science education. Soil Sci Soc Am J 71(3):639

Hussein AH (2009) Modeling of sea-level rise and deforestation in submerging coastal Ultisols of Chesapeake Bay. Soil Sci Soc Am J 73(1):185–196

Jackson T (2014) EPA awards funds to fight Lake Erie algal blooms. Sandusky Register. http://www.sanduskyregister.com/news/environment/6265991. Accessed 23 Feb 2015

Jogelekar A (2012) Why we need to stop comparing every Big Science Project to the Manhattan Project. Scientific American http://blogs.scientificamerican.com/the-curious-wavefunction/2013/11/25/why-we-need-to-stop-comparing-every-big-science-project-to-the-manhattan-project/. Accessed 23 Feb 2015

Johnson LC (1987) Soil loss tolerance: fact or myth. J Soil Water Conserv 42(3):155–160

Kahn PH, Severson RL, Ruckert JH (2009) The human relation with nature and technological nature. Curr Dir Psychol Sci 18(1):37–42

Karl, T R, Melillo JM, Peterson TC (eds) (2009) Global climate change impacts in the United States. http://www.globalchange.gov/browse/reports/global-climate-change-impacts-united-states. Accessed 23 Feb 2015

Kates RW, Travis WR, Wilbanks TJ (2012) Transformational adaptation when incremental adaptations to climate change are insufficient. Proc Natl Acad Sci 109(19):7156–7161

Kellogg CE (1939) Soil and society. US Department of Agriculture

Kildow JT, Colgan CS, Scorse JD (2009) State of the U.S. Ocean and Coastal Economies 2009. http://cbe.miis.edu/noep_publications/4. Accessed 23 Feb 2015

Kiley K (2013) The pupil cliff. https://www.insidehighered.com/news/2013/01/11/wiche-report-highlights-decline-high-school-graduates-and-growing-diversity. Accessed 23 Feb 2015

Laidlaw MA, Zahran S, Mielke HW et al (2012) Re-suspension of lead contaminated urban soil as a dominant source of atmospheric lead in Birmingham, Chicago, Detroit and Pittsburgh, USA. Atmos Environ 49:302–310

Lee J (2014) Driven by climate change, algae blooms behind Ohio water scare are new normal. http://news.nationalgeographic.com/news/2014/08/140804-harmful-algal-bloom-lake-erie-climate-change-science/. Accessed 23 Feb 2015

Lindbo DL, Kozlowski DA, Robinson C (eds) (2012) Know soil, know life. Soil Science Society of America, Madison

Lucas WC, Sample DJ (2015) Reducing combined sewer overflows by using outlet controls for Green Stormwater Infrastructure: case study in Richmond, Virginia. J Hydrol 520:473–488

Lund A, McMillan J, Kelly R et al (2014) Long term impacts of combined sewer overflow remediation on water quality and population dynamics of *Culex quinquefasciatus*, the main urban West Nile virus vector in Atlanta, GA. Environ Res 129:20–26

McBratney A, Field DJ, Koch A (2014) The dimensions of soil security. Geoderma 213:203–213

McLellan SL, Hollis EJ, Depas MM et al (2007) Distribution and fate of *Escherichia coli* in Lake Michigan following contamination with urban stormwater and combined sewer overflows. J Great Lakes Res 33(3):566–580

Megonigal JP, Stauffer B, Starrs S et al (2010) "Dig it!": how an exhibit breathed life into soils education. Soil Sci Soc Am J 74(3):706–716

Meyfroidt P (2013) Environmental cognitions, land change, and social–ecological feedbacks: an overview. J Land Use Sci 8(3):341–367

Miilar CM, Aduomih O, Ama A, Still B, Stolt MH (2015) Estuarine subaqueous soil organic carbon accounting: sequestration and storage. Soil Sci Soc Am J 79(2):389–397

Mitsch WJ, Day JW, Zhang L et al (2005) Nitrate-nitrogen retention in wetlands in the Mississippi River Basin. Ecol Eng 24(4):267–278

Mitsch WJ, Zhang L, Fink DF et al (2008) Ecological engineering of floodplains. Ecohydrol Hydrobiol 8(2):139–147

Monger HC, Grossman RB, Ahrens RJ et al. (2007) A 50th anniversary guidebook for the Desert Project. US Department of Agriculture, Natural Resources Conservation Service, National Soil Survey Center. http://jornada.nmsu.edu/biblio/50th-anniversary-guidebook-desert-project. Accessed 23 Feb 2015

Mortensen DA, Egan JF, Maxwell BD et al (2012) Navigating a critical juncture for sustainable weed management. Bioscience 62(1):75–84

Myrick TE, Berven BA, Haywood FF (1983) Determination of concentrations of selected radionuclides in surface soil in the US. Health Phys 45(3):631–642

National Agriculture and Biotechnology Committee (2013) NABC Report 24: Water Sustainability in Agriculture (2012). http://nabc.cals.cornell.edu/Publications/Reports/nabc_24/24_0_3_Workshops.pdf. Accessed 23 Feb 2015

National Agriculture Statistics Service (2014) Land values 2014 summary. USDA National Agricultural Statistics Service. http://www.nass.usda.gov/Publications/Todays_Reports/reports/land0814.pdf. Accessed 23 Feb 2015

Natural Resources Conservation Service (1998) NRCS Soil Survey of City of Baltimore, Maryland (Soil Survey Report) Natural Resource Conservation Service, Washington, DC

Nauman TW, Thompson JA (2014) Semi-automated disaggregation of conventional soil maps using knowledge driven data mining and classification trees. Geoderma 213:385–399

Newcomer ME, Gurdak JJ, Sklar LS et al (2014) Urban recharge beneath low impact development and effects of climate variability and change. Water Resour Res 50(2):1716–1734

Odgers NP, Sun W, McBratney AB et al (2014) Disaggregating and harmonising soil map units through resampled classification trees. Geoderma 214:91–100

Penn State University (2014) Net Price Calculator. http://collegecostestimate.ais.psu.edu/. Accessed 23 Feb 2015

Peters CJ, Bills NL, Lembo AJ et al (2009) Mapping potential foodsheds in New York State: a spatial model for evaluating the capacity to localize food production. Renewable Agric Food Syst 24(01):72–84

Prensky M (2005) Computer games and learning: Digital game-based learning. Handb Comput Game Stud 18:97–122

Pyke C, Warren MP, Johnson T et al (2011) Assessment of low impact development for managing stormwater with changing precipitation due to climate change. Landsc Urban Plan 103(2):166–173

Rabenhorst MC, Stolt MH (2012) Subaqueous soils: pedogenesis, mapping, and applications. In: Lin H (ed) Hydropedology: synergistic integration of soil science and hydrology. Elsevier, Oxford, pp 173–204

Robertson J, Howells C (2008) Computer game design: opportunities for successful learning. Comput Educ 50(2):559–578

Rosenzweig C, Solecki W (2014) Hurricane sandy and adaptation pathways in New York: lessons from a first-responder city. Glob Environ Change 28:395–408

Ruby MV, Schoof R, Brattin W et al (1999) Advances in evaluating the oral bioavailability of inorganics in soil for use in human health risk assessment. Environ Sci Technol 33(21):3697–3705

Russett BM (1964) Inequality and instability: the relation of land tenure to politics. World Polit 16(03):442–454

Salisbury A, Stolt MH (2011) Estuarine subaqueous soil temperature. Soil Sci Soc Am J 75(4):1584–1587

Scheyer JM, Hipple KW (2005) Urban soil primer. United States Department of Agriculture, Natural Resources Conservation Service, Washington, DC. http://www.nrcs.usda.gov/wps/portal/nrcs/detail/soils/use/urban/?cid=nrcs142p2_053993. Accessed 23 Feb 2015

Shuster WD, Garmestani AS (2014) Adaptive exchange of capitals in urban water resources management: an approach to sustainability? Clean Technol Environ Policy 1–8

Shuster WD, Dadio S, Drohan P et al (2014) Residential demolition and its impact on vacant lot hydrology: implications for the management of stormwater and sewer system overflows. Landsc Urban Plan 125:48–56

Sims JT, Bergström L, Bowman BT et al (2005) Nutrient management for intensive animal agriculture: policies and practices for sustainability. Soil Use Manag 21(1):141–151

Smil V (2000) Feeding the world: a challenge for the 21st century. The MIT Press, Cambridge

Soil Survey Staff (2013) Hazelton official series description. https://soilseries.sc.egov.usda.gov/OSD_Docs/H/HAZLETON.html. Accessed 4 Feb 2012

Stephens P (2013) Washington monthly. International students: separate but profitable. http://www.washingtonmonthly.com/magazine/september_october_2013/features/international_students_separat046454.php?page=all. Accessed 23 Feb 2015

Stevenson CM, Puleston CO, Vitousek PM et al (2015) Variation in Rapa Nui (Easter Island) land use indicates production and population peaks prior to European contact. Proc Natl Acad Sci. doi:10.1073/pnas.1420712112

Still BM, Stolt MH (2015) Subaqueous soils and coastal acidification: a hydropedology perspective with implications for calcifying organisms. Soil Sci Soc Am J. doi:10.2136/sssaj2014.09.0366

Townsend AR, Howarth RW, Bazzaz FA et al (2003) Human health effects of a changing global nitrogen cycle. Front Ecol Environ 1 (5):240–246

Trienekens JH, Wognum PM, Beulens AJ et al (2012) Transparency in complex dynamic food supply chains. Adv Eng Inform 26(1):55–65

U.S. Environmental Protection Agency (2013) Combined sewer overflows CSO control policy. http://cfpub.epa.gov/npdes/cso/cpolicy.cfm?program_id=5. Accessed 23 Feb 2015

USDA-SCS (1992.) Historical notes 1. Readings in the history of the soil conservation service. http://www.nrcs.usda.gov/Internet/FSE_DOCUMENTS/stelprdb1043484.pdf. Accessed 23 Feb 2015

Verheijen FG, Jones RJ, Rickson RJ et al (2009) Tolerable versus actual soil erosion rates in Europe. Earth Sci Rev 94(1):23–38

Wagner M, Chhetri N, Sturm M (2014) Adaptive capacity in light of Hurricane Sandy: the need for policy engagement. Appl Geogr 50:15–23

Webb NP, Herrick JE, Duniway MC (2014) Ecological site-based assessments of wind and water erosion: informing accelerated soil erosion management in rangelands. Ecol Appl 24(6):1405–1420

Wognum PN, Bremmers H, Trienekens JH et al (2011) Systems for sustainability and transparency of food supply chains—current status and challenges. Adv Eng Inform 25(1):65–76

Wolman MG (1967) A cycle of sedimentation and erosion in urban river channels. Geografiska Annaler Ser A Phys Geogr 385–395

Woodward EE, Andrews DM, Williams CF et al (2014) Vadose zone transport of natural and synthetic estrogen hormones at penn state's "living filter" wastewater irrigation site. J Environ Qual 43(6):1933–1941

Index

A
Aa lava, 307, 313
Ablation till, 265, 343
Acid sulfate soil(s), 90, 172, 276, 351, 363
Acrudoxs, 293
Active layer, 332, 333, 363
Additions, 1, 13–15, 78, 101, 110, 146, 199, 201, 344, 356, 359
Acid sulfate soils, 90, 172, 276, 351, 363
Aeolian, *see* eolian
Aerial photo, 19
Agriculture/Agricultural, 1, 2, 5, 30, 49, 74, 77, 79, 87, 89, 103–106, 109, 113, 115, 119, 125, 127, 137, 140, 146, 157, 160, 167, 172, 174, 185–187, 195, 201, 203, 205, 206, 209, 224, 225, 240, 252, 257, 259, 261, 262, 266, 269, 271, 273, 275, 276, 278, 281, 282, 285, 286, 290, 291, 293, 311, 312, 316, 317, 319, 323, 324, 326, 351, 353–359, 363–365, 374–379, 381
A horizon, 13, 14, 32, 101, 103, 110, 112, 143, 146, 148, 158, 186, 200, 204, 230, 248, 251–253, 256, 257, 281, 284, 296, 300, 303, 332, 345, 348, 360
Al, *see* aluminum
Alaquods, 253, 254, 285, 303
Alaska, 1, 11, 14, 34, 43, 45, 73–75, 329–332, 334, 335, 337–340, 342–344, 346, 348, 363
Albaqualfs, 79, 258
Albic, 267, 268, 284
Albic horizon(s), 37
Alfalfa, 105, 127, 140, 213
Alfisols, 43, 45, 49, 75, 83, 90, 106, 108, 110, 111, 113, 124, 126–128, 135, 144, 155, 167, 168, 171, 173, 174, 179, 187, 191, 202–204, 206, 209, 210, 214, 215, 218, 219, 221, 223, 230, 234, 235, 238, 240, 243, 251, 255, 257–259, 266, 268, 271, 275, 285, 292, 301, 303, 324, 356
Algal blooms, 313
Alkalinization, 339
Allophane, 107, 268
Alluvial fan(s), 13, 16, 73, 87, 90, 113, 115, 116, 119, 122, 128, 146, 149, 155, 215, 270, 302, 334, 336, 337, 343
Alluvial material(s), 255, 285, 313, 314, 322
Alluvial soil(s), 146, 149, 153, 199, 235, 246, 265, 277, 323, 339
Alluvial terrace(s), 2, 14, 87, 148, 157, 182, 199
Alluvium, 9, 14, 23, 68–74, 77, 78, 81, 84, 86, 88, 90, 91, 99, 101, 107, 127, 128, 132, 146–149, 153, 154, 179, 182, 186, 195, 198, 199, 201, 211, 214, 215, 218, 221, 230, 234, 235, 238, 243, 252, 257–259, 276, 292, 307, 317, 322, 330, 337, 343
Alpine, 14, 73, 82, 83, 99, 104, 111, 119, 126, 273, 339, 342, 344, 361
Al-silicates, 348
Altitude(s), 12, 14, 133, 363
Aluminum, 9, 15, 49, 107, 108, 110, 111, 203, 204, 245, 268, 313, 314, 322, 324, 344, 348, 360
Amorphous silica, 17, 234
Amphibians, 156
Amphibole(s), 93
Anaerobic, 15, 199, 344, 358, 363
Andesite, 82, 95, 127, 128
Andic, 78, 107, 108, 110, 313, 314
Andic horizon(s), 107
Andic properties, *see* andic soil properties
Andic soil properties, 33
Andisolization, 53, 344
Andisols, 45, 49, 75, 83, 106–110, 125–127, 305, 307, 308, 311–313, 316, 318, 324, 326, 345–348
Animal(s), 13, 14, 17, 365
Annual precipitation, 2, 53, 68, 74, 78, 81–84, 101, 104, 106, 118, 124, 135, 140, 143, 148, 193, 209, 227, 232, 236, 239, 243, 308, 314, 343, 344, 355
Anthropogenic, 1, 6, 87, 88, 94, 160, 229, 323, 356, 359, 360, 363, 365
Ants, 13, 14, 17, 32
Aqualfs, 43, 70–73, 203
Aquents, 68, 201, 300
Aquepts, 43, 68, 71–73, 201, 202, 273, 300
Aquerts, 45, 69, 71, 302
Aquic, 11, 13, 35, 73, 77, 87, 89, 90, 125, 192, 209, 214, 218, 230, 234, 240, 243, 261, 281, 285, 292, 303, 344, 346, 347
Aquic soil moisture regime, 78, 192
Aquifer, 131, 156, 257, 291, 317, 355, 379
Aquiturbels, 332, 334, 339
Aquods, 45, 72, 73, 253, 273, 347
Aquolls, 69, 71, 74, 204, 300
Aquorthels, 332, 346
Aquults, 43, 73
Argialbolls, 79, 156, 157
Argiaquolls, 156, 211, 218, 219, 224, 256, 290
Argids, 68–70, 104, 105, 115
Argidurids, 128
Argillic horizon, 15, 45, 87, 104, 110–112, 120, 122, 139, 140, 144, 148, 155, 157, 167, 168, 181, 183, 187, 230, 248, 252, 255, 257, 258
Argillite, 278
Argilluviation, 49, 75, 144
Argiudolls, 139, 140, 148, 149, 218, 221
Argiustolls, 128, 139, 140, 143, 153, 175, 176, 179, 185, 186
Argixerolls, 78, 79, 83, 86, 90, 93, 94, 96, 103, 104, 119, 124, 128
Aridic, 35, 77, 83, 87, 91, 92, 101, 104, 106, 115, 118, 124, 127, 129, 140, 143, 144, 147, 149, 184, 185, 302, 307, 314, 316, 361
Aridic soil moisture regime, 11

L.T. West et al. (eds.), *The Soils of the USA*, World Soils Book Series,
DOI 10.1007/978-3-319-41870-4

Aridisols, 45, 68, 75, 101, 104–106, 108, 111, 113, 115, 119, 121, 127–129, 135, 181, 182, 185, 357
Asbestos, 93, 95
Ash, 2, 14, 17, 53, 68, 72–74, 78, 82, 83, 91, 101, 105, 107, 108, 110, 128, 220, 281, 303, 307, 308, 313, 314, 316, 318, 322, 324
Aspect, 12, 29, 77, 91, 95, 96, 124, 261, 337–340, 375, 376, 378, 381
Association(s), 20, 21
Atmospheric carbon, 1, 6
Available water, *see* available water holding capacity
Available water capacity, *see* available water holding capacity
Available water holding capacity, 10, 33, 34, 78, 95, 104, 107, 108, 112, 185, 186, 218, 220–222, 224, 234, 270, 275, 278, 284, 313, 339, 364

B

Backswamp(s), 82, 257, 258
Backwater, 146, 148, 246
Bacteria, 13
Balds, 238
Barley, 101, 105, 127, 140, 278, 342
Basalt, 9, 79–83, 91, 101, 103, 104, 122, 127, 128, 281, 293, 296, 303, 307, 314, 322, 346
Base-cation cycling, 49
Base map, 19
Base saturation, 29, 30, 110, 132, 143, 173, 230, 245, 246, 250–253, 273, 318, 344, 348
Basin and range, 115, 117, 125
Basin(s), 16, 45, 74, 104, 115, 117, 124, 129, 160, 227, 238, 248, 337
Batholith(s), 178, 179
Bayou(s), 255
BC horizon(s), 246, 249, 250
Bedrock, 18, 35, 91, 113, 133, 149, 153, 182, 191, 192, 197, 214, 227, 245, 265, 276, 297, 314, 343, 347
Beef, 109, 218, 270, 273
Bh horizon(s), 15, 253, 268, 344
Biocycling, 13
Biologic activity, *see* biological activity
Biological activity, 9, 10, 115, 363
Biological crusts, 351, 362
Biome(s), 168, 217
Biotite, 245, 246, 265
Bioturbation, 17, 195
Bk horizon(s), 15, 153
Bkkm horizon(s), 151
Block diagram, 19, 22
Bluff(s), 214
Bog(s), 195, 199, 206, 240, 265, 268, 276, 339, 344, 345, 347, 338
Boreal, 131, 195, 261, 278, 339, 340, 363
Boreal forest, 262, 331, 363
Breccia, 92, 95, 98
Brunification, 338
Bs horizon(s), 268, 348
Bt horizon(s), 15, 112, 139, 140, 172, 202, 247, 250
Btk horizon(s), 140, 153, 212
Bulk density, 34, 104, 107, 128, 265, 270, 351, 355, 358, 361
Bunchgrass steppes, 135
Burrowing rodents, 13

C

Ca, *see* calcium
Ca:Mg ratio(s), 9
$CaCO_3$, *see* calcium carbonate
Calciaquolls, 140, 211
Calciargids, 122, 127, 129
Calcic horizon, 15, 16, 34, 87, 124, 153, 181, 182
Calcids, 34, 68, 69, 115
Calcification, 49
Calcite, 181
Calciudolls, 211
Calcium, 9, 13, 15, 33, 84, 110, 120–122, 148, 153, 157, 171, 201, 211, 313, 317, 318, 364
Calcium carbonate/Calcium carbonate equivalent, 11, 15, 16, 30, 32, 33, 35, 37, 87, 101, 103–105, 109, 110, 113, 120, 121, 148, 195, 211–213, 216
Calciustepts, 140, 145, 186
Calciusterts, 175
Calciustolls, 140, 149, 153, 159, 175, 176, 186, 211
Caliche, 151, 186
Cambic, *see* cambic horizon
Cambic horizon, 36, 45, 120
Cambids, 105
Cambisolization, 43, 49, 75
Cambrian, 178, 179, 198, 209, 238, 244
Capillary rise, 16
Carbon, 1, 15, 77, 94, 129, 264, 312, 332, 340
Carbonate(s), 15, 17, 33, 49, 99, 118, 122, 140, 152, 196, 201, 224, 340
Carbon dioxide, 14–16, 33, 334, 358, 363
Carbon sequestration, 77, 115, 311
Caribbean, 45, 50, 75, 281, 303
Category, 32, 35, 38, 40, 172
Cation-exchange, 15, 31, 32, 87
Cation exchange capacity, 110, 230, 245, 313, 334, 364
Cattle, 109, 127, 172, 218, 222, 271, 285, 318, 361
Caves, 16, 87, 118, 214, 227, 233
Cd horizon(s), 265, 272, 275
Cedar, 80–82, 250, 281, 344
Cenozoic, 110, 116, 133, 265
Chalk(s), 165, 174, 250, 251
Chemical dispersion, 15
Chemical weathering, 93, 120, 121, 124, 126, 128, 153, 155, 340
Chihuahuan desert, 118, 119, 121
C horizon(s), 3, 15, 16, 30, 31, 34, 43, 88, 89, 103, 120, 122, 140, 158, 181, 202, 284
Chronosequence, 122, 148
Cicada casts, 13
Cinder(s), 74, 82, 117, 128, 307, 309, 313, 322, 324, 346
Cinder cones, 127, 307, 308, 313, 316
Class(es), 29, 30, 32, 35, 38, 40, 144, 277, 347, 376
Clast(s), 149, 181
Clayey soils, 10, 17, 98, 172, 258
Clay film(s), 29
Clay formation, 121, 122
Clay illuviation, 167
Clay pan, 71, 168, 171–173, 185, 203
Clay translocation, 49, 172, 196, 200, 202, 248
Clay(s), 9, 11, 15, 17, 70, 81, 105, 111, 140, 153, 156, 160, 239, 246, 248, 250, 251, 255, 257, 259, 268, 272, 290, 314, 340
Climate, 1, 3–5, 9, 11–16, 19, 21, 23, 27, 43, 45, 72, 73, 75, 77, 80, 83, 87, 89, 92–94, 96, 98, 99, 101, 103–105, 111, 113, 115, 117–120, 124–126, 129, 133, 135, 140, 143, 153, 155, 156, 160, 171, 174, 180, 192, 193, 198, 203, 204, 206, 209, 216, 218, 220, 222, 224, 225, 227, 232, 236, 239, 240, 243, 247, 253, 259, 261, 268, 273, 278, 282, 292, 300, 302, 303, 305, 307, 314, 322, 329, 331, 333–335, 338, 340, 343, 344, 351, 353, 354, 363, 373, 379, 380
Climate change, 15, 188, 250, 262, 264, 278, 314, 334, 357, 359, 360, 373, 377–380
Climatic regime, 106, 118
Closed depressions, 2
CO_2, *see* carbon dioxide

Coastal marshes, 2, 73, 359
Colloidal particles, 15
Colloids, 14, 107
Colluvial parent material(s), 160, 198, 337
Colluvial soil(s), 198
Colluvium, 9, 68, 72–74, 77, 80, 83, 86, 91, 107, 154, 155, 182, 238
Complex(es), 15, 20, 107, 117, 176, 182, 253, 268, 313, 346, 348
Component(s), 20, 104, 266, 268, 275, 312, 332
Concave, 13, 176, 256, 257, 270, 316, 344
Concretions, 29, 146, 148, 153, 156
Conifer forest(s), *see* coniferous forest(s)
Coniferous, 74, 99, 108, 119, 194, 253, 278, 344
Coniferous forest(s), 13, 68, 109, 111, 118, 129, 243, 261, 262, 273, 278, 344, 347
Conifers, 69, 73, 74, 111, 195, 204, 344
Conservation, 75, 156, 209, 355, 364, 375
Conservation tillage, 273
Consociation(s), 20
Continental, 2, 45, 117, 121, 138, 237, 335, 338, 343
Continental glaciers, 191
Convex, 13, 151, 245, 273, 342
Coral, 72, 307, 323, 326, 356
Cover crop(s), 104, 313, 323, 355
Cracks, 15, 104, 157, 332
Crayfish, 17
Cretaceous, 117, 147, 165, 177, 179, 182, 248, 259, 343
Crevices, 214
Cropland, 101, 104, 115, 127, 146, 191, 252, 351, 354
Crop rotation, 103, 324, 355
Crops, 22, 73, 80, 137, 203, 224, 243, 247, 275, 285, 316, 378
Cryalfs, 69, 111
Cryands, 45, 69, 74
Cryaquands, 345
Cryaquents, 332, 339
Cryaquepts, 83, 339, 344
Cryaquods, 344
Cryepts, 69, 73, 339
Cryic, 53, 82, 83, 113, 273, 347
Cryods, 45, 73, 348
Cryofibrists, 332, 339, 346, 347
Cryofluvents, 333, 339, 346
Cryohemists, 83, 347
Cryolls, 68, 69, 106
Cryopsamments, 346
Cryorthents, 332, 346
Cryosaprists, 85, 346, 347
Cryoturbation, 332, 363
Cuesta(s), 174, 270
Cultivation, 103, 143
Cumulization, 146

D

Dairy, 209, 210, 224, 270, 273
Deciduous forest, 49, 238, 261, 262, 268, 269, 273
Deciduous hardwood, *see* deciduous forest
Decomposers, 14
Decomposition, 11–13, 109, 204, 358
Degradation, 243, 252, 303, 323, 361, 373
Delineation(s), 9, 20, 185, 186
Delta(s), 2, 88, 89, 92, 224, 265, 332, 358
Denitrification, 340
Densic contact, 272
Densic horizon, 265
Depression(s), 12, 70, 140, 188, 198, 217, 219, 224, 270, 285, 339
Desertification, 108
Desert landform(s), 117
Desert pavement(s), 122, 123
Desert shrub(s), 68
Devonian, 209, 219, 223, 269
Diabase, 246, 278
Diagnostic horizon(s), 31, 32, 120
Digital elevation models, 19
Diorite, 140, 278, 344
Dispersion, 15, 252
Diversity, 3, 4, 43, 95, 97, 99, 127, 131, 156, 206, 243
Dolomite, 127, 223, 272, 274
Dolostone, 128, 228, 230, 240
Drainage, 21, 77, 89, 141, 206, 233, 252, 264, 269, 286, 331, 339, 357, 365
Drained, 12, 79, 89, 129, 158, 218, 234, 238, 245, 252, 257, 265, 270, 273, 288, 296, 339, 344, 357
Drift, 71, 138, 139, 157, 221, 223, 332
Driftless, 215
Drought(s), 88, 135, 181, 257
Drumlin(s), 192, 265
Dunes, 13, 90, 187, 191, 200, 240, 254, 256
Duraquods, 81
Durids, 68
Duripan, 82, 87, 101, 105
Durixeralfs, 87, 90
Durudands, 81, 82
Durustepts, 186
Dust, 14, 88, 121, 181, 316, 363
Dust bowl, 355
Dynamic soil properties, 339, 361
Dystrocryepts, 339, 344
Dystrogelepts, 332, 339
Dystroxerepts, 78, 82, 86, 88
Dystrudepts, 80–82, 250, 270, 275, 276
Dystruderts, 251

E

Earthworms, 13, 17, 360
Ecological sections, 3
Ecoregions, 363
Ecosystem, 2, 15, 115, 124, 126, 262, 351, 353, 360, 373, 374, 377
Ecosystem function, 352
Ecosystem services, 115, 353, 357, 375, 379
E horizon, 15, 108, 250, 252, 253, 266
Electrical conductivity, 157, 185
Eluvial, 167, 173, 230
Endoaqualfs, 219, 223, 272
Endoaquepts, 82, 88, 223, 266, 271, 277
Endoaquerts, 87
Endoaquolls, 88, 212, 219, 221, 223
Endoaquults, 254
Endosaturation, 273
Entisols, 43, 75, 108, 113, 121, 128, 171, 200, 238, 243, 283, 324
Eocene, 117, 307
Eolian, 9, 121, 122, 141, 180, 265, 281, 283, 289, 291, 303, 326
Epiaqualfs, 219, 235, 258, 272
Epiaquepts, 223, 258
Epiaquerts, 156, 251, 258
Epiaquolls, 146
Epipedon, 29, 30, 32, 106, 111, 148, 314
Episaturation, 172
Erodibility, 248, 251
Erosion, 12, 14, 91, 98, 107, 133, 138, 153, 176, 199, 209, 219, 221, 225, 238, 247, 264, 270, 311, 314, 316, 318, 374
Escarpment(s), 129, 179, 324

Esker(s), 192, 270, 276
Estuarine, 81, 272, 380
Eutrophication, 351
Eutrudepts, 80, 230, 251, 272
Evaporation, 16, 94, 118
Evergreen forest(s), 2
Exchangeable Al, 313, 348
Extragrades, 35

F
Factors of soil formation, 2, 19, 243
Fallow, 103, 317
Family, 35, 276, 340
Farallon plate, 117
Farmland, 140, 146, 210, 221, 378
Faulting, 94, 116, 228, 264
Fauna, 195
Fe, 9, 15, 107
  concentrations, 140
  oxidation, 120
  oxide, 245, 294
  oxyhydroxides, 126, 316
Fe oxyhydroxides, 126
Fe-oxyhydroxides, *see* Fe oxyhydroxides
Feldspars, 18, 307
Felsic, 117, 245
Fen(s), 199, 338, 345
Fern(s), 308, 316, 323
Ferralitization, 49
Ferrihydrite, 107, 268, 348
Ferrolysis, 252
Fertility, 30, 32, 201, 253, 288, 312, 317, 323, 354
Fertilization/Fertilizer, 104, 225, 312, 313, 318, 323, 354, 361, 374, 377
Fibristels, 332
Fibrists, 74
Fir, 72, 86, 106, 194, 238
Fire, 88, 91, 98, 105, 143, 229, 322, 337, 339, 361
Flatwoods, 73, 252, 254, 286
Floodplain(s), 14, 82, 146, 147, 160, 176, 235, 246, 266, 338
Fluvaquents, 81, 87, 215, 252
Fluvents, 115
Fluvial, 122, 170, 246
Fluviomarine, 240
Fog drip, 9, 82, 90
Folistels, 339
Folists, 73, 310
Footslope(s), 68, 155, 238, 343
Forage species, 22
Forest, 3, 12, 71, 78, 106, 115, 191, 210, 227, 240, 271, 331, 338, 344, 352
Forest community, 108, 195
Forestland, 125
Forest soils, 13
Formative elements, 32, 34, 38
Fragiaqualfs, 269
Fragiaquepts, 270, 271
Fragiaquods, 271
Fragic properties, 234
Fragiorthods, 271
Fragipan(s), 104, 230, 252
Fragiudalfs, 203, 231, 269
Fragiudepts, 266, 270
Fragiudults, 234
Fragixeralfs, 104
Fraglossudalfs, 203, 252, 258
Frigid, 80, 83, 125, 135, 344
Frost
  churning, 363
  -free, 106, 287, 335
  heave, 332, 342
  mounds, 339
Frost free, 106, 193, 268, 287, 335
Frost-free, *see* frost free
Fulvicryands, 345
Fulvudands, 80–82
Fungi, 14

G
Gabbro, 84, 92, 93
Gelaquents, 332
Gelepts, 332
Geliaquepts, 339
Gelic, 101, 106, 329, 332
Gelifluction, 329
Gelifluvents, 332
Gelisols, 332, 334
Geloll, 332
Gelorthents, 332, 339
General soil map(s), 20, 21
Genesis, 2–4, 157, 243, 307, 353
Geology, 3, 43, 131, 180, 209
Gibbsite, 245, 276
Gilgai, 176
Glacial–interglacial cycles, 118
Glacial outwash, 12, 112, 153, 211
Glacial till, 84, 107, 138, 139, 240
Glaciation, 13, 14, 117, 209, 211, 262
Glacic layer, 332
Glacier(s), 2, 14, 108, 157, 198, 209, 265, 343
Glaciofluvial, 101, 139, 151, 224, 265
Glaciolacustrine, 70, 71, 140, 276
Glacistels, 332
Glass, 87, 348
Gleization, 49
Global carbon, 6
Glossaqualfs, 252
Glossic, 202, 203
Glossocryalfs, 111, 112
Glossudalfs, 203
Gneiss, 178, 238, 240, 272–274
Goethite, 245, 348
Gophers, 13, 17
Graben, 117
Grand Canyon, 117, 127
Granite, 2, 11, 92, 128, 133, 272
Granitic, 86, 117, 209
Grassland(s), 13, 32, 70, 128, 204, 235
Gravity, 10, 235
Grazingland, 167
Great group(s), 20, 32, 34, 35, 168, 240, 254, 257, 275
Great Plains, 4, 131, 135, 146, 147, 153, 160
Greenhouse gas(es), 6, 15, 278
Ground ice, 330, 332, 336
Ground squirrels, 13, 17
Groundwater, 14, 118, 129, 158, 324
Growing degree days, 264, 329
Growing season, 110, 113, 227, 235, 257
Gully, 318
Gypsic horizon(s), 157

Gypsum, 154

H
Halaquepts, 157
Halophytic, 136
Haplaquepts, 251
Haplargids, 93, 113, 128
Haplocalcids, 113, 119, 129, 182
Haplocambids, 113, 119, 182
Haplocryands, 82, 83
Haplocryepts, 82, 83, 340
Haplocryods, 82, 344, 347
Haplodurids, 128
Haplogelepts, 332, 339
Haplogelods, 339
Haplohemists, 81, 88
Haplohumults, 79, 83, 125, 293, 297
Haplorthels, 339
Haplorthods, 81, 266, 271–273, 278
Haplosaprists, 88, 90, 220, 224
Haploturbels, 339
Haploxeralfs, 79, 83, 84, 86, 87, 90, 92–94, 96, 125
Haploxerands, 78, 82, 84, 125
Haploxerepts, 83, 87, 92, 95, 99
Haploxererts, 87, 90, 95
Haploxerolls, 83, 86, 90, 94, 98, 103, 124
Haploxerults, 86, 87
Hapludalfs, 35, 171, 203, 218, 221, 230, 234, 235, 245, 258
Hapludands, 80
Hapluderts, 251, 256
Hapludolls, 140, 149, 152, 211, 212, 215, 221, 230
Hapludoxs, 293
Hapludults, 230, 234, 238, 245, 248, 250, 303
Haplustalfs, 82, 94, 127, 171, 179, 295
Haplustepts, 149, 155, 159, 171, 176, 295
Haplusterts, 128, 153, 174, 176, 178, 256
Haplustolls, 94, 124, 140, 148, 153, 155, 174, 215, 295
Hardwood forest(s), 2, 111, 220, 222, 261, 275
Hawaii, 4, 11, 14, 43, 50, 305–307, 313, 318, 321, 326
Hawaiian Islands, 2, 305, 307, 313, 317
Heat flux, 13
Hematite, 10, 12, 245
Hemistels, 332
Hemists, 70
Hemlock, 81–83, 275, 276, 344
High plains, 69, 131, 135, 146, 149, 156, 160, 355
Histels, 74, 329, 339, 340
Historthels, 332, 339, 340
Histosols, 9, 34, 45, 75, 85, 124, 191, 204, 206, 209, 219, 223, 238, 255, 271, 281, 310, 346
Histoturbels, 332, 339, 342
Holocene, 81, 107, 122, 143, 148, 149, 200, 243
Horizon, 1, 12, 15, 31, 34, 107, 139, 152, 181, 184, 195, 202
Horst, 117
Human-altered, 35, 268, 276, 277
Human-transported, 35, 276, 277
Humaquepts, 81, 278
Humicryods, 82, 85, 344, 345, 347, 348
Humification, 49
Humudepts, 82, 83
Humults, 53, 73, 74
Humus, 15, 18, 107, 249, 332, 348
Humustepts, 124
Hydraquents, 255
Hydraulic properties, 124
Hydric soils, 49, 247, 319
Hydrocryands, 345
Hydromorphic, 344
Hydromorphism, 331, 339, 344
Hydrophobic, 91, 99
Hydrothermal, 95
Hydrous, 308
Hydroxy-interlayered, 245, 265
Hydruands, 312
Hyperthermic, 184, 243, 281, 285, 293, 299, 307

I
Ice age, 106, 140
Igneous rocks, 127, 227, 228, 243, 274, 337
Illinoian, 138–140, 221
Illuvial, 15, 179
Imogolite, 107, 268, 348
Inceptisols, 43, 49, 108, 111, 113, 127, 131, 186, 191, 201, 209, 243, 259, 266, 268, 275, 307, 316
Insects, 14, 262
Interglacial, 138
Intergrades, 35, 110
Interpretations, 2, 22, 25
Invertebrates, 17, 87, 156
Ions, 14–16
Iron, 11, 12, 18, 167, 348
Iron-mangnaese, 29
Iron oxides, 11, 13, 18
Ironstone, 249
Iron sulfide, 275, 363
Irrigation, 87, 89, 109, 317, 324, 354
Irrigation water, 87, 88, 357
ISEE, 4
Isohyperthermic, 292, 295, 300
Isostatic rebound, 272
Isotic, 268
Isotopic, 104

J
Jurassic, 127, 244

K
K, *see* potassium
Kame(s), 192, 221, 224, 270, 276
Kandic horizon(s), 10, 15, 49, 248, 314
Kandiudults, 248, 249
Kanhapludults, 248
Kaolinite/Kaolinitic, 245, 246, 248, 265
Kaolins, 126
Karst, 227, 233, 234
Kettle(s), 191, 265
Kettle lakes, 191, 200

L
Lacustrine deposits, 128, 139, 140, 153, 266, 337
Lacustrine plains, 192, 197
Lacustrine sediments, 101, 108, 219, 268, 276
Lake plains, 71, 141, 191, 192, 199, 201, 271
Lamellae, 144, 202
Landform(s), 2, 4, 93, 131, 146, 149, 191, 230, 257, 270, 337
Land resource region, 43, 74, 99, 165, 224, 301, 342

Landscape(s), 1, 3, 14, 19, 21, 23, 94, 111, 138, 165, 179, 182, 195, 217, 234, 248, 250, 257, 275, 307, 314, 316, 323, 324, 354, 361, 377
Landslide(s), 80, 98, 316
Land use, 2, 20, 74, 101, 185, 239, 224, 265, 271, 334, 356, 377
Lateral flow, 104
Latitude(s), 12, 14, 236
Laurentide glacier, 137, 138, 140, 153, 265, 272
Laurentide ice sheet, *see* Laurentide glacier
Lava, 2, 127, 308, 313, 314
Lentils, 103
Lessivage, 202
Levee(s), 88, 89, 91, 257, 258
Limestone, 2, 70, 127, 175, 176, 179, 182, 221, 230, 233, 238, 273, 287, 289, 301, 307, 322, 324
Lithic, 128, 179, 288
Lithologic discontinuity(ies), 234
Lithosequence, 128
Livestock grazing, 101, 110, 129
Lodgment till, 275
Loess, 2, 68, 71, 104, 105, 132, 143, 200, 211, 215, 220, 221, 235, 252, 266, 332
Losses, 1, 104, 319
Low activity clay, 250
Lunettes, 146

M
Macropores, 15, 18, 156
Mafic, 92, 244
Magnesic, 95
Magnesium, 9, 13, 33, 84, 110, 171, 201, 317, 322
Management, 2, 5, 29, 101, 213, 270, 311, 316, 319
Manganese, 84, 146
Mangrove, 73, 302
Mangrove swamps, 2
Map scale, 4, 20
Map unit, 9, 20
Map unit description(s), 19, 20
Marine deposit(s), 78, 240, 271, 272
Marine sediment(s), 16, 79, 281, 346
Marine terrace(s), 74, 81
Marl(s), 176, 248
Marshes, 73, 88, 206, 239, 359
Marshlands, 34
Masses, 15, 153
Mass movement, 96, 200
Mass wasting, 13, 96
Mean annual air temperature, 243, 261, 329
Mean annual precipitation, 69, 101, 103, 126, 329
Mean annual soil temperature, 126
Medial, 347
Melanic, 49
Melanic epipedon, 47
Melanization, 49, 112
Melanocryands, 345, 348
Mesic, 77, 79, 83, 129, 234, 271
Mesozoic, 83, 110, 343
Metamorphic, 71, 72, 77, 83, 228, 240, 272
Metasedimentary, 87
Metavolcanic, 92
Methane, 334
Methyl mercury, 90
Mg, *see* magnesium
Microbe(s), 13, 14
Microbial activity, 12, 13
Microbial diversity, 355
Micronutrients, 110
Microphytic crust, 105
Micropores, 15
Middens, 118, 119
Mineralogy class, 316
Minimal tillage, *see* minimum tillage
Minimum tillage, 104, 320
Mining, 5, 87, 91, 239, 364
Miocene, 78, 83, 248
Mirabilite, 158
Mississippian, 217, 219, 228, 234, 268, 269
Missoula, 78
Missoula flood, 79
Mixed-grass prairies, 135
Mixed hardwoods, 71, 72, 250
Mixed pines, 70
Moderately-well drained, 246
Moisture regime, 11, 53, 210, 292, 322
Moisture regimes, 90
Mojave desert, 2, 123
Mollic, 43, 49, 109, 112, 140
Mollic epipedon, 35, 45, 109, 153
Mollisols, 43, 69, 106, 110, 113, 124, 127, 149, 175, 191, 211
Moraines, 13, 140, 191, 211, 214, 239
Morphogenetic, 31
Mosses, 308, 331
Mountain blocks, 116
Mountain meadow, 2, 125
Mountain slopes, 2, 108, 332, 343
Mount Mazama, 83, 107
Muck, 83
Mulch, 313

N
Na, *see* sodium
NaF, 268
National Cooperative Soil Survey, 2, 4, 29, 30, 40, 305, 379–381
Native prairie, 104, 354
Native vegetation, 2, 112, 268, 285, 348
Natraqualfs, 256
Natric, 49
Natrustalfs, 179
NCSS, *see* National Cooperative Soil Survey
Nitrate(s), 17, 318
Nitrogen, 13, 360, 374
North American plate, 84
Northern deciduous hardwood(s), 59, 60, 63, 64, 70–72, 195, 261, 262, 269
Northern hardwood(s) *see* northern deciduous hardwoods
No-till, 356
Nutrient(s), 6, 202, 230, 313, 324, 361, 374

O
Oak-hickory forests, 71, 229
Oak woodlands, 126, 229
Oak-savannah, 165, 167
Ochric epipedon, 230
Ogallala aquifer, 131
O horizon, 10, 13, 261
Ophiolites, 95
Order(s), 3, 15, 32, 35, 43, 110, 209, 307
Ordovician, 209, 210, 228, 235
Organic carbon(C), *see* soil organic carbon

Organic horizon, 89, 266, 334, 339, 342
Organic matter, *see* soil organic carbon
Organic-metal complexes, 15
Organic soil(s), 35, 307
Organisms, 9, 13, 308
Orographic, 99, 261
Orthels, 74, 332
Orthents, 43, 69
Orthods, 45, 72
Ortstein, 204
Outcrop(s), 73, 127, 178
Outwash, 70, 71, 138, 153, 178, 212, 221, 238, 265
Outwash plain(s), 71, 219, 221, 223, 239, 270
Over grazing, 12
Oxalate-extractable, 107
Oxbow(s), 146, 257
Oxidic, 73, 314, 324
Oxidized, 18, 363
Oxisols, 45, 49, 113, 313
Oxygen, 13, 310

P

Pachic, 140, 149
Pacific Islands, 2, 14, 322
Pacific Northwest, 3, 15, 107
Pahoehoe lava, 305, 313
Paleaquults, 254
Paleargids, 122
Palecryolls, 111
Palehumults, 79
Paleosol(s), 104, 138, 186
Paleozoic, 127, 259, 337
Paleudalfs, 140, 221, 251
Paleudults, 248, 250, 283
Paleustalfs, 168, 171–173, 186
Paleustolls, 128, 152
Palexeralfs, 90, 93
Palexerolls, 90, 104
Paludification, 53, 339
Paralithic, 93
Parent material(s), 1, 4, 9, 43, 68, 71, 73, 107, 153, 175, 176, 197, 203, 243, 252, 271, 313, 332
Particle-size class, 129
Particulate organic matter, 355
Pasture, 49, 172, 247, 271, 299, 361
Pastureland, 125, 293
Patterned ground, 329, 330
Peat, 88, 199, 206, 339
Pedogenesis, 122, 146, 200
Pedologists, 5
Pedology, 3, 376
Pedon, 265, 360
Pedoturbation, 17
Pegmatite, 133
Peneplain, 244
Pennsylvanian, 217, 228, 238
Perched water table, 78, 141
Percolation, 10, 14, 202
Pergelic, 329
Periglacial, 101, 138
Permafrost, 73, 74, 329, 330, 332, 334, 336, 339, 348, 363
Permeability, 154, 178, 212, 230
Permian, 146, 153, 178, 337
Perox, 45
Pesticides, 146, 224
Petrocalcic horizon, 151, 180–182
Petrocalcids, 151, 182
P fixation, 312, 318
pH, 30, 39
Phosphate(s), 348, 364
Phosphorous, *see* phosphorus
Phosphorus, 33, 312, 324
Photosynthesis, 13
Phyllosilicates, 348
Physiography, 43, 106, 111, 128
Pine savanna, 68
Pingos, 329
Placaquands, 81
Placer deposits, 94
Plant community(ies), 101, 174
Plant debris, 9
Plants, 1, 22, 84, 107, 319
Playa(s), 12, 128, 156
Pleistocene, 16, 124, 141, 187, 248, 259, 330
Plinthite, 248
Pliocene, 131
Plowing, 12, 31, 137, 357
Pluton, 237
POD Index, 204
Podzolization, 53, 203, 344
Podzols, 253
Polygenesis, 105
Polygons, 19
Poorly crystalline, 50, 110, 312
Poorly drained, 79, 108, 146, 234, 248, 257, 264, 270, 281, 289, 344
Potatoes, 141, 204, 273
Potential Evapotranspiration, 181
Pothole(s), 220, 214
Prairie, 32, 68, 69, 71, 73, 172, 175, 183, 211, 230
Prairie pothole(s), 157, 158, 211
Prairie vegetation, 212, 220, 225
Precambrian, 111, 178, 228, 243, 264
Pre-Illinoian, 138, 140
Pressure faces, 146
P retention, 107
Primary mineral(s), 9, 196, 245
Primary producers, 13
Profile(s), 19, 31, 32, 99, 200, 203, 276, 334
Proglacial, 265, 272
Proterozoic, 116
Psammaquents, 187, 256
Psamments, 69, 70, 73
Psammorthels, 332
Psammoturbels, 332
Puerto Rico, 74, 301

Q

Quartz, 230, 246, 249, 267
Quartzipsamments, 240, 254
Quartzite, 140, 228
Quaternary, 104, 127, 131, 252

R

Rain, 9, 82, 101
Range, 1, 14, 39, 43, 53, 68, 69, 73, 74, 79, 82, 91, 104, 109, 124, 227, 234, 248, 266, 307, 329, 337
Rangeland, 105, 108, 127, 361
Recharge, 129, 156, 379
Red beds, 147, 153, 257, 274

Redbeds, *see* Red beds
Redox concentrations, 12, 249
Redox depletions, 11, 12, 29
Redox features, *see* Redoximorphic features
Redoximorphic features, 31, 37, 38, 139, 140, 173, 257, 344
Redox processes, 18, 250
Reducing conditions, 11, 18
Reduction-oxidation, 338
Regolith, 200
Relief, 1, 176, 192, 227, 237, 246, 252, 257, 273, 343
Rendolls, 297
Residual material(s), 73, 186
Residual soil(s), 168, 174
Residuum, 69–71, 91, 216, 234
Resilient, 6
Resistant, 9, 98, 176, 238, 265
Resolution, 4
Rhodoxeralfs, 93
Rhodustalfs, 182
Rhyolite, 95, 128
Rock, 2, 82, 84, 93, 217
Rock plow, 356
Roots, 12, 14, 29, 106
Runoff, 10, 13, 33, 89, 98, 99, 108, 146, 153, 155, 192, 200, 224, 240, 251, 270, 272, 283, 311, 316–319, 324, 355, 360, 374, 375, 377
Rural, 22

S
Salic, 185
Salic horizon, 17
Salids, 68
Salinas, 157, 158
Saline, 33, 37, 136, 158
Salinity, 39, 91, 212
Salts, 14, 16, 122
San Andreas Fault, 94
Sand, 14, 70, 132, 187, 214, 250
Sand dunes, 132, 195, 337, 343
Sand grains, 35
Sand prairies, 135
Sandy soils, 10, 137, 197
Sapristels, 339
Saprists, 70, 73
Saprock, 11, 18
Saprolite, 244–246, 314
SAR, *see* sodium adsorption ratios
Saturated hydraulic conductivity(ies), 235, 251, 259, 265, 317
Saturated-paste extract, 158
Savannah, 2, 165
Schist, 2, 96, 178, 272
Seasonal water table(s), 248, 250, 289
Sediment(s), 2, 70, 72, 79, 87, 107, 141, 230, 248
Sedimentary rocks, 69, 79, 83, 128, 264
Semiarid, 73, 129, 292
Series, 3, 20, 21, 39, 43, 45, 239
Serpentine, 95
Serpentinite, 81, 84
Sesquioxides, 246
Shale, 2, 69, 154, 171, 221, 266
Shape, 13, 83, 344
Short-grass prairie, *see* shortgrass prairie
Shortgrass prairie, 68–70, 135, 137, 217, 218, 220
Shrink-swell, 35, 122, 173, 251
Shrink-swell potential, 139, 245
Shrub(s), 22, 129, 250, 297, 341, 344
Si, 9, 49
Sierra Nevada, 68, 87, 91, 92, 124
Silica, 17, 37, 101, 103, 105, 313
Silicate clay(s), 108
Silicate minerals, 13
Siliceous, 230, 266, 330
Silicification, 50
Silicious, 330
Silt, 70, 79, 139, 149, 155, 218, 250
Siltstone, 79, 83, 217, 234, 246, 268
Silurian, 219
Silvicultural production, 2
Sinkhole(s), 158, 160, 214, 283
Sinks, 99, 233
SiO_2, *see* Silica
Site productivity, 108, 109
Slate, 228, 238, 245
Slickensides, 156
Slope aspect, 13
Slope shape, 13
Smectite, 245, 251
Smectitic, 129, 156, 316
Snowfall, 81, 193, 227, 239
Snowmelt, 9, 112, 339
Snowpack(s), 192, 204
SO_4, *see* sulfate
Sodium, 15, 33, 157, 171, 201, 252, 256, 322
Sodium adsorption ratios, 157
Sodium chloride, 16
Soil aggregation, 13
Soil behavior, 2, 14
Soil change, 5, 351, 353, 360, 364
Soil classification, 3, 4, 43, 356
Soil color, 18
Soil description(s), 18, 25
Soil development, 13, 32, 124, 146, 258
Soil erosion, 135, 217, 360, 374
Soil family(ies), 35
Soil feature(s), 4, 35
Soil formation, 1, 5, 13, 149
Soil forming factors, 5, 20, 83
Soil-forming process(es), 22, 31, 98
Soil function(s), 351
Soil genesis, 4, 195
Soil health, 291, 313
Soil landscape(s), 99, 197
Soil-landscape unit(s), 19
Soil map(s), 21, 22
Soil moisture regime, 11, 77, 104, 344
Soil organic C, *see* Soil organic carbon
Soil organic carbon, 1, 11–15, 17–19, 30–32, 34–39, 77–79, 87, 89, 91–93, 95, 96, 101, 103, 106, 110–112, 124, 140, 143, 146–148, 150, 156, 158, 199, 200, 203–206, 209, 216–218, 220–222, 224, 225, 230, 240, 245, 248, 251, 253, 255–257, 261, 264, 281, 290, 297, 303, 307, 308, 310, 311, 314, 316, 317, 319, 322, 331, 333, 334, 339, 340, 344, 347, 348, 352–356, 358, 364, 365
Soil organic matter, *see* Soil organic carbon
Soil organisms, 1, 6, 32, 264
Soil polygon, 19
Soil quality, 34, 353, 361, 362, 364
Soil reaction, 212, 216, 220, 332, 339
Soil science, 4, 373, 375–377, 380
Soil science education, 5, 380
Soil science research, 373, 379
Soil series, 4, 20, 32, 43
Soil solution(s), 15

Soil structure, 13, 247
Soil survey, 18, 20, 22, 380
Soil taxonomy, 1, 15, 29, 348
Soil temperature, 11, 53, 81, 89, 329, 346
Soil temperature regime, 32, 34, 53, 79, 82, 90, 96, 110, 111, 113, 129, 210, 238, 243, 307
Soil type, 99
SoilWeb, 380
Sola, 153
Solar radiation, 9, 13
Solifluction, 200
Soluble salts, 16, 339
Solum, 149, 252, 313
Solution, 5, 15, 16, 377
Sonoran desert, 68, 118, 119, 122
Sorghum, 143, 147
Spatial variability, 195
Spodic horizon, 16, 253, 254, 268
Spodic materials, 203, 268
Spodosols, 45, 106, 203, 223, 243, 266, 278, 344
STATSGO2, 20
Steppes, 9
Storms, 252, 305
Stream channel(s), 14, 149, 247
Stream incision, 13
Stream valley(s), 78, 131, 214, 227, 265
Stromatolitic fabric, 105
Subaqueous, 276, 380
Subglacial, 265, 270
Subgroup modifiers
- aeric, 40
- andic, 33, 36
- aquertic, 148
- aquic, 13
- arenic, 40
- arenic aridic, 145
- argiaquic, 157
- aridic, 35, 39
- calcic, 35, 184
- cumulic, 40
- entic, 140
- grossarenic, 40
- gypsic, 35
- humic, 34
- kandic, 35
- leptic vertic, 153, 154
- lithic, 40, 182
- natric, 35
- pachic, 107
- plinthic, 250
- ruptic-histic, 332
- spodic, 37
- torric, 185
- torriorthentic, 155
- typic, 40
- udertic, 172
- udic, 53, 82, 106
- ultic, 103, 172
- ustic, 35, 39
- vertic, 33, 156
- vitrandic, 108
- xeric, 92, 101

Subgroup(s), 13, 19, 32–40, 45, 106, 108, 156, 173, 179, 184, 186, 248, 250, 257, 274, 275
Suborder(s), 5, 20, 43
Subsidence, 77, 89, 92
Subsoil, 15, 17, 153
Substrate, 144, 337
Sugar beets, 147
Sugar cane, *see* Sugarcane
Sugarcane, 186, 311
Sulfaquents, 90
Sulfate(s), 90, 158, 276, 363
Sulfide(s), 90
Sulfihemists, 240, 276
Sulfuric horizon(s), 363
Sulfuricization, 172
Supraglacial, 265, 266
Surface horizon textures, 252
Surface runoff, 99, 155, 270, 374
Suspension, 10, 14, 15
Swamps, 34, 199, 200, 243, 252, 281, 287, 300

T

Tabular data, 19
Tall-grass prairie, *see* Tallgrass prairie
Tallgrass prairie, 69, 135, 137, 148, 168, 212, 228, 250, 258
Taxa, 4, 5, 19, 20, 25, 43, 184, 186
Taxon, 4, 20
Temperature(s), 1, 15, 32, 35, 78, 80, 82–84, 90, 91, 94, 103, 106, 111, 116, 135, 160, 171, 181, 193, 209, 227, 232, 236, 240, 261, 281, 282, 287, 290, 292, 308, 322, 330, 339, 343, 355, 363
Temperature regime, *see* soil temperature regime
Tephra, 9, 82, 84, 99, 341, 343, 344, 346
Termites, 17
Terrace(s), 13, 78, 81, 83, 87, 88, 90, 91, 94, 143, 146, 148, 149, 153, 160, 171, 174, 175, 209, 224, 234, 240, 248, 250, 257, 258, 270, 300, 302, 339, 347, 360
Tertiary, 79, 81, 83, 127, 165, 173, 174, 184–188, 243, 250, 252, 255, 295, 300, 330, 337, 343, 379
Texture(s), 9–11, 22, 31, 32, 39, 79, 87, 101, 104, 124, 139, 140, 143, 144, 146, 147, 149, 153, 155, 160, 165, 172, 178, 182, 188, 197, 212, 218, 221, 224, 230, 234, 235, 238, 240, 245, 246, 248, 249, 251–253, 255, 256, 261, 266, 270, 271, 275, 277, 281, 333, 337, 340, 343, 344, 347, 356, 357, 364, 365
Thermic, 35, 77, 87, 89, 90, 92, 96, 126, 127, 229, 243
Thermokarsting, 333, 336
Tidal marsh, 239, 240, 265, 276, 359
Tile drains, 217, 225
Tile drainage, 234
Till, 9, 68–71, 83, 107, 138–140, 167, 191, 197, 203, 206, 209, 211, 214–217, 221, 224, 265, 270, 271, 275, 278, 347, 353
Tillage, 88, 104, 225, 245, 247, 311, 312, 316, 319, 323, 354–356, 358, 364, 365
Tillage pan, 311, 312, 316
Till plains, 71, 78, 138, 140, 191, 211, 215, 217, 219, 221, 223, 269, 278
Tilth, 32
Timber production, 106, 109, 110, 113, 114, 155
Time, 1, 5, 9, 11, 14, 15, 19, 20, 23, 27, 32, 43, 45, 75, 99, 101, 131, 146, 160, 198, 199, 289, 303, 351, 353–355
Topographic map, 19, 159
Topography, 1, 3, 4, 9, 13, 14, 23, 73, 78, 87, 104, 113, 124, 129, 135, 153, 174, 176, 191, 192, 195, 197, 227, 229, 233, 234, 238, 243, 248, 259, 270, 281, 282, 290, 301, 303, 313, 333, 338, 344, 351, 363
Torrifluvents, 129
Torriorthents, 124, 127–129, 181
Torripsamments, 93
Trade wind, 305, 308, 314
Transformation(s), 1, 14, 18, 128

Translocated, 15, 17, 200, 202, 204
Translocation(s), 14, 15, 112, 144, 200, 201, 203, 268
Tree(s), 12, 13, 17, 22, 80, 83, 94, 135, 172, 187, 195, 204, 229–231, 234, 238, 253, 262, 268, 271, 278, 282, 287, 290, 292, 293, 295, 300, 308, 356
Tree throw, 12, 278
Triassic, 238, 244, 246, 274, 276
Tropical rainforest(s), 2
Tundra, 1, 73, 74, 82, 83, 262, 330, 333, 334, 339, 342, 363, 364
Turbels, 74, 332
Turbidites, 83

U
Udalfs, 43, 70–72, 203, 301
Udands, 45, 53, 72, 73, 312
Udepts, 43, 53, 69–74, 201, 202, 273
Uderts, 71, 73
Udic, 77, 80–82, 92, 126, 170, 171, 173, 209, 210, 230, 234, 238, 281, 283, 309, 346
Udic soil moisture regime, 11, 53, 83, 106, 192, 215, 218, 240, 261, 344, 347
Udifluvents, 81, 82, 215, 218, 221, 222, 230, 252, 257, 258, 277
Udipsamments, 81, 224, 256, 257, 269, 271, 277
Udolls, 43, 69–72, 140, 205, 301
Udorthents, 251, 271, 276, 277, 288
Udox, 45, 73, 74, 300
Udults, 71–74, 300
Ultisols, 14, 43, 45, 49, 75, 90, 125, 127, 170, 171, 173, 199, 209, 230, 234, 238, 240, 243, 245, 255, 259, 268, 269, 281, 283, 303, 313, 314, 316, 318, 356, 364
Ultra-mafic, *see* Ultramafic
Ultramafic, 9, 84, 93
Umbric, 34, 49
Umbric epipedon, 33
Upland landforms, 2, 337
Urban, 22, 90, 96, 174, 210, 221, 231, 236, 240, 247, 262, 275, 277, 281, 282, 292, 297, 300, 302, 320, 324, 342, 348, 374, 376, 377, 379, 381
Ustalfs, 43, 68–70, 72
Ustands, 45, 72, 73
Ustepts, 43, 69, 70, 72, 74
Usterts, 69, 70, 74, 302
Ustic soil moisture regime, 94, 101, 127, 128, 300, 302, 316
Ustifluvents, 147, 171, 172
Ustipsamments, 144, 145, 154, 159, 160, 187, 256
Ustolls, 43, 69, 70, 72–74, 302
Ustorthents, 127, 128, 143, 144, 158, 174–176, 179

V
Vegetables, 224, 240, 268, 270, 281, 285, 288, 316, 358
Vegetation, 1, 9, 21, 43, 45, 68, 70, 71, 74, 83, 85, 90, 111, 124–126, 165, 174, 200, 216, 217, 221, 225, 230, 238, 261, 271, 276, 284, 289, 290, 300, 331, 337, 344, 347, 351, 361, 363
Vegetation community(ies), 340, 344, 361
Vermiculite, 245, 248, 265, 348
Vertisols, 18, 35, 43, 45, 49, 75, 87, 106, 111, 167, 174, 176, 185, 187, 191, 193, 238, 243, 255, 259, 302, 316, 317
Vertization, 53
Very poorly drained, 195, 211, 253, 254, 259, 271, 272, 276, 285, 291, 346
Vesicular horizons, 122
Vesicular pores, 122
Vitrandic, 108, 110
Vitrands, 69
Vitric, 33, 309
Vitricryands, 82, 83, 346
Vitrixerands, 83, 84
Volcanic ash, 14, 69, 73, 74, 78, 82, 84, 107, 110, 138, 307, 314, 316, 317, 322, 324, 326, 330, 344, 348
Volcanic eruptions, 9, 101, 307, 313, 343, 347
Volcanic material, 72, 281, 299, 303, 314, 322, 324
Volcanic rocks, 82, 92, 127, 128, 299, 337, 343
Volcanic terrain, 124
Volcano(s), 2, 14, 72, 82, 188, 299, 305, 307, 313, 316, 322, 324, 337, 346

W
Water, 1, 9–11, 13, 15, 16, 73, 78, 82, 89, 91, 95, 98, 104, 105, 139, 173, 195, 202, 218, 224, 234, 249, 257, 266, 287, 317, 324, 339, 379
Water balance, 9, 11, 14, 94
Water erosion, 103, 104, 222, 316, 356
Water flow, 13
Water-holding capacity, *see* available water holding capacity
Water quality, 104, 217, 219, 221, 291, 357, 359, 374, 375, 377
Water quantity, 357
Water resources, 3
Water retention, 107, 175, 268, 312
Watershed, 13, 86, 87, 109, 110, 113, 125, 249, 316, 348, 351, 355, 359, 360, 377
Water-stable aggregates, 312
Water storage capacity, *see* available water holding capacity
Weathering, 9, 11, 15, 18, 50, 87, 95, 110, 126, 171, 196, 250, 322, 357
Weathering product(s), 14, 107, 110, 246, 347
Weathering reaction(s), 10, 128
Website, 4
Web soil survey, 4, 380
Well drained, 79, 125, 129, 141, 178, 180, 184, 187, 211, 215, 218, 229, 234, 235, 246, 248, 250, 252, 256, 259, 269, 270, 293, 296, 298, 300, 302, 340
Wetland(s), 18, 136, 156, 157, 191, 192, 206, 224, 247, 273, 313, 320, 334, 340, 343, 354, 357, 358, 365, 375, 380
Wheat belt, 132, 137
Wildlife, 106, 110, 188, 326, 340, 341
Wildlife habitat, 22, 109, 113, 125, 221, 276, 277, 334, 357, 361
Wind, 9, 14, 132, 188, 213, 219, 256, 305, 314, 318, 337
Wind erosion, 12, 89, 105
Wisconsin, 70, 138, 140, 153, 191, 193, 198
Wisconsinan, 138, 238, 239
Woodlands, 127, 129, 165, 229, 276, 338

X
Xeralfs, 68
Xerands, 53, 68
Xerepts, 43, 53, 68
Xeric, 11, 53, 79, 82, 89, 101, 129
Xeric soil moisture regime, 79, 90, 125
Xerofluvents, 87, 90
Xerolls, 43, 53, 68, 75, 87
Xeropsamments, 90
Xerorthents, 90, 91, 93, 99

Y
Younger Dryas Chronozone, 149

The manufacturer's authorised representative in the EU is Springer Nature Customer Service Centre GmbH, Europaplatz 3, 69115 Heidelberg, Germany. If you have any concerns regarding our products, please contact ProductSafety@springernature.com

Printed and bound by CPI Group (UK) Ltd, Croydon, CR0 4YY
05/10/2026
02228166-0001